Wnt Signaling in Development and Disease

Wnt Signaling in Development and Disease

Molecular Mechanisms and Biological Functions

Edited by

Stefan Hoppler
University of Aberdeen, Aberdeen, Scotland

Randall T. Moon
Howard Hughes Medical Institute, The University of Washington, Seattle, WA, USA

WILEY Blackwell

Published by John Wiley & Sons, Inc., Hoboken, New Jersey
Published simultaneously in Canada

Library of Congress Cataloging-in-Publication Data:

Wnt signaling in development and disease : molecular mechanisms and biological functions / [edited by] Stefan Hoppler, Randall T. Moon.
p. ; cm.
Includes bibliographical references and index.
ISBN 978-1-118-44416-0 (cloth)
I. Hoppler, Stefan, 1965– editor of compilation. II. Moon, Randall T., editor of compilation.
[DNLM: 1. Wnt Proteins–physiology. QU 55.2]
QP552.W58
572′.64–dc23

2013042745

Cover images: First and third images from iStockphoto.com. Middle image courtesy of Randall T. Moon and colleagues.
Cover design by Modern Alchemy LLC

Printed and bound in Malaysia by Vivar Printing Sdn Bhd

10 9 8 7 6 5 4 3 2 1

Contents

Contributors vii

Preface xi

Part 1 Molecular Signaling Mechanisms: From Pathways to Networks 1

1 Wnt Signal Production, Secretion, and Diffusion 3
Madelon M. Maurice and Hendrik C. Korswagen

2 Wnt Signaling at the Membrane 15
Gary Davidson and Christof Niehrs

3 Wnt Signal Transduction in the Cytoplasm: an Introduction to the Destruction Complex 33
Tony W. Chen, Heather A. Wallace, Yashi Ahmed, and Ethan Lee

4 An Overview of Gene Regulation by Wnt/β-Catenin Signaling 51
Chen U. Zhang and Ken M. Cadigan

5 Finding a Needle in a Genomic Haystack: Genome-Wide Approaches to Identify Wnt/TCF Transcriptional Targets 73
Chandan Bhambhani and Ken M. Cadigan

6 Introduction to β-Catenin-Independent Wnt Signaling Pathways 89
Susanne Kühl and Michael Kühl

7 Molecular Mechanisms of Wnt Pathway Specificity 101
Alexandra Schambony, Guido J.R. Zaman, and Folkert Verkaar

8 Modulation of Wnt Signaling by Endocytosis of Receptor Complexes 113
Akira Kikuchi, Shinji Matsumoto, Katsumi Fumoto, and Akira Sato

9 New Insights from Proteomic Analysis of Wnt Signaling 125
Matthew P. Walker, Dennis Goldfarb, and Michael B. Major

10 New Insights about Wnt/β-Catenin Pathway Mechanisms from Global siRNA Screens 137
Tenzin Gocha and Ramanuj DasGupta

11 Mathematical Models of Wnt Signaling Pathways 153
Michael Kühl, Barbara Kracher, Alexander Groß, and Hans A. Kestler

12 The Wnt's Tale: On the Evolution of a Signaling Pathway 161
Jenifer C. Croce and Thomas W. Holstein

Part 2 Selected Key Molecules in Wnt Signaling 177

13 Secreted Wnt Inhibitors or Modulators 179
Paola Bovolenta, Anne-Kathrin Gorny, Pilar Esteve, and Herbert Steinbeisser

14 Frizzleds as G Protein-Coupled Receptors 195
Gunnar Schulte

15 Dishevelled at the Crossroads of Pathways 207
Vítězslav Bryja and Ondřej Bernatík

16 β-Catenin: a Key Player in Both Cell Adhesion and Wnt Signaling 217
Jonathan Pettitt

17 Evolutionary Diversification of Vertebrate TCF/LEF Structure, Function, and Regulation 225
Stefan Hoppler and Marian L. Waterman

18 Insights from Structural Analysis of Protein–Protein Interactions by Wnt Pathway Components and Functional Multiprotein Complex Formation 239
Zhihong Cheng and Wenqing Xu

Part 3 Wnt Signaling in Embryonic Development and Adult Tissue Homeostasis 251

19 Wnt Signaling in Early Vertebrate Development: From Fertilization to Gastrulation 253
Eliza Zylkiewicz, Sergei Y. Sokol, and Stefan Hoppler

20 Wnt/β-Catenin Signaling in Embryonic Stem Cells: Insights into Early Mammalian Development 267
Kathryn C. Davidson

21 Wnt Signaling in Neural Development 279
Richard I. Dorsky

22 Wnt Signaling in Heart Organogenesis 293
Stefan Hoppler, Silvia Mazzotta, and Michael Kühl

23 Wnt Signaling in Kidney Organogenesis 303
Kimmo Halt and Seppo Vainio

24 Wnt Signaling Regulation of Tissue Architecture (EMT and MET) and Morphogenesis: Consequences for Colorectal and Liver Cancer 315
Theodora Fifis, Bang M. Tran, Renate H.M. Schwab, Timothy M. Johanson, Nadia Warner, Nick Barker, and Elizabeth Vincan

25 Wnt Signaling in Adult Stem Cells: Tissue Homeostasis and Regeneration 329
Frank J.T. Staal and Riccardo Fodde

26 Restoring Tissue Homeostasis: Wnt Signaling in Tissue Regeneration After Acute Injury 339
Günes Özhan and Gilbert Weidinger

Part 4 Wnt Signaling in Chronic Disease 357

27 Wnt Signaling and Colorectal Cancer 359
Kevin Myant and Owen J. Sansom

28 Wnt Signaling in Melanoma 369
Jamie N. Anastas and Andy J. Chien

29 Wnt Signaling in Mood and Psychotic Disorders 379
Stephen J. Haggarty, Karun Singh, Roy H. Perlis, and Rakesh Karmacharya

30 Neuropsychiatric Disease-Associated Genetic Variation in the Wnt Pathway 393
Stephen J. Haggarty, Karun Singh, Roy H. Perlis, and Rakesh Karmacharya

31 Wnt Signaling in Dementia 411
Stephen J. Haggarty

32 Therapeutic Targeting of the Wnt Signaling Network 421
Felicity Rudge and Trevor Dale

Index 445

Contributors

Yashi Ahmed. Department of Genetics and the Norris Cotton Cancer Center, Geisel School of Medicine at Dartmouth, Hanover, NH, USA

Jamie N. Anastas. The University of Washington School of Medicine, Seattle, WA, USA

Nick Barker. Institute of Medical Biology, Immunos, Singapore

Ondřej Bernatík. Faculty of Science, Institute of Experimental Biology, Masaryk University & Institute of Biophysics, Academy of Sciences of the Czech Republic, Brno, Czech Republic

Chandan Bhambhani. Department of Molecular, Cellular and Developmental Biology, University of Michigan, Ann Arbor, MI, USA

Paola Bovolenta. Centro de Biología Molecular "Severo Ochoa," CSIC-UAM; and CIBER de Enfermedades Raras (CIBERER), Madrid, Spain

Vítězslav Bryja. Faculty of Science, Institute of Experimental Biology, Masaryk University & Institute of Biophysics, Academy of Sciences of the Czech Republic, Brno, Czech Republic

Ken M. Cadigan. Department of Molecular, Cellular and Developmental Biology, University of Michigan, Ann Arbor, MI, USA

Tony W. Chen. Department of Cell and Developmental Biology and Program in Developmental Biology, Vanderbilt University Medical Center, Nashville, TN, USA; and Vanderbilt Ingram Cancer Center, Vanderbilt University School of Medicine, Nashville, TN, USA

Zhihong Cheng. Department of Biological Structure, University of Washington, Seattle, WA, USA

Andy J. Chien. The Group Health Research Institute; and The University of Washington School of Medicine, Seattle, WA, USA

Jenifer C. Croce. CNRS, UMR7009, Sorbonne Universités, UPMC Univ Paris 06, Laboratoire de Biologie du Développement de Villefranche-sur-mer, EvoInSiDe Team, Observatoire Océanographique, 06230, Villefranche-sur-mer, France

Trevor Dale. School of Biosciences, Cardiff University, Cardiff, UK

Ramanuj DasGupta. Department of Biochemistry and Molecular Pharmacology, and the NYU Cancer Institute, New York University Langone Medical Center, New York, NY, USA

Gary Davidson. Institute of Toxicology and Genetics, Karlsruhe Institute of Technology, Karlsruhe, Germany

Kathryn C. Davidson. Department of Pharmacology, Institute for Stem Cell and Regenerative Medicine, Howard Hughes Medical Institute, University of Washington School of Medicine, Seattle, WA, USA; and Centre for Eye Research Australia, University of Melbourne, and Royal Victorian Eye and Ear Hospital, East Melbourne, Victoria, Australia

Richard I. Dorsky. Department of Neurobiology and Anatomy, University of Utah, Salt Lake City, UT, USA

Pilar Esteve. Centro de Biología Molecular "Severo Ochoa," CSIC-UAM; and CIBER de Enfermedades Raras (CIBERER), Madrid, Spain

Theodora Fifis. Austin Hospital, University of Melbourne, Heidelberg, Melbourne, Victoria, Australia

Riccardo Fodde. Josephine Nefkens Institute, Erasmus MC, Rotterdam, The Netherlands

Katsumi Fumoto. Department of Molecular Biology and Biochemistry, Graduate School of Medicine, Osaka University, Osaka, Japan

Tenzin Gocha. Department of Biochemistry and Molecular Pharmacology, and the NYU Cancer Institute, New York University Langone Medical Center, New York, NY, USA

Dennis Goldfarb. Department of Computer Science, Lineberger Comprehensive Cancer Center, University of North Carolina at Chapel Hill, Chapel Hill, NC, USA

Anne-Kathrin Gorny. Institute of Human Genetics, University of Heidelberg, Heidelberg, Germany

Alexander Groß. Research Group Bioinformatics and Systems Biology, Institute of Neural Information Processing, Ulm University, Ulm, Germany

Stephen J. Haggarty. Department of Neurology, Department of Psychiatry, and MGH Psychiatry Center for Experimental Drugs & Diagnostics, Center for Human Genetic Research, Massachusetts General Hospital & Harvard Medical School, Boston, MA, USA

Kimmo Halt. The Centre of Excellence in Cell-Extracellular Matrix Research, Biocenter Oulu; and Laboratory of Developmental Biology, Department of Medical Biochemistry and Molecular Biology, Institute of Biomedicine, University of Oulu, Oulu, Finland

Thomas W. Holstein. Department of Molecular Evolution and Genomics, Centre for Organismal Studies (COS), Heidelberg University, Heidelberg, Germany

Stefan Hoppler. Institute of Medical Sciences, University of Aberdeen, Aberdeen, Scotland

Timothy M. Johanson. University of Melbourne and Victorian Infectious Diseases Reference Laboratories, Melbourne, Australia

Rakesh Karmacharya. Department of Psychiatry, and MGH Psychiatry Center for Experimental Drugs & Diagnostics, Center for Human Genetic Research, Massachusetts General Hospital & Harvard Medical School, Boston; Chemical Biology Program, Broad Institute of Harvard & MIT, Cambridge; and Schizophrenia and Bipolar Disorder Program, McLean Hospital, Belmont, MA, USA

Hans A. Kestler. Research Group Bioinformatics and Systems Biology, Institute of Neural Information Processing, Ulm University, Ulm, Germany

Akira Kikuchi. Department of Molecular Biology and Biochemistry, Graduate School of Medicine, Osaka University, Osaka, Japan

Hendrik C. Korswagen. Hubrecht Institute, Royal Netherlands Academy of Arts and Sciences and University Medical Center Utrecht, Utrecht, The Netherlands

Barbara Kracher. Department of Plant-Microbe Interactions, Max Planck Institute for Plant Breeding Research, Cologne, Germany

Michael Kühl. Institute of Biochemistry and Molecular Biology, Ulm University, Ulm, Germany

Susanne Kühl. Institute of Biochemistry and Molecular Biology, Ulm University, Ulm, Germany

Ethan Lee. Department of Cell and Developmental Biology and Program in Developmental Biology, Vanderbilt University Medical Center; and Vanderbilt Ingram Cancer Center, Vanderbilt University School of Medicine, Nashville, TN, USA

Michael B. Major. Department of Cell Biology and Physiology and Department of Computer Science, Lineberger Comprehensive Cancer Center, University of North Carolina at Chapel Hill, Chapel Hill, NC, USA

Shinji Matsumoto. Department of Molecular Biology and Biochemistry, Graduate School of Medicine, Osaka University, Osaka, Japan

Madelon M. Maurice. Department of Cell Biology, University Medical Center Utrecht, Utrecht, The Netherlands

Silvia Mazzotta. Institute of Medical Sciences, University of Aberdeen, Aberdeen, Scotland

Kevin Myant. Beatson Institute for Cancer Research, Garscube Estate, Glasgow, UK

Christof Niehrs. Molecular Embryology, Deutsches Krebsforschungszentrum, Heidelberg; and Faculty of Biology, Institute of Molecular Biology (IMB), University of Mainz, Mainz, Germany

Günes Özhan. Biotechnology Center, Technische Universität Dresden, Dresden, Germany

Roy H. Perlis. Department of Psychiatry, and MGH Psychiatry Center for Experimental Drugs & Diagnostics, Center for Human Genetic Research, Massachusetts General Hospital & Harvard Medical School, Boston, MA, USA

Jonathan Pettitt. Institute of Medical Sciences, University of Aberdeen, Aberdeen, Scotland

Felicity Rudge. School of Biosciences, Cardiff University, Cardiff, UK

Owen J. Sansom. Beatson Institute for Cancer Research, Garscube Estate, Glasgow, UK

Akira Sato. Department of Molecular Biology and Biochemistry, Graduate School of Medicine, Osaka University, Osaka, Japan

Alexandra Schambony. Developmental Biology, Biology Department, Friedrich-Alexander University Erlangen-Nuremberg, Erlangen, Germany

Gunnar Schulte. Department of Physiology & Pharmacology, Section for Receptor Biology & Signaling, Karolinska Institutet, Stockholm, Sweden

Renate H.M. Schwab. University of Melbourne and Victorian Infectious Diseases Reference Laboratories, Melbourne, Australia

Karun Singh. Department of Biochemistry and Biomedical Sciences, McMaster Stem Cell & Cancer Research Institute, McMaster University, Hamilton, Ontario, Canada

Sergei Y. Sokol. Department of Developmental and Regenerative Biology, Icahn School of Medicine at Mount Sinai, New York, NY, USA

Frank J.T. Staal. Leiden University Medical Center, Leiden, The Netherlands

Herbert Steinbeisser. Institute of Human Genetics, University of Heidelberg, Heidelberg, Germany

Bang M. Tran. University of Melbourne and Victorian Infectious Diseases Reference Laboratories, Melbourne, Australia

Seppo Vainio. The Centre of Excellence in Cell-Extracellular Matrix Research, Biocenter Oulu; and Laboratory of Developmental Biology, Department of Medical Biochemistry and Molecular Biology, Institute of Biomedicine, University of Oulu, Oulu, Finland

Folkert Verkaar. Department of Chemistry and Pharmaceutical Sciences, VU University Amsterdam, Amsterdam, The Netherlands

Elizabeth Vincan. University of Melbourne and Victorian Infectious Diseases Reference Laboratories, Melbourne, Australia

Matthew P. Walker. Department of Cell Biology and Physiology, Lineberger Comprehensive Cancer Center, University of North Carolina at Chapel Hill School of Medicine, Chapel Hill, NC, USA

Heather A. Wallace. Department of Genetics and the Norris Cotton Cancer Center, Geisel School of Medicine at Dartmouth, Hanover, NH, USA

Nadia Warner. University of Melbourne and Victorian Infectious Diseases Reference Laboratories, Melbourne, Australia

Marian L. Waterman. Department of Microbiology and Molecular Genetics, University of California, Irvine, CA, USA

Gilbert Weidinger. Institute for Biochemistry and Molecular Biology, Ulm University, Ulm, Germany

Wenqing Xu. Department of Biological Structure, University of Washington, Seattle, WA, USA

Guido J.R. Zaman. Netherlands Translational Research Center B.V. (NTRC), Oss, The Netherlands

Chen U. Zhang. Department of Molecular, Cellular and Developmental Biology, University of Michigan, Ann Arbor, MI, USA

Eliza Zylkiewicz. Department of Developmental and Regenerative Biology, Icahn School of Medicine at Mount Sinai, New York, NY, USA

Preface

It is 30 years since what is generally considered the discovery of Wnt signaling: the realization that the *Drosophila* developmental selector gene wingless and the mouse oncogene int-1 are homologs. This Dr. Jekyll and Mr. Hyde split personality (good and bad) of Wnt signaling has endured: Wnt signaling is considered benevolent in embryonic development and in regulating stem cell differentiation for tissue homeostasis in the adult yet is also implicated in unpleasant diseases, particularly cancer. Wnt signaling has come a long way in the last three decades and is now recognized among the most important signaling mechanisms in development and disease. This 30-year anniversary provides an ideal opportunity to review the remarkable progress we have made in understanding the mechanisms and functions of Wnt signaling but also to reflect upon the still unresolved important questions about fundamental molecular pathway mechanisms and their biological roles.

Core topics in Wnt signaling will be explored by expert reviewers, providing first clear access to the core foundations before advancing to some of the very cutting edge of current scientific research. In this book, we start by discussing the molecular pathway mechanisms and their integration into the cell's regulatory networks and then home in on a select few molecules considered to be key players before reviewing some of the benevolent roles of Wnt signaling in embryonic development and adult tissue homeostasis and ending up considering the roles of Wnt signaling mechanisms in chronic disease. Each of these parts of the book will be briefly introduced to facilitate independent access to individual chapters of interest to the reader.

This book aims to focus on biological insight and current scientific questions about Wnt signaling that are likely widely applicable. Advantages of different model systems and application of novel methods for ingenious experimental approaches of course give access to and drive this scientific discovery and are therefore embraced in the individual chapters, which however focus on what is revealed about the fundamental biology of Wnt signaling.

Part 1

Molecular Signaling Mechanisms: From Pathways to Networks

Multicellular organisms need to coordinate gene expression in cells of their tissues. Wnt signaling represents one of the most important molecular cell-to-cell signaling mechanisms in animal and human cells. In its basic form, it can be described as a linear pathway where a gene encoding a Wnt signal is transcribed in the nucleus of some Wnt-sending cells, a signal protein is synthesized including necessary post-translational modifications (Chapter 1) and secreted from that cell. The Wnt signal is not cell permeable; Wnt binding to cell membrane receptor proteins triggers biochemical mechanisms inside the Wnt-responding cell (often called signal transduction), which ultimately changes gene expression in the nucleus of Wnt-responding cells. Wnt signaling was discovered about 30 years ago after the discovery of the MapK signal transduction pathway, which was described as a bucket brigade or cascade of positive interactions (sequential phosphorylation) involving proteins that could clearly be recognized as enzymes.

Genetic analysis in Drosophila soon revealed that Wnt signal transduction is fundamentally different since it involves many proteins that are not obvious enzymes and in being a double negative pathway, whereby upstream signaling mechanisms inhibit the function of proteins that would otherwise restrict further downstream Wnt signal transduction mechanisms. This double negative pathway architecture has fundamental implications in that loss of function mutations in some Wnt signaling components will lead to constitutively active downstream signal transduction activity (e.g., APC in colorectal cancer; see Chapter 27) and that small molecule inhibitors may lead to increased signal transduction activity if their molecular target normally restricts pathway activity (e.g., GSK3 inhibitors; see Chapter 32). Studying the function and cell biology of Wnt signaling components that are not obvious enzymes revealed that Wnt signal transduction is characterized by alternate assembly and disassembly of multi-protein complexes at the membrane (see Chapter 2), in the cytoplasm (see Chapter 3), and in the nucleus (see Chapter 4). Enzymes such as kinases are now recognized as important regulators that assemble and disassemble these protein complexes, and cutting-edge comprehensive analysis (e.g., see Chapters 9 and 10) provides further insight into additional modulators of these processes. Context-specific

Wnt Signaling in Development and Disease: Molecular Mechanisms and Biological Functions, First Edition.
Edited by Stefan Hoppler and Randall T. Moon.

modulators enable Wnt signaling to regulate gene expression in a cell-type-specific or stage-specific manner in animals and humans (e.g., see Chapter 5).

However, multicellular organisms also coordinate tissue polarity and other aspects of cell biology in addition to gene expression. Wnt signaling also turned out to be a potent regulator in this context but often through molecular mechanisms that diverged from the above-described linear pathway. Alternative linear Wnt pathways are being identified and described (see Chapter 6), which also reveal that the various Wnt signaling pathways are interconnected and influence each other at several levels. While thinking in terms of linear signaling pathways is still clearly very useful for studying the functional roles of Wnt signaling (see Parts 3 and 4), the concept of a Wnt signaling network of molecular mechanisms will become more prominent. Insight into the nonlinear complexity of such a Wnt signaling network will come from systems level analysis with mathematical tools (see Chapter 11).

A network of interactions also exists between Wnt signaling mechanisms and biochemical mechanisms regulated by other cell-to-cell signaling pathways. They are increasingly recognized as regulating context-dependent modulators of Wnt signaling as mentioned above, and Wnt signaling in turn changes the cellular and molecular context for other cell-to-cell signaling pathways. Such combinatorial signaling emerges as paramount for the cell-type- and context-specific functional roles of Wnt signaling in Embryonic Development and Adult Tissue Homeostasis (Part 3) and – no doubt related to that – in Chronic Disease (Part 4).

1 Wnt Signal Production, Secretion, and Diffusion

Madelon M. Maurice[1] and Hendrik C. Korswagen[2]

[1]Department of Cell Biology, University Medical Center Utrecht, Utrecht, The Netherlands
[2]Hubrecht Institute, Royal Netherlands Academy of Arts and Sciences and University Medical Center Utrecht, Utrecht, The Netherlands

Introduction

Wnt proteins are members of an evolutionarily conserved family of secreted signaling proteins that play a central role in the development of metazoan organisms (Willert and Nusse, 2012) (see Chapter 12). Wnts are lipid-modified glycoproteins that can signal in a short-range manner to target cells that are directly adjacent to Wnt-producing cells (Sato *et al.*, 2011). Importantly, Wnts can also form long-range concentration gradients that provide positional information to cells in developing tissues (Zecca, Basler, and Struhl, 1996). The formation and regulation of such morphogenic gradients is one of the major enigmas in the Wnt field, raising questions on how the hydrophobic Wnt protein is efficiently released from producing cells and on how it spreads in the aqueous extracellular environment of the tissue. In this chapter, we will briefly discuss the lipid and sugar modification of Wnt proteins and then focus on the specialized secretion machinery that mediates the release of Wnt from producing cells and the mechanisms that facilitate and control the spreading of Wnt in morphogen gradient formation.

Posttranslational modification of Wnt

Biosynthesis of Wnt proteins is initiated by their translation and translocation in the rough endoplasmic reticulum (ER), after which the Wnt proteins undergo a number of maturation and modification steps (Figure 1.1, step 1). First, all Wnt proteins harbor a large number of conserved cysteines (23–25 on average), which participate in the formation of intramolecular disulfide bonds (Janda *et al.*, 2012). In addition, Wnts undergo two major types of posttranslational modification, N-glycosylation and lipidation. Although the addition of *N*-glycans may facilitate the secretion of a subset of Wnts, they appear generally dispensable for the activity of mature Wnt proteins, as glycosylation-deficient mutants exhibit no major defects in signaling (Doubravska *et al.*, 2011; Komekado *et al.*, 2007; Kurayoshi *et al.*, 2007; Tang *et al.*, 2012). In agreement, the two glycan groups attached to *Xenopus* Wnt8 (XWnt8) did not contribute to the Wnt–Frizzled (Fz) interaction in the recently solved crystal structure (Janda *et al.*, 2012).

A number of studies have reported on the modification of both vertebrate and invertebrate Wnts with two acyl groups: a palmitate at an

Wnt Signaling in Development and Disease: Molecular Mechanisms and Biological Functions, First Edition.
Edited by Stefan Hoppler and Randall T. Moon.

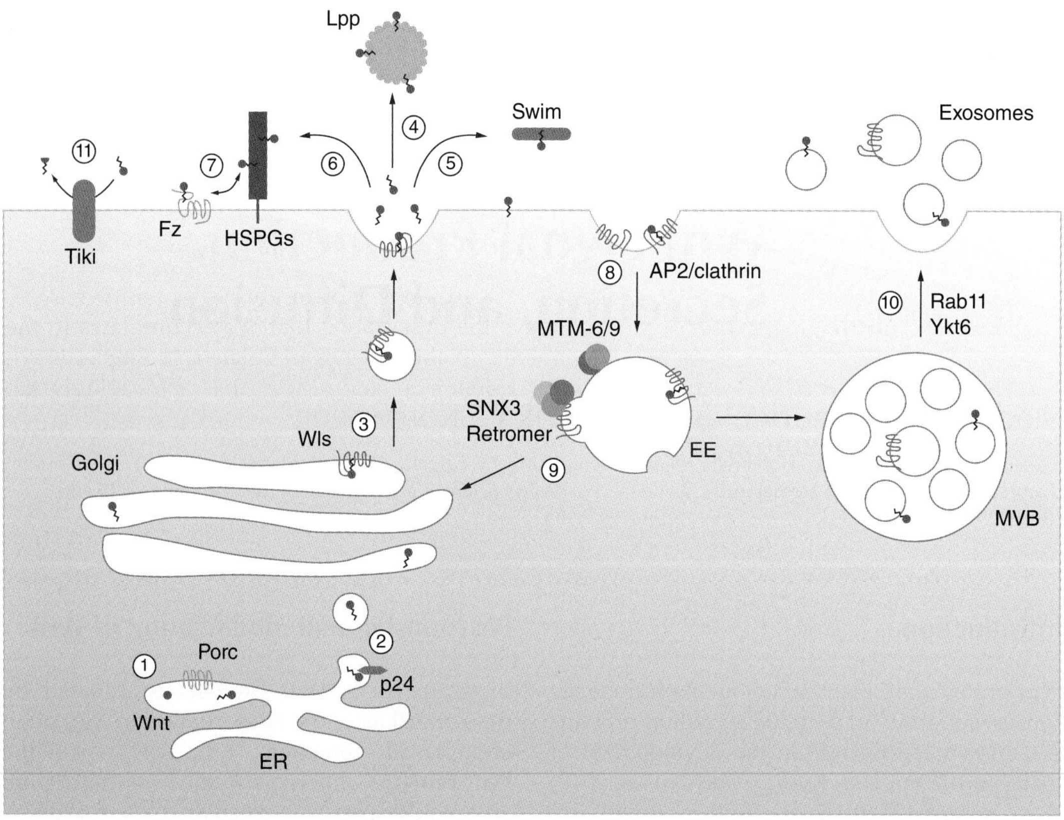

Figure 1.1 Wnt production, secretion, and spreading mechanisms. Wnt is lipid modified by Porc in the ER (1) and is transported to the Golgi through a p24-dependent mechanism (2). Next, Wnt binds the sorting receptor Wls, which transports Wnt to the cell surface (3). Release and diffusion of Wnt is facilitated by binding to lipoprotein particles (4), the lipocalin Swim (5), or HSPGs such as Dally and Dlp (6). HSPGs may also function as coreceptors that promote binding of Wnt to the Fz receptor (7). Wls is recycled from the plasma membrane through AP2-/clathrin-mediated endocytosis (8) and retromer-dependent endosome to Golgi retrieval (9). Wls and Wnt can also be internalized on intraluminal vesicles and be secreted on exosomes (10). The activity of secreted Wnt is modulated by Tiki (11). ER, endoplasmic reticulum; EE, early endosome; MVB, multivesicular body.

N-terminal cysteine and a palmitoleic acid at an internal serine, exemplified by the Cys93 and Ser239 residues, respectively, in the *Drosophila* Wnt family member Wingless (Wg) (Doubravska *et al.*, 2011; Galli *et al.*, 2007; Kurayoshi *et al.*, 2007; Miura and Treisman, 2006; Takada *et al.*, 2006; Willert *et al.*, 2003). Genetic evidence strongly suggests that the ER-resident multi-span *O*-acyltransferase Porcupine (Porc) is responsible for the acylation of Wnts (Zhai, Chaturvedi, and Cumberledge, 2004). *Porc* mutants show ER accumulation of Wnts, disrupted secretion, and reduced hydrophobicity of Wnt proteins (Zhai, Chaturvedi, and Cumberledge, 2004). Thus, Porc is required for the lipidation of Wnt proteins and subsequently drives ER exit and entry of Wnts into the secretory pathway. To understand how Wnt acylation impacts on protein function, Wnt mutants that lack the cysteine and serine acyl attachment sites were used in both cell culture and developmental studies (Tang *et al.*, 2012). While palmitoleic acid modification at Ser239 was found essential for Wnt secretion and signaling (Tang *et al.*, 2012), the palmitate at Cys93 appeared of less importance in the regulation of Wnt signaling *in vivo* (Tang *et al.*, 2012). Intriguingly, the analogous *N*-terminal cysteine (Cys55) in the XWnt8 crystal structure is engaged in the formation of a disulfide bond that is predicted to be conserved across all Wnts (Janda *et al.*, 2012). Thus, phenotypes observed for Cys-to-Ala mutants may have resulted from conformational alterations in the protein, due to the absence of this important disulfide bond. The essential signaling role of the acyl modification of Ser was confirmed by the structure as this lipid moiety directly engaged a groove on the extracellular cysteine-rich domain (CRD) of Fz8 (Janda *et al.*, 2012).

The Wnt secretion pathway

Once the Wnt protein is lipid modified and glycosylated, it is transported to the cell surface for release. Current evidence suggests that this is mediated through a specialized trafficking pathway (Lorenowicz and Korswagen, 2009; Port and Basler, 2010) and that different release mechanisms may contribute to the formation of distinct pools of Wnt that have different signaling activities in the tissue (Beckett *et al.*, 2013; Gross *et al.*, 2012; Panakova *et al.*, 2005).

p24 proteins mediate ER to Golgi transport of Wnt

The first leg in the journey of Wnt to the cell surface is transport from the ER to the Golgi network (Figure 1.1, step 2). It has recently been shown that members of the p24 cargo adaptor family play a central role in this trafficking step (Buechling *et al.*, 2011; Port, Hausmann, and Basler, 2011). Using large-scale RNA interference (RNAi) screens, it was found that the knockdown of the p24 family members Éclair, Emp24, and Opossum leads to a reduction in Wg secretion from Schneider 2 (S2) cells. Further experiments showed that these p24 family members are also required for Wg secretion *in vivo*. Thus, the knockdown of Éclair, Emp24, and Opossum in the wing imaginal disc resulted in the accumulation of Wg in producing cells, a reduction in target gene expression, and defects in the formation of wing margin tissue, a hallmark of defective Wg signaling. Importantly, Éclair, Emp24, and Opossum were not required for general protein secretion, as the ER export of the heparan sulfate proteoglycans (HSPGs) Dally and Dally-like (Dlp) and the Wnt secretion factor Wntless (Wls) (see succeeding text) were not affected.

p24 proteins have been proposed to function as cargo-specific adaptors that facilitate the sorting of cargo proteins into COPII vesicles, a class of transport carriers that mediate the trafficking of cargo proteins from the ER to the Golgi (Castillon *et al.*, 2011). Emp24 and Opossum interact with Wg in coimmunoprecipitation experiments, indicating that they may have a similar cargo adaptor function towards Wnt proteins. In support of such a role is the observation that Wg accumulates in the ER in the absence of Éclair and Emp24 (Port, Hausmann, and Basler, 2011). Interestingly, the secretion of the non-lipid-modified *Drosophila* Wnt protein WntD is dependent on Opossum as well (Buechling *et al.*, 2011). The role of Opossum in WntD secretion may, however, be indirect, as WntD does not bind to Opossum in coimmunoprecipitation experiments.

The Wnt binding protein Wntless is essential for Wnt secretion

The second stage in the transport of Wnt to the cell surface is mediated by Wls (Banziger *et al.*, 2006; Bartscherer *et al.*, 2006; Goodman *et al.*, 2006), a protein that is also known as Evi or Sprinter in *Drosophila*, MIG-14 in *Caenorhabditis elegans* (Banziger *et al.*, 2006), and GPR177 in the mouse (Fu *et al.*, 2009). Like Porc, Wls is essential for Wnt secretion. In the *Drosophila* wing imaginal disc, for example, the loss of *wls* leads to the accumulation of Wg in producing cells and a strong reduction in Wg signaling (Banziger *et al.*, 2006; Bartscherer *et al.*, 2006), and also in the mouse and in *C. elegans*, mutation of *Gpr177* and *mig-14* disrupts Wnt signaling (Fu *et al.*, 2009; Harris *et al.*, 1996; Thorpe *et al.*, 1997; Yang *et al.*, 2008). This function appears to be specific to Wnt, as the loss of Wls does not affect general protein secretion or the release of the related lipid-modified morphogen Hedgehog (Hh) (Banziger *et al.*, 2006). Wls encodes a highly conserved seven-pass transmembrane protein that binds Wnt in coimmunoprecipitation experiments (Banziger *et al.*, 2006; Bartscherer *et al.*, 2006). This interaction requires Porc and the conserved lipid-modified internal serine residue in Wnt proteins (Ser239 in Wg and Ser209 in Wnt3a), indicating that the addition of a palmitoleic acid chain at this position is essential for binding to Wls (Coombs *et al.*, 2010; Herr and Basler, 2012). This is consistent with the observation that the non-lipid-modified Wnt protein WntD does not bind to Wls and also does not require Wls for its secretion (Ching, Hang, and Nusse, 2008; Herr and Basler, 2012). The region of Wls that is necessary for Wnt binding is contained within the first extracellular loop (Fu *et al.*, 2009). Interestingly, this region is predicted to share structural similarities with proteins of the lipocalin family, which bind lipid moieties of proteins to enable their extracellular transport. An analogous interaction mechanism thus may be involved in the Wls–Wnt interaction (Coombs *et al.*, 2010). What is the function of Wls in the Wnt secretion pathway? Endogenous as well as fluorescently tagged versions of Wls show a prominent localization to the Golgi network, the endosomes, and the plasma membrane (Belenkaya *et al.*, 2008; Franch-Marro *et al.*, 2008; Port *et al.*, 2008; Yang *et al.*, 2008). Taken together with the observation that Wg accumulates in the Golgi of *wls* mutant cells (Port *et al.*, 2008), these findings led to the hypothesis that Wls functions as a sorting receptor that facilitates trafficking of Wnt from the Golgi to the cell surface for release (Lorenowicz and Korswagen, 2009; Port and Basler, 2010).

Wntless is recycled to maintain efficient Wnt secretion

Once Wls reaches the plasma membrane, it is internalized and retrieved back to the Golgi to take part in further rounds of Wnt secretion (Belenkaya *et al.*, 2008; Franch-Marro *et al.*, 2008; Pan *et al.*, 2008; Port *et al.*, 2008; Yang *et al.*, 2008). Mutations that interfere with this recycling induce strong defects in Wnt signaling, indicating that Wls is a limiting component in the pathway that needs to be recycled to maintain efficient Wnt secretion. In *Drosophila*, the expression of Wls is independent of Wnt signaling (Herr and Basler, 2012). Interestingly, the mouse Wls ortholog *Gpr177* is a direct Wnt target gene, indicating that mammalian Wnt proteins may stimulate their own secretion by upregulating Wls expression (Fu *et al.*, 2009). However, also in mammalian cells, interfering with Wls retrieval induces defects in Wnt secretion (Belenkaya *et al.*, 2008), indicating that despite this potential positive feedback, Wls recycling is still necessary for efficient Wnt secretion.

The first step in the recycling of Wls is internalization from the plasma membrane, which is mediated through AP2 adaptin- and clathrin-dependent endocytosis (Figure 1.1, step 8) and requires a conserved YXXΦ sorting motif that is present in the third intracellular loop of Wls (Gasnereau *et al.*, 2011; Pan *et al.*, 2008; Yang *et al.*, 2008).

Next, Wls is retrieved from the endosomal system and is transported back to the Golgi (Figure 1.1, step 9) through a retromer-dependent trafficking pathway (Belenkaya *et al.*, 2008; Franch-Marro *et al.*, 2008; Port *et al.*, 2008; Yang *et al.*, 2008). The retromer is a multisubunit membrane coat complex that mediates the retrograde transport of cargo proteins such as the cation-independent mannose 6-phosphate receptor (CI-MPR) from endosomes to the *trans*-Golgi network (TGN) (Cullen and Korswagen, 2012; Seaman, 2005). The retromer consists of a stable trimer of the subunits Vps26, Vps29, and Vps35 that binds to a loosely defined sorting signal in the cargo protein and a membrane-bound heterodimer of the SNX–BAR sorting nexins SNX1/2 and SNX5/6. The SNX–BAR sorting nexins contain a membrane curvature-sensing BAR domain that drives the formation of membrane tubules into which cargo proteins such as the CI-MPR are sorted. Scission of these tubules generates transport carriers that deliver the cargo back to the TGN.

A role of the retromer in Wnt signaling was first discovered in *C. elegans*, where mutations in the cargo-selective subunits were found to induce a range of Wnt-related phenotypes (Coudreuse *et al.*, 2006; Prasad and Clark, 2006). Subsequent studies showed that this function is evolutionarily conserved and that the retromer is required in Wnt-producing cells for Wls retrieval. The Vps26, Vps29, and Vps35 trimer binds to Wls in coimmunoprecipitation experiments (Belenkaya *et al.*, 2008; Franch-Marro *et al.*, 2008), and in the absence of retromer function, Wls fails to be retrieved from the endosomal system and is degraded in lysosomes (Belenkaya *et al.*, 2008; Franch-Marro *et al.*, 2008; Port *et al.*, 2008; Yang *et al.*, 2008). Interestingly, it was found that the endosome to TGN transport of Wls is independent of the

SNX–BAR sorting nexins (Harterink *et al.*, 2011). Instead, Wls retrieval requires the unrelated sorting nexin SNX3, which sorts Wls into vesicular transport carriers that are morphologically distinct from the tubular carriers formed by the SNX–BAR sorting nexins. SNX3 is recruited to endosomal membranes through a phosphatidylinositol 3-monophosphate (PI3P)-binding PX domain. In *C. elegans*, this endosomal association is regulated by the myotubularin lipid phosphatases MTM-6 and MTM-9, and MIG-14/Wls retrieval is strongly disrupted in their absence (Silhankova *et al.*, 2010). Why Wls retrieval is mediated through a specialized retromer pathway remains to be established.

Release of Wnt from producing cells

The first step in the release of Wnt is dissociation from Wls. This was shown to be dependent on vacuolar acidification, most likely of the secretory vesicles that transport the Wnt–Wls complex to the cell surface (Coombs *et al.*, 2010). When vacuolar acidification is blocked, Wnt3a still reaches the cell surface, but is not released from Wls into the medium. Interestingly, a decrease in pH is not sufficient for the dissociation of Wnt from Wls, indicating that additional mechanisms are required for Wnt release. Such mechanisms may involve binding of Wnt to HSPGs on the surface of Wnt-producing cells or may require the presence of specific carriers.

A role for carriers in Wnt secretion and diffusion was first proposed based on the observation that Wg colocalizes with punctate structures in the *Drosophila* wing imaginal disc (Greco, Hannus, and Eaton, 2001). These so-called argosomes, which derive from the Wg-producing cells, were proposed to act as vehicles for Wg diffusion (Figure 1.1, step 4). Although the exact nature of these argosomes remains unknown, subsequent studies have provided evidence that they may represent lipoprotein particles or exosomes.

Density gradient centrifugation experiments with *Drosophila* larval extracts showed that Wg and Hh cofractionate with lipophorin (Panakova *et al.*, 2005). Lipophorin is a component of lipoprotein particles, structures that consist of apolipoproteins and a phospholipid monolayer that surround a core of esterified cholesterol and triglycerides. Lipoprotein particles act as lipid carriers and would therefore be ideally suited to facilitate diffusion of Wg and Hh, which may bind to the particles through insertion of their fatty acid and cholesterol tails into the lipid core. Consistent with such a role, it was found that Wg colocalizes with lipophorin in the wing disc and that the knockdown of lipophorin interferes with Wg gradient formation, resulting in reduced expression of the long-range target gene *distal-less (dll)* (Panakova *et al.*, 2005). In addition to promoting diffusion, lipoprotein particles may also have a role in the release of Wnt from mouse fibroblast L cells, a cell line commonly used for mammalian Wnt secretion (Willert *et al.*, 2003; Neumann *et al.*, 2009). Thus, the secretion of Wnt3a from L cells into the medium requires the presence of low-density lipoprotein (LDL) and especially high-density lipoprotein (HDL) particles (Neumann *et al.*, 2009). Interestingly, it was found that the release of Wnt3a from LDL receptor mutant Chinese hamster ovary (CHO) cells was strongly stimulated by the expression of the HDL receptor SR-BI/II, indicating that the SR-BI/II receptor may stimulate Wnt secretion by binding and releasing HDL particles. However, this does not appear to be a general mechanism, as the knockdown of SR-BI/II did not interfere with Wnt secretion from L cells (Neumann *et al.*, 2009).

A study on Wg signaling in the neuromuscular junction of *Drosophila* revealed that Wls and Wg are present on small vesicles that traverse the synaptic cleft (Korkut *et al.*, 2009). Interestingly, it was found that these vesicles are also formed by S2 cells and can be transferred between cells in tissue culture. Recently, three separate studies have shown that these vesicles are exosomes (Figure 1.1, step 10) (Beckett *et al.*, 2013; Gross *et al.*, 2012; Koles *et al.*, 2012), but their function in Wnt release and signaling remains unclear.

Exosomes are small vesicles that are secreted from cells when multivesicular bodies fuse with the plasma membrane and release their content of intraluminal vesicles into the extracellular space (Simons and Raposo, 2009). Both Wls and Wg can be purified together with exosomes from the culture medium of Wg-expressing S2 cells (Beckett *et al.*, 2013; Gross *et al.*, 2012; Koles *et al.*, 2012). This exosome fraction of Wg is active but represents only part of the total amount of Wg present in the medium, indicating that secretion on exosomes acts in parallel to other release mechanisms. What is the function of this exosome-associated pool of Wnt? On this topic, disagreement is apparent between the three studies. The Vincent group found no evidence for the secretion of Wls on exosomes in the *Drosophila* wing imaginal disc and also found that blocking the formation of Wls-containing exosomes by interfering with the small GTPase Rab11 (Beckett *et al.*, 2013; Koles *et al.*, 2012) had no effect on Wg signaling in this tissue (Beckett *et al.*, 2013). In contrast, the Boutros group did observe colocalization of Wls and Wg with punctate structures labeled with the exosomal marker CD63/GFP in the wing disc (Gross *et al.*, 2012). Furthermore, they found that the inhibition of exosome secretion by knocking down the SNARE Ykt6 resulted in a reduction in Wg target gene expression and Wg loss of function phenotypes such as the loss of wing margin tissue. Taken together, these studies clearly show that Wnt proteins can be secreted on exosomes, but the *in vivo* role of this secretion mechanism in Wnt signaling needs to be further established.

How is the secretion of Wnt and Wls on exosomes related to the Golgi retrieval of Wls that we discussed in the previous section? An interesting possibility is that the exosome pathway acts in parallel to other release mechanisms. In such a scenario, Wnt binding may determine whether Wls is recycled or secreted through the exosome pathway. Thus, the pool of Wls that has released Wnt at the plasma membrane will be recycled, while Wnt-bound Wls may be shunted into the exosome pathway to generate a pool of Wnt with potential long-range signaling activity (Beckett *et al.*, 2013; Gross *et al.*, 2012).

Mechanisms that promote and control the diffusion of Wnt

Wnts mediate short- and long-range signaling

Wnt-producing cells can signal to directly neighboring cells but also to cells that are located at a distance. Short-range signaling occurs via direct cell–cell contact between the Wnt-producing cell and the signal-receiving cell. This type of Wnt-mediated cell communication is exemplified in the crypts of Lieberkühn of the small intestine, where differentiated Paneth cells directly present Wnt3 and other growth factors to sustain the adjacent stem cells (Sato *et al.*, 2011). During embryonic development, Wnt signals are also communicated over longer distances to mediate tissue pattern formation. In these processes, Wnts act as morphogens by forming a gradient of extracellular protein to drive the activation of specific gene programs and cellular responses in a concentration-dependent manner (Strigini and Cohen, 2000; Vincent and Briscoe, 2001). The question of how the lipid-modified Wnts can be transported over long distances has been an intensely debated subject. Accumulating evidence suggests that the mechanisms of secretion and extracellular transport of Wnts differ between short- and long-range signaling (Bartscherer and Boutros, 2008; Coudreuse and Korswagen, 2007).

A number of factors that act at the interplay of proteins and lipids were implicated in long-range rather than short-range signaling. First, the association of secreted Wnts with lipoprotein particles promotes long-range signaling but leaves the expression of short-range target genes unaffected (Panakova *et al.*, 2005). Of note, only a minor fraction of secreted Wnts was associated with lipoprotein particles in these studies (Panakova *et al.*, 2005). These findings suggest that a small pool of Wnts destined for long-range signaling may require selective packaging.

A second factor implicated in the secretion and spreading of Wg in *Drosophila* wing discs is the membrane microdomain-forming component Reggie-1/flotillin-2 (Katanaev *et al.*, 2008). Both Reggie-1/flotillin-2 and Reggie-2/flotillin-1 isoforms tightly bind the inner leaflet of the plasma membrane where they associate and polymerize to define specific microdomains in the plasma membrane (Otto and Nichols, 2011). A number of activities have been assigned to these proteins, including the regulation of endocytosis, signal transduction, and modulation of the cortical cytoskeleton. In Wnt-producing cells, Reggie-1 appears to be specifically required to generate and release a mobile form of Wnt that spreads efficiently into the tissue to mediate long-range target gene expression (Katanaev *et al.*, 2008). Reggie-1 also promoted secretion and spreading of the lipid-modified morphogen Hh, while other secreted factors such as Dpp and a GPI-linked form of GFP remained unaffected. The mechanism by which Reggie-1 activity contributes to the generation and secretion of Wnts remains unresolved. It is plausible that Reggie-1 facilitates trafficking or perhaps incorporation of Wg proteins in lipoprotein particles or exosomes. The level of conservation of this mechanism needs further investigation. As Reggie-1 or Reggie-2 homologs are absent in *C. elegans*, it will be interesting to determine if other proteins with similar microdomain-organizing activity may facilitate the extracellular mobility of Wnts in this organism.

Another protein called secreted Wingless-interacting molecule (Swim) was recently shown to promote long-range Wnt signaling (Mulligan *et al.*, 2012) (Figure 1.1, step 5). Swim was identified in direct association with secreted Wg and significantly potentiated cellular responses to Wg. *Swim* RNAi experiments in *Drosophila* wing discs revealed no effect of Swim on the levels of Wg secretion or short-range signaling but demonstrated its involvement in the spreading of Wg and long-range target gene activation. How does Swim facilitate Wg mobility and signaling? The Swim protein shares a motif with members of the lipocalin family, which commonly facilitate the extracellular transport of hydrophobic proteins by shielding their lipid components (Flower, 2000; Ganfornina *et al.*, 2000). Indeed, Swim binds Wg with high affinity and the interaction can be disrupted by palmitate in a dose-dependent manner, suggesting that Swim directly interacts with the lipid moiety on Wg (Mulligan *et al.*, 2012). Together, these results lead to a model in which Swim maintains extracellular Wnt in soluble form by binding to and shielding its lipid tail. Once the Swim–Wnt complex reaches its target cells, the lipid tail of Wnt will need to be transferred to a groove in the Fz extracellular domain for a productive interaction (Janda *et al.*, 2012). In agreement with this notion, Swim and Fz CRD were shown to compete for binding to Wg (Mulligan *et al.*, 2012). The question whether Swim plays a part in the formation of Wg-containing lipoprotein particles or exosomes or perhaps represents another parallel pathway for Wnt transport remains unknown.

In conclusion, Wnt proteins destined for distant signaling require interactions with selective membrane microdomains as well as lipid-binding transport proteins. Thus, accumulating evidence suggests that packing of Wnts in specialized carriers is necessary for their long-range transport in extracellular space. Additional investigation is required to solve the question of whether the identified factors act in parallel or in sequential molecular steps that involve production, packing, release, and transport of Wnts.

Roles of HSPG in Wnt gradient formation

Wnt morphogens form gradients of extracellular protein that trigger concentration-dependent cellular responses during tissue patterning (Strigini and Cohen, 2000; Vincent and Briscoe, 2001). How these gradients are formed, shaped, and maintained is of fundamental interest but remains poorly understood (Lander, 2007). *Drosophila* 3rd instar larval wing imaginal discs have provided a powerful model for studying Wnt/Wg gradient formation. In this tissue, Wg secretion is confined to a narrow strip of cells at the dorsoventral (DV) boundary. Secreted, extracellular Wg proteins subsequently diffuse through the adjacent tissue to form a concentration gradient. High Wg concentrations close to the producing cells induce high-threshold target genes such as *senseless* (*sens*), while lower Wg concentrations farther away from the source induce low-threshold genes such as *dll* (Cadigan *et al.*, 1998; Neumann and Cohen, 1997; Zecca, Basler, and Struhl, 1996).

The formation of a robust and stable morphogen gradient depends on an array of regulatory parameters, including rates of production, diffusion, retention, and endocytosis (Lander, 2007). Genetic approaches have identified HSPGs as major regulators of Wnt gradient formation and target gene activation in developing tissues (Hacker, Nybakken, and Perrimon, 2005; Lin, 2004) (Figure 1.1, step 6). HSPGs consist of a core protein that is heavily modified with heparan sulfate (HS), a type of glycosaminoglycan (GAG) (Sarrazin, Lamanna, and Esko, 2011). The highly negatively charged HS biopolymers can undergo an endless number of alterations in number, length, and modification of the sugar chains, giving rise to an enormous diversity. Secreted HSPGs perform roles in the extracellular matrix and in secretory vesicles, while membrane-bound HSPGs were implicated in the formation of Hh, bone morphogenetic protein (BMP), fibroblast growth factor (FGF), and Wnt morphogen gradients (Hacker, Nybakken, and Perrimon, 2005; Lin, 2004; Sarrazin, Lamanna, and Esko, 2011).

Membrane HSPGs are subdivided in two families, called glypicans and syndecans. Glypicans are anchored to the cell membrane via glycosylphosphatidylinositol (GPI). Mammals carry six glypican genes, *Drosophila* has two, and *C. elegans* at least one. Sequence homology between family members is limited, but all glypicans share 14 Cys residues and 2–3 GAG attachment sites to membrane-proximal regions of the protein. Syndecans are type I transmembrane proteins that carry up to five GAG attachment sites that mainly contain HS chains. In mammals, four syndecan genes have been identified while invertebrates carry only one.

What is the evidence for the role of membrane HSPGs in Wg gradient formation and signaling? The initial identification of HSPGs in Wnt gradient formation came from genetic screens that searched for genes involved in embryonic segment polarity in *Drosophila* (Binari *et al.*, 1997; Hacker, Nybakken, and Perrimon, 2005; Haerry *et al.*, 1997; Lin and Perrimon, 1999; Luders *et al.*, 2003; Selva *et al.*, 2001). All of the identified genes in these studies encoded for enzymes or nucleotide sugar transporters involved in GAG biosynthesis (Hacker, Lin and Perrimon, 1997). Striking similarities in phenotypes between genes in HS biosynthesis and those of Hh and Wg pathways were found. As the Hh and Wg pathways are strongly interlinked through a positive feedback loop at this developmental stage, it has remained difficult to distinguish selective effects of HSPGs on the individual pathways. Subsequent studies in larval stage *Drosophila* wing discs provided conclusive evidence for a role of HS biosynthesis in Wg signaling and distribution. Mutations in genes involved in the cytosol-to-Golgi transport of GAG building blocks (*slalom, sll*), the transfer of sulfate groups to GAGs (*sulfateless, sfl*), or the assembly of the GAG backbone (*Ext* class of genes: *ttv, botv, sotv*) led to a decrease in levels of extracellular Wg and an abrogation of high-threshold Wnt target gene expression (Baeg *et al.*, 2001; Han *et al.*, 2004; Luders *et al.*, 2003; Takei *et al.*, 2004).

These findings clearly implicate HS biosynthesis in Wg signaling, but what HSPG core proteins are involved? Two HSPGs of the glypican family, Dally and Dlp, were placed centrally to Wg signaling events in *Drosophila* embryonic epidermis and developing wing discs (Baeg *et al.*, 2001; Lin and Perrimon, 1999; Tsuda *et al.*, 1999). The expression of Dally is positively regulated by Wg signaling, yielding highest levels close to the DV boundary in wing discs, where Wg is produced (Fujise *et al.*, 2001; Han *et al.*, 2005). Dlp, on the other hand, is negatively regulated by Wg signaling yielding low levels in a 7–10 cell-wide strip spanning the expression domain of Wg and increasing expression towards the tail end of the gradient (Han *et al.*, 2005).

Dally binds and maintains Wg at the surface of cells within the range of the Wg gradient and shows genetic interaction with Wg signaling pathway components, and *dally* mutants display reduced extracellular Wg protein and wing margin defects (Han *et al.*, 2005; Lin and Perrimon, 1999). The combined evidence suggests that Dally acts as a classical coreceptor that binds Wg and facilitates its interaction with Fz receptors (Figure 1.1, step 7), leading to the activation of signaling and rapid degradation of the complex (Franch-Marro *et al.*, 2005; Han *et al.*, 2005; Lin and Perrimon, 1999).

In contrast, the role of Dlp has puzzled researchers due to its biphasic activity in Wg signaling. In wing discs, Dlp expression promotes Wg activity in the tail end of the gradient where Wg ligands are low and reduces Wg activity close to the DV boundary where ligands are high (Franch-Marro *et al.*, 2005; Hufnagel *et al.*, 2006; Kirkpatrick *et al.*, 2004; Kreuger *et al.*, 2004; Yan *et al.*, 2009). A number of studies have come up with explanations for this phenomenon. A consistent view is that Dlp captures Wg at the cell surface, prevents its degradation, and passes it on to neighboring cells, facilitating a unidirectional flow of Wg along the epithelial sheet to promote long-range signaling (Baeg *et al.*, 2001; Franch-Marro *et al.*, 2005; Hufnagel *et al.*, 2006; Yan *et al.*, 2009). But how does overexpressed Dlp inhibit short-range gene expression when overexpressed near the Wg-producing cells? Recent work suggests that the biphasic activity of Dlp does not depend on its GPI anchor and does not involve shedding of Wg-bound Dlp from the cell surface, as suggested previously (Gallet, Staccini-Lavenant, Therond, 2008; Kreuger *et al.*, 2004). Instead, Dlp either may compete with the Fz receptor for Wg binding or may retain Wg at the cell surface to promote its interaction with Fz, depending on the concentration ratio of Dlp, Wg, and Fz (Yan *et al.*, 2009). Thus, the net flow of Wg is determined by the relative levels of ligand, receptor, and Dlp. Interestingly, a Dlp variant that lacks HS chains (DlpΔHS) showed a similar biphasic activity as the wild-type protein, suggesting involvement of the core protein (Yan *et al.*, 2009). Indeed, the Dlp core protein interacted with Wg, and its modification with GAG chains further enhanced this interaction (Yan *et al.*, 2009). These findings suggest that besides the evident contributions of HS chains, the core proteins enhance specificity to the roles of HSPG in different signaling pathways.

Tiki abrogates Wnt activity via cleavage of the Wnt N-terminus

A number of well-described secreted and membrane proteins can antagonize Wnt activity in the extracellular space, either by preventing productive Wnt-receptor interactions or by inhibiting Wnt receptor maturation (Cruciat and Niehrs, 2013) (see Chapter 13). Recently, a unique and highly conserved novel negative feedback mechanism was identified in Wnt signaling that involves the membrane-bound metalloprotease Tiki (Zhang *et al.*, 2012). Tiki is a type I membrane protein that is induced by maternal Wnt signaling in the *Xenopus* head organizer region to drive anterior patterning via selective inhibition of the Wnt pathway. Tiki1 overexpression induced head enlargement (its name refers to the large-headed humanoid in Polynesian mythology), and knockdown led to anterior defects and loss of forebrain structures (Zhang *et al.*, 2012).

Important mechanistic insight came from experiments in which Tiki was coexpressed with Wnt3a in mouse L cells. While Wnt3a was secreted normally from Tiki-expressing L cells, the protein showed faster electrophoretic migration, exhibited strongly impaired activity, and failed to bind its cognate receptors Fz and Lrp6 (Zhang *et al.*, 2012). Edman amino acid sequencing revealed that Tiki induced cleavage of the eight most amino-terminal residues of Wnt3a (Figure 1.1, step 11). The purified Tiki ectodomain cleaved Wnt3a *in vitro* and its activity depended on bivalent metal ions, suggesting that Tiki is a metalloprotease. In phase-separation assays, the wild-type Wnt3a protein resided in hydrophobic detergent-solubilized fraction, while Tiki-modified Wnt3a partitioned exclusively in the aqueous phase (Zhang *et al.*, 2012). Thus, by cleaving the Wnt N-terminus, Tiki alters the hydrophobicity of the Wnt protein. Strikingly, Tiki did not hamper lipidation of Wnt3a, suggesting that the enhanced Wnt3a solubility is mediated via conformational rearrangements. Indeed, Tiki-cleaved Wnt3a (as well as ΔN-Wnt3a) formed large soluble oligomeric complexes that were brought about via oxidation-mediated formation of intermolecular disulfide bonds (Zhang *et al.*, 2012). How does the N-terminus of Wnt prevent oxidation–oligomerization? The recently solved structure of XWnt8 in complex with the XFz8 Cys-rich domain unfortunately does not reveal information on the orientation of the most N-terminal (and likely flexible) part of Wnt (Janda *et al.*, 2012). Possibly, the N-terminus folds back onto the secreted Wnt protein to stabilize disulfide bonds in a conformation that allows exposure of the lipid tail. N-terminal cleavage by Tiki protease would drive the formation of alternative oligomeric conformations of Wnt that bury the lipid inside and render the complex hydrophilic. Interestingly, Tiki displays specificity for a number of different Wnts but fails to cleave Wnt11 (Zhang *et al.*, 2012). This raises the question of how specificity of Tiki is regulated. Moreover, to what extent Wnts bound to the cell surface, exosomes, or lipoprotein particles are susceptible to Tiki cleavage and what is the role of Tiki in Wnt gradient formation remain important issues that await elucidation.

Conclusions and perspectives

The identification of specific cellular components that assist Wnt maturation and secretion has delivered essential new insights into the mechanisms that underlie the generation of active Wnt proteins. Clearly, the lipid moiety on Wnts is a critical factor in the regulation of ER exit, Golgi-to-plasma membrane trafficking, release from the cell surface, extracellular spreading, and signaling of Wnts. During its intra- and extracellular journey, a number of regulatory proteins (including Wls, Swim, lipoproteins, Fz) bind and control Wnt activity in a lipid-dependent manner. These findings raise important new questions. How are lipidated Wnts transferred between cellular membranes, lipoprotein particles, and Fz receptors? How does Wls facilitate these events? Emerging evidence further suggests that distinct pools of extracellular Wnts exist. How different Wnt pools composed of exosomes, lipoproteins, or Swim-bound complexes contribute to Wnt gradient formation has yet to be solved. In addition, what are the regulatory mechanisms that control the activity of these Wnt subsets? Do these Wnt pools interact equally well with extracellular factors such as HSPGs, Tiki, and Fz receptors? A complete understanding of these issues will require the integration of genetic, cell biological, and biochemical approaches.

References

Baeg, G.H., Lin, X., Khare, N. *et al.* (2001) Heparan sulfate proteoglycans are critical for the organization of the extracellular distribution of Wingless. *Development*, **128**, 87–94.

Banziger, C., Soldini, D., Schutt, C. *et al.* (2006) Wntless, a conserved membrane protein dedicated to the secretion of Wnt proteins from signaling cells. *Cell*, **125**, 509–522.

Bartscherer, K. and Boutros, M. (2008) Regulation of Wnt protein secretion and its role in gradient formation. *EMBO Reports*, **9**, 977–982.

Bartscherer, K., Pelte, N., Ingelfinger, D., and Boutros, M. (2006) Secretion of Wnt ligands requires Evi, a conserved transmembrane protein. *Cell*, **125**, 523–533.

Beckett, K., Monier, S., Palmer, L. *et al.* (2013) Drosophila S2 cells secrete wingless on exosome-like vesicles but the wingless gradient forms independently of exosomes. *Traffic*, **14**, 82–96.

Belenkaya, T.Y., Wu, Y., Tang, X. *et al.* (2008) The retromer complex influences Wnt secretion by recycling Wntless from endosomes to the trans-Golgi network. *Developmental Cell*, **14**, 120–131.

Binari, R.C., Staveley, B.E., Johnson, W.A. *et al.* (1997) Genetic evidence that heparin-like glycosaminoglycans are involved in wingless signaling. *Development*, **124**, 2623–2632.

Buechling, T., Chaudhary, V., Spirohn, K. *et al.* (2011) p24 proteins are required for secretion of Wnt ligands. *EMBO Reports*, **12**, 1265–1272.

Cadigan, K.M., Fish, M.P., Rulifson, E.J., and Nusse, R. (1998) Wingless repression of Drosophila frizzled 2 expression shapes the Wingless morphogen gradient in the wing. *Cell*, **93**, 767–777.

Castillon, G.A., Aguilera-Romero, A., Manzano-Lopez, J. *et al.* (2011) The yeast p24 complex regulates GPI-anchored protein transport and quality control by monitoring anchor remodeling. *Molecular Biology of the Cell*, **22**, 2924–2936.

Ching, W., Hang, H.C., and Nusse, R. (2008) Lipid-independent secretion of a Drosophila Wnt protein. *The Journal of Biological Chemistry*, **283**, 17092–17098.

Coombs, G.S., Yu, J., Canning, C.A. *et al.* (2010) WLS-dependent secretion of WNT3A requires Ser209 acylation and vacuolar acidification. *Journal of Cell Science*, **123**, 3357–3367.

Coudreuse, D. and Korswagen, H.C. (2007) The making of Wnt: new insights into Wnt maturation, sorting and secretion. *Development*, **134**, 3–12.

Coudreuse, D.Y., Roel, G., Betist, M.C. *et al.* (2006) Wnt gradient formation requires retromer function in Wnt-producing cells. *Science*, **312**, 921–924.

Cruciat, C.M. and Niehrs, C. (2013) Secreted and transmembrane wnt inhibitors and activators. *Cold Spring Harbor Perspectives in Biology*, **5** (3), a015081.

Cullen, P.J. and Korswagen, H.C. (2012) Sorting nexins provide diversity for retromer-dependent trafficking events. *Nature Cell Biology*, **14**, 29–37.

Doubravska, L., Krausova, M., Gradl, D. *et al.* (2011) Fatty acid modification of Wnt1 and Wnt3a at serine is prerequisite for lipidation at cysteine and is essential for Wnt signalling. *Cell Signal*, **23**, 837–848.

Flower, D.R. (2000) Beyond the superfamily: the lipocalin receptors. *Biochimica et Biophysica Acta*, **1482**, 327–336.

Franch-Marro, X., Marchand, O., Piddini, E. *et al.* (2005) Glypicans shunt the Wingless signal between local signalling and further transport. *Development*, **132**, 659–666.

Franch-Marro, X., Wendler, F., Guidato, S. *et al.* (2008) Wingless secretion requires endosome-to-Golgi retrieval of Wntless/Evi/Sprinter by the retromer complex. *Nature Cell Biology*, **10**, 170–177.

Fu, J., Jiang, M., Mirando, A.J. *et al.* (2009) Reciprocal regulation of Wnt and Gpr177/mouse Wntless is required for embryonic axis formation. *Proceedings of the National Academy of Sciences of the United States of America*, **106**, 18598–18603.

Fujise, M., Izumi, S., Selleck, S.B., and Nakato, H. (2001) Regulation of Dally, an integral membrane proteoglycan, and its function during adult sensory organ formation of Drosophila. *Developmental Biology*, **235**, 433–448.

Gallet, A., Staccini-Lavenant, L., and Therond, P.P. (2008) Cellular trafficking of the glypican Dally-like is required for full-strength Hedgehog signaling and Wingless transcytosis. *Developmental Cell*, **14**, 712–725.

Galli, L.M., Barnes, T.L., Secrest, S.S. *et al.* (2007) Porcupine-mediated lipid-modification regulates the activity and distribution of Wnt proteins in the chick neural tube. *Development*, **134**, 3339–3348.

Ganfornina, M.D., Gutierrez, G., Bastiani, M., and Sanchez, D. (2000) A phylogenetic analysis of the lipocalin protein family. *Molecular Biology and Evolution*, **17**, 114–126.

Gasnereau, I., Herr, P., Chia, P.Z. *et al.* (2011) Identification of an endocytosis motif in an intracellular loop of Wntless protein, essential for its recycling and the control of Wnt protein signaling. *The Journal of Biological Chemistry*, **286**, 43324–43333.

Goodman, R.M., Thombre, S., Firtina, Z. *et al.* (2006) Sprinter: a novel transmembrane protein required for Wg secretion and signaling. *Development*, **133**, 4901–4911.

Greco, V., Hannus, M., and Eaton, S. (2001) Argosomes: a potential vehicle for the spread of morphogens through epithelia. *Cell*, **106**, 633–645.

Gross, J.C., Chaudhary, V., Bartscherer, K., and Boutros, M. (2012) Active Wnt proteins are secreted on exosomes. *Nature Cell Biology*, **14**, 1036–1045.

Hacker, U., Lin, X., and Perrimon, N. (1997) The Drosophila sugarless gene modulates Wingless signaling and encodes an enzyme involved in polysaccharide biosynthesis. *Development*, **124**, 3565–3573.

Hacker, U., Nybakken, K., and Perrimon, N. (2005) Heparan sulphate proteoglycans: the sweet side of development. *Nature Reviews Molecular Cell Biology*, **6**, 530–541.

Haerry, T.E., Heslip, T.R., Marsh, J.L., and O'Connor, M.B. (1997) Defects in glucuronate biosynthesis disrupt Wingless signaling in Drosophila. *Development*, **124**, 3055–3064.

Han, C., Belenkaya, T.Y., Khodoun, M. *et al.* (2004) Distinct and collaborative roles of Drosophila EXT family proteins in morphogen signalling and gradient formation. *Development*, **131**, 1563–1575.

Han, C., Yan, D., Belenkaya, T.Y., and Lin, X. (2005) Drosophila glypicans Dally and Dally-like shape the extracellular Wingless morphogen gradient in the wing disc. *Development*, **132**, 667–679.

Harris, J., Honigberg, L., Robinson, N., and Kenyon, C. (1996) Neuronal cell migration in C. elegans: regulation of Hox gene expression and cell position. *Development*, **122**, 3117–3131.

Harterink, M., Port, F., Lorenowicz, M.J. *et al.* (2011) A SNX3-dependent retromer pathway mediates retrograde transport of the Wnt sorting receptor Wntless and is required for Wnt secretion. *Nature Cell Biology*, **13**, 914–923.

Herr, P. and Basler, K. (2012) Porcupine-mediated lipidation is required for Wnt recognition by Wls. *Developmental Biology*, **361**, 392–402.

Hufnagel, L., Kreuger, J., Cohen, S.M., and Shraiman, B.I. (2006) On the role of glypicans in the process of morphogen gradient formation. *Developmental Biology*, **300**, 512–522.

Janda, C.Y., Waghray, D., Levin, A.M. *et al.* (2012) Structural basis of Wnt recognition by Frizzled. *Science*, **337**, 59–64.

Katanaev, V.L., Solis, G.P., Hausmann, G. *et al.* (2008) Reggie-1/flotillin-2 promotes secretion of the long-range signalling forms of Wingless and Hedgehog in Drosophila. *The EMBO Journal*, **27**, 509–521.

Kirkpatrick, C.A., Dimitroff, B.D., Rawson, J.M., and Selleck, S.B. (2004) Spatial regulation of Wingless morphogen distribution and signaling by Dally-like protein. *Developmental Cell*, **7**, 513–523.

Koles, K., Nunnari, J., Korkut, C. *et al.* (2012) Mechanism of evenness interrupted (Evi)-exosome release at synaptic boutons. *The Journal of Biological Chemistry*, **287**, 16820–16834.

Komekado, H., Yamamoto, H., Chiba, T., and Kikuchi, A. (2007) Glycosylation and palmitoylation of Wnt-3a are coupled to produce an active form of Wnt-3a. *Genes to Cells*, **12**, 521–534.

Korkut, C., Ataman, B., Ramachandran, P. *et al.* (2009) Trans-synaptic transmission of vesicular Wnt signals through Evi/Wntless. *Cell*, **139**, 393–404.

Kreuger, J., Perez, L., Giraldez, A.J., and Cohen, S.M. (2004) Opposing activities of Dally-like glypican at high and low levels of Wingless morphogen activity. *Developmental Cell*, **7**, 503–512.

Kurayoshi, M., Yamamoto, H., Izumi, S., and Kikuchi, A. (2007) Post-translational palmitoylation and glycosylation of Wnt5a are necessary for its signalling. *The Biochemical Journal*, **402**, 515–523.

Lander, A.D. (2007) Morpheus unbound: reimagining the morphogen gradient. *Cell*, **128**, 245–256.

Lin, X. (2004) Functions of heparan sulfate proteoglycans in cell signaling during development. *Development*, **131**, 6009–6021.

Lin, X. and Perrimon, N. (1999) Dally cooperates with Drosophila Frizzled 2 to transduce Wingless signalling. *Nature*, **400**, 281–284.

Lorenowicz, M.J. and Korswagen, H.C. (2009) Sailing with the Wnt: charting the Wnt processing and secretion route. *Experimental Cell Research*, **315**, 2683–2689.

Luders, F., Segawa, H., Stein, D. *et al.* (2003) Slalom encodes an adenosine 3′-phosphate 5′-phosphosulfate transporter essential for development in Drosophila. *The EMBO Journal*, **22**, 3635–3644.

Miura, G.I. and Treisman, J.E. (2006) Lipid modification of secreted signaling proteins. *Cell Cycle*, **5**, 1184–1188.

Mulligan, K.A., Fuerer, C., Ching, W. *et al.* (2012) Secreted Wingless-interacting molecule (Swim) promotes long-range signaling by maintaining Wingless solubility. *Proceedings of the National Academy of Sciences of the United States of America*, **109**, 370–377.

Neumann, C.J. and Cohen, S.M. (1997) Long-range action of Wingless organizes the dorsal-ventral axis of the Drosophila wing. *Development*, **124**, 871–880.

Neumann, S., Coudreuse, D.Y., van der Westhuyzen, D.R. *et al.* (2009) Mammalian Wnt3a is released on lipoprotein particles. *Traffic*, **10**, 334–343.

Otto, G.P. and Nichols, B.J. (2011) The roles of flotillin microdomains—endocytosis and beyond. *Journal of Cell Science*, **124**, 3933–3940.

Pan, C.L., Baum, P.D., Gu, M. *et al.* (2008) Caenorhabditis elegans AP-2 and retromer control Wnt signaling by regulating mig-14/Wntless. *Developmental Cell*, **14**, 132–139.

Panakova, D., Sprong, H., Marois, E. *et al.* (2005) Lipoprotein particles are required for Hedgehog and Wingless signalling. *Nature*, **435**, 58–65.

Port, F. and Basler, K. (2010) Wnt trafficking: new insights into Wnt maturation, secretion and spreading. *Traffic*, **11**, 1265–1271.

Port, F., Hausmann, G., and Basler, K. (2011) A genome-wide RNA interference screen uncovers two p24 proteins as regulators of Wingless secretion. *EMBO Reports*, **12**, 1144–1152.

Port, F., Kuster, M., Herr, P. *et al.* (2008) Wingless secretion promotes and requires retromer-dependent cycling of Wntless. *Nature Cell Biology*, **10**, 178–185.

Prasad, B.C. and Clark, S.G. (2006) Wnt signaling establishes anteroposterior neuronal polarity and requires retromer in C. elegans. *Development*, **133**, 1757–1766.

Sarrazin, S., Lamanna, W.C., and Esko, J.D. (2011) Heparan sulfate proteoglycans. *Cold Spring Harbor Perspectives in Biology*, **3** (7), pii, a004952.

Sato, T., van Es, J.H., Snippert, H.J. *et al.* (2011) Paneth cells constitute the niche for Lgr5 stem cells in intestinal crypts. *Nature*, **469**, 415–418.

Seaman, M.N. (2005) Recycle your receptors with retromer. *Trends in Cell Biology*, **15**, 68–75.

Selva, E.M., Hong, K., Baeg, G.H. *et al.* (2001) Dual role of the fringe connection gene in both heparan sulphate and fringe-dependent signalling events. *Nature Cell Biology*, **3**, 809–815.

Silhankova, M., Port, F., Harterink, M. *et al.* (2010) Wnt signalling requires MTM-6 and MTM-9 myotubularin lipid-phosphatase function in Wnt-producing cells. *The EMBO Journal*, **29**, 4094–4105.

Simons, M. and Raposo, G. (2009) Exosomes—vesicular carriers for intercellular communication. *Current Opinion in Cell Biology*, **21**, 575–581.

Strigini, M. and Cohen, S.M. (2000) Wingless gradient formation in the Drosophila wing. *Current Biology*, **10**, 293–300.

Takada, R., Satomi, Y., Kurata, T. *et al.* (2006) Monounsaturated fatty acid modification of Wnt protein: its role in Wnt secretion. *Developmental Cell*, **11**, 791–801.

Takei, Y., Ozawa, Y., Sato, M. *et al.* (2004) Three Drosophila EXT genes shape morphogen gradients through synthesis of heparan sulfate proteoglycans. *Development*, **131**, 73–82.

Tang, X., Wu, Y., Belenkaya, T.Y. *et al.* (2012) Roles of N-glycosylation and lipidation in Wg secretion and signaling. *Developmental Biology*, **364**, 32–41.

Thorpe, C.J., Schlesinger, A., Carter, J.C., and Bowerman, B. (1997) Wnt signaling polarizes an early C. elegans blastomere to distinguish endoderm from mesoderm. *Cell*, **90**, 695–705.

Tsuda, M., Kamimura, K., Nakato, H. *et al.* (1999) The cell-surface proteoglycan Dally regulates Wingless signalling in Drosophila. *Nature*, **400**, 276–280.

Vincent, J.P. and Briscoe, J. (2001) Morphogens. *Current Biology*, **11**, R851–R854.

Willert, K. and Nusse, R. (2012) Wnt proteins. *Cold Spring Harbor Perspectives in Biology*, **4**, a007864.

Willert, K., Brown, J.D., Danenberg, E. *et al.* (2003) Wnt proteins are lipid-modified and can act as stem cell growth factors. *Nature*, **423**, 448–452.

Yan, D., Wu, Y., Feng, Y. *et al.* (2009) The core protein of glypican Dally-like determines its biphasic activity in wingless morphogen signaling. *Developmental Cell*, **17**, 470–481.

Yang, P.T., Lorenowicz, M.J., Silhankova, M. *et al.* (2008) Wnt signaling requires retromer-dependent recycling of MIG-14/Wntless in Wnt-producing cells. *Developmental Cell*, **14**, 140–147.

Zecca, M., Basler, K., and Struhl, G. (1996) Direct and long-range action of a Wingless morphogen gradient. *Cell*, **87**, 833–844.

Zhai, L., Chaturvedi, D., and Cumberledge, S. (2004) Drosophila Wnt-1 undergoes a hydrophobic modification and is targeted to lipid rafts, a process that requires porcupine. *The Journal of Biological Chemistry*, **279**, 33220–33227.

Zhang, X., Abreu, J.G., Yokota, C. *et al.* (2012) Tiki1 is required for head formation via Wnt cleavage-oxidation and inactivation. *Cell*, **149**, 1565–1577.

2 Wnt Signaling at the Membrane

Gary Davidson[1] and Christof Niehrs[2,3]

[1]Institute of Toxicology and Genetics, Karlsruhe Institute of Technology, Karlsruhe, Germany
[2]Molecular Embryology, Deutsches Krebsforschungszentrum, Heidelberg, Germany
[3]Faculty of Biology, Institute of Molecular Biology (IMB), University of Mainz, Mainz, Germany

Introduction

After Wnts have undergone synthesis, post-translational modification, packaging, and release from secreting cells, they must engage with specific cell surface receptors on nearby cells to elicit a cellular response. The 19 highly related Wnt ligands engage a diverse set of receptors and coreceptors to control a vast range of cellular functions. Wnt binding receptors and coreceptors include members of the Frizzled (FZD), LRP, Ror, Ryk, MuSK, and PTK7 families (Figure 2.1). The recent discovery of additional receptor classes for the R-spondin family of Wnt agonists adds to the already complex set of mechanisms that regulates Wnt signaling on the membrane of responding cells.

Because Wnt ligands associate with a variety of receptors, they can activate different downstream pathways. Wnt signaling pathways have been classified as either β-catenin dependent (Wnt/β-catenin signaling) or β-catenin independent, which includes Wnt/PCP and Wnt/Ca^{2+} (see Chapter 6). Although specific Wnts preferentially activate β-catenin-dependent or β-catenin-independent pathways, they act according to cellular context as well as receptor profiles and hence cannot be rigorously classified. It is nevertheless an approximation that, for example, Wnt1, Wnt3a, and Wnt8 activate Wnt/β-catenin signaling, while Wnt5a and Wnt11 are predominantly associated with β-catenin-independent signaling. At least 15 different receptors direct the downstream Wnt pathway engaged, and deciphering this combinatorial code of Wnt-receptor coupling remains a major challenge in the field. In general, for β-catenin-dependent signaling, Wnt/FZD forms a ternary complex with LRP5/6, while for β-catenin-independent signaling, Ror or PTK7 is used in place of LRP5/6. The reader is referred to Chapter 7 for a more detailed account of how Wnt signals are routed to either the β-catenin-dependent or β-catenin-independent pathways.

Here, we provide a detailed overview of the principal receptors that transduce Wnt/β-catenin signals, namely, the FZD family and two specialized members of the low-density lipoprotein receptor family, LRP5 and LRP6. We focus on their interaction with Wnt ligands; their regulation by extracellular, transmembrane, as well as intracellular modifiers; and their aggregation into activated receptor platforms referred to as

Wnt Signaling in Development and Disease: Molecular Mechanisms and Biological Functions, First Edition.
Edited by Stefan Hoppler and Randall T. Moon.

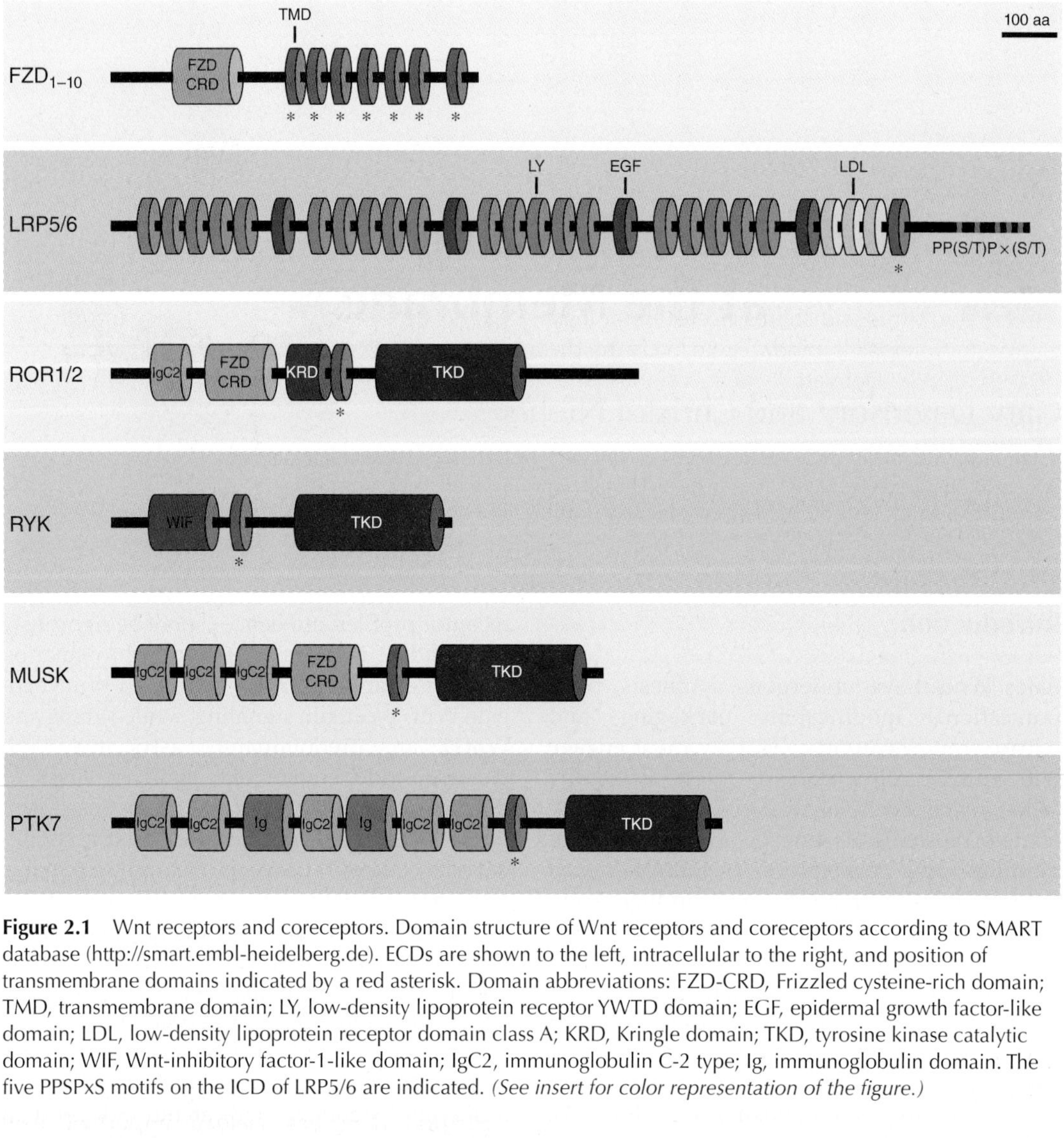

Figure 2.1 Wnt receptors and coreceptors. Domain structure of Wnt receptors and coreceptors according to SMART database (http://smart.embl-heidelberg.de). ECDs are shown to the left, intracellular to the right, and position of transmembrane domains indicated by a red asterisk. Domain abbreviations: FZD-CRD, Frizzled cysteine-rich domain; TMD, transmembrane domain; LY, low-density lipoprotein receptor YWTD domain; EGF, epidermal growth factor-like domain; LDL, low-density lipoprotein receptor domain class A; KRD, Kringle domain; TKD, tyrosine kinase catalytic domain; WIF, Wnt-inhibitory factor-1-like domain; IgC2, immunoglobulin C-2 type; Ig, immunoglobulin domain. The five PPSPxS motifs on the ICD of LRP5/6 are indicated. *(See insert for color representation of the figure.)*

signalosomes. Extracellular regulators include six types of secreted Wnt antagonists (sFRP, Dkk, WIF, Wise/SOST, Cerberus, IGFBP) that play important roles in modulating Wnt signaling under various physiological and pathological settings. Norrin is the only non-Wnt extracellular agonist known to interact with Wnt receptors, whereas the R-spondins are a distinct family of secreted Wnt agonists that signal through a different set of receptors to activate Wnt signaling. For intracellular modification of Wnt receptors, phosphorylation events appear to be the key mediators of receptor activation, at least for LRP5/6. Ubiquitinylation (Ub) of receptors is, however, increasingly recognized as an important regulatory modification. We start to build a consensus overview of how all these molecules interact and are regulated at the level of the membrane and which pathways they engage.

The Frizzled family of Wnt receptors

General background

FZDs are seven-pass-transmembrane proteins initially identified as regulators of tissue polarity in *Drosophila* (Vinson and Adler, 1987). In 1996, it was demonstrated that FZD overexpression conferred responsiveness to Wg signaling in otherwise unresponsive *Drosophila* cells (Bhanot *et al.*, 1996) and soon after a vertebrate

FZD was also identified as a Wnt receptor (Yang-Snyder *et al.*, 1996). This was a milestone in the history of Wnt signaling and led to an explosion of interest and new research in the field. In addition to Wnt proteins, the only other ligand so far known to bind with high affinity to FZD proteins is the product of the Norrie disease-linked gene, Norrin. Like Wnt proteins, Norrin is a secreted ligand with limited extracellular diffusion properties but is otherwise unrelated to Wnts and shares closest homology to TGFb proteins. It binds exclusively to the CRD of FZD4 to activate β-catenin signaling in an LRP5/6-dependent manner and is required for correct retinal vascularization (Smallwood *et al.*, 2007; Xu *et al.*, 2004). Although FZD proteins were the first Wnt receptors identified, are the most widely expressed in tissues, and likely contribute to all known Wnt signaling, many questions remain with respect to the biochemistry of how they are activated upon Wnt binding. In this respect, LRP5/6 are better understood (discussed later). There are 10 FZD proteins in the human and mouse genome (FZD 1–10), four in *Drosophila* (Fz, Dfz2, Dfz3, Dfz4), and four in *Caenorhabditis elegans* (MOM-5, LIN-17, CFZ-2, MIG-1) (Schulte, 2010). The Hedgehog (Hh) pathway receptor Smoothened (Smo) is highly related to FZD and is classified as the 11th member of the FZD family, although it does not function as a receptor for Wnt proteins (Schulte, 2010).

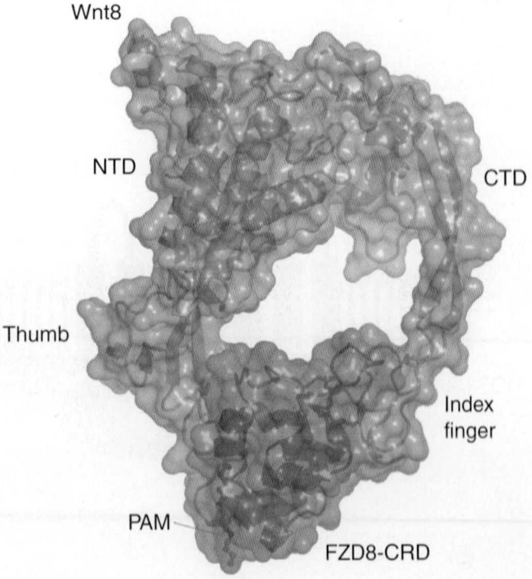

Figure 2.2 Structure of Wnt8 in complex with the extracellular CRD of Frizzled 8. Surface representation model developed from crystal structure data of *Xenopus* Wnt8 in complex with a CRD of mouse Frizzled 8 (FZD8-CRD). The palmitoleic acid (PAM) group is shown in red, extending from the tip of the Wnt thumb and passing through a hydrophobic groove in the FZD-CRD. Wnt8 is shown in pink with its N-terminal domain (NTD) and C-terminal domain (CTD), from which the thumb and index finger regions protrude. The figure was kindly supplied by, and with permission to publish from, Chris Garcia and Claudia Janda, Howard Hughes Medical Institute, Stanford University School of Medicine, Stanford. *(See insert for color representation of the figure.)*

Structural and functional features of Frizzled proteins

The topology of the transmembrane domains in FZD proteins resembles G protein-coupled receptors, and there is indeed evidence that FZD proteins can dimerize and couple to heterotrimeric G proteins (Koval *et al.*, 2011; see Chapter 14 for more details on this subject). In addition to the conserved TM domains, FZDs have a highly conserved cysteine-rich domain (FZD-CRD) that is located extracellularly at the N-terminus and binds Wnt (Figure 2.1 and Figure 2.3). Although the hydrophobic nature of Wnt molecules has hampered analysis of Wnt/FZD interaction, the long-awaited structure of Wnt in complex with its receptor has recently been solved. This was achieved after purification of a mouse FZD8-CRD-Fc fusion protein that was coexpressed with, and therefore bound to, *Xenopus* Wnt8 (Janda *et al.*, 2012). It was worth waiting for, since analysis of the structural data has yielded important new mechanistic insights into Wnt signaling at the membrane. First, there is a 1:1 interaction between a Wnt8 monomer and FZD8-CRD. Second, all cysteine residues in Wnt8 participate in intramolecular disulfide bonds, precluding additional posttranslational modifications such as acylation that could further regulate interaction. The third and most exciting aspect reveals a novel two-domain Wnt structure that forms "thumb and index fingerlike" extensions to grasp the FZD8-CRD at two sites (Figure 2.2). The hydrophobic palmitoyl group on Wnt extends from the tip of the N-terminal thumb domain and fits into a hydrophobic groove in the FZD8-CRD. The conserved nature of residues on both Wnts and FRZs that contribute to this

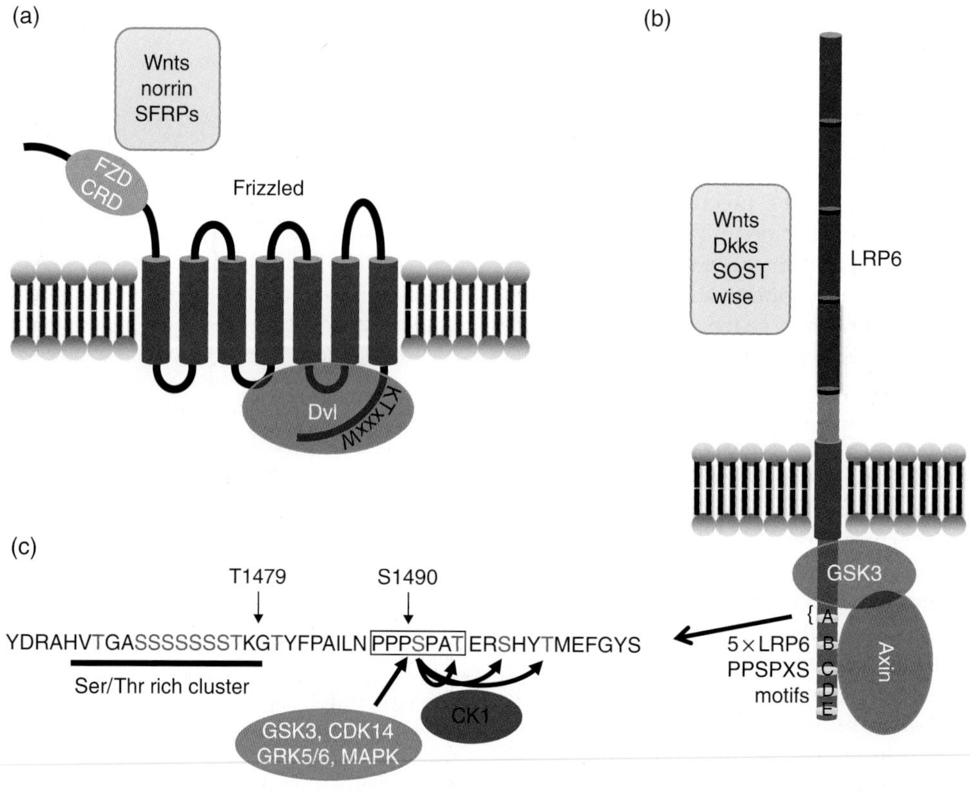

Figure 2.3 FZD and LRP6 and their interaction partners. (a) Extracellularly, Wnt, Norrin, and sFRP proteins bind the FZD-CRD of FZD receptors, whereas intracellularly, and in a Wnt-dependent manner, Dvl interacts with a KTxxxW motif on the C-Terminal domain and two additional motifs within the third intracellular loop of FZD. (b) Wnts, Dkks (Dkk1, 2, and 4), Wise, and SOST bind to the extracellular EGF/YWTD repeats of LRP6, whereas Axin/GSK3 interact with phosphorylated motifs on the ICD of LRP6. (c) Short region of LRP6 ICD spanning the Ser/Thr-rich cluster (important for GSK3 interaction) and PPSPxS motif A (boxed). The dual nature of PPSPxS site phosphorylation is illustrated, where both PPSP site kinases (GSK3, CDK14, GRK5/6, MAPK) and CK1 contribute to phosphorylation of the motifs to create Axin binding sites. Within PPSPxS motif A, the priming of CK1 site phosphorylation by prior PPSP site phosphorylation is also illustrated. Phosphospecific antibodies recognizing S1490 and T1479 are commonly used to detect PPSP and CK1 site phosphorylation of LRP6, respectively. *(See insert for color representation of the figure.)*

binding interface, as well as the conserved nature of Wnt palmitoylation itself, provides an explanation for the promiscuous nature of Wnt/FZD interactions between family members. The C-terminal "index finger" domain of Wnt also engages in hydrophobic contacts, although without lipid groups, and contact residues are highly conserved in all Wnts. The Wnt finger, however, touches a depression in FZD8-CRD that harbors incompletely conserved residues between different FZDs and may therefore contribute to specifying distinct Wnt/FZD binding preferences. Indeed, since Wnts form ternary complexes consisting of Wnt, FZD, and a coreceptor such as LRP5/6, ROR1/2, and PTK7, the protruding surface patch on the index finger of Wnt8 may govern such additional interactions. In this respect, the overall flexibility of the Wnt index finger domain, when Wnt is engaged with FZD solely by the lipid thumb interaction, is important.

Frizzled CRD in non-Frizzled proteins

Apart from FZD receptors themselves, the five secreted FZD-related proteins (sFRP1 5) represent the largest family of proteins containing FZD-CRD domains. The founding member, frzb (sFRP3), was shown to function primarily as a secreted Wnt inhibitor because its CRD domain resembles that of FZD and therefore competes for Wnt binding in the extracellular environment (Leyns *et al.*, 1997; Lin *et al.*, 1997; Wang *et al.*, 1997). However, the CRDs of FZD and sFRP can also interact to form dimers, suggesting an additional mechanism of Wnt inhibition (Bafico *et al.*, 1999; Dann *et al.*, 2001; Rodriguez *et al.*, 2005). In addition to their FZD-CRD, sFRPs have Netrin-like domains at the C-terminus which are found in a number of unrelated proteins (Bányai and Patthy, 1999). There is evidence that both the FZD-CRD and the NTR domains of sFRPs are involved in Wnt binding (Bhat *et al.*, 2007). Interestingly, association of Wnt ligands to sFRPs can also promote Wnt signaling by enhancing their extracellular diffusion (Mii and Taira, 2009). The reader is referred to Chapter 13 for a more comprehensive overview of the sFRP family.

A FZD-CRD is also found on the extracellular region of ROR1/2 and MuSK (Xu and Nusse, 1998; Figure 2.1), Wnt receptors that are reported to interact with Wnt5a and Wnt11/Wnt4, respectively (Gordon *et al.*, 2012; Jing *et al.*, 2009; Oishi *et al.*, 2003). The only other proteins known to contain FZD-CRDs are the metalloprotease carboxypeptidase Z (CPZ) and collagen XVIII (Xu and Nusse, 1998). CPZ removes basic amino acids, particularly arginine residues, from the carboxyl terminus of proteins, and there is evidence that the FZD-CRD of CPZ associates with Wnt and regulates Wnt signaling (Moeller *et al.*, 2003; Reznik and Fricker, 2001). It may also remove the C-terminal arginine from Wnt4 in a thyroid hormone-dependent manner to activate Wnt signaling and chondrocyte differentiation (Wang, Shao, and Ballock, 2009). A more recent study has implicated carboxypeptidase E (CPE) in Wnt signaling, where it has been shown to form a complex with Wnt3a and the CRD of FZD1 and to function in intracellular sorting of endocytosed receptor complexes (Skalka *et al.*, 2012; see also Chapter 8). This is intriguing because, unlike CPZ, CPE does not contain an FZD-CRD. There is also evidence that the FZD-CRD module of collagen XVIII can bind to and inhibit Wnt signaling (Hendaoui *et al.*, 2012; Quélard *et al.*, 2008). It will be interesting to study these proteins in different cellular contexts to determine if they play more general roles in regulating Wnt signaling and the mechanisms employed.

Biochemical regulation of FZD proteins

FZD proteins were identified as high-affinity Wnt receptors almost two decades ago; however, many aspects of their biochemistry remain unclear. They have four intracellular domains (ICDs) that can potentially interact with the cytoplasmic machinery to regulate Wnt signaling. Three of these ICDs link successive transmembrane domains, and the fourth is located at the C-terminus which varies in length between different members. The best understood mechanism by which FZD transduces Wnt signals is Dvl recruitment. Dvl interacts with the intracellular region of FZD receptors upon Wnt activation, and this interaction is mediated by a highly conserved KTxxxW motif located in the C-terminal domain close to the membrane; however, additional residues in other ICDs may also be involved (Cong, Schweizer, and Varmus, 2004). Indeed, a recent report has shown that two Dvl domains (DEP and C-terminal, termed DEP-C) interact with a discontinuous structure in FZD consisting of two motifs within the third intracellular loop domain as well as the C-terminal KTxxxW motif (Tauriello *et al.*, 2012; Figure 2.3a).

There are putative Ser/Thr phosphorylation sites in the FZD ICD, although evidence linking specific phosphorylation events to activation is limited. The FZD-related Hh pathway receptor Smo is better characterized in this respect: Hh stimulation results in Smo phosphorylation at several cytoplasmic residues by protein kinase A (PKA) and casein kinase 1 (CK1), and phosphorylation at these sites is required for Smo activity (Jia *et al.*, 2004; Zhang *et al.*, 2004). It is therefore likely that similar phosphorylation events are triggered on FZD residues, and there is indeed evidence that Wnt/PCP signaling is negatively regulated by phosphorylation of *Drosophila* FZD1 and *Xenopus* FZD3 (Djiane, Yogev, and Mlodzik, 2005; Yanfeng *et al.*, 2006).

Ub and deubiquitinylation are known to regulate the balance between cell surface localization of receptors and their trafficking to lysosomes for degradation (Gruenberg and Stenmark, 2004; Haglund and Dikic, 2005; see also Chapter 8). FZDs are subject to regulatory Ub modifications (Mukai *et al.*, 2010), and the first FZD ubiquitinases were recently identified as the transmembrane RING finger E3 Ub ligases ZNRF3 and RNF43 (Hao *et al.*, 2012; Koo *et al.*, 2012; MacDonald and He, 2012). As predicted, ZNRF3 and RNF43 promote the removal of FZD receptors from the plasma membrane, followed by targeted degradation in the lysosomal compartment, leading to the inhibition of both β-catenin-dependent and β-catenin-independent Wnt signaling. Interestingly, and in contrast to ligand-induced negative feedback Ub that mediates endocytosis of most receptors, FZD appears to be constitutively ubiquitinylated by ZNRF3/RNF43, resulting in a dampened state of Wnt signaling that is relieved by the R-Spondin family of Wnt agonists (discussed in more detail later and in Chapter 8). However, the expression of ZNRF3/RNF43 is also strongly induced by Wnt signaling, implying they operate via a negative feedback loop to prevent overactivation of Wnt signaling.

Regulated maturation of Frizzled receptors

FZD receptors, like most cell surface proteins, undergo a series of posttranslational modifications in the endoplasmic reticulum (ER) en route to the cell surface. A predominant secretory pathway modification is glycosylation, and indeed, mature cell surface FZD (also LRP6) can be distinguished from immature ER-localized FZD by decreased electrophoretic mobility of the more heavily glycosylated mature form (Yamamoto *et al.*, 2005; Yanfeng *et al.*, 2006). Glycosylation and maturation of FZD proteins are inhibited by Shisa family proteins, resulting in attenuation of Wnt signaling (Furushima *et al.*, 2007; Yamamoto *et al.*, 2005). Shisa proteins have a single transmembrane domain with an N-terminal cysteine-rich and a C-terminal proline-rich domain. Shisa is required for correct anteroposterior development and somitogenesis in *Xenopus*, processes that are dependent on Wnt and FGF signaling (Nagano *et al.*, 2006). Biochemically, Shisa proteins retain FZD and FGFR proteins in the ER, thus preventing receptor maturation. Shisa, however, does not possess an obvious ER retention signal and indeed is also reported to be secreted (He, 2005). Surprisingly, Shisa proteins are not required for anterior development or somitogenesis in mice (Furushima *et al.*, 2007), suggesting additional Wnt inhibitors, such as DKKs, may compensate.

The LRP family of Wnt coreceptors

General background

The next major breakthrough in understanding Wnt reception came with the simultaneous identification, by three different research groups, of low-density lipoprotein receptor-related proteins (LRPs) as Wnt coreceptors (Pinson *et al.*, 2000; Tamai *et al.*, 2000; Wehrli *et al.*, 2000). Using different approaches and model organisms, these groups convincingly demonstrated that LRP5/6 and the *Drosophila* homolog, Arrow, function as Wnt coreceptors (reviewed in He *et al.*, 2004). LRP5/6 are divergent members of the LRP family and appear to function specifically in β-catenin-dependent signaling (He *et al.*, 2004). The extracellular domain (ECD) of LRP6 associates robustly with Wnt in a ternary complex together with FZD (Tamai *et al.*, 2000). Indeed, the LRP6 ECD contains several Wnt binding sites that mediate the interaction with Wnt/FZD (Bourhis *et al.*, 2010). Analogous to the recruitment of Dvl to the cytoplasmic domains of activated FZD receptors, recruitment of Axin to LRP6 is a key mechanism for signaling and is regulated by phosphorylation events on the ICD of LRP6 (Liu *et al.*, 2003; Mao *et al.*, 2001a; Tamai *et al.*, 2004; Tolwinski *et al.*, 2003) (Figure 2.3b and c, detailed later in this chapter). Recruitment of Axin to LRP5 was shown to occur within 4 min after exposure to Wnt3a (Mao *et al.*, 2001a).

Lrp6$^{-/-}$ mice display phenotypes characteristic of *Wnt1*, *Wnt3a*, and *Wnt7a* mutants (Pinson *et al.*, 2000), whereas *Lrp5*$^{-/-}$ mice develop normally but have osteoporosis and metabolic abnormalities as adults (Fujino *et al.*, 2003; Kato *et al.*, 2002; Magoori *et al.*, 2003). LRP6 therefore functions as the principal Wnt coreceptor; however, LRP5 and LRP6 function redundantly to some degree (Kelly, Pinson, and Skarnes, 2004). The inferred link between Wnt signaling and lipoprotein biology has puzzled researchers and, despite the fact that Wnts are lipid-modified proteins and have even been reported to associate with lipoproteins (Panáková *et al.*, 2005), an unambiguous connection between

Wnt signaling and lipoprotein metabolism has yet to emerge. Nevertheless, the cholesterol homeostasis and lipid metabolism defects associated with *Lrp5*$^{-/-}$ and *Lrp6*$^{-/-}$ mice as well as human mutations hint at a possible connection that needs to be addressed in more detail.

In addition to LRP5 and LRP6, LRP4 (also called Megf7) has been implicated in Wnt signaling (Johnson, Hammer, and Herz, 2005; Ohazama *et al.*, 2010). LRP4$^{-/-}$ mice display defects in limb and tooth development, and LRP4 appears to function as a Wnt and BMP antagonist in these tissues during development. The mechanisms used by LRP4 to inhibit BMP and Wnt signaling are incompletely understood but involve the secreted Wnt/BMP antagonist, Wise. LRP4$^{-/-}$ mice also have defective neuromuscular junctions (NMJs) because LRP4 is required for postsynaptic assembly of acetylcholine receptors (AChRs) (Weatherbee, Anderson, and Niswander, 2006), a process that also depends on the Wnt receptor, MuSK (see later).

Structural features of LRP6

The ECD of LRP5/6/Arrow (termed simply LRP6 here) shares common features with other lipoprotein receptors, such as Cys-rich LDLR type A repeats and Cys-rich EGF repeats with their associated YWTD motif spacer domains, although overall arrangement differs (Brown *et al.*, 1998). The ICD is, however, unrelated to other members of the family and notably lacks the NPxY and YxxL motifs that mediate internalization of classical LDL receptors via clathrin-coated pits. LRP6 does, however, have dileucine (IL or LL) and YRxY motifs that can potentially interact with the endocytotic machinery, although it is unknown if these play a role in endocytosis for LRP6 (Schneider and Nimpf, 2003). Unlike other LDL receptors, the ICD of LRP6 is rich in proline, serine, and threonine residues, which are concentrated in and around repetitive PPP[S/T]Px[S/T] motifs. Although either serine or threonine ([S/T]) can act as the phospho-acceptor residue within these motifs, they are commonly referred to as "PPSPxS" motifs for simplicity. There are five PPSPxS motifs in total, termed A to E, distributed mostly in the C-terminal half of the ICD (Figure 2.3b). The motif itself contains two phosphorylation sites, the PPSP site that is targeted by proline-dependent kinases and the adjacent SPxS site that is targeted by CK1 members (Figure 2.3c). The PPSP site residue is serine for PPSPxS motifs A, D, and E, whereas it is threonine for motifs B and C. The adjacent CK1 site residue is threonine if the PPSP site residue is serine and vice versa, so that there is always a Ser/Thr or Thr/Ser motif pairing. The unique exception to this rule is PPSPxS motif D in *Drosophila* Arrow, which harbors two serine residues. Progressive removal of PPSPxS motifs from the C-terminus of LRP6 results in a progressive inhibition of Axin binding and signaling activity (Davidson *et al.*, 2005; Mao *et al.*, 2001a), and, in agreement, these motifs function cooperatively to activate Wnt/β-catenin signaling (Macdonald *et al.*, 2008). Highlighting their importance, transplantation of a single PPSPxS motif to the short cytoplasmic tail of LDLRΔN, which is otherwise inert, is sufficient to transduce Wnt signals.

Regulatory phosphorylation of the LRP6 intracellular domain

Phosphorylation of LRP6 at PPSPxS motifs is not only required for signaling, but it represents one of the most highly regulated nodes within the Wnt/β-catenin signaling pathway. Mutation of the PPSP sites in LRP6 to PPAP (LRP6m5) prevents phosphorylation and blocks signaling (Tamai *et al.*, 2004) and indeed produces a dominant-negative form of LRP6 similar to LRP6ΔC which lacks the entire ICD (Tamai *et al.*, 2000). Importantly, the prediction that Wnt stimulates LRP6 phosphorylation of PPSPxS motifs to create docking sites for Axin (Mao *et al.*, 2001a) was confirmed when antibodies recognizing phosphorylated LRP6 became available, providing invaluable tools for the biochemical readout of activated Wnt signaling (Davidson *et al.*, 2005; Macdonald *et al.*, 2008; Tamai *et al.*, 2004). The most widely used antibody recognizes phosphoserine at position 1490 in PPSP site A (Figure 2.3c), although, due to the conserved nature of the five PPSP sites, some degree of cross-specificity likely exists.

The identification of kinases that phosphorylate LRP6 has revealed an unexpected level of complexity with respect to both the number of kinases involved and the different mechanisms used to regulate phosphorylation. We first give an overview of the proline-dependent kinases that phosphorylate PPSP sites before discussing the adjacent CK1 site phosphorylation.

LRP6 PPSP site proline-dependent kinases

GSK3

The first PPSP site kinases identified were GSK3α/β (Zeng *et al.*, 2005). This is surprising because, on the one hand, GSK3 functions negatively in Wnt signaling as the principal mediator of β-catenin destruction but, on the other hand, functions positively to destabilize the same destruction complex by promoting Axin binding to LRP6. It is likely that different subcellular compartments of GSK3 are responsible for their opposing activities in Wnt/β-catenin signaling, with membrane-bound GSK3 acting positively on LRP6 and cytosolic GSK3 acting negatively on β-catenin. In support of this idea, forced targeting of GSK3 to the membrane switches its activity from a negative to a positive regulator and results in more efficient phosphorylation of LRP6 (Zeng *et al.*, 2005). Another twist to the GSK3–LRP6 connection came when it was shown that phosphorylated PPSPxS motifs themselves function to inhibit the kinase activity of GSK3 (Cselenyi *et al.*, 2008; Piao *et al.*, 2008; Wu *et al.*, 2009). It is therefore likely that the pool of GSK3 associated with β-catenin in the destruction complex is inactivated when it is brought to the LRP6 ICD together with Axin. GSK3 substrates are commonly phosphorylated within a [S/T]xxx[S/T] motif that is first primed by phosphorylation at the +4 (C-terminal) S/T site (Cohen and Frame, 2001). GSK3 phosphorylation of LRP6 PPSPxS sites is, however, unusual because the priming residue at +4 is absent. Consistent with evidence that GSK3 alone does not account for all PPSP sites phosphorylation (Davidson *et al.*, 2009), several additional LRP6 PPSP site kinases have since been identified.

CDK14

A genome-wide kinase RNAi screen in *Drosophila* cells using a PPSP site phosphospecific antibody (Sp1490) identified CDK14 (called L63 in *Drosophila*) as an LRP6 kinase (Davidson *et al.*, 2009). CDK14, also known as Pftk1, is a relatively uncharacterized member of the Cyclin-dependent kinase (CDK) family; however, Cyclin Y was identified as its Cyclin partner in this study, confirming it is a "classical" CDK (Davidson *et al.*, 2009). Nevertheless, Cyclin Y has the unusual property of being associated with the plasma membrane, and this explains why CDK14, which alone localizes to the cytoplasm and nucleus, can phosphorylate LRP6 as Cyclin Y/CDK14 complexes. Indeed, Cyclin Y is a G2/M-type Cyclin and thus results in preferential phosphorylation of LRP6 during G2/M. This unexpected link of LRP6 phosphorylation with the cell cycle predicts that Wnt signaling itself should be regulated by the cell cycle, with maximal signaling at G2/M, and this indeed seems to be the case, at least under some conditions (Davidson *et al.*, 2009).

GRK5/6

G protein-coupled receptor kinases (GRKs) were the next class of PPSP site kinases to be identified (Chen *et al.*, 2009). GRKs are best known for their role in phosphorylation and desensitization of G protein-coupled receptors. They specifically associate with and phosphorylate activated receptors, which leads to interaction of the receptors with β-arrestins and subsequent endocytosis and degradation (Mushegian, Gurevich, and Gurevich, 2012). Membrane association is essential for their activity, and this is mediated by palmitoylation in the case of GRK6, whereas GRK5 is thought to bind membrane lipids via its carboxyl-terminal domain (Stoffel *et al.*, 1994; Stoffel, Pitcher, and Lefkowitz, 1997). In contrast to the β-arrestin-mediated negative regulation of most receptors by GRKs, GRK5/6 act positively on LRP6 because they phosphorylate domains that promote Axin binding. It is interesting to note that β-arrestin-2 is recruited to FZD4 receptors in a Dvl-dependent manner upon activation by Wnt (Chen *et al.*, 2003). It is therefore tempting to speculate that GRK's function in Wnt signaling may not be limited to LRP6 phosphorylation but may include FZD modification via a more classical GRK-mediated mechanism involving β-arrestin.

PKA

PKA is a cAMP-dependent Ser/Thr kinase that phosphorylates many targets (Kirschner *et al.*, 2009). It has been reported that parathyroid hormone (PTH) can signal through LRP6 by forming a complex with its receptor PTH1R and LRP6 to promote osteoblast differentiation (Wan *et al.*, 2008). PTH treatment induces rapid phosphorylation of LRP6 PPSP sites that is blocked by pharmacological PKA inhibitors, suggesting that this kinase is involved in LRP6 phosphorylation. Interestingly, PTH1R is regulated by the LRP6 kinase GRK5 (Dicker *et al.*, 1999), raising the possibility that PTH and GRK5 may act together to regulate PTH1R–LRP6 receptor complexes.

MAPK

The most recent addition to the list of LRP6 PPSP site kinases are the MAPKs p38, ERK1/2, and JNK1 that were initially identified from a human kinome-wide siRNA screen (Cervenka *et al.*, 2010). JNK and p38 phosphorylation of LRP6 PPSP sites is Wnt dependent; however, ERK1/2-mediated LRP6 phosphorylation is dependent on extracellular stimuli that activate ERK1/2, such as FGF2. This suggests that cells can select different kinases for LRP6 PPSP site phosphorylation, depending on the environmental/mitogenic state they are exposed to.

Considering that there are multiple PPSPxS motifs that function cooperatively as well as multiple PPSP site kinases, it seems likely that cells use a combination of these proline-dependent kinases in a context-dependent manner to control the precise activation state of LRP6.

CK1 site phosphorylation of LRP6

CK1 family members are well known to positively and negatively regulate Wnt signaling by phosphorylating cytoplasmic pathway components such as Dvl, Axin, APC, and β-catenin (Polakis, 2002) (see Chapter 3). The CK1 family, however, also phosphorylates LRP6 at the membrane (Davidson *et al.*, 2005; Zeng *et al.*, 2005). Within the PPPSPxS motifs of LRP6, PPSP site phosphorylation by the aforementioned proline-dependent kinases serves as a priming event for subsequent phosphorylation at the adjacent SPxS sites by members of the casein kinase I (CKI) family, notably CKIγ (Figure 2.3c). This dual phosphorylation of PPSPxS motifs is required for Wnt/β-catenin signaling (Davidson *et al.*, 2005; Zeng *et al.*, 2005). CKIγ is uniquely located to the inner leaflet of the plasma membrane due to palmitoylation, facilitating LRP6 interaction (Davidson *et al.*, 2005). Deletion of a short C-terminal region containing the palmitoylation site results in a CK1γ protein unable to either associate with or phosphorylate LRP6. In addition to CK1γ, CK1ε and CK1α have also been shown to play a role in phosphorylation of LRP6 (Zeng *et al.*, 2005; Zhang *et al.*, 2006), likely due to their association at the membrane with E-cadherin/p120 catenin or Axin, respectively (del Valle-Perez *et al.*, 2011).

CK1γ also phosphorylates an additional Ser/Thr-rich motif N-terminal to the first PPSP motif (Davidson *et al.*, 2005; Figure 2.3c). Upon phosphorylation, this motif has been proposed to act as a docking site for GSK3 and to promote the phosphorylation of the PPSP site (Piao *et al.*, 2008; Yum *et al.*, 2009). Although this is in contradiction to the generally accepted idea that PPSP site phosphorylation promotes CK1 site phosphorylation, the Ser/Thr-rich cluster is not directly adjacent to PPSP sites like the other CK1 sites in LRP6. It is likely that the phosphorylated Ser/Thr-rich cluster helps to maintain the GSK3–Axin complex at the membrane, because Axin recruitment to LRP6 is GSK3 dependent (Figure 2.3b).

Wnt-dependent versus Wnt-independent LRP6 phosphorylation

Considering Axin is recruited to LRP6 in a Wnt signal-dependent manner and that Axin binding to LRP6 requires phosphorylation, it is logical to assume that LRP6 kinases, or at least their ability to phosphorylate LRP6, are activated via Wnt ligand-dependent mechanisms.

CK1-mediated phosphorylation of LRP6 indeed requires Wnt ligand activation. Exactly how Wnt activates CK1-dependent LRP6 phosphorylation is unresolved, although it does not appear to involve direct activation of kinase activity. A likely mechanism is accumulation of

Table 2.1 Protein kinases participating in LRP6 phosphorylation.

Kinase	Phosphorylated site	Inducer	Role
GSK3	PPSP	Wnt	Primes
GRK5/6	PPSP	Wnt	Primes
PKA	PPSP	PTH	Primes
MAPK	PPSP	Phorbol ester, FGF	Primes
CycY/CDK14	PPSP	Cell cycle, primarily G2/M	Primes
CKIγ	PPSPxS; Ser/Thr cluster	Wnt	Activates

The indicated kinases phosphorylate the underlined residues within the ICD of LRP6. How these phosphorylation events are induced and whether they prime or activate LRP6 is shown.

LRP6 within CK1γ-enriched lipid rafts in response to Wnt stimulation (Sakane, Yamamoto, and Kikuchi, 2010; Yamamoto *et al.*, 2008). Full activation of LRP6 by CKI-mediated phosphorylation also depends on several additional regulatory mechanisms such as oligomerized Dvl and LRP signalosome formation (Bilic *et al.*, 2007; see in the following), phosphatidylinositol 4,5-bisphosphate (Pan *et al.*, 2008), transmembrane protein 198 (Liang *et al.*, 2011), and Amer1/WTX (Tanneberger *et al.*, 2011).

The situation is more complex for regulation of PPSP site phosphorylation, due to the fact that more kinases are involved (Table 2.1). There are conflicting reports on the degree to which PPSP sites are constitutively phosphorylated versus Wnt stimulated (Nusse, 2005). The reason for such differences is unclear but may result from the use of different phosphospecific antibodies. GSK3, GRK5/6, and the MAP kinases p38 and JNK1 likely account for Wnt-dependent increases in LRP6 PPSP site phosphorylation; however, what is the nature of the constitutive PPSP site phosphorylation?

First, CDK14/Cyclin Y phosphorylates LRP6 in a cell cycle-dependent manner, independent of Wnt signaling. CDK14/Cyclin Y mediates phosphorylation of LRP6 predominantly, although not exclusively, at G2/M and will therefore contribute to "constitutive" PPSP site phosphorylation detected in cultured cells. In agreement with the PPSP site priming model, CDK14/Cyclin Y phosphorylation of PPSP sites appears to sensitize LRP6 at G2/M for subsequent Wnt-dependent CK1 phosphorylation. This is because CK1-mediated phosphorylation is more strictly Wnt-dependent than PPSP phosphorylation events but nevertheless depends on prior PPSP site phosphorylation (Davidson *et al.*, 2005). It is thought that CDK14/Cyclin Y phosphorylates LRP6 at G2/M to induce Wnt/β-catenin signaling for orchestrating a mitotic program beyond transcriptional regulation (reviewed in Davidson and Niehrs, 2010).

Second, as mentioned earlier, PKA mediates LRP6 PPSP site phosphorylation in response to PTH, and the MAP kinase ERK1/2 phosphorylate PPSP sites in response to non-Wnt extracellular stimuli such as fibroblast growth factors, both of which will contribute, to varying degrees, to constitutive PPSP site phosphorylation.

Regulation of LRP6 by the Dickkopf family of secreted Wnt antagonists

Dickkopf (DKK) proteins are a small family of evolutionarily conserved secreted glycoproteins that function as general and potent inhibitors of the Wnt/β-catenin pathway. The founding member of the family, DKK1, was identified as an embryonic head inducer (hence the name "Dickkopf," which means "big head" in German) in *Xenopus* (Glinka *et al.*, 1998). The family comprises four members, DKK1–4, all of which share two conserved domains, termed DKK-N and DKK-C for the N-terminal and C-terminal domains, respectively. DKK1, 2, and 4 bind with high affinity to the ECDs of LRP5/6 and inhibit Wnt signaling (Mao *et al.*, 2001b), whereas DKK3 is a divergent family member that neither binds LRP6 nor regulates Wnt signaling (Cruciat and Niehrs, 2012). DKK-N is unique to the DKKs, whereas DKK-C shows homology with the colipase fold found in a wide range of functionally unrelated proteins (Niehrs, 2006). Binding of DKK1 to LRP6 is mediated by synergistic interactions of LRP6 EGF repeats 3–4 with DKK-C and EGF repeats 1–2 with DKK1-N (Ahn *et al.*, 2011; Bourhis *et al.*, 2010). DKK1 therefore competitively inhibits the interactions of LRP6 with multiple Wnt ligands, accounting for its broad antagonism of Wnt signaling. The importance of DKK1 in physiological regulation of Wnt/β-catenin-dependent anteroposterior embryonic patterning is demonstrated by the headless phenotype of *dkk1*$^{-/-}$ mice (Mukhopadhyay *et al.*, 2001). For a comprehensive overview on the DKK proteins, the reader is referred to dedicated reviews (Niehrs, 2006).

In addition to LRP6, DKK1 also interacts with the single-pass TM proteins Kremen 1/2 (Mao *et al.*, 2002). Although not absolutely required for DKK1-mediated inhibition, Kremen proteins form ternary complexes with DKK1 and LRP6 and significantly enhance the inhibitory activity of DKK1 by rapidly removing LRP6 from the cell surface (Mao *et al.*, 2002). Kremen protein can, however, also act as activators of Wnt signaling in the absence of DKK (Hassler *et al.*, 2007).

Regulation of LRP6 by additional factors: Wise/SOST, IGFBP-4, Waif1, and C1q

DKK is not the only secreted antagonist of Wnt signaling that competes with Wnt ligands for binding to LRP6. Wise and SOST (Sclerostin) are closely related secreted Wnt inhibitors belonging to the CAN subfamily of cysteine knot-containing BMP antagonists (Avsian-Kretchmer and Hsueh, 2004; Ellies *et al.*, 2006; Lintern *et al.*, 2009). Wise binds to LRP6 and competes with Wnt binding to inhibit signaling; however, it can also act positively on Wnt/β-catenin in some contexts (Itasaki *et al.*, 2003; Lintern *et al.*, 2009). Interestingly, Wise has also been demonstrated to inhibit Wnt/β-catenin signaling by functioning in the ER to prevent cell surface accumulation of LRP6 (Guidato and Itasaki, 2007). SOST is highly expressed in osteoblasts and osteocytes where it inhibits Wnt/β-catenin signaling by binding to the EGF repeats 1–2 of LRP5 (Li *et al.*, 2005; Semenov, Tamai, and He, 2005). SOST is the product of the gene mutated in sclerosteosis, a disease characterized by high bone mass, a phenotype also seen in gain-of-function mutations in LRP5 (Johnson *et al.*, 2004). Indeed, these gain-of-function mutations in LRP5 (e.g., G171V) disrupt the SOST binding site on LRP5, strongly suggesting that SOST functions *in vivo* by inhibiting LRP5 signaling to regulate bone mass (Semenov and He, 2006).

Wnt-activated inhibitory factor 1 (Waif1/5T4) is a metastasis-associated transmembrane protein that binds to LRP6 and inhibits Wnt-induced LRP6 internalization, a process that is required for pathway activation (Kagermeier-Schenk *et al.*, 2011). Waif1 inhibition of Wnt/β-catenin signaling also results in the activation of noncanonical Wnt pathways.

A newly identified and somewhat surprising mode of LRP6 regulation appears to involve the serum complement system protein C1q, which is a component of the mammalian innate immune system response. C1q activates Wnt/β-catenin signaling by binding to FZD via the FZD-CRD and subsequent induction of C1s-mediated cleavage of the LRP6 ECD to yield a constitutively active receptor (Naito *et al.*, 2012). However, because relatively high amounts of C1q are needed for efficient Wnt activation *in vitro* (50–100 μg/ml), the physiological relevance of this phenomenon remains unclear (Naito *et al.*, 2012). The authors postulate that the activation of a small fraction of LRP6 by C1s cleavage, as seen by increased LRP6 ICD phosphorylation, may lead to some form of amplification mechanism that phosphorylates uncleaved LRP6. In addition, a direct correlation between age-related increases in levels of C1q and Wnt signaling was observed in mice (Naito *et al.*, 2012).

Regulation of LRP6 cell surface localization by Mesd

As is the case for FZD receptors, LRP6 coreceptors require plasma membrane localization for Wnt signaling, and this is mediated by the ER-resident chaperone Mesd (Culi and Mann, 2003; Hsieh *et al.*, 2003) (called Boca in *Drosophila*). Mice or flies that lack functional Mesd/Boca have LRP6 trapped within the ER of their cells and display Wnt/β-catenin mutant phenotypes. Within the ER, Mesd helps to fold the large EGF/YWTD domain repeats of the

LRP6 ECD and traffic it efficiently through the ER and Golgi to the cell surface. Separate "chaperone" and "escort" domains on Mesd have been proposed to initiate proper folding within the ER and prevent premature ligand binding during transport to the cell surface, respectively (Chen *et al.*, 2011). It is, however, not clear to what extent Mesd facilitated transport of LRP6 to the cell surface is a regulated process.

Additional Wnt coreceptors

RYK

RYK (derailed (Drl) in *Drosophila*) is a single-pass transmembrane receptor tyrosine kinase that binds Wnts (He, 2004; Inoue *et al.*, 2004; Lu *et al.*, 2004). RYK contains an extracellular leucine-rich motif with similarity to the secreted Wnt antagonist Wif-1 and an intracellular tyrosine kinase domain that is inactive. The kinase domain, however, associates with SRC, and Wnt/RYK signaling plays important roles in axon guidance and neuronal cell fate determination (Wouda *et al.*, 2008). There is evidence that RYK acts as a Wnt/FZD coreceptor to activate both β-catenin-dependent and β-catenin-independent signaling (Lu *et al.*, 2004; Macheda *et al.*, 2012). The E3 ubiquitin ligase Mindbomb 1 (MIB1) ubiquitylates RYK and reduces its plasma membrane levels, leading to β-catenin activation, possibly by inhibiting RYK/PCP signaling (Berndt *et al.*, 2011). The RYK ICD is also reported to be cleaved by gamma-secretase and translocated to the nucleus in a Wnt-dependent manner to directly regulate transcription, likely independent of β-catenin (Lyu, Yamamoto, and Lu, 2008).

MuSK

MuSK is a receptor tyrosine kinase that is involved in the initiation of NMJ formation and has an ECD with homology to the FZD-CRD. In zebrafish, Wnt11 binds to the MuSK (called unplugged) ectodomain during neuromuscular development and activates a PCP-like pathway. Mechanistically, MuSK undergoes Wnt-induced localized endocytosis that somehow restricts postsynaptic AChRs and aids alignment of pre- and postsynaptic components (Gordon *et al.*, 2012; Jing *et al.*, 2009). Similarly, in mammalian muscle cells, Wnt4, Wnt9a, and Wnt11 proteins regulate AChR clustering, which involves MuSK dimerization and tyrosine phosphorylation (Strochlic *et al.*, 2012; Zhang *et al.*, 2012). As mentioned earlier, MuSK and AChR clustering also requires LRP4 (Weatherbee, Anderson, and Niswander, 2006; Zhang *et al.*, 2012), suggesting that Wnt may mediate NMJ formation through a variety of receptors and/or mechanisms.

PTK7 and ROR1/2

The single-pass transmembrane protein PTK7 binds Wnts in combination with FZD and recruits Dvl to the plasma membrane domain. Like many PCP signaling activators, PTK7 is a negative regulator of β-catenin-dependent Wnt signaling (Peradziryi, Tolwinski, and Borchers, 2012). More recently, however, PTK7 has been implicated in regulating Wnt/β-catenin signaling in *Xenopus* and to interact with β-catenin (Puppo *et al.*, 2011).

ROR1 and ROR2 (receptor tyrosine kinase-like orphan receptor 1/2) are single transmembrane receptor tyrosine kinases containing an FRZ-CRD and bind Wnt5a to transmit Wnt/PCP signaling in vertebrates (Ho *et al.*, 2012; Oishi *et al.*, 2003; Schambony and Wedlich, 2007). More information on the function of these primarily PCP pathway receptors is provided in Chapters 6 and 7.

Heparan sulfate proteoglycans

Heparan sulfate proteoglycans (HSPGs) are macromolecules containing a protein core with sulfated glycosaminoglycan (GAG) chains that have been implicated as coreceptors for a variety of growth factors such as Hh, FGF, and Wnts and mainly promote ligand receptor interactions (Perrimon and Bernfield, 2000). Early work showed that Wnt1 is a heparin-binding protein and glypicans and syndecans have been shown to bind Wnt/Wg and Fz proteins to enhance Wnt signaling (Fuerer, Habib, and Nusse, 2010). The *Drosophila* glypicans Dally and Dlp promote β-catenin signaling, although evidence from vertebrates suggests they function in PCP signaling (Kikuchi *et al.*, 2011). Glypicans can also be regulated by the lipase Notum, releasing them from the cell surface to inhibit Wnt/β-catenin signaling (Flowers, Topczewska, and Topczewski, 2012; Giráldez, Copley, and Cohen, 2002). The

HSPG biglycan is a MuSK ligand that is required for proper AChR cluster formation (Amenta *et al.*, 2012).

Wnt receptor clustering and signalosomes

Upon binding of Wnt ligands to FZD receptors, LRP6 is recruited and a ternary complex of FZD/Wnt/LRP6 is formed. Activated receptors recruit Dvl and Axin to the plasma membrane, possibly together with associated proteins of the β-catenin destruction complex, promoting the formation of large Wnt receptor complexes termed "signalosomes" (Bilic *et al.*, 2007). Exactly how the formation of these high-molecular-weight receptor-ligand complexes is initiated and the order in which the various components are added to form a fully activated signalosome is unclear, but the propensity of Dvl to polymerize is thought to be a key mediator for aggregation (Bilic *et al.*, 2007; Schwarz-Romond *et al.*, 2007). Recently, it was proposed that Dvl may function to inhibit Axin via Dvl–Axin interaction through their respective DIX domains, leading to suppression of cytoplasmic Axin complexes in favor of membrane-localized Dvl–Axin complexes in response to Wnt (Fiedler *et al.*, 2011). It has also been proposed that Dvl–Axin interactions progressively bring more GSK3 to the membrane to amplify LRP6 phosphorylation, resulting in a feedforward amplification process (Zeng *et al.*, 2008). Interestingly, CK1-mediated phosphorylation of LRP6 appears to require prior receptor clustering, indicating this is a relatively late event in the formation and/or activation of signalosomes (Bilic *et al.*, 2007). Wnt receptor signalosomes are presumed to inhibit GSK3-mediated phosphorylation of β-catenin, and therefore activate Wnt/β-catenin signaling, by two distinct mechanisms. First, as mentioned earlier, phosphorylated LRP6 binds and directly inhibits GSK3 (Cselenyi *et al.*, 2008; Piao *et al.*, 2008; Wu *et al.*, 2009), and second, GSK3 is sequestered in multivesicular bodies (MVBs) (Taelman *et al.*, 2010). To what extent MVB localization of GSK3 contributes to accumulation of β-catenin signaling relative to direct inhibition of GSK3 by phosphorylated LRP6 PPSPxS motifs remains unclear. Intriguingly, β-catenin itself appears to be essential for the sequestration of GSK3 in MVBs. Endocytosis of receptor signalosomes, their endosomal acidification by vacuolar ATPases, and sequestration in MVBs are all essential for Wnt signaling and covered in more detail in Chapter 8.

R-Spondin family of secreted Wnt activators

R-Spondins function as Wnt agonists by derepressing FZD. As mentioned earlier, ubiquitination of FZD by ZRNF and RNF43 removes FZD from the cell surface, and it seems R-Spondins prevent this by tethering ZRNF3 to the R-Spondin receptor LGR4, resulting in internalization of the Ub ligases (Hao *et al.*, 2012; see also Chapter 8). Although R-Spondin1 binds directly to LGR4, it remains to be confirmed whether this is also true for ZNRF3. The mechanisms used by R-Spondins to regulate Wnt signaling appear to be more complex however. In stark contrast to their ability to stabilize FZD at the plasma membrane upon engagement with LGR4/ZRNF3, R-Spondins also promote internalization of FZD upon engagement with the cell surface proteoglycan Syndecan 4 (Ohkawara, Glinka, and Niehrs, 2011). How these opposing functions of R-Spondin are regulated is unclear but will be an important issue to address. Considering R-Spondins can interact with three different receptors (LGR4, Syndecan 4, and possibly ZNRF3), the final outcome on FZD is likely to be context dependent.

The discovery of how R-Spondins signal via non-Wnt binding receptors to activate Wnt signaling is a clear reminder of how little we may actually know about how Wnt signaling is regulated at the membrane and that despite decades of intensive research by many dedicated groups spanning different fields, our understanding of Wnt signaling remains far from complete.

Acknowledgments

We would like to thank Claudia Janda for providing the surface model representation of the Wnt/FZD crystal structure shown in Figure 2.2. We apologize to authors we could not cite due to space limitation.

References

Ahn, V.E., Chu, M.L., Choi, H.J. *et al.* (2011) Structural basis of Wnt signaling inhibition by Dickkopf binding to LRP5/6. *Developmental Cell*, **21**, 862–873.

Amenta, A.R., Creely, H.E., Mercado, M.L. *et al.* (2012) Biglycan is an extracellular MuSK binding protein important for synapse stability. *The Journal of Neuroscience*, **32**, 2324–2334.

Avsian-Kretchmer, O. and Hsueh, A.J.W. (2004) Comparative genomic analysis of the eight-membered ring cystine knot-containing bone morphogenetic protein antagonists. *Molecular Endocrinology*, **18**, 1–12.

Bafico, A., Gazit, A., Pramila, T. *et al.* (1999) Interaction of frizzled related protein (FRP) with Wnt ligands and the frizzled receptor suggests alternative mechanisms for FRP inhibition of Wnt signaling. *The Journal of Biological Chemistry*, **274**, 16180–16187.

Bányai, L. and Patthy, L. (1999) The NTR module: domains of netrins, secreted frizzled related proteins, and type I procollagen C-proteinase enhancer protein are homologous with tissue inhibitors of metalloproteases. *Protein Science*, **8**, 1636–1642.

Berndt, J.D., Aoyagi, A., Yang, P. *et al.* (2011) Mindbomb 1, an E3 ubiquitin ligase, forms a complex with RYK to activate Wnt/β-catenin signaling. *The Journal of Cell Biology*, **194**, 737–750.

Bhanot, P., Brink, M., Samos, C.H. *et al.* (1996) A new member of the frizzled family from Drosophila functions as a Wingless receptor. *Nature*, **382**, 225–230.

Bhat, R.A., Stauffer, B., Komm, B.S., and Bodine, P.V.N. (2007) Structure–function analysis of secreted frizzled-related protein-1 for its Wnt antagonist function. *Journal of Cellular Biochemistry*, **102**, 1519–1528.

Bilic, J., Huang, Y-L., Davidson, G. *et al.* (2007) Wnt induces LRP6 signalosomes and promotes dishevelled-dependent LRP6 phosphorylation. *Science*, **316**, 1619–1622.

Bourhis, E., Tam, C., Franke, Y. *et al.* (2010) Reconstitution of a frizzled8.Wnt3a.LRP6 signaling complex reveals multiple Wnt and Dkk1 binding sites on LRP6. *The Journal of Biological Chemistry*, **285**, 9172–9179.

Brown, S.D., Twells, R.C., Hey, P.J. *et al.* (1998) Isolation and characterization of LRP6, a novel member of the low density lipoprotein receptor gene family. *Biochemical and Biophysical Research Communications*, **248**, 879–888.

Cervenka, I., Wolf, J., Mašek, J. *et al.* (2010) Mitogen-activated protein kinases promote WNT/β-catenin signaling via phosphorylation of LRP6. *Molecular and Cellular Biology*, **31**, 179–189.

Chen, W., ten Berge, D., Brown, J. *et al.* (2003) Dishevelled 2 recruits beta-arrestin 2 to mediate Wnt5A-stimulated endocytosis of Frizzled 4. *Science*, **301**, 1391–1394.

Chen, M., Philipp, M., Wang, J. *et al.* (2009) G protein-coupled receptor kinases phosphorylate LRP6 in the Wnt pathway. *The Journal of Biological Chemistry*, **284**, 35040–35048.

Chen, J., Liu, CC., Li, Q. *et al.* (2011) Two structural and functional domains of MESD required for proper folding and trafficking of LRP5/6. *Structure*, **19**, 313–323.

Cohen, P. and Frame, S. (2001) The renaissance of GSK3. *Nature Reviews Molecular Cell Biology*, **2**, 769–776.

Cong, F., Schweizer, L., and Varmus, H. (2004) Wnt signals across the plasma membrane to activate the beta-catenin pathway by forming oligomers containing its receptors, Frizzled and LRP. *Development*, **131**, 5103–5115.

Cruciat, C-M. and Niehrs, C. (2012) Secreted and transmembrane Wnt inhibitors and activators. *Cold Spring Harbor Perspectives in Biology*, **5**, a015081.

Cselenyi, C.S., Jernigan, K.K., Tahinci, E. *et al.* (2008) LRP6 transduces a canonical Wnt signal independently of Axin degradation by inhibiting GSK3's phosphorylation of beta-catenin. *Proceedings of the National Academy of Sciences of the United States of America*, **105**, 8032–8037.

Culi, J. and Mann, R.S. (2003) Boca, an endoplasmic reticulum protein required for wingless signaling and trafficking of LDL receptor family members in Drosophila. *Cell*, **112**, 343–354.

Dann, C.E., Hsieh, J.C., Rattner, A. *et al.* (2001) Insights into Wnt binding and signalling from the structures of two Frizzled cysteine-rich domains. *Nature*, **412**, 86–90.

Davidson, G. and Niehrs, C. (2010) Emerging links between CDK cell cycle regulators and Wnt signaling. *Trends in Cell Biology*, **20**, 453–460.

Davidson, G., Wu, W., Shen, J. *et al.* (2005) Casein kinase 1 gamma couples Wnt receptor activation to cytoplasmic signal transduction. *Nature*, **438**, 867–872.

Davidson, G., Shen, J., Huang, Y.L. *et al.* (2009) Cell cycle control of wnt receptor activation. *Developmental Cell*, **17**, 788–799.

.del Valle-Perez, B., Arques, O., Vinyoles, M. *et al.* (2011) Coordinated action of CK1 isoforms in canonical Wnt signaling. *Molecular and Cellular Biology*, **31**, 2877–2888.

Dicker, F., Quitterer, U., Winstel, R. *et al.* (1999) Phosphorylation-independent inhibition of parathyroid hormone receptor signaling by G protein-coupled receptor kinases. *Proceedings of the National Academy of Sciences of the United States of America*, **96**, 5476–5481.

Djiane, A., Yogev, S., and Mlodzik, M. (2005) The apical determinants aPKC and dPatj regulate

Frizzled-dependent planar cell polarity in the Drosophila eye. *Cell*, **121**, 621–631.

Ellies, D.L. Viviano, B., McCarthy, J. *et al.* (2006) Bone density ligand, Sclerostin, directly interacts with LRP5 but not LRP5G171V to modulate Wnt activity. *Journal of Bone and Mineral Research*, **21**, 1738–1749.

Fiedler, M., Mendoza-Topaz, C., Rutherford, T.J. *et al.* (2011) Dishevelled interacts with the DIX domain polymerization interface of Axin to interfere with its function in down-regulating β-catenin. *Proceedings of the National Academy of Sciences of the United States of America*, **108**, 1937–1942.

Flowers, G.P., Topczewska, J.M., and Topczewski, J. (2012) A zebrafish Notum homolog specifically blocks the Wnt/β-catenin signaling pathway. *Development*, **139**, 2416–2425.

Fuerer, C., Habib, S.J., and Nusse, R. (2010) A study on the interactions between heparan sulfate proteoglycans and Wnt proteins. *Developmental Dynamics*, **239**, 184–190.

Fujino, T., Hiroshi, A., Man-Jong, K. *et al.* (2003) Low-density lipoprotein receptor-related protein 5 (LRP5) is essential for normal cholesterol metabolism and glucose-induced insulin secretion. *Proceedings of the National Academy of Sciences of the United States of America*, **100**, 229–234.

Furushima, K., Yamamoto, A., Nagano, T. *et al.* (2007) Mouse homologues of Shisa antagonistic to Wnt and Fgf signalings. *Developmental Biology*, **306**, 480–492.

Giráldez, A.J., Copley, R.R., and Cohen, S.M. (2002) HSPG modification by the secreted enzyme Notum shapes the Wingless morphogen gradient. *Developmental Cell*, **2**, 667–676.

Glinka, A., Wu, W., Delius, H. *et al.* (1998) Dickkopf-1 is a member of a new family of secreted proteins and functions in head induction. *Nature*, **391**, 357–362.

Gordon, L.R., Gribble, K.D., Syrett, C.M., and Granato, M. (2012) Initiation of synapse formation by Wnt-induced MuSK endocytosis. *Development*, **139**, 1023–1033.

Gruenberg, J. and Stenmark, H. (2004) The biogenesis of multivesicular endosomes. *Nature Reviews Molecular Cell Biology*, **5**, 317–323.

Guidato, S. and Itasaki, N. (2007) Wise retained in the endoplasmic reticulum inhibits Wnt signaling by reducing cell surface LRP6. *Developmental Biology*, **310**, 250–263.

Haglund, K. and Dikic, I. (2005) Ubiquitylation and cell signaling. *The EMBO Journal*, **24**, 3353–3359.

Hao, H-X., Xie, Y., Zhang, Y. *et al.* (2012) ZNRF3 promotes Wnt receptor turnover in an R-spondin-sensitive manner. *Nature*, **485**, 195–200.

Hassler, C., Cruciat, C.M., Huang, Y.L. *et al.* (2007) Kremen is required for neural crest induction in *Xenopus* and promotes LRP6-mediated Wnt signaling. *Development*, **134**, 4255–4263.

He, X. (2004) Wnt signaling went derailed again: a new track via the LIN-18 receptor? *Cell*, **118**, 668–670.

He, X. (2005) Antagonizing Wnt and FGF receptors: an enemy from within (the ER). *Cell*, **120**, 156–158.

He, X., Semenov, M., Tamai, K., and Zeng, X. (2004) LDL receptor-related proteins 5 and 6 in Wnt/beta-catenin signaling: arrows point the way. *Development*, **131**, 1663–1677.

Hendaoui, I., Lavergne, E., Lee, H.S. *et al.* (2012) Inhibition of Wnt/β-catenin signaling by a soluble collagen-derived frizzled domain interacting with Wnt3a and the receptors frizzled 1 and 8. *PLoS One*, **7**, e30601.

Ho, H-Y.H., Susman, M.W., Bikoff, J.B. *et al.* (2012) Wnt5a-Ror-Dishevelled signaling constitutes a core developmental pathway that controls tissue morphogenesis. *Proceedings of the National Academy of Sciences of the United States of America*, **109**, 4044–4051.

Hsieh, J-C., Lee, L., Zhang, L. *et al.* (2003) Mesd encodes an LRP5/6 chaperone essential for specification of mouse embryonic polarity. *Cell*, **112**, 355–367.

Inoue, T., Oz, H.S., Wiland, D. *et al.* (2004) *C. elegans* LIN-18 is a Ryk ortholog and functions in parallel to LIN-17/Frizzled in Wnt signaling. *Cell*, **118**, 795–806.

Itasaki, N., Jones, C.M., Mercurio, S. *et al.* (2003) Wise, a context-dependent activator and inhibitor of Wnt signalling. *Development*, **130**, 4295–4305.

Janda, C.Y., Waghray, D., Levin, A.M. *et al.* (2012) Structural basis of Wnt recognition by Frizzled. *Science*, **337**, 59–64.

Jia, J., Tong, C., Wang, B. *et al.* (2004) Hedgehog signalling activity of Smoothened requires phosphorylation by protein kinase A and casein kinase I. *Nature*, **432**, 1045–1050.

Jing, L., Lefebvre, J.L., Gordon, L.R., and Granato, M. (2009) Wnt signals organize synaptic prepattern and axon guidance through the zebrafish unplugged/MuSK receptor. *Neuron*, **61**, 721–733.

Johnson, E.B., Hammer, R.E., and Herz, J. (2005) Abnormal development of the apical ectodermal ridge and polysyndactyly in Megf7-deficient mice. *Human Molecular Genetics*, **14**, 3523–3538.

Johnson, M.L., Harnish, K., Nusse, R., and Van Hul, W. (2004) LRP5 and Wnt signaling: a union made for bone. *Journal of Bone and Mineral Research*, **19**, 1749–1757.

Kagermeier-Schenk, B., Wehner, D., Ozhan-Kizil, G. *et al.* (2011) Waif1/5T4 inhibits Wnt/β-catenin signaling and activates noncanonical Wnt pathways by modifying LRP6 subcellular localization. *Developmental Cell*, **21**, 1129–1143.

Kato, M., Patel, M.S., Levasseur, R. *et al.* (2002) Cbfa1-independent decrease in osteoblast proliferation, osteopenia, and persistent embryonic eye vascularization in mice deficient in Lrp5, a Wnt coreceptor. *The Journal of Cell Biology*, **157**, 303–314.

Kelly, O.G., Pinson, K.I., and Skarnes, W.C. (2004) The Wnt co-receptors Lrp5 and Lrp6 are essential for gastrulation in mice. *Development*, **131**, 2803–2815.

Kikuchi, A., Yamamoto, H., Sato, A., and Matsumoto, S. (2011) New insights into the mechanism of Wnt signaling pathway activation. *International Review of Cell and Molecular Biology*, **291**, 21–71.

Kirschner, L.S., Yin, Z., Jones, G.N., and Mahoney, E. (2009) Mouse models of altered protein kinase A signaling. *Endocrine-Related Cancer*, **16**, 773–793.

Koo, B-K., Spit, M., Jordens, I. *et al.* (2012) Tumour suppressor RNF43 is a stem-cell E3 ligase that induces endocytosis of Wnt receptors. *Nature*, **488**, 665–669.

Koval, A., Purvanov, V., Egger-Adam, D., and Katanaev, V.L. (2011) Yellow submarine of the Wnt/Frizzled signaling: submerging from the G protein harbor to the targets. *Biochemical Pharmacology*, **82**, 1311–1319.

Leyns, L., Bouwmeester, T., Kim, S.H. *et al.* (1997) Frzb-1 is a secreted antagonist of Wnt signaling expressed in the Spemann organizer. *Cell*, **88**, 747–756.

Li, X., Zhang, Y., Kang, H. *et al.* (2005) Sclerostin binds to LRP5/6 and antagonizes canonical Wnt signaling. *The Journal of Biological Chemistry*, **280**, 19883–19887.

Liang, J., Fu, Y., Cruciat, C.M. *et al.* (2011) Transmembrane protein 198 promotes LRP6 phosphorylation and Wnt signaling activation. *Molecular and Cellular Biology*, **31**, 2577–2590.

Lin, K., Wang, S., Julius, M.A. *et al.* (1997) The cysteine-rich frizzled domain of Frzb-1 is required and sufficient for modulation of Wnt signaling. *Proceedings of the National Academy of Sciences of the United States of America*, **94**, 11196–11200.

Lintern, K.B., Guidato, S., Rowe, A. *et al.* (2009) Characterization of wise protein and its molecular mechanism to interact with both Wnt and BMP signals. *The Journal of Biological Chemistry*, **284**, 23159–23168.

Liu, G., Bafico, A., Harris, V.K., and Aaronson, S.A. (2003) A novel mechanism for Wnt activation of canonical signaling through the LRP6 receptor. *Molecular Cell Biology*, **23**, 5825–5835.

Lu, W., Yamamoto, V., Ortega, B., and Baltimore, D. (2004) Mammalian Ryk is a Wnt coreceptor required for stimulation of neurite outgrowth. *Cell*, **119**, 97–108.

Lyu, J., Yamamoto, V., and Lu, W. (2008) Cleavage of the Wnt receptor Ryk regulates neuronal differentiation during cortical neurogenesis. *Developmental Cell*, **15**, 773–780.

MacDonald, B.T. and He, X. (2012) A finger on the pulse of Wnt receptor signaling. *Cell Research*, **22**, 1410–1412.

MacDonald, B.T., Yokota, C., Tamai, K. *et al.* (2008) Wnt signal amplification: activity, cooperativity and regulation of multiple intracellular PPPSP motifs in the Wnt coreceptor LRP6. *The Journal of Biological Chemistry*, **283**, 16115–16123.

Macheda, M.L., Sun, W.W., Kugathasan, K. *et al.* (2012) The Wnt receptor Ryk plays a role in mammalian planar cell polarity signaling. *The Journal of Biological Chemistry*, **287**, 29312–29323.

Magoori, K., Kang, M.J., Ito, M.R. *et al.* (2003) Severe hypercholesterolemia, impaired fat tolerance, and advanced atherosclerosis in mice lacking both low density lipoprotein receptor-related protein 5 and apolipoprotein E. *The Journal of Biological Chemistry*, **278**, 11331–11336.

Mao, J., Wang, J., Liu, B. *et al.* (2001a) Low-density lipoprotein receptor-related protein-5 binds to Axin and regulates the canonical Wnt signaling pathway. *Molecular Cell*, **7**, 801–809.

Mao, B., Wu, W., Li, Y. *et al.* (2001b) LDL-receptor-related protein 6 is a receptor for Dickkopf proteins. *Nature*, **411**, 321–325.

Mao, B., Wu, W., Davidson, G. *et al.* (2002) Kremen proteins are Dickkopf receptors that regulate Wnt/beta-catenin signalling. *Nature*, **417**, 664–667.

Mii, Y. and Taira, M. (2009) Secreted Frizzled-related proteins enhance the diffusion of Wnt ligands and expand their signalling range. *Development*, **136**, 4083–4088.

Moeller, C., Swindell, E.C., Kispert, A., and Eichele, G. (2003) Carboxypeptidase Z (CPZ) modulates Wnt signaling and regulates the development of skeletal elements in the chicken. *Development*, **130**, 5103–5111.

Mukai, A., Yamamoto-Hino, M., Awano, W. *et al.* (2010) Balanced ubiquitylation and deubiquitylation of Frizzled regulate cellular responsiveness to Wg/Wnt. *The EMBO Journal*, **29**, 2114–2125.

Mukhopadhyay, M., Shtrom, S., Rodriguez-Esteban, C. *et al.* (2001) Dickkopf1 is required for embryonic head induction and limb morphogenesis in the mouse. *Developmental Cell*, **1**, 423–434.

Mushegian, A., Gurevich, V.V., and Gurevich, E.V. (2012) The origin and evolution of G protein-coupled receptor kinases. *PLoS One*, **7**, e33806.

Nagano, T., Takehara, S., Takahashi, M. *et al.* (2006) Shisa2 promotes the maturation of somitic precursors and transition to the segmental fate in *Xenopus* embryos. *Development*, **133**, 4643–4654.

Naito, A.T., Sumida, T., Nomura, S. *et al.* (2012) Complement C1q activates canonical Wnt signaling and promotes aging-related phenotypes. *Cell*, **149**, 1298–1313.

PtdIns(4,5)P2 to LRP6 phosphorylation. *The EMBO Journal*, **30**, 1433–1443.

Tauriello, D.V.F., Jordens, I., Kirchner, K. *et al.* (2012) Wnt/β-catenin signaling requires interaction of the Dishevelled DEP domain and C terminus with a discontinuous motif in Frizzled. *Proceedings of the National Academy of Sciences of the United States of America*, **109**, E812–E820.

Tolwinski, N.S., Wehrli, M., Rives, A. *et al.* (2003) Wg/Wnt signal can be transmitted through arrow/LRP5,6 and Axin independently of Zw3/Gsk3beta activity. *Developmental Cell*, **4**, 407–418.

Vinson, C.R. and Adler, P.N. (1987) Directional non-cell autonomy and the transmission of polarity information by the frizzled gene of Drosophila. *Nature*, **329**, 549–551.

Wan, M., Yang, C., Li, J. *et al.* (2008) Parathyroid hormone signaling through low-density lipoprotein-related protein 6. *Genes & Development*, **22**, 2968–2979.

Wang, L., Shao, Y.Y., and Ballock, R.T. (2009) Carboxypeptidase Z (CPZ) links thyroid hormone and Wnt signaling pathways in growth plate chondrocytes. *Journal of Bone and Mineral Research*, **24**, 265–273.

Wang, S., Krinks, M., Lin, K. *et al.* (1997) Frzb, a secreted protein expressed in the Spemann organizer, binds and inhibits Wnt-8. *Cell*, **88**, 757–766.

Weatherbee, S.D., Anderson, K.V., and Niswander, L.A. (2006) LDL-receptor-related protein 4 is crucial for formation of the neuromuscular junction. *Development*, **133**, 4993–5000.

Wehrli, M., Dougan, S.T., Caldwell, K. *et al.* (2000) Arrow encodes an LDL-receptor-related protein essential for Wingless signalling. *Nature*, **407**, 527–530.

Wouda, R.R., Bansraj, M.R.K.S., de Jong, A.W.M. *et al.* (2008) Src family kinases are required for WNT5 signaling through the Derailed/RYK receptor in the Drosophila embryonic central nervous system. *Development*, **135**, 2277–2287.

Wu, G., Huang, H., Garcia Abreu, J., and He, X. (2009) Inhibition of GSK3 phosphorylation of beta-catenin via phosphorylated PPPSPXS motifs of Wnt coreceptor LRP6. *PLoS One*, **4**, e4926.

Xu, Y.K. and Nusse, R. (1998) The Frizzled CRD domain is conserved in diverse proteins including several receptor tyrosine kinases. *Current Biology*, **8**, R405–R406.

Xu, Q., Wang, Y., Dabdoub, A. *et al.* (2004) Vascular development in the retina and inner ear: control by Norrin and Frizzled-4, a high-affinity ligand–receptor pair. *Cell*, **116**, 883–895.

Yamamoto, A., Nagano, T., Takehara, S. *et al.* (2005) Shisa promotes head formation through the inhibition of receptor protein maturation for the caudalizing factors, Wnt and FGF. *Cell*, **120**, 223–235.

Yamamoto, H., Sakane, H., Yamamoto, H. *et al.* (2008) Wnt3a and Dkk1 regulate distinct internalization pathways of LRP6 to tune the activation of β-catenin signaling. *Developmental Cell*, **15**, 37–48.

Yanfeng, W.A., Tan, C., Fagan, R.J., and Klein, P.S. (2006) Phosphorylation of frizzled-3. *The Journal of Biological Chemistry*, **281**, 11603–11609.

Yang-Snyder, J., Miller, J.R., Brown, J.D. *et al.* (1996) A frizzled homolog functions in a vertebrate Wnt signaling pathway. *Current Biology*, **6**, 1302–1306.

Yum, S., Lee, S-J., Piao, S. *et al.* (2009) The role of the Ser/Thr cluster in the phosphorylation of PPPSP motifs in Wnt coreceptors. *Biochemical and Biophysical Research Communications*, **381**, 345–349.

Zeng, X., Tamai, K., Doble, B. *et al.* (2005) A dual-kinase mechanism for Wnt co-receptor phosphorylation and activation. *Nature*, **438**, 873–877.

Zeng, X., Huang, H., Tamai, K. *et al.* (2008) Initiation of Wnt signaling: control of Wnt coreceptor Lrp6 phosphorylation/activation via frizzled, dishevelled and axin functions. *Development*, **135**, 367–375.

.Zhang, C., Williams, E.H., Guo, Y. *et al.* (2004) Extensive phosphorylation of Smoothened in Hedgehog pathway activation. *Proceedings of the National Academy of Sciences of the United States of America*, **101**, 17900–17907.

Zhang, L., Jia, J., Wang, B. *et al.* (2006) Regulation of wingless signaling by the CKI family in Drosophila limb development. *Developmental Biology*, **299**, 221–237.

Zhang, B., Liang, C., Bates, R. *et al.* (2012) Wnt proteins regulate acetylcholine receptor clustering in muscle cells. *Molecular Brain*, **5**, 7.

Niehrs, C. (2006) Function and biological roles of the Dickkopf family of Wnt modulators. *Oncogene*, **25**, 7469–7481.

Nusse, R. (2005) Cell biology: relays at the membrane. *Nature*, **438**, 747–749.

Ohazama, A., Porntaveetus, T., Ota, M.S. *et al.* (2010) Lrp4: a novel modulator of extracellular signaling in craniofacial organogenesis. *American Journal of Medical Genetics A*, **152A**, 2974–2983.

Ohkawara, B., Glinka, A., and Niehrs, C. (2011) Rspo3 binds Syndecan 4 and induces Wnt/PCP signaling via clathrin-mediated endocytosis to promote morphogenesis. *Developmental Cell*, **20**, 303–314.

Oishi, I., Suzuki, H., Onishi, N. *et al.* (2003) The receptor tyrosine kinase Ror2 is involved in non-canonical Wnt5a/JNK signalling pathway. *Genes to Cells*, **8**, 645–654.

Pan, W., Choi, S-C., Wang, H. *et al.* (2008) Wnt3a-mediated formation of phosphatidylinositol 4,5-bisphosphate regulates LRP6 phosphorylation. *Science*, **321**, 1350–1353.

Panáková, D., Sprong, H., Marois, E. *et al.* (2005) Lipoprotein particles are required for Hedgehog and Wingless signalling. *Nature*, **435**, 58–65.

Peradziryi, H., Tolwinski, N.S., and Borchers, A. (2012) The many roles of PTK7: a versatile regulator of cell-cell communication. *Archives of Biochemistry and Biophysics*, **524**, 71–76.

Perrimon, N. and Bernfield, M. (2000) Specificities of heparan sulphate proteoglycans in developmental processes. *Nature*, **404**, 725–728.

Piao, S., Lee, S-H., Kim, H. *et al.* (2008) Direct inhibition of GSK3β by the phosphorylated cytoplasmic domain of LRP6 in Wnt/β-catenin signaling. *PLoS One*, **3**, e4046.

Pinson, K.I., Brennan, J., Monkley, S. *et al.* (2000) An LDL-receptor-related protein mediates Wnt signalling in mice. *Nature*, **407**, 535–538.

Polakis, P. (2002) Casein kinase 1: a Wnt'er of disconnect. *Current Biology*, **12**, R499–R501.

Puppo, F., Thomé, V., Lhoumeau, A.C. *et al.* (2011) Protein tyrosine kinase 7 has a conserved role in Wnt/β-catenin canonical signalling. *EMBO Report*, **12**, 43–49.

Quélard, D., Lavergne, E., Hendaoui, I. *et al.* (2008) A cryptic frizzled module in cell surface collagen 18 inhibits Wnt/beta-catenin signaling. *PLoS One*, **3**, e1878.

Reznik, S.E. and Fricker, L.D. (2001) Carboxypeptidases from A to z: implications in embryonic development and Wnt binding. *Cellular and Molecular Life Sciences*, **58**, 1790–1804.

Rodriguez, J., Esteve, P., Weinl, C. *et al.* (2005) SFRP1 regulates the growth of retinal ganglion cell axons through the Fz2 receptor. *Nature Neuroscience*, **8**, 1301–1309.

Sakane, H., Yamamoto, H., and Kikuchi, A. (2010) LRP6 is internalized by Dkk1 to suppress its phosphorylation in the lipid raft and is recycled for reuse. *Journal of Cell Science*, **123**, 360–368.

Schambony, A. and Wedlich, D. (2007) Wnt-5A/Ror2 regulate expression of XPAPC through an alternative noncanonical signaling pathway. *Developmental Cell*, **12**, 779–792.

Schneider, W.J. and Nimpf, J. (2003) LDL receptor relatives at the crossroad of endocytosis and signaling. *Cellular and Molecular Life Sciences*, **60**, 892–903.

Schulte, G. (2010) International Union of Basic and Clinical Pharmacology. LXXX. The class Frizzled receptors. *Pharmacological Reviews*, **62**, 632–667.

Schwarz-Romond, T., Fiedler, M., Shibata, N. *et al.* (2007) The DIX domain of Dishevelled confers Wnt signaling by dynamic polymerization. *Nature Structural and Molecular Biology*, **14**, 484–492.

Semenov, M.V. and He, X. (2006) LRP5 mutations linked to high bone mass diseases cause reduced LRP5 binding and inhibition by SOST. *The Journal of Biological Chemistry*, **281**, 38276–38284.

Semenov, M., Tamai, K., and He, X. (2005) SOST is a ligand for LRP5/LRP6 and a Wnt signaling inhibitor. *The Journal of Biological Chemistry*, **280**, 26770–26775.

Skalka, N., Caspi, M., Caspi, E. *et al.* (2012) Carboxypeptidase E: a negative regulator of the canonical Wnt signaling pathway. *Oncogene*, **32**, 2836–2847.

Smallwood, P.M., Williams, J., Xu, Q. *et al.* (2007) Mutational analysis of Norrin-Frizzled4 recognition. *The Journal of Biological Chemistry*, **282**, 4057–4068.

Stoffel, R.H., Pitcher, J.A., and Lefkowitz, R.J. (1997) Targeting G protein-coupled receptor kinases to their receptor substrates. *Journal of Membrane Biology*, **157**, 1–8.

Stoffel, R.H., Randall, R.R., Premont, R.T. *et al.* (1994) Palmitoylation of G protein-coupled receptor kinase, GRK6. Lipid modification diversity in the GRK family. *The Journal of Biological Chemistry*, **269**, 27791–27794.

Strochlic, L., Falk, J., Goillot, E. *et al.* (2012) Wnt4 participates in the formation of vertebrate neuromuscular junction. *PLoS One*, **7**, e29976.

Taelman, V.F., Dobrowolski, R. Plouhinec, J.L. *et al.* (2010) Wnt signaling requires sequestration of glycogen synthase kinase 3 inside multivesicular endosomes. *Cell*, **143**, 1136–1148.

Tamai, K., Semenov, M., Kato, Y. *et al.* (2000) LDL-receptor-related proteins in Wnt signal transduction. *Nature*, **407**, 530–535.

Tamai, K., Zeng, X., Liu, C. *et al.* (2004) A mechanism for Wnt coreceptor activation. *Molecular Cell*, **13**, 149–156.

Tanneberger, K., Pfister, A., Brauburger, K. *et al.* (2011) Amer1/WTX couples Wnt-induced formation of

3 Wnt Signal Transduction in the Cytoplasm: an Introduction to the Destruction Complex

Tony W. Chen[1,2], Heather A. Wallace[3], Yashi Ahmed[3], and Ethan Lee[1,2]

[1]Department of Cell and Developmental Biology and Program in Developmental Biology, Vanderbilt University Medical Center, Nashville, TN, USA
[2]Vanderbilt Ingram Cancer Center, Vanderbilt University School of Medicine, Nashville, TN, USA
[3]Department of Genetics and the Norris Cotton Cancer Center, Geisel School of Medicine at Dartmouth, Hanover, NH, USA

Introduction: A central role for control of β-catenin degradation in the Wnt pathway

Wnt/β-catenin signal transduction, at its core, is a pathway that regulates steady-state cytoplasmic β-catenin levels to modulate its transcriptional activity. β-Catenin is continually synthesized and degraded in the cytoplasm by the β-catenin destruction complex when the Wnt pathway is inactive to maintain a low steady-state concentration of β-catenin. The destruction complex is composed of the scaffold proteins Axin and adenomatous polyposis coli (APC), the kinases GSK3 and CK1α, and β-catenin (Figure 3.1). The activation of Wnt signaling results in the removal of APC from the destruction complex and relocalization of the other components to the plasma membrane via the adaptor protein Dishevelled (Dsh), thereby disrupting destruction complex formation (see Chapter 2); this disruption of the destruction complex results in the stabilization and accumulation of cytoplasmic β-catenin, which enters the nucleus to mediate transcription (see Chapter 4).

Components of the β-catenin destruction complex

The β-catenin destruction complex is an efficient, highly regulated, macromolecular complex that, in the absence of a Wnt signal, efficiently maintains low steady-state levels of β-catenin within the cell. In spite of intense investigation utilizing cellular, biochemical, and crystallographic approaches, little is known about the detailed molecular mechanisms by which β-catenin is recruited, phosphorylated, ubiquitinated, and degraded by this remarkable machine. Herein, we will describe the major players involved in the formation of the β-catenin destruction complex.

The transcriptional coactivator β-catenin

β-Catenin was identified in *Drosophila* as the product of the segment polarity gene *armadillo* and in *Xenopus* also as a component of the adherens junctions (McCrea, Turck, and Gumbiner, 1991; Nusslein-Volhard and Wieschaus, 1980; see

Wnt Signaling in Development and Disease: Molecular Mechanisms and Biological Functions, First Edition.
Edited by Stefan Hoppler and Randall T. Moon.

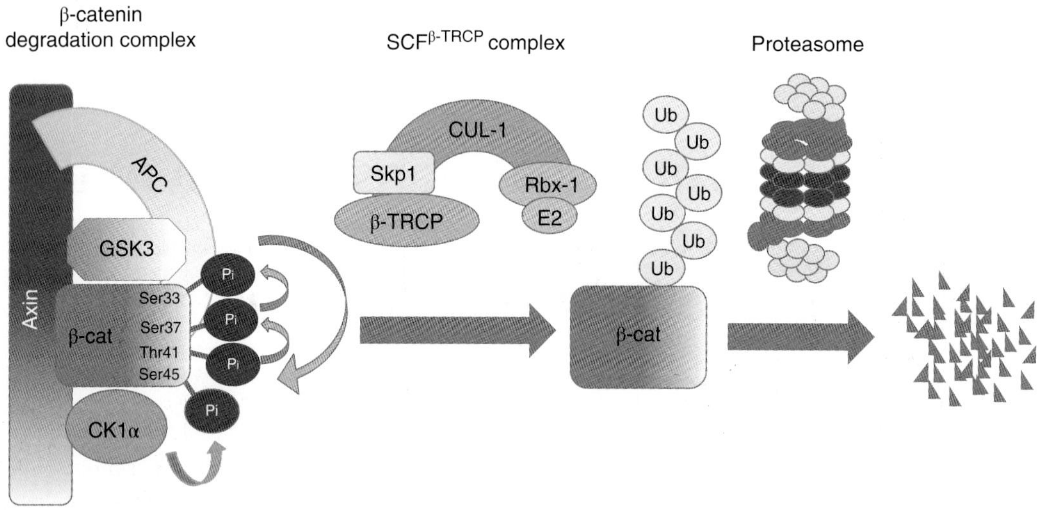

Figure 3.1 The core β-catenin destruction complex. The core β-catenin destruction complex consists of Axin, APC, GSK3, CK1α, and β-catenin. In the absence of a Wnt signal, these proteins form a complex that promotes β-TrCP-mediated ubiquitination of β-catenin, which is ultimately degraded by the proteasome. SCF$^{β\text{-}TrCP}$ consists of the F-box protein β-TrCP that binds phosphorylated β-catenin and the core catalytic complex containing Cul1, Skp1, and the Rbx1 RING finger protein. Binding of the E2 ubiquitin-conjugating enzyme to the SCF complex occurs via interaction with the Rbx1 protein. *(See insert for color representation of the figure.)*

Chapter 16). Elucidation of the crystal structure of β-catenin revealed a central superhelical structure consisting of 12 helical 42-amino acid armadillo repeats (Huber, Nelson, and Weis, 1997), which form a positively charged groove that mediates interactions between β-catenin and other Wnt pathway components (Graham *et al.*, 2000; Huber, Nelson, and Weis, 1997; Xing *et al.*, 2003, 2004). The N- and C-terminal domains of β-catenin are less structured and form dynamic intramolecular interactions with the armadillo repeats (Xing *et al.*, 2008; see Chapter 16).

In the absence of Wnt signaling, β-catenin is targeted for ubiquitin-mediated proteasomal degradation in the cytoplasm through its incorporation into a "destruction complex" (Aberle *et al.*, 1997; Hart *et al.*, 1999; Liu *et al.*, 1999; Orford *et al.*, 1997; Thomas *et al.*, 1999; Winston *et al.*, 1999). In the presence of Wnt signaling, β-catenin degradation is inhibited, its cytoplasmic levels rise, and it enters the nucleus to mediate a Wnt-specific transcriptional program.

Axin: The limiting component of the destruction complex

Axin, initially identified as the product of the *fused* gene in mice (Zeng *et al.*, 1997), is the scaffold protein that assembles components of the destruction complex to promote efficient phosphorylation and ubiquitin-mediated degradation of β-catenin (Figure 3.1 and Figure 3.2). The crystal structures of domains of Axin in complex with APC, β-catenin, or GSK3β have been solved (Dajani *et al.*, 2003; Spink, Polakis, and Weis, 2000; Xing *et al.*, 2003), although a structure containing multiple components of the destruction complex remains to be determined. Such a structure would provide insight into how β-catenin phosphorylation (and ubiquitination) is precisely regulated within this complex. In addition to its role in assembling the β-catenin degradation complex, studies in *Drosophila* suggest that Axin may also act as a cytoplasmic anchor to restrict Armadillo/β-catenin import into the nucleus (Peterson-Nedry *et al.*, 2008; Tolwinski and Wieschaus, 2001).

Axin is present at low concentrations and is the limiting component of the β-catenin destruction complex in *Xenopus* embryos (Lee *et al.*, 2003). The low concentration of Axin has been proposed to be a design feature that isolates the Wnt pathway from other signal transduction pathways because fluctuations in the concentration of the assembled Axin complex would minimally alter the concentrations of Wnt components that

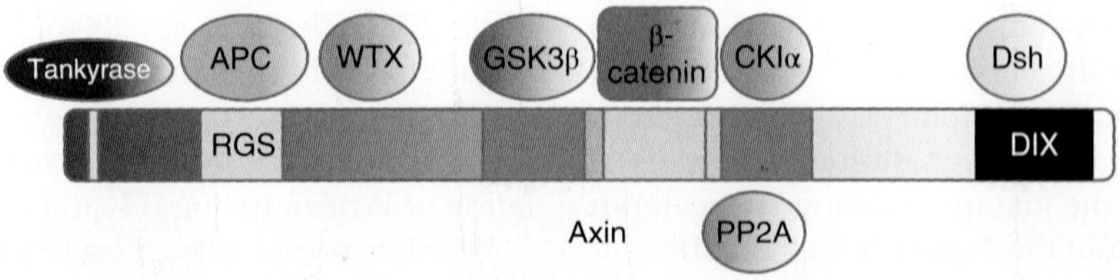

Figure 3.2 Axin is a multidomain scaffold protein that nucleates assembly of the β-catenin destruction complex. Axin contains binding sites for the core components of the destruction complex as well as Tankyrase, PP2A, WTX, and Dsh. The capacity of Axin to bind with high affinity to other components of the destruction complex and its limiting concentration make it a critical regulator of Wnt/β-catenin signaling. *(See insert for color representation of the figure.)*

play roles in other cellular pathways (e.g., GSK3, CK1α, and APC) (Lee *et al.*, 2003). The low intracellular level of Axin is due, in part, to its ubiquitin-mediated turnover, which is promoted by the actions of the enzymes Tankyrase and RNF146 (Callow *et al.*, 2011; Chen *et al.*, 2009; Huang *et al.*, 2009a; Zhang *et al.*, 2011).

Interestingly, given the positive role of both APC and Axin in promoting β-catenin degradation, APC appears to be required for Axin degradation (Lee *et al.*, 2003; Takacs *et al.*, 2008). The regulation of Axin turnover by APC has been proposed to represent a mechanism by which low levels of β-catenin are maintained in cells regardless of fluctuations in intracellular levels of APC (Lee *et al.*, 2003; Takacs *et al.*, 2008). According to this model, a reduction in APC levels, which would normally inhibit β-catenin degradation, would result in a compensatory increase in Axin concentration due to its decreased rate of turnover. Because Axin is limiting, this would presumably lead to an increase in the number of degradation complexes assembled, thereby mitigating the effect of reduced APC on β-catenin degradation. Axin turnover has also been shown to be promoted by LRP5/6 and inhibited by GSK3 (Cselenyi *et al.*, 2008; Mao *et al.*, 2001; Tolwinski *et al.*, 2003; Yamamoto *et al.*, 1999), although the biological significance of these proteins in this process is not clear. Axin degradation and its role in the β-catenin destruction complex will be discussed in more detail later in this review.

The Ser/Thr kinase GSK3

Initially identified as a regulator of glucose metabolism that phosphorylates muscle glycogen synthase (Embi, Rylatt, and Cohen, 1980), GSK3 (*Shaggy/Zeste White* 3 in *Drosophila*) is a serine/threonine protein kinase that regulates many critical cellular pathways (Forde and Dale, 2007; Siegfried, Chou, and Perrimon, 1992). β-Catenin phosphorylated by GSK3 (at Ser33, Ser37, and Thr41) acts as the substrate for its subsequent ubiquitination (Peifer, Pai, and Casey, 1994; Yost *et al.*, 1996) (Figure 3.1). Antagonizing the capacity of GSK3 to phosphorylate β-catenin is central to almost all mechanistic models of Wnt signaling. GSK3 also phosphorylates other major Wnt pathway components including APC, Axin, and LRP6, in addition to β-catenin (Rubinfeld *et al.*, 1996). In mammals, there are two distinct GSK3 genes, α and β, that have redundant functions in the Wnt pathway (Doble *et al.*, 2007). The crystal structure of GSK3β indicates a classic protein kinase bilobed structure topology consisting of an amino-terminal β-sheet domain linked to a carboxy-terminal α-helical domain (Dajani *et al.*, 2001; Haar *et al.*, 2001).

The consensus GSK3 phosphorylation site is Ser/Thr-X-X-X-Ser/Thr, where the first Ser/Thr is the site of GSK3 phosphorylation. GSK3 has a preference for substrates that are prephosphorylated ("primed") carboxyl terminal to its site of phosphorylation (last Ser/Thr of the consensus site). Priming of GSK3 substrates dramatically increases (up to 1000-fold) the efficiency of substrate phosphorylation by GSK3 compared to the unphosphorylated substrate (Thomas *et al.*, 1999). Phosphorylation of GSK3β at Ser9 (Ser21 in GSK3α) greatly reduces its kinase activity (Cross *et al.*, 1995; Sutherland, Leighton, and Cohen, 1993) and is predicted to form a primed pseudosubstrate that acts in an autoinhibitory fashion to block access to the catalytic site (Cohen and Frame, 2001; Cross *et al.*, 1995; Dajani *et al.*, 2001). Binding of the Ser9/Ser21 phosphorylated forms of GSK3 to the degradation complex has the potential to inhibit β-catenin degradation by acting in a dominant-negative fashion to sequester non-productive β-catenin degradation complexes. It is not clear if the juxtapositioning of β-catenin and GSK3 within the Axin complex is sufficient to overcome intramolecular inhibition by the phosphorylated pseudosubstrate of GSK3, and there may exist a mechanism by which the inhibited, phosphorylated form of GSK3 is "reactivated" – possibly via the action of a phosphatase.

The Ser/Thr kinase CK1α

As with GSK3, CK1 is a ubiquitously expressed family of Ser/Thr kinases (encoded by seven distinct genes in mammals (α, β, γ1, γ2, γ3, δ, and ε)) with a large number of substrates (Knippschild *et al.*, 2005; Price, 2006). All the CK1 family members have conserved catalytic domains, but they differ significantly in their C-terminal domains; it is not clear if the C-terminal domains of CK1 family members determine their substrate specificity and/or activity within the Wnt pathway. CK1 family members have been implicated as both positive and negative regulators of the Wnt pathway through their phosphorylation of various Wnt pathway components, including Dsh, LRP5, β-catenin, APC, Axin, and TCF/LEF (Cong, Schweizer, and Varmus, 2004; Gao *et al.*, 2002; Hammerlein, Weiske, and Huber, 2005; Kishida *et al.*, 2001; Lee, Salic, and Kirschner, 2001; Liu *et al.*, 2002; Peters *et al.*, 1999; Rubinfeld, Tice, and Polakis, 2001; Sakanaka *et al.*, 1999; Swiatek *et al.*, 2004; Yanagawa *et al.*, 1995; Zeng *et al.*, 2005; Zhang *et al.*, 2006).

The clearest involvement of a CK1 family member in regulating Wnt signaling is the negative regulation of Wnt signaling by CK1α, the *in vivo* priming kinase for GSK3 (Figure 3.1). CK1α phosphorylates β-catenin at Ser45, which acts as a priming site for subsequent phosphorylation by GSK3 (Liu *et al.*, 2002). Consistent with a primarily negative role for CK1α in Wnt signaling, two independent genome-wide *Drosophila S2* cell RNA interference (RNAi) screens and two *in vivo* loss-of-function studies revealed that CK1α is essential for suppressing Wnt/Wg signaling in unstimulated cells (DasGupta *et al.*, 2005; Legent *et al.*, 2012; Lum *et al.*, 2003; Zhang *et al.*, 2006). Evidence for a negative role in vertebrates comes from studies of CK1α in the murine gut in which conditional knockout of CK1α resulted in massive upregulation of Wnt signaling (Elyada *et al.*, 2011).

Whether or not phosphorylation of β-catenin by CK1α is regulated by Wnt signaling or is a constitutively active event is controversial. The initial identification of CK1α as the priming kinase required for subsequent GSK3-mediated phosphorylation of β-catenin suggested that its activity was not regulated in a Wnt-dependent manner (Liu *et al.*, 2002). Demonstration that Wnt signaling leads to a decrease in the levels of β-catenin phosphorylated by CK1α supports the hypothesis that CK1 activity is regulated upon Wnt pathway activation (Hernandez, Klein, and Kirschner, 2012). Indeed, a recent study shows that the RNA helicase, DDX3, regulates CK1 activity and is required for Wnt signaling and that this mechanism is evolutionarily conserved across phyla (Cruciat *et al.*, 2013).

The tumor suppressor APC

Studies of familial adenomatous polyposis (FAP), a familial form of colon cancer, revealed a causative mutation in the gene encoding APC (Kinzler *et al.*, 1991; Nishisho *et al.*, 1991). A clear connection between APC and Wnt signaling was subsequently established in a series of studies that showed that APC binds to β-catenin and that *APC* mutant cancer cells exhibit elevated levels of β-catenin (Munemitsu *et al.*, 1995; Rubinfeld *et al.*, 1993; Su, Vogelstein, and Kinzler, 1993). These studies provided clear evidence that APC acts as a negative regulator of the Wnt pathway.

APC is a large protein consisting of 2843 amino acids and with a predicted size of ~310 kDa. There are two distinct *APC* genes in humans, mice, and flies; *Drosophila Apc1* and *Apc2* are functionally redundant in regulating Wnt signaling in many cells (Ahmed, Nouri, and Wieschaus, 2002; Akong *et al.*, 2002). Several non-Wnt signaling cellular functions have been attributed to APC via distinct APC pools, including migration, microtubule dynamics, and apoptosis (Faux *et al.*, 2008; Nathke, 2006). Notably, APC has been shown to be involved in microtubule dynamics to regulate mitotic spindle alignment and cell migration that is mediated by its binding to EB1 and Discs large proteins (Matsumine *et al.*, 1996; Nathke, 2006; Su *et al.*, 1995).

It has been proposed that APC is regulated by its phosphorylation and ubiquitination, which alters its capacity to regulate Wnt signaling. APC is phosphorylated by the kinases CK1, protein kinase A (PKA), and GSK3 (Morin *et al.*, 1997; Rubinfeld *et al.*, 1996). Phosphorylation of APC by CK1 and GSK3 has been shown to enhance the affinity of APC for β-catenin by nearly 140-fold (Ha *et al.*, 2004; Rubinfeld *et al.*, 1996; Salic *et al.*, 2000). At least two deubiquitinases and one ubiquitin ligase have been shown to regulate APC activity. Trabid, which removes K63-linked ubiquitin chains from APC, positively regulates Wnt signaling, consistent with the hypothesis that K63-linked ubiquitination of APC potentiates APC activity (Tran *et al.*, 2008). In addition, the deubiquitinase USP15, which is associated with the β-catenin degradation complex, has been shown to stabilize APC (Huang *et al.*, 2009b). The HECT D1 E3 ligase ubiquitinates APC, which promotes its interaction with Axin (Tran *et al.*, 2013).

Similar to Axin, APC binds multiple components of the β-catenin degradation complex, including β-catenin, GSK3, and Axin (Eklof Spink, Fridman, and Weis, 2001; Fagotto *et al.*, 1999; Ikeda *et al.*, 1998; Itoh, Krupnik, and Sokol, 1998; Rubinfeld *et al.*, 1996). There remains intense speculation about exactly how the interaction between APC and multiple components of the β-catenin degradation machinery facilitates phosphorylation/ubiquitination of β-catenin or promotes assembly of the degradation complex. There have been numerous studies on APC (due, in part, to the importance of APC as a tumor suppressor in the genesis of colorectal cancer); yet a clear role for APC in controlling β-catenin degradation remains disappointingly elusive, and multiple models have been proposed for how APC functions within the Axin complex to control the destruction of β-catenin as well as β-catenin transcriptional activity (Cadigan and Peifer, 2009; MacDonald, Tamai, and He, 2009).

Other destruction complex factors

Our understanding of the molecular machinery involved in β-catenin destruction is incomplete. Preliminary molecular studies have implicated several additional components that have not received a thorough structural or biochemical analysis. These "noncore" components include the protein phosphatase PP2A, the tumor suppressor Amer1/WTX, the structural protein Diversin, and the protease Presenilin 1 (PS1).

The current body of work indicates that the phosphatase PP2A can act as either a negative or positive regulator of Wnt signaling. Thus, similar to that of CK1 and GSK3, its activity in the Wnt pathway (not surprisingly) may be context dependent. The PP2A catalytic subunit (C) has been shown to act on Axin and APC *in vitro* and potentiates the capacity of Dsh to induce secondary body axis formation in *Xenopus* embryos (Hsu, Zeng, and Costantini, 1999; Ratcliffe, Itoh, and Sokol, 2000; Willert, Shibamoto, and Nusse, 1999). In contrast, the PP2A subunits A, B56α, and C exhibit ventralizing activity in *Xenopus* embryos and inhibit Wnt signaling when overexpressed in cultured human cells (Gao *et al.*, 2002; Seeling *et al.*, 1999). Consistent with the biochemical activity of PP2A acting on the β-catenin degradation complex as a negative regulator of Wnt signaling, PP2A and its subunit, B56, have been shown to be required for β-catenin degradation in *Xenopus* egg extract (Gao *et al.*, 2002; Li *et al.*, 2001). The protein phosphatases PP2C and PP1 have also been implicated as positive regulators of Wnt signaling by opposing the phosphorylation of Axin by CK1, possibly leading to decreased association between Axin and GSK3 (Luo *et al.*, 2007; Strovel, Wu, and Sussman, 2000).

Both Amer1/WTX and Diversin have been shown to be associated with the β-catenin degradation complex to promote β-catenin turnover. Amer1/WTX stimulates β-catenin turnover through its interaction with the degradation complex in cultured mammalian cells and *Xenopus* egg extract (Major *et al.*, 2007). Diversin, an ankyrin repeat-containing protein, promotes the efficient priming of β-catenin (for subsequent GSK3 phosphorylation) via recruitment of CK1ε to Axin, possibly at centrosomes (Itoh *et al.*, 2009; Schwarz-Romond *et al.*, 2002).

PS1, the catalytic subunit of the protease γ-secretase that has been implicated in Notch signaling and Alzheimer's disease, also inhibits Wnt signaling (Kang *et al.*, 2002; Killick *et al.*, 2001). PS1 is thought to regulate β-catenin stability by acting as a scaffold protein similar to Axin to promote phosphorylation of β-catenin by GSK3, thereby targeting it for ubiquitin-mediated degradation (Kang *et al.*, 2002). In this model, PKA serves as the priming kinase (equivalent to the role of CK1α in the Axin complex), and degradation of β-catenin is dependent on E-cadherin and independent of APC (Kang *et al.*, 2002; Serban *et al.*, 2005).

Formation of the β-catenin destruction complex

Axin serves as the scaffold protein that nucleates the formation of the β-catenin destruction complex. It binds with high affinity to the two kinases of the complex, GSK3 and CK1α (Figure 3.1). GSK3 binds to a central region of Axin, whereas CK1 binds to a C-terminal region of Axin (Figure 3.2) (Dajani *et al.*, 2003; Sobrado *et al.*, 2005). The RGS domain of Axin, an N-terminal region that has structural homology to domains found in regulators of G protein signaling, binds to APC (Spink, Polakis, and Weis, 2000). This scaffold function of Axin brings together the components of the destruction complex that act in concert to bind, phosphorylate, and target β-catenin for degradation.

It is not clear whether binding of components of the β-catenin degradation complex is ordered or stochastic (Lee *et al.*, 2003). Regardless, β-catenin enters the complex by binding the 15-amino acid repeats of APC and a region of Axin that is C-terminal to the GSK3 binding site. The phosphorylation of Axin (e.g., by GSK3) increases its affinity for β-catenin (Willert, Shibamoto, and Nusse, 1999). Binding of β-catenin to Axin results in its phosphorylation by CK1 at Ser45 (Figure 3.1). This phosphorylated species of β-catenin is recognized by Axin-bound GSK3, which subsequently phosphorylates β-catenin at Thr41, Ser37, and Ser33 (Amit *et al.*, 2002; Liu *et al.*, 2002). APC is also phosphorylated by CK1 and GSK3, increasing its affinity for β-catenin (Ha *et al.*, 2004). Based on these observations, it has been proposed that phosphorylation of APC promotes a cycle in which phosphorylated β-catenin undergoes ubiquitin-mediated degradation, and nonphosphorylated β-catenin is recruited to the destruction complex (Kimelman and Xu, 2006).

β-Catenin degradation by the destruction complex

This cycle of Axin binding, GSK3 and CK1 phosphorylation, APC binding, and subsequent removal of phosphorylated/ubiquitinated β-catenin and incorporation of a nonphosphorylated/nonubiquitinated β-catenin molecules into the destruction complex occurs continuously in cells when Wnt/β-catenin signaling is not active (Figure 3.1). Once β-catenin is phosphorylated at Ser33 and Ser37, it is recognized by β-TrCP/Slimb, the specificity subunit of the Skp1–Cullin–F-box (SCF) E3 ubiquitin ligase complex (Jiang and Struhl, 1998; Kitagawa *et al.*, 1999; Lagna *et al.*, 1999; Liu *et al.*, 1999; Marikawa and Elinson, 1998; Winston *et al.*, 1999; Zu *et al.*, 2011). $SCF^{β\text{-}TrCP}$ binds to β-catenin and catalyzes its polyubiquitination (via K48 linkages that occur when a lysine residue of ubiquitin at position 48 is linked to the carboxy-terminal glycine of another ubiquitin) and subsequent proteasomal degradation (Figure 3.1). This process ensures that the steady-state levels of newly synthesized, free cytosolic pools of β-catenin are kept at a level below the threshold necessary for nuclear translocation and β-catenin-mediated gene regulation.

In contrast to the $SCF^{β\text{-}TrCP}$ complex, which ubiquitinates β-catenin via K48 linkages and targets β-catenin for degradation, β-catenin has also been shown to be ubiquitinated by the HECT domain-containing E3 ligase, EDD (identified by differential display), which modifies β-catenin via K29 or K11 linkages (Hay-Koren *et al.*, 2010). Ubiquitination of β-catenin by EDD, unlike ubiquitination by the $SCF^{β\text{-}TrCP}$ complex, inhibits β-catenin degradation via a mechanism that is not well understood. It will be interesting to determine whether the modification by EDD represents a significant mechanism for regulating β-catenin degradation and whether the sites of ubiquitination utilized by the $SCF^{β\text{-}TrCP}$ complex and EDD are the same or different.

The role of Axin turnover in regulating β-catenin turnover

The regulation of Axin protein turnover is a major question in the Wnt field given the critical role of Axin in Wnt signaling. Axin is thought to be tightly regulated due to its concentration-limiting role in the formation of the destruction complex. The importance of Axin turnover was uncovered by the identification of IWR-1 and XAV939, small molecule inhibitors of Tankyrase that potently inhibit Wnt signaling by increasing steady-state Axin levels (Chen *et al.*, 2009; Huang *et al.*, 2009a). Tankyrase was shown to promote the degradation of Axin through PARsylation of Axin (covalent addition of poly(ADP-ribose) moieties) (Huang *et al.*, 2009a). Poly(ADP-ribosylated) Axin is recognized by the poly(ADP-ribose)-directed E3 ubiquitin ligase RNF146, which ubiquitinates Axin, targeting it for degradation and maintaining low steady-state levels of Axin (Callow *et al.*, 2011; Zhang *et al.*, 2011). Another ubiquitin ligase, Smad ubiquitin regulatory factor 2 (Smurf2), has also been proposed to target Axin for degradation to promote Wnt signaling (Kim and Jho, 2010). Conversely, the deubiquitinating enzyme, USP34, has been shown to catalyze the removal of ubiquitin from Axin and to increase steady-state levels of Axin in cultured cells (Lui *et al.*, 2011). In contrast to PARsylation, SUMOylation (covalent attachment of a small ubiquitin-related modifier or SUMO) of Axin at its C-terminal end has been shown to confer stability by inhibiting Axin ubiquitination (Kim, Chia, and Costantini, 2008). Recently, quantitative measurements of Axin concentration in a variety of mammalian cells suggest that its levels vary significantly to alter the dynamics of Wnt signaling (Tan *et al.*, 2012). Thus, the regulation of Axin levels and stability is likely a major mechanism by which cells control the response to Wnt signals.

Potential roles of APC in the β-catenin destruction complex

The mechanisms by which APC negatively regulate Wnt/β-catenin signaling are seemingly shrouded in mystery. The proposed diverse roles of APC within the Wnt pathway have made it particularly difficult to interpret results of mechanistic studies of APC mutations and their effects on the Wnt pathway. The majority of models of APC action focus on its role for maintaining low steady-state levels of β-catenin through its association with the β-catenin degradation complex and include (i) localization of β-catenin to the cell cortex for ubiquitin-mediated degradation (McCartney *et al.*, 1999; Nathke *et al.*, 1996; Yu, Waltzer, and Bienz, 1999), (ii) release of β-catenin from the destruction complex for subsequent ubiquitination (Kimelman and Xu, 2006), (iii) protection of the phosphorylated form of β-catenin from the action of the phosphatase PP2A (Su *et al.*, 2008), and (iv) shielding the activity the β-catenin destruction complex from inhibition by Dsh (Mendoza-Topaz, Mieszczanek, and Bienz, 2011). Other proposed mechanisms of action for APC in the Wnt pathway include cytoplasmic anchoring of β-catenin (Ahmed, Nouri, and Wieschaus, 2002; Tolwinski and Wieschaus, 2001), nuclear export of β-catenin (Hamada and Bienz, 2004) (although recent studies indicate that APC does not need to shuttle into the nucleus in order to regulate Wnt signaling (Roberts *et al.*, 2012)), and Wnt target gene repression (Sierra *et al.*, 2006). Thus, it appears that APC participates at multiple levels in the β-catenin destruction complex and in other aspects of the Wnt pathway to regulate the activity of β-catenin. It is important to keep in mind, however, that some of the proposed activities of APC need to be more fully vetted. For example, although the action of APC has also been proposed to prevent the dephosphorylation of β-catenin by PP2A that would be predicted to block ubiquitin-mediated proteolysis of β-catenin (Su *et al.*, 2008), PP2A has also been shown to be required for β-catenin degradation (Gao *et al.*, 2002; Li *et al.*, 2001).

Models for inhibition of β-catenin degradation

Inhibition of β-catenin degradation and its subsequent cytoplasmic accumulation represent the central dogma of currently accepted models of Wnt/β-catenin pathway activation. The detailed mechanism by which canonical Wnt pathway activation leads to inhibition of the β-catenin destruction complex has been under intense investigation, although it remains controversial (Figure 3.3). The basis for β-catenin destruction complex inhibition can be categorized into three general mechanisms: (i) disruption of the β-catenin destruction complex (via complex dissociation and/or Axin degradation), (ii) inhibition of GSK3 activity (via its inhibitory

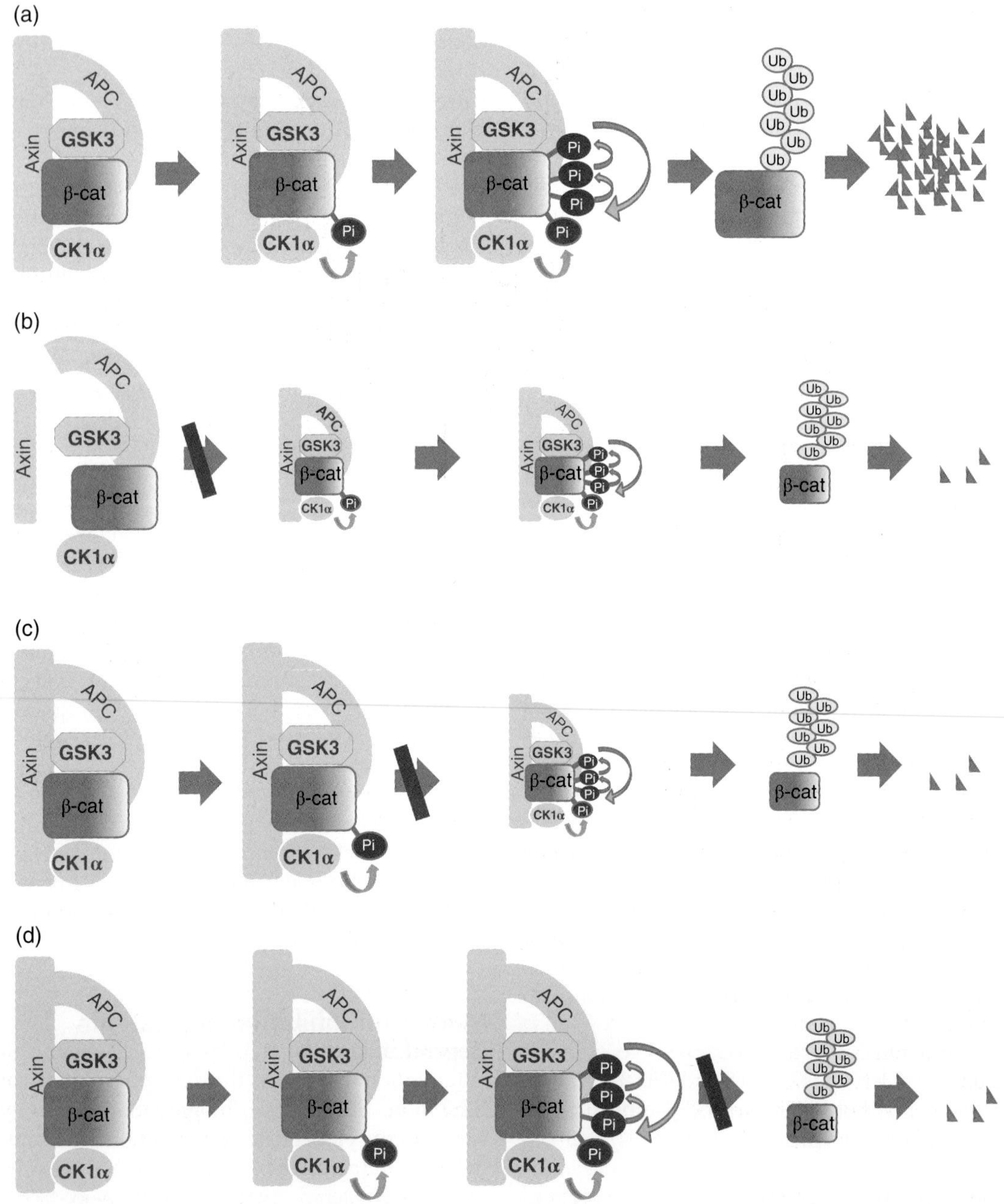

Figure 3.3 Proposed models for inhibition of the β-catenin degradation complex upon Wnt pathway activation. (a) In the absence of a Wnt signal, β-catenin is assembled into a complex where it is phosphorylated by CK1α and GSK3 and targeted for ubiquitin-mediated proteasomal degradation. (b) Wnt signaling inhibits β-catenin turnover by blocking its phosphorylation by CK1α and GSK3 (possibly due to disruption of complex assembly via subunit dissociation and/or Axin degradation). (c) Wnt signaling inhibits β-catenin phosphorylation by GSK3 only. (d) Wnt signaling does not inhibit β-catenin phosphorylation, but rather its ubiquitination. In b–d, blocking the degradation of β-catenin ultimately results in its increased cytoplasmic concentration, thereby promoting its entry into the nucleus and initiation of a Wnt-specific transcriptional program (see Chapter 4). *(See insert for color representation of the figure.)*

phosphorylation, sequestration, or the action of LRP6), and (iii) inhibition of β-catenin ubiquitination by the $SCF^{\beta\text{-TrCP}}$ complex. We discuss specific models in more detail in the following.

Dissociation of the β-catenin destruction complex

There are conflicting results from studies that assess the integrity of the β-catenin degradation complex and whether or not the degradation complex is disrupted upon Wnt pathway activation. Several studies have suggested that the β-catenin degradation complex remains intact and localizes to Frizzled (Fz) and LRP6 at the membrane upon Wnt stimulation (Bilić *et al.*, 2007; Hendriksen *et al.*, 2008; Li *et al.*, 2012; Mao *et al.*, 2001; Yamamoto, Komekado, and Kikuchi, 2006). Most models invoking dissociation of the complex have focused on the critical interaction between Axin and GSK3 (Gao *et al.*, 2002; Liu *et al.*, 2002; Logan and Nusse, 2004; Luo *et al.*, 2007). The basis of GSK3 dissociation from Axin was initially attributed to the action of GSK3 binding protein (GBP) in studies using *Xenopus laevis* embryos and egg extract (Farr *et al.*, 2000; Rubinfeld *et al.*, 1996; Salic *et al.*, 2000; Yost *et al.*, 1998). The apparent lack of requirement for GBP in Wnt signaling in mice and the lack of a GBP ortholog in *Drosophila* suggest that inhibition of GSK3 within the β-catenin degradation complex via the action of GBP is not a general mechanism by which the β-catenin degradation complex is inhibited (van Amerongen *et al.*, 2005). Regardless of the exact mechanism by which disruption of the β-catenin degradation complex occurs, recent studies by Hernandez, Klein, and Kirschner (2012) showed that both GSK3 and CK1 phosphorylation activities are regulated by Wnt pathway activation, further lending support to the model that disruption of the entire complex occurs by Wnt-mediated inhibition of β-catenin degradation (reviewed by Tacchelly-Benites *et al.*, 2013).

Degradation of the scaffold protein Axin in response to Wnt stimulation

Several studies from cell culture, *Xenopus* embryos and egg extract, and *Drosophila* suggest that a conserved and critical event of the Wnt pathway is the degradation of Axin upon ligand stimulation (Cselenyi *et al.*, 2008; Yamamoto *et al.*, 1999; Willert, Shibamoto, and Nusse, 1999). Because Axin is a concentration-limiting factor, the regulation of its stability is likely to have a dramatic effect on signaling (Lee *et al.*, 2003). *Drosophila* embryos genetically null for GSK3 have elevated levels of β-catenin (as expected), yet some aspects of Wg-dependent patterning remain intact when β-catenin activity is simultaneously reduced to more physiological levels in these embryos. These results suggest that modulated degradation of β-catenin is not required for Wnt signaling; instead, Axin inhibits the Wg pathway by acting as a "cytoplasmic anchor" to prevent nuclear translocation of Armadillo/β-catenin (Tolwinski *et al.*, 2003). Thus, Wnt-induced degradation of Axin likely acts as a mechanism by which β-catenin is liberated from the destruction complex to enter the nucleus.

Other studies have shown that Axin degradation appears to temporally lag behind β-catenin stabilization and, therefore, is not required for the initial stabilization of β-catenin in response to Wnt stimulation (Cselenyi *et al.*, 2008; Hino *et al.*, 2005; Willert, Shibamoto, and Nusse, 1999). These studies support a model in which the initial block in β-catenin destruction that occurs upon Wnt exposure does not require Axin turnover, and degradation of Axin is a distinct event that may modulate the nature or duration of the response.

Phosphorylation of GSK3 at Ser9/Ser21

Several studies have demonstrated that Wnt stimulation promotes the phosphorylation of GSK3 at Ser9/Ser21, inhibiting its activity within the β-catenin degradation complex (Ding, Chen, and McCormick, 2000; Yokoyama and Malbon, 2007). Although inhibitory phosphorylation of GSK3 at Ser9/Ser21 has been attributed to the activity of several kinases (e.g., AKT, S6 kinase, and ERK2) (Cross *et al.*, 1995; Sutherland, Leighton, and Cohen, 1993), the kinase that phosphorylates GSK3 at Ser9/Ser21 within the β-catenin degradation complex is unknown. Furthermore, the physiological relevance of GSK3 phosphorylation at Ser9/Ser21 is not clear, as there were no observable developmental defects and/or perturbation in Wnt signaling in mice in which nonphosphorylatable forms of GSK3 were "knocked in" (McManus *et al.*, 2005). Given the critical role of GSK3 in regulating multiple important biological pathways, it is possible that there are several ways the GSK3 activity could be inhibited. Further studies on the effects of inhibitory Ser9/Ser21 phosphorylation of GSK3 will be required to further elucidate its role (if any) in regulating the Wnt pathway.

Global inhibition of GSK3 by sequestration

Studies by Taelman *et al.* (2010) suggest that Wnt activation results in global inhibition of GSK3 activity via its sequestration into the lumen of multivesicular bodies (MVBs). Sequestration of GSK3 into MVBs is predicted to block its capacity to phosphorylate many of its physiological substrates (including β-catenin). Thus, Wnt signaling is predicted to have broad and substantial biological effects as GSK3 is involved in a large number of biological functions. Demonstration that Wnt signaling via inhibition of GSK3 alters the turnover of ~20% of all intracellular proteins (as assessed by pulse-chase experiments) has been provided as evidence for such a model (Taelman *et al.*, 2010), although these events were observed hours after actual Wnt pathway activation. If this model is correct, Wnt activation would be expected to alter many diverse cellular activities in which GSK3 plays a critical role (other than activating a β-catenin-mediated transcriptional program), including insulin signaling and glucose metabolism, translational control, vesicular transport, cytoskeletal dynamics, and apoptosis. Further confirmation for a model involving such profound effects of Wnt activation on a multitude of cell metabolism and signaling processes (especially their timing relative to Wnt pathway activation) is needed.

Inhibition of GSK3 by LRP6

A number of mammalian cell culture and biochemical studies suggest that the β-catenin destruction complex relocalizes to the cell surface upon Wnt stimulation (Bilić *et al.*, 2007; Cselenyi *et al.*, 2008; Hendriksen *et al.*, 2008; Piao *et al.*, 2008; Wu *et al.*, 2009; Yamamoto, Komekado, and Kikuchi, 2006; see Chapter 2). Translocation of Axin is initiated by the activation of LRP6 that occurs, in part, via its phosphorylation by Axin-bound GSK3 and is initially Fz and Dsh dependent (Mao *et al.*, 2001; Schwarz-Romond, Metcalfe, and Bienz, 2007; Zeng *et al.*, 2008). LRP6 phosphorylation (on its five PPPSPXS motifs) has been proposed to create multiple binding sites for Axin (see Chapter 2). Because Axin (bound to GSK3) is required for phosphorylation of LRP6, and phosphorylated LRP6 is required for Axin binding, it has been proposed that a positive feedback mechanism exists in which the initial stimulus is amplified (in a mechanism independent of Fz and Dsh) via phosphorylation of all five PPPSPxS motifs of LRP6 (Baig-Lewis, Peterson-Nedry, and Wehrli, 2007; Zeng *et al.*, 2008). It is not clear if there is successive stepwise recruitment of Axin–GSK3 and subsequent LRP6 phosphorylation or whether the initial recruitment of Axin–GSK3 by LRP6 results in a rapid processive phosphorylation of all five PPPSPXS motifs.

Paradoxically, the phosphorylated PPPSPXS motif of LRP6 has been shown to act as a direct competitive inhibitor of GSK3 (Cselenyi *et al.*, 2008; Piao *et al.*, 2008; Wu *et al.*, 2009) (see Chapter 2). Thus, the product of LRP6 phosphorylation by GSK3 (phosphorylated PPPSPXS) would potentially inhibit the kinase activity of GSK3-bound Axin. Such a mechanism would be necessarily limiting and would be inconsistent with a proposed amplification step in Wnt pathway activation. A careful biochemical study to more finely dissect this part of the pathway is clearly warranted in order to resolve this discrepancy.

Inhibition of β-catenin ubiquitination

Recently, by carefully analyzing Axin complexes immunoprecipitated from Wnt-stimulated cells, Li *et al.* (2012) provided evidence that Wnt activation does not inhibit β-catenin phosphorylation *per se*, but rather its ubiquitination by β-TrCP. In this new model of Wnt pathway activation, phosphorylated β-catenin accumulates within the complex because the complex is no longer able to target β-catenin for ubiquitination and subsequent degradation. Thus, Wnt activation serves to "tie up" the complex so as to prevent the degradation complex from catalyzing multiple cycles of β-catenin binding and ubiquitin-mediated degradation (Li *et al.*, 2012). This model is very different from the majority of proposed models of Wnt pathway activation in that inhibition of GSK3 activity (directly or indirectly) is not involved in pathway activation, and elevated intracellular levels of GSK3 phosphorylated β-catenin are predicted to be indicative of an "activated state" for the pathway.

β-Catenin turnover as a futile cycle

As previously mentioned, β-catenin is continually synthesized and rapidly degraded. This "futile" cycle of constitutive β-catenin synthesis–degradation is energetically expensive. Why would a cell utilize such a seemingly inefficient system to regulate the Wnt pathway? Such expenditure of energy may serve as a major mechanism of regulation and has been shown to exist in diverse biochemical circuits including GTPase cycles (Donovan, Shannon, and Bollag, 2002), protein phosphorylation–dephosphorylation in the MAPK pathway (Chang *et al.*, 2003), cell cycle regulation (Sulis and Parsons, 2003), actin assembly–disassembly (Chen, Bernstein, and Bamburg, 2000), and metabolic regulation (Newsholme and Crabtree, 1976). The characteristic switch-like behavior associated with futile cycles is thought to allow the system to exhibit complex behavioral patterns in such a way that even small changes in enzymatic activities and/or concentrations of constituent components could potentially be exploited to significantly alter pathway behavior (Cohen and Frame, 2001; Samoilov, Plyasunov, and Arkin, 2005). In the case of the β-catenin destruction complex, such small changes (e.g., in GSK3 and/or CKIα activity) would allow β-catenin levels to rapidly overcome the threshold required for transcriptional activation in response to stimulus. Finally, the maintenance of futile cycles appears to be critical for systems in which low concentrations and limited affinities might render them susceptible to cellular noise. In these systems, a futile cycle may impart critical control on the system to increase its sensitivity, specificity, and robustness (Cohen and Frame, 2001; Qian, 2003; Qian and Reluga, 2005; Zhu *et al.*, 2004). Further experimental and mathematical exploration of how the futile cycle of β-catenin turnover may contribute to its complex behavior and regulation will likely yield critical insights into how this pathway may be fine-tuned to promote the complex patterning of an organism during development as well as its misregulation in pathological conditions (e.g., cancer).

Future directions

Although much is known about the activities of the individual major core members of the β-catenin destruction complex, the molecular basis by which they are assembled and how the individual components are coordinated to drive phosphorylation and ubiquitin-mediated proteolysis of β-catenin remain elusive. Specifically, the role of APC in the β-catenin destruction complex remains unclear with many potential functions that individually require a deeper level of analysis. The roles of other β-catenin destruction complex components (e.g., PP2A, PP1, and WTX) must also be elucidated because, in spite of their designation as "noncore" components, their activities are critical for efficient functioning of the complex.

The second major question that remains to be resolved is the mechanism by which the β-catenin degradation complex is inhibited upon Wnt pathway activation. Most of the models that have been proposed suggest that phosphorylation of β-catenin (primarily by GSK3) is the critical point at which upstream Wnt pathway components impinge on the β-catenin degradation machinery. It remains to be determined how this is mechanistically achieved (e.g., via dissociation of the assembled complex or direct kinase inhibition).

Although genetic and cell-based studies have greatly contributed to our knowledge of the functions and mechanistic underpinnings of the β-catenin destruction complex, a complete understanding of this intricate macromolecular machine will require its *in vitro* assembly. Thus, one near-term goal is to biochemically reconstitute the phosphorylation and ubiquitination of β-catenin using purified components with the long-term goal of reconstituting receptor-mediated inhibition of the complex. A similar approach was successfully used in pioneering studies of the heterotrimeric G protein system and thus is not merely a "pipe dream" (Gilman, 1987). If successful, the results from such experiments would significantly advance our understanding of this critical step in the Wnt signaling pathway.

Acknowledgments

The authors would like to thank A. Hanson, C. Thorne, and L. Lee for their input and help in editing the manuscript. This work was supported by NIH R01GM081635 and R01GM103926 (E.L); Emerald Foundation, National Cancer Institute (RO1CA105038 to Y.A); NIH 5T32CA11992504 (H.A.W); and American Heart Association 12 PRE 6590007 (T.W.C).

References

Aberle, H., Bauer, A., Stappert, J. *et al.* (1997) Beta-catenin is a target for the ubiquitin–proteasome pathway. *The EMBO Journal*, **16**, 3797–3804.

Ahmed, Y., Nouri, A., and Wieschaus, E. (2002) Drosophila Apc1 and Apc2 regulate Wingless transduction throughout development. *Development*, **129**, 1751–1762.

Akong, K., Grevengoed, E.E., Price, M.H. *et al.* (2002) Drosophila APC2 and APC1 play overlapping roles in wingless signaling in the embryo and imaginal discs. *Developmental Biology*, **250**, 91–100.

Amit, S., Hatzubai, A., Birman, Y. *et al.* (2002) Axin-mediated CKI phosphorylation of beta-catenin at Ser 45: a molecular switch for the Wnt pathway. *Genes & Development*, **16**, 1066–1076.

Baig-Lewis, S., Peterson-Nedry, W., and Wehrli, M. (2007) Wingless/Wnt signal transduction requires distinct initiation and amplification steps that both depend on Arrow/LRP. *Developmental Biology*, **306**, 94–111.

Bilić, J., Huang, Y-L., Davidson, G. *et al.* (2007) Wnt Induces LRP6 signalosomes and promotes dishevelled-dependent LRP6 phosphorylation. *Science*, **316**, 1619–1622.

Cadigan, K.M. and Peifer, M. (2009) Wnt signaling from development to disease: insights from model systems. *Cold Spring Harbor Perspectives in Biology*, **1**, a002881.

Callow, M.G., Tran, H., Phu, L. *et al.* (2011) Ubiquitin ligase RNF146 regulates tankyrase and Axin to promote Wnt signaling. *PLoS One*, **6**, e22595.

Chang, F., Steelman, L.S., Lee, J.T. *et al.* (2003) Signal transduction mediated by the Ras/Raf/MEK/ERK pathway from cytokine receptors to transcription factors: potential targeting for therapeutic intervention. *Leukemia*, **17**, 1263–1293.

Chen, H., Bernstein, B.W., and Bamburg, J.R. (2000) Regulating actin-filament dynamics in vivo. *Trends in Biochemical Sciences*, **25**, 19–23.

Chen, B., Dodge, M.E., Tang, W. *et al.* (2009) Small molecule-mediated disruption of Wnt-dependent signaling in tissue regeneration and cancer. *Nature Chemical Biology*, **5**, 100–107.

Cohen, P. and Frame, S. (2001) The renaissance of GSK3. *Nature Reviews Molecular Cell Biology*, **2**, 769–776.

Cong, F., Schweizer, L., and Varmus, H. (2004) Casein kinase Iepsilon modulates the signaling specificities of dishevelled. *Molecular Cell Biology*, **24**, 2000–2011.

Cross, D.A., Alessi, D.R., Cohen, P. *et al.* (1995) Inhibition of glycogen synthase kinase-3 by insulin mediated by protein kinase B. *Nature*, **378**, 785–789.

Cruciat, C.M., Dolde, C., de Groot, R.E. *et al.* (2013) RNA helicase DDX3 is a regulatory subunit of casein kinase 1 in Wnt-β-catenin. *Science*, **339**, 1436–1441.

Cselenyi, C.S., Jernigan, K.K., Tahinci, E. *et al.* (2008) LRP6 transduces a canonical Wnt signal independently of Axin degradation by inhibiting GSK3's phosphorylation of β-catenin. *Proceedings of the National Academy of Sciences of the United States of America*, **105**, 8032–8037.

Dajani, R., Fraser, E., Roe, S.M. *et al.* (2001) Crystal structure of glycogen synthase kinase 3 beta: structural basis for phosphate-primed substrate specificity and autoinhibition. *Cell*, **105**, 721–732.

Dajani, R., Fraser, E., Roe, S.M. *et al.* (2003) Structural basis for recruitment of glycogen synthase kinase 3beta to the axin–APC scaffold complex. *The EMBO Journal*, **22**, 494–501.

DasGupta, R., Kaykas, A., Moon, R.T., and Perrimon, N. (2005) Functional genomic analysis of the Wnt-wingless signaling pathway. *Science*, **308**, 826–833.

Ding, V.W., Chen, R.H., and McCormick, F. (2000) Differential regulation of glycogen synthase kinase 3beta by insulin and Wnt signaling. *The Journal of Biological Chemistry*, **275**, 32475–32481.

Doble, B.W., Patel, S., Wood, G.A. *et al.* (2007) Functional redundancy of GSK-3alpha and GSK-3beta in Wnt/beta-catenin signaling shown by using an allelic series of embryonic stem cell lines. *Developmental Cell*, **12**, 957–971.

Donovan, S., Shannon, K.M., and Bollag, G. (2002) GTPase activating proteins: critical regulators of intracellular signaling. *Biochimica et Biophysica Acta*, **1602**, 23–45.

Eklof Spink, K., Fridman, S.G., and Weis, W.I. (2001) Molecular mechanisms of beta-catenin recognition by adenomatous polyposis coli revealed by the structure of an APC-beta-catenin complex. *The EMBO Journal*, **20**, 6203–6212.

Elyada, E., Pribluda, A., Goldstein, R.E. *et al.* (2011) CKIalpha ablation highlights a critical role for p53 in invasiveness control. *Nature*, **470**, 409–413.

Embi, N., Rylatt, D.B., and Cohen, P. (1980) Glycogen synthase kinase-3 from rabbit skeletal muscle. Separation from cyclic-AMP-dependent protein kinase and phosphorylase kinase. *European Journal of Biochemistry*, **107**, 519–527.

Fagotto, F., Jho, E., Zeng, L. *et al.* (1999) Domains of axin involved in protein–protein interactions, Wnt pathway inhibition, and intracellular localization. *The Journal of Cell Biology*, **145**, 741–756.

Farr, G.H., 3rd, Ferkey, D.M., Yost, C. *et al.* (2000) Interaction among GSK-3, GBP, axin, and APC in *Xenopus* axis specification. *The Journal of Cell Biology*, **148**, 691–702.

Faux, M.C., Coates, J.L., Catimel, B. *et al.* (2008) Recruitment of adenomatous polyposis coli and beta-catenin to axin-puncta. *Oncogene*, **27**, 5808–5820.

Forde, J.E. and Dale, T.C. (2007) Glycogen synthase kinase 3: a key regulator of cellular fate. *Cellular and Molecular Life Sciences*, **64**, 1930–1944.

Gao, Z.H., Seeling, J.M., Hill, V. *et al.* (2002) Casein kinase I phosphorylates and destabilizes the beta-catenin degradation complex. *Proceedings of the National Academy of Sciences of the United States of America*, **99**, 1182–1187.

Gilman, A.G. (1987) G proteins: transducers of receptor-generated signals. *Annual Reviews of Biochemistry*, **56**, 615–649.

Graham, T.A., Weaver, C., Mao, F. *et al.* (2000) Crystal structure of a beta-catenin/Tcf complex. *Cell*, **103**, 885–896.

Ha, N-C., Tonozuka, T., Stamos, J.L. *et al.* (2004) Mechanism of phosphorylation-dependent binding of APC to [beta]-catenin and its role in [beta]-catenin degradation. *Molecular Cell*, **15**, 511–521.

Haar, E.t., Coll, J.T., Austen, D.A. *et al.* (2001) Structure of GSK3beta reveals a primed phosphorylation mechanism. *Nature Structural Biology*, **8**, 593–596.

Hamada, F. and Bienz, M. (2004) The APC tumor suppressor binds to C-terminal binding protein to divert nuclear beta-catenin from TCF. *Developmental Cell*, **7**, 677–685.

Hammerlein, A., Weiske, J., and Huber, O. (2005) A second protein kinase CK1-mediated step negatively regulates Wnt signalling by disrupting the lymphocyte enhancer factor-1/beta-catenin complex. *Cellular and Molecular Life Sciences*, **62**: 606–618.

Hart, M., Concordet, J.P., Lassot, I. *et al.* (1999) The F-box protein beta-TrCP associates with phosphorylated beta-catenin and regulates its activity in the cell. *Current Biology*, **9**, 207–210.

Hay-Koren, A., Caspi, M., Zilberberg, A., and Rosin-Arbesfeld, R. (2010) The EDD E3 ubiquitin ligase ubiquitinates and up-regulates beta-catenin. *Molecular Biology of the Cell*, **22**, 399–411.

Hendriksen, J., Jansen, M., Brown, C.M. *et al.* (2008) Plasma membrane recruitment of dephosphorylated beta-catenin upon activation of the Wnt pathway. *Journal of Cell Science*, **121**, 1793–1802.

Hernandez, A.R., Klein, A.M., and Kirschner, M.W. (2012) Kinetic responses of beta-catenin specify the sites of Wnt control. *Science*, **338**, 1337–1340.

Hino, S., Tanji, C., Nakayama, K.I., and Kikuchi, A. (2005) Phosphorylation of beta-catenin by cyclic AMP-dependent protein kinase stabilizes beta-catenin through inhibition of its ubiquitination. *Molecular Cell Biology*, **25**, 9063–9072.

Hsu, W., Zeng, L., and Costantini, F. (1999) Identification of a domain of Axin that binds to the serine/threonine protein phosphatase 2A and a self-binding domain. *The Journal of Biological Chemistry*, **274**, 3439–3445.

Huang, S-M.A., Mishina, Y.M., Liu, S. *et al.* (2009a) Tankyrase inhibition stabilizes axin and antagonizes Wnt signalling. *Nature*, **461**, 614–620.

Huang, X., Langelotz, C., Hetfeld-Pechoc, B.K. *et al.* (2009b) The COP9 signalosome mediates beta-catenin degradation by deneddylation and blocks adenomatous polyposis coli destruction via USP15. *Journal of Molecular Biology*, **391**, 691–702.

Huber, A.H., Nelson, W.J., and Weis, W.I. (1997) Three-dimensional structure of the armadillo repeat region of beta-catenin. *Cell*, **90**, 871–882.

Ikeda, S., Kishida, S., Yamamoto, H. *et al.* (1998) Axin, a negative regulator of the Wnt signaling pathway, forms a complex with GSK-3[beta] and [beta]-catenin and promotes GSK-3[beta]-dependent phosphorylation of [beta]-catenin. *The EMBO Journal*, **17**, 1371–1384.

Itoh, K., Krupnik, V.E., and Sokol, S.Y. (1998) Axis determination in *Xenopus* involves biochemical interactions of axin, glycogen synthase kinase 3 and [beta]-catenin. *Current Biology*, **8**, 591–594.

Itoh, K., Jenny, A., Mlodzik, M., and Sokol, S.Y. (2009) Centrosomal localization of Diversin and its relevance to Wnt signaling. *Journal of Cell Science*, **122**, 3791–3798.

Jiang, J. and Struhl, G. (1998) Regulation of the Hedgehog and Wingless signalling pathways by the F-box/WD40-repeat protein Slimb. *Nature*, **391**, 493–496.

Kang, D.E., Soriano, S., Xia, X. *et al.* (2002) Presenilin couples the paired phosphorylation of beta-catenin independent of axin: implications for beta-catenin activation in tumorigenesis. *Cell*, **110**, 751–762.

Killick, R., Pollard, C.C., Asuni, A.A. *et al.* (2001) Presenilin 1 independently regulates beta-catenin stability and transcriptional activity. *The Journal of Biological Chemistry*, **276**, 48554–48561.

Kim, S. and Jho, E-H. (2010) The protein stability of Axin, a negative regulator of Wnt signaling, is regulated by Smad ubiquitination regulatory factor 2 (Smurf2). *The Journal of Biological Chemistry*, **285**, 36420–36426.

Kim, M.J., Chia, I.V., and Costantini, F. (2008) SUMOylation target sites at the C terminus protect Axin from ubiquitination and confer protein stability. *The FASEB Journal*, **22**, 3785–3794.

Kimelman, D. and Xu, W. (2006) Beta-catenin destruction complex: insights and questions from a structural perspective. *Oncogene*, **25**, 7482–7491.

Kinzler, K.W., Nilbert, M.C., Su, L.K. *et al.* (1991) Identification of FAP locus genes from chromosome 5q21. *Science*, **253**, 661–665.

Kishida, M., Hino, S., Michiue, T. *et al.* (2001) Synergistic activation of the Wnt signaling pathway by Dvl and casein kinase Iepsilon. *The Journal of Biological Chemistry*, **276**, 33147–33155.

Kitagawa, M., Hatakeyama, S., Shirane, M. *et al.* (1999) An F-box protein, FWD1, mediates ubiquitin-dependent proteolysis of beta-catenin. *The EMBO Journal*, **18**, 2401–2410.

Knippschild, U., Gocht, A., Wolff, S. *et al.* (2005) The casein kinase 1 family: participation in multiple cellular processes in eukaryotes. *Cellular Signalling*, **17**, 675–689.

Lagna, G., Carnevali, F., Marchioni, M., and Hemmati-Brivanlou, A. (1999) Negative regulation of axis formation and Wnt signaling in *Xenopus* embryos by the F-box/WD40 protein beta TrCP. *Mechanisms of Development*, **80**, 101–106.

Lee, E., Salic, A., and Kirschner, M.W. (2001) Physiological regulation of [beta]-catenin stability by Tcf3 and CK1epsilon. *The Journal of Cell Biology*, **154**, 983–993.

Lee, E., Salic, A., Krüger, R. *et al.* (2003) The roles of APC and Axin derived from experimental and theoretical analysis of the Wnt pathway. *PLoS Biology*, **1**, E10.

Legent, K., Steinhauer, J., Richard, M., and Treisman, J.E. (2012) A screen for X-linked mutations affecting *Drosophila* photoreceptor differentiation identifies casein kinase 1alpha as an essential negative regulator of wingless signaling.

Li, X., Yost, H.J., Virshup, D.M., and Seeling, J.M. (2001) Protein phosphatase 2A and its B56 regulatory subunit inhibit Wnt signaling in *Xenopus*. *The EMBO Journal*, **20**, 4122–4131.

Li, V.S., Ng, S.S., Boersema, P.J. *et al.* (2012) Wnt signaling through inhibition of beta-catenin degradation in an intact Axin1 complex. *Cell*, **149**, 1245–1256.

Liu, C., Kato, Y., Zhang, Z. *et al.* (1999) beta-Trcp couples beta-catenin phosphorylation–degradation and regulates *Xenopus* axis formation. *Proceedings of the National Academy of Sciences of the United States of America*, **96**, 6273–6278.

Liu, C., Li, Y., Semenov, M. *et al.* (2002) Control of beta-catenin phosphorylation/degradation by a dual-kinase mechanism. *Cell*, **108**, 837–847.

Logan, C.Y. and Nusse, R. (2004) The Wnt signaling pathway in development and disease. *Annual Review of Cell and Developmental Biology*, **20**, 781–810.

Lui, T.T.H., Lacroix, C., Ahmed, S.M. *et al.* (2011) The ubiquitin-specific protease USP34 regulates Axin stability and Wnt/β-catenin signaling. *Molecular and Cellular Biology*, **31**, 2053–2065.

Lum, L., Yao, S., Mozer, B. *et al.* (2003) Identification of Hedgehog pathway components by RNAi in Drosophila cultured cells. *Science*, **299**, 2039–2045.

Luo, W., Peterson, A., Garcia, B.A. *et al.* (2007) Protein phosphatase 1 regulates assembly and function of the beta-catenin degradation complex. *The EMBO Journal*, **26**, 1511–1521.

MacDonald, B.T., Tamai, K., and He, X. (2009) Wnt/β-catenin signaling: components, mechanisms, and diseases. *Developmental Cell*, **17**, 9–26.

Major, M.B., Camp, N.D., Berndt, J.D. *et al.* (2007) Wilms tumor suppressor WTX negatively regulates WNT/beta-catenin signaling. *Science*, **316**, 1043–1046.

Mao, J., Wang, J., Liu, B. *et al.* (2001) Low-density lipoprotein receptor-related protein-5 binds to Axin and regulates the canonical Wnt signaling pathway. *Molecular Cell*, **7**, 801–809.

Marikawa, Y. and Elinson, R.P. (1998) Beta-TrCP is a negative regulator of Wnt/beta-catenin signaling pathway and dorsal axis formation in *Xenopus* embryos. *Mechanisms of Development*, **77**, 75–80.

Matsumine, A., Ogai, A., Senda, T. *et al.* (1996) Binding of APC to the human homolog of the Drosophila discs large tumor suppressor protein. *Science*, **272**, 1020–1023.

McCartney, B.M., Dierick, H.A., Kirkpatrick, C. *et al.* (1999) Drosophila APC2 is a cytoskeletally-associated protein that regulates wingless signaling in the embryonic epidermis. *The Journal of Cell Biology*, **146**, 1303–1318.

McCrea, P.D., Turck, C.W., and Gumbiner, B. (1991) A homolog of the armadillo protein in Drosophila (plakoglobin) associated with E-cadherin. *Science*, **254**, 1359–1361.

McManus, E.J., Sakamoto, K., Armit, L.J. *et al.* (2005) Role that phosphorylation of GSK3 plays in insulin and Wnt signalling defined by knockin analysis. *The EMBO Journal*, **24**, 1571–1583.

Mendoza-Topaz, C., Mieszczanek, J., and Bienz, M. (2011) The adenomatous polyposis coli tumour suppressor is essential for Axin complex assembly and function and opposes Axin's interaction with Dishevelled. *Open Biology*, **1**, 110013.

Morin, P.J., Sparks, A.B., Korinek, V. *et al.* (1997) Activation of beta-catenin-Tcf signaling in colon cancer by mutations in beta-catenin or APC. *Science*, **275**, 1787–1790.

Munemitsu, S., Albert, I., Souza, B. *et al.* (1995) Regulation of intracellular beta-catenin levels by the adenomatous polyposis coli (APC) tumor-suppressor protein. *Proceedings of the National Academy of Sciences*, **92**, 3046–3050.

Nathke, I. (2006) Cytoskeleton out of the cupboard: colon cancer and cytoskeletal changes induced by loss of APC. *Nature Reviews Cancer*, **6**, 967–974.

Nathke, I.S., Adams, C.L., Polakis, P. *et al.* (1996) The adenomatous polyposis coli tumor suppressor protein localizes to plasma membrane sites involved in active cell migration. *The Journal of Cell Biology*, **134**, 165–179.

Newsholme, E.A. and Crabtree, B. (1976) Substrate cycles in metabolic regulation and in heat generation. *Biochemical Society Symposium*, 61–109.

Nishisho, I., Nakamura, Y., Miyoshi, Y. *et al.* (1991) Mutations of chromosome 5q21 genes in FAP and colorectal cancer patients. *Science*, **253**, 665–669.

Nusslein-Volhard, C. and Wieschaus, E. (1980) Mutations affecting segment number and polarity in Drosophila. *Nature*, **287**, 795–801.

Orford, K., Crockett, C., Jensen, J.P. *et al.* (1997) Serine phosphorylation-regulated ubiquitination and degradation of beta-catenin. *The Journal of Biological Chemistry*, **272**, 24735–24738.

Peifer, M., Pai, L.M., and Casey, M. (1994) Phosphorylation of the Drosophila adherens junction protein Armadillo: roles for wingless signal and zeste-white 3 kinase. *Developmental Biology*, **166**, 543–556.

Peters, J.M., McKay, R.M., McKay, J.P., and Graff, J.M. (1999) Casein kinase I transduces Wnt signals. *Nature*, **401**, 345–350.

Peterson-Nedry, W., Erdeniz, N., Kremer, S. *et al.* (2008) Unexpectedly robust assembly of the Axin destruction complex regulates Wnt/Wg signaling in Drosophila as revealed by analysis in vivo. *Developmental Biology*, **320**, 226–241.

Piao, S., Lee, S-H., Kim, H. *et al.* (2008) Direct inhibition of GSK3β by the phosphorylated cytoplasmic domain of LRP6 in Wnt/β-catenin signaling. *PLoS One*, **3**, e4046.

Price, M.A. (2006) CKI, there's more than one: casein kinase I family members in Wnt and Hedgehog signaling. *Genes & Development*, **20**, 399–410.

Qian, H. (2003) Thermodynamic and kinetic analysis of sensitivity amplification in biological signal transduction. *Biophysical Chemistry*, **105**, 585–593.

Qian, H. and Reluga, T.C. (2005) Nonequilibrium thermodynamics and nonlinear kinetics in a cellular signaling switch. *Physical Review Letters*, **94**, 028101.

Ratcliffe, M.J., Itoh, K., and Sokol, S.Y. (2000) A positive role for the PP2A catalytic subunit in Wnt signal transduction. *The Journal of Biological Chemistry*, **275**, 35680–35683.

Roberts, D.M., Pronobis, M.I., Poulton, J.S. *et al.* (2012) Regulation of Wnt signaling by the tumor suppressor adenomatous polyposis coli does not require the ability to enter the nucleus or a particular cytoplasmic localization. *Molecular Biology of the Cell*, **23**, 2041–2056.

Rubinfeld, B., Tice, D.A., and Polakis, P. (2001) Axin-dependent phosphorylation of the adenomatous polyposis coli protein mediated by casein kinase 1epsilon. *The Journal of Biological Chemistry*, **276**, 39037–39045.

Rubinfeld, B., Souza, B., Albert, I. *et al.* (1993) Association of the APC gene product with beta-catenin. *Science*, **262**, 1731–1734.

Rubinfeld, B., Albert, I., Porfiri, E. *et al.* (1996) Binding of GSK3β to the APC-β-catenin complex and regulation of complex assembly. *Science*, **272**, 1023–1026.

Sakanaka, C., Leong, P., Xu, L. *et al.* (1999). Casein kinase iepsilon in the wnt pathway: regulation of beta-catenin function. *Proceedings of the National Academy of Sciences of the United States of America*, **96**, 12548–12552.

Salic, A., Lee, E., Mayer, L., and Kirschner, M.W. (2000) Control of beta-catenin stability: reconstitution of the cytoplasmic steps of the wnt pathway in *Xenopus* egg extracts. *Molecular Cell*, **5**, 523–532.

Samoilov, M., Plyasunov, S., and Arkin, A.P. (2005). Stochastic amplification and signaling in enzymatic futile cycles through noise-induced bistability with oscillations. *Proceedings of the National Academy of Sciences of the United States of America*, **102**, 2310–2315.

Schwarz-Romond, T., Metcalfe, C., and Bienz, M. (2007) Dynamic recruitment of axin by Dishevelled protein assemblies. *Journal of Cell Science*, **120**, 2402–2412.

Schwarz-Romond, T., Asbrand, C., Bakkers, J. *et al.* (2002) The ankyrin repeat protein Diversin recruits Casein kinase Iepsilon to the beta-catenin degradation complex and acts in both canonical Wnt and Wnt/JNK signaling. *Genes & Development*, **16**, 2073–2084.

Seeling, J.M., Miller, J.R., Gil, R. *et al.* (1999) Regulation of beta-catenin signaling by the B56 subunit of protein phosphatase 2A. *Science*, **283**, 2089–2091.

Serban, G., Kouchi, Z., Baki, L. *et al.* (2005) Cadherins mediate both the association between PS1 and beta-catenin and the effects of PS1 on beta-catenin stability. *The Journal of Biological Chemistry*, **280**, 36007–36012.

Siegfried, E., Chou, T.B., and Perrimon, N. (1992) wingless signaling acts through zeste-white 3, the Drosophila homolog of glycogen synthase kinase-3, to regulate engrailed and establish cell fate. *Cell*, **71**, 1167–1179.

Sierra, J., Yoshida, T., Joazeiro, C.A., and Jones, K.A. (2006) The APC tumor suppressor counteracts beta-catenin activation and H3K4 methylation at Wnt target genes. *Genes & Development*, **20**, 586–600.

Sobrado, P., Jedlicki, A., Bustos, V.H. *et al.* (2005) Basic region of residues 228–231 of protein kinase CK1alpha is involved in its interaction with axin: binding to axin does not affect the kinase activity. *Journal of Cell Biochemistry*, **94**, 217–224.

Spink, K.E., Polakis, P., and Weis, W.I. (2000) Structural basis of the Axin-adenomatous polyposis coli interaction. *The EMBO Journal*, **19**, 2270–2279.

Strovel, E.T., Wu, D., and Sussman, D.J. (2000) Protein phosphatase 2Calpha dephosphorylates axin and activates LEF-1-dependent transcription. *The Journal of Biological Chemistry*, **275**, 2399–2403.

Su, L.K., Vogelstein, B., and Kinzler, K.W. (1993) Association of the APC tumor suppressor protein with catenins. *Science*, **262**, 1734–1737.

Su, L-K., Burrell, M., Hill, D.E. *et al.* (1995) APC binds to the novel protein EB1. *Cancer Research*, **55**, 2972–2977.

Su, Y., Fu, C., Ishikawa, S. *et al.* (2008) APC is essential for targeting phosphorylated beta-catenin to the SCFbeta-TrCP ubiquitin ligase. *Molecular Cell*, **32**, 652–661.

Sulis, M.L. and Parsons, R. (2003) PTEN: from pathology to biology. *Trends in Cell Biology*, **13**, 478–483.

Sutherland, C., Leighton, I.A., and Cohen, P. (1993) Inactivation of glycogen synthase kinase-3 beta by phosphorylation: new kinase connections in insulin and growth-factor signalling. *Biochemical Journal*, **296** (Pt 1), 15–19.

Swiatek, W., Tsai, I.C., Klimowski, L. *et al.* (2004) Regulation of casein kinase I epsilon activity by Wnt signaling. *The Journal of Biological Chemistry*, **279**, 13011–13017.

Tacchelly-Benites, O., Wang, Z., Yang, E. *et al.* (2013) Toggling a conformational switch in Wnt/β-catenin signaling: regulation of Axin phosphorylation. *Bioessays*, **35**, 1063–1070.

Taelman, V.F., Dobrowolski, R., Plouhinec, J-L. *et al.* (2010) Wnt signaling requires sequestration of glycogen synthase kinase 3 inside multivesicular endosomes. *Cell*, **143**, 1136–1148.

Takacs, C.M., Baird, J.R., Hughes, E.G. *et al.* (2008) Dual positive and negative regulation of wingless signaling by adenomatous polyposis coli. *Science*, **319**, 333–336.

Tan, C.W., Gardiner, B.S., Hirokawa, Y. *et al.* (2012) Wnt signalling pathway parameters for mammalian cells. *PLoS One*, **7**, e31882.

Thomas, G.M., Frame, S., Goedert, M. *et al.* (1999) A GSK3-binding peptide from FRAT1 selectively inhibits the GSK3-catalysed phosphorylation of axin and beta-catenin. *FEBS Letters*, **458**, 247–251.

Tolwinski, N.S. and Wieschaus, E. (2001). Armadillo nuclear import is regulated by cytoplasmic anchor Axin and nuclear anchor dTCF/Pan. *Development*, **128**, 2107–2117.

Tolwinski, N.S., Wehrli, M., Rives, A. *et al.* (2003) Wg/Wnt signal can be transmitted through arrow/LRP5,6 and Axin independently of Zw3/Gsk3beta activity. *Developmental Cell*, **4**, 407–418.

Tran, H., Hamada, F., Schwarz-Romond, T., and Bienz, M. (2008) Trabid, a new positive regulator of Wnt-induced transcription with preference for binding and cleaving K63-linked ubiquitin chains. *Genes & Development*, **22**, 528–542.

Tran, H., Bustos, D., Yeh, R. *et al.* (2013) HectD1 E3 ligase modifies adenomatous polyposis coli (APC) with polyubiquitin to promote the APC–axin interaction. *Journal of Biological Chemistry*, **288**, 3753–3767.

van Amerongen, R., Nawijn, M., Franca-Koh, J. *et al.* (2005) Frat is dispensable for canonical Wnt signaling in mammals. *Genes & Development*, **19**, 425–430.

Willert, K., Shibamoto, S., and Nusse, R. (1999) Wnt-induced dephosphorylation of axin releases beta-catenin from the axin complex. *Genes & Development*, **13**, 1768–1773.

Winston, J.T., Strack, P., Beer-Romero, P. *et al.* (1999). The SCFbeta-TRCP-ubiquitin ligase complex associates specifically with phosphorylated destruction motifs in IkappaBalpha and beta-catenin and stimulates IkappaBalpha ubiquitination in vitro. *Genes & Development*, **13**, 270–283.

Wu, G., Huang, H., Garcia Abreu, J., and He, X. (2009) Inhibition of GSK3 phosphorylation of beta-catenin via phosphorylated PPPSPXS motifs of Wnt coreceptor LRP6. *PLoS One*, **4**, e4926.

Xing, Y., Clements, W.K., Kimelman, D., and Xu, W. (2003) Crystal structure of a beta-catenin/axin complex suggests a mechanism for the beta-catenin destruction complex. *Genes & Development*, **17**, 2753–2764.

Xing, Y., Clements, W.K., Le Trong, I. *et al.* (2004) Crystal structure of a beta-catenin/APC complex reveals a critical role for APC phosphorylation in APC function. *Molecular Cell*, **15**, 523–533.

Xing, Y., Takemaru, K., Liu, J. *et al.* (2008) Crystal structure of a full-length beta-catenin. *Structure*, **16**, 478–487.

Yamamoto, H., Komekado, H., and Kikuchi, A. (2006) Caveolin is necessary for Wnt-3a-dependent internalization of LRP6 and accumulation of beta-catenin. *Developmental Cell*, **11**, 213–223.

Yamamoto, H., Kishida, S., Kishida, M. *et al.* (1999) Phosphorylation of axin, a Wnt signal negative regulator, by glycogen synthase kinase-3beta regulates its stability. *The Journal of Biological Chemistry*, **274**, 10681–10684.

Yanagawa, S., van Leeuwen, F., Wodarz, A. *et al.* (1995) The dishevelled protein is modified by wingless signaling in Drosophila. *Genes & Development*, **9**, 1087–1097.

Yokoyama, N. and Malbon, C.C. (2007) Phosphoprotein phosphatase-2A docks to Dishevelled and counterregulates Wnt3a/beta-catenin signaling. *Journal of Molecular Signalling*, **2**, 12.

Yost, C., Torres, M., Miller, J.R. *et al.* (1996) The axis-inducing activity, stability, and subcellular distribution of beta-catenin is regulated in *Xenopus* embryos by glycogen synthase kinase 3. *Genes & Development*, **10**, 1443–1454.

Yost, C., Farr, G.H., 3rd, Pierce, S.B. *et al.* (1998) GBP, an inhibitor of GSK-3, is implicated in *Xenopus* development and oncogenesis. *Cell*, **93**, 1031–1041.

Yu, X., Waltzer, L., and Bienz, M. (1999). A new Drosophila APC homologue associated with adhesive zones of epithelial cells. *Nature Cell Biology*, **1**, 144–151.

Zeng, L., Fagotto, F., Zhang, T. *et al.* (1997) The mouse Fused locus encodes Axin, an inhibitor of the Wnt signaling pathway that regulates embryonic axis formation. *Cell*, **90**, 181–192.

Zeng, X., Tamai, K., Doble, B. *et al.* (2005) A dual-kinase mechanism for Wnt co-receptor phosphorylation and activation. *Nature*, **438**, 873–877.

Zeng, X., Huang, H., Tamai, K. *et al.* (2008) Initiation of Wnt signaling: control of Wnt coreceptor Lrp6 phosphorylation/activation via frizzled, dishevelled and axin functions. *Development*, **135**, 367–375.

Zhang, L., Jia, J., Wang, B. *et al.* (2006) Regulation of wingless signaling by the CKI family in Drosophila limb development. *Developmental Biology*, **299**, 221–237.

Zhang, Y., Liu, S., Mickanin, C. *et al.* (2011) RNF146 is a poly(ADP-ribose)-directed E3 ligase that regulates axin degradation and Wnt signalling. *Nature Cell Biology*, **13**, 623–629.

Zhu, X.M., Yin, L., Hood, L., and Ao, P. (2004) Robustness, stability and efficiency of phage lambda genetic switch: dynamical structure analysis. *Journal of Bioinformatics and Computational Biology*, **2**, 785–817.

Zu, T., Gibbens, B., Doty, N.S. *et al.* (2011) Non-ATG–initiated translation directed by microsatellite expansions. *Proceedings of the National Academy of Sciences of the United States of America*, **108**, 260–265.

4 An Overview of Gene Regulation by Wnt/β-Catenin Signaling

Chen U. Zhang and Ken M. Cadigan

Department of Molecular, Cellular and Developmental Biology, University of Michigan, Ann Arbor, MI, USA

Introduction

In addition to playing an essential role in cadherin-based cell adhesion (Stepniak, Radice, and Vasioukhin, 2009), β-catenin is a key player in Wnt/β-catenin signaling, often referred to as "canonical Wnt signaling" (see Chapter 16). Without Wnt stimulation, the cytoplasmic pool of β-catenin is phosphorylated and ubiquitinated by a complex containing Axin, adenomatous polyposis coli (APC) protein, glycogen synthase kinase 3 (GSK3), casein kinase I, and the F-box protein βTrCP (Cadigan and Piefer, 2009; Kennell and Cadigan, 2009; see Chapter 3). Wnt signaling compromises the activity of this "destructive complex," leading to accumulation of β-catenin (Cadigan and Piefer, 2009; MacDonald, Tamai, and He, 2009). The Armadillo (Arm) repeats of β-catenin resemble those of β-importin, and several lines of evidence indicate that β-catenin has the intrinsic ability to translocate across the nuclear pore complex (Henderson and Fagotto, 2002; Sharma *et al.*, 2012). Thus, stabilized β-catenin enters the nucleus, a process that can be influenced by the concentrations of cytoplasmic tethers such as Axin and APC and transcription factors (TFs) that bind β-catenin (Jamieson, Sharma, and Henderson, 2012; Tolwinski and Wieschaus, 2004; Zhang *et al.*, 2011).

This chapter provides an overview of the regulation of Wnt target gene expression which occurs when nuclear β-catenin binds to TFs. T-cell factors (TCFs) are the best-understood TFs that recruit β-catenin to target gene chromatin. TCFs often act as transcriptional switches, repressing transcription in concert with corepressors in the absence of signaling and activating the same targets when complexed with β-catenin and other coactivators. The differences between this switch in invertebrate and vertebrate systems will be discussed. An overview of the large number of coregulators that influence TCF and β-catenin/TCF gene regulation will be provided, with some discussion of how Wnt researchers might deal with the growing complexity of this transcriptional switch machinery going forward. Finally, other TFs that recruit β-catenin to chromatin or promote TCF binding will also be covered.

Wnt Signaling in Development and Disease: Molecular Mechanisms and Biological Functions, First Edition.
Edited by Stefan Hoppler and Randall T. Moon.

TCFs and transcriptional switches

TCFs were first discovered as sequence-specific DNA-binding proteins in lymphocytes (Laudet, Stehelin, and Clevers, 1993). The N-terminus of TCFs can bind to β-catenin and expressions of TCFs lacking this domain are potent inhibitors of Wnt/β-catenin signaling (e.g., Molenaar *et al.*, 1996; van de Wetering *et al.*, 1997). All TCFs contain a High Mobility Group (HMG) domain followed by a stretch of basic residues that contribute to DNA binding (Giese, Amsterdam, and Grosschedl, 1991; Love *et al.*, 1995) and nuclear import (Prieve *et al.*, 1998; see Chapter 17). Synthetic reporters containing TCF-binding sites are activated by the pathway in many *in vitro* and *in vivo* contexts (Barolo, 2006). Genetic analysis of TCF genes in several invertebrate and vertebrate organisms cemented their position as important physiological regulators of Wnt/β-catenin signaling (Archbold *et al.*, 2012; Cadigan and Waterman, 2012).

The current working model for TCFs acting as transcriptional switches was derived from several lines of evidence. For example, in *Drosophila*, mutations in the single *TCF* gene, *TCF/Pangolin* (*TCF/Pan*), resulted in a reduction in Wingless (Wg, a fly Wnt) signaling (Brunner *et al.*, 1997; van de Wetering *et al.*, 1997). However, the embryonic phenotype of *wg* mutants was partially rescued by *TCF/Pan* mutants (Cavallo *et al.*, 1998). A similar suppression was also observed in *wg*, *groucho* (*gro*) double mutants (Cavallo *et al.*, 1998). Gro belongs to the Transducin-like Enhancer of Split (TLE) family of corepressors, which directly bind to TCFs (Roose *et al.*, 1998). In addition, while mutation of TCF-binding sites in reporters with Wnt Response Elements (WREs) reduces their expression in transgenic fly tissues, sometimes, there is an accompanying expansion of reporter gene expression (Lee and Frasch, 2000; Yang *et al.*, 2000). A model summarizing these data is shown in Figure 4.1. In the absence of Wg/Wnt signaling, TCF/Pan and Gro repress (along with other factors discussed in the following sections) target gene expression (Figure 4.1a). When signaling promotes nuclear accumulation of β-catenin, it binds to TCF/Pan, displacing Gro and recruiting coactivators to activate Wnt targets (Figure 4.1b).

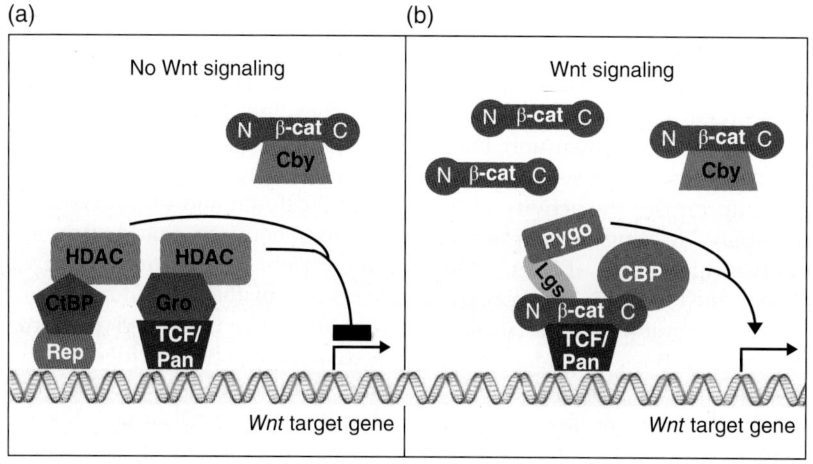

Figure 4.1 Summary of the TCF transcriptional switch in *Drosophila* cells. Depiction of a Wnt target gene and surrounding nucleoplasm in the absence or presence of Wnt/β-catenin signaling. (a) Under conditions of low nuclear β-catenin (β-cat), TCF/Pan is complexed with corepressors such as Gro, which recruits HDACs to inhibit transcription. Other factors act in concert with TCF/Pan to silence Wnt targets, such as CtBP, which is recruited to WREs by other, as yet unidentified transcriptional repressors (Rep). In addition, β-catenin binding proteins such as Cby can associate with β-catenin and block its interaction with TCF/Pan. (b) Wnt signaling causes higher levels of nuclear β-catenin, which overcomes the binding proteins and TCF corepressors and binds to TCF/Pan on target gene chromatin. β-catenin recruits a variety of coactivators, such as Lgs and Pygo through Arm repeats in the N-terminal half of β-catenin and CBP through its C-terminal transactivation domain, leading to activation of transcription. *(See insert for color representation of the figure.)*

While the classic switch model outlined earlier has been a useful paradigm for understanding Wnt target gene regulation, it does not universally apply to all targets in all organisms. In *Caenorhabditis elegans*, there is abundant evidence that a distinct type of switch occurs involving POP-1 (the single worm TCF) and Sys-1 (one of the four worm β-catenins). This alternative switch is often referred to as the "Wnt/β-catenin asymmetry pathway" due to its prevalence in regulating asymmetric cell divisions in worm development. This pathway has been reviewed in detail elsewhere (Phillips and Kimble, 2009; Sawa, 2012) and is briefly summarized here. In addition to nuclear accumulation of Sys-1, this pathway also requires nuclear efflux of POP-1 for Wnt targets to be activated. This efflux is mediated by binding of the β-catenin Wrm-1 to POP-1, along with phosphorylation by Lit-1, a Nemo-like kinase (NLK) (Phillips and Kimble, 2009; Sawa, 2012). This POP-1 efflux is required to enhance the formation of POP-1–Sys-1 complexes on Wnt target gene chromatin (Figure 4.2). The genetic evidence in worms indicates that this asymmetry pathway plays several important roles in *C. elegans* cell fate specification, but it is not yet clear how important this pathway is in other animals (Cadigan, 2012; Phillips and Kimble, 2009).

Flies and worms have only one *TCF* gene each with little isoform diversity (Archbold *et al.*, 2012; Cadigan and Waterman, 2012). This implies that a single species of TCF acts in both sides of the transcriptional switches outlined earlier. The situation is more complicated in vertebrates, where amphibians and mammals have four TCF genes, *TCF1*, *LEF1*, *TCF3*, and *TCF4*, also known as *TCF7*, *LEF1*, *TCF7L1*, and *TCF7L2*, respectively (see Chapter 17). Bony fish have two closely related *TCF3* genes in addition to the other three TCFs (Dorsky *et al.*, 2003). Loss of function analysis suggests that the vertebrate TCFs are more specialized for repression or activation than invertebrate TCFs. TCF3 appears to function solely as a repressor (Kim *et al.*, 2000; Liu *et al.*, 2005; Merrill *et al.*, 2004), while LEF1 appears to be an activator (Kratochwil *et al.*, 2002; Reya *et al.*, 2000; van Genderen *et al.*, 1994). The data for

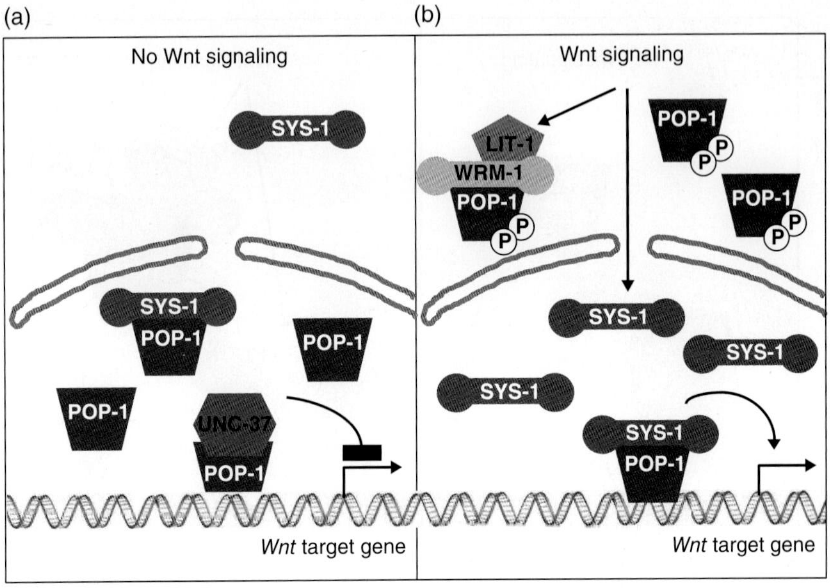

Figure 4.2 Summary of the TCF transcriptional switch in the Wnt/β-catenin asymmetry pathway in *C. elegans*. (a) In unstimulated cells, there are low levels of the worm β-catenin Sys-1 and high levels of POP-1 in the nucleus. The Gro homolog Unc-37 contributes to repression (Calvo *et al.*, 2001). (b) Wnt signaling increases the nuclear concentration of Sys-1 and lowers the level of POP-1 through its phosphorylation via the Lit-1 kinase, which promotes nuclear efflux. This shifts the equilibrium on WRE chromatin from POP-1 (or POP-1-Unc-37) to POP-1-Sys-1 heterodimers, resulting in transcriptional activation. *(See insert for color representation of the figure.)*

TCF1 and TCF4 suggest that these TCFs retain both functions (Galceran *et al.*, 1999; Korinek *et al.*, 1998; Nguyen *et al.*, 2009; Roose *et al.*, 1999; Tang *et al.*, 2008). However, these genes can produce truncated isoforms lacking the β-catenin binding domain, which can function as inhibitors of the pathway (Vacik and Lemke, 2011). Indeed, in colorectal tumor cells, which possess elevated Wnt/β-catenin signaling (Polakis, 2012; see Chapter 28), there is an enrichment of "full-length" TCF1 isoforms, where normal tissue expresses mostly the truncated TCF1 (Najdi *et al.*, 2009). This may explain why TCF1 behaves as an intestinal tumor suppressor in mouse knockouts (Roose *et al.*, 1999). A morpholino specific for the dominant-negative TCF4 isoform revealed a biologically important role for this truncated TCF4 in antagonizing Wnt/β-catenin during *Xenopus* embryogenesis (Vacik, Stubbs, and Lemke, 2011). Development of more isoform-specific inhibitors of TCF1 and TCF4 will be needed to better understand how these TCFs regulate the Wnt/β-catenin pathway.

Mutagenesis of some vertebrate WREs clearly suggests that they are both negatively and positively regulated by TCFs (Brannon *et al.*, 1997; Hikasa and Sokol, 2011). This raises the possibility that multiple TCFs are involved in a transcriptional switch. Wnt/β-catenin signaling stimulates the phosphorylation of TCF3 in frog embryos, which inhibits its ability to associate with target gene chromatin (Hikasa and Sokol, 2011; Hikasa *et al.*, 2010). This phosphorylation occurs through homeodomain-interacting kinase 2 (HIPK2). TCF1 lacks HIPK2 phosphorylation sites (Hikasa *et al.*, 2010), supporting the "TCF exchange" model outlined in Figure 4.3.

In mouse embryonic stem cells (mESCs), TCF3 promotes differentiation by repressing pluripotency genes (Merrill, 2012; Sokol, 2011). TCF1 functions antagonistically with TCF3 in this process (Yi *et al.*, 2011), suggesting that some Wnt targets may undergo a TCF exchange as described in frogs, though this remains to be demonstrated. TCF3 expression is also inhibited by Wnt/β-catenin signaling in mESCs (Atlasi *et al.*, 2013), providing another variation

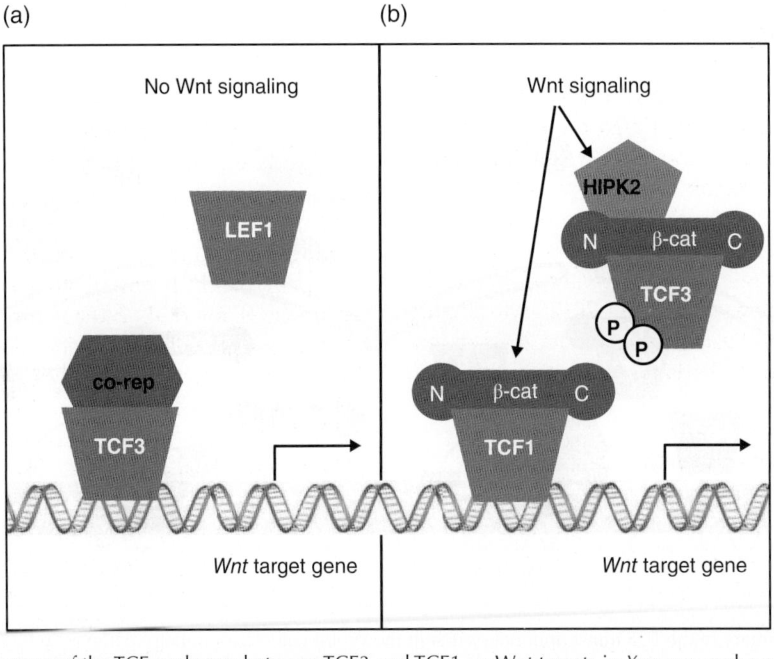

Figure 4.3 Summary of the TCF exchange between TCF3 and TCF1 on Wnt targets in *Xenopus* embryos. (a) In cells with low nuclear β-catenin, TCF3 represses Wnt target gene transcription. (b) Wnt signaling activates HIPK2, which acts with β-catenin to phosphorylate TCF3, removing it from target gene chromatin, where it is replaced by TCF1, which activates target gene transcription (Hikasa and Sokol, 2011; Hikasa *et al.*, 2010). *(See insert for color representation of the figure.)*

on how multiple TCFs can regulate Wnt targets in vertebrate cells.

As described earlier, many Wnt targets and WREs are both repressed and activated by TCFs in the absence and presence of Wnt signaling, respectively. However, it should be pointed out that many WREs have little detectable TCF repression in the absence of signaling, based on TCF site mutagenesis (e.g., Chang *et al.*, 2008b; Galceran *et al.*, 2004; Lam, Chesney, and Kimble, 2006; Yamaguchi *et al.*, 1999). Most likely, these WREs lack sequences for general activators, so that they are inactive whether TCF is bound or not, but require TCF and β-catenin for activation (Archbold *et al.*, 2012). Wnt targets can also have multiple WREs controlling their transcription, as has been found for the *naked cuticle* gene in flies (Chang *et al.*, 2008a) and *c-myc* in humans (He *et al.*, 1998; Pomerantz *et al.*, 2009; Wright, Brown, and Cole, 2010; Yochum, Cleland, and Goodman, 2008). Due to these complexities, the generic situations outlined in Figure 4.1, Figure 4.2, and Figure 4.3 may be an oversimplification for many Wnt targets.

Coregulators of β-catenin/TCF transcription

There are a large number of nuclear factors that negatively or positively influence Wnt target gene regulation. The intention of the following section is to provide a brief outline of some of the mechanisms by which these coregulators operate. A list of many of these factors are shown in Table 4.1 and Table 4.2 and more comprehensive reviews can be found elsewhere (Cadigan, 2012; Mosimann, Hausmann, and Basler, 2009; Valenta, Hausmann, and Basler, 2012; Willert and Jones, 2006).

Factors and mechanisms repressing Wnt target genes

In general, the described mechanisms for repressing Wnt target gene transcription fall into two main categories: recruitment of corepressors and histone deacetylases (HDACs) to WREs and factors that block TCF binding to β-catenin or DNA. Examples are discussed subsequently.

Several Wnt target gene corepressors act by direct binding to TCFs. The best-studied case is members of the Gro/TLE family, which repress Wnt target gene expression in flies, worms, and vertebrate systems (reviewed in Cadigan, 2012). Gro/TLEs bind to many other TFs besides TCFs and repress transcription by recruiting HDACs to target gene chromatin (Turki-Judeh and Courey, 2012). Other TCF corepressors include Corepressor of Pangolin (Coop) in *Drosophila* (Song *et al.*, 2010) and members of the Myeloid Translocation Gene (MTG) family in mammals (Barrett *et al.*, 2012; Moore *et al.*, 2008). Like Gro/TLE, these proteins compete with β-catenin for binding to TCFs (Arce, Pate, and Waterman, 2009; Daniels and Weis, 2005; Moore *et al.*, 2008; Song *et al.*, 2010). This means in addition to their role as corepressors, they may prevent low levels of nuclear β-catenin from inappropriately activating Wnt targets.

After β-catenin recruitment to WREs, corepressors can still influence Wnt target expression. For example, Reptin/TIP49b directly binds β-catenin and represses gene expression via its DNA helicase activity (Rottbauer *et al.*, 2002). The nuclear receptor corepressor (NCoR) and silencing mediator of retinoic acid and thyroid hormone receptor (SMRT) corepressors can bind to the β-catenin/TCF heterodimer and recruit HDACs to WREs (Song and Gelmann, 2008). These factors could act to dampen the amplitude of target gene expression levels in cells receiving Wnt stimulation.

In addition, some corepressors act on WRE chromatin in parallel with TCF. In *Drosophila* cells, C-terminal binding protein (CtBP) is recruited to Wnt targets independently of TCF/Pan, presumably through other transcriptional repressors (Fang *et al.*, 2006; Figure 4.1a). In vertebrate systems, CtBP has been reported to bind directly to TCFs (Brannon *et al.*, 1999; Valenta, Lukas, and Korinek, 2003), but other reports do not see this interaction (Hamada and Bienz, 2004; Valenta *et al.*, 2006). In *Xenopus* embryos, Kaiso and TCF bind to WREs in close proximity and repress transcription (Hong *et al.*, 2010; Park *et al.*, 2005, 2006). The importance of Kaiso in regulating Wnt targets has been questioned (Ruzov *et al.*, 2009a, b) and Kaiso has also been reported to activate Wnt targets in *Xenopus* (Iioka, Doerner, and Tamai, 2009). Likewise, CtBP can also directly activate Wnt targets in flies (Fang *et al.*, 2006). This bimodal regulation is not uncommon for Wnt regulators (see Table 4.1 and Table 4.2),

Table 4.1 List of factors that repress β-catenin/TCF transcription. Factors are grouped according to general mechanism of action

Category	Factor	System	Binding partners and interaction domains	Potential mechanisms	References
TCF bound co-repressors which can also disrupt β-catenin/TCF interactions	Coop	*Drosophila*	TCF/Pan: a motif containing HMG domain.	Recruit HDACs. Do not disrupt TCF/DNA interaction.	Song *et al.* (2010).
	Groucho/ TLE	Vertebrate, *Drosophila* and *C. elegans*	TCFs: central and HMG domains. Gro: Q domain.		Roose *et al.* (1998), Daniels and Weis (2005), Sierra *et al.* (2006), Arce, Pate, and Waterman (2009), Cadigan (2012)[a].
	HIC-5	Vertebrate	TCFs: a motif in the central domain (absent in xLEF1).	Excludes β-catenin on *c-myc* promoter.	Ghogomu *et al.* (2006), Li *et al.* (2011a).
	MTGs	Vertebrate	A N-terminal motif and the HMG domain on TCF4 contribute to interaction.	MTGR1 interacts with TCF1, TCF4 and LEF1. MTG8/16 could be recruited by Kaiso.	Moore *et al.* (2008), Barrett *et al.* (2012).
Bind to β-catenin and disrupt β-catenin/TCF interactions	Chibby	Vertebrate and *Drosophila*	β-catenin: Arm C-terminus.	Besides directly disrupt β-catenin binding, Chibby and 14-3-3 together sequester β-catenin in the cytosol.	Takemaru *et al.* (2003), Li *et al.* (2008), Love *et al.* (2010), Enjolras *et al.* (2012).
	Duplin/CHD8	Vertebrate	β-catenin: Arm repeats 1–7 shows strongest interaction.	Unclear.	Sakamoto *et al.* (2000), Thompson *et al.* (2008), Nishiyama, Skoultchi, and Nakayama (2012).
	ICAT	Vertebrate	β-catenin: Arm repeats 10–12.	ICAT masks Arm repeats 5–10 and competes for β-catenin binding. TCF4 and LEF1 were tested.	Tago *et al.* (2000), Daniels and Weis (2002), Graham *et al.* (2002), Hasegawa *et al.* (2007).
	p15RS	Mammalian cell culture	TCF: N-terminus. β-catenin: Arm repeats 6–8.	Also binds to TCF with the same domain (RPR domain).	Wu *et al.* (2010).
	Sox6, Sox9, Sox17	Vertebrate	β-catenin: Arm repeats 4–10 for interaction with Sox9, 1–6 with Sox17 and 1–4 with Sox6. TCF3 and TCF4: HMG domains.	Sox9 can also promote β-catenin degradation.	Akiyama *et al.* (2004), Kan *et al.* (2004), Sinner *et al.* (2007), Topol *et al.* (2009), Kormish, Sinner, and Zorn (2010)[a].

β-catenin bound co-repressors	Reptin/Tip49b	Vertebrate and *Drosophila*	β-catenin: unspecified.	Inibitory effect requires its DNA-dependent ATPase activity. Might antagonize Pontin.	Bauer *et al*. (2000), Rottbauer *et al*. (2002), Kim *et al*. (2005), Olson *et al*. (2006).
	TIS7[b]	Mammalian cell culture	β-catenin: unspecified.	Unclear.	Vietor *et al*. (2005).
Modify TCF and inhibit its function	CBP/P300[b]	Mammalian cell culture, *Drosophila* and *C. elegans*	TCF/Pan: HMG domain. Acetylation site: TCF/Pan K25; TCF/POP-1 K185, K187 and/or K188.	Acetylation of TCF/Pan reduces its affinity to β-catenin. Acetylation of TCF/POP-1 enhances its nuclear retention.	Waltzer and Bienz (1998), Gay *et al*. (2003), Li *et al*. (2007).
	PIASy[b]	Vertebrate	LEF1: a motif containing HMG domain. SUMOylation site: K25 and K267 of LEF1.	SUMOylates LEF1, which is then sequestered into nuclear bodies and inhibited.	Sachdev *et al*. (2001), Roth *et al*. (2004).
Recruited to WREs in parallel of TCF	CtBP[b]	Vertebrate and *Drosophila*	APC: 15 aa repeats	Recruited to WREs in a TCF-independent manner, functioning as a homo-oligomer; can also divert β-catenin/APC complexes away from TCF.	Hamada and Bienz (2004), Fang *et al*. (2006), Bhambhani *et al*. (2011).
	ISWI[b]/ SNF2H and 2L, SNF5	Mammalian cell culture and *Drosophila*	Unspecified	Interacts with ACF1 and antagonizes histone acetylation on Wnt targets.	Liu *et al*. (2008), Eckey *et al*. (2012), Mora- Blanco *et al*. (2013).
	Kaiso[b]	Vertebrate	Unspecified	Recruit co-repressors to WREs. May also disrupt TCF/DNA interaction. Can activate Wnt targets in *Xenopus*.	Park *et al*. (2005, 2006), Ruzov *et al*. (2009a, b), Iioka, Doerner, and Tamai (2009), Hong *et al*. (2012).
Misc	Eaf1, Eaf2	Vertebrate	β-catenin: Arm repeats 1-12 and C- terminus. TCF4: unspecified.	Interact with both TCF4 and β-catenin.	Liu *et al*. (2013).
	NCoR and SMRT	Mammalian cell culture	β-catenin: Arm repeats1-12. TCF4: a motif containing HMG domain.	Known chromatin modifiers that interact with both TCF and β-catenin.	Song and Gelmann (2008).
	Osterix	Vertebrate	Unspecified.	Disrupt TCF/DNA interaction.	Zhang *et al*. (2008), Chen *et al*. (2012).

[a]Reviews.
[b]Factors that can both inhibit and activate Wnt/β-catenin signaling.
For the "System" column, Vertebrate denotes factors where positive results were obtained in one or more vertebrate species *in vivo* and can also include mammalian cell culture. For the column indicating the interaction domains, the Arm repeats denote the 12 motifs forming central domain of β-catenin (see Valenta *et al*., 2012). Note that some factors can also promote β-catenin/TCF transcription in some contexts (see Table 4.2). Abbreviations used that are not defined in the text are as follows: HIC-5, hydrogen peroxide-inducible clone; Hint1, histidine triad protein; ISWI, imitation switch; p15RS, p15Ink4b-related protein; PIAS, protein inhibitor of activated STAT; TIS7, 12-O-tetradecanoylphorbol-13-acetate-induced sequence 7.

Table 4.2 List of coactivators of β-catenin/TCF transcription

Category	Factor	System	Binding partners and interaction domains	Potential mechanisms	References
Interacts with N-terminal half of β-catenin	BCL9/Lgs and Pygo proteins	Vertebrate and *Drosophila*	β-catenin: Arm repeat 1 interacts with BCL9/Lgs; D172 (fly) and D164 (mouse) are cruicial for the interactions.	BCL9/Lgs recruits Pygo, which in turn recruits many other transcription co-activators. They also help to retain β-catenin in the nucleus.	Kramps *et al.*, (2002), Parker, Jemison, and Cadigan (2002), Thompson *et al.* (2002), Brembeck *et al.* (2004), Jessen *et al.* (2008)[a], Valenta *et al.* (2011).
	Pontin/ TIP49	Vertebrate and *Drosophila*	Arm repeats 1–4.	DNA-dependent helicase that can complex with histone acetyltransferases (HATs). Also binds to TBP, another co-activator.	Bauer *et al.* (2000), Rottbauer *et al.* (2002), Feng, Lee, and Fearon (2003).
Interacts with C-terminal half of β-catenin	CBP and P300	Vertebrate and *Drosophila*	Arm repeats 10–12 and C-terminus.	HATs.	Hecht *et al.* (2000), Sun *et al.* (2000),Takemaru and Moon (2000), Kioussi *et al.* (2002), Ma *et al.* (2005), Sierra *et al.* (2006), Li *et al.* (2007).
	Brg-1/ Brm	Vertebrate and *Drosophila*	Arm repeats 7–12.	ATPase-dependent chromatin remodeller.	Barker *et al.* (2001), Major *et al.* (2008), Mahmoudi *et al.* (2010).
	ISWI[a]/ SNF2H and 2L	Mammalian cell culture and *Drosophila*	Arm repeats 11–12 and C-terminus.	ATPase-dependent chromatin remodeler.	Sierra *et al.* (2006), Song, Spichiger-Haeusermann, and Basle (2009).
	Hyrax/ Para-fibromin	Mammalian cell culture and *Drosophila*	Arm repeat 12 and C-terminus.	Member of the PAF1 complex, which is involved in transcription initiation and elongation. Target activation is dependent on Pygo. Activity is regulated by SHP2.	Mosimann, Hausmann, and Basler (2006), Takahashi *et al.* (2011).
	Med12 and Med13	*Vertebrate and Drosophila*	Arm repeats 11–12 and C-terminus.	Subunit of the mediator complex. Also found to be recruited by Pygo.	Kim *et al.* (2006), Carrera *et al.* (2008), Rocha *et al.* (2010).
	TAF complex	Mammalian cell culture and *Drosophila*	TBP interacts with Arm repeats 11–12 and C-terminus.	TBP is recruited by β-catenin. TAF4 is recruited by Pygo. These are members of the TFIID complex.	Hecht *et al.* (1999), Wright *et al.* (2009), Simoneau *et al.* (2011).

Other histone modifiers	MLL1/ MLL2	Mammalian cell culture	Arm repeats 11–12 and C-terminus.	MLL2 was also shown to be recruited by Pygo2. MLL1/2 are Histone methyltransferases (HMTs). Catalyzes H3K4 mono-, di- and tri-methylation.	Sierra *et al.* (2006), Chen *et al.* (2010).
	Dot1L (Dot1)	Vertebrate and *Drosophila*	Might be recruited through TRRAP/Tip60.	Found in several complexes with MLL partners and has HMT activity. Catalyzes H3K79 methylation.	Mohan *et al.* (2010), Mahmoudi *et al.* (2010).
	SET8	Vertebrate	Interacts with a TCF4 fragment spanning from N-terminus to the end of HMG domain.	HMT. Catalyzes H4K20 mono-methylation.	Li *et al.* (2011b).
	PRMT2	Vertebrate	Interacts with β-catenin.	HMT. Catalyzes H3R8 methylation.	Blythe *et al.* (2010).
Other histone modifiers	Carm1	Mammalian cell culture	Interacts with β-catenin but not LEF1.	HMT. Catalyzes H3R17me2.	Ou *et al.* (2011).
	TRRAP p400 and TIP60	Mammalian cell culture	Arm repeats 11–12 and C-terminus.	HAT complex. Might also mediate β-catenin ubiquitination through Skp1/SCF.	Sierra *et al.* (2006), Sustmann *et al.* (2008).
Facilitating β-catenin/ TCF interaction	Jerky/ Ebd1	Mammalian cell culture and *Drosophila*	Interacts with β-catenin, LEF1 and Pygo2.	The localization of Ebd1 on polytene chromosomes requires a DNA-binding protein called NRF-1/Ewg.	Benchabane *et al.* (2011), Xin *et al.* (2011).
	RNF14	Vertebrate	Interaction requires the N-terminal half of TCF	Contributes to β-catenin recruitment on the chromatin.	Wu *et al.* (2013).
	TBL1/ TBLR1	Mammalian cell culture	TBL1 interacts with TCF4. TBL1 and TBLR1 both interact with β-catenin.	TBL1 and TBLR1 are SUMOylated in response to Wnt signaling, which releases these factors from the NCoR complex, increasing recruitment to WREs.	Li and Wang (2008), Choi *et al.* (2011).
Misc	APPL1 and APPL2	Mammalian cell culture	Interact with Reptin.	Interact with the co-repressor Reptin and remove it from the chromatin.	Rashid *et al.* (2009).
	CtBP[a]	*Drosophila*	Unspecified.	CtBP monomers activate some Wg targets downstream of Pygo.	Fang *et al.* (2006), Bhambhani *et al.* (2011).
	PIASy[a]	Mammalian cell culture	SUMOylation site: K297 of TCF4.	SUMOylates TCF4 and increases its activity. SUMOylation of PIASy is required for PIASy activity.	Yamamoto *et al.* (2003), Ihara *et al.* (2005).
	TIS7[a]	Mammalian cell culture	Unspecified.	Unclear.	Nakamura *et al.* (2003).

[a]Factors that can both activate and inhibit Wnt/β-catenin signaling.
[b]Reviews.
Factors are grouped according to general mechanism of action. Vertebrate systems and Arm repeats of β-catenin are defined as in Table 4.1. Note that some factors listed can also repress β-catenin/TCF transcription (see Table 4.1). Abbreviations used that are not defined in the text are as follows: Brm, Brahma; MED12, Mediator 12; TAF, TBP-associated factor; TBP, TATA-box binding protein; TRRAP, transcription/transformation domain-associated protein.

though the molecular mechanism is usually unknown. In the case of CtBP, its oligomerization status determines whether it will repress (oligomeric) or activate (monomeric) Wnt targets (Bhambhani *et al.*, 2011).

CtBP can also inhibit Wnt signaling by diverting β-catenin away from TCF, in a complex with APC (Hamada and Bienz, 2004). APC binds β-catenin and has been proposed to promote its nuclear efflux (Brocardo and Henderson, 2008) and is thought to act on WRE chromatin to remove β-catenin from TCF activating complexes (Sierra *et al.*, 2006). Other proteins that bind β-catenin and prevent it from associating with TCFs include inhibitor of β-catenin and TCF4 (ICAT) (Hasegawa *et al.*, 2007; Tago *et al.*, 2000), Sry-type HMG box containing protein 9 (Sox9) (Akiyama *et al.*, 2004; Topol *et al.*, 2009), and Chibby (Cby; Li *et al.*, 2008; Love *et al.*, 2010; Takemaru *et al.*, 2003). Depletion of *Cby* in *Drosophila* embryos via RNAi can partially rescue *wg* but not *armadillo* (*arm*, the fly β-catenin) mutants (Takemaru *et al.*, 2003), suggesting that even in the absence of Wnt signaling, there is some β-catenin in the nucleus with the potential for activating Wnt targets. The significance of Cby in fly development has recently been challenged, since *Cby* null mutants do not have Wg-related phenotypes (Enjolras *et al.*, 2012). Nonetheless, it seems likely that there are several "β-catenin/TCF buffers" in the nucleoplasm that help set the threshold for how much β-catenin is required to convert TCFs to transcriptional activators (Figure 4.1b).

Factors and mechanisms activating target genes

Once β-catenin binds to TCF and displaces or otherwise overcomes the aforementioned negative regulators, it serves as a "landing platform" for a variety of transcriptional coactivators (Table 4.2). Many of these factors can be roughly divided into four categories: (i) factors that facilitate β-catenin/TCF interaction, (ii) coactivators bound to the N-terminal transactivation domain of β-catenin, (iii) coactivators bound to the C-terminal transactivation domain, and (iv) chromatin modifying complexes recruited to WREs by β-catenin. Here, we briefly describe the basic features of these proteins.

Although the N-terminus of TCF is sufficient for interaction with the Arm repeats of β-catenin *in vitro* (Behrens *et al.*, 1996; Graham *et al.*, 2000; Poy *et al.*, 2001; van de Wetering *et al.*, 1997), there are additional factors that are necessary for association of these proteins *in vivo*. Transducin β-like protein 1 (TBL1) and TBL1-related protein (TBLR1), which are subunits of the SMRT–NCoR corepressor complex, have a distinct function in recruiting β-catenin to WREs, with TBL1 binding to both TCFs and β-catenin (Li and Wang, 2008). TBL1 and TBLR1 are SUMOylated upon Wnt signaling, which releases them from the SMRT–NCoR complex, allowing them to promote β-catenin recruitment to Wnt targets (Choi *et al.*, 2011). The RING Finger Protein 14 (RNF14) binds to TCFs and is required for β-catenin recruitment to several vertebrates WREs (Wu *et al.*, 2013). While TBL1, TBLR1, and RNF14 appear to be general promoters of Wnt/β-catenin signaling, the Centromere Binding Protein B (CENPB) domain protein Jerky/Earthbound 1 (Jerky/Ebd1) also functions as a β-catenin/TCF adaptor but only in specific *Drosophila* tissues (Benchabane *et al.*, 2011).

β-catenin contains at least two domains capable of activating transcription when fused to DNA-binding domains (reviewed in Mosimann, Hausmann, and Basler, 2009; see Chapter 16). Several factors that bind to the N-terminal transactivation domain have been reported (Table 4.2) but the best characterized is known as Legless (Lgs) in flies and B-cell lymphoma 9 (Bcl9) and Bcl9-2 in mammals (Mosimann, Hausmann, and Basler, 2009). These proteins bind to the first Arm repeat in β-catenin (Hoffmans and Basler, 2004; Valenta *et al.*, 2011) serving as an adaptor between β-catenin and Pygopus (Pygo) proteins (fly Pygo and mammalian Pygo1 and Pygo2) (Kramps *et al.*, 2002). Lgs and Pygo are essential for Wg/Wnt signaling in flies (Belenkaya *et al.*, 2002; Kramps *et al.*, 2002; Parker, Jemison, and Cadigan, 2002; Thompson *et al.*, 2002) and are also significant contributors to the pathway in mice (Brack *et al.*, 2009; Gu *et al.*, 2009; Schwab *et al.*, 2007). The Pygo proteins (fly Pygo and the mammalian Pygo 1 and Pygo 2) are thought to activate transcription by interacting with subunits of the mediator complex (Carrera *et al.*, 2008), basal TFs (Wright and Tjian, 2009),

CREB-binding protein (CBP) (Andrews *et al.*, 2009), and the mixed lineage leukemia 2 (MLL2) histone methyltransferase (Chen *et al.*, 2010). While clearly a major mediator of N-terminal transactivation by β-catenin, mouse embryos carrying a point mutation in β-catenin (D164A), which abolishes BCL9/BCL9-2 binding, display more severe defects than BCL9/BCL9-2 double mutants (Schwab *et al.*, 2007; Valenta *et al.*, 2011). These results suggest that additional coactivators utilize this region of β-catenin to activate Wnt targets.

The C-terminal transactivation domain consists of the last three Arm repeats and the adjacent C-terminus of β-catenin (Mosimann, Hausmann, and Basler, 2009; Willert and Jones, 2006). Several coactivators have been found to bind directly with this domain (Table 4.2), among them the histone acetyltransferases (HATs) CBP and p300 (Hecht *et al.*, 2000; Li *et al.*, 2007; Sun *et al.*, 2000; Takemaru and Moon, 2000) as well as Brahma-related protein 1 (Brg-1), the ATPase subunit of the Swi/Snf chromatin remodeling complex (Barker *et al.*, 2001). Consistent with a role for HATs in Wnt target gene activation, Wnt/β-catenin signaling promotes an increase in acetylated histones at Wnt targets (Kioussi *et al.*, 2002; Parker, Jemison, and Cadigan, 2008; Sierra *et al.*, 2006). Studying the contribution of factors such as CBP, p300, and Brg-1 to Wnt gene activation is difficult, since they are general coactivators involved in the regulation of many genes (Goodman and Smolik, 2000; Sudarsanam and Winston, 2000). However, clonal analysis in flies and partial knockdown by RNAi have demonstrated specific loss of Wg/Wnt signaling for CBP (Li *et al.*, 2007). Likewise, siRNA of Brg-1 and p300 in mammalian cell culture results in loss of regulation of Wnt targets, as well as many non-Wnt targets (Mahmoudi *et al.*, 2010). In addition, a small molecule (ICG-001) that blocks the interaction between β-catenin and CBP inhibits several Wnt/β-catenin readouts (Henderson *et al.*, 2010; Ma *et al.*, 2005).

In addition to histone acetylation, several other chromatin marks and the enzymes that catalyze them have been linked to gene activation by the Wnt/β-catenin pathway (Table 4.3). These include MLL1/2 (H3K4me3), protein arginine methyltransferase 2 (PMRT2, H3R8me), SET8 (H3K20me), CARM1 (H3R17me2), and DOTL1/DOT1 (H3K79me3) (Blythe *et al.*, 2010; Chen *et al.*, 2010; Li *et al.*, 2011b; Mahmoudi *et al.*, 2010; Mohan *et al.*, 2010; Ou *et al.*, 2011; Sierra *et al.*, 2006). It is not clear how consistently chromatin modifications occur among different Wnt targets, for example, for some targets, there is no change in histone acetylation upon Wnt signaling (Blythe *et al.*, 2010; Wohrle, Wallmen, and Hecht, 2007). In contrast, a microarray study in HEK293T cells demonstrated that most genes that are activated by Wnt3a treatment required DOTL1 for this regulation (Mahmoudi *et al.*, 2010).

With so many factors connected with β-catenin and TCF on either the *on* or *off* side of the transcriptional switch, it is difficult to envision them all working simultaneously. This has led to suggestions of coactivator cycling on and off β-catenin (Mosimann, Hausmann, and Basler, 2009; Valenta, Hausmann, and Basler, 2012). Indeed, there is some evidence for cycling of negative and positive regulators on the *c-myc* WRE (Sierra *et al.*, 2006). While coregulator dynamics is likely occurring on WREs, another consideration is whether all identified factors act on every Wnt target. There are some clear examples of tissue-specific regulators, for example, Osterix, an osteoblast-specific TF that binds to TCFs and inhibits their ability to bind DNA (Zhang *et al.*, 2008) and Jerky/Ebd which is only required for Wg/Wnt signaling in a few fly cell types (Benchabane *et al.*, 2011). For most Wnt coregulators, their involvement in other pathways or possible redundancy with related proteins makes it more difficult to assess whether they are general or gene-/cell-specific Wnt factors.

Other TFs that mediate Wnt/β-catenin signaling

The genetic data in *Drosophila* suggests that TCF/Pan mediates most Wnt/β catenin signaling in this organism, at least during embryonic and larval development (Brunner *et al.*, 1997; van de Wetering *et al.*, 1997). However, the overall importance of TCFs for the pathway in vertebrate is much less clear. Conditional deletion of β-catenin has revealed numerous developmental phenotypes in mice

Table 4.3 List of other TFs that bind to β-catenin

Factors	System	Binding partners and interaction domains	Potential mechanism	References
Prop1 and PitX2	Vertebrate	LEF1: can interact with PITX2, domain unspecified. β-catenin: Arm repeats 5–9 with Prop1.	LEF1 could also be involved.	Kioussi *et al.* (2002), Olson *et al.* (2006), Amen *et al.* (2007).
FOXO proteins	Mammalian cell culture and *C.elegans*	β-catenin: unspecified.	Could reduce interaction between β-catenin and TCF4.	Essers *et al.* (2005), Almeida *et al.* (2007), Hoogeboom *et al.* (2008).
Sox17[a]	Vertebrate	β-catenin: Arm repeats 1–6.	TCF4 may also be involved.	Sinner *et al.* (2004), Kormish *et al.* (2010)[b].
MyoD	Mammalian cell culture	β-catenin: Arm repeats 1–9.	Unclear.	Kim *et al.* (2008)
Tbx5 and YAP1	Vertebrate	The co-activator YAP1 interacts with β-catenin while Tbx5 binds DNA.	Tbx5, YAP1 and β-catenin interact with each other. YAP1 and β-catenin colocalize on chromatin. These interactions require phosphorylation of YAP1 by YES1.	Rosenbluh *et al.* (2012).
HIF-1	Mammalian cell culture	β-catenin: Arm repeats 9–12 and C-terminus.	TCFs could also be involved. β-catenin and HIF-1 can form a ternary complex with androgen receptor, activating androgen-dependent targets.	Kaidi, Williams, and Paraskeva (2007), Mazumdar *et al.* (2010), Mitani *et al.* (2012).
Androgen receptor	Vertebrate	β-catenin: Arm repeats 2–7.	Can compete with TCF for β-catenin binding.	Song *et al.* (2003), Cronauer *et al.* (2005), Wang *et al.* (2008), Mitani *et al.* (2012).
Vitamin D receptor	Vertebrate	β-catenin: C-terminus; acetylation in K671 and K672 regulates its specificity.	Acetylation of K671/672 on β-catenin promotes TCF targets while inhibits VDR target.	Shah *et al.* (2006), Palmer *et al.* (2008).
LRH-1	Mammalian cell culture	β-catenin: Arm repeats, key residues are Y306, K345 and W383.	Unclear.	Botrugno *et al.* (2004), Yumoto *et al.* (2012).
PPARγ	Mammalian cell culture	β-catenin and TCF4: domain unspecified.	TCF/β-catenin interactions can antagonize PPARg targets through chromatin loops.	Jansson *et al.* (2005), Hwang *et al.* (2012).
Other nuclear receptors	Mainly the RAR, RXR and LXR proteins.			Mulholland *et al.* (2005)[b], Beildeck, Gelmann, and Byers (2010)[b].

[a]Factors that can both activate and inhibit Wnt/β-catenin signaling.
[b]Reviews
The vertebrate system is defined as in Table 4.1. Abbreviations used that are not defined in the text are as follows: PitX2, paired-like homeodomain transcription factor 2; Prop1, homeobox protein prophet of PIT-1; LRH-1, liver receptor homolog 1; PPARγ, peroxisome proliferator-activated receptor gamma.

(Grigoryan *et al.*, 2008), but only a limited number can be unambiguously linked to TCFs (e.g., Galceran *et al.*, 1999; Korinek *et al.*, 1998; Kratochwil *et al.*, 2002; van Genderen *et al.*, 1994). While this may be due to redundancy and the repressive properties of some TCFs (see Chapter 17), the other possibility is that additional TFs can also recruit β-catenin to their respective enhancers. Indeed, the list of TFs with this function is large and diverse (Table 4.3) and has been reviewed in detail elsewhere (Archbold *et al.*, 2012; Beildeck, Gelmann, and Byers, 2010; Cadigan and Waterman, 2012; Valenta, Hausmann, and Basler, 2012). In one recent report, RNAi-based screens in human cancer cells with elevated Wnt/β-catenin signaling identified the T-box protein Tbx5 and the coactivator Yes-associated protein 1 (YAP1) as β-catenin binding proteins. When YAP1 is phosphorylated by the tyrosine kinase YES1, the YAP1–β-catenin–Tbx5 complex associated with and activated antiapoptotic genes (Rosenbluh *et al.*, 2012). A YES1 inhibitor dramatically reduced growth of β-catenin-dependent cancer cells and tumors (Rosenbluh *et al.*, 2012), providing a dramatic example of how β-catenin can act through non-TCFs to affect cell behavior.

In addition to recruiting β-catenin to their respective target genes, the aforementioned TFs can also divert β-catenin away from TCFs, inhibiting TCF-dependent gene expression. This appears to be a biologically important function of hypoxia-induced factor 1α (HIF1α) and Forkhead box (FOX) proteins during hypoxia and oxidative or nutritional stress (Almeida *et al.*, 2007; Hoogeboom *et al.*, 2008; Kaidi, Williams, and Paraskeva, 2007; Liu *et al.*, 2011).

There is also a growing list of TFs that interact with TCFs on chromatin. In some cases, this appears to be a mechanism for enhancers to integrate information from Wnt and other signaling pathways, for example, serum growth factor signaling via c-Jun–TCF interactions (Nateri, Spencer-Dene, and Behrens, 2005; Yochum, Cleland, and Goodman, 2008) or bone morphogenetic protein (BMP) signaling via Smad–TCF binding (Eivers, Demagny, and De Robertis, 2009; Itasaki and Hoppler, 2010). Recent ChIP-seq data suggests that TCF occupancy is heavily influenced in distinct cell types by colocalization with other TFs, some of which bind directly to TCFs (Bottomly *et al.*, 2010; Frietze *et al.*, 2012; Junion *et al.*, 2012; Trompouki *et al.*, 2011).

Adding to the complex nature of transcriptional responses to Wnt signaling, β-catenin is not the only transcriptional regulator whose stability is controlled by the β-catenin destruction complex. The transcriptional repressor Snail is phosphorylated by GSK3 and undergoes β-TrCP ubiquitination and proteasomal degradation (Yook *et al.*, 2005; Zhou *et al.*, 2004). Downregulation of this process by Wnt leads to increased Snail levels, which can promote epithelial–mesenchymal transitions (Yook *et al.*, 2005). More recently, transcriptional activator with PDZ-binding motif (TAZ), a relative of YAP1 and an important transcriptional coactivator in the Hippo signaling pathway that controls cell proliferation and survival (Pan, 2010), has been reported to be targeted for degradation by the β-catenin destruction complex (Azzolin *et al.*, 2012). Wnt stimulation leads to accumulation of nuclear TAZ, and transcriptome analysis revealed that the majority of Wnt targets in a human breast cancer cell line were TAZ dependent (Azzolin *et al.*, 2012). These examples make it clear that Wnt researchers have to look beyond the classic Wnt/β-catenin/TCF axis when considering how Wnts affect gene expression.

Concluding comments

Despite the extensive literature and relative maturity of the Wnt signaling field, the immense complexity of the system has ensured that many basic questions remain to be answered. One challenge for vertebrate researchers interested in TCF targets is how to deal with the large number of isoforms produced by the four/five *TCF* genes (see Chapter 17). A better catalog of which isoforms are endogenously expressed in various cells is needed, unless one settles for misexpression of individual TCF isoforms to study their function. For non-TCF TFs that mediate Wnt/β-catenin signaling, do they also operate as transcriptional switches, and is the coactivator cohort recruited to β-catenin/TCF complexes utilized in other TF–β-catenin contexts? These are issues that can be addressed with current technology and reagents.

One important issue that will require new approaches is the dynamics that likely occurs on Wnt target chromatin during induction. How can the large number of coregulators be studied in living cells at high resolution? This is a challenging problem, and perhaps approaches like single molecular FRET (Deindl *et al.*, 2013; Padilla-Parra and Tramier, 2012) can be utilized as this technology continues to be refined.

Finally, it should be noted that many basic features of basic Wnt gene regulation have largely been unexplored. Is RNA Pol II pausing an important feature of Wnt gene regulation? Is nucleosome positioning altered during Wnt target gene activation? Hopefully, these questions and others will be addressed in the near future.

Acknowledgments

Apologies to researchers whose work we could not discuss/cite due to space limits. This work was supported by AHA predoctoral fellowship 9520018 to CUZ and NIH grant GM082994 and NSF grant 0950348 to KMC.

References

Akiyama, H., Lyons, J.P., Mori-Akiyama, Y. *et al.* (2004) Interactions between Sox9 and beta-catenin control chondrocyte differentiation. *Genes & Development*, **18**, 1072–1087.

Almeida, M., Han, L., Martin-Millan, M. *et al.* (2007) Oxidative stress antagonizes Wnt signaling in osteoblast precursors by diverting beta-catenin from T cell factor- to forkhead box O-mediated transcription. *The Journal of Biological Chemistry*, **282**, 27298–27305.

Amen, M., Liu, X., Vadlamudi, U. *et al.* (2007) PITX2 and beta-catenin interactions regulate Lef-1 isoform expression. *Molecular and Cell Biology*, **27**, 7560–7573.

Andrews, P.G., He, Z., Popadiuk, C., and Kao, K.R. (2009) The transcriptional activity of Pygopus is enhanced by its interaction with cAMP-response-element-binding protein (CREB)-binding protein. *Biochemistry Journal*, **422**, 493–501.

Arce, L., Pate, K.T., and Waterman, M.L. (2009) Groucho binds two conserved regions of LEF-1 for HDAC-dependent repression. *BMC Cancer*, **9**, 159.

Archbold, H.C., Yang, Y.X., Chen, L., and Cadigan, K.M. (2012) How do they do Wnt they do? Regulation of transcription by the Wnt/beta-catenin pathway. *Acta Physiologica*, **204**, 74–109.

Atlasi, Y., Noori, R., Gaspar, C. *et al.* (2013) Wnt signaling regulated the lineage differentiation potential of mouse embryonic stem cells through TCF3 down-regulation. *PLoS Genetics*, **9**, e1003424.

Azzolin, L., Zanconato, F., Bresolin, S. *et al.* (2012) Role of TAZ as mediator of Wnt signaling. *Cell*, **151**, 1443–1456.

Barker, N., Hurlstone, A., Musisi, H. *et al.* (2001) The chromatin remodelling factor Brg-1 interacts with beta-catenin to promote target gene activation. *The EMBO Journal*, **20**, 4935–4943.

Barolo, S. (2006) Transgenic Wnt/TCF pathway reporters: all you need is Lef? *Oncogene*, **25**, 7505–7511.

Barrett, C.W., Smith, J.J., Lu, L.C. *et al.* (2012) Kaiso directs the transcriptional corepressor MTG16 to the Kaiso binding site in target promoters. *PLoS One*, **7**, e51205.

Bauer, A., Chauvet, S., Huber, O. *et al.* (2000) Pontin52 and reptin52 function as antagonistic regulators of beta-catenin signalling activity. *The EMBO Journal*, **19**, 6121–6130.

Behrens, J., vonKries, J.P., Kuhl, M. *et al.* (1996) Functional interaction of beta-catenin with the transcription factor LEF-1. *Nature*, **382**, 638–642.

Beildeck, M.E., Gelmann, E.P., and Byers, S.W. (2010) Cross-regulation of signaling pathways: an example of nuclear hormone receptors and the canonical Wnt pathway. *Experimental Cell Research*, **316**, 1763–1772.

Belenkaya, T.Y., Han, C., Standley, H.J. *et al.* (2002) pygopus Encodes a nuclear protein essential for wingless/Wnt signaling. *Development*, **129**, 4089–4101.

Benchabane, H., Xin, N., Tian, A. *et al.* (2011) Jerky/Earthbound facilitates cell-specific Wnt/Wingless signalling by modulating beta-catenin-TCF activity. *The EMBO Journal*, **30**, 1444–1458.

Bhambhani, C., Chang, J.L., Akey, D.L., and Cadigan, K.M. (2011) The oligomeric state of CtBP determines its role as a transcriptional co-activator and co-repressor of Wingless targets. *The EMBO Journal*, **30**, 2031–2043.

Blythe, S.A., Cha, S.W., Tadjuidje, E. *et al.* (2010) Beta-catenin primes organizer gene expression by recruiting a histone H3 arginine 8 methyltransferase, Prmt2. *Developmental Cell*, **19**, 220–231.

Botrugno, O.A., Fayard, E., Annicotte, J.S. *et al.* (2004) Synergy between LRH-1 and beta-catenin induces G1 cyclin-mediated cell proliferation. *Molecular Cell*, **15**, 499–509.

Bottomly, D., Kyler, S.L., McWeeney, S.K., and Yochum, G.S. (2010) Identification of β-catenin binding regions in colon cancer cells using ChIP-Seq. *Nucleic Acids Research*, **38**, 5735–5745.

Brack, A.S., Murphy-Seiler, F., Hanifi, J. *et al.* (2009) BCL9 is an essential component of canonical Wnt signaling that mediates the differentiation of myogenic progenitors during muscle regeneration. *Developmental Biology*, **335** (1), 93–105.

Brannon, M., Gomperts, M., Sumoy, L. *et al.* (1997) A beta-catenin/XTcf-3 complex binds to the siamois promoter to regulate dorsal axis specification in Xenopus. *Genes & Development*, **11**, 2359–2370.

Brannon, M., Brown, J.D., Bates, R. *et al.* (1999) XCtBP is a XTcf-3 co-repressor with roles throughout Xenopus development. *Development*, **126**, 159–170.

Brembeck, F.H., Schwarz-Romond, T., Bakkers, J. *et al.* (2004) Essential role of BCL9-2 in the switch between beta-catenin's adhesive and transcriptional functions. *Genes & Development*, **18**, 2225–2230.

Brocardo, M. and Henderson, B.R. (2008) APC shuttling to the membrane, nucleus and beyond. *Trends in Cell Biology*, **18**, 587–596.

Brunner, E., Peter, O., Schweizer, L., and Basler, K. (1997) pangolin Encodes a Lef-1 homologue that acts downstream of Armadillo to transduce the Wingless signal in *Drosophila*. *Nature*, **385**, 829–833.

Cadigan, K.M. (2012) TCFs and Wnt/beta-catenin signaling: more than one way to throw the switch. *Current Topics in Developmental Biology*, **98**, 1–34.

Cadigan, K.M. and Peifer, M. (2009) Wnt signaling from development to disease: insights from model systems. *Cold Spring Harbor Perspectives in Biology*, **1**, a002881.

Cadigan, K.M. and Waterman, M.L. (2012) TCF/LEFs and Wnt signaling in the nucleus. *Cold Spring Harbor Perspectives in Biology*, **4**, a007906.

Calvo, D., Victor, M., Gay, F. *et al.* (2001) A POP-1 repressor complex restricts inappropriate cell type-specific gene transcription during *Caenorhabditis elegans* embryogenesis. *The EMBO Journal*, **20**, 7197–7208.

Carrera, I., Janody, F., Leeds, N. *et al.* (2008) Pygopus activates Wingless target gene transcription through the mediator complex subunits Med12 and Med13. *Proceedings of the National Academy of Sciences of the United States of America*, **105**, 6644–6649.

Cavallo, R.A., Cox, R.T., Moline, M.M. *et al.* (1998) *Drosophila* Tcf and Groucho interact to repress Wingless signalling activity. *Nature*, **395**, 604–608.

Chang, J.L., Chang, M.V., Barolo, S., and Cadigan, K.M. (2008a) Regulation of the feedback antagonist naked cuticle by Wingless signaling. *Developmental Biology*, **321**, 446–454.

Chang, M.V., Chang, J.L., Gangopadhyay, A. *et al.* (2008b) Activation of wingless targets requires bipartite recognition of DNA by TCF. *Current Biology*, **18**, 1877–1881.

Chen, J., Luo, Q., Yuan, Y. *et al.* (2010) Pygo2 associates with MLL2 histone methyltransferase and GCN5 histone acetyltransferase complexes to augment Wnt target gene expression and breast cancer stem-like cell expansion. *Molecular and Cell Biology*, **30**, 5621–5635.

Chen, D., Tian, W., Li, Y. *et al.* (2012) Osteoblast-specific transcription factor Osterix (Osx) and HIF-1alpha cooperatively regulate gene expression of vascular endothelial growth factor (VEGF). *Biochemical and Biophysical Research Communications*, **424**, 176–181.

Choi, H.K., Choi, K.C., Yoo, J.Y. *et al.* (2011) Reversible SUMOylation of TBL1-TBLR1 regulates beta-catenin-mediated Wnt signaling. *Molecular Cell*, **43**, 203–216.

Cronauer, M.V., Schulz, W.A., Ackermann, R., and Burchardt, M. (2005) Effects of WNT/beta-catenin pathway activation on signaling through T-cell factor and androgen receptor in prostate cancer cell lines. *International Journal of Oncology*, **26**, 1033–1040.

Daniels, D.L. and Weis, W.I. (2002) ICAT inhibits beta-catenin binding to Tcf/Lef-family transcription factors and the general coactivator p300 using independent structural modules. *Molecular Cell*, **10** (3), 573–584.

Daniels, D.L. and Weis, W.I. (2005) Beta-catenin directly displaces Groucho/TLE repressors from Tcf/Lef in Wnt-mediated transcription activation. *Nature Structural & Molecular Biology*, **12**, 364–371.

Deindl, S., Hwang, W.L., Hota, S.K. *et al.* (2013) ISWI remodelers slide nucleosomes with coordinated multi-base-pair entry steps and single-base-pair exit steps. *Cell*, **152**, 442–452.

Dorsky, R.I., Itoh, M., Moon, R.T., and Chitnis, A. (2003) Two tcf3 genes cooperate to pattern the zebrafish brain. *Development*, **130**, 1937–1947.

Eckey, M., Kuphal, S., Straub, T. *et al.* (2012) Nucleosome remodeler SNF2L suppresses cell proliferation and migration and attenuates Wnt signaling. *Molecular and Cell Biology*, **32**, 2359–2371.

Eivers, E., Demagny, H., and De Robertis, E.M. (2009) Integration of BMP and Wnt signaling via vertebrate Smad1/5/8 and *Drosophila* Mad. *Cytokine & Growth Factor Reviews*, **20**, 357–365.

Enjolras, C., Thomas, J., Chhin, B. *et al.* (2012) *Drosophila* chibby is required for basal body formation and ciliogenesis but not for Wg signaling. *Journal of Cell Biology*, **197**, 313–325.

Essers, M.A., de Vries-Smits, L.M., Barker, N. *et al.* (2005) Functional interaction between beta-catenin and FOXO in oxidative stress signaling. *Science*, **308**, 1181–1184.

Fang, M., Li, J., Blauwkamp, T. *et al.* (2006) C-terminal-binding protein directly activates and represses Wnt transcriptional targets in *Drosophila*. *The EMBO Journal*, **25**, 2735–2745.

Feng, Y., Lee, N., and Fearon, E.R. (2003) TIP49 regulates beta-catenin-mediated neoplastic transformation and T-cell factor target gene induction via effects on chromatin remodeling. *Cancer Research*, **63**, 8726–8734.

Frietze, S., Wang, R., Yao, L. *et al.* (2012) Cell type-specific binding patterns reveal that TCF7L2 can be tethered to the genome by association with GATA3. *Genome Biology*, **13**, R52.

Galceran, J., Farinas, I., Depew, M.J. *et al.* (1999) Wnt3a−/−-like phenotype and limb deficiency in Lef1(−/−)Tcf1(−/−) mice. *Genes & Development*, **13**, 709–717.

Galceran, J., Sustmann, C., Hsu, S.C. *et al.* (2004) LEF1-mediated regulation of Delta-like1 links Wnt and Notch signaling in somitogenesis. *Genes & Development*, **18**, 2718–2723.

Gay, F., Calvo, D., Lo, M.C. *et al.* (2003) Acetylation regulates subcellular localization of the Wnt signaling nuclear effector POP-1. *Genes & Development*, **17** (6), 717–722.

Ghogomu, S.M., van Venrooy, S., Ritthaler, M. *et al.* (2006) HIC-5 is a novel repressor of lymphoid enhancer factor/T-cell factor-driven transcription. *The Journal of Biological Chemistry*, **281**, 1755–1764.

Giese, K., Amsterdam, A., and Grosschedl, R. (1991) DNA-binding properties of the HMG domain of the lymphoid-specific transcriptional regulator LEF-1. *Genes & Development*, **5**, 2567–2578.

Goodman, R.H. and Smolik, S. (2000) CBP/p300 in cell growth, transformation, and development. *Genes & Development*, **14**, 1553–1577.

Graham, T.A., Weaver, C., Mao, F. *et al.* (2000) Crystal structure of a beta-catenin/Tcf complex. *Cell*, **103**, 885–896.

Graham, T.A., Clements, W.K., Kimelman, D., and Xu, W. (2002) The crystal structure of the beta-catenin/ICAT complex reveals the inhibitory mechanism of ICAT. *Molecular Cell*, **10** (3), 563–571.

Grigoryan, T., Wend, P., Klaus, A., and Birchmeier, W. (2008) Deciphering the function of canonical Wnt signals in development and disease: conditional loss- and gain-of-function mutations of beta-catenin in mice. *Genes & Development*, **22** (17), 2308–2341.

Gu, B., Sun, P., Yuan, Y. *et al.* (2009) Pygo2 expands mammary progenitor cells by facilitating histone H3 K4 methylation. *The Journal of Cell Biology*, **185**, 811–826.

Hamada, F. and Bienz, M. (2004) The APC tumor suppressor binds to C-terminal binding protein to divert nuclear beta-catenin from TCF. *Developmental Cell*, **7**, 677–685.

Hasegawa, Y., Satoh, K., Iizuka-Kogo, A. *et al.* (2007) Loss of ICAT gene function leads to arrest of ureteric bud branching and renal agenesis. *Biochemical & Biophysical Research Communications*, **362**, 988–994.

He, T.C., Sparks, A.B., Rago, C. *et al.* (1998) Identification of c-MYC as a target of the APC pathway. *Science*, **281**, 1509–1512.

Hecht, A., Litterst, C.M., Huber, O., and Kemler, R. (1999) Functional characterization of multiple transactivating elements in beta-catenin, some of which interact with the TATA-binding protein in vitro. *The Journal of Biological Chemistry*, **274**, 18017–18025.

Hecht, A., Vleminckx, K., Stemmler, M.P. *et al.* (2000) The p300/CBP acetyltransferases function as transcriptional coactivators of beta-catenin in vertebrates. *The EMBO Journal*, **19**, 1839–1850.

Henderson, B.R. and Fagotto, F. (2002) The ins and outs of APC and beta-catenin nuclear transport. *EMBO Reports*, **3**, 834–839.

Henderson, W.R., Jr., Chi, E.Y., Ye, X. *et al.* (2010) Inhibition of Wnt/beta-catenin/CREB binding protein (CBP) signaling reverses pulmonary fibrosis. *Proceedings of the National Academy of Sciences of the United States of America*, **107**, 14309–14314.

Hikasa, H. and Sokol, S.Y. (2011) Phosphorylation of TCF proteins by homeodomain-interacting protein kinase 2. *The Journal of Biological Chemistry*, **286**, 12093–12100.

Hikasa, H., Ezan, J., Itoh, K. *et al.* (2010) Regulation of TCF3 by Wnt-dependent phosphorylation during vertebrate axis specification. *Developmental Cell*, **19**, 521–532.

Hoffmans, R. and Basler, K. (2004) Identification and in vivo role of the Armadillo-Legless interaction. *Development*, **131**, 4393–4400.

Hong, J.Y., Park, J.I., Cho, K. *et al.* (2010) Shared molecular mechanisms regulate multiple catenin proteins: canonical Wnt signals and components modulate p120-catenin isoform-1 and additional p120 subfamily members. *Journal of Cell Science*, **123**, 4351–4365.

Hong, J.Y., Park, J.I., Lee, M. *et al.* (2012) Down's-syndrome-related kinase Dyrk1A modulates the p120-catenin-Kaiso trajectory of the Wnt signaling pathway. *Journal of Cell Science*, **125** (Pt.3), 561–569.

Hoogeboom, D., Essers, M.A., Polderman, P.E. *et al.* (2008) Interaction of FOXO with beta-catenin inhibits beta-catenin/T cell factor activity. *The Journal of Biological Chemistry*, **283**, 9224–9230.

Hwang, I., Kim, J., and Jeong, S. (2012) Beta-catenin and peroxisome proliferator-activated receptor-delta coordinate dynamic chromatin loops for the transcription of vascular endothelial growth factor A gene in colon cancer cells. *The Journal of Biological Chemistry*, **287**, 41364–41373.

Ihara, M., Yamamoto, H., and Kikuchi, A. (2005) SUMO-1 modification of PIASy, an E3 ligase, is necessary for PIASy-dependent activation of Tcf-4. *Molecular and Cell Biology*, **25**, 3506–3518.

Iioka, H., Doerner, S.K., and Tamai, K. (2009) Kaiso is a bimodal modulator for Wnt/beta-catenin signaling. *FEBS Letters*, **583**, 627–632.

Itasaki, N. and Hoppler, S. (2010) Crosstalk between Wnt and bone morphogenic protein signaling: a turbulent relationship. *Developmental Dynamics*, **239**, 16–33.

Jamieson, C., Sharma, M., and Henderson, B.R. (2012) Wnt signaling from membrane to nucleus: β-catenin caught in a loop. *The International Journal of Biochemistry & Cell Biology*, **44**, 847–850.

Jansson, E.A., Are, A., Greicius, G. *et al.* (2005) The Wnt/beta-catenin signaling pathway targets PPARgamma activity in colon cancer cells. *Proceedings of the National Academy of Sciences of the United States of America*, **102**, 1460–1465.

Jessen, S., Gu, B., and Dai, X. (2008) Pygopus and the Wnt signaling pathway: a diverse set of connections. *Bioessays*, **30**, 448–456.

Junion, G., Spivakov, M., Girardot, C. *et al.* (2012) A transcription factor collective defines cardiac cell fate and reflects lineage history. *Cell*, **148**, 473–486.

Kaidi, A., Williams, A.C., and Paraskeva, C. (2007) Interaction between beta-catenin and HIF-1 promotes cellular adaptation to hypoxia. *Nature Cell Biology*, **9**, 210–217.

Kan, L., Israsena, N., Zhang, Z. *et al.* (2004) Sox1 acts through multiple independent pathways to promote neurogenesis. *Developmental Biology*, **269** (2), 580–594.

Kennell, J. and Cadigan, K.M. (2009) APC and beta-catenin degradation. *Advances in Experimental Medicine and Biology*, **656**, 1–12.

Kim, C.H., Oda, T., Itoh, M. *et al.* (2000) Repressor activity of Headless/Tcf3 is essential for vertebrate head formation. *Nature*, **407**, 913–916.

Kim, J.H., Kim, B., Cai, L. *et al.* (2005) Transcriptional regulation of a metastasis suppressor gene by Tip60 and beta-catenin complexes. *Nature*, **434**, 921–926.

Kim, S., Xu, X., Hecht, A., and Boyer, T.G. (2006) Mediator is a transducer of Wnt/beta-catenin signaling. *The Journal of Biological Chemistry*, **281**, 14066–14075.

Kim, C.H., Neiswender, H., Baik, E.J. *et al.* (2008) Beta-catenin interacts with MyoD and regulates its transcription activity. *Molecular and Cell Biology*, **28**, 2941–2951.

Kioussi, C., Briata, P., Baek, S.H. *et al.* (2002) Identification of a Wnt/Dvl/beta-Catenin → Pitx2 pathway mediating cell-type-specific proliferation during development. *Cell*, **111**, 673–685.

Korinek, V., Barker, N., Moerer, P. *et al.* (1998) Depletion of epithelial stem-cell compartments in the small intestine of mice lacking Tcf-4. *Nature Genetics*, **19**, 379–383.

Kormish, J.D., Sinner, D., and Zorn, A.M. (2010) Interactions between SOX factors and Wnt/beta-catenin signaling in development and disease. *Developmental Dynamics*, **239**, 56–68.

Kramps, T., Peter, O., Brunner, E. *et al.* (2002) Wnt/wingless signaling requires BCL9/legless-mediated recruitment of pygopus to the nuclear beta-catenin-TCF complex. *Cell*, **109**, 47–60.

Kratochwil, K., Galceran, J., Tontsch, S. *et al.* (2002) FGF4, a direct target of LEF1 and Wnt signaling, can rescue the arrest of tooth organogenesis in Lef1(−/−) mice. *Genes & Development*, **16**, 3173–3185.

Lam, N., Chesney, M.A., and Kimble, J. (2006) Wnt signaling and CEH-22/tinman/Nkx2.5 specify a stem cell niche in *C. elegans*. *Current Biology*, **16**, 287–295.

Laudet, V., Stehelin, D., and Clevers, H. (1993) Ancestry and diversity of the HMG box superfamily. *Nucleic Acids Research*, **21**, 2493–2501.

Lee, H.H. and Frasch, M. (2000) Wingless effects mesoderm patterning and ectoderm segmentation events via induction of its downstream target sloppy paired. *Development*, **127**, 5497–5508.

Li, J. and Wang, C.Y. (2008) TBL1-TBLR1 and beta-catenin recruit each other to Wnt target-gene promoter for transcription activation and oncogenesis. *Nature Cell Biology*, **10**, 160–169.

Li, J., Sutter, C., Parker, D.S. *et al.* (2007) CBP/p300 are bimodal regulators of Wnt signaling. *The EMBO Journal*, **26**, 2284–2294.

Li, F.Q., Mofunanya, A., Harris, K., and Takemaru, K. (2008) Chibby cooperates with 14-3-3 to regulate beta-catenin subcellular distribution and signaling activity. *Journal of Cell Biology*, **181**, 1141–1154.

Li, X., Martinez-Ferrer, M., Botta, V. *et al.* (2011a) Epithelial Hic-5/ARA55 expression contributes to prostate tumorigenesis and castrate responsiveness. *Oncogene*, **30**, 167–177.

Li, Z., Nie, F., Wang, S., and Li, L. (2011b) Histone H4 Lys 20 monomethylation by histone methylase SET8 mediates Wnt target gene activation. *Proceedings of the National Academy of Sciences of the United States of America*, **108**, 3116–3123.

Liu, F., van den Broek, O., Destree, O., and Hoppler, S. (2005) Distinct roles for Xenopus Tcf/Lef genes in mediating specific responses to Wnt/beta-catenin signalling in mesoderm development. *Development*, **132**, 5375–5385.

Liu, Y.I., Chang, M.V., Li, H.E. *et al.* (2008) The chromatin remodelers ISWI and ACF1 directly repress Wingless transcriptional targets. *Developmental Biology*, **323**, 41–52.

Liu, H., Fergusson, M.M., Wu, J.J. *et al.* (2011) Wnt signaling regulates hepatic metabolism. *Science Signaling*, **4**, ra6.

Liu, J.X., Zhang, D., Xie, X. *et al.* (2013) Eaf1 and Eaf2 negatively regulate canonical Wnt/beta-catenin signaling. *Development*, **140**, 1067–1078.

Love, J.J., Li, X., Case, D.A. *et al.* (1995) Structural basis for DNA bending by the architectural transcription factor LEF-1. *Nature*, **376**, 791–795.

Love, D., Li, F.Q., Burke, M.C. *et al.* (2010) Altered lung morphogenesis, epithelial cell differentiation and mechanics in mice deficient in the Wnt/beta-catenin antagonist Chibby. *PLoS One*, **5**, e13600.

Ma, H., Nguyen, C., Lee, K.S., and Kahn, M. (2005) Differential roles for the coactivators CBP and p300 on TCF/beta-catenin-mediated survivin gene expression. *Oncogene*, **24**, 3619–3631.

MacDonald, B.T., Tamai, K., and He, X. (2009) Wnt/beta-catenin signaling: components, mechanisms, and diseases. *Developmental Cell*, **17**, 9–26.

Mahmoudi, T., Boj, S.F., Hatzis, P. *et al.* (2010) The leukemia-associated Mllt10/Af10-Dot1l are Tcf4/beta-catenin coactivators essential for intestinal homeostasis. *PLoS Biology*, **8**, e1000539.

Major, M.B., Roberts, B.S., Berndt, J.D. *et al.* (2008) New regulators of Wnt/beta-catenin signaling revealed by integrative molecular screening. *Science Signaling*, **1**, ra12.

Mazumdar, J., O'Brien, W.T., Johnson, R.S. *et al.* (2010) O_2 regulates stem cells through Wnt/β-catenin signalling. *Nature Cell Biology*, **12** (10), 1007–1013.

Merrill, B.J. (2012) Wnt pathway regulation of embryonic stem cell self-renewal. *Cold Spring Harbor Perspectives in Biology*, **4**, a007971.

Merrill, B.J., Pasolli, H.A., Polak, L. *et al.* (2004) Tcf3: a transcriptional regulator of axis induction in the early embryo. *Development*, **131**, 263–274.

Mitani, T., Harada, N., Nakano, Y. *et al.* (2012) Coordinated action of hypoxia-inducible factor-1alpha and beta-catenin in androgen receptor signaling. *The Journal of Biological Chemistry*, **287**, 33594–33606.

Mohan, M., Herz, H.M., Takahashi, Y.H. *et al.* (2010) Linking H3K79 trimethylation to Wnt signaling through a novel Dot1-containing complex (DotCom). *Genes & Development*, **24** (6), 574–589.

Molenaar, M., van de Wetering, M., Oosterwegel, M. *et al.* (1996) XTcf-3 transcription factor mediates beta-catenin-induced axis formation in Xenopus embryos. *Cell*, **86**, 391–399.

Moore, A.C., Amann, J.M., Williams, C.S. *et al.* (2008) Myeloid translocation gene family members associate with T-cell factors (TCFs) and influence TCF-dependent transcription. *Molecular and Cell Biology*, **28**, 977–987.

Mora-Blanco, E.L., Mishina, Y., Tillman, E.J. *et al.* (2013) Activation of beta-catenin/TCF targets following loss of the tumor suppressor SNF5. *Oncogene*. doi: 10.1038/onc.2013.37

Mosimann, C., Hausmann, G., and Basler, K. (2006) Parafibromin/Hyrax activates Wnt/Wg target gene transcription by direct association with beta-catenin/Armadillo. *Cell*, **125**, 327–341.

Mosimann, C., Hausmann, G., and Basler, K. (2009) Beta-catenin hits chromatin: regulation of Wnt target gene activation. *Nature Reviews Molecular Cell Biology*, **10**, 276–286.

Mulholland, D.J., Dedhar, S., Coetzee, G.A., and Nelson, C.C. (2005) Interaction of nuclear receptors with the Wnt/beta-catenin/Tcf signaling axis: Wnt you like to know? *Endocrine Reviews*, **26** (7), 898–915.

Najdi, R., Syed, A., Arce, L. *et al.* (2009) A Wnt kinase network alters nuclear localization of TCF-1 in colon cancer. *Oncogene*, **28**, 4133–4146.

Nakamura, Y., Hinoi, E., Iezaki, T. *et al.* (2013) Repression of adipogenesis through promotion of Wnt/β-catenin signaling by TIS7 up-regulated in adipocytes under hypoxia. *Biochimica et Biophysica Acta*, **1832** (8), 1117–1128.

Nateri, A.S., Spencer-Dene, B., and Behrens, A. (2005) Interaction of phosphorylated c-Jun with TCF4 regulates intestinal cancer development. *Nature*, **437**, 281–285.

Nguyen, H., Merrill, B.J., Polak, L. *et al.* (2009) Tcf3 and Tcf4 are essential for long-term homeostasis of skin epithelia. *Nature Genetics*, **41**, 1068–1075.

Nishiyama, M., Skoultchi, A.I., and Nakayama, K.I. (2012) Histone H1 recruitment by CHD8 is essential for suppression of the Wnt-beta-catenin signaling pathway. *Molecular and Cell Biology*, **32**, 501–512.

Olson, L.E., Tollkuhn, J., Scafoglio, C. *et al.* (2006) Homeodomain-mediated beta-catenin-dependent switching events dictate cell-lineage determination. *Cell*, **125**, 593–605.

Ou, C.Y., LaBonte, M.J., Manegold, P.C. *et al.* (2011) A coactivator role of CARM1 in the dysregulation of beta-catenin activity in colorectal cancer cell growth and gene expression. *Molecular Cancer Research*, **9**, 660–670.

Padilla-Parra, S. and Tramier, M. (2012) FRET microscopy in the living cell: different approaches, strengths and weaknesses. *Bioessays*, **34**, 369–376.

Palmer, H.G., Anjos-Afonso, F., Carmeliet, G. *et al.* (2008) The vitamin D receptor is a Wnt effector that controls hair follicle differentiation and specifies tumor type in adult epidermis. *PLoS One*, **3**, e1483.

Pan, D. (2010) The hippo signaling pathway in development and cancer. *Developmental Cell*, **19**, 491–505.

Park, J.I., Kim, S.W., Lyons, J.P. *et al.* (2005) Kaiso/p120-catenin and TCF/beta-catenin complexes coordinately regulate canonical Wnt gene targets. *Developmental Cell*, **8**, 843–854.

Park, J.I., Ji, H., Jun, S. *et al.* (2006) Frodo links Dishevelled to the p120-catenin/Kaiso pathway: distinct catenin subfamilies promote Wnt signals. *Developmental Cell*, **11**, 683–695.

Parker, D.S., Jemison, J., and Cadigan, K.M. (2002) Pygopus, a nuclear PHD-finger protein required for Wingless signaling in *Drosophila*. *Development*, **129**, 2565–2576.

Parker, D.S., Ni, Y.Y., Chang, J.L. *et al.* (2008) Wingless signaling induces widespread chromatin remodeling of target loci. *Molecular and Cell Biology*, **28**, 1815–1828.

Phillips, B.T. and Kimble, J. (2009) A new look at TCF and beta-catenin through the lens of a divergent *C. elegans* Wnt pathway. *Developmental Cell*, **17**, 27–34.
Polakis, P. (2012) Wnt signaling in cancer. *Cold Spring Harbor Perspectives in Biology*, **4**, a008052.
Pomerantz, M.M., Ahmadiyeh, N., Jia, L. *et al.* (2009) The 8q24 cancer risk variant rs6983267 shows long-range interaction with MYC in colorectal cancer. *Nature Genetics*, **41**, 882–884.
Poy, F., Lepourcelet, M., Shivdasani, R.A., and Eck, M.J. (2001) Structure of a human Tcf4-beta-catenin complex. *Nature Structural & Molecular Biology*, **8**, 1053–1057.
Prieve, M.G., Guttridge, K.L., Munguia, J., and Waterman, M.L. (1998) Differential importin-alpha recognition and nuclear transport by nuclear localization signals within the high-mobility-group DNA binding domains of lymphoid enhancer factor 1 and T-cell factor 1. *Molecular and Cell Biology*, **18**, 4819–4832.
Rashid, S., Pilecka, I., Torun, A. *et al.* (2009) Endosomal adaptor proteins APPL1 and APPL2 are novel activators of beta-catenin/TCF-mediated transcription. *The Journal of Biological Chemistry*, **284**, 18115–18128.
Reya, T., O'Riordan, M., Okamura, R. *et al.* (2000) Wnt signaling regulates B lymphocyte proliferation through a LEF-1 dependent mechanism. *Immunity*, **13**, 15–24.
Rocha, P.P., Scholze, M., Bleiss, W., and Schrewe, H. (2010) Med12 is essential for early mouse development and for canonical Wnt and Wnt/PCP signaling. *Development*, **137**, 2723–2731.
Roose, J., Molenaar, M., Peterson, J. *et al.* (1998) The Xenopus Wnt effector XTcf-3 interacts with Groucho-related transcriptional repressors. *Nature*, **395**, 608–612.
Roose, J., Huls, G., van Beest, M. *et al.* (1999) Synergy between tumor suppressor APC and the beta-catenin-Tcf4 target Tcf1. *Science*, **285**, 1923–1926.
Rosenbluh, J., Nijhawan, D., Cox, A.G. *et al.* (2012) beta-Catenin-driven cancers require a YAP1 transcriptional complex for survival and tumorigenesis. *Cell*, **151**, 1457–1473.
Roth, W., Sustmann, C., Kieslinger, M. *et al.* (2004) PIASy-deficient mice display modest defects in IFN and Wnt signaling. *Journal of Immunology*, **173** (10), 6189–6199.
Rottbauer, W., Saurin, A.J., Lickert, H. *et al.* (2002) Reptin and pontin antagonistically regulate heart growth in zebrafish embryos. *Cell*, **111**, 661–672.
Ruzov, A., Hackett, J.A., Prokhortchouk, A. *et al.* (2009a) The interaction of xKaiso with xTcf3: a revised model for integration of epigenetic and Wnt signalling pathways. *Development*, **136**, 723–727.
Ruzov, A., Savitskaya, E., Hackett, J.A. *et al.* (2009b) The non-methylated DNA-binding function of Kaiso is not required in early Xenopus laevis development. *Development*, **136**, 729–738.
Sachdev, S., Bruhn, L., Sieber, H. *et al.* (2001) PIASy, a nuclear matrix-associated SUMO E3 ligase, represses LEF1 activity by sequestration into nuclear bodies. *Genes & Development*, **15** (23), 3088–3103.
Sakamoto, I., Kishida, S., Fukui, A. *et al.* (2000) A novel beta-catenin-binding protein inhibits beta-catenin-dependent Tcf activation and axis formation. *The Journal of Biological Chemistry*, **275**, 32871–32878.
Sawa, H. (2012) Control of cell polarity and asymmetric division in *C. elegans*. *Current Topics in Developmental Biology*, **101**, 55–76.
Schwab, K.R., Patterson, L.T., Hartman, H.A. *et al.* (2007) Pygo1 and Pygo2 roles in Wnt signaling in mammalian kidney development. *BMC Biology*, **5**, 15.
Shah, S., Islam, M.N., Dakshanamurthy, S. *et al.* (2006) The molecular basis of vitamin D receptor and beta-catenin crossregulation. *Molecular Cell*, **21**, 799–809.
Sharma, M., Jamieson, C., Johnson, M. *et al.* (2012) Specific armadillo repeat sequences facilitate beta-catenin nuclear transport in live cells via direct binding to nucleoporins Nup62, Nup153, and RanBP2/Nup358. *The Journal of Biological Chemistry*, **287**, 819–831.
Sierra, J., Yoshida, T., Joazeiro, C.A., and Jones, K.A. (2006) The APC tumor suppressor counteracts beta-catenin activation and H3K4 methylation at Wnt target genes. *Genes & Development*, **20**, 586–600.
Simoneau, M., Coulombe, G., Vandal, G. *et al.* (2011) SHP-1 inhibits beta-catenin function by inducing its degradation and interfering with its association with TATA-binding protein. *Cell Signal*, **23**, 269–279.
Sinner, D., Rankin, S., Lee, M., and Zorn, A.M. (2004) Sox17 and beta-catenin cooperate to regulate the transcription of endodermal genes. *Development*, **131**, 3069–3080.
Sinner, D., Kordich, J.J., Spence, J.R. *et al.* (2007) Sox17 and Sox4 differentially regulate beta-catenin/T-cell factor activity and proliferation of colon carcinoma cells. *Molecular and Cell Biology*, **27**, 7802–7815.
Sokol, S.Y. (2011) Maintaining embryonic stem cell pluripotency with Wnt signaling. *Development*, **138**, 4341–4350.
Song, L.N. and Gelmann, E.P. (2008) Silencing mediator for retinoid and thyroid hormone receptor and nuclear receptor corepressor attenuate transcriptional activation by the beta-catenin-TCF4 complex. *The Journal of Biological Chemistry*, **283**, 25988–25999.
Song, H., Spichiger-Haeusermann, C., and Basler, K. (2009) The ISWI-containing NURF complex regulates the output of the canonical Wingless pathway. *EMBO Reports*, **10**, 1140–1146.

Song, L.N., Herrell, R., Byers, S. *et al.* (2003) Beta-catenin binds to the activation function 2 region of the androgen receptor and modulates the effects of the N-terminal domain and TIF2 on ligand-dependent transcription. *Molecular and Cell Biology*, **23**, 1674–1687.

Song, H., Goetze, S., Bischof, J. *et al.* (2010) Coop functions as a corepressor of Pangolin and antagonizes Wingless signaling. *Genes & Development*, **24**, 881–886.

Stepniak, E., Radice, G.L., and Vasioukhin, V. (2009) Adhesive and signaling functions of cadherins and catenins in vertebrate development. *Cold Spring Harbor Perspectives in Biology*, **1**, a002949.

Sudarsanam, P. and Winston, F. (2000) The Swi/Snf family nucleosome-remodeling complexes and transcriptional control. *Trends in Genetics*, **16**, 345–351.

Sun, Y., Kolligs, F.T., Hottiger, M.O. *et al.* (2000) Regulation of beta-catenin transformation by the p300 transcriptional coactivator. *Proceedings of the National Academy of Sciences of the United States of America*, **97**, 12613–12618.

Sustmann, C., Flach, H., Ebert, H. *et al.* (2008) Cell-type-specific function of BCL9 involves a transcriptional activation domain that synergizes with beta-catenin. *Molecular and Cell Biology*, **28**, 3526–3537.

Tago, K., Nakamura, T., Nishita, M. *et al.* (2000) Inhibition of Wnt signaling by ICAT, a novel beta-catenin-interacting protein. *Genes & Development*, **14** (14), 1741–1749.

Takahashi, A., Tsutsumi, R., Kikuchi, I. *et al.* (2011) SHP2 tyrosine phosphatase converts parafibromin/Cdc73 from a tumor suppressor to an oncogenic driver. *Molecular Cell*, **43**, 45–56.

Takemaru, K.I. and Moon, R.T. (2000) The transcriptional coactivator CBP interacts with beta-catenin to activate gene expression. *The Journal of Cell Biology*, **149**, 249–254.

Takemaru, K., Yamaguchi, S., Lee, Y.S. *et al.* (2003) Chibby, a nuclear beta-catenin-associated antagonist of the Wnt/Wingless pathway. *Nature*, **422**, 905–909.

Tang, W., Dodge, M., Gundapaneni, D. *et al.* (2008) A genome-wide RNAi screen for Wnt/beta-catenin pathway components identifies unexpected roles for TCF transcription factors in cancer. *Proceedings of the National Academy of Sciences of the United States of America*, **105**, 9697–9702.

Thompson, B., Townsley, F., Rosin-Arbesfeld, R. *et al.* (2002) A new nuclear component of the Wnt signalling pathway. *Nature Cell Biology*, **4**, 367–373.

Thompson, B.A., Tremblay, V., Lin, G., and Bochar, D.A. (2008) CHD8 is an ATP-dependent chromatin remodeling factor that regulates beta-catenin target genes. *Molecular and Cell Biology*, **28**, 3894–3904.

Tolwinski, N.S. and Wieschaus, E. (2004) Rethinking WNT signaling. *Trends in Genetics*, **20**, 177–181.

Topol, L., Chen, W., Song, H. *et al.* (2009) Sox9 inhibits Wnt signaling by promoting beta-catenin phosphorylation in the nucleus. *The Journal of Biological Chemistry*, **284**, 3323–3333.

Trompouki, E., Bowman, T.V., Lawton, L.N. *et al.* (2011) Lineage regulators direct BMP and Wnt pathways to cell-specific programs during differentiation and regeneration. *Cell*, **147**, 577–589.

Turki-Judeh, W. and Courey, A.J. (2012) Groucho: a corepressor with instructive roles in development. *Current Topics in Developmental Biology*, **98**, 65–96.

Vacik, T. and Lemke, G. (2011) Dominant-negative isoforms of Tcf/Lef proteins in development and disease. *Cell Cycle*, **10**, 4199–4200.

Vacik, T., Stubbs, J.L., and Lemke, G. (2011) A novel mechanism for the transcriptional regulation of Wnt signaling in development. *Genes & Development*, **25**, 1783–1795.

Valenta, T., Lukas, J., and Korinek, V. (2003) HMG box transcription factor TCF-4's interaction with CtBP1 controls the expression of the Wnt target Axin2/Conductin in human embryonic kidney cells. *Nucleic Acids Research*, **31**, 2369–2380.

Valenta, T., Hausmann, G., and Basler, K. (2012) The many faces and functions of beta-catenin. *The EMBO Journal*, **31**, 2714–2736.

Valenta, T., Lukas, J., Doubravska, L. *et al.* (2006) HIC1 attenuates Wnt signaling by recruitment of TCF-4 and beta-catenin to the nuclear bodies. *The EMBO Journal*, **25**, 2326–2337.

Valenta, T., Gay, M., Steiner, S. *et al.* (2011) Probing transcription-specific outputs of beta-catenin in vivo. *Genes & Development*, **25**, 2631–2643.

van de Wetering, M., Cavallo, R., Dooijes, D. *et al.* (1997) Armadillo coactivates transcription driven by the product of the *Drosophila* segment polarity gene dTCF. *Cell*, **88**, 789–799.

van Genderen, C., Okamura, R.M., Farinas, I. *et al.* (1994) Development of several organs that require inductive epithelial-mesenchymal interactions is impaired in LEF-1-deficient mice. *Genes & Development*, **8**, 2691–2703.

Vietor, I., Kurzbauer, R., Brosch, G., and Huber, L.A. (2005) TIS7 regulation of the beta-catenin/Tcf-4 target gene osteopontin (OPN) is histone deacetylase-dependent. *The Journal of Biological Chemistry*, **280**, 39795–39801.

Wang, G., Wang, J., and Sadar, M.D. (2008) Crosstalk between the androgen receptor and beta-catenin in castrate-resistant prostate cancer. *Cancer Research*, **68**, 9918–9927.

Willert, K. and Jones, K.A. (2006) Wnt signaling: is the party in the nucleus? *Genes & Development*, **20**, 1394–1404.

Wohrle, S., Wallmen, B., and Hecht, A. (2007) Differential control of Wnt target genes involves epigenetic mechanisms and selective promoter occupancy by T-cell factors. *Molecular and Cell Biology*, **27**, 8164–8177.

Wright, K.J. and Tjian, R. (2009) Wnt signaling targets ETO coactivation domain of TAF4/TFIID in vivo. *Proceedings of the National Academy of Sciences of the United States of America*, **106**, 55–60.

Wright, J.B., Brown, S.J., and Cole, M.D. (2010) Upregulation of c-MYC in cis through a large chromatin loop linked to a cancer risk-associated single-nucleotide polymorphism in colorectal cancer cells. *Molecular and Cell Biology*, **30**, 1411–1420.

Wu, Y., Zhang, Y., Zhang, H. *et al.* (2010) p15RS attenuates Wnt/{beta}-catenin signaling by disrupting {beta}-catenin TCF4 interaction. *The Journal of Biological Chemistry*, **285**, 34621–34631.

Wu, B., Piloto, S., Zeng, W. *et al.* (2013) Ring Finger Protein 14 is a new regulator of TCF/beta-catenin-mediated transcription and colon cancer cell survival. *EMBO Reports*, **14**, 347–355.

Xin, N., Benchabane, H., Tian, A. *et al.* (2011) Erect Wing facilitates context-dependent Wnt/Wingless signaling by recruiting the cell-specific Armadillo-TCF adaptor Earthbound to chromatin. *Development*, **138**, 4955–4967.

Yamaguchi, T.P., Takada, S., Yoshikawa, Y. *et al.* (1999) T (Brachyury) is a direct target of Wnt3a during paraxial mesoderm specification. *Genes & Development*, **13**, 3185–3190.

Yamamoto, H., Ihara, M., Matsuura, Y., and Kikuchi, A. (2003) Sumoylation is involved in beta-catenin-dependent activation of Tcf-4. *The EMBO Journal*, **22**, 2047–2059.

Yang, X., van Beest, M., Clevers, H. *et al.* (2000) Decapentaplegic is a direct target of dTcf repression in the *Drosophila* visceral mesoderm. *Development*, **127**, 3695–3702.

Yi, F., Pereira, L., Hoffman, J.A. *et al.* (2011) Opposing effects of Tcf3 and Tcf1 control Wnt stimulation of embryonic stem cell self-renewal. *Nature Cell Biology*, **13**, 762–770.

Yochum, G.S., Cleland, R., and Goodman, R.H. (2008) A genome-wide screen for beta-catenin binding sites identifies a downstream enhancer element that controls c-Myc gene expression. *Molecular and Cell Biology*, **28**, 7368–7379.

Yook, J.I., Li, X.Y., Ota, I. *et al.* (2005) Wnt-dependent regulation of the E-cadherin repressor snail. *The Journal of Biological Chemistry*, **280**, 11740–11748.

Yumoto, F., Nguyen, P., Sablin, E.P. *et al.* (2012) Structural basis of coactivation of liver receptor homolog-1 by beta-catenin. *Proceedings of the National Academy of Sciences of the United States of America*, **109**, 143–148.

Zhang, C., Cho, K., Huang, Y. *et al.* (2008) Inhibition of Wnt signaling by the osteoblast-specific transcription factor Osterix. *Proceedings of the National Academy of Sciences of the United States of America*, **105**, 6936–6941.

Zhang, N., Wei, P., Gong, A. *et al.* (2011) FoxM1 promotes beta-catenin nuclear localization and controls Wnt target-gene expression and glioma tumorigenesis. *Cancer Cell*, **20**, 427–442.

Zhou, B.P., Deng, J., Xia, W. *et al.* (2004) Dual regulation of Snail by GSK-3beta-mediated phosphorylation in control of epithelial–mesenchymal transition. *Nature Cell Biology*, **6**, 931–940.

5 Finding a Needle in a Genomic Haystack: Genome-Wide Approaches to Identify Wnt/TCF Transcriptional Targets

Chandan Bhambhani and Ken M. Cadigan

Department of Molecular, Cellular and Developmental Biology, University of Michigan Ann Arbor, MI, USA

TCFs: Major nuclear mediators of Wnt/β-catenin signaling

T-cell factor 1 (TCF1, also known as TCF7) and lymphoid enhancer factor 1 (LEF1) were originally identified in screens for transcriptional regulators required for specification of various lymphoid lineages (Oosterwegel *et al.*, 1991; Travis *et al.*, 1991; van de Wetering *et al.*, 1991; Waterman and Jones, 1990). Most vertebrates have two additional *TCF* genes (*TCF3* and *TCF4*; also known as *TCF7L* and *TCF7L2*, respectively) (Cadigan and Waterman, 2012; see also Chapter 17). Association of TCFs with Wnt signaling initially came from a yeast two-hybrid screen where they were found to interact with β-catenin (Behrens *et al.*, 1996; Huber *et al.*, 1996; Molenaar *et al.*, 1996; see also Chapter 16). Requirement of TCF for mediation of Wnt/β-catenin signaling in several organisms established its role as a crucial nuclear mediator of the pathway (reviewed in (Archbold *et al.*, 2012; Cadigan and Waterman, 2012; Logan and Nusse, 2004).

TCFs belong to the High-Mobility Group (HMG) family of transcription factors (TFs) with conserved residues in the HMG domain making key DNA contacts (Giese, Amsterdam, and Grosschedl, 1991; Love *et al.*, 1995; Figure 5.1a and b). These HMG domains bind to a consensus of 5′-CCTTTGATS-3′ (S = G/C) with high affinity (Hallikas *et al.*, 2006; van Beest *et al.*, 2000; van de Wetering *et al.*, 1997), and several studies have shown that multimerization of this consensus upstream of a heterologous promoter (e.g., TOPFLASH) responds to Wnt/β-catenin signaling in many contexts (DasGupta and Fuchs, 1999; Denayer, Van Roy, and Vleminckx, 2006; Dorsky, Sheldahl, and Moon, 2002; Green, Inoue, and Sternberg, 2008; Korinek *et al.*, 1997; Maretto *et al.*, 2003). However, concatemerized TCF-binding sites are not typically found in naturally occurring Wnt Response Elements (WREs) (reviewed in Barolo, 2006). This chapter will review DNA recognition by TCFs and how computational and genome-wide surveys of TCF occupancy are being used to identify WREs and the transcriptional targets that are regulated by the Wnt/β-catenin pathway.

DNA binding by TCFs: Considerable flexibility in what constitutes a TCF-binding site

The initial consensus for TCF-binding sites was found through sequential enrichment of preferred binding sites from random mixtures

Wnt Signaling in Development and Disease: Molecular Mechanisms and Biological Functions, First Edition.
Edited by Stefan Hoppler and Randall T. Moon.

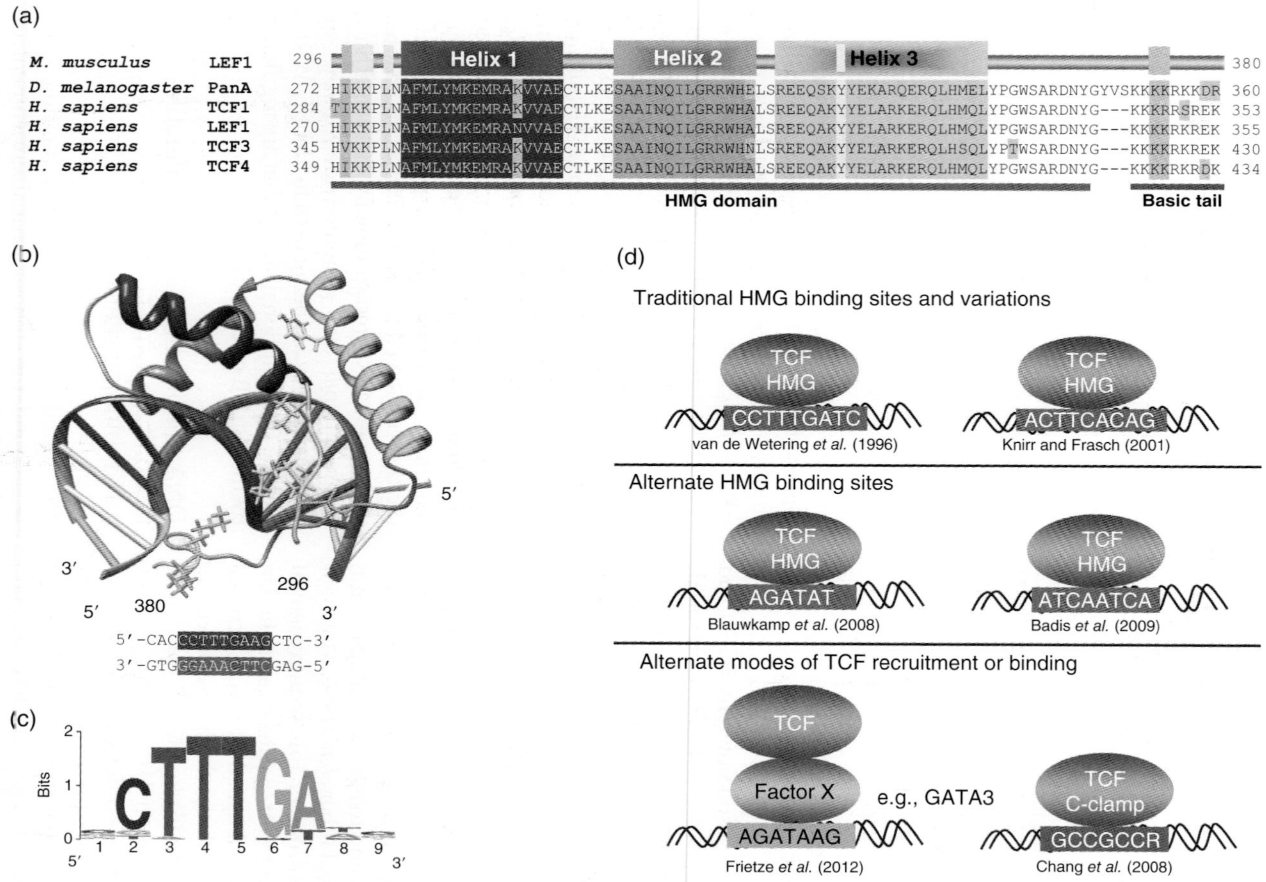

Figure 5.1 (a) Sequence alignment of the HMG domain (*red bar*) and basic tail (*brown bar*) of the *Drosophila* TCF/Pan (isoform A; NP_726522), human TCF1E (EAW62279.1), human LEF1 (NP_001124185), human TCF3 (NP_112573.1), and human TCF4E (CAB97213.1). Above them is a cartoon showing the corresponding region of mouse LEF1 (mLEF1; NP_034833.2) for which the solution structure is available (Love *et al.*, 1995), with the three α-helices in *blue*, *orange*, and *cyan*. Residues not conserved with mLEF1in the sequence alignment are shown in gray. Residues in yellow are essential for DNA binding and the ones in green are required for optimal binding (Giese, Amsterdam, and Grosschedl, 1991). (b) Ribbon diagram based on the solution structure of mLef1 (modified from Love *et al.*, 1995; PDB id 2LEF). Shown is the sequence of the double-stranded DNA motif used to determine the structure in complex with mLEF1, with nine base pairs of the consensus shown in *red* and *dark gray*. (c) Sequence logo of TCF functional binding sites derived from 32 WREs of various species, highlighting the CTTTGA consensus. (d) Models representing the diversity in HMG binding sites or alternative mechanisms of TCF recruitment to WREs. See text for further explanation. *(See insert for color representation of the figure.)*

of oligonucleotides, with a bias for detecting the highest affinity binding sequences (Atcha *et al.*, 2007; van Beest *et al.*, 2000; van de Wetering *et al.*, 1997). A systematic analysis of variations from the consensus revealed a graded reduction in affinity for TCF, with some positions allowing more degeneracy than others and the CTTTGA core being most invariant (Hallikas *et al.*, 2006). This consensus can also be observed in a collection of functional TCF sites from several species studied in the Cadigan Lab (Figure 5.1c). However, many functional sites in WREs have substitutions in this core sequence, for example, WREs from the human *c-Myc* locus (CCTTTCATC) (Pomerantz *et al.*, 2009; Tuupanen *et al.*, 2009; Wright, Brown, and Cole, 2010) and *Caenorhabditis elegans ceh-22* (CTTTTGAAG) (Lam, Chesney, and Kimble, 2006). Even more divergent sites (e.g., CCTTTTTTC, TTTTTGTGT, or ACTTCACAG) have been suggested to contribute to Wnt responsiveness of other WREs (Knirr and Frasch, 2001; Lee and Frasch, 2000; Wisniewska *et al.*, 2010). In *Drosophila*, there are Wnt targets that bear no resemblance to the traditional consensus (e.g., AGATAT), which mediate Wnt/β-catenin-dependent repression (Blauwkamp, Chang, and Cadigan, 2008). These studies indicate that relevant TCF-binding sites come in many distinct flavors (Figure 5.1d).

The full extent of binding site degeneracy for TCFs can be seen in a study where tiled arrays were used to monitor TF binding to every possible eight-base-pair sequence (Badis *et al.*, 2009; Berger and Bulyk, 2009). For the four mouse TCFs, more than 10 000 of the 24 000 sites assayed showed higher than background affinity (e.g., Figure 5.1d). In total, 104 mouse TFs from several different families were analyzed, and almost all displayed binding site flexibility to a similar degree as the TCFs (Badis *et al.*, 2009). This work demonstrates that TFs as a general rule have a large number of secondary sites, and the challenge is to determine the biological relevance of their promiscuous DNA-binding properties.

The tremendous flexibility of TCF/DNA binding has huge implications for the identification of WREs in genomes. The greater the diversity of functional TCF sites, the more difficult it is to pick them out from excess DNA, even with sophisticated algorithms that factor in some degeneracy and utilize phylogenetic conservation (e.g., Hallikas *et al.*, 2006). Genome-wide occupancy studies have found that a significant number of regions occupied by TCFs do not contain recognizable TCF sites (Blahnik *et al.*, 2010; Frietze *et al.*, 2012; Hatzis *et al.*, 2008; Junion *et al.*, 2012; Wu *et al.*, 2012). One possibility is that protein–protein interactions are responsible for TCF recruitment (Frietze *et al.*, 2012; Junion *et al.*, 2012). However, it is also likely that some of the TCF-bound regions lacking classic sites contain divergent TCF sites, and these caveats should be given more weight as the field moves forward.

Bipartite binding by TCFs: One mechanism to increase specificity

Many TCFs contain a second DNA-binding domain, known as the C-clamp, which can increase the selectivity of the sites bound by the HMG domain (Atcha *et al.*, 2007; Chang *et al.*, 2008; Hoverter *et al.*, 2012). This 30 amino acid domain contains four conserved cysteines and is located at the C-terminal of the HMG domain. The C-clamp binds to a GC-rich motif known as a helper site. Helper sites are critical for activation of several WREs by Wnt signaling (Atcha *et al.*, 2007; Chang *et al.*, 2008; Hoverter *et al.*, 2012). Nearly all invertebrate TCFs contain a C-clamp (Archbold *et al.*, 2012; Cadigan and Waterman, 2012), as do many "E box" isoforms of vertebrate TCF1 and TCF4 (Atcha *et al.*, 2007; Weise *et al.*, 2010). While the helper site consensus has been defined (RCCGCCR; R = A/G), it seems likely that the C-clamp will display the same degeneracy found for most other DNA-binding domains (Badis *et al.*, 2009). Nonetheless, the presence of a helper site almost doubles the amount of DNA sequence that this class of TCFs can recognize, which is likely critical for location of some WREs (Chang *et al.*, 2008; Hoverter *et al.*, 2012).

Identification of Wnt target genes

Many Wnt/β-catenin target genes have been identified through non-systems-level approaches. The expression patterns of many *Wnt* genes are well characterized in developmental systems such as *Drosophila* and mouse (Baker, 1988a, b; Couso, Bate, and Martinez-Arias, 1993; Gavin, McMahon, and McMahon, 1990; McMahon *et al.*, 1992) so that genes with similar expression patterns quickly became candidates for target genes, for example, *engrailed* in flies (DiNardo *et al.*, 1988; Martizez Arias, Baker, and Ingham, 1988) and *brachyury* in mice (Yamaguchi *et al.*, 1999). In other systems where the localized *Wnt* expression has not been unambiguously identified, targets have been identified on the basis of their loss or gain-of-function phenotypes sharing strong similarity with Wnt/β-catenin signaling components, for example, *ceh-22* in *C. elegans* (Lam, Chesney, and Kimble, 2006) and *siamois* in *Xenopus* (Brannon *et al.*, 1997; Brannon and Kimelman, 1996; see also list of Wnt targets on the Wnt home page; http://www.stanford.edu/group/nusselab/cgi-bin/wnt/). While this candidate gene approach has been successful in many systems and contexts, for example, planaria regeneration (Petersen and Reddien, 2011) or vertebrate neural crest induction (Elkouby *et al.*, 2010; Li *et al.*, 2009), it is obviously limited in scope.

Microarrays provide a systems-level snapshot of changes in cellular transcript levels and have been successfully used to find Wnt targets important for development, adult homeostasis and disease (reviewed in (Vlad *et al.*, 2008). For example, regulation of the oncogene *c-Myc* by Wnt signaling was discovered from a microarray screen of a colorectal cancer (CRC) cell line with constitutive Wnt signaling (He *et al.*, 1998). Perhaps the most intensive transcriptome analysis comes from Clevers and colleagues (van de Wetering *et al.*, 2002; Van der Flier *et al.*, 2007; van Es *et al.*, 2005), who have identified a list of genes they termed "the intestinal Wnt/TCF signature" (Van der Flier *et al.*, 2007). This is a set of 208 genes that are regulated by Wnt/β-catenin signaling in CRC cell lines and upregulated in primary adenomas and/or carcinomas (Van der Flier *et al.*, 2007). Several of these Wnt targets have been followed up in detail (reviewed in (Barker *et al.*, 2012), the most prominent of which is leucine-rich-repeat-containing G protein-coupled receptor 5 (LGR5), which is a specific marker for intestinal stem cells (Barker *et al.*, 2007; Sato *et al.*, 2009) and colon cancer stem cells (Barker *et al.*, 2009; Kemper *et al.*, 2012). A key in the work from Clevers and coworkers was the use of several cell lines where the Wnt/β-catenin pathway could be manipulated, combined with human tumor samples (Van der Flier *et al.*, 2007). The use of several microarray data sets facilitated the identification of robust and biologically relevant Wnt transcriptional targets and provides a blueprint for further studies in other tissues/systems.

Determining direct from indirect Wnt targets

The studies cited earlier do not distinguish whether the Wnt target is directly or indirectly regulated by the pathway. Our definition of a direct target is one that does not require synthesis of an intermediate for its regulation by Wnt signaling. Ideally, pretreatment of cultured cells with a protein synthesis inhibitor, for example, cycloheximide (CHX), prior to Wnt stimulation could allow the identification of direct targets. However, Wnt stimulated stabilization of β-catenin to activate signaling requires *de novo* protein synthesis, and hence, it has not been a preferred method to find direct targets (Willert *et al.*, 2002). For other signaling pathways, an optimal dose of CHX (10–20 μg/ml) has been used to block protein synthesis and identify direct targets (e.g., Kang, Chen, and Massague, 2003; Willert *et al.*, 2002). However, limiting the dose of CHX to low and probably suboptimal levels can block activation of some Wnt targets, demonstrating that they are indirect (Buttitta *et al.*, 2003; Sanchez-Ferras *et al.*, 2012). An alternative is to fuse TCF or β-catenin to a steroid hormone receptor and conditionally activate the pathway with steroid agonists (Blauwkamp, Chang, and Cadigan, 2008; Chamorro *et al.*, 2005; Elkouby *et al.*, 2010; Garnett, Square, and Medeiros, 2012; Leung *et al.*, 2002; Li *et al.*, 2009). This strategy circumvents the requirement of *de novo* protein synthesis, and a higher dose of CHX (10 μg/ml) can be used to distinguish between direct and indirect targets (Blauwkamp, Chang, and Cadigan, 2008; Garnett, Square, and Medeiros, 2012).

Given the limitations of using protein synthesis inhibitors, the most common strategy for identifying direct targets of Wnt/β-catenin signaling is to demonstrate the presence of functional TCF-binding sites in WREs (reviewed in Archbold *et al.*, 2012). This can be done in any system where WRE reporter constructs can be introduced, including *C. elegans* (e.g., Arata *et al.*, 2006; Lam, Chesney, and Kimble, 2006; Maduro *et al.*, 2005; Shetty *et al.*, 2005), *Drosophila* (e.g., (Blauwkamp, Chang, and Cadigan, 2008; Chang *et al.*, 2008; Knirr and Frasch, 2001; Lee and Frasch, 2000), zebrafish (e.g., (Dorsky, Raible, and Moon, 2000; Ryu *et al.*, 2001; Weidinger *et al.*, 2005), *Xenopus* (e.g., Brannon *et al.*, 1997; Laurent *et al.*, 1997; McKendry *et al.*, 1997), mouse (e.g., Kuwabara *et al.*, 2009; Wisniewska *et al.*, 2010; Yamaguchi *et al.*, 1999), and mammalian cell culture (e.g., He *et al.*, 1998; Jho *et al.*, 2002; Yochum, Cleland, and Goodman, 2008). In many cases, TCFs have been shown to associate with the WRE at the endogenous loci via chromatin immunoprecipitation (ChIP) (e.g., Blauwkamp, Chang, and Cadigan, 2008; Fang *et al.*, 2006; Hikasa *et al.*, 2010; Parker *et al.*, 2008; Wright, Brown, and Cole, 2010). While the combination of reporter gene mutagenesis and ChIP with TCF and/or β-catenin is considered the gold standard for determining whether a Wnt target is direct, this approach is predicated on knowledge of the location of the WRE(s) at the target loci.

The *c-Myc* locus offers an example of the complexity of WRE organization in a Wnt target. This gene is activated by Wnt/β-catenin signaling in many CRC cell lines (e.g., He *et al.*, 1998; Van der Flier *et al.*, 2007) and mouse intestine, where it is genetically required for epithelial hyperplasia (Myant and Sansom, 2011). A WRE upstream of the c-Myc transcription start site (TSS) was initially identified and fulfills the aforementioned criteria for direct regulation by TCF and β-catenin (He *et al.*, 1998; Sierra *et al.*, 2006). Subsequently, an additional WRE was identified 3′ of the *c-Myc* transcript that is activated by the pathway (Yochum, Cleland, and Goodman, 2008). More recently, a SNP linked to increased risk of colorectal and prostate cancers in humans led to the identification of a *c-Myc* WRE located 335 kb upstream of the *c-Myc* TSS. The risk allele contains a higher affinity TCF site that is preferentially bound by TCF (Pomerantz *et al.*, 2009; Tuupanen *et al.*, 2009; Wright, Brown, and Cole, 2010). Both the 3′ and distal 5′ WREs interact with the *c-Myc* promoter via chromatin looping (Pomerantz *et al.*, 2009; Wright, Brown, and Cole, 2010; Yochum *et al.*, 2010). Deletion of the distal upstream WRE reduces *c-Myc* expression and the ability of constitutive Wnt/β-catenin signaling to form tumors in the mouse intestine (Sur *et al.*, 2012). Paradoxically, deletion of the 3′ WRE results in heightened expression of *c-Myc* in the intestine (Konsavage, Jin, and Yochum, 2012), presumably due to removal of repressive elements. No other Wnt target gene has been as extensively studied as *c-Myc*, so the complexity of regulation of this locus by the pathway may also be typical for other targets.

Computational approaches

There are several computational approaches which have been implemented to identify *cis*-regulatory modules (reviewed in Aerts, 2012; Elnitski *et al.*, 2006). One such approach to identify WREs used an algorithm known as the Enhancer Element Locator (EEL) (Hallikas *et al.*, 2006). This program takes into account distance-based clustering of two or more high-affinity binding sites of TCF at genomic loci conserved in mice and humans. Novel WREs for some of the known Wnt targets such as *Axin2* and *c-Myc* were identified using this approach, in addition to WREs for novel targets (Hallikas *et al.*, 2006). EEL was instrumental in one group's identification of the 5′ distal *c-Myc* WRE mentioned in the previous section (Tuupanen *et al.*, 2009). While this bioinformatics approach is capable of identifying *bona fide* WREs, many will escape detection, given the vast excess of DNA sequence in mammalian genomes.

Genome-wide studies of TCF binding

Over the past 5 years, there have been a growing number of genome-wide surveys measuring TCF binding. The initial studies used ChIP to enrich for TCF bound regions, which were then hybridized to slides containing tiled arrays of nonrepetitive genomic sequence (Cole *et al.*, 2008; Hatzis *et al.*, 2008; Tam *et al.*, 2008). This "ChIP on chip" technology has been replaced by ChIP coupled with high-throughput sequencing (ChIP-seq). While there is one report carried out in *Drosophila* embryos (Junion *et al.*, 2012), all others have been in mammalian cell culture. TCF4 occupancy has been examined by several groups in human CRC cell lines (Blahnik *et al.*, 2010; Frietze *et al.*, 2012; Hatzis *et al.*, 2008; Mokry *et al.*, 2010, 2012; Zhao *et al.*, 2010) and several other cell types (Frietze *et al.*, 2012; Norton *et al.*, 2011; Trompouki *et al.*, 2011; Verzi *et al.*, 2010; Zhou *et al.*, 2012). The other vertebrate TCFs have received less attention, with comparatively fewer studies for TCF3 (Cole *et al.*, 2008; Marson *et al.*, 2008; Tam *et al.*, 2008) and TCF1 (Germar *et al.*, 2011; Wu *et al.*, 2012).

General features of genome-wide TCF binding

One general feature of surveys of TCF binding in genomes is large number of high-confidence bound regions. In murine embryonic stem (ES) cells, TCF3 was enriched in ~1200–14 000 regions (Cole *et al.*, 2008; Marson *et al.*, 2008; Tam *et al.*, 2008). Over 9600 TCF1 bound regions were identified in hematopoietic precursors (Wu *et al.*, 2012). For TCF4, the number of bound regions ranges from ~1100 (Zhao *et al.*, 2010) to tens of thousands (Frietze *et al.*, 2012). Frietze and coworkers probably provide the best picture for TCF4, given that they examined six different cell lines with the same TCF4 antibody and used a sophisticated computational method for calling specific TCF4 peaks (Blahnik *et al.*, 2010). They found between 24 000 and 53 000 TCF4 peaks in six different human cell lines (Frietze *et al.*, 2012). These large numbers are typical for many other TFs in vertebrate and invertebrate genomes (Frietze *et al.*, 2012; Junion *et al.*, 2012; Trompouki *et al.*, 2011; Yip *et al.*, 2012).

Another hallmark of TCF4 occupancy in genomes is cell specificity. In two studies using eight different human cell lines, the overlap between TCF4-enriched regions ranged from 12% to 46% in pair-wise comparisons (Frietze *et al.*, 2012; Trompouki *et al.*, 2011). There were only 1864 peaks that were common to six cell lines in one study, which is less than 2% of the total peaks identified (Frietze *et al.*, 2012). The binding pattern of TCF4 also shifts dramatically during differentiation from multipotent hematopoietic progenitor cells to erythroid cells (Trompouki *et al.*, 2011). These results are consistent with cell-specific regulation of Wnt target genes, suggested by transcriptome analysis (Vlad *et al.*, 2008) as well as more focused studies on Wnt transcriptional programs in several systems (reviewed in Archbold *et al.*, 2012). While there are likely multiple *mechanisms* influencing cell-type-specific TCF-binding patterns, interactions between TCFs and other TFs likely play a major role in this level of regulation, as discussed below.

Another general rule for TCF4 is that its binding is not confined to promoter regions just upstream of the TSS, but can also be found at more distal locations (Frietze *et al.*, 2012; Hatzis *et al.*, 2008; Mokry *et al.*, 2010; Trompouki *et al.*, 2011; Zhao *et al.*, 2010). For example, in LS174T cells, a CRC cell line, greater than 70% of the TCF4 enrichment peaks were found at distances more than 10 kb from the nearest TSS (Hatzis *et al.*, 2008; Mokry *et al.*, 2010). In HCT116 cells, another CRC cell line, 80% of the TCF4 peaks were found at a distance more than 5 kb from the TSS (Zhao *et al.*, 2010). A similar pattern was seen for TCF3 in murine ES cells, where 73% of the peaks were found at distances more than 8 kb from the TSS (Marson *et al.*, 2008). Functional WREs can be found in introns or downstream of Wnt target loci (Chang *et al.*, 2008; Theisen *et al.*, 2007; Yochum, Cleland, and Goodman, 2008) or several hundred kb removed (Pomerantz *et al.*, 2009; Sur *et al.*, 2012; Tuupanen *et al.*, 2009; Wright, Brown, and Cole, 2010). This is consistent with other genome-wide studies, demonstrating the existence of functional enhancers at great distances from TSSs (Blow *et al.*, 2010; May *et al.*, 2012; Visel *et al.*, 2009), and suggests that many of the distal TCF-bound regions could be functional WREs.

Examination of regions bound by TCF1 and TCF4 for consensus TCF-binding sites reveals that many of these regions do not contain such motifs. For example, only 44% of the TCF1-bound regions contain predicted TCF sites (Wu *et al.*, 2012). For TCF4, this number ranges between 45% and 70% depending on the study and the criteria for what constitutes a TCF consensus site (Blahnik *et al.*, 2010; Frietze *et al.*, 2012; Hatzis *et al.*, 2008; Mokry *et al.*, 2010; Norton *et al.*, 2011). In one study, restricting the analysis to the top 1000 TCF4 peaks (of over 100 000) resulted in a higher incidence (80%) of predicted TCF sites (Frietze *et al.*, 2012). These results suggest that a considerable portion of TCF recruitment in genomes is not dependent on the consensus TCF-binding site. It is possible that more divergent sites are playing a significant role in TCF recruitment, such as those found in specific WREs (Blauwkamp, Chang, and Cadigan, 2008; Knirr and Frasch, 2001; Wisniewska *et al.*, 2010) or the secondary sites defined by Badis and coworkers (Badis *et al.*, 2009) (Figure 5.1d). Another alternative is that TCFs may be recruited by protein–protein interactions with other TFs. Supporting this, several other TF binding sites were also enriched in TCF-bound regions (Blahnik *et al.*, 2010; Bottomly *et al.*, 2010; Frietze *et al.*, 2012; Hatzis *et al.*, 2008; Wu *et al.*, 2012).

Although many of the data described earlier lend validation to genome-wide surveys of TCF binding, some caution is warranted when considering false-positives due to non-TCF binding of α-TCF antisera. The use of null mutants to control for nonspecific antibody binding in ChiP-seq studies in *Arabidopsis* seedlings demonstrated that many peaks are false-positives (Wierzbicki *et al.*, 2012; Zheng *et al.*, 2012). Efficient siRNA would be required for this approach to be adopted in mammalian cell culture. An alternative would be to perform the analysis with multiple antisera, to concentrate on the common regions that were immunoprecipitated.

What fraction of TCF-bound regions are functional?

With thousands to tens of thousands of TCF-bound locations in genomes, it is difficult to imagine that they are all functional. In many cases, there are several TCF peaks within or surrounding gene loci (Frietze *et al.*, 2012; Hatzis *et al.*, 2008; Mokry *et al.*, 2010; Wu *et al.*, 2012). Many TCF4 peaks colocalize with the chromatin marks H3K4me and H3K27ac (Frietze *et al.*, 2012; Mokry *et al.*, 2010; Trompouki *et al.*, 2011), which have been linked to enhancer activity (reviewed in Heintzman and Ren, 2009). In some cells, up to 40% of the H3K4me/H3K27ac peaks overlap with TCF4 (Frietze *et al.*, 2012). Whether this is indicative of TCF4 contributing to global enhancer activity or simply reflects TCF4 enrichment in more accessible areas of the genome requires further analysis.

One obvious method for determining which TCF-bound regions might be functional is to cross-reference occupancy data with transcriptome analysis designed to identify TCF or Wnt/β-catenin signaling targets. For example, in erythroblasts, ~31% of loci with associated TCF4 peaks were regulated by Wnt signaling (Trompouki *et al.*, 2011). In CRC cells, ~20–22% of the TCF4-bound regions were within 100 kb of loci that were dependent on TCF4 or upregulated in primary adenomas (Hatzis *et al.*, 2008; Zhao *et al.*, 2010). In the breast cancer cell line MCF7, only 4.5% of the genes associated with TCF4 were altered upon TCF4 knockdown (Frietze *et al.*, 2012). These studies relied on microarrays to measure the steady-state levels of transcripts, and it has been suggested that monitoring RNA polymerase II occupancy at promoters is a more sensitive approach to determining the transcriptional state of loci (Mokry *et al.*, 2012). However, this does not change the overall picture: The majority of TCF4-occupied regions do not have a detectable effect on gene expression. It should be noted here that there is evidence that TCF4-bound regions may also correspond to areas of non-mRNA synthesis that are not represented on microarrays (Hatzis *et al.*, 2008), which could be short or long noncoding RNAs. Indeed, in murine ES cells, a small yet significant fraction of TCF3 peaks were associated with micro-RNA promoters (Marson *et al.*, 2008; Tam *et al.*, 2008).

An alternative approach for finding out which peaks may be functional WREs is to monitor the genomic distribution of β-catenin. In HCT116 CRC cells, over 2000 regions bound by β-catenin have been identified (Bottomly *et al.*, 2010; Yochum *et al.*, 2007) and a third of these regions overlap with TCF4 (Bottomly *et al.*, 2010). It is not clear whether these regions are enriched for *bona fide* WREs.

As mentioned before, several studies have found that a majority of TCF4 occupancy peaks are located distally from the core promoter. In order to test if there was a correlation between location of TCF4 peaks and Wnt regulation of the nearest target, we chose a small biased data

Table 5.1 TCF4 occupancy peak locations in LS174T cells relative to the TSS of the intestinal Wnt/TCF signature gene set.

Genes with peak(s) proximal to the TSS		Genes with peak(s) distal from the TSS		Genes with peaks proximal to and distal from the TSS	
Gene ID	Gene name	Gene ID	Gene name	Gene ID	Gene name
ENSG00000133216	***EPHB2***	ENSG00000117479	*SLC19A2*	ENSG00000162981	*NSE1*
ENSG00000117318	***ID3***[a]	ENSG00000196950	*SLC39A10*	ENSG00000115946	*LOC56902*[a]
ENSG00000067533	*CGI-115*	ENSG00000173638	***SLC19A1***[a]	ENSG00000144354	***CDCA7***
ENSG00000175701	*LOC205251*	ENSG00000112996	*MRPS30*	ENSG00000138795	***LEF1***
ENSG00000144485	***HES6***	ENSG00000118513	***MYB***	ENSG00000171617	***ENC1***
ENSG00000182580	***EPHB3***[a]	ENSG00000016402	*IL20RA*[a]	ENSG00000164379	***FOXQ1***[b]
ENSG00000163735	*CXCL5*	ENSG00000086289	***UCC1***	ENSG00000124766	***SOX4***[a]
ENSG00000109534	***NOLA1***	ENSG00000106615	*RHEB*[a]	ENSG00000183779	*ZNF703*
ENSG00000005073	*HOXA11*	ENSG00000138134	*AMSH-LP*	ENSG00000147642	*FLJ20366*
ENSG00000136997	***MYC***	ENSG00000168003	***SLC3A2***	ENSG00000150347	***ARID5B***
ENSG00000165802	***NELF***	ENSG00000139289	*PHLDA1*[a]	ENSG00000165655	*ZNF503*
ENSG00000099194	***SCD***[a]	ENSG00000043355	*ZIC2*	ENSG00000026508	***CD44***
ENSG00000183734	***ASCL2***[a]	ENSG00000100526	***CDKN3***	ENSG00000137693	*YAP1*[a]
ENSG00000173465	*MTVR1*	ENSG00000103888	***KIAA1199***	ENSG00000139292	***LGR5***
ENSG00000175575	*FLJ11848*	ENSG00000103544	*MGC16824*[a]	ENSG00000135111	*TBX3*[a]
ENSG00000175538	*KCNE3*	ENSG00000175832	***ETV4***	ENSG00000103257	*SLC7A5*[a]
ENSG00000196371	*FUT4*[a]	ENSG00000125398	*SOX9*	ENSG00000108375	*FLJ20315*[a]
ENSG00000186184	*POLR1D*	ENSG00000088970	*C20orf19*	ENSG00000168646	***AXIN2***
ENSG00000177426	*TGIF*	ENSG00000183579	***ZNRF3***[a]	ENSG00000101384	*JAG1*
ENSG00000100105	*PATZ1*	ENSG00000169756	*LIMS1*	ENSG00000101230	*C20orf82*
ENSG00000156697	*SDCCAG16*[a]			ENSG00000157557	***ETS2***
				ENSG00000164125	*DKFZp434I142*

Analysis of peak locations of TCF4 relative to the TSS of the intestinal Wnt signature genes suggests that there is no bias for TCF occupancy relative to the nearest core promoters. This is a refined list of 62 targets from the 208 that show an overlap for regulation in primary adenomas, primary carcinomas, and colorectal cancer lines LS174T and DLD1. The same is true for the ones overlapping for regulation by β-cat/TCF in LS174T cells. This is different from the occupancy profile of TCF4 relative to the TSS in the context of the whole genome, where TCF4 seems to be enriched at sites distal from the core promoter (see text for details).

"Proximal" corresponds to a TCF4 peak < 10 kb, while "distal" corresponds to a TCF4 peak > 10 kb from the TSS based on ChIP-on-chip analysis in LS174T cells (Hatzis *et al.*, 2008). Genes in bold are the ones downregulated by misexpression of dnTCF4 or depletion of β-cat, based on microarray analysis in LS174T cells (Mokry *et al.*, 2010).

[a]Genes expressed in intestinal crypts of wild-type mice and adenomas of *APCmin* mice (Van der Flier *et al.*, 2007).

[b]*FoxQ1* was positively regulated based on microarray analyses in colorectal cancer cells, primary human adenomas and carcinomas (Van der Flier *et al.*, 2007), but negatively regulated based on analysis in LS174T cells (Mokry *et al.*, 2010).

set for further analysis. This is the previously mentioned collection of 208 genes, termed "the intestinal Wnt/TCF signature," which are regulated by TCF4 and TCF1 in CRC cell lines and upregulated in primary adenomas and/or carcinomas (Van der Flier *et al.*, 2007). Cross-referencing TCF4 distribution in LS174T cells to these genes (Hatzis *et al.*, 2008), we found that there were a total of 63 targets that had one or more high-confidence TCF4 peaks associated with them (197 peaks total). The distribution of this subset of peaks in relation to the TSS of the nearest gene was very similar to the distribution of the entire collection of 6868 peaks (Hatzis *et al.*, 2008; data not shown). Out of the 63 intestinal Wnt signature targets, 33% have one or more peaks exclusively within 10 kb of their TSS (proximal), 32% have one or more peaks at distances greater than 10 kb from their TSS (distal), and 35% have peaks in both classes (proximal and distal) (Table 5.1). To further bias the target loci for those that are relevant for LS174T cells, we examined the 27 targets that were significantly downregulated upon reduction of Wnt/β-catenin signaling in these cells (Mokry *et al.*, 2010). The distributions of the TCF4 peaks were similar to the larger collection (33% proximal, 30% distal, 37% both proximal and distal (Table 5.1; in bold)). While it is not known which of these TCF4-bound regions correspond to functional WREs, these analyses suggest no obvious bias in TCF4 distribution at Wnt-regulated loci compared to the entire genome.

Colocalization of TCFs with other TFs

One emerging theme from the growing number of genome-wide studies is the existence of cell-specific groups of TFs that colocalize with TCFs. For example, approximately one third of the regions bound by caudal-related homeodomain protein 2 (CDX2) in an intestinal and CRC cell line are also bound by TCF4 (Verzi *et al.*, 2010). TCF4 and c-Jun, a basic helix–loop–helix protein that binds c-Fos to form AP-1 (Shaulian and Karin, 2002), colocalize at a subset of genomic loci occupied by β-catenin (Bottomly *et al.*, 2010). Significant overlap was also observed when TCF4 and c-Jun occupancy patterns were cross-referenced from different cell lines (Blahnik *et al.*, 2010). In liver-derived HepG2 cells, TCF4 was found to co-occupy genomic loci with liver-specific factors HNF4 and FoxA2 and with GATA3 in the breast cancer cell line MCF7 (Frietze *et al.*, 2012). An impressive overlap of TCF4 was also found with Smad1, a mediator of TGF-β signaling (Moustakas and Heldin, 2009), in different hematopoietic lineages (Trompouki *et al.*, 2011). In erythroid cells, TCF4 and Smad1 colocalized with GATA1 and GATA2, while in myeloid cells, they were enriched at regions also bound with C/EBPα (Trompouki *et al.*, 2011). TCF3 has also been shown to colocalize with Oct4, Nanog, and Sox2 in murine ES cells (Cole *et al.*, 2008; Marson *et al.*, 2008; Tam *et al.*, 2008). In murine hematopoietic multipotent cells, almost half of the loci occupied by TCF1 were also bound by Runx1 (Wu *et al.*, 2012). These results are summarized in Table 5.2 and suggest a picture where TCFs operate in the context of other cell-type-specific TFs to regulate distinct targets.

The mechanistic relationships between many of the factors implicated with TCFs in the aforementioned studies remain to be determined, but in other cases, there is good evidence that they cooperate with TCFs in binding to WREs. There is a large body of literature linking TCFs, β-catenin, and Smads in several systems. These studies favor a model where TCFs, Smads, and β-catenin form complexes in a signaling-dependent manner on enhancers (Hussein, Duff, and Sirard, 2003; Labbe, Letamendia, and Attisano, 2000; Nakano *et al.*, 2010; Nishita *et al.*, 2000; Shafer and Towler, 2009). A similar picture is emerging for TCFs and AP-1, where Wnt and mitogenic signals converge (Bottomly *et al.*, 2010; Domenzain-Reyna *et al.*, 2009; Nateri, Spencer-Dene, and Behrens, 2005). Other examples are Cdx proteins and TCFs, where Cdx1 associates with LEF1 (Beland *et al.*, 2004), TCF4 binding to chromatin is reduced a Cdx2 mutant cells (Verzi *et al.*, 2010) and functional TCF and Cdx binding sites are found in close proximity in a chick WRE (Castillo *et al.*, 2010).

While we typically think of protein–DNA interactions when considering the factors that influence TCF occupancy in genomes, some recent data indicate that protein–protein interactions may be more important than currently appreciated. In MCF7 cells, where there is considerable overlap between TCF4 and GATA3 binding, there was enrichment for GATA binding sites at regions bound by both factors but little enrichment for TCF-binding sites (Frietze *et al.*, 2012). GATA3 and TCF4 proteins could be coimmunoprecipitated, and TCF4 binding to many co-occupied regions was reduced upon depletion of *GATA3* via siRNA (Frietze *et al.*, 2012). In another study in *Drosophila* embryos, areas where TCF colocalized with several other TFs were strikingly devoid of predicted TCF-binding sites (Junion *et al.*, 2012), suggesting that protein–protein interactions may be a significant factor in determining TCF chromatin occupancy.

One important aspect of the work of Junion and coworkers is that it provides a blueprint for the efficient identification of functional enhancers, including WREs. This study determined the genomic locations of five TFs known to be genetically required for cardiogenesis and/or visceral mesoderm development in the fly embryo (Junion *et al.*, 2012). Thirty-four of the TCF-bound regions were tested for enhancer activity, with an astonishing 82% exhibiting expression in mesoderm and/or its derivatives. Interestingly, even though most of these functional enhancers were bound by all five TFs, no consistent binding site grammar could be discerned (Junion *et al.*, 2012). These *cis*-regulatory

Table 5.2 TF collectives for β-catenin and/or TCF identified by genome-wide occupancy analyses with other transcriptional regulators in various cell types.

Cellular origin/ differentiated tissue	Species	Cell type	General mediator of Wnt signaling	Co-occupancy with	References
Colon	Human	Caco-2 and LS174T	TCF4	CDX2	Verzi *et al.* (2010)
	Human	HCT116	β-catenin and TCF4[a]	c-Jun[a]	Bottomly *et al.* (2010)
	Human	LS174T	β-catenin and TCF4	c-Myc and E2A	Mokry *et al.* (2012)
Liver	Human	HepG2	TCF4	HNF4α and FoxA2	Frietze *et al.* (2012)
Mammary	Human	MCF7	TCF4	GATA3	Frietze *et al.* (2012)
Blood	Human	K562	TCF4	SMAD1, GATA1 and GATA2	Trompouki *et al.* (2011)
	Human	$CD34^+$ Progenitors	TCF4	SMAD1 and GATA2	Trompouki *et al.* (2011)
	Human	U937	TCF4	C/EBPα	Trompouki *et al.* (2011)
	Mouse	Lin^-CD34^+	TCF1	RUNX1	Wu *et al.* (2012)
Embryo	Mouse	ES	TCF3	Oct4 and Nanog	Cole *et al.* (2008), Tam *et al.* (2008)
	Mouse	ES	TCF3	Oct4, Nanog, and Sox2	Marson *et al.* (2008)
	Drosophila	Cardiac mesoderm and its precursors	TCF	Tin, pMad, Doc, and Pnr	Junion *et al.* (2012)
	Drosophila	Visceral mesodermal precursors	TCF	Tin, pMad, Doc, Pnr, Slp, and Biniou[a]	Junion *et al.* (2012)

List of context-specific factors that are enriched at loci occupied by β-cat and/or TCFs. There are several cell-type-specific factors and comparatively general transcriptional regulators which co-occupy loci with TCFs supporting a TF collective model in each case. The vertebrate homologs of the fly TFs from the Junion *et al.* study are listed below, after their respective fly counterparts: Tinman (Tin): Nkx2.5; phosphorylated Mothers against Dpp (pMad): pSmad 1/5/8; Dorsalcross (Doc): Tbx5/6; Pannier (Pnr): Gata4/5/6; Sloppy Paired (Slp): FoxG1 and Biniou: FoxF1/2.
[a]Occupancy for this factor was determined on a subset of genomic loci and not genome-wide.

modules may belong to the "flexible billboard" class of enhancers, where the spacing and orientation of essential TF binding sites can be quite variable (Arnosti and Kulkarni, 2005). In any case, the strategy of identifying *bona fide* enhancers by cross-referencing multiple TFs, which has also been highly successful for identifying enhancers involved in dorsoventral patterning in *Drosophila* (Zeitlinger *et al.*, 2007), is likely to continue to be a productive method for identifying novel WREs.

Conclusions

Further research is needed to enhance our ability to identify direct Wnt transcriptional targets and WREs in a more efficient manner. This information is necessary to understand how Wnt/β-catenin signaling affects cell behavior in normal and pathological states. Genome-wide studies combined with intensive investigation of WREs that fall into distinct classes (e.g., ones with or without recognizable TCF-binding sites) may shed light on how the Wnt pathway can achieve a diverse array of transcriptional outputs. Towards this end, more attention needs to be directed at understanding the interactions between different TCFs in a given cell type and whether their genomic distributions are altered by Wnt stimulation. For example, is the Wnt signaling-dependent removal of TCF from WREs (Hikasa and Sokol, 2011; Hikasa *et al.*, 2010) a general feature of vertebrate systems, given that individual TCFs are known to be more specialized for activation or repression (reviewed in Cadigan and Waterman, 2012; see also Chapter 17).

Clearly, genome-wide ChIP-seq studies will continue to play a major role in the identification of novel WREs, with cross-referencing of several TFs serving as a model (Junion *et al.*, 2012). However, this strategy depends on knowing which TFs are linked with TCFs, and it seems likely that the current list is likely to grow, given that some of these TCF cofactors are cell-type-specific (Table 5.2). Continued investigation of what (besides TCF-binding sites) defines various WREs, coupled with continued development of bioinformatic tools, will complement the data provided by genome-wide surveys of TCFs and potential cofactors. This multipronged approach will be necessary to further unravel the complexities of Wnt/β-catenin-dependent gene regulation and understand how TCFs and other TFs "read" the genome to identify WREs.

Acknowledgments

We thank Dr Andrzej T. Wierzbicki and M. Jordan Rowley for helpful comments and Hilary C. Archbold and Pete E. Burby for critical reading of the manuscript. This work was supported by NIH grant GM082994 and NSF grant 0950348 to KMC.

References

Aerts, S. (2012) Computational strategies for the genome-wide identification of cis-regulatory elements and transcriptional targets. *Current Topics in Developmental Biology*, **98**, 121–145.

Arata, Y., Kouike, H., Zhang, Y. *et al.* (2006) Wnt signaling and a Hox protein cooperatively regulate psa-3/Meis to determine daughter cell fate after asymmetric cell division in C. elegans. *Developmental Cell*, **11**, 105–115.

Archbold, H.C., Yang, Y.X., Chen, L., and Cadigan, K.M. (2012) How do they do Wnt they do? Regulation of transcription by the Wnt/beta-catenin pathway. *Acta Physiologica (Oxford)*, **204**, 74–109.

Arnosti, D.N. and Kulkarni, M.M. (2005) Transcriptional enhancers: intelligent enhanceosomes or flexible billboards? *The Journal of Cellular Biochemistry*, **94**, 890–898.

Atcha, F.A., Syed, A., Wu, B. *et al.* (2007) A unique DNA binding domain converts T-cell factors into strong Wnt effectors. *Molecular and Cell Biology*, **27**, 8352–8363.

Badis, G., Berger, M.F., Philippakis, A.A. *et al.* (2009) Diversity and complexity in DNA recognition by transcription factors. *Science*, **324**, 1720–1723.

Baker, N.E. (1988a) Localization of transcripts from the wingless gene in whole Drosophila embryos. *Development*, **103**, 289–298.

Baker, N.E. (1988b) Transcription of the segment-polarity gene wingless in the imaginal discs of Drosophila, and the phenotype of a pupal-lethal wg mutation. *Development*, **102**, 489–497.

Barker, N., van Es, J.H., Kuipers, J. *et al.* (2007) Identification of stem cells in small intestine and colon by marker gene Lgr5. *Nature*, **449**, 1003–1007.

Barker, N., Ridgway, R.A., van Es, J.H. *et al.* (2009) Crypt stem cells as the cells-of-origin of intestinal cancer. *Nature*, **457**, 608–611.

Barker, N., Rookmaaker, M.B., Kujala, P. *et al.* (2012) Lgr5(+ve) stem/progenitor cells contribute to nephron formation during kidney development. *Cell Reports*, **2**, 540–552.

Barolo, S. (2006) Transgenic Wnt/TCF pathway reporters: all you need is Lef? *Oncogene*, **25**, 7505–7511.

Behrens, J., vonKries, J.P., Kuhl, M. *et al.* (1996) Functional interaction of beta-catenin with the transcription factor LEF-1. *Nature*, **382**, 638–642.

Beland, M., Pilon, N., Houle, M. *et al.* (2004) Cdx1 autoregulation is governed by a novel Cdx1-LEF1 transcription complex. *Molecular and Cellular Biology*, **24**, 5028–5038.

Berger, M.F. and Bulyk, M.L. (2009) Universal protein-binding microarrays for the comprehensive characterization of the DNA-binding specificities of transcription factors. *Nature Protocols*, **4**, 393–411.

Blahnik, K.R., Dou, L., O'Geen, H. *et al.* (2010) Sole-Search: an integrated analysis program for peak detection and functional annotation using ChIP-seq data. *Nucleic Acids Research*, **38**, e13.

Blauwkamp, T.A., Chang, M.V., and Cadigan, K.M. (2008) Novel TCF-binding sites specify transcriptional repression by Wnt signalling. *The EMBO Journal*, **27**, 1436–1446.

Blow, M.J., McCulley, D.J., Li, Z. *et al.* (2010) ChIP-Seq identification of weakly conserved heart enhancers. *Nature Genetics*, **42**, 806–810.

Bottomly, D., Kyler, S.L., McWeeney, S.K., and Yochum, G.S. (2010) Identification of β-catenin binding regions in colon cancer cells using ChIP-Seq. *Nucleic Acids Research*, **38**, 5735–5745.

Brannon, M. and Kimelman, D. (1996) Activation of Siamois by the Wnt pathway. *Developmental Biology*, **180**, 344–347.

Brannon, M., Gomperts, M., Sumoy, L. *et al.* (1997) A beta-catenin/XTcf-3 complex binds to the siamois promoter to regulate dorsal axis specification in Xenopus. *Genes & Development*, **11**, 2359–2370.

Buttitta, L., Tanaka, T.S., Chen, A.E. *et al.* (2003) Microarray analysis of somitogenesis reveals novel targets of different WNT signaling pathways in the somitic mesoderm. *Developmental Biology*, **258**, 91–104.

Cadigan, K.M. and Waterman, M.L. (2012) TCF/LEFs and Wnt signaling in the nucleus. *Cold Spring Harbor Perspectives in Biology*, **4**, a007906.

Castillo, H.A., Cravo, R.M., Azambuja, A.P. *et al.* (2010) Insights into the organization of dorsal spinal cord pathways from an evolutionarily conserved raldh2 intronic enhancer. *Development*, **137**, 507–518.

Chamorro, M.N., Schwartz, D.R., Vonica, A. *et al.* (2005) FGF-20 and DKK1 are transcriptional targets of beta-catenin and FGF-20 is implicated in cancer and development. *The EMBO Journal*, **24**, 73–84.

Chang, M.V., Chang, J.L., Gangopadhyay, A. *et al.* (2008) Activation of wingless targets requires bipartite recognition of DNA by TCF. *Current Biology*, **18**, 1877–1881.

Cole, M.F., Johnstone, S.E., Newman, J.J. *et al.* (2008) Tcf3 is an integral component of the core regulatory circuitry of embryonic stem cells. *Genes & Development*, **22**, 746–755.

Couso, J.P., Bate, M., and Martinez-Arias, A. (1993) A wingless-dependent polar coordinate system in Drosophila imaginal discs. *Science*, **259**, 484–489.

DasGupta, R. and Fuchs, E. (1999) Multiple roles for activated LEF/TCF transcription complexes during hair follicle development and differentiation. *Development*, **126**, 4557–4568.

Denayer, T., Van Roy, F., and Vleminckx, K. (2006) In vivo tracing of canonical Wnt signaling in Xenopus tadpoles by means of an inducible transgenic reporter tool. *FEBS Letters*, **580**, 393–398.

DiNardo, S., Sher, E., Heemskerk-Jongens, J. *et al.* (1988) Two-tiered regulation of spatially patterned engrailed gene expression during Drosophila embryogenesis. *Nature*, **332**, 604–609.

Domenzain-Reyna, C., Hernandez, D., Miquel-Serra, L. *et al.* (2009) Structure and regulation of the versican promoter: the versican promoter is regulated by AP-1 and TCF transcription factors in invasive human melanoma cells. *The Journal of Biological Chemistry*, **284**, 12306–12317.

Dorsky, R.I., Raible, D.W., and Moon, R.T. (2000) Direct regulation of nacre, a zebrafish MITF homolog required for pigment cell formation, by the Wnt pathway. *Genes & Development*, **14**, 158–162.

Dorsky, R.I., Sheldahl, L.C., and Moon, R.T. (2002) A transgenic Lef1/beta-catenin-dependent reporter is expressed in spatially restricted domains throughout zebrafish development. *Developmental Biology*, **241**, 229–237.

Elkouby, Y.M., Elias, S., Casey, E.S. *et al.* (2010) Mesodermal Wnt signaling organizes the neural plate via Meis3. *Development*, **137**, 1531–1541.

Elnitski, L., Jin, V.X., Farnham, P.J., and Jones, S.J. (2006) Locating mammalian transcription factor binding sites: a survey of computational and experimental techniques. *Genome Research*, **16**, 1455–1464.

Fang, M., Li, J., Blauwkamp, T. *et al.* (2006) C-terminal-binding protein directly activates and represses Wnt transcriptional targets in Drosophila. *The EMBO Journal*, **25**, 2735–2745.

Frietze, S., Wang, R., Yao, L. *et al.* (2012) Cell type-specific binding patterns reveal that TCF7L2 can be tethered to the genome by association with GATA3. *Genome Biology*, **13**, R52.

Garnett, A.T., Square, T.A., and Medeiros, D.M. (2012) BMP, Wnt and FGF signals are integrated through evolutionarily conserved enhancers to achieve robust expression of Pax3 and Zic genes at the zebrafish neural plate border. *Development*, **139**, 4220–4231.

Gavin, B.J., McMahon, J.A., and McMahon, A.P. (1990) Expression of multiple novel Wnt-1/int-1-related genes during fetal and adult mouse development. *Genes & Development*, **4**, 2319–2332.

Germar, K., Dose, M., Konstantinou, T. *et al.* (2011) T-cell factor 1 is a gatekeeper for T-cell specification in response to Notch signaling. *Proceedings of the National Academy of Sciences of the United States of America*, **108**, 20060–20065.

Giese, K., Amsterdam, A., and Grosschedl, R. (1991) DNA-binding properties of the HMG domain of the lymphoid-specific transcriptional regulator LEF-1. *Genes & Development*, **5**, 2567–2578.

Green, J.L., Inoue, T., and Sternberg, P.W. (2008) Opposing Wnt pathways orient cell polarity during organogenesis. *Cell*, **134**, 646–656.

Hallikas, O., Palin, K., Sinjushina, N. *et al.* (2006) Genome-wide prediction of mammalian enhancers based on analysis of transcription-factor binding affinity. *Cell*, **124**, 47–59.

Hatzis, P., van der Flier, L.G., van Driel, M.A. *et al.* (2008) Genome-wide pattern of TCF7L2/TCF4 chromatin occupancy in colorectal cancer cells. *Molecular and Cell Biology*, **28**, 2732–2744.

He, T.C., Sparks, A.B., Rago, C. *et al.* (1998) Identification of c-MYC as a target of the APC pathway. *Science*, **281**, 1509–1512.

Heintzman, N.D. and Ren, B. (2009) Finding distal regulatory elements in the human genome. *Current Opinion in Genetics & Development*, **19**, 541–549.

Hikasa, H. and Sokol, S.Y. (2011) Phosphorylation of TCF proteins by homeodomain-interacting protein kinase 2. *The Journal of Biological Chemistry*, **286**, 12093–12100.

Hikasa, H., Ezan, J., Itoh, K. *et al.* (2010) Regulation of TCF3 by Wnt-dependent phosphorylation during vertebrate axis specification. *Developmental Cell*, **19**, 521–532.

Hoverter, N.P., Ting, J.H., Sundaresh, S. *et al.* (2012) A WNT/p21 circuit directed by the C-clamp, a sequence-specific DNA binding domain in TCFs. *Molecular and Cellular Biology*, **32**, 3648–3662.

Huber, O., Korn, R., McLaughlin, J. *et al.* (1996) Nuclear localization of beta-catenin by interaction with transcription factor LEF-1. *Mechanisms of Development*, **59**, 3–10.

Hussein, S.M., Duff, E.K., and Sirard, C. (2003) Smad4 and beta-catenin co-activators functionally interact with lymphoid-enhancing factor to regulate graded expression of Msx2. *The Journal of Biological Chemistry*, **278**, 48805–48814.

Jho, E.H., Zhang, T., Domon, C. *et al.* (2002) Wnt/beta-catenin/Tcf signaling induces the transcription of Axin2, a negative regulator of the signaling pathway. *Molecular and Cellular Biology*, **22**, 1172–1183.

Junion, G., Spivakov, M., Girardot, C. *et al.* (2012) A transcription factor collective defines cardiac cell fate and reflects lineage history. *Cell*, **148**, 473–486.

Kang, Y., Chen, C.R., and Massague, J. (2003) A self-enabling TGFbeta response coupled to stress signaling: Smad engages stress response factor ATF3 for Id1 repression in epithelial cells. *Molecular Cell*, **11**, 915–926.

Kemper, K., Prasetyanti, P.R., De Lau, W. *et al.* (2012) Monoclonal antibodies against lgr5 identify human colorectal cancer stem cells. *Stem Cells*, **30**, 2378–2386.

Knirr, S. and Frasch, M. (2001) Molecular integration of inductive and mesoderm-intrinsic inputs governs even-skipped enhancer activity in a subset of pericardial and dorsal muscle progenitors. *Developmental Biology*, **238**, 13–26.

Konsavage, W.M., Jr., Jin, G., and Yochum, G.S. (2012) The Myc 3′ Wnt-responsive element regulates homeostasis and regeneration in the mouse intestinal tract. *Molecular and Cellular Biology*, **32**, 3891–3902.

Korinek, V., Barker, N., Morin, P.J. *et al.* (1997) Constitutive transcriptional activation by a beta-catenin-Tcf complex in APC−/− colon carcinoma. *Science*, **275**, 1784–1787.

Kuwabara, T., Hsieh, J., Muotri, A. *et al.* (2009) Wnt-mediated activation of NeuroD1 and retro-elements during adult neurogenesis. *Nature Neuroscience*, **12**, 1097–1105.

Labbe, E., Letamendia, A., and Attisano, L. (2000) Association of Smads with lymphoid enhancer binding factor 1/T cell-specific factor mediates cooperative signaling by the transforming growth factor-beta and wnt pathways. *Proceedings of the National Academy of Sciences of the United States of America*, **97**, 8358–8363.

Lam, N., Chesney, M.A., and Kimble, J. (2006) Wnt signaling and CEH-22/tinman/Nkx2.5 specify a stem cell niche in C. elegans. *Current Biology*, **16**, 287–295.

Laurent, M.N., Blitz, I.L., Hashimoto, C. *et al.* (1997) The *Xenopus* homeobox gene twin mediates Wnt induction of goosecoid in establishment of Spemann's organizer. *Development*, **124**, 4905–4916.

Lee, H.H. and Frasch, M. (2000) Wingless effects mesoderm patterning and ectoderm segmentation events via induction of its downstream target sloppy paired. *Development*, **127**, 5497–5508.

Leung, J.Y., Kolligs, F.T., Wu, R. *et al.* (2002) Activation of AXIN2 expression by beta-catenin-T cell factor. A feedback repressor pathway regulating Wnt signaling. *The Journal of Biological Chemistry*, **277**, 21657–21665.

Li, B., Kuriyama, S., Moreno, M., and Mayor, R. (2009) The posteriorizing gene Gbx2 is a direct target of Wnt signalling and the earliest factor in neural crest induction. *Development*, **136**, 3267–3278.

Logan, C.Y. and Nusse, R. (2004) The Wnt signaling pathway in development and disease. *Annual Review of Cell and Developmental Biology*, **20**, 781–810.

Love, J.J., Li, X., Case, D.A. *et al.* (1995) Structural basis for DNA bending by the architectural transcription factor LEF-1. *Nature*, **376**, 791–795.

Maduro, M.F., Kasmir, J.J., Zhu, J., and Rothman, J.H. (2005) The Wnt effector POP-1 and the PAL-1/Caudal homeoprotein collaborate with SKN-1 to activate C. elegans endoderm development. *Developmental Biology*, **285**, 510–523.

Maretto, S., Cordenonsi, M., Dupont, S. *et al.* (2003) Mapping Wnt/beta-catenin signaling during mouse development and in colorectal tumors. *Proceedings of the National Academy of Sciences of the United States of America*, **100**, 3299–3304.

Marson, A., Levine, S.S., Cole, M.F. *et al.* (2008). Connecting microRNA genes to the core transcriptional regulatory circuitry of embryonic stem cells. *Cell*, **134**, 521–533.

Martizez Arias, A., Baker, N.E., and Ingham, P.W. (1988) Role of segment polarity genes in the definition and maintenance of cell states in the Drosophila embryo. *Development*, **103**, 157–170.

May, D., Blow, M.J., Kaplan, T. *et al.* (2012) Large-scale discovery of enhancers from human heart tissue. *Nature Genetics*, **44**, 89–93.

McKendry, R., Hsu, S.C., Harland, R.M. *et al.* (1997) LEF-1/TCF proteins mediate wnt-inducible transcription from the *Xenopus* nodal-related 3 promoter. *Developmental Biology*, **192**, 420–431.

McMahon, A.P., Joyner, A.L., Bradley, A., and McMahon, J.A. (1992) The midbrain-hindbrain phenotype of Wnt-1-/Wnt-1- mice results from stepwise deletion of engrailed-expressing cells by 9.5 days postcoitum. *Cell*, **69**, 581–595.

Mokry, M., Hatzis, P., de Bruijn, E. *et al.* (2010) Efficient double fragmentation ChIP-seq provides nucleotide resolution protein-DNA binding profiles. *PLoS One*, **5**, e15092.

Mokry, M., Hatzis, P., Schuijers, J. *et al.* (2012) Integrated genome-wide analysis of transcription factor occupancy, RNA polymerase II binding and steady-state RNA levels identify differentially regulated functional gene classes. *Nucleic Acids Research*, **40**, 148–158.

Molenaar, M., van de Wetering, M., Oosterwegel, M. *et al.* (1996) XTcf-3 transcription factor mediates beta-catenin-induced axis formation in Xenopus embryos. *Cell*, **86**, 391–399.

Moustakas, A. and Heldin, C.H. (2009) The regulation of TGFbeta signal transduction. *Development*, **136**, 3699–3714.

Myant, K. and Sansom, O.J. (2011) Wnt/Myc interactions in intestinal cancer: partners in crime. *Experimental Cell Research*, **317**, 2725–2731.

Nakano, N., Itoh, S., Watanabe, Y. *et al.* (2010) Requirement of TCF7L2 for TGF-beta-dependent transcriptional activation of the TMEPAI gene. *The Journal of Biological Chemistry*, **285**, 38023–38033.

Nateri, A.S., Spencer-Dene, B., and Behrens, A. (2005) Interaction of phosphorylated c-Jun with TCF4 regulates intestinal cancer development. *Nature*, **437**, 281–285.

Nishita, M., Hashimoto, M.K., Ogata, S. *et al.* (2000) Interaction between Wnt and TGF-beta signalling pathways during formation of Spemann's organizer. *Nature*, **403**, 781–785.

Norton, L., Fourcaudot, M., Abdul-Ghani, M.A. *et al.* (2011) Chromatin occupancy of transcription factor 7-like 2 (TCF7L2) and its role in hepatic glucose metabolism. *Diabetologia*, **54**, 3132–3142.

Oosterwegel, M., van de Wetering, M., Dooijes, D. *et al.* (1991) Cloning of murine TCF-1, a T cell-specific transcription factor interacting with functional motifs in the CD3-epsilon and T cell receptor alpha enhancers. *Journal of Experimental Medicine*, **173**, 1133–1142.

Parker, D.S., Ni, Y.Y., Chang, J.L. *et al.* (2008) Wingless signaling induces widespread chromatin remodeling of target loci. *Molecular and Cellular Biology*, **28**, 1815–1828.

Petersen, C.P. and Reddien, P.W. (2011) Polarized notum activation at wounds inhibits Wnt function to promote planarian head regeneration. *Science*, **332**, 852–855.

Pomerantz, M.M., Ahmadiyeh, N., Jia, L. *et al.* (2009) The 8q24 cancer risk variant rs6983267 shows long-range interaction with MYC in colorectal cancer. *Nature Genetics*, **41**, 882–884.

Ryu, S.L., Fujii, R., Yamanaka, Y. *et al.* (2001). Regulation of dharma/bozozok by the Wnt pathway. *Developmental Biology*, **231**, 397–409.

Sanchez-Ferras, O., Coutaud, B., Djavanbakht Samani, T. *et al.* (2012) Caudal-related homeobox (Cdx) protein-dependent integration of canonical Wnt signaling on paired-box 3 (Pax3) neural crest enhancer. *The Journal of Biological Chemistry*, **287**, 16623–16635.

Sato, T., Vries, R.G., Snippert, H.J. *et al.* (2009) Single Lgr5 stem cells build crypt-villus structures in vitro without a mesenchymal niche. *Nature*, **459**, 262–265.

Shafer, S.L. and Towler, D.A. (2009) Transcriptional regulation of SM22alpha by Wnt3a: convergence with TGFbeta(1)/Smad signaling at a novel regulatory element. *Journal of Molecular and Cellular Cardiology*, **46**, 621–635.

Shaulian, E. and Karin, M. (2002) AP-1 as a regulator of cell life and death. *Nature Cell Biology*, **4**, E131–136.

Shetty, P., Lo, M.C., Robertson, S.M., and Lin, R. (2005) C. elegans TCF protein, POP-1, converts from repressor to activator as a result of Wnt-induced lowering of nuclear levels. *Developmental Biology*, **285**, 584–592.

Sierra, J., Yoshida, T., Joazeiro, C.A., and Jones, K.A. (2006) The APC tumor suppressor counteracts beta-catenin activation and H3K4 methylation at Wnt target genes. *Genes & Development*, **20**, 586–600.

Sur, I.K., Hallikas, O., Vaharautio, A. *et al.* (2012) Mice lacking a Myc enhancer that includes human SNP rs6983267 are resistant to intestinal tumors. *Science*, **338**, 1360–1363.

Tam, W.L., Lim, C.Y., Han, J. *et al.* (2008) T-cell factor 3 regulates embryonic stem cell pluripotency and self-renewal by the transcriptional control of multiple lineage pathways. *Stem Cells*, **26**, 2019–2031.

Theisen, H., Syed, A., Nguyen, B.T. *et al.* (2007). Wingless directly represses DPP morphogen expression via an armadillo/TCF/Brinker complex. *PLoS One*, **2**, e142.

Travis, A., Amsterdam, A., Belanger, C., and Grosschedl, R. (1991) LEF-1, a gene encoding a lymphoid-specific protein with an HMG domain, regulates T-cell receptor alpha enhancer function [corrected]. *Genes & Development*, **5**, 880–894.

Trompouki, E., Bowman, T.V., Lawton, L.N. *et al.* (2011) Lineage regulators direct BMP and Wnt pathways to cell-specific programs during differentiation and regeneration. *Cell*, **147**, 577–589.

Tuupanen, S., Turunen, M., Lehtonen, R. *et al.* (2009) The common colorectal cancer predisposition SNP rs6983267 at chromosome 8q24 confers potential to enhanced Wnt signaling. *Nature Genetics*, **41**, 885–890.

van Beest, M., Dooijes, D., van De Wetering, M. *et al.* (2000) Sequence-specific high mobility group box factors recognize 10–12-base pair minor groove motifs. *The Journal of Biological Chemistry*, **275**, 27266–27273.

van de Wetering, M., Oosterwegel, M., Dooijes, D., and Clevers, H. (1991) Identification and cloning of TCF-1, a T lymphocyte-specific transcription factor containing a sequence-specific HMG box. *The EMBO Journal*, **10**, 123–132.

van de Wetering, M., Cavallo, R., Dooijes, D. *et al.* (1997) Armadillo coactivates transcription driven by the product of the Drosophila segment polarity gene dTCF. *Cell*, **88**, 789–799.

van de Wetering, M., Sancho, E., Verweij, C. *et al.* (2002) The beta-catenin/TCF-4 complex imposes a crypt progenitor phenotype on colorectal cancer cells. *Cell*, **111**, 241–250.

Van der Flier, L.G., Sabates-Bellver, J., Oving, I. *et al.* (2007) The intestinal Wnt/TCF signature. *Gastroenterology*, **132**, 628–632.

van Es, J.H., Jay, P., Gregorieff, A. *et al.* (2005) Wnt signalling induces maturation of Paneth cells in intestinal crypts. *Nature Cell Biology*, **7**, 381–386.

Verzi, M.P., Hatzis, P., Sulahian, R. *et al.* (2010) TCF4 and CDX2, major transcription factors for intestinal function, converge on the same cis-regulatory regions. *Proceedings of the National Academy of Sciences of the United States of America*, **107**, 15157–15162.

Visel, A., Blow, M.J., Li, Z. *et al.* (2009) ChIP-seq accurately predicts tissue-specific activity of enhancers. *Nature*, **457**, 854–858.

Vlad, A., Rohrs, S., Klein-Hitpass, L., and Muller, O. (2008) The first five years of the Wnt targetome. *Cellular Signalling*, **20**, 795–802.

Waterman, M.L. and Jones, K.A. (1990) Purification of TCF-1 alpha, a T-cell-specific transcription factor that activates the T-cell receptor C alpha gene enhancer in a context-dependent manner. *New Biology*, **2**, 621–636.

Weidinger, G., Thorpe, C.J., Wuennenberg-Stapleton, K. *et al.* (2005) The Sp1-related transcription factors sp5 and sp5-like act downstream of Wnt/beta-catenin signaling in mesoderm and neuroectoderm patterning. *Current Biology*, **15**, 489–500.

Weise, A., Bruser, K., Elfert, S. *et al.* (2010) Alternative splicing of Tcf7l2 transcripts generates protein variants with differential promoter-binding and transcriptional activation properties at Wnt/beta-catenin targets. *Nucleic Acids Research*, **38**, 1964–1981.

Wierzbicki, A.T., Cocklin, R., Mayampurath, A. *et al.* (2012) Spatial and functional relationships among Pol V-associated loci, Pol IV-dependent siRNAs, and cytosine methylation in the Arabidopsis epigenome. *Genes & Development*, **26**, 1825–1836.

Willert, J., Epping, M., Pollack, J.R. *et al.* (2002) A transcriptional response to Wnt protein in human embryonic carcinoma cells. *BMC Developmental Biology*, **2**, 8.

Wisniewska, M.B., Misztal, K., Michowski, W. *et al.* (2010) LEF1/beta-catenin complex regulates transcription of the Cav3.1 calcium channel gene (Cacna1g) in thalamic neurons of the adult brain. *The Journal of Neuroscience*, **30**, 4957–4969.

Wright, J.B., Brown, S.J., and Cole, M.D. (2010) Upregulation of c-MYC in cis through a large chromatin loop linked to a cancer risk-associated single-nucleotide polymorphism in colorectal cancer cells. *Molecular and Cellular Biology*, **30**, 1411–1420.

Wu, J.Q., Seay, M., Schulz, V.P. *et al.* (2012) Tcf7 is an important regulator of the switch of self-renewal and differentiation in a multipotential hematopoietic cell line. *PLoS Genetics*, **8**, e1002565.

Yamaguchi, T.P., Takada, S., Yoshikawa, Y. *et al.* (1999) T (Brachyury) is a direct target of Wnt3a during paraxial mesoderm specification. *Genes & Development*, **13**, 3185–3190.

Yip, K.Y., Cheng, C., Bhardwaj, N. *et al.* (2012) Classification of human genomic regions based on experimentally determined binding sites of more than 100 transcription-related factors. *Genome Biology*, **13**, R48.

Yochum, G.S., Cleland, R., and Goodman, R.H. (2008). A genome-wide screen for beta-catenin binding sites identifies a downstream enhancer element that controls c-Myc gene expression. *Molecular and Cellular Biology* **28**, 7368–7379.

Yochum, G.S., McWeeney, S., Rajaraman, V. *et al.* (2007) Serial analysis of chromatin occupancy identifies beta-catenin target genes in colorectal carcinoma cells. *Proceedings of the National Academy of Sciences of the United States of America*, **104**, 3324–3329.

Yochum, G.S., Sherrick, C.M., Macpartlin, M., and Goodman, R.H. (2010) A beta-catenin/TCF-coordinated chromatin loop at MYC integrates 5′ and 3′ Wnt responsive enhancers. *Proceedings of the National Academy of Sciences of the United States of America*, **107**, 145–150.

Zeitlinger, J., Zinzen, R.P., Stark, A. *et al.* (2007) Whole-genome ChIP-chip analysis of Dorsal, Twist, and Snail suggests integration of diverse patterning processes in the Drosophila embryo. *Genes & Development*, **21**, 385–390.

Zhao, J., Schug, J., Li, M. *et al.* (2010) Disease-associated loci are significantly over-represented among genes bound by transcription factor 7-like 2 (TCF7L2) in vivo. *Diabetologia*, **53**, 2340–2346.

Zheng, Q., Rowley, M.J., Bohmdorfer, C. *et al.* (2012) RNA polymerase V targets transcriptional silencing components to promoters of protein-coding genes. *The Plant Journal*, **73**, 179–189.

Zhou, Y., Zhang, E., Berggreen, C. *et al.* (2012) Survival of pancreatic beta cells is partly controlled by a TCF7L2-p53-p53INP1-dependent pathway. *Human Molecular Genetics*, **21**, 196–207.

6 Introduction to β-Catenin-Independent Wnt Signaling Pathways

Susanne Kühl and Michael Kühl
Institute of Biochemistry and Molecular Biology, Ulm University, Ulm, Germany

Introduction

In 2012, the scientific community was celebrating 30 years of Wnt research. The initial starting point was in 1982 when Roel Nusse and Harald Varmus identified the founding member of the Wnt family, int1, today named Wnt1 by studying the mouse mammary tumor virus (Nusse and Varmus, 1982). Nowadays, we know 19 members of the Wnt family of extracellular ligands, 10 members of the Frizzled (Fzd) family of 7 transmembrane Wnt receptors (Chapter 14), as well as several Wnt coreceptors such as LRP5/6, Ror2, or Ryk (Chapters 2 and 7) (Niehrs, 2012). Given the high number of ligands and receptors as well as possible ligand/receptor combinations, it was tempting to speculate that Wnt proteins are able to activate more than only one intracellular signaling cascade.

The best described Wnt signaling cascade is still the Wnt/β-catenin pathway, which is characterized by the stabilization of cytoplasmic β-catenin (see Figure 6.1; Chapters 3 and 4). For quite some time, this pathway has been called canonical Wnt pathway. In contrast, noncanonical Wnt signaling pathways are per definition independent of β-catenin. Today, these Wnt signaling cascades are more commonly referred to as β-catenin-dependent or independent Wnt signaling pathways, respectively.

Today, we are astonished by the high diversity and complexity of Wnt signaling pathways. Moreover, these pathways are not isolated acting entities but are interwoven to form a complex Wnt signaling network. For a more detailed discussion of this concept of a Wnt signaling network, we refer the reader to Kestler and Kuhl (2008). This chapter overviews the architecture of the currently known β-catenin-independent Wnt signaling pathways and highlights their biological relevance.

Wnt/calcium signaling

Roughly half the way from 1982, it became obvious that Wnt proteins are indeed able to activate different intracellular signaling cascades. Earlier work already indicated that not all Wnt proteins cause axis duplication in *Xenopus* embryos upon overexpression at the ventral body side, nowadays a very well established assay for analyzing the Wnt/β-catenin pathway (see Chapters 3 and 4 for a

Wnt Signaling in Development and Disease: Molecular Mechanisms and Biological Functions, First Edition.
Edited by Stefan Hoppler and Randall T. Moon.

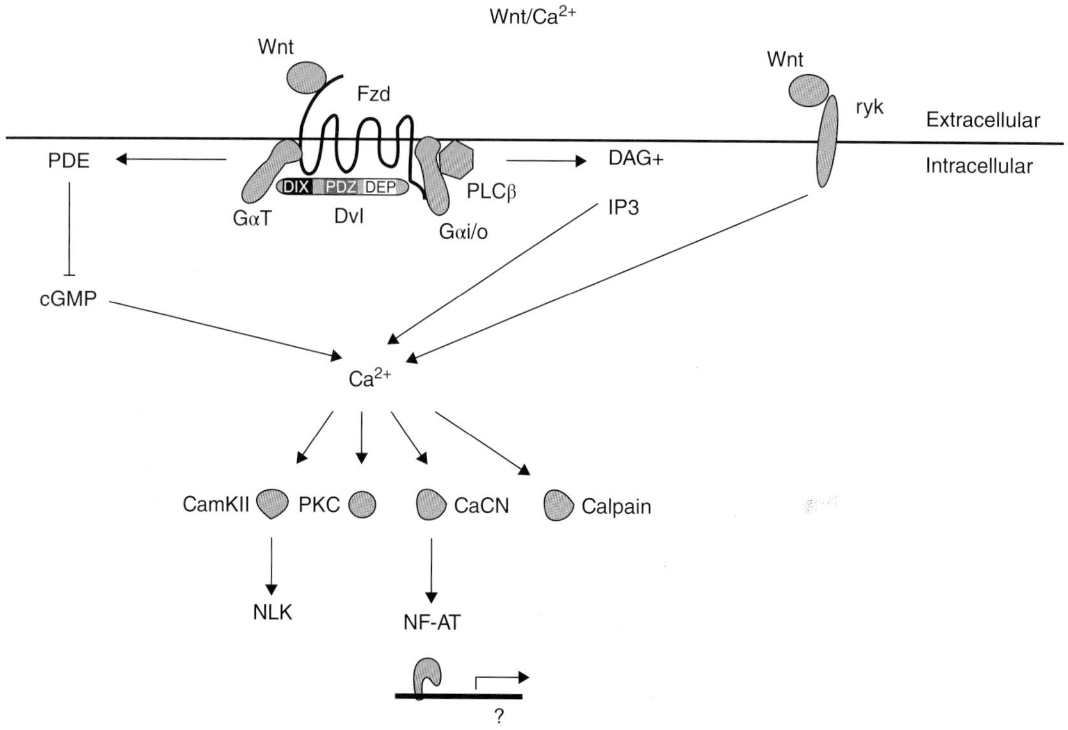

Figure 6.1 Architecture of Wnt/Ca^{2+} signaling pathways. Different branches of Wnt signaling are given as graph presentations as indicated. Please refer to the main text for details. ? indicates yet unknown direct target genes.

detailed description of this pathway). Though some Wnts did not stimulate axis duplication, when overexpressed on the dorsal side of *Xenopus* embryos, some Wnts such as Wnt5a interfered with cell movements during gastrulation upon overexpression (Du *et al.*, 1995). This was the first hint of the existence of β-catenin-independent Wnt signaling.

But how does one make the leap from altered gastrulation movements to the identification of downstream signaling? In a fortuitous conversation with William Busa at Woods Hole, Randall Moon learned that the Wnt5a-mediated phenotype of an altered gastrulation was also observed by Busa who injected the serotonin 5HT1c receptor in *Xenopus*. This receptor was known to stimulate calcium release; hence, Moon and Busa decided to investigate whether Wnt5a can also induce calcium release. In 1997, the group of Randall Moon showed that Wnt5a is really able to trigger an intracellular calcium release upon overexpression in zebrafish embryos but not of double axis formation (Slusarski, Corces, and Moon, 1997; Slusarski *et al.*, 1997). Noteworthy, these key experiments also indicated the involvement of heterotrimeric G proteins and suggested Fzd receptors being G protein coupled.

The characteristic feature of the Wnt/calcium pathway is an increase of the intracellular calcium ion concentration upon stimulation of cells by Wnt ligands (Figure 6.1). So far, three different possible mechanisms have been described of how Wnt stimulation results in calcium release. The first publication about Wnt/calcium signaling in 1997 also revealed that the overexpression of certain Fzd receptors, such as Fzd2 in zebrafish embryos, triggers intracellular calcium release (Slusarski, Corces, and Moon, 1997; Slusarski *et al.*, 1997). These data also indicated the involvement of further factors such as phospholipase Cβ (PLCβ), inositoltrisphosphate (IP_3), and diacylglycerol (DAG). The activation of PLCβ occurs in a G protein-dependent manner and likely involves members of the $GQ_{i/o}$ family. Five years later, another pathway eliciting calcium release was described (Ahumada *et al.*, 2002). This pathway includes a specific α-subunit of heterotrimeric G proteins named transducin ($G\alpha_t$) and a cyclic guanosine monophosphate (cGMP)-specific phospho diesterase (PDE). PDE lowers cGMP levels, resulting in an increase of intracellular calcium levels. The exact molecular mechanism, however, has not yet been clarified. More recently, the Wnt coreceptor Ryk (see subsequent sections) has also been implicated in Wnt/calcium signaling (Hutchins, Li, and Kalil, 2012). Till now, it remains unclear whether these pathway mechanisms act independently of each other, overlap, or even synergize.

For quite some time, the scientific community debated whether the calcium release is a direct response to a Wnt/Fzd interaction or rather reflects an indirect response to an unknown intermediate. A direct response would likely be a quick response of cells upon Wnt ligand treatment. Indeed, several independent studies showed that Wnt triggers calcium release very rapidly within seconds or minutes in a characteristic kinetik (Dejmek *et al.*, 2006; Jenei *et al.*, 2009).

Wnt ligands that have originally been shown to cause intracellular calcium release are Wnt5a and Wnt11 (Kuhl *et al.*, 2000b). More recently, also Wnt3a was added to the list of calcium releasing Wnt proteins (Nalesso *et al.*, 2011). Intriguingly, the same study revealed that Wnt3a additionally activates the Wnt/β-catenin pathway in the same cell. Whether Wnt3a simulates calcium signaling or β-catenin stabilization seems to depend on its concentration (Nalesso *et al.*, 2011). Furthermore, Wnt5a or Wnt11 can not only cause intracellular calcium release but also activate β-catenin signaling in particular cellular settings. Thus, the concentration-dependent activation of different pathways could be a general motif that requires additional work in the future (Kestler and Kuhl, 2011).

The intracellular calcium release subsequently results in the activation of calcium-sensitive enzymes. These include the protein kinase C (PKC; Sheldahl *et al.*, 1999), the calcium/calmodulin-dependent kinase II (CamKII) (Kuhl *et al.*, 2000a), the calcium-sensitive phosphatase calcineurin (CaCN) (Saneyoshi *et al.*, 2002), and the calcium-sensitive protease calpain (Li and Iyengar, 2002). Likely, other calcium-sensitive proteins are regulated as well, which, however, awaits further experimental confirmation.

The downstream events of these pathways have also been elucidated in more detail. Downstream of CamKII, Nemo-like kinase (NLK) is regulated by Wnt proteins (Ishitani *et al.*, 1999, 2003). Activated NLK in turn acts as an inhibitor of the Wnt/β-catenin pathway by influencing TCF/LEF transcription factors. Calcineurin can regulate the transcription factor NF-AT, thereby also linking this pathway to transcriptional effects. So far, no direct target gene of NF-AT has been described in this context.

The Wnt/calcium pathway is of relevance for diverse biological processes including dorso-ventral patterning of zebrafish and *Xenopus* embryos (Kuhl *et al.*, 2000a; Saneyoshi *et al.*, 2002; Westfall, Hjertos, and Slusarski, 2003; Westfall *et al.*, 2003), regulation of cell migration during gastrulation (Kuhl *et al.*, 2001), migration of neural crest cells (Garriock and Krieg, 2006), cartilage development (Nalesso *et al.*, 2011), axon growth and guidance (Hutchins, Li, and Kalil, 2012), tubulogenesis in the developing kidney (Burn *et al.*, 2011; Tanigawa *et al.*, 2011), and T-cell development (Liang *et al.*, 2007). The Wnt/calcium pathway has also been discussed to function during tumor formation, in particular metastasis, but also in tumor suppression (Kremenevskaja *et al.*, 2005; Liang *et al.*, 2003; Weeraratna *et al.*, 2002).

Planar cell polarity signaling

When Fzd proteins were identified as Wnt receptors (Bhanot *et al.*, 1996), it was also suggested that Fzd-regulated planar cell polarity (PCP) might also be Wnt regulated. PCP is an important biological phenomenon describing the polarity of epithelial cells within the plane of cells perpendicular to the apical–basal axis. Well-known examples for this phenomenon are the polarities of *Drosophila* wing cells. Each of these cells has a single hair, and all hairs of the wing point into the same direction (Vinson and Adler, 1987). Similarly, the hairs on a mouse appendage point distally (Guo, Hawkins, and Nathans, 2004). Other processes such as the arrangement of ommatidia in the *Drosophila* eye (Adler, 2002) and the orientation of stereocilia in the vertebrate inner ear (Dabdoub *et al.*, 2003; Montcouquiol and Kelley, 2003; Montcouquiol *et al.*, 2003) are also regulated by PCP signaling.

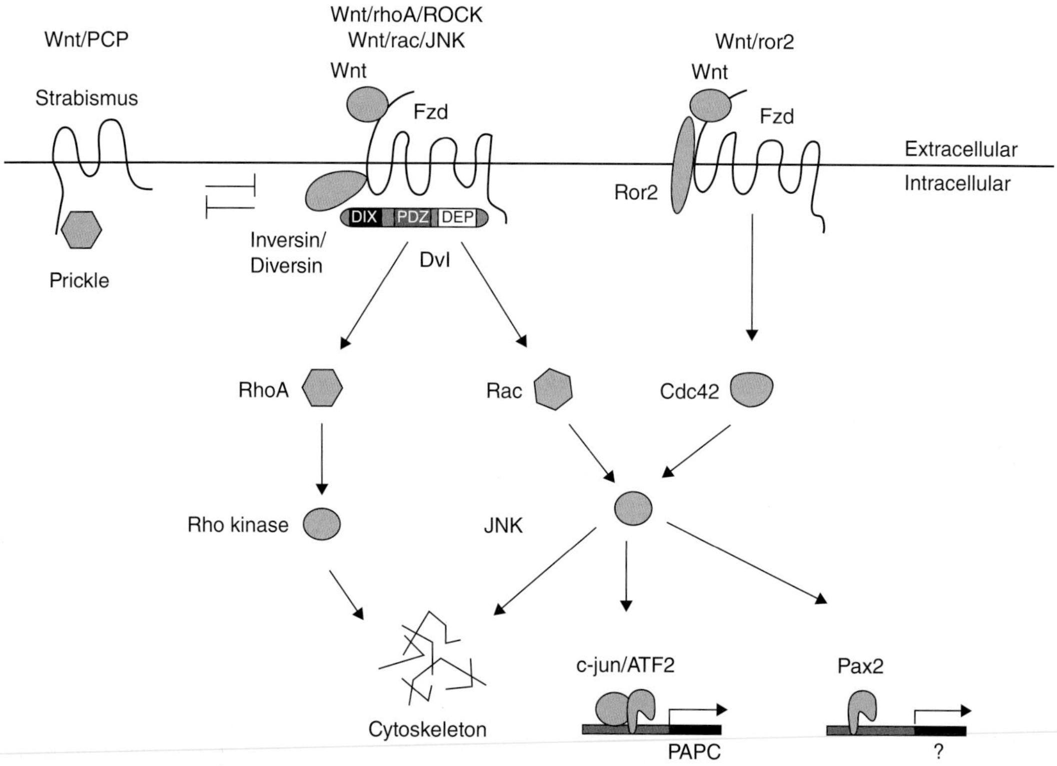

Figure 6.2 Architecture of Wnt/PCP-related signaling pathways. Different branches of Wnt signaling are given as graph presentations as indicated. Please refer to the main text for details. Negative arrows indicate that strabismus and prickle as well as Fzd and dvl negatively influence each other's cellular localization. ? indicates yet unknown direct target genes.

Further aspects regulated by PCP signaling include gastrulation movements, neural tube closure, and morphogenesis of the proximal tubule during pronephros development in *Xenopus* (Simons and Mlodzik, 2008).

Components of PCP signaling can be classified into core components, regulators, and effectors of PCP signaling (Seifert and Mlodzik, 2007). Core components in *Drosophila* include Fzd, strabismus, prickle, diego, and Dishevelled (dvl). Today, also vertebrate homologs of these proteins are well known. The function of these proteins is the cross-regulation of each other's intracellular localization, resulting in an asymmetric cellular distribution. As an effect, strabismus and prickle on the one hand and Fzd and dvl on the other hand are localized on opposite sites of cells, thus resulting in cell polarization of epithelial cells (Figure 6.2). Whereas this pathway in *Drosophila* does not require any Wnt ligand, convergent extension movements in vertebrates are regulated by Wnts such as Wnt5a or Wnt11 (Du *et al.*, 1995). Both Wnt ligands therefore are considered as regulators of PCP signaling.

Cytoplasmic proteins that are regulated by PCP signaling are called effectors of the pathway. These are, among others, the small GTPases rhoA and rac, rho kinase (ROCK), and PKCδ. For a more detailed list of core components, regulators, and effectors of PCP signaling, see a recent review (Seifert and Mlodzik, 2007). The activation of rhoA and rac1 as well as further downstream components is discussed in subsequent sections in more detail.

Cilia formation and function: Components of PCP signaling

Key components of PCP signaling are especially important for cilia formation and function. Cilia are cellular structures that are based on microtubules. They consist of a basal body containing two centrioles and one axonem forming a cellular, membrane-covered extrusion. Cilia serve different functions. First, they can send extracellular signals of chemical or mechanical nature. Second, they can function to generate either moving forces or extracellular fluid flow, for example, in the airway epithelium, in the nephric tubules, or in the node of early embryos. It is currently thought that only certain cell types of invertebrates but most cell types of vertebrates can generate a cilium. One distinguishes mono- and multiciliated cells. Monociliated cells have one single cilium that is positioned asymmetrically on the apical surface of epithelial cells. This polarized position reflects the planar polarization of an epithelial cell sheet. Monociliated cells undergo rotational beating to generate a fluid flow (Wallingford, 2010). In contrast, multiciliated cells have multiple cilia on the apical surface. They undergo directional beating to generate fluid flow. Multiciliated cells have accessory structures linked to the basal body named basal foot and rootlet. They are aligned in a parallel manner and with respect to cellular geometry, a feature called rotational polarity (Mirzadeh *et al.*, 2010; Wallingford, 2010). In addition, these multiple cilia are in general generated at the center of a cell but are translocated after formation to the posterior side of cells, a feature called translational planar polarity (Mirzadeh *et al.*, 2010; Wallingford, 2010).

Interestingly, different proteins involved in PCP signaling are required for cilia formation and function, including Fzd receptors, dvl, inversin, or Vangl2. This is based on localization studies and loss-of-function analyses. Detailed discussions of this issue have recently been published elsewhere (Lienkamp, Ganner, and Walz, 2012; Wallingford, 2010; Wallingford and Mitchell, 2011). Of note, the misregulation of PCP signaling can lead to cilia dysfunction and different diseases called ciliopathies (Hildebrandt, Benzing, and Katsanis, 2011; Oh and Katsanis, 2012).

Moreover, it has been demonstrated that cilia can receive extracellular signals such as hedgehog ligands. These signals can furthermore be translated in intracellular signals. Earlier reports that indicate a particular role of cilia in triggering Wnt signaling (Corbit *et al.*, 2008; Gerdes *et al.*, 2007) could recently not be confirmed using genetic mouse and zebrafish models with mutations in cilia-associated genes (Huang and Schier, 2009; Ocbina, Tuson, and Anderson, 2009). This suggests that those previous reports might rather reflect an indirect role of the cilium on Wnt signaling. Furthermore, a role of cilia in the spatial regulation of Wnt signaling has been discussed (Lancaster, Schroth, and Gleeson, 2011).

Wnt/rho/ROCK and Wnt/rac/JNK signaling

Wnt proteins are also able to activate monomeric small G proteins of the rho protein family, namely, rhoA, rac, and cdc42 (Figure 6.2). This section focuses only on the activation of rhoA and rac1 (the activation of cdc42 will be discussed in a later section). RhoA and rac1 are induced by the scaffolding phosphoprotein dvl. Dvl contains multiple domains including the DEP, DIX, and PDZ domains that are differentially used for further downstream signaling (Chapter 15). The activation of rhoA requires the PDZ domain of dvl, and the activation of rac1 the DEP domain (Habas, Kato, and He, 2001).

Moreover, the activation of rhoA involves the formin homology protein Daam1 that can interact with rhoA and dvl. Furthermore, the weak-similarity guanine nucleotide-exchange factor (WGEF) interacts with both proteins and is required for the activation of rhoA (Tanegashima, Zhao, and Dawid, 2008). Upon rhoA activation, the downstream factor ROCK is activated to regulate cytoskeletal rearrangement (Marlow *et al.*, 2002). In contrast, the activation of rac1 is not yet fully elucidated but requires β-arrestin and CK1/2 (Bryja *et al.*, 2008). Rac1 subsequently activates jun N-terminal kinase (JNK). This results either in cytoskeletal reorganization (Oishi *et al.*, 2006) or in activation of transcription factors such as c-jun, ATF2, or Pax2 (Cai *et al.*, 2002; Rosso *et al.*, 2005; Schambony and Wedlich, 2007).

Taken together, the activation of rhoA and rac1 represents two distinct signaling branches downstream of dvl. Another argument for this branching comes with the observation of different kinetics of enzyme activation. Rac1 is activated rapidly, whereas rhoA activation occurs much slower (Habas, Dawid, and He, 2003).

The most prominent functional example for Wnt/rhoA/ROCK and Wnt/rac/JNK signaling is likely the regulation of convergent extension movements during vertebrate gastrulation or neural tube closure (Seifert and Mlodzik, 2007). Other examples are the requirement of Wnt11/JNK signaling during cardiac (Garriock *et al.*, 2005; Pandur *et al.*, 2002) or eye development (Maurus *et al.*, 2005; Rasmussen *et al.*, 2001).

It is noteworthy that some Wnt proteins that have been described earlier to be specific for Wnt/β-catenin signaling are also able to activate rhoA or rac1 in certain settings. Wnt3A, for example, has been shown to activate rhoA and ROCK, thereby regulating cell migration and retraction of neurites (Endo *et al.*, 2005; Kishida, Yamamoto, and Kikuchi, 2004). Of general importance is the observation that rac1 is also required for Wnt/β-catenin signaling (Wu *et al.*, 2008), suggesting that Wnt3a is able to activate different signaling branches in parallel. This finding further supports the idea that the Wnt signaling branches are highly interconnected to form a complex Wnt signaling network (Kestler and Kuhl, 2008).

Wnt/Ror signaling

The single-span transmembrane tyrosine receptor kinases Ror1/2 have been described to bind Wnt ligands through their N-terminal cysteine-rich domain (Liu *et al.*, 2008; Minami *et al.*, 2010). Experimental evidence indicates that Wnt binding results in dimerization of Ror1/2, resulting in the typical activation pattern of tyrosine kinases: kinase activation, cross receptor autophosphorylation, and subsequent activation of downstream intracellular factors (Figure 6.2). This includes the activation of PI3K, cdc42, JNK, and finally ATF2/c-jun as transcription factors (Oishi *et al.*, 2003; Schambony and Wedlich, 2007; Unterseher *et al.*, 2004). It remains under debate whether activation of this pathway involves or requires Fzd receptors. A secreted glycoprotein named collagen triple helix repeat containing 1 (CTHRC1) has been shown to stabilize the interaction between Wnt, Ror2, and Fzd, suggesting the involvement of Fzd receptors (Yamamoto *et al.*, 2008).

Wnt/Ror signaling has been shown to be essential to regulate convergent extension movements during gastrulation in *Xenopus* (Schambony and Wedlich, 2007). Mutations in Ror2 are linked to Robinow syndrome that is characterized by face and limb malformations. Patients with Robinow syndrome, in addition, often have heart defects (Al-Ata, Paquet, and Teebi, 1998). Heart development also depends on β-catenin-independent Wnt signaling (see Chapter 22). Furthermore, brachydactyly has been linked to mutations in Ror2 (Minami *et al.*, 2010). Important to note is also the ability of Ror2 to modulate and inhibit Wnt/β-catenin signaling (Billiard *et al.*, 2005; Mikels and Nusse, 2006; Witte *et al.*, 2010).

Wnt/Ryk signaling

Ryk is a single-span transmembrane protein with an intracellular kinase domain. Its intracellular part contains a PDZ protein–protein interaction domain and an S/T domain. Wnt binding to the extracellular region occurs

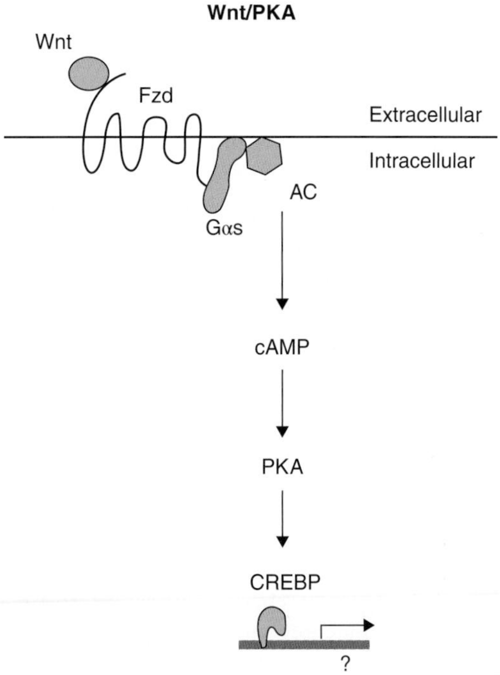

Figure 6.3 Architecture of Wnt/cAMP/PKA signaling. Please refer to the main text for details. ? indicates yet unknown direct target genes.

through a Wnt inhibitory factor homology domain (Fradkin, Dura, and Noordermeer, 2010; Patthy, 2000). Homologs of this gene are also found in *Drosophila* and in *Caenorhabditis elegans* named derailed or LIN-18, respectively (Inoue *et al.*, 2004; Yoshikawa *et al.*, 2001). Interestingly, the kinase domain of Ryk seems not to be catalytically active, but Ryk signaling requires the cytoplasmic src kinase (Hovens *et al.*, 1992; Katso, Russell, and Ganesan, 1999; Wouda *et al.*, 2008; Yoshikawa *et al.*, 2001). Important *in vivo* functions of ryk are given by axon guidance and neurite outgrowth (Lu *et al.*, 2004; Yoshikawa *et al.*, 2003).

Wnt/cAMP/protein kinase A

Fzd have been predicted to interact with Gαs subunits (Wang, Liu, and Malbon, 2006). A recent study describes Fzd as G protein-coupled receptors (Koval and Katanaev, 2011; Koval *et al.*, 2011). For a detailed description, see the discussion in Chapter 14 of this book. Several pieces of evidences indeed suggest that Wnt proteins can activate adenylate cyclase that is normally modulated by Gαs. Thereby, the intracellular concentrations of cAMP increases. This in turn results in an activation of protein kinase A (PKA) (Figure 6.3). Initially, Wnt1 and Wnt7a were shown to regulate the expression of muscle-specific genes through PKA (Chen, Ginty, and Fan, 2005). This pathway was also described to function in dermal fibroblasts (Torii *et al.*, 2008) and breast cancer cells (Hansen *et al.*, 2009). In dermal fibroblasts, this pathway was considered to be antiapoptotic (Torii *et al.*, 2008). Clearly, additional experimental approaches will be required to firmly establish the Wnt/cAMP/PKA signaling cascade.

Wnt/TOR signaling

Cell proliferation is a key feature regulated by Wnt/β-catenin signaling. This occurs through transcriptional upregulation of TCF/LEF target

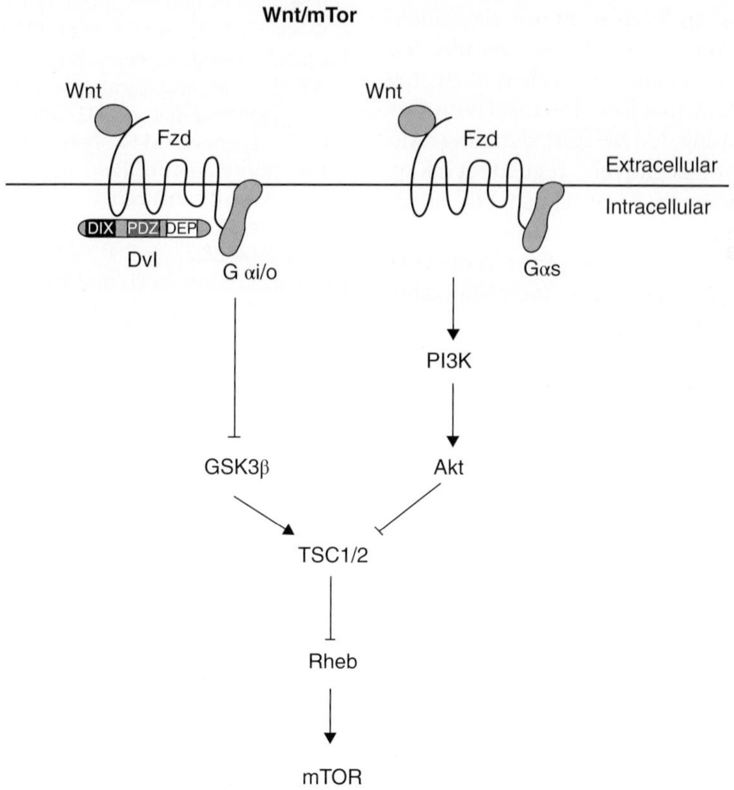

Figure 6.4 Architecture of Wnt/mTOR signaling pathways. Different branches of Wnt signaling are given as graph presentations as indicated. Please refer to the main text for details.

genes such as *c-myc* and *Cyclin D1* that are responsible for driving the cell cycle. In most cases, a prerequisite for cell cycle progression is an increase in cell size. Cell growth requires the synthesis of proteins that in turn depends on an increased ribosome amount. In this scenario, the Wnt/β-catenin target gene, *c-myc*, regulates cell growth through upregulating the transcriptional activity of all three RNA polymerases (RNA pol I–III), resulting in an increase of ribosome number. Another major signaling pathway regulating cell growth represents mTOR (Zoncu, Efeyan, and Sabatini, 2011). Interestingly, Wnt signaling has been shown to activate mTOR signaling by two different mechanisms. In both cases, this occurs through a β-catenin-independent signaling branch (Figure 6.4).

GSK3β turned out to be an inhibitor of mTOR signaling. In this process, GSK3β phosphorylates tuberous sclerosis 2, leading to mTOR inhibition through the small GTPase Rheb. Thus, Wnt-stimulated inhibition of GSK3β activates mTOR and its downstream targets. This Wnt-mediated activation of mTOR results in cell growth as demonstrated by an increase in cell size (Inoki *et al.*, 2006). Of note, this does not require the involvement of β-catenin. In another study, the authors could show that Fzd7 interacts with Gαs and PI3K, activating Akt/mTOR signaling. Akt is also known to negatively regulate GSK3β. Whether these described mechanisms therefore overlap mechanistically remains to be analyzed. In differentiated myofibers, the Wnt/TOR pathway results in hypertrophy (von Maltzahn, Bentzinger, and Rudnicki, 2012).

Outlook

This brief introduction into β-catenin-independent Wnt signaling branches indicates the diversity of different intracellular signal transduction pathways initiated by Wnt proteins. It remains unclear, however, how these signaling branches overlap and how they are interconnected. A major unsolved question relates to how signaling specificity is achieved, which likely occurs at the cell membrane. Further work will also have to identify Wnt ligand/receptor combinations and how these combinations relate to different intracellular signaling events. Clearly, much work and exciting discoveries are ahead of us. It will be interesting to see which of these questions will be solved until we celebrate 50 years of Wnt research. Even more exciting will be the answers to those questions.

References

Adler, P.N. (2002) Planar signaling and morphogenesis in Drosophila. *Developmental Cell*, **2**, 525–535.

Ahumada, A., Slusarski, D.C., Liu, X. *et al.* (2002) Signaling of rat Frizzled-2 through phosphodiesterase and cyclic GMP. *Science*, **298**, 2006–2010.

Al-Ata, J., Paquet, M., and Teebi, A.S. (1998) Congenital heart disease in Robinow syndrome. *American Journal of Medical Genetics*, **77**, 332–333.

Bhanot, P., Brink, M., Samos, C.H. *et al.* (1996) A new member of the frizzled family from Drosophila functions as a Wingless receptor. *Nature*, **382**, 225–230.

Billiard, J., Way, D.S., Seestaller-Wehr, L.M. *et al.* (2005) The orphan receptor tyrosine kinase Ror2 modulates canonical Wnt signaling in osteoblastic cells. *Molecular Endocrinology*, **19**, 90–101.

Bryja, V., Schambony, A., Cajanek, L. *et al.* (2008) Beta-arrestin and casein kinase 1/2 define distinct branches of non-canonical WNT signalling pathways. *EMBO Report*, **9**, 1244–1250.

Burn, S.F., Webb, A., Berry, R.L. *et al.* (2011) Calcium/NFAT signalling promotes early nephrogenesis. *Developmental Biology*, **352**, 288–298.

Cai, Y., Lechner, M.S., Nihalani, D. *et al.* (2002) Phosphorylation of Pax2 by the c-Jun N-terminal kinase and enhanced Pax2-dependent transcription activation. *The Journal of Biological Chemistry*, **277**, 1217–1222.

Chen, A.E., Ginty, D.D., and Fan, C.M. (2005) Protein kinase A signalling via CREB controls myogenesis induced by Wnt proteins. *Nature*, **433**, 317–322.

Corbit, K.C., Shyer, A.E., Dowdle, W.E. *et al.* (2008) Kif3a constrains beta-catenin-dependent Wnt signalling through dual ciliary and non-ciliary mechanisms. *Nature Cell Biology*, **10**, 70–76.

Dabdoub, A., Donohue, M.J., Brennan, A. *et al.* (2003) Wnt signaling mediates reorientation of outer hair cell stereociliary bundles in the mammalian cochlea. *Development*, **130**, 2375–2384.

Dejmek, J., Safholm, A., Kamp Nielsen, C. *et al.* (2006) Wnt-5a/Ca^{2+}-induced NFAT activity is counteracted by Wnt-5a/Yes-Cdc42-casein kinase 1alpha signaling in human mammary epithelial cells. *Molecular Cell Biology*, **26**, 6024–6036.

Du, S.J., Purcell, S.M., Christian, J.L. *et al.* (1995) Identification of distinct classes and functional domains of Wnts through expression of wild-type and chimeric proteins in Xenopus embryos. *Molecular Cell Biology*, **15**, 2625–2634.

Endo, Y., Wolf, V., Muraiso, K. *et al.* (2005) Wnt-3a-dependent cell motility involves RhoA activation and is specifically regulated by dishevelled-2. *The Journal of Biological Chemistry*, **280**, 777–786.

Fradkin, L.G., Dura, J.M., and Noordermeer, J.N. (2010) Ryks: new partners for Wnts in the developing and regenerating nervous system. *Trends in Neuroscience*, **33**, 84–92.

Garriock, R.J. and Krieg, P.A. (2006) Wnt11-R signaling regulates a calcium sensitive EMT event essential for dorsal fin development of Xenopus. *Developmental Biology*, **304**, 127–140.

Garriock, R.J., D'Agostino, S.L., Pilcher, K.C., and Krieg, P.A. (2005) Wnt11-R, a protein closely related to mammalian Wnt11, is required for heart morphogenesis in Xenopus. *Developmental Biology*, **279**, 179–192.

Gerdes, J.M., Liu, Y., Zaghloul, N.A. *et al.* (2007) Disruption of the basal body compromises proteasomal function and perturbs intracellular Wnt response. *Nature Genetics*, **39**, 1350–1360.

Guo, N., Hawkins, C., and Nathans, J. (2004) Frizzled6 controls hair patterning in mice. *Proceedings of the National Academy of Sciences of the United States of America*, **101**, 9277–9281.

Habas, R., Kato, Y., and He, X. (2001) Wnt/Frizzled activation of Rho regulates vertebrate gastrulation and requires a novel Formin homology protein Daam1. *Cell*, **107**, 843–854.

Habas, R., Dawid, I.B., and He, X. (2003) Coactivation of Rac and Rho by Wnt/Frizzled signaling is required for vertebrate gastrulation. *Genes & Development*, **17**, 295–309.

Hansen, C., Howlin, J., Tengholm, A. *et al.* (2009) Wnt-5a-induced phosphorylation of DARPP-32 inhibits breast cancer cell migration in a CREB-dependent manner. *The Journal of Biological Chemistry*, **284**, 27533–27543.

Hildebrandt, F., Benzing, T., and Katsanis, N. (2011) Ciliopathies. *The New England Journal of Medicine*, **364**, 1533–1543.

Hovens, C.M., Stacker, S.A., Andres, A.C. *et al.* (1992) RYK, a receptor tyrosine kinase-related molecule with unusual kinase domain motifs. *Proceedings of the National Academy of Sciences of the United States of America*, **89**, 11818–11822.

Huang, P. and Schier, A.F. (2009) Dampened Hedgehog signaling but normal Wnt signaling in zebrafish without cilia. *Development*, **136**, 3089–3098.

Hutchins, B.I., Li, L., and Kalil, K. (2012) Wnt-induced calcium signaling mediates axon growth and guidance in the developing corpus callosum. *Science Signaling*, **5**, pt 1.

Inoki, K., Ouyang, H., Zhu, T. *et al.* (2006) TSC2 integrates Wnt and energy signals via a coordinated phosphorylation by AMPK and GSK3 to regulate cell growth. *Cell*, **126**, 955–968.

Inoue, T., Oz, H.S., Wiland, D. *et al.* (2004) C. elegans LIN-18 is a Ryk ortholog and functions in parallel to LIN-17/Frizzled in Wnt signaling. *Cell*, **118**, 795–806.

Ishitani, T., Ninomiya-Tsuji, J., Nagai, S.-I. *et al.* (1999) The TAK1-NLK-MAPK-related pathway antagonizes signalling between beta-catenin and transcription factor TCF. *Nature*, **399**, 798–802.

Ishitani, T., Kishida, S., Hyodo-Miura, J. *et al.* (2003) The TAK1-NLK mitogen-activated protein kinase cascade functions in the Wnt-5a/Ca(2+) pathway to antagonize Wnt/beta-catenin signaling. *Molecular Cell Biology*, **23**, 131–139.

Jenei, V., Sherwood, V., Howlin, J. *et al.* (2009) A t-butyloxycarbonyl-modified Wnt5a-derived hexapeptide functions as a potent antagonist of Wnt5a-dependent melanoma cell invasion. *Proceedings of the National Academy of Sciences of the United States of America*, **106**, 19473–19478.

Katso, R.M., Russell, R.B., and Ganesan, T.S. (1999) Functional analysis of H-Ryk, an atypical member of the receptor tyrosine kinase family. *Molecular Cell Biology*, **19**, 6427–6440.

Kestler, H.A. and Kuhl, M. (2008) From individual Wnt pathways towards a Wnt signalling network. *Philosophical Transactions of the Royal Society of London. Series B, Biological Sciences*, **363**, 1333–1347.

Kestler, H.A. and Kuhl, M. (2011) Generating a Wnt switch: it's all about the right dosage. *The Journal of Cell Biology*, **193**, 431–433.

Kishida, S., Yamamoto, H., and Kikuchi, A. (2004) Wnt-3a and Dvl induce neurite retraction by activating Rho-associated kinase. *Molecular Cell Biology*, **24**, 4487–4501.

Koval, A. and Katanaev, V.L. (2011) Wnt3a stimulation elicits G-protein-coupled receptor properties of mammalian Frizzled proteins. *The Biochemical Journal*, **433**, 435–440.

Koval, A., Purvanov, V., Egger-Adam, D., and Katanaev, V.L. (2011) Yellow submarine of the Wnt/Frizzled signaling: submerging from the G protein harbor to the targets. *Biochemical Pharmacology*, **82**, 1311–1319.

Kremenevskaja, N., von Wasielewski, R., Rao, A.S. *et al.* (2005) Wnt-5a has tumor suppressor activity in thyroid carcinoma. *Oncogene*, **24**, 2144–2154.

Kuhl, M., Sheldahl, L.C., Malbon, C.C., and Moon, R.T. (2000a) Ca(2+)/calmodulin-dependent protein kinase II is stimulated by Wnt and Frizzled homologs and promotes ventral cell fates in Xenopus. *The Journal of Biological Chemistry*, **275**, 12701–12711.

Kuhl, M., Sheldahl, L.C., Park, M. *et al.* (2000b) The Wnt/Ca^{2+} pathway: a new vertebrate Wnt signaling pathway takes shape. *Trends in Genetics*, **16**, 279–283.

Kuhl, M., Geis, K., Sheldahl, L.C. *et al.* (2001) Antagonistic regulation of convergent extension movements in Xenopus by Wnt/beta-catenin and Wnt/Ca^{2+} signaling. *Mechanisms of Development*, **106**, 61–76.

Lancaster, M.A., Schroth, J., and Gleeson, J.G. (2011) Subcellular spatial regulation of canonical Wnt signalling at the primary cilium. *Nature Cell Biology*, **13**, 700–707.

Li, G. and Iyengar, R. (2002) Calpain as an effector of the Gq signaling pathway for inhibition of Wnt/beta-catenin-regulated cell proliferation. *Proceedings of the National Academy of Sciences of the United States of America*, **99**, 13254–13259.

Liang, H., Chen, Q., Coles, A.H. *et al.* (2003) Wnt5a inhibits B cell proliferation and functions as a tumor suppressor in hematopoietic tissue. *Cancer Cell*, **4**, 349–360.

Liang, H., Coles, A.H., Zhu, Z. *et al.* (2007) Noncanonical Wnt signaling promotes apoptosis in thymocyte development. *Journal of Experimental Medicine*, **204**, 3077–3084.

Lienkamp, S., Ganner, A., and Walz, G. (2012) Inversin, Wnt signaling and primary cilia. *Differentiation*, **83**, S49–S55.

Liu, Y., Rubin, B., Bodine, P.V., and Billiard, J. (2008) Wnt5a induces homodimerization and activation of Ror2 receptor tyrosine kinase. *Journal of Cell Biochemistry*, **105**, 497–502.

Lu, W., Yamamoto, V., Ortega, B., and Baltimore, D. (2004) Mammalian Ryk is a Wnt coreceptor required for stimulation of neurite outgrowth. *Cell*, **119**, 97–108.

Marlow, F., Topczewski, J., Sepich, D., and Solnica-Krezel, L. (2002) Zebrafish Rho kinase 2 acts downstream of Wnt11 to mediate cell polarity and effective convergence and extension movements. *Current Biology*, **12**, 876–884.

Maurus, D., Heligon, C., Burger-Schwarzler, A. *et al.* (2005) Noncanonical Wnt-4 signaling and EAF2 are required for eye development in Xenopus laevis. *The EMBO Journal*, **24**, 1181–1191.

Mikels, A.J. and Nusse, R. (2006) Purified Wnt5a protein activates or inhibits beta-catenin-TCF signaling depending on receptor context. *PLoS Biology*, **4**, e115.

Minami, Y., Oishi, I., Endo, M., and Nishita, M. (2010) Ror-family receptor tyrosine kinases in noncanonical Wnt signaling: their implications in developmental morphogenesis and human diseases. *Developmental Dynamics*, **239**, 1–15.

Mirzadeh, Z., Han, Y.G., Soriano-Navarro, M. *et al.* (2010) Cilia organize ependymal planar polarity. *The Journal of Neuroscience*, **30**, 2600–2610.

Montcouquiol, M. and Kelley, M.W. (2003) Planar and vertical signals control cellular differentiation and patterning in the mammalian cochlea. *The Journal of Neuroscience*, **23**, 9469–9478.

Montcouquiol, M., Rachel, R.A., Lanford, P.J. *et al.* (2003) Identification of Vangl2 and Scrb1 as planar polarity genes in mammals. *Nature*, **423**, 173–177.

Nalesso, G., Sherwood, J., Bertrand, J. *et al.* (2011) WNT-3A modulates articular chondrocyte phenotype by activating both canonical and noncanonical pathways. *The Journal of Cell Biology*, **193**, 551–564.

Niehrs, C. (2012) The complex world of WNT receptor signalling. *Nature Reviews Molecular Cell Biology*, **13**, 767–779.

Nusse, R. and Varmus, H.E. (1982) Many tumors induced by the mouse mammary tumor virus contain a provirus integrated in the same region of the host genome. *Cell*, **31**, 99–109.

Ocbina, P.J., Tuson, M., and Anderson, K.V. (2009) Primary cilia are not required for normal canonical Wnt signaling in the mouse embryo. *PLoS One*, **4**, e6839.

Oh, E.C. and Katsanis, N. (2012) Cilia in vertebrate development and disease. *Development*, **139**, 443–448.

Oishi, I., Suzuki, H., Onishi, N. *et al.* (2003) The receptor tyrosine kinase Ror2 is involved in noncanonical Wnt5a/JNK signalling pathway. *Genes Cells*, **8**, 645–654.

Oishi, I., Kawakami, Y., Raya, A. *et al.* (2006) Regulation of primary cilia formation and left-right patterning in zebrafish by a noncanonical Wnt signaling mediator, duboraya. *Nature Genetics*, **38**, 1316–1322.

Pandur, P., Lasche, M., Eisenberg, L.M., and Kuhl, M. (2002) Wnt-11 activation of a non-canonical Wnt signalling pathway is required for cardiogenesis. *Nature*, **418**, 636–641.

Patthy, L. (2000) The WIF module. *Trends in Biochemical Sciences*, **25**, 12–13.

Rasmussen, J.T., Deardorff, M.A., Tan, C. *et al.* (2001) Regulation of eye development by frizzled signaling in Xenopus. *Proceedings of the National Academy of Sciences of the United States of America*, **98**, 3861–3866.

Rosso, S.B., Sussman, D., Wynshaw-Boris, A., and Salinas, P.C. (2005) Wnt signaling through Dishevelled, Rac and JNK regulates dendritic development. *Nature Neuroscience*, **8**, 34–42.

Saneyoshi, T., Kume, S., Amasaki, Y., and Mikoshiba, K. (2002) The Wnt/calcium pathway activates NF-AT and promotes ventral cell fate in Xenopus embryos. *Nature*, **417**, 295–299.

Schambony, A. and Wedlich, D. (2007) Wnt-5A/Ror2 regulate expression of XPAPC through an alternative noncanonical signaling pathway. *Developmental Cell*, **12**, 779–792.

Seifert, J.R. and Mlodzik, M. (2007) Frizzled/PCP signalling: a conserved mechanism regulating cell polarity and directed motility. *Nature Reviews Genetics*, **8**, 126–138.

Sheldahl, L.C., Park, M., Malbon, C.C., and Moon, R.T. (1999) Protein kinase C is differentially stimulated by Wnt and Frizzled homologs in a G-protein-dependent manner. *Current Biology*, **9**, 695–698.

Simons, M. and Mlodzik, M. (2008) Planar cell polarity signaling: from fly development to human disease. *Annual Review of Genetics*, **42**, 517–540.

Slusarski, D.C., Corces, V.G., and Moon, R.T. (1997) Interaction of Wnt and a Frizzled homologue triggers G-protein-linked phosphatidylinositol signalling. *Nature*, **390**, 410–413.

Slusarski, D.C., Yang-Snyder, J., Busa, W.B., and Moon, R.T. (1997) Modulation of embryonic intracellular Ca^{2+} signaling by Wnt-5A. *Developmental Biology*, **182**, 114–120.

Tanegashima, K., Zhao, H., and Dawid, I.B. (2008) WGEF activates Rho in the Wnt-PCP pathway and controls convergent extension in Xenopus gastrulation. *The EMBO Journal*, **27**, 606–617.

Tanigawa, S., Wang, H., Yang, Y. *et al.* (2011) Wnt4 induces nephronic tubules in metanephric mesenchyme by a non-canonical mechanism. *Developmental Biology*, **352**, 58–69.

Torii, K., Nishizawa, K., Kawasaki, A. *et al.* (2008) Anti-apoptotic action of Wnt5a in dermal fibroblasts is mediated by the PKA signaling pathways. *Cell Signaling*, **20**, 1256–1266.

Unterseher, F., Hefele, J.A., Giehl, K. *et al.* (2004) Paraxial protocadherin coordinates cell polarity during convergent extension via Rho A and JNK. *The EMBO Journal*, **23**, 3259–3269.

Vinson, C.R. and Adler, P.N. (1987) Directional non-cell autonomy and the transmission of polarity information by the frizzled gene of Drosophila. *Nature*, **329**, 549–551.

von Maltzahn, J., Bentzinger, C.F., and Rudnicki, M.A. (2012) Wnt7a–Fzd7 signalling directly activates the Akt/mTOR anabolic growth pathway in skeletal muscle. *Nature Cell Biology*, **14**, 186–191.

Wallingford, J.B. (2010) Planar cell polarity signaling, cilia and polarized ciliary beating. *Current Opinion in Cell Biology*, **22**, 597–604.

Wallingford, J.B. and Mitchell, B. (2011) Strange as it may seem: the many links between Wnt signaling, planar cell polarity, and cilia. *Genes & Development*, **25**, 201–213.

Wang, H.Y., Liu, T., and Malbon, C.C. (2006) Structure–function analysis of Frizzleds. *Cell Signaling*, **18**, 934–941.

Weeraratna, A.T., Jiang, Y., Hostetter, G. *et al.* (2002) Wnt5a signaling directly affects cell motility and invasion of metastatic melanoma. *Cancer Cell*, **1**, 279–288.

Westfall, T.A., Brimeyer, R., Twedt, J. *et al.* (2003) Wnt-5/pipetail functions in vertebrate axis formation as a negative regulator of Wnt/beta-catenin activity. *The Journal of Cell Biology*, **162**, 889–898.

Westfall, T.A., Hjertos, B., and Slusarski, D.C. (2003) Requirement for intracellular calcium modulation in zebrafish dorsal-ventral patterning. *Developmental Biology*, **259**, 380–391.

Witte, F., Bernatik, O., Kirchner, K. *et al.* (2010) Negative regulation of Wnt signaling mediated by CK1-phosphorylated Dishevelled via Ror2. *The FASEB Journal*, **24**, 2417–2426.

Wouda, R.R., Bansraj, M.R., de Jong, A.W. *et al.* (2008) Src family kinases are required for WNT5 signaling through the Derailed/RYK receptor in the Drosophila embryonic central nervous system. *Development*, **135**, 2277–2287.

Wu, X., Tu, X., Joeng, K.S. *et al.* (2008) Rac1 activation controls nuclear localization of beta-catenin during canonical Wnt signaling. *Cell*, **133**, 340–353.

Yamamoto, S., Nishimura, O., Misaki, K. *et al.* (2008) Cthrc1 selectively activates the planar cell polarity pathway of Wnt signaling by stabilizing the Wnt–receptor complex. *Developmental Cell*, **15**, 23–36.

Yoshikawa, S., Bonkowsky, J.L., Kokel, M. *et al.* (2001) The derailed guidance receptor does not require kinase activity in vivo. *The Journal of Neuroscience*, **21**, RC119.

Yoshikawa, S., McKinnon, R.D., Kokel, M., and Thomas, J.B. (2003) Wnt-mediated axon guidance via the Drosophila Derailed receptor. *Nature*, **422**, 583–588.

Zoncu, R., Efeyan, A., and Sabatini, D.M. (2011) mTOR: from growth signal integration to cancer, diabetes and ageing. *Nature Reviews Molecular Cell Biology*, **12**, 21–35.

7 Molecular Mechanisms of Wnt Pathway Specificity

Alexandra Schambony[1], Guido J.R. Zaman[2], and Folkert Verkaar[3]

[1]Developmental Biology, Biology Department, Friedrich-Alexander University Erlangen-Nuremberg, Erlangen, Germany
[2]Netherlands Translational Research Center B.V. (NTRC), Oss, The Netherlands
[3]Department of Chemistry and Pharmaceutical Sciences, VU University Amsterdam, Amsterdam, The Netherlands

Early Wnt classifications

Previous chapters have described the signaling components of the canonical (Chapters 2–5) and noncanonical (Chapter 6) Wnt signaling pathways. But what determines whether a Wnt protein will activate canonical or noncanonical signaling? This chapter will focus on the molecular determinants behind Wnt pathway specificity.

Early classifications of Wnt proteins were based on their ability to elicit formation of a secondary axis upon overexpression in *Xenopus laevis* embryos (Du *et al.*, 1995; McMahon and Moon, 1989) or to induce morphological transformation when transfected into murine mammary tumor or fibroblast cells (Bradbury *et al.*, 1994; Brown *et al.*, 1986; Wong, Gavin, and McMahon, 1994). Such studies led to a rough division into a class of transforming, axis-inducing Wnts, spearheaded by the prototypical members Wnt1 and Wnt3a, and a group including Wnt4, Wnt5a, and Wnt11 that were inactive in these assays. The activity of Wnts in these assays was traced down to the ability to activate the Wnt/β-catenin signaling pathway (Shimizu *et al.*, 1997). Wnts that were inactive in the said assays have since been found to initiate noncanonical signaling (Chapter 6). We know now that such crude classifications of Wnts into a canonical and noncanonical class bear only limited predictive power for their signaling behavior. For instance, Wnt11 governs dorsoventral specification in early *Xenopus* embryogenesis through β-catenin signaling, while mediating noncanonical convergent extension movements at later developmental stages (Cha *et al.*, 2008, 2009; Tao *et al.*, 2005). Likewise, the prototypical "noncanonical" Wnt5a has been implicated in multiple signaling cascades. We will briefly describe the role of Wnt5a in these pathways to illustrate that cellular responses to a given Wnt protein are by no means set in stone.

Wnt5a exemplifies the ability of Wnts to activate multiple signaling cascades

Wnt5a was initially implicated in calcium signaling when it was found that injection of Wnt5a mRNA induced transient release of intracellular calcium ions in zebrafish embryos overexpressing rat Frizzled2 (Fz2) (Slusarski,

Wnt Signaling in Development and Disease: Molecular Mechanisms and Biological Functions, First Edition.
Edited by Stefan Hoppler and Randall T. Moon.

Corces, and Moon, 1997). Wnt5a-induced Ca^{2+} signaling led to the activation of protein kinase C (PKC) and Ca^{2+}/calmodulin-dependent kinase (CamKII) (Kühl *et al.*, 2000), cell division control protein 42 (cdc42) (Choi and Han, 2002), calcineurin, and nuclear factor of activated T-cells (NFAT) (Saneyoshi *et al.*, 2002). Direct pharmacological evidence for Wnt5a/Fz2-mediated Ca^{2+} signaling was provided by Ma and Wang, who noted that recombinant Wnt5a-elicited Ca^{2+} mobilization and NFAT reporter gene activity in murine F9 cells overexpressing rat Fz2 (Ma and Wang, 2006).

Wnt5a is not only a ligand of Frizzled (Fz) receptors but also of the receptor tyrosine kinase Ror2. Like Fz, Ror2 contains a cysteine-rich domain (CRD) that binds Wnt5a. Initial evidence for this interaction was based on genetic studies in mice and subsequent functional assays in cell lines (Nishita *et al.*, 2006, 2010; Oishi *et al.*, 2003; Schambony and Wedlich, 2007). Additionally, mutations in both Wnt5a and Ror2 can cause Robinow syndrome, a disorder characterized by short-limbed dwarfism, hemivertebrae, and genital defects (Afzal *et al.*, 2000; Person *et al.*, 2010; van Bokhoven *et al.*, 2000).

The signaling pathways downstream of Wnt5a/Ror2 have not yet been well defined, but current evidence suggests at least two pathways are activated. First, Wnt5a and Ror2 are part of what appears to be a *bona fide* receptor tyrosine kinase-activated mitogen-activated protein (MAP) kinase cascade that is required for the transcriptional regulation of the paraxial protocadherin XPAPC in *Xenopus* (Feike *et al.*, 2010; Schambony and Wedlich, 2007). Wnt5a-induced homodimerization and autophosphorylation of Ror2, which are typical prerequisites for receptor tyrosine kinase signaling, have also been observed in cell culture models (Liu *et al.*, 2007; Mikels, Minami, and Nusse, 2009), further supporting these findings. Second, Wnt5a/Ror2 signaling might be implicated in planar cell polarity (PCP) (Gao *et al.*, 2011; Yamamoto *et al.*, 2008), a pathway that directs cellular polarization over a common axis (Chapter 6). It should be noted, though, that by themselves Ror2- and Wnt5a null mice do not display the strong defects typically associated with knockout of core PCP components (Gao *et al.*, 2011; Qian *et al.*, 2007; R. van Amerongen, personal communication), and the contribution of Ror2 and Wnt5a to mammalian PCP remains incompletely understood.

Finally, Wnt5a has also been shown to activate Wnt/β-catenin signaling. For instance, Wnt5a activates β-catenin signaling in HEK293 cells overexpressing Fz4 and LRP5 (Mikels and Nusse, 2006; Verkaar *et al.*, 2009) and in *Xenopus* embryos coexpressing Wnt5a and *Xenopus* Fz4, Fz7, or human Fz5 (He *et al.*, 1997; Holmen *et al.*, 2002; Umbhauer *et al.*, 2000). Furthermore, fusion proteins composed of human Fz5 N-terminally extended with either Wnt5a or Wnt11 activated β-catenin-dependent genes and induced a secondary body axis in *Xenopus* embryos (Holmen *et al.*, 2002). In fact, maternal (endogenously expressed) Wnt5a mediates early body axis specification through stabilization of β-catenin (Cha *et al.*, 2008, 2009). There is also evidence for activated β-catenin signaling in the calvarial mesenchyme of mice conditionally overexpressing Wnt5a (Van Amerongen *et al.*, 2012). Finally, primary cells from acute myeloid leukemia patients have been shown to respond to Wnt5a treatment with a rise in Fz4-mediated β-catenin signaling (Despeaux *et al.*, 2012).

Whereas the aforementioned might give the impression that distinct branches of Wnt signaling exist relatively autonomously, Wnt pathways are intimately intertwined. This is perhaps best illustrated by the seminal study by Moon and colleagues, who demonstrated that the Wnt5a class of noncanonical Wnts could inhibit β-catenin signaling through canonical Wnts (Torres *et al.*, 1996). In the following paragraph, we will summarize our current understanding of the mechanisms behind this.

Wnt5a-mediated inhibition of β-catenin signaling has been suggested to be induced by the upregulation of the E3 ubiquitin ligase Siah2, leading to GSK3-independent proteasomal degradation of β-catenin (MacLeod, Hayes, and Pacheco, 2007; Topol *et al.*, 2003). Others provided evidence that Wnt5a inhibits Wnt3a-dependent β-catenin accumulation through competitive binding to the CRD of Fz (Grumolato *et al.*, 2010; Sato *et al.*, 2010). Indeed, Wnt3a and Wnt5a bind to many of the same Fz CRDs *in vitro* (Bourhis *et al.*, 2010; Grumolato *et al.*, 2010; Sato *et al.*, 2010; Yamamoto *et al.*, 2008). However, Wnt5a has also been demonstrated to inhibit β-catenin in colon cancer cell lines that harbor activating mutations in β-catenin (Li *et al.*, 2010; Smit *et al.*, 2004; Topol *et al.*, 2003; Ying *et al.*, 2008), a phenomenon competitive Fz binding cannot account for.

Some evidence suggests that Wnt5a inhibits Wnt/β-catenin signaling through noncanonical signaling pathways. Wnt5a-induced Ca^{2+} signaling has been suggested to lead to the activation of a MAP kinase pathway consisting of CamKII, TGF-β-activated kinase 1 (TAK1), and Nemo-like kinase (NLK) in HEK293 cells (Ishitani, Ninomiya-Tsuji, and Matsumoto, 2003; Ishitani *et al.*, 1999, 2003). Activated NLK can phosphorylate TCF family members, thereby inhibiting association of β-catenin/TCF to TCF target sequences on DNA (Ishitani, Ninomiya-Tsuji, and Matsumoto, 2003; Ishitani *et al.*, 1999) (see Chapter 17). This model is enticing given a similar regulatory role of TAK1 and NLK homologs in β-catenin signaling in *C. elegans* (Meneghini *et al.*, 1999; Rocheleau *et al.*, 1999; Shin *et al.*, 1999). Furthermore, it has been shown that CamKII can directly phosphorylate lymphoid enhancer factor 1 (LEF1) *in vitro*, potentially inhibiting Wnt/β-catenin signaling downstream of the β-catenin destruction complex in *Xenopus* embryos (Kühl *et al.*, 2001). In contrast, other investigators found no evidence of Wnt5a-mediated TAK1/NLK activation in HEK293 cells (Smit *et al.*, 2004). Alternatively, Wnt5a/Ca^{2+} signaling might interfere with β-catenin signaling *via* PKC. PKCα can activate retinoic acid-related orphan receptor α (RORα), which was shown to compete with coactivator proteins for binding to β-catenin, thereby inhibiting β-catenin-dependent transcription (Lee *et al.*, 2010).

Arguably, the best genetic evidence exists for Wnt5a/Ror2-mediated inhibition of Wnt/β-catenin signaling. Both *Wnt5a*$^{-/-}$ and *Ror2*$^{-/-}$ mice possess regions of elevated β-catenin activity, indicative of derepression of β-catenin signaling through the genetic ablation of these proteins (Mikels, Minami, and Nusse, 2009; Sato *et al.*, 2010; Topol *et al.*, 2003). *In vitro*, activation of Ror2 by Wnt5a has been suggested to inhibit Wnt3a-induced β-catenin signaling downstream of nuclear translocation of β-catenin (Kurayoshi *et al.*, 2007; Mikels and Nusse, 2006; Verkaar *et al.*, 2010), but the intracellular pathway mediating this inhibition is incompletely understood. Ror2 does not seem to be implicated in Wnt5a-mediated inhibition of β-catenin signaling in all instances. For instance, Wnt5a-induced inhibition of β-catenin signaling appeared unaffected in HeLa cells transfected with siRNAs targeting Ror1/Ror2 (Sato *et al.*, 2010) as well as in *Ror1/Ror2*-deficient mouse embryonic fibroblasts (Ho *et al.*, 2012).

In summary, the verdict is not yet out on how Wnt5a inhibits β-catenin. Rather, we may have to consider the existence of multiple parallel or partly overlapping mechanisms of Wnt5a-mediated inhibition of Wnt/β-catenin signaling.

Multiple signaling routes through single Frizzled family members

Like Wnts, Fzs do not appear to be uniquely dedicated to a single signaling cascade. *Drosophila* Fz1 provides a striking example of such pathway bifunctionality, because it can mediate both β-catenin signaling and PCP (Bhanot *et al.*, 1999; Boutros *et al.*, 2000; Rulifson, Wu, and Nusse, 2000). An equally compelling vertebrate example is Fz7, which regulates convergent extension movements and dorsoventral patterning during *Xenopus* embryogenesis (Djiane *et al.*, 2000; Medina, Reintsch, and Steinbeisser, 2000; Sumanas *et al.*, 2000; Winklbauer *et al.*, 2001), processes that have been attributed to noncanonical and Wnt/β-catenin signaling, respectively. Furthermore, there is evidence for both noncanonical ROR2-mediated signaling and β-catenin signaling through Fz7 in cell lines (Carron *et al.*, 2003; Li *et al.*, 2008; Liu, Bafico, and Aaronson, 2005; Nishita *et al.*, 2010). Similarly, mammalian Fz2 has been implicated in Wnt/Ca^{2+} signaling (Slusarski, Corces, and Moon, 1997) and ROR2-dependent JNK signaling (Sato *et al.*, 2010), but it can also activate β-catenin signaling in heterologous overexpression studies (Cong, Schweizer, and Varmus, 2004a; Liu, Bafico, and Aaronson, 2005; Verkaar *et al.*, 2009). Additionally, zebrafish Fz2 morphants display phenotypes reminiscent of defects in convergent extension, whereas Fz2 overexpression induces hyperdorsalization and secondary axis formation in zebrafish embryos, indicative of

Wnt/β-catenin signaling (Kilian *et al.*, 2003; Sumanas *et al.*, 2001).

These examples illustrate a common theme, namely, that evidence for signaling through multiple pathways seems to exist for nearly all mammalian Fz receptor family members.

Determinants of Wnt pathway specificity

The involvement of single Wnt and Fz family members in multiple signaling cascades begs the question of how signaling specificity between pathways is attained. Recent years have seen the continual refinement of models to explain this fascinating phenomenon (Angers and Moon, 2009; Van Amerongen, 2012; Van Amerongen and Nusse, 2009; Verkaar and Zaman, 2010).

A first layer of signaling specificity is imposed by the tight spatial and temporal control on the expression of Wnts, receptors, and accessory proteins *in vivo*. Due to the near-infinite potential interactions governed by the Wnt signaling system, it has proven difficult to unambiguously deduce Wnt–receptor pairs from expression studies, at least in vertebrates, and Wnt researchers have had to resort to cell systems to build models for Wnt pathway specificity (Yu *et al.*, 2012).

From such studies, it has been argued that signaling specificity is, in part, determined by the specificity of Fz–Wnt binding (Figure 7.1a). The recently published crystal structure of *Xenopus* Wnt8 in complex with Fz8 CRD sheds some light on how these proteins interact by revealing that Wnt binds to the Fz CRD through two distinct sites. One amino-terminal site on Wnt8, containing a serine-linked lipid modification, appears to mainly ensure sufficient affinity, whereas a carboxy-terminal site determines Wnt/Fz binding specificity. Accordingly, a "mini-Wnt" consisting only of this second binding region is able to discriminate between different Fz CRDs (Janda *et al.*, 2012). The high sequence conservation of Fzs and Wnts in this second site is unlikely to allow for highly restrictive Wnt/Fz pairing (Figure 7.1b), which is in line with earlier biochemical studies that overall noted only a very limited degree of specificity in Wnt/Fz binding (Bourhis *et al.*, 2010; Hsieh *et al.*, 1999; Rulifson, Wu, and Nusse, 2000; Wu and Nusse, 2002). Clearly, there are additional determinants of Wnt pathway specificity.

The propensity of a given Fz protein to signal *via* a certain pathway is most likely a combination of its ability to bind Wnt proteins (or accessory proteins) with specific signaling preferences (see subsequent sections) and intrinsic features of the Fz proteins themselves. The ability of *Drosophila* Fz1 and Fz2 to preferentially activate PCP and β-catenin (Armadillo in *Drosophila*) signaling, respectively, has been attributed to their C-termini, which direct these receptors to different plasma membrane compartments in the fly imaginal disc (apical vs. basolateral for DFz1 and DFz2, respectively) (Boutros *et al.*, 2000; Wu, Klein, and Mlodzik, 2004). However, another study has suggested that pathway specificity of these receptors is explained by their differential ability to bind the *Drosophila* Wnt ligand Wingless (Rulifson, Wu, and Nusse, 2000). Umbhauer and colleagues argued that a C-terminal motif in Fzs (KTxxxW, where x is any amino acid) is essential for the propagation of β-catenin signaling (Umbhauer *et al.*, 2000). However, since all human Fzs contain this sequence, it is unlikely to direct Fz signaling specificity. A recent article describes a motif in the third intracellular loop of Fz receptors that is essential for β-catenin signaling and is only present in Fzs that preferentially activate β-catenin signaling (Tauriello *et al.*, 2012). Future studies will have to determine whether this motif is involved in directing signaling to a given pathway. Finally, individual Fz family members have been demonstrated to form homo- and hetero-oligomers (Kaykas *et al.*, 2004). Heterodimerization of related seven-transmembrane (G protein-coupled) receptors has been shown to markedly influence receptor signaling (Vischer *et al.*, 2011). In this light, it is enticing to speculate that Fz heterodimerization may direct pathway specificity, although direct evidence of this is lacking.

Wnt pathway specificity appears to be in large part determined by the differential ability of Wnts to bind to coreceptors. In this view, formation of a Wnt–Fz–LRP5/6 complex would lead to initiation of β-catenin signaling, whereas recruitment of other coreceptors to Wnt/Fz complexes would trigger noncanonical signaling pathways (Figure 7.1c, e, and g). Such receptor complexes have been described for Wnt, Fz, and LRP6 during β-catenin signaling

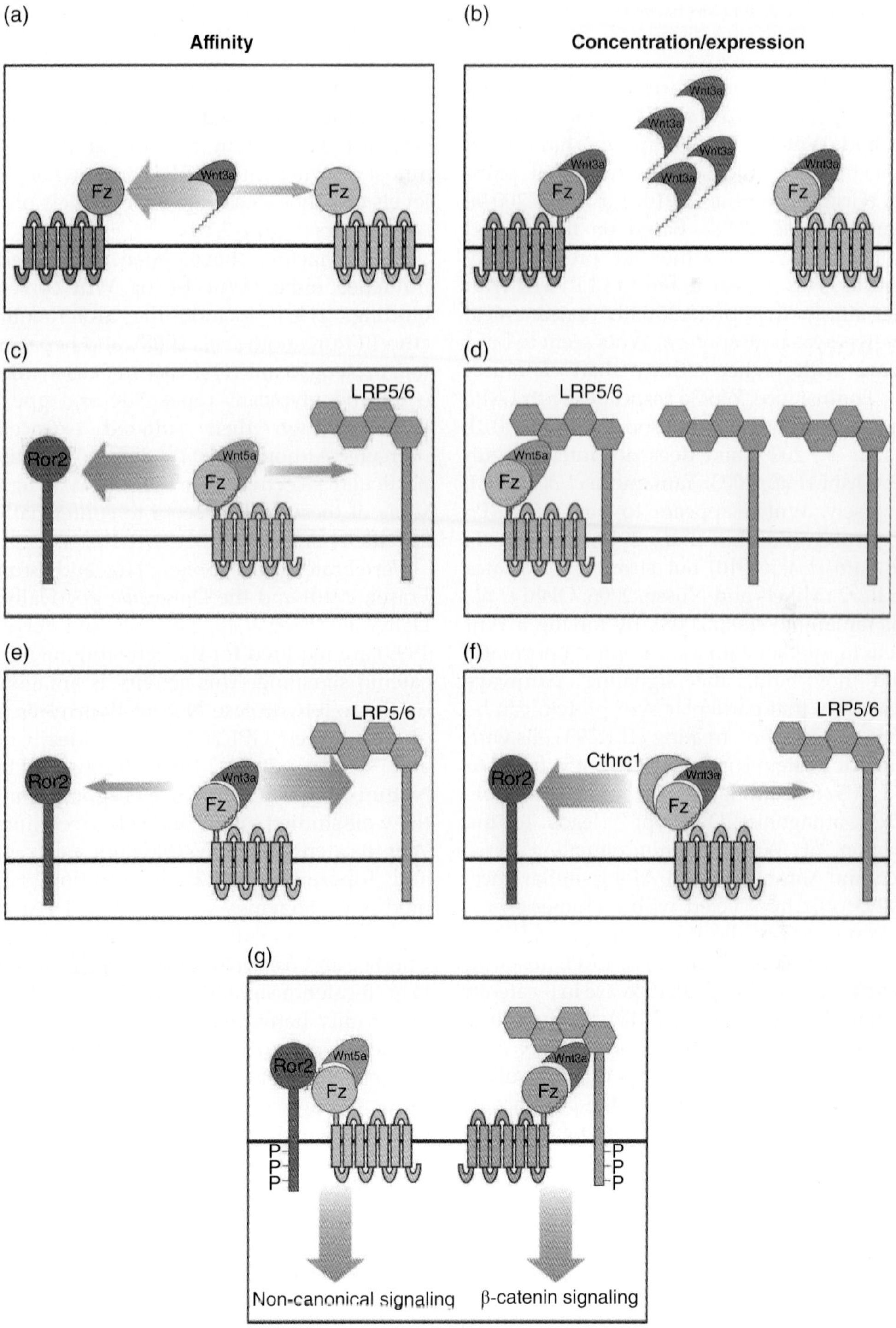

Figure 7.1 Multiple layers of Wnt pathway specificity. Wnt pathway specificity is dependent on affinities between the Wnt signaling components (left panels, i.e., a, c, and e) and the expression levels or concentrations of these components (right panels, i.e., b, d, and f). Wnt proteins differentially associate with Fz receptor family members depending on Wnt–Fz affinities (a) and Fz expression levels and Wnt concentrations (b). Furthermore, Wnt proteins bind coreceptor molecules with different affinity. For instance, Wnt5a preferentially interacts with Ror2 (c), unless LRP5/6 levels greatly exceed Ror2 levels (d). Conversely, Wnt3a has greater affinity for LRP5/6 than for Ror2 (e) but can interact with Ror2 in the presence of additional factors, such as Cthrc1 (f). In general (g), Wnt–Fz–LRP5/6 complexes transduce β-catenin signaling, whereas complexes between Wnt, Fz, and Ror2 mediate noncanonical signaling pathways. *(See insert for color representation of the figure.)*

(Cong, Schweizer, and Varmus, 2004a; Tamai *et al.*, 2000; Yamamoto *et al.*, 2008), and Wnt–Fz–Ror2 and Wnt–Fz–Ryk complexes have been demonstrated to mediate noncanonical pathways (Grumolato *et al.*, 2010; Lu *et al.*, 2004a; Yamamoto *et al.*, 2008). Based on the limited amount of Wnts tested thus far, preferentially canonical Wnts appear to bind to LRP5/6 with higher affinity than preferentially noncanonical Wnts, whereas noncanonical Wnts seem to bind to Ror2 with higher affinity than canonical Wnts. For instance, Wnt3a associates with LRP6 with nanomolar affinities (Bourhis *et al.*, 2010; Tamai *et al.*, 2000) and does not interact with Ror2 (Oishi *et al.*, 2003; Yamamoto *et al.*, 2008). Conversely, Wnt5a appears to bind to LRP6 very weakly (Bourhis *et al.*, 2010; Bryja *et al.*, 2009; Sato *et al.*, 2010) but strongly associates with Ror2 (Mikels and Nusse, 2006; Oishi *et al.*, 2003; Yamamoto *et al.*, 2008). By forcing a Wnt protein to interact with a coreceptor it normally would not bind, the signaling pathways activated by that particular Wnt protein can be altered. For instance, treating HEK293 cells with a chimeric protein consisting of Wnt5a fused to the LRP5/6 binding region of the soluble LRP5/6 antagonist Dickkopf-2 leads to the activation of Wnt/β-catenin signaling (Liu, Bafico, and Aaronson, 2005). Along similar lines, HEK293 cells transfected with a chimeric protein, composed of the Wnt5a-binding CRD of Ror2 and the intracellular tail of LRP6, respond to Wnt5a treatment with an increase in β-catenin signaling (Grumolato *et al.*, 2010). Furthermore, the otherwise canonical Wnt protein Wnt3a activates a PCP-like pathway that involves activation of Rac1 and RhoA in the presence of collagen triple helix repeat-containing protein 1 (cthrc1), a protein that enables Wnt3a to bind to Ror2 (Figure 7.1f) (Yamamoto *et al.*, 2008).

Importantly, whether a Wnt protein activates canonical or noncanonical signaling is not just dependent on its affinities for LRP5/6 and Ror2 but also on the expression levels of these coreceptors (Figure 7.1d). This point is best exemplified by the fact that HEK293 cells stably expressing Fz4 only activate β-catenin signaling upon Wnt5a treatment when they are transiently transfected with LRP5 (Mikels and Nusse, 2006; Verkaar *et al.*, 2009). This is striking, because endogenous levels of LRP5/6 suffice for Wnt/β-catenin signaling through Wnt3a in these cells but apparently are insufficient to allow Wnt5a-mediated β-catenin signaling. The simplest explanation for this is that Wnt5a binds and activates LRP5 only when LRP5 levels greatly exceed expression levels of other coreceptors (Figure 7.1d).

Many factors have been suggested to influence either Wnt/Fz or Wnt–coreceptor binding. These include the aforementioned cthrc1 (Yamamoto *et al.*, 2008), and heparin sulfate proteoglycans (HSPGs), such as syndecans (sdc) and glypicans (gpc). Sdc and gpc bind Wnts through their sulfated extracellular domains. Although *in vitro* they do not display particular selectivity for specific Wnt ligands, some of these HSPGs seem to confer pathway specificity *in vivo*.

Vertebrate gpc-1 (Shiau, Hu, and Bronner-Fraser, 2010) and the *Drosophila* gpc Dally and Dally-like (Baeg *et al.*, 2001; Lin and Perrimon, 1999) are required for the activation of Wnt/β-catenin signaling. This activity is antagonized by the α/β-hydrolase Notum that releases the gpc from their GPI anchor (Giráldez, Copley, and Cohen, 2002). Interestingly, zebrafish Notum-1a specifically cleaves gpc-1, but not the well-studied gpc-4, and selectively inhibits Wnt/β-catenin signaling (Flowers, Topczewska, and Topczewski, 2012). In addition, HSPG-modifying enzymes such as Qsulf1 (Ai *et al.*, 2003) and 2-*O*-sulfotransferase (Cadwalader, Condic, and Yost, 2012) specifically influence Wnt/β-catenin signaling *in vivo* by modulating the affinity between HSPGs and Wnts. Gpc-3 has been reported to play a role in both canonical and noncanonical Wnt signaling, although it needs to be processed to act in the Wnt/PCP pathway (Capurro *et al.*, 2005; De Cat *et al.*, 2003; Gao and Ho, 2011). In contrast, gpc-4 apparently exclusively promotes Wnt/PCP signaling in zebrafish and *Xenopus* development (Caneparo *et al.*, 2007; Ohkawara *et al.*, 2003; Topczewski *et al.*, 2001). Likewise, sdc-4 specifically activates noncanonical Wnt pathways including the PCP and Wnt/Ca^{2+} pathway in many vertebrates (Matthews *et al.*, 2008; Muñoz *et al.*, 2006; Ohkawara, Glinka, and Niehrs, 2011; Saoncella *et al.*, 2004). In summary, HSPGs add to the specificity of Wnt signaling *in vivo* by binding Wnts and facilitating or inhibiting ligand interaction with Fz receptors, coreceptors, and intracellular effector proteins.

Some earlier work has suggested that pathway selectivity not only derives from Wnt–receptor interactions but also from downstream events. For instance, Wnt signaling has been proposed to be determined by clathrin- versus caveolin-mediated endocytosis of Wnt–receptor complexes (Kikuchi, Yamamoto, and Sato, 2009) (see Chapter 8). This was based on evidence that Wnt/β-catenin signaling requires caveolin-dependent endocytosis of the Wnt–Fz–LRP5/6 receptor complex (Bilic *et al.*, 2007; Yamamoto, Komekado, and Kikuchi, 2006; Yamamoto *et al.*, 2008), whereas clathrin-mediated endocytosis mediates noncanonical Wnt pathways (Kim *et al.*, 2008; Yu *et al.*, 2007). More recent data suggest that the type of coreceptor, namely, LRP6 and Ror2, determines the mode of receptor endocytosis (Sakane *et al.*, 2012; Sato *et al.*, 2010), thus strengthening the model of coreceptors as major determinants of Wnt signaling specificity.

Others have suggested that the scaffolding protein Dishevelled (Dvl), which binds to Fz, forms a molecular switch between pathways (see Chapter 15). Indeed, Dvl has a prominent role in both canonical and noncanonical signaling pathways (Sheldahl *et al.*, 2003; Wang *et al.*, 2006; Zeng *et al.*, 2008). Phosphorylation of Dvl by casein kinase Iε (CKIε) has been hypothesized to direct signaling towards the β-catenin pathway at the expense of noncanonical signaling (Cong, Schweizer, and Varmus, 2004b; Witte *et al.*, 2010). On the other hand, an equally stimulatory role for CKIε in *Drosophila* PCP and β-catenin signaling has been noted in other studies (Klein *et al.*, 2006; Strutt, Price, and Strutt, 2006). The prominent role of Dvl in multiple Wnt signaling pathways can also be explained by assuming that canonical and noncanonical receptor complexes employ common activation mechanisms, including Dvl recruitment and glycogen synthase kinase 3-mediated coreceptor phosphorylation, as has recently been put forth (Grumolato *et al.*, 2010).

Perspectives

While the models discussed in this chapter are in agreement with most of the current data, there are notable exceptions. For instance, coreceptors do not appear uniquely dedicated to a single signaling pathway in all instances. "Noncanonical" Wnt receptors such as Ryk, Ror2, and protein tyrosine kinase 7 (PTK7) have all been shown to activate β-catenin signaling in some contexts (Billiard *et al.*, 2005; Li *et al.*, 2008; Lu *et al.*, 2004a; Puppo *et al.*, 2011; Winkel *et al.*, 2008). Furthermore, Ryk, Ror2, and PTK7 have all been implicated in multiple noncanonical pathways (Lin *et al.*, 2010; Lu *et al.*, 2004b; Macheda *et al.*, 2012; Wehner *et al.*, 2011). This suggests either that noncanonical pathways overlap considerably or that these receptors are not uniquely dedicated to a single noncanonical pathway.

Another anomalous finding is that the otherwise prototypically noncanonical Wnts Wnt5a and Wnt11 have been reported to induce β-catenin signaling as dimers or multimers during *Xenopus* axis specification (Cha *et al.*, 2008, 2009). Clearly, there are additional determinants of signaling specificity that are not captured by our models yet.

A defining step in our understanding of Wnt signaling may be heralded by the recent elucidation of the Wnt8–Fz8 cocrystal structure (Janda *et al.*, 2012), which provided structural information on Wnt/Fz binding specificity. Future crystallization efforts may one day structurally define complete Wnt–Fz–coreceptor complexes.

Despite 30 years of intense research, we have only begun to understand the mechanics behind Wnt pathway specificity. Much of the challenge in upcoming years will be to translate the findings from heterologous overexpression systems to the intact organism and to verify whether the current models we have based on a limited set of Wnt proteins and receptors hold true for the whole Wnt/Fz signaling system.

Acknowledgments

The authors thank Dr. Renée van Amerongen (NKI-AVL, the Netherlands) for critical reading of the manuscript.

References

Afzal, A.R., Rajab, A., Fenske, C.D. *et al.* (2000) Recessive Robinow syndrome, allelic to dominant brachydactyly type B, is caused by mutation of ROR2. *Nature Genetics*, **25**, 419–422.

Ai, X., Do, A.-T., Lozynska, O. *et al.* (2003) QSulf1 remodels the 6-O sulfation states of cell surface heparan sulfate proteoglycans to promote Wnt signaling. *The Journal of Cell Biology*, **162**, 341–351.

Angers, S. and Moon, R.T. (2009) Proximal events in Wnt signal transduction. *Nature Reviews. Molecular Cell Biology*, **10**, 468–477.

Baeg, G.H., Lin, X., Khare, N. *et al.* (2001) Heparan sulfate proteoglycans are critical for the organization of the extracellular distribution of Wingless. *Development*, **128**, 87–94.

Bhanot, P., Fish, M., Jemison, J.A. *et al.* (1999) Frizzled and Dfrizzled-2 function as redundant receptors for Wingless during Drosophila embryonic development. *Development*, **126**, 4175–4186.

Bilic, J., Huang, Y.L., Davidson, G. *et al.* (2007) Wnt induces LRP6 signalosomes and promotes dishevelled-dependent LRP6 phosphorylation. *Science*, **316**, 1619–1622.

Billiard, J., Way, D.S., Seestaller-Wehr, L.M. *et al.* (2005) The orphan receptor tyrosine kinase Ror2 modulates canonical Wnt signaling in osteoblastic cells. *Molecular Endocrinology*, **19**, 90–101.

Bourhis, E., Tam, C., Franke, Y. *et al.* (2010) Reconstitution of a frizzled8.Wnt3a.LRP6 signaling complex reveals multiple Wnt and Dkk1 binding sites on LRP6. *The Journal of Biological Chemistry*, **285**, 9172–9179.

Boutros, M., Mihaly, J., Bouwmeester, T., and Mlodzik, M. (2000) Signaling specificity by Frizzled receptors in Drosophila. *Science*, **288**, 1825–1828.

Bradbury, J.M., Niemeyer, C.C., Dale, T.C., and Edwards, P.A. (1994) Alterations of the growth characteristics of the fibroblast cell line C3H 10T1/2 by members of the Wnt gene family. *Oncogene*, **9**, 2597–2603.

Brown, A.M., Wildin, R.S., Prendergast, T.J., and Varmus, H.E. (1986) A retrovirus vector expressing the putative mammary oncogene int-1 causes partial transformation of a mammary epithelial cell line. *Cell*, **46**, 1001–1009.

Bryja, V., Andersson, E.R., Schambony, A. *et al.* (2009) The extracellular domain of Lrp5/6 inhibits noncanonical Wnt signaling in vivo. *Molecular Biology of the Cell*, **20**, 924–936.

Cadwalader, E.L., Condic, M.L., and Yost, H.J. (2012) 2-O-Sulfotransferase regulates Wnt signaling, cell adhesion and cell cycle during zebrafish epiboly. *Development*, **139**, 1296–1305.

Caneparo, L., Huang, Y.-L., Staudt, N. *et al.* (2007) Dickkopf-1 regulates gastrulation movements by coordinated modulation of Wnt/beta catenin and Wnt/PCP activities, through interaction with the Dally-like homolog Knypek. *Genes & Development*, **21**, 465–480.

Capurro, M.I., Shi, W., Sandal, S., and Filmus, J. (2005) Processing by convertases is not required for glypican-3-induced stimulation of hepatocellular carcinoma growth. *The Journal of Biological Chemistry*, **280**, 41201–41206.

Carron, C., Pascal, A., Djiane, A. *et al.* (2003) Frizzled receptor dimerization is sufficient to activate the Wnt/beta-catenin pathway. *Journal of Cell Science*, **116**, 2541–2550.

Cha, S.W., Tadjuidje, E., Tao, Q. *et al.* (2008) Wnt5a and Wnt11 interact in a maternal Dkk1-regulated fashion to activate both canonical and non-canonical signaling in Xenopus axis formation. *Development*, **135**, 3719–3729.

Cha, S.W., Tadjuidje, E., White, J. *et al.* (2009) Wnt11/5a complex formation caused by tyrosine sulfation increases canonical signaling activity. *Current Biology*, **19**, 1573–1580.

Choi, S.C. and Han, J.K. (2002) Xenopus Cdc42 regulates convergent extension movements during gastrulation through Wnt/Ca^{2+} signaling pathway. *Developmental Biology*, **244**, 342–357.

Cong, F., Schweizer, L., and Varmus, H. (2004a) Wnt signals across the plasma membrane to activate the beta-catenin pathway by forming oligomers containing its receptors, Frizzled and LRP. *Development*, **131**, 5103–5115.

Cong, F., Schweizer, L., and Varmus, H. (2004b) Casein kinase Iepsilon modulates the signaling specificities of dishevelled. *Molecular and Cellular Biology*, **24**, 2000–2011.

De Cat, B., Muyldermans, S.-Y., Coomans, C. *et al.* (2003) Processing by proprotein convertases is required for glypican-3 modulation of cell survival, Wnt signaling, and gastrulation movements. *The Journal of Cell Biology*, **163**, 625–635.

Despeaux, M., Chicanne, G., Rouer, E. *et al.* (2012) Focal adhesion kinase splice variants maintain primitive acute myeloid leukemia cells through altered Wnt signaling. *Stem Cells*, **30**, 1597–1610.

Djiane, A., Riou, J., Umbhauer, M. *et al.* (2000) Role of frizzled 7 in the regulation of convergent extension movements during gastrulation in Xenopus laevis. *Development*, **127**, 3091–3100.

Du, S.J., Purcell, S.M., Christian, J.L. *et al.* (1995) Identification of distinct classes and functional domains of Wnts through expression of wild-type and chimeric proteins in Xenopus embryos. *Molecular and Cellular Biology*, **15**, 2625–2634.

Feike, A.C., Rachor, K., Gentzel, M., and Schambony, A. (2010) Wnt5a/Ror2-induced upregulation of xPAPC requires xShcA. *Biochemical and Biophysical Research Communications*, **400**, 500–506.

Flowers, G.P., Topczewska, J.M., and Topczewski, J. (2012) A zebrafish Notum homolog specifically blocks the Wnt/β-catenin signaling pathway. *Development*, **139**, 2416–2425.

Gao, W. and Ho, M. (2011) The role of glypican-3 in regulating Wnt in hepatocellular carcinomas. *Journal of Molecular Recognition*, **1**, 14–19.

Gao, B., Song, H., Bishop, K. *et al.* (2011) Wnt signaling gradients establish planar cell polarity by inducing Vangl2 phosphorylation through Ror2. *Developmental Cell*, **20**, 163–176.

Giráldez, A.J., Copley, R.R., and Cohen, S.M. (2002). HSPG modification by the secreted enzyme Notum shapes the Wingless morphogen gradient. *Developmental Cell*, **2**, 667–676.

Grumolato, L., Liu, G., Mong, P. *et al.* (2010) Canonical and noncanonical Wnts use a common mechanism to activate completely unrelated coreceptors. *Genes & Development*, **24**, 2517–2530.

He, X., Saint-Jeannet, J.P., Wang, Y. *et al.* (1997) A member of the Frizzled protein family mediating axis induction by Wnt-5A. *Science*, **275**, 1652–1654.

Ho, H.Y., Susman, M.W., Bikoff, J.B. *et al.* (2012) Wnt5a-Ror-Dishevelled signaling constitutes a core developmental pathway that controls tissue morphogenesis. *Proceedings of the National Academy of Sciences United States of America*, **109**, 4044–4051.

Holmen, S.L., Salic, A., Zylstra, C.R. *et al.* (2002) A novel set of Wnt–Frizzled fusion proteins identifies receptor components that activate beta-catenin-dependent signaling. *The Journal of Biological Chemistry*, **277**, 34727–34735.

Hsieh, J.C., Rattner, A., Smallwood, P.M., and Nathans, J. (1999) Biochemical characterization of Wnt–Frizzled interactions using a soluble, biologically active vertebrate Wnt protein. *Proceedings of the National Academy of Sciences of the United States of America*, **96**, 3546–3551.

Ishitani, T., Ninomiya-Tsuji, J., and Matsumoto, K. (2003) Regulation of lymphoid enhancer factor 1/T-cell factor by mitogen-activated protein kinase-related Nemo-like kinase-dependent phosphorylation in Wnt/beta-catenin signaling. *Molecular and Cellular Biology*, **23**, 1379–1389.

Ishitani, T., Ninomiya-Tsuji, J., Nagai, S. *et al.* (1999) The TAK1-NLK-MAPK-related pathway antagonizes signalling between beta-catenin and transcription factor TCF. *Nature*, **399**, 798–802.

Ishitani, T., Kishida, S., Hyodo-Miura, J. *et al.* (2003) The TAK1-NLK mitogen-activated protein kinase cascade functions in the Wnt-5a/Ca(2+) pathway to antagonize Wnt/beta-catenin signaling. *Molecular and Cellular Biology*, **23**, 131–139.

Janda, C.Y., Waghray, D., Levin, A.M. *et al.* (2012) Structural basis of Wnt recognition by Frizzled. *Science*, **337**, 59–64.

Kaykas, A., Yang-Snyder, J., Heroux, M. *et al.* (2004) Mutant Frizzled-4 associated with vitreoretinopathy traps wild-type Frizzled in the endoplasmic reticulum by oligomerization. *Nature Cell Biology*, **6**, 52–58.

Kikuchi, A., Yamamoto, H., and Sato, A. (2009) Selective activation mechanisms of Wnt signaling pathways. *Trends in Cell Biology*, **19**, 119–129.

Kilian, B., Mansukoski, H., Barbosa, F.C. *et al.* (2003) The role of Ppt/Wnt5 in regulating cell shape and movement during zebrafish gastrulation. *Mechanisms of Development*, **120**, 467–476.

Kim, G.-H., Her, J.-H., and Han, J.-K. (2008) Ryk cooperates with Frizzled 7 to promote Wnt11-mediated endocytosis and is essential for *Xenopus laevis* convergent extension movements. *Journal of Cell Biology*, **182**, 1073–1082.

Klein, T.J., Jenny, A., Djiane, A., and Mlodzik, M. (2006) CKIepsilon/discs overgrown promotes both Wnt–Fz/beta-catenin and Fz/PCP signaling in Drosophila. *Current Biology*, **16**, 1337–1343.

Kühl, M., Sheldahl, L.C., Malbon, C.C., and Moon, R.T. (2000) Ca(2+)/calmodulin-dependent protein kinase II is stimulated by Wnt and Frizzled homologs and promotes ventral cell fates in Xenopus. *The Journal of Biological Chemistry*, **275**, 12701–12711.

Kühl, M., Geis, K., Sheldahl, L.C. *et al.* (2001) Antagonistic regulation of convergent extension movements in Xenopus by Wnt/beta-catenin and Wnt/Ca^{2+} signaling. *Mechanisms of Development*, **106**, 61–76.

Kurayoshi, M., Yamamoto, H., Izumi, S., and Kikuchi, A. (2007) Post-translational palmitoylation and glycosylation of Wnt-5a are necessary for its signaling. *The Biochemical Journal*, **402**, 515–523.

Lee, J.M., Kim, I.S., Kim, H. *et al.* (2010) RORalpha attenuates Wnt/beta-catenin signaling by PKCalpha-dependent phosphorylation in colon cancer. *Molecular Cell*, **37**, 183–195.

Li, C., Chen, H., Hu, L. *et al.* (2008) Ror2 modulates the canonical Wnt signaling in lung epithelial cells through cooperation with Fzd2. *BMC Molecular Biology*, **9**, 11.

Li, J., Ying, J., Fan, Y. *et al.* (2010) WNT5A antagonizes WNT/beta-catenin signaling and is frequently silenced by promoter CpG methylation in esophageal squamous cell carcinoma. *Cancer Biology and Therapy*, **10**, 617–624.

Lin, X. and Perrimon, N. (1999) Dally cooperates with Drosophila Frizzled 2 to transduce Wingless signalling. *Nature*, **400**, 281–284.

Lin, S., Baye, L.M., Westfall, T.A., and Slusarski, D.C. (2010) Wnt5b-Ryk pathway provides directional signals to regulate gastrulation movement. *Journal of Cell Biology*, **190**, 263–278.

Liu, G., Bafico, A., and Aaronson, S.A. (2005) The mechanism of endogenous receptor activation functionally distinguishes prototype canonical and noncanonical Wnts. *Molecular and Cellular Biology*, **25**, 3475–3482.

Liu, Y., Ross, J.F., Bodine, P.V.N., and Billiard, J. (2007) Homodimerization of Ror2 tyrosine kinase receptor induces 14-3-3(beta) phosphorylation and promotes osteoblast differentiation and bone formation. *Molecular Endocrinology*, **21**, 3050–3061.

Witte, F., Bernatik, O., Kirchner, K. *et al.* (2010) Negative regulation of Wnt signaling mediated by CK1-phosphorylated Dishevelled via Ror2. *The FASEB Journal*, **24**, 2417–2426.

Wong, G.T., Gavin, B.J., and McMahon, A.P. (1994) Differential transformation of mammary epithelial cells by Wnt genes. *Molecular and Cellular Biology*, **14**, 6278–6286.

Wu, C.H. and Nusse, R. (2002) Ligand receptor interactions in the Wnt signaling pathway in Drosophila. *The Journal of Biological Chemistry*, **277**, 41762–41769.

Wu, J., Klein, T.J., and Mlodzik, M. (2004) Subcellular localization of Frizzled receptors, mediated by their cytoplasmic tails, regulates signaling pathway specificity. *PLoS Biology*, **2**, e158.

Yamamoto, S., Komekado, H., and Kikuchi, A. (2006) Caveolin is necessary for Wnt-3a-dependent internalization of LRP6 and accumulation of beta-catenin. *Developmental Cell*, **11**, 213–223.

Yamamoto, S., Nishimura, O., Misaki, K. *et al.* (2008) Cthrc1 selectively activates the planar cell polarity pathway of Wnt signaling by stabilizing the Wnt–receptor complex. *Developmental Cell*, **15**, 23–36.

Ying, J., Li, H., Yu, J. *et al.* (2008) WNT5A exhibits tumor-suppressive activity through antagonizing the Wnt/beta-catenin signaling, and is frequently methylated in colorectal cancer. *Clinical Cancer Research*, **14**, 55–61.

Yu, A., Rual, J.F., Tamai, F. *et al.* (2007) Association of Dishevelled with the clathrin AP-2 adaptor is required for Frizzled endocytosis and planar cell polarity signaling. *Developmental Cell*, **12**, 129–141.

Yu, H., Ye, X., Guo, N., and Nathans, J. (2012) Frizzled 2 and frizzled 7 function redundantly in convergent extension and closure of the ventricular septum and palate: evidence for a network of interacting genes. *Development*, **139**, 4383–4394.

Zeng, X., Huang, H., Tamai, K. *et al.* (2008) Initiation of Wnt signaling: control of Wnt coreceptor Lrp6 phosphorylation/activation via frizzled, dishevelled and axin functions. *Development*, **135**, 367–375.

8 Modulation of Wnt Signaling by Endocytosis of Receptor Complexes

Akira Kikuchi, Shinji Matsumoto, Katsumi Fumoto, and Akira Sato

Department of Molecular Biology and Biochemistry, Graduate School of Medicine, Osaka University, Osaka, Japan

Introduction

At least 19 Wnt members have been shown to be present in humans and mice to date, and they exhibit unique expression patterns and distinct functions during development (Logan and Nusse, 2004). In humans and mice, the 10 members of the Frizzled (Fz) family comprise a series of 7-pass transmembrane receptors that have been identified as Wnt receptors (Chapter 14; Schulte, 2010). In addition to Fz proteins, single-pass transmembrane proteins, such as low-density lipoprotein receptor-related protein 5 (LRP5), LRP6, receptor tyrosine kinase-like orphan receptor 1 (Ror1), Ror2, and receptor-like tyrosine kinase (Ryk), have been shown to function as coreceptors for Wnt signaling (Chapter 7; Fradkin, Dura, and Noordermeer, 2009; Green, Kuntz, and Sternberg, 2008; Kikuchi, Yamamoto, and Sato, 2009; Kikuchi *et al.*, 2011). Therefore, it has been assumed traditionally that the binding of different Wnts to their specific receptors selectively triggers different outcomes via distinct intracellular pathways, including the β-catenin-dependent and β-catenin-independent pathways (see Chapters 2–7). However, current evidence suggests that in addition to the specific combination of Wnt ligands and receptors, endocytosis of the Wnt–receptor complex, binding of Wnt to heparan sulfate proteoglycans (HSPGs), and ubiquitination of Wnt receptors are involved in the regulation of the Wnt signaling pathway. This chapter highlights new insight into the mechanisms of Wnt signaling pathway activation, which occurs on and inside the cell surface membrane.

Endocytic pathways

Endocytosis and signaling

Endocytosis, a process by which eukaryotic cells internalize plasma membrane along with cell surface receptors and diverse soluble molecules, is used for a number of different cellular functions. In cellular signaling, it plays critical roles in initiating and spreading signals and in determining the specific signaling subpathways to activate or terminate. Originally, endocytosis was simply thought of as a process to downregulate signaling. There are at least two routes: clathrin- and caveolin-mediated endocytosis (Lajoie and Nabi, 2010; Le Roy and Wrana, 2005; Razani, Woodman, and Lisanti, 2002; Sorkin and von

Wnt Signaling in Development and Disease: Molecular Mechanisms and Biological Functions, First Edition.
Edited by Stefan Hoppler and Randall T. Moon.

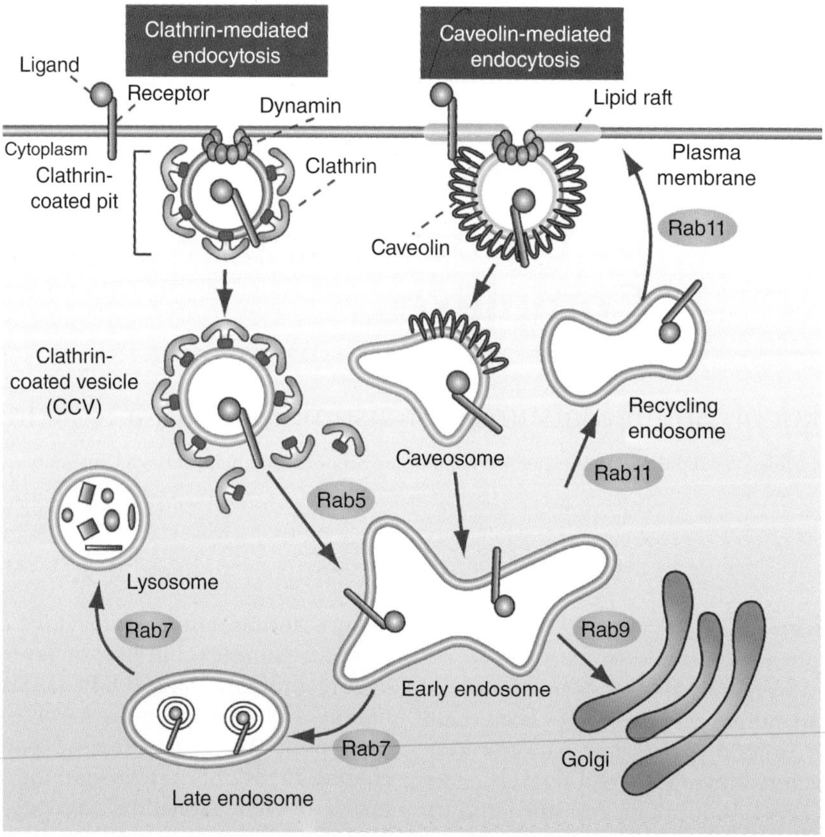

Figure 8.1 Endocytic pathways. There are two different types of endocytosis, clathrin-mediated endocytosis and caveolin-mediated endocytosis. Clathrin and caveolin play a major role in the respective pathways.

Zastrow, 2009; Traub, 2009; Figure 8.1). Clathrin-mediated endocytosis targets plasma membrane proteins and lipids to the early endosome and is an important pathway for downregulating many receptors through ubiquitin-dependent sorting processes involving ubiquitin-binding proteins resident in the clathrin pathway (Mousavi *et al.*, 2004). As non-clathrin-dependent endocytosis, lipid raft microdomains, which are enriched in cholesterol and sphingolipids, and caveolae function in vesicular and cholesterol trafficking as well as in the internalization of toxins and SV40 virus, and they regulate the internalization of receptors for autocrine motility factor, transforming growth factor β (TGF-β), and epidermal growth factor (EGF) (Lajoie and Nabi, 2010).

While lower doses of EGF are internalized through clathrin-mediated endocytosis, which is necessary for EGF dependent phosphorylation of the EGF receptor, higher doses of EGF are associated with ubiquitination of EGFR in lipid raft microdomains, followed by caveolin-mediated endocytosis and degradation (Sigismund *et al.*, 2005). The TGF-β receptor is endocytosed in a clathrin-dependent manner into early endosome antigen 1-positive early endosomes, which promotes TGF-β signaling. TGF-β receptor is also internalized in a caveolin-dependent manner into caveosomes, which leads to receptor turnover and signal downregulation (Di Guglielmo *et al.*, 2003). These results suggest that ligands can regulate different signaling pathways selectively within cells by inducing different endocytic pathways. In addition, it is becoming increasingly evident that specific endocytic pathways are essential for the appropriate cellular responses to different cues.

Clathrin-mediated endocytosis

In clathrin-mediated endocytosis (Figure 8.1, left side), the ligand-receptor complex is concentrated into "coated pits" on the plasma membrane, which is formed by the assembly of the cytosolic coat protein, clathrin. Coated pits are encapsulated by a polygonal clathrin coat and carry ligand-receptor complexes, including clathrin adaptor protein-2 (AP-2), into the cell. Then, dynamin severs the invaginated clathrin-coated pits to release clathrin-coated vesicles. After internalization from the plasma membranes, the vesicles are uncoated and then fuse with the early endosomes. Early endosomes are multifunctional organelles that regulate membrane transport between the plasma membrane and various intracellular organelles (Zerial and McBride, 2001). After fusing with the early endosomes, vesicles including ligands and receptors are returned to the plasma membrane for recycling, transported to late endosomes and lysosomes for degradation, or connected to the trans-Golgi network. Ras-related GTP-binding proteins (Rabs), including Rab4, Rab5, Rab7, Rab9, and Rab11, exert a regulatory function in vesicle trafficking.

Caveolin-mediated endocytosis

Non-clathrin-dependent endocytosis through lipid raft microdomains and the caveolar pathway have emerged as another important trafficking pathway (Figure 8.1, right side). Caveolae are flask-shaped invaginations of the plasma membrane. The shape and structural organization of caveolae are conferred by caveolin, which binds directly to cholesterol in the membrane. Caveolin self-associates to form a striated caveolin coat on the cytoplasmic side of the distortion and invagination of membranes. Caveolin-mediated endocytosis shares common features with clathrin-dependent endocytosis. Although the mechanism of caveolae formation is becoming clearer, the trafficking pathway after the internalization of caveolae has not yet been clarified.

Selective activation of Wnt signaling pathways by receptor-mediated endocytosis

Receptor-mediated endocytosis in the β-catenin-dependent pathway

Wnt/β-catenin signaling stabilizes cytoplasmic β-catenin. In the absence of Wnt, β-catenin is phosphorylated and ubiquitinated in the Axin complex and thereby degraded in the proteasome (Kikuchi *et al.*, 2011). The binding of Wnt3a to Fz/LRP6 causes the phosphorylation of LRP6. Phosphorylated LRP6 has been shown to form intracellular protein assemblies that are termed signalosomes in response to Wnt3a (Bilic *et al.*, 2007; Chapter 2). These assemblies do not colocalize with endocytic markers except for caveolin. Consistent with this observation, it has been shown that LRP6 is internalized through a caveolin-mediated route in cultured cells in response to Wnt3a and that manipulations to inhibit caveolin-mediated endocytosis blocks Wnt3a-dependent β-catenin accumulation and T-cell factor (Tcf) transcriptional activation (Sakane, Yamamoto, and Kikuchi, 2010; Yamamoto, Komekado, and Kikuchi, 2006; Yamamoto *et al.*, 2008; Figure 8.2a). Therefore, caveolin-mediated endocytosis is required for Wnt3a-dependent activation of the β-catenin pathway. However, the inhibition of internalization of LRP6 does not suppress Wnt3a-dependent phosphorylation of LRP6, and the internalization of LRP6 occurs independently of its phosphorylation, suggesting that phosphorylation and internalization of LRP6 are independent events and both are required for the activation of the β-catenin-dependent pathway. Rab5 is involved in the internalization of LRP6 to early endosomes, and internalized LRP6 is recycled to the cell surface membrane from recycling endosomes (Yamamoto, Komekado, and Kikuchi, 2006). These results were confirmed by the findings that transmembrane protein Wnt-activated inhibitory factor 1 (WAIF1), which was identified in a negative regulator of Wnt signaling in zebrafish screening, inhibits Wnt3a-dependent internalization of LRP6 but not phosphorylation of LRP6, resulting in the suppression of β-catenin signaling (Kagermeier-Schenk *et al.*, 2011).

Although the precise mechanism by which caveolin-mediated endocytosis promotes the β-catenin pathway is unclear, two models have been proposed (Figure 8.2a). When Axin binds to phosphorylated LRP6 upon stimulation with Wnt3a and the complex is internalized with caveolin, caveolin affects the Axin complex to interfere with the binding of β-catenin to Axin, thereby reducing the phosphorylation of β-catenin by glycogen synthase kinase 3 (GSK3) (Yamamoto, Komekado, and Kikuchi, 2006).

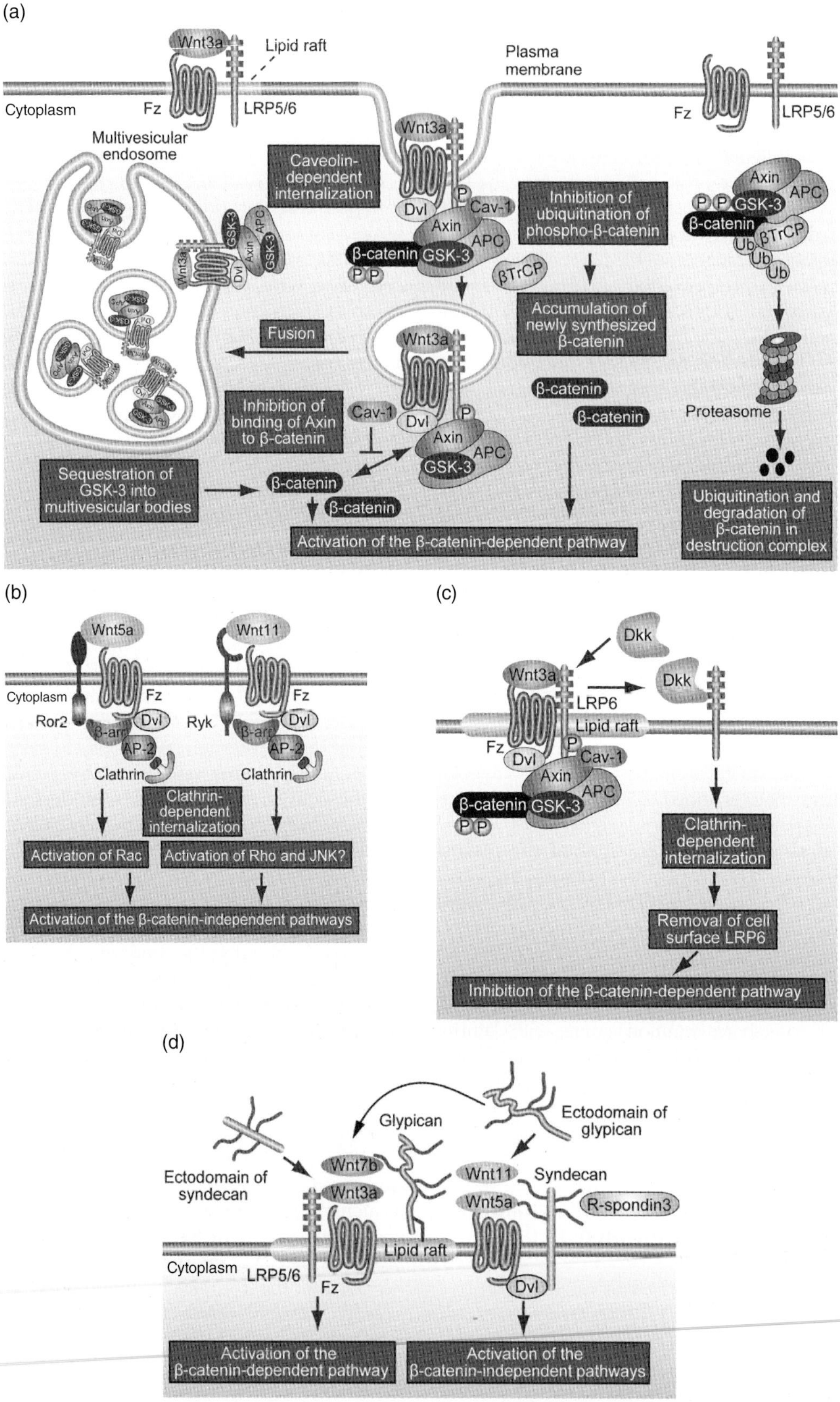
(a)
Wnt3a
Lipid raft
Plasma membrane
Cytoplasm
Fz
LRP5/6
Multivesicular endosome
Caveolin-dependent internalization
Wnt3a
P
Cav-1
Dvl
Axin
APC
β-catenin
GSK-3
P P
Inhibition of ubiquitination of phospho-β-catenin
Fz
LRP5/6
Axin
APC
P P GSK-3
β-catenin
βTrCP
Ub
Ub
Ub
GSK-3
APC
Axin
GSK-3
Wnt3a
Dvl
βTrCP
Accumulation of newly synthesized β-catenin
Fusion
Wnt3a
β-catenin
β-catenin
Proteasome
Inhibition of binding of Axin to β-catenin
Cav-1
Dvl
P
Axin
APC
GSK-3
Sequestration of GSK-3 into multivesicular bodies
β-catenin
β-catenin
Activation of the β-catenin-dependent pathway
Ubiquitination and degradation of β-catenin in destruction complex
(b)
Wnt5a
Wnt11
Cytoplasm
Fz
Fz
Ror2
Dvl
Ryk
Dvl
β-arr
β-arr
AP-2
AP-2
Clathrin
Clathrin
Clathrin-dependent internalization
Activation of Rac
Activation of Rho and JNK?
Activation of the β-catenin-independent pathways
(c)
Dkk
Wnt3a
Dkk
LRP6
Lipid raft
Fz
P
Dvl
Cav-1
Axin
APC
β-catenin
GSK-3
P P
Clathrin-dependent internalization
Removal of cell surface LRP6
Inhibition of the β-catenin-dependent pathway
(d)
Ectodomain of glypican
Glypican
Ectodomain of syndecan
Wnt7b
Wnt3a
Wnt11
Syndecan
Wnt5a
R-spondin3
Lipid raft
Cytoplasm
LRP5/6
Fz
Dvl
Activation of the β-catenin-dependent pathway
Activation of the β-catenin-independent pathways

Another model suggests that Wnt3a causes the sequestration of GSK3 from the cytosol into multivesicular bodies (MVBs) so that the enzyme becomes separated from its cytosolic substrates. In such a model, LRP6/GSK3-containing signalosomes are internalized, presumably in caveolin-containing vesicles, and may recruit the endosomal sorting complex required for transport (ESCRT) complex. The internalized vesicles are sorted to vesicles destined for the formation of MVBs, wherein GSK3 is sequestered from cytosolic β-catenin, resulting in the stabilization of β-catenin (Taelman *et al.*, 2010).

It has also been proposed that clathrin-mediated endocytosis is involved in the activation of the β-catenin pathway. The inhibition of clathrin-mediated endocytosis with pharmacological reagents suppressed the Wnt3a-induced β-catenin stabilization in mouse fibroblast L cells (Blitzer and Nusse, 2006). Consistent with the positive role of clathrin in the β-catenin pathway, it was shown that β-arrestin, a clathrin adaptor, binds to Dvl, and these proteins activate Wnt-dependent Tcf transcriptional activity synergistically (Chen *et al.*, 2001). In addition, knockdown of β-arrestin suppresses Wnt3a-dependent phosphorylation of Dvl and stabilization of β-catenin (Bryja *et al.*, 2007; Chapter 15). These results suggest that clathrin and β-arrestin play a role in the β-catenin pathway, but whether β-arrestin is involved in the endocytosis of Fz and LRP6 through clathrin remains to be clarified. Thus, there are contradictory reports on the role of endocytosis in the β-catenin pathway.

In addition, the new mechanism that underlies Wnt-dependent stabilization of β-catenin has been proposed (Li *et al.*, 2012). In this model, Wnt causes the interaction of phosphorylated LRP6 and the Axin complex, resulting in the inhibition of the ubiquitination of phosphorylated β-catenin, thereby stabilizing β-catenin (Figure 8.2a). However, it remains to be clarified whether receptor endocytosis is involved for this mechanism.

Clathrin-mediated endocytosis in the β-catenin-independent pathway

Compared with the controversial roles of endocytic pathways in the β-catenin-dependent pathway, the importance of endocytosis in the β-catenin-independent pathway is well documented. Because Fz belongs to the G protein-coupled receptor family (Schulte, 2010; Chapter 14), β-arrestin is expected to be involved in the internalization of Fz. Wnt5a triggers internalization of Fz4, which requires the recruitment of Dvl2 and β-arrestin-2 to the plasma membrane in HEK293 cells (Chen *et al.*, 2003). The reduction of β-arrestin-2 levels in *Xenopus* embryos leads to defects in convergent extension, which is regulated by β-catenin-independent pathway (Kim and Han, 2007). Dvl2 also interacts with μ2-adaptin of AP-2, and this interaction is required for the internalization of Fz4 and for convergent extension (Yu *et al.*, 2007).

It has also been demonstrated that Wnt5a induces the internalization of Fz2 and Ror2 in a clathrin-dependent manner and that knockdown of Dvl2, β-arrestin-2, and Ror2 suppresses Wnt5a-dependent internalization of receptors and the activation of Rac (Sakane *et al.*, 2012; Sato *et al.*, 2010; Figure 8.2a). In addition, Ryk interacts with β-arrestin-2 and promotes Wnt11-induced internalization of Fz7 and Dvl in *Xenopus* (Kim, Her, and Han, 2008). It is notable that Ryk plays a role in the activation of RhoA and c-Jun N-terminal kinase (JNK) and that it

Figure 8.2 Selective activation of Wnt signaling pathways by receptor-mediated endocytosis. (a) Activation of the β-catenin-dependent pathway by receptor-mediated endocytosis. After the Axin complex is translocated to the Fz and LRP6 complex in response to Wnt3a, the receptor complex may be internalized via caveolin-mediated routes to stabilize β-catenin. (b) Activation of the β-catenin-independent pathway by receptor-mediated endocytosis. Wnt5a induces the internalization of Fz and Ror2 in a clathrin-dependent manner, thereby activating Rac. Ryk interacts with β-arrestin-2 and promotes Wnt11-induced internalization of Fz7 and Dvl, leading to the activation of the β-catenin-independent pathway in *Xenopus*. (c) Negative regulation of the β-catenin-dependent pathways by Dkk. Dkk removes LRP6 from the lipid raft microdomains to nonlipid raft microdomains and internalizes LRP6 through a clathrin-dependent route, thereby inhibiting the β-catenin-dependent pathway. (d) Cell surface HSPGs that regulate Wnt signaling. Glypican and syndecan bind to Wnt and concentrate it on the cell surface, making Wnt–receptor interaction more efficient. β-arr, β-arrestin; Cav-1, caveolin-1; Dkk, Dickkopf.

is necessary for convergent extension. Taken together, it is possible that Wnt5a and Wnt11 induce the internalization of Fzs with Ror2 or Ryk in cooperation with Dvl2 and β-arrestin-2 via a clathrin-mediated route, thereby activating the β-catenin-independent pathway (Figure 8.2a). Using purified Wnt11, similar findings were observed in mammalian cells. Wnt11 induces the internalization of Fz7 and Ryk, which is required for the activation of Rac in HeLa and HEK293 cells (Yamamoto, H. *et al.*, unpublished data).

Wnt5a was shown to be involved in modulating the aggressiveness of human cancers, including gastric, prostate, and lung cancers and melanoma (Kikuchi and Yamamoto, 2008), through the enhancement of focal adhesion turnover and gene expression (Kurayoshi *et al.*, 2006; Matsumoto *et al.*, 2010; Yamamoto *et al.*, 2009, 2010). An anti-Wnt5a antibody suppressed Wnt5a-dependent Rac1 activation through inhibition of the internalization of Fz2 and Ror2 in KKLS cells, a gastric cancer cell line that expresses Wnt5a highly (Hanaki *et al.*, 2012). KKLS cells exhibits liver metastatic ability in nude mice, and Wnt5a is required for this ability. When an anti-Wnt5a antibody was injected intraperitoneally, the metastatic ability of KKLS cells in nude mice was suppressed (Hanaki *et al.*, 2012). Taken together, receptor-mediated endocytosis might be a therapeutic target for diseases caused by abnormal activation of Wnt signaling.

Negative regulation of the β-catenin-dependent pathway by Dickkopf and Dab2

Family members of the secreted protein Dickkopf (Dkk) antagonize the β-catenin-dependent pathway (Niehrs, 2006). Dkk1 binds specifically to LRP5/6 but not to Wnt ligands and Fzs, interfering with the binding between LRP6 and Fzs induced by Wnt ligands (Bafico *et al.*, 2001; Semënov, Zhang, and He, 2008; Figure 8.2c). This causes the internalization of LRP5/6 through the clathrin-mediated route, and the removal of LRP5/6 from the plasma membrane attenuates β-catenin-dependent signaling (Figure 8.2c; Yamamoto *et al.*, 2008). In fact, knockdown of clathrin blocks Dkk1-dependent inhibition of the Wnt3a-induced β-catenin accumulation (Yamamoto *et al.*, 2008). Taken together with the finding that the knockdown of caveolin inhibits Wnt3a-dependent β-catenin accumulation (Yamamoto, Komekado, and Kikuchi, 2006), Wnt3a and Dkk1 may shunt LRP6 to distinct internalization pathways in order to activate and inhibit, respectively, the β-catenin-dependent pathway. Because Dkk1 can convert the distribution of LRP6 in the lipid raft fraction into the nonlipid raft fraction (Yamamoto *et al.*, 2008), Wnt3a and Dkk1 may determine the distribution of LRP6 in lipid raft microdomains containing caveolin on the cell surface membrane.

Disabled-2 (Dab2) is a widely expressed endocytic adaptor protein shown to be involved in several receptor-mediated signaling pathways (Hocevar *et al.*, 2003). Dab2 expression levels regulate whether LRP6 is internalized through the caveolin or clathrin endocytic pathways (Jiang, He, and Howe, 2012). In this model, Wnt stimulates the casein kinase 2-dependent phosphorylation of LRP6 at Ser1595, thereby inducing the binding of LRP6 and Dab2 and its internalization via a clathrin route. Consistent with previous observations (Yamamoto, Komekado, and Kikuchi, 2006), in the absence of Dab2, LRP6 interacts with caveolin in a Wnt-dependent manner, and in its presence, Wnt stimulation leads to sequestration of LRP6 with clathrin. Therefore, the tumor suppressor functions of Dab2 may involve modulation of Wnt/β-catenin signaling by regulating the endocytic fate of LRP6.

Modulation of receptor-mediated endocytosis by HSPGs

HSPGs are involved in extracellular regulation of many signaling pathways such as EGF, FGFs, TGF-β, Hedgehog, and Wnt signaling (Hacker, Nybakken, and Perrimon, 2005; Figure 8.2d; Chapter 1). To date, six glypicans (GPCs), cell surface HSPGs, have been identified in mammals (GPC1–6) (Song and Filmus, 2002). GPCs are linked to the exocytoplasmic surface of the plasma membrane through a covalent glycosylphosphatidyl inositol (GPI) linkage modified with heparan sulfates at sites near to a GPI anchor. GPC3 knockout mice embryos exhibit a reduction in JNK activity (Song *et al.*, 2005). GPC3 binds to Wnt5a and enhances Wnt5a-dependent JNK activation in mammalian mesothelioma cells, and interference with GPC3 expression disrupts gastrulation movements in zebrafish (De Cat *et al.*, 2003; Song *et al.*, 2005). Furthermore, zebrafish Knypek (GPC4/6) binds to Dkk1 and potentiates the activity of Dkk1 to activate JNK (Caneparo *et al.*, 2007). *Xenopus* GPC4 interacts with Wnt11 and enhances Wnt-stimulated cell migration in convergent extension (Ohkawara *et al.*, 2003). In addition to the functional relevance of GPCs in the β-catenin-independent pathway, it has also been reported that GPC3 knockout mice embryos contain elevated levels of β-catenin in the cytoplasm and that GPC3 binds to Wnt3a and Wnt7b and enhances the Wnt-dependent Tcf activities in hepatocellular carcinoma cell lines, suggesting that GPCs are also able to potentiate the β-catenin pathway (Capurro *et al.*, 2005). However, it is not clear whether these actions of GPCs on Wnt signaling are linked to receptor endocytosis.

The transmembrane-type HSPG syndecan (SDC) also regulates Wnt signaling. SDC1–4 have a region near the N-terminus that bears heparan sulfate chains and a transmembrane domain with a C-terminal cytoplasmic region (Couchman, 2003). Some SDCs are also modified with chondroitin sulfates. Mice with mammary gland-specific transgenic expression of Wnt1 develop mammary tumors, and SDC1 is required for Wnt1-induced tumorigenesis (Alexander *et al.*, 2000), suggesting that SDC1 is involved in the regulation of the β-catenin-dependent pathway. In contrast, SDC4 was shown to form a complex with Fz7 and Dvl and regulate convergent extension in *Xenopus* (Munoz *et al.*, 2006). In addition, SDC4 interacts with R-spondin3 and is involved in convergent extension movement in cooperation with Wnt5a and Fz7 in *Xenopus* embryos (Ohkawara, Glinka, and Niehrs, 2011). In this scenario, Wnt5a and R-spondin3 induce the internalization of Fz and SDC4 through a clathrin-mediated endocytic route. These results suggest that SDC is also involved in the regulation of the β-catenin-independent pathway. Therefore, although HSPG has a critical role for modulating Wnt signaling by binding to Wnt ligands and receptors, whether the action of HSPG is positive or negative for β-catenin-dependent or β-catenin-independent pathway might be cell-type and tissue specific.

GPCs are present in both lipid raft and nonlipid raft microdomains, and SDCs are predominantly in nonlipid raft microdomains (Sakane *et al.*, 2012; Figure 8.2d). GPC4 in lipid raft microdomains activates Wnt3a-dependent activation of the β-catenin pathway, and GPC4 in nonlipid raft microdomains activates the Wnt5a-dependent β-catenin-independent pathway. GPC3 is subjected to endoproteolytic processing by convertases in Chinese hamster ovary (CHO) cells, and processed GPC3 modulates both the β-catenin-dependent and β-catenin-independent pathways (De Cat *et al.*, 2003). The GPC4 ectodomain showed inhibitory actions for both Wnt signaling pathways (Sakane *et al.*, 2012). It is known that the ectodomains of HSPGs are shed into the extracellular space. The SDC1 ectodomain was shown to activate the β-catenin pathway in cultured cells, and this may explain why the knockout of SDC1 suppresses Wnt1-induced mammary tumorigenesis (Alexander *et al.*, 2000). Taken together, the localization (lipid raft or nonlipid raft microdomains) and form (membrane anchored structure or soluble structure) of HSPGs could also be important for the regulation of Wnt signaling.

Regulation of Wnt signaling by receptor ubiquitination

Ubiquitination and deubiquitination of receptors

Receptor presentation on the cell surface membrane is controlled by the level of expression and by subcellular localization and is critical for detection of extracellular ligands and for regulation of the duration of downstream signal activation. For example, EGFR is mainly located on the cell surface membrane because its recycling rate is about 10 times higher than its internalization rate. After EGF binds to EGFR, the complex is sorted to different cellular destinations, such as the cell surface and the lysosomes through MVBs. It has been shown that ubiquitination functions as the sorting signal for EGFR to direct lysosomes (Duan *et al.*, 2003).

Ubiquitination is one of the major posttranslational modifications. Ubiquitin is conjugated to target proteins by ubiquitin-activating (E1), ubiquitin-conjugating (E2), and ubiquitin-ligating (E3) enzymes (Kerscher, Felberbaum, and Hochstrasser, 2006). The removal of ubiquitin from modified proteins is catalyzed by deubiquitination enzymes (Komander, Clague, and Urbe, 2009). There are two types of ubiquitinations: polyubiquitination and monoubiquitination. Polyubiquitination is involved in selected degradation of target proteins in 26S proteasomes, the activation of signaling cascades such as Nuclear factor-kappa B (NF-kB) pathway, the orchestration of different steps during DNA repair, and degradation of target proteins in lysozomes, whereas monoubiquitination serves to incorporate ubiquitinated proteins into vesicles at different stages of the endocytic pathway (Kerscher, Felberbaum, and Hochstrasser, 2006; Mukhopadhyay and Riezman, 2007). It was speculated that there are 600–1000 E3 enzymes and more than 100 deubiquitination enzymes and that different enzymes have different spectrums of substrates. Therefore, the balance of ubiquitination and deubiquitination activities may play essential roles in each protein's functions.

Deubiquitination of Frizzled by UBPY

Genome-wide *Drosophila* RNA interference library screening revealed that depletion of dUBPY, an ortholog of the mammalian deubiquitinating enzyme ubiquitin-specific protease Y (UBPY)/USP8, causes loss of sensory bristles at the wing margin (Mukai *et al.*, 2010). The amount of Armadillo (an ortholog of β-catenin) and expression of Sense, a target gene of Armadillo, are reduced in dUBPY knockout clones in the wing disc. Human UBPY facilitated Wnt3a-dependent β-catenin pathway activation in mammalian cells (Figure 8.3a). Fz4 undergoes monoubiquitination at multiple sites constitutively and is internalized to the early endosome where UBPY is localized (Figure 8.3a). Fz4 not deubiquitinated by UBPY is transported to the lysosome for degradation through the late endosome. When Fz4 is deubiquitinated by UBPY, Fz4 is recycled back to the cell surface membrane through the

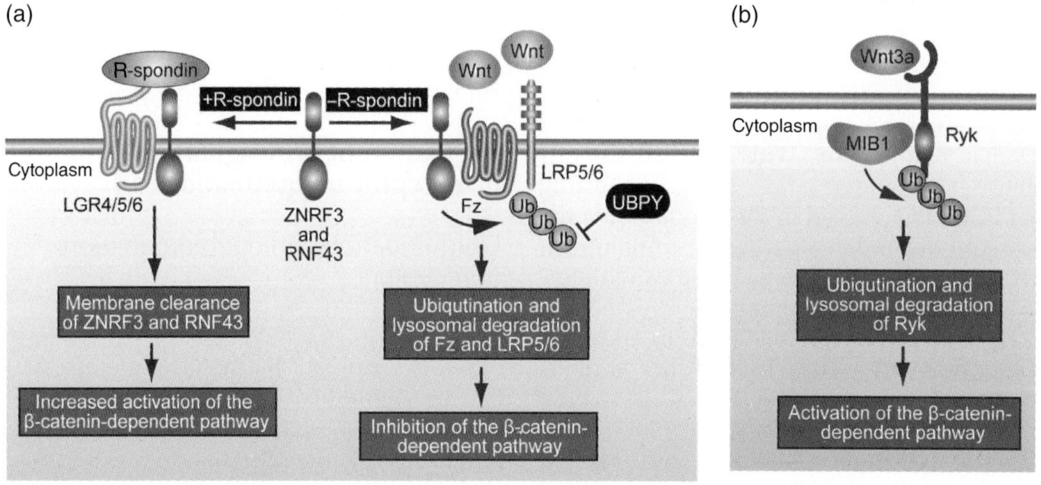

Figure 8.3 Regulation of Wnt signaling by receptor ubiquitination. (a) The R-spondin and ZNFR3/RNF43 mechanism. The transmembrane E3 ubiquitin ligase proteins ZNRF3 or RNF43 and deubiquitination enzyme UBPY regulates the ubiquitination of Fz its cell surface levels. R-spondin causes the clearance of ZNRF3 or RNF43 from the cell surface membrane, thereby leading to reduced ubiquitination of Fz and enhanced β-catenin-dependent signaling. (b) The MIB1 mechanism. An E3 ligase MIB1 interacts with Ryk and promotes ubiquitination of Ryk and its degradation through proteasomes and lysosomes. LGR, leucine-rich repeat-containing G protein-coupled receptor; MIB1, Mindbomb E3 ubiquitin protein ligase 1; RNF43; ring finger protein 43; Ub, ubiquitin; ZNRF3, zinc and ring finger 3.

recycling endosome. dUBPY indeed increases cell surface levels of *Drosophila* Fz2 and enhances Wg target gene expression in *Drosophila* wing disc. Therefore, balanced ubiquitination and deubiquitination of Fz regulates cellular responses to Wg/Wnt signaling by the tuning of the cell surface level of Fz. It has been shown that UBPY mRNA is upregulated in chronic lymphocytic leukemia (CLL) and Fz is also elevated in CLL (Lu *et al.*, 2004; Mukai *et al.*, 2010). Deubiquitination of Fz by elevated activity of UBPY may contribute to tumorigenesis in CLL.

Ubiquitination of Fz by ZNRF3/RNF43

ZNRF3 and RNF43 were identified by the screening of new β-catenin target genes and potential negative regulators of Wnt signaling (Hao *et al.*, 2012; Koo *et al.*, 2012). They are closely related RING finger proteins that act as an E3 ubiquitin ligase localized to the plasma membrane. ZNRF3 promotes the ubiquitination-mediated endocytosis and lysosomal targeting of Fz4 and reduces cell surface levels of Fz4 and LRP6, thereby suppressing the β-catenin-dependent pathway (Figure 8.3a). The R-spondin family exhibits synergistic activity regulating the β-catenin pathway with Wnts (Kim *et al.*, 2005) through direct interaction with LGR4/5/6 (de Lau *et al.*, 2011; Glinka *et al.*, 2011), which are seven transmembrane proteins expressed in tissue-specific adult stem cells. R-spondin also interacts with the extracellular domain of ZNRF3 and reduces its cell surface level in a LGR4-dependent manner by endocytosis (Figure 8.3a). Therefore, it is conceivable that R-spondin induces membrane clearance of ZNRF3 through LGR4, leading to the accumulation of Fz and LRP6 on the cell surface by suppressing their ubiquitination, resulting in the enhancement of the β-catenin-dependent pathway activity.

LGR5 is also a protein that is expressed by activation of the β-catenin pathway (Barker *et al.*, 2007), and ZNRF3 and RNF43 are expressed in LGR5-positive stem cells resided at the crypt bottom. *ZNRF3/RNF43* compound mutant mice undergo a marked expansion of the proliferative compartment in the mutant intestine. Intestinal organoids from wild-type mice require R-spondin1 and die within 5 days after the removal of R-spondin, whereas *ZNRF3/RNF43* compound mutant organoids are maintained for 4 weeks without R-spondin (Koo *et al.*, 2012). These results suggest that ZNRF3 and RNF43 function as a tumor suppressor.

Ubiquitination of Ryk by MIB1

Ryk is a single transmembrane protein that binds to Wnt3a and Wnt11 and is involved in both the β-catenin-dependent and β-catenin-independent pathways (Fradkin, Dura, and Noordermeer, 2009). An E3 ligase Mindbomb 1 (MIB1) was originally identified as an Ryk-binding protein using affinity purification followed by mass spectrometry (Berndt *et al.*, 2011). MIB1 promotes ubiquitination and turnover of Ryk and reduces cell surface levels of Ryk in the absence of degradation (Figure 8.3b). Internalized Ryk is colocalized with MIB1 in the cytoplasmic puncta where Rab5 is positive. In addition, MIB and Ryk are required for the activation of the β-catenin-dependent pathway. Therefore, MIB1 regulates the cell surface level of Ryk to sense extracellular Wnt. It has been shown that MIB family proteins promote the ubiquitination and endocytosis of the Notch ligand Delta, resulting in the activation of Notch signaling (Itoh *et al.*, 2003; Zhang, Li, and Jiang, 2007). In Wnt signaling, it seems that MIB acts with Ryk in a cell autonomous manner, but it still remains unclear how the internalization of Ryk by MIB leads to the activation of the β-catenin-dependent pathway. Although there are no data showing that Wnt3a induces the internalization of Ryk, it remains formally possible that MIB1 regulates the endocytosis of Ryk after Wnt binding.

Concluding remarks and future perspectives

It is becoming clear that Wnt signaling is modulated at the cell surface at different levels. However, there are still unresolved issues that we should elucidate in order to draw a complete picture of the regulatory mechanisms of Wnt signaling at the ligand and receptor levels. For example, although caveolin-mediated and clathrin-mediated endocytoses are required for the β-catenin-dependent and β-catenin-independent pathways, respectively, whether caveolae- or clathrin-coated vesicles are structurally required for their signaling pathways is unclear. It is also unclear where Wnt signaling is activated within the cell. Because there are few antibodies to detect endogenous Wnt receptors immunocytochemically, it is still hard to observe that endogenous Wnt receptors are internalized. If such antibodies became available in the future, detailed observation of the structure and subcellular localization of Wnt receptors would become feasible by ultrastructural and immunoelectron microscopy.

Intensive studies are needed to characterize the biochemical nature of Wnts to understand the mechanism of their secretion and interaction with receptors. Lipid modification at conserved cysteine and serine residues of Wnt3a by Porcupine is critical for Wnt3a action and secretion (Takada *et al.*, 2006; Willert *et al.*, 2003). It is necessary to determine lipid modification of other Wnts. It was reported that Wnts are secreted on extracellular vesicles, exosomes, in *Drosophila* (Gross *et al.*, 2012). This may explain why hydrophobic Wnts can act as a morphogen gradient targeting Wnt-responsive cells far from Wnt-producing cells. In addition, cleavage of N-terminal residues of Wnt3a by Tiki, a single-pass transmembrane protein, inactivates its activity owing to a loss of hydrophobicity by oxidation–oligomerization (Zhang *et al.*, 2012). Therefore, changes in the hydrophobic character of Wnt proteins must occur in various biological situations. In addition to lipid modification, Wnts are well known to be modified with glycans. But, their glycan profiles and the functions of glycosylation of Wnts are unknown. Glycosylation sites of Wnts need to be identified by mass spectrometric analysis, and functional characteristics should be clarified in terms of exocytic and endocytic trafficking.

Acknowledgments

We thank all of our laboratory members for reading this manuscript carefully. Studies in our laboratory were supported by Grants-in-Aid for Scientific Research and for Scientific Research on priority areas from the Ministry of Education, Science, and Culture of Japan.

References

Alexander, C.M., Reichsman, F., Hinkes, M.T. *et al.* (2000) Syndecan-1 is required for Wnt-1-induced mammary tumorigenesis in mice. *Nature Genetics*, **25**, 329–332.

Bafico, A., Liu, G., Yaniv, A. *et al.* (2001) Novel mechanism of Wnt signalling inhibition mediated by Dickkopf-1 interaction with LRP6/Arrow. *Nature Cell Biology*, **3**, 683–686.

Barker, N., van Es, J.H., Kuipers, J. *et al.* (2007) Identification of stem cells in small intestine and colon by marker gene Lgr5. *Nature*, **449**, 1003–1007.

Berndt, J.D., Aoyagi, A., Yang, P. *et al.* (2011) Mindbomb 1, an E3 ubiquitin ligase, forms a complex with RYK to activate Wnt/β-catenin signaling. *Journal of Cell Biology*, **194**, 737–750.

Bilic, J., Huang, Y.L., Davidson, G. *et al.* (2007) Wnt induces LRP6 signalosomes and promotes dishevelled-dependent LRP6 phosphorylation. *Science*, **316**, 1619–1622.

Blitzer, J.T. and Nusse, R. (2006) A critical role for endocytosis in Wnt signaling. *BMC Cell Biology*, **7**, 28.

Bryja, V., Gradl, D., Schambony, A. *et al.* (2007) β-Arrestin is a necessary component of Wnt/β-catenin signaling in vitro and in vivo. *Proceedings of the National Academy of Sciences of the United States of America*, 104, 6690–6695.

Caneparo, L., Huang, Y.L., Staudt, N. *et al.* (2007) Dickkopf-1 regulates gastrulation movements by coordinated modulation of Wnt/βcatenin and Wnt/PCP activities, through interaction with the Dally-like homolog Knypek. *Genes & Development*, **21**, 465–480.

Capurro, M.I., Xiang, Y.Y., Lobe, C., and Filmus, J. (2005) Glypican-3 promotes the growth of hepatocellular carcinoma by stimulating canonical Wnt signaling. *Cancer Research*, **65**, 6245–6254.

Chen, W., Hu, L.A., Semenov, M.V. *et al.* (2001) β-Arrestin1 modulates lymphoid enhancer factor transcriptional activity through interaction with phosphorylated dishevelled proteins. *Proceedings of the National Academy of Sciences of the United States of America*, **98**, 14889–14894.

Chen, W., ten Berge, D., Brown, J. *et al.* (2003) Dishevelled 2 recruits β-arrestin 2 to mediate Wnt5A-stimulated endocytosis of Frizzled 4. *Science*, **301**, 1391–1394.

Couchman, J.R. (2003) Syndecans: proteoglycan regulators of cell-surface microdomains? *Nature Reviews Molecular Cell Biology*, **4**, 926–937.

De Cat, B., Muyldermans, S.Y., Coomans, C. *et al.* (2003) Processing by proprotein convertases is required for glypican-3 modulation of cell survival, Wnt signaling, and gastrulation movements. *Journal of Cell Biology*, **163**, 625–635.

de Lau, W., Barker, N., Low, T.Y. *et al.* (2011) Lgr5 homologues associate with Wnt receptors and mediate R-spondin signalling. *Nature*, **476**, 293–297.

Di Guglielmo, G.M., Le Roy, C., Goodfellow, A.F., and Wrana, J.L. (2003) Distinct endocytic pathways regulate TGF-β receptor signalling and turnover. *Nature Cell Biology*, 5, 410–421.

Duan, L., Miura, Y., Dimri, M. *et al.* (2003) Cbl-mediated ubiquitinylation is required for lysosomal sorting of epidermal growth factor receptor but is dispensable for endocytosis. *The Journal of Biological Chemistry*, **278**, 28950–28960.

Fradkin, L.G., Dura, J.M., and Noordermeer, J.N. (2009) Ryks: new partners for Wnts in the developing and regenerating nervous system. *Trends in Neuroscience*, **33**, 84–92.

Glinka, A., Dolde, C., Kirsch, N. *et al.* (2011) LGR4 and LGR5 are R-spondin receptors mediating Wnt/β-catenin and Wnt/PCP signalling. *EMBO Report*, **12**, 1055–1061.

Green, J.L., Kuntz, S.G., and Sternberg, P.W. (2008) Ror receptor tyrosine kinases: orphans no more. *Trends in Cell Biology*, **18**, 536–544.

Gross, J.C., Chaudhary, V., Bartscherer, K., and Boutros, M. (2012) Active Wnt proteins are secreted on exosomes. *Nature Cell Biology*, **14**, 1036–1045.

Hacker, U., Nybakken, K., and Perrimon, N. (2005) Heparan sulphate proteoglycans: the sweet side of development. *Nature Reviews Molecular Cell Biology*, **6**, 530–541.

Hanaki, H., Yamamoto, H., Sakane, H. *et al.* (2012) An anti-Wnt5a antibody suppresses metastasis of gastric cancer cells in vivo by inhibiting receptor-mediated endocytosis. *Molecular Cancer Therapeutics*, **11**, 298–307.

Hao, H.X., Xie, Y., Zhang, Y. *et al.* (2012) ZNRF3 promotes Wnt receptor turnover in an R-spondin-sensitive manner. *Nature*, **485**, 195–200.

Hocevar, B.A., Mou, F., Rennolds, J.L. *et al.* (2003) Regulation of the Wnt signaling pathway by disabled-2 (Dab2). *The EMBO Journal*, **22**, 3084–3094.

Itoh, M., Kim, C.H., Palardy, G. *et al.* (2003) Mind bomb is a ubiquitin ligase that is essential for efficient activation of Notch signaling by Delta. *Developmental Cell*, **4**, 67–82.

Jiang, Y., He, X., and Howe, P.H. (2012) Disabled-2 (Dab2) inhibits Wnt/β-catenin signalling by binding LRP6 and promoting its internalization through clathrin. *The EMBO Journal*, **31**, 2336–2349.

Kagermeier-Schenk, B., Wehner, D., Ozhan-Kizil, G. *et al.* (2011) Waif1/5T4 inhibits Wnt/β-catenin signaling and activates noncanonical Wnt pathways by modifying LRP6 subcellular localization. *Developmental Cell*, **21**, 1129–1143.

Kerscher, O., Felberbaum, R., and Hochstrasser, M. (2006) Modification of proteins by ubiquitin and ubiquitin-like proteins. *Annual Review of Cell and Developmental Biology*, **22**, 159–180.

Kikuchi, A. and Yamamoto, H. (2008) Tumor formation due to abnormalities in the β-catenin-independent pathway of Wnt signaling. *Cancer Science*, **99**, 202–208.

Kikuchi, A., Yamamoto, H., and Sato, A. (2009) Selective activation mechanisms of Wnt signaling pathways. *Trends in Cell Biology*, **19**, 119–129.

Kikuchi, A., Yamamoto, H., Sato, A., and Matsumoto, S. (2011) New insights into the mechanism of Wnt signaling pathway activation. *International Review of Cell and Molecular Biology*, **291**, 21–71.

Kim, G.H. and Han, J.K. (2007) Essential role for β-arrestin 2 in the regulation of Xenopus convergent extension movements. *The EMBO Journal*, **26**, 2513–2526.

Kim, G.H., Her, J.H., and Han, J.K. (2008) Ryk cooperates with Frizzled 7 to promote Wnt11-mediated endocytosis and is essential for Xenopus laevis convergent extension movements. *Journal of Cell Biology*, **182**, 1073–1082.

Kim, K.A., Kakitani, M., Zhao, J. *et al.* (2005) Mitogenic influence of human R-spondin1 on the intestinal epithelium. *Science*, **309**, 1256–1259.

Komander, D., Clague, M.J., and Urbe, S. (2009) Breaking the chains: structure and function of the deubiquitinases. *Nature Reviews Molecular Cell Biology*, **10**, 550–563.

Koo, B.K., Spit, M., Jordens, I. *et al.* (2012) Tumour suppressor RNF43 is a stem-cell E3 ligase that induces endocytosis of Wnt receptors. *Nature*, **488**, 665–669.

Kurayoshi, M., Oue, N., Yamamoto, H. *et al.* (2006) Expression of Wnt-5a is correlated with aggressiveness of gastric cancer by stimulating cell migration and invasion. *Cancer Research*, **66**, 10439–10448.

Lajoie, P. and Nabi, I.R. (2010) Lipid rafts, caveolae, and their endocytosis. *International Review of Cell and Molecular Biology*, **282**, 135–163.

Le Roy, C. and Wrana, J.L. (2005) Clathrin- and non-clathrin-mediated endocytic regulation of cell signalling. *Nature Reviews Molecular and Cell Biology*, **6**, 112–126.

Li, V.S., Ng, S.S., Boersema, P.J. *et al.* (2012) Wnt signaling through inhibition of β-catenin degradation in an intact Axin1 complex. *Cell*, **149**, 1245–1256.

Logan, C.Y. and Nusse, R. (2004) The Wnt signaling pathway in development and disease. *Annual Review of Cell and Developmental Biology*, **20**, 781–810.

Lu, D., Zhao, Y., Tawatao, R. *et al.* (2004) Activation of the Wnt signaling pathway in chronic lymphocytic leukemia. *Proceedings of the National*

Academy of Sciences of the United States of America, **101**, 3118–3123.

Matsumoto, S., Fumoto, K., Okamoto, T. *et al.* (2010) Binding of APC and dishevelled mediates Wnt5a-regulated focal adhesion dynamics in migrating cells. *The EMBO Journal*, **29**, 1192–1204.

Mousavi, S.A., Malerod, L., Berg, T., and Kjeken, R. (2004) Clathrin-dependent endocytosis. *Biochemical Journal*, **377**, 1–16.

Mukai, A., Yamamoto-Hino, M., Awano, W. *et al.* (2010) Balanced ubiquitylation and deubiquitylation of Frizzled regulate cellular responsiveness to Wg/Wnt. *The EMBO Journal*, **29**, 2114–2125.

Mukhopadhyay, D. and Riezman, H. (2007) Proteasome-independent functions of ubiquitin in endocytosis and signaling. *Science*, **315**, 201–205.

Munoz, R., Moreno, M., Oliva, C. *et al.* (2006) Syndecan-4 regulates non-canonical Wnt signalling and is essential for convergent and extension movements in Xenopus embryos. *Nature Cell Biology*, **8**, 492–500.

Niehrs, C. (2006) Function and biological roles of the Dickkopf family of Wnt modulators. *Oncogene*, **25**, 7469–7481.

Ohkawara, B., Glinka, A., and Niehrs, C. (2011) Rspo3 binds syndecan 4 and induces Wnt/PCP signaling via clathrin-mediated endocytosis to promote morphogenesis. *Developmental Cell*, **20**, 303–314.

Ohkawara, B., Yamamoto, T.S., Tada, M., and Ueno, N. (2003) Role of glypican 4 in the regulation of convergent extension movements during gastrulation in Xenopus laevis. *Development*, **130**, 2129–2138.

Razani, B., Woodman, S.E., and Lisanti, M.P. (2002) Caveolae: from cell biology to animal physiology. *Pharmacological Reviews*, **54**, 431–467.

Sakane, H., Yamamoto, H., and Kikuchi, A. (2010) LRP6 is internalized by Dkk1 to suppress its phosphorylation in the lipid raft and is recycled for reuse. *Journal of Cell Science*, **123**, 360–368.

Sakane, H., Yamamoto, H., Matsumoto, S. *et al.* (2012) Localization of glypican-4 in different membrane microdomains is involved in the regulation of Wnt signaling. *Journal of Cell Science*, **125**, 449–460.

Sato, A., Yamamoto, H., Sakane, H. *et al.* (2010) Wnt5a regulates distinct signalling pathways by binding to Frizzled2. *The EMBO Journal*, **29**, 41–54.

Schulte, G. (2010) International Union of Basic and Clinical Pharmacology. LXXX. The class Frizzled receptors. *Pharmacological Reviews*, **62**, 632–667.

Semënov, M.V., Zhang, X., and He, X. (2008) DKK1 antagonizes Wnt signaling without promotion of LRP6 internalization and degradation. *The Journal of Biological Chemistry*, **283**, 21427–21432.

Sigismund, S., Woelk, T., Puri, C. *et al.* (2005) Clathrin-independent endocytosis of ubiquitinated cargos. *Proceedings of the National Academy of Sciences of the United States of America*, **102**, 2760–2765.

Song, H.H. and Filmus, J. (2002) The role of glypicans in mammalian development. *Biochimica et Biophysica Acta*, **1573**, 241–246.

Song, H.H., Shi, W., Xiang, Y.Y., and Filmus, J. (2005) The loss of glypican-3 induces alterations in Wnt signaling. *The Journal of Biological Chemistry*, **280**, 2116–2125.

Sorkin, A. and von Zastrow, M. (2009) Endocytosis and signalling: intertwining molecular networks. *Nature Reviews Molecular Cell Biology*, **10**, 609–622.

Taelman, V.F., Dobrowolski, R., Plouhinec, J.L. *et al.* (2010) Wnt signaling requires sequestration of glycogen synthase kinase 3 inside multivesicular endosomes. *Cell*, **143**, 1136–1148.

Takada, R., Satomi, Y., Kurata, T. *et al.* (2006) Monounsaturated fatty acid modification of Wnt protein: its role in Wnt secretion. *Developmental Cell*, **11**, 791–801.

Traub, L.M. (2009) Tickets to ride: selecting cargo for clathrin-regulated internalization. *Nature Reviews Molecular Cell Biology*, **10**, 583–596.

Willert, K., Brown, J.D., Danenberg, E. *et al.* (2003) Wnt proteins are lipid-modified and can act as stem cell growth factors. *Nature*, **423**, 448–452.

Yamamoto, H., Komekado, H., and Kikuchi, A. (2006) Caveolin is necessary for Wnt-3a-dependent internalization of LRP6 and accumulation of β-catenin. *Developmental Cell*, **11**, 213–223.

Yamamoto, H., Sakane, H., Yamamoto, H. *et al.* (2008) Wnt3a and Dkk1 regulate distinct internalization pathways of LRP6 to tune the activation of β-catenin signaling. *Developmental Cell*, **15**, 37–48.

Yamamoto, H., Kitadai, Y., Yamamoto, H. *et al.* (2009) Laminin γ2 mediates Wnt5a-induced invasion of gastric cancer cells. *Gastroenterology*, **137**, 242–252.

Yamamoto, H., Oue, N., Sato, A. *et al.* (2010) Wnt5a signaling is involved in the aggressiveness of prostate cancer and expression of metalloproteinase. *Oncogene*, **29**, 2036–2046.

Yu, A., Rual, J.F., Tamai, K. *et al.* (2007) Association of Dishevelled with the clathrin AP-2 adaptor is required for Frizzled endocytosis and planar cell polarity signaling. *Developmental Cell*, **12**, 129–141.

Zerial, M. and McBride, H. (2001) Rab proteins as membrane organizers. *Nature Reviews. Molecular Cell Biology*, **2**, 107–117.

Zhang, C., Li, Q., and Jiang, Y.J. (2007) Zebrafish Mib and Mib2 are mutual E3 ubiquitin ligases with common and specific delta substrates. *The Journal of Molecular Biology*, **366**, 1115–1128.

Zhang, X., Abreu, J.G., Yokota, C. *et al.* (2012) Tiki1 is required for head formation via Wnt cleavage-oxidation and inactivation. *Cell*, **149**, 1565–1577.

9 New Insights from Proteomic Analysis of Wnt Signaling

Matthew P. Walker[1,*], Dennis Goldfarb[2,*], and Michael B. Major[1,2]

[1]Department of Cell Biology and Physiology, Lineberger Comprehensive Cancer Center, University of North Carolina at Chapel Hill School of Medicine, Chapel Hill, NC, USA
[2]Department of Computer Science, Lineberger Comprehensive Cancer Center, University of North Carolina at Chapel Hill, Chapel Hill, NC, USA

A brief introduction to protein mass spectrometry

Over the past several decades, major advances in engineering, experimental approaches, and computational algorithms have propelled protein mass spectrometry (MS) to the forefront of biological discovery. Through its application, peptides and proteins as well as their post-translational modifications (PTMs) are identified and quantified within complex biological samples. Fundamentally, mass spectrometers measure the mass to charge ratio (m/z) of ions, whether they be metabolites, nucleic acids, or proteins. In the powerful and popular experimental workflow of liquid chromatography-shotgun MS, a protein sample is digested into peptides, which are then chromatographically separated under high pressure and high voltage in liquid phase. The fractionated peptides emerge from the chromatography column as charged droplets. Subsequent evaporation yields ionized peptides suitable for mass and charge determination. The m/z of a given peptide ion is referred to as the precursor ion; its intensity is used to determine peptide abundance. Peptide identity, on the other hand, is revealed through dissociation of the precursor ion into a diagnostic series of fragment ions, referred to as the MS2 spectrum. As peptides elute from the chromatography column, an ion detector continually measures the m/z ratio of the precursor ions, the most intense of which at a given retention time are selected for fragmentation. Collision-induced dissociation (CID), which employs neutral atoms to bombard and fragment an isolated parent ion, is a commonly used fragmentation method due to its ability to create sequence-determining ions. The fragment m/z values are then recorded into an MS2 spectrum. The precursor's amino acid sequence is most often determined computationally by a database search in which the observed MS2 spectrum is compared to *in silico* predicted MS2 spectra for peptides within the known proteome; if the peptide is not known or predicted from available genomic sequences, the spectrum will be assigned to an incorrect peptide – usually with a poor score. A single experiment commonly generates tens of thousands of spectra, of which ~15% are confidently assigned to peptides.

*Matthew P. Walker and Dennis Goldfarb contributed equally to this chapter.

Wnt Signaling in Development and Disease: Molecular Mechanisms and Biological Functions, First Edition.
Edited by Stefan Hoppler and Randall T. Moon.

Protein–protein interaction networks and activity-based proteomics

Arguably, the most valuable and informative application of proteomics to the biological sciences has been the definition of co-complexed proteins, including protein interaction dynamics and PTMs. Interacting proteins are most often revealed through affinity purification–mass spectrometry (APMS), which isolates a protein of interest – known as the bait – and its associated proteins from a protein lysate. Proteins identified from an APMS experiment are referred to as prey and are assumed to have been physically bound to the bait – directly or indirectly – and therefore constitute one or more of the bait's various protein complexes. Great care must be taken when interpreting APMS data as most identified proteins are false-positives. The identities of these contaminants vary considerably and depend partly on the tissue origin and the purification approach. Control experiments are frequently employed to filter these candidate interaction lists, as well as hand picking based on a prey's abundance, reproducibility, and uniqueness. Several computational methods are available to automate this process, such as SAINT, CompPASS, MiST, and HGSCore (Choi *et al.*, 2011; Guruharsha *et al.*, 2011; Jager *et al.*, 2012; Sowa *et al.*, 2009).

Subsets of the proteome can also be purified by virtue of varying biochemical properties within the proteome. Activity-based proteomics selectively isolates proteins based upon their enzymatic activities (Nomura, Dix, and Cravatt, 2010). Active kinases, for example, can be easily purified and quantitated using kinobead technology coupled to MS (Duncan *et al.*, 2012). Other activity-based proteomics approaches involve comparing APMS data between enzymatic-dead and wild-type proteins, such as protease traps, which aides in the identification of substrates (Dix, Simon, and Cravatt, 2008). Beyond activity-based fractionation of the proteome, recent methodological advancements have propelled our ability to simplify the proteome through depletion of high-copy proteins, such as is required for biomarker discovery platforms, which seek low-abundant proteins from complex mixtures. Similarly, peptide enrichment strategies are frequently employed to isolate low-abundant peptides of functional interest. For example, TiO_2 can be easily employed to enrich for phosphorylated peptides (Olsen *et al.*, 2006). In addition, a recently developed antibody recognizing the diglycine tryptic remnant on ubiquitinated lysines provides an enrichment approach to profile the ubiquitinated proteome (Kim *et al.*, 2011). These methods have given rise to numerous studies on proteome-wide ubiquitination and phosphorylation site identification.

Quantitative proteomics

In the often employed shotgun APMS approach, identified proteins are represented by varying numbers of identified peptides, where specific peptides may be observed multiple times. With spectral counting approaches, the number of MS2 spectra mapping to a given protein serves as a semiquantitative metric of protein abundance. Though there are many examples of its successful use, spectral counting suffers from missing low-abundant proteins due to the standard data-dependent acquisition strategy of only fragmenting the most intense precursor peaks, as well as run-to-run variability due to the stochastic nature of the MS pipeline. Protein and peptide labeling methods, on the other hand, remove run-to-run variability by directly comparing samples simultaneously and concomitantly reduce machine time by multiplexing the experiment. Proteins belonging to different samples are differentiated by some form of isotopic or chemical labeling, such as SILAC, which uses cells grown in media containing heavy amino acids (Ong *et al.*, 2002). The ratios of the heavy and light peptide abundances provide relative quantification. Both labeled and label-free approaches are powerful due to their ability to quantify proteins in an unbiased, discovery-based fashion.

Selected reaction monitoring (SRM, also referred to as multiple reaction monitoring) is an exquisitely sensitive, targeted approach that provides accurate and precise abundance measurements of a predetermined set of proteins. It requires not only *a priori* selection of the proteins to be quantified but also their corresponding proteotypic peptides – peptides whose sequence is unique to the protein and with physicochemical properties ideal for mass spectrometric analysis – as well as the most common fragment ions produced from these peptides, known as transitions, and the peptide's expected charge state. Several databases and software applications are now available to facilitate SRM-based discovery (MacLean *et al.*, 2010; Picotti *et al.*, 2013). As stated, typical data-dependent acquisition selects the most intense m/z peaks for fragmentation. With SRM, peaks are chosen only if they match one of the predetermined ions – regardless of their signal strength. They are then subjected to fragmentation, and if the observed transitions match those expected, the ion is determined to be the preselected peptide and its signal intensity is recorded. This strategy allows for very low-abundant proteins to be quantified, with the proteotypic peptides serving as surrogates for the proteins of interest. When coupled with spiked-in stable isotope dilution of synthetic peptides to serve as internal standards, absolute quantification can be determined (Gerber *et al.*, 2003). Using SRM, we can now study protein dynamics in a hypothesis-driven manner.

Network analysis of β-catenin-dependent WNT signaling

Protein MS has enabled significant advancements within the WNT field, primarily through the discovery of novel regulatory proteins and PTMs. Here, we will break down and summarize the results of previous studies, beginning at the membrane and concluding within the nucleus. Because MS-based proteomics is a relatively new technology for the biological sciences, a major goal of this review is not only to summarize the experiments that have been done but to call attention to areas in need of further study. To better frame this goal, we surveyed the literature and compiled 26 published studies that used proteomics to tackle some aspect of WNT signaling (Table 9.1).

Proximal components of WNT signaling

The proximal components of the WNT signaling pathway have received very little proteomic attention to date; MS analysis of only WNT3a and the RYK protein complex has been reported. MS figured prominently in the discovery of WNT ligand PTMs, a few of which control WNT protein secretion and activity, including glycosylation, acylation, and sulfation (Takada *et al.*, 2006; Willert *et al.*, 2003; see also Chapter 1). MS analysis of purified WNT3a revealed the type and site of two acetyl groups: a palmitic acid linked to a cysteine and a palmitoleic acid on a serine. Acylation of serine 209 (murine WNT3a) is required for WNT3a secretion and activity (Takada *et al.*, 2006). This knowledge has greatly improved WNT ligand purification, as well as contributed to the identification of porcupine, an ER-localized protein that is required for WNT processing. Small molecule screens identified a number of compounds that impede WNT secretion through porcupine inhibition, thus blocking autocrine WNT pathway activity (Proffitt *et al.*, 2013). These compounds are important research tools and have shown some success in preclinical studies. Whether additional PTMs exist on WNT3a, how PTMs vary between different WNT ligands, and whether PTMs show tissue and context-specific patterns await future proteomic study. Furthermore, the analysis of the WNT ligand protein–protein interaction network is sadly lacking; such APMS studies promise to reveal new insight into WNT–Frizzled (Fz)–LRP dynamics, possibly through new effector proteins and PTMs.

Beyond WNT3a, the atypical receptor tyrosine kinase RYK is the only other proximal WNT component to be examined by MS (Berndt *et al.*, 2011; see also Chapter 7). Proteomic analysis of the RYK protein complex revealed 66 high-confidence interactions belonging to seven functional groups: cell cycle/apoptosis, cell adhesion, desmosome, cell–cell signaling, planar cell polarity (PCP), actin-remodeling, and ubiquitin trafficking. Expanding on these data, MIB1, an E3 ubiquitin ligase famed for its control of Delta/Notch signaling, was shown to control WNT signaling by ubiquitinating RYK. Precisely how MIB1 controls WNT pathway activity and whether it constitutes a node of pathway cross talk remains unknown. More recently, the RYK protein interaction network, with an obvious enrichment of PCP proteins, served to support functional studies into the role of RYK in establishing PCP (Macheda *et al.*,

Table 9.1 APMS experiments performed on members of the Wnt/β-catenin pathway.

Baits	Cell line/tissue	Purification	Treatment	Filtering	References
WNT3A	L-cell	FLAG	—	NA	Takada *et al.* (2006)
WNT3A	L-cell	Chromatography	—	NA	Willert *et al.* (2003)
RYK	HEK293T	TAP	—	Hand filtered	Berndt *et al.* (2011)
DVL3	HEK293	Streptavadin, FLAG, IP	—	NA	Wu *et al.* (2012)
DVL3	HEK293	HA	—	Controls	Bikkavilli and Malbon (2010)
DVL (*Drosophila*)	HEK293T	IP Western/Gel excision	—	NA	Yanfeng *et al.* (2011)
DVL2, DVL3	HEK293T	TAP	—	Hand filtered	Angers *et al.* (2006)
APC, AXIN1, CTBP, DVL2	HeLa	GFP	±WNT3A	SILAC ratio	Hilger and Mann (2012)
GSK3B, DKK1	HEK293	FLAG	—	Regression model	Ewing *et al.* (2007)
CSNK1A1	Rat brain lysates	IP	—	NA	Dubois *et al.* (2002)
PP2A complex	HEK293	Streptavadin + HA	—	Controls	Glatter *et al.* (2009)
AXIN1	Ls174T	IP	—	Controls	Sue Ng *et al.* (2010)
AXIN1	HEK293T	IP	±WNT3A	Hand filtered	Li *et al.* (2012)
AXIN1, AXIN2	HEK293T	TAP	—	Controls	Lui *et al.* (2011)
CDC73, BTK	HEK293T	FLAG	—	Hand filtered	James *et al.* (2009)
APC	HEK293T	FLAG	—	Controls	Breitman *et al.* (2008)
CTNNB1, AXIN1, BTRC, FBXW11, WTX	HEK293T	TAP	—	Hand filtered	Major *et al.* (2007)
APC, CDC73, AGGF1	DLD1	TAP, IP	—	siRNA	Major *et al.* (2008)
CTNNB1	SW480	IP Western/Gel excision	—	NA	Sierra *et al.* (2006)
CTNNB1	Ls174T	IP	—	SILAC ratio	Selbach and Mann (2006)
AMER1, AMER2, AMER3	HEK293T	TAP	—	Hand filtered	Siesser *et al.* (2012)
CUL1, CUL3	HEK293T	HA	±Neddylation inhibitor	CompPASS	Bennett *et al.* (2010)
BTRC	HEK293T	HA	—	CompPASS	Sowa *et al.* (2009)
BTRC	HEK293T	HA	—	CompPASS	Gao *et al.* (2011)
CREBBP	C2C12	IP Western/Gel excision	Phospho-enrichment	NA	Xia *et al.* (2011)
RUVBL1	HEK293T	FLAG	—	NA	Kim *et al.* (2007)

TAP, tandem affinity purification; IP, immunoprecipitation; GFP, green fluorescent protein; SILAC, stable isotope labeling by amino acids in cell culture.

2012; see also Chapter 6). The insight gained from APMS studies of RYK underscores the pressing need to solve the LRP5, LRP6, and the Fz receptor protein interaction networks, as well as their PTMs.

Dishevelled proteomics

Given their central yet enigmatic role in both β-catenin-dependent and β-catenin-independent WNT signaling, the DVL family of proteins has been the topic of several proteomic studies, both with regard to PTMs and protein–protein interactions (see also Chapter 15). The analysis of the DVL proteins themselves revealed numerous phosphorylation and methylation PTMs, but to date, none of which have been shown to affect DVL function on β-catenin signaling (Wu *et al.*, 2012). This contrasts to tyrosine phosphorylation at position 473 of *Drosophila* Dsh, which is required for Dsh-driven PCP signaling (Yanfeng *et al.*, 2011). Of the 11 reported PTMs, dimethylation at R698 is transiently increased following WNT stimulation; again, the functional consequence of this awaits further study.

The Moon, Malbon, and Mann laboratories have independently defined the protein interaction networks for DVL2 and/or DVL3. From these studies, two novel DVL-interacting proteins were functionally examined: KLHL12 and KSRP (Angers *et al.*, 2006; Bikkavilli and Malbon, 2010). KLHL12 is an E3 ubiquitin ligase that controls the stability of DVL2 and DVL3. Consequently, KLHL12 regulates both β-catenin-dependent and β-catenin-independent WNT signaling in cell lines, *Xenopus* and zebrafish. Through APMS study of DVL3, the Malbon laboratory discovered a novel interaction with the RNA-binding protein KSRP. Functional studies showed that when complexed with DVL3, KSRP negatively regulates β-catenin-dependent WNT signaling by destabilizing the β-catenin mRNA. Interestingly, this suppression was transiently alleviated following WNT3a stimulation, thereby establishing a regulator loop contributing to WNT-driven β-catenin activation. In a third study by the Mann laboratory, the DVL2 protein complex was interrogated as part of a larger APMS study on WNT signaling proteins. Although novel interactions were identified, functional assessments were not reported.

It is important to note that despite using the same cell line and similar instrumentation, minimal overlap was observed between the DLV3 APMS analyses from the Moon and Malbon studies (Angers *et al.*, 2006; Bikkavilli and Malbon, 2010). Of the proteins reported, only Cullin3 was identified by both groups. On the other hand, there was moderate overlap between the DVL2 runs from the Moon and Mann laboratories despite the use of different cell lines, HEK293T and HeLa cells, respectively (Angers *et al.*, 2006; Hilger and Mann, 2012). Although both studies reported many known interactors, there was little overlap between the novel binding proteins, including KLHL12. This variability could be the result of differences in affinity protocols, lysis conditions (detergent), cell lines, or data filtration strategies; it is unlikely to be due to problems with the mass spectrometer as sufficient bait protein was observed in all experiments. Despite their differences, these APMS studies revealed new facets of DVL biology within the WNT pathway and provided numerous candidate proteins for future work.

The β-catenin destruction complex

Cytosolic protein complexes, often nucleated by a scaffolding protein, typically yield strong and highly reproducible protein interaction networks. Consequently, APMS characterization of the β-catenin destruction complex has been very informative, revealing new signaling components as well as insight into the mechanics of established components (see also Chapter 3). To date, proteomic analysis of the destruction complex comprises 23 publications on 11 unique bait proteins (Table 9.1).

In stark contrast to the relatively stable protein interactions of cytosolic scaffolding proteins, APMS analysis of catalytic proteins often fails to capture many of the associated proteins (Ewing *et al.*, 2007). For this and other reasons, negative data within APMS experiments should be interpreted with little or no confidence, particularly when dealing with enzymes and substrates. Indeed, although proteomic characterization of the CSNK1A (CKIα) protein kinase identified a number of potential interacting proteins, including PP2A, neither β-catenin nor destruction complex components were detected (Dubois *et al.*, 2002). More recently, reciprocal APMS runs on various PP2A subunits similarly failed to reveal WNT pathway components, suggesting that, as expected, the interaction is likely transient and/or comprises a small fraction of total cellular PP2A (Glatter *et al.*, 2009). That said, in keeping with their central role within the destruction complex, the casein kinases are among the more commonly observed proteins from APMS experiments employing the following baits: AXIN1, CTNNB1, APC, and DVLs (Table 9.1).

In a comprehensive study by the Mann laboratories, SILAC-based quantitative analysis of the AXIN1, APC, CTBP2, and DVL2 protein interaction networks revealed known and novel associated proteins, including dynamic remodeling following WNT ligand stimulation (Hilger and Mann, 2012). While molecular follow-up studies remain incomplete, a number of interacting proteins were identified. The most notable interactor was CCDC88A (Girdin), which bound APC, AXIN1, and DVL2 in the presence of WNT3a and in the absence of WNT3a only bound DVL2. Girdin is a relatively unstudied actin binding protein with reported functional roles in the AKT-mTOR signaling pathway and in actin cytoskeletal dynamics. Interestingly, Girdin mRNA expression positively correlates with an increased risk of metastasis in colorectal cancer (Liu *et al.*, 2012). Given its reported role in an important cancer signaling cascade, its correlation with increases metastasis, and its interaction with multiple WNT pathway components, Girdin is an obvious target for further study.

Unlike the DVL2 and DVL3 APMS runs discussed in the previous section, there is considerable overlap between AXIN1 APMS experiments performed in the Angers, Mann, Clevers, and Moon laboratories (Hilger and Mann, 2012; Li *et al.*, 2012; Lui *et al.*, 2011; Major *et al.*, 2007). Over 50% of the interactions were shared between the groups despite different purification techniques and cell lines. Furthermore, a comparative study of the AXIN1 and AXIN2 protein interaction networks revealed considerable redundancy (Lui *et al.*, 2011). Of the 17 reported interactions, 12 were observed within both AXIN1 and AXIN2 protein complexes. Of these, the ubiquitin-specific protease USP34 and WDR26 represented novel discoveries. USP34 removes ubiquitin from AXIN, resulting in its stabilization and, unexpectedly, its nuclear accumulation. Through an unknown mechanism, an increased AXIN level within the nuclear compartment promotes WNT signaling. Therefore, USP34 presents as an attractive drug target because it is catalytic and functions to promote WNT signaling downstream of the β-catenin destruction complex.

Through integration of AXIN1, APC, and β-catenin APMS experiments, we previously identified the AMER1 (WTX) tumor suppressor protein as a novel component of the destruction complex (Major *et al.*, 2007, 2008). In agreement with the interaction data, WTX expression increases CTNNB1 ubiquitination and is critical for proper stasis of WNT signaling in cell lines, *Xenopus*, zebrafish, and mice. Inactivating mutations in AMER1 are common in Wilms' tumor, and it has been hypothesized that these mutations lead to activation of WNT signaling. This work represents an important breakthrough in the understanding of WNT signaling and demonstrated the power of APMS in the identification of clinically significant modulators of WNT signaling. Further work expanded upon this initial discovery to include APMS analysis on all AMER family members. Both AMER1 and AMER2 copurify with APC and FBXW11 and inhibit WNT signaling but curiously only AMER1 binds β-catenin (Siesser *et al.*, 2012). Unlike AMER1, AMER2 has a distinct role in the regulation of microtubule stability, a discovery that was made through comparative APMS analyses of the protein family. Finally, AMER3 has no effect on WNT signaling and does not bind any of the core components of WNT signaling. Because the initial discovery and much of the molecular/mechanistic analysis was deeply enhanced through APMS, the AMER family of proteins is a shining example of the power of MS-mediated proteomics.

Although not directly used to study WNT signaling, extensive studies by Wade Harper and colleagues defined the protein interaction networks for Cullin-RING ubiquitin ligases, ubiquitin-specific proteases, autophagy pathway members, and β-TRCP1/2 (Bennett *et al.*, 2010;

Gao *et al.*, 2011; Sowa *et al.*, 2009). While there were no WNT-specific advances, this data contains a number of proteins that are important for WNT signaling and can be mined for hypothesis generation. In similar fashion, an automated proteome scale yeast two-hybrid screen of ~5000 baits revealed numerous protein interactions of immediate interest to the WNT field. For example, ANP32a and CRMP1 were identified as AXIN1-interacting proteins and were shown to repress WNT signaling via TOPFLASH assays (Stelzl *et al.*, 2005).

Comprehensive analysis has been done on most of the core components of the destruction complex, but these studies were done in either HEK293T or HeLa cells, cell lines that do not represent clinically significant tissue. There is a pressing need to expand the range of cell lines in which APMS experiments are done. For example, proteomic analysis of AXIN1 and APC from the colon cancer line Ls174T revealed MAP3K1 and FLII as novel discoveries. Interestingly, we identified both MAP3K1 and FLI1 as WNT pathway-interacting proteins using HCT116 colon cancer cells, but not in any APMS experiments employing HEK293T cells (Song *et al.*, 2012; Sue Ng *et al.*, 2010). MAP3K1 inhibition abrogates WNT activity and represents a novel therapeutic target for WNT pathway inhibition (Sue Ng *et al.*, 2010). Given the relative lack of information on FLII, any insight into its molecular role could be beneficial. These possible tissue-specific discoveries indicate a strong need for a more diverse set of cell lines used in APMS studies.

Proteomic analysis of nuclear β-catenin

Similar to the proximal signaling complex, the nuclear events of WNT signaling have untapped potential in regard to potential APMS impact. A few studies have been reported. First, although nuclear fraction was not employed, APMS experiments on S37A stabilized β-catenin did not reveal any novel nuclear interactors (Major *et al.*, 2007, 2008). Second, using the Armadillo repeats 11–12 of β-catenin as bait, TRRAP, p400, MLL1, SNF2H, and TIP49 were identified as novel nuclear interactors of β-catenin (Sierra *et al.*, 2006). Interestingly, the association of these chromatin remodelers with β-catenin depends on APC. Third, small molecule and small interfering RNA (siRNA)-based functional screening identified Bruton's tyrosine kinase (BTK) as a negative regulator of WNT signaling. BTK associates with and functions through the parafibromin complex to control β-catenin-driven transcription (James *et al.*, 2009). Finally, a recent yeast two-hybrid screen on the DNA-binding domain of TCF7L2 identified RNF14 as a novel nuclear WNT activator (Wu *et al.*, 2013). RNF14 enhances WNT signaling, but cannot activate in the absence of TCF4 or β-catenin. In regard to proteomics, this work demonstrated that yeast two-hybrid remains a valuable tool for identifying direct protein–protein interactions, and there are still as of yet undiscovered regulators of WNT signaling in the nucleus, reemphasizing the need for APMS studies on nuclear WNT components.

Integration of previously described APMS experiments

Historically, APMS experiments were generally performed and analyzed with little regard to previous studies and/or other proteins. In an effort to compile the WNT-relevant protein interactions, we have begun to create a merged network (Figure 9.1). Through the combination of previously published work, we illustrate proteins that bind multiple core components of the WNT pathway. While large in scale, this figure best represents the contribution of MS-based proteomics to our understanding of β-catenin-dependent WNT signaling. To complement these APMS-based interactions, we integrated the network with yeast two-hybrid screen discoveries. In contrast to the many well-established protein interactions represented within the yeast two-hybrid dataset, we observe no overlap in novel interacting proteins between the APMS and global yeast two-hybrid studies. For example, although DPPA2, BAHD1, PRPF3, and C16ORF48 interact with DVL2 and DVL3 by yeast two-hybrid, these proteins were not identified by APMS with any analyzed bait protein. Like APMS, false-positives within yeast two-hybrid datasets are common; defining and managing the noise is required for data interpretation.

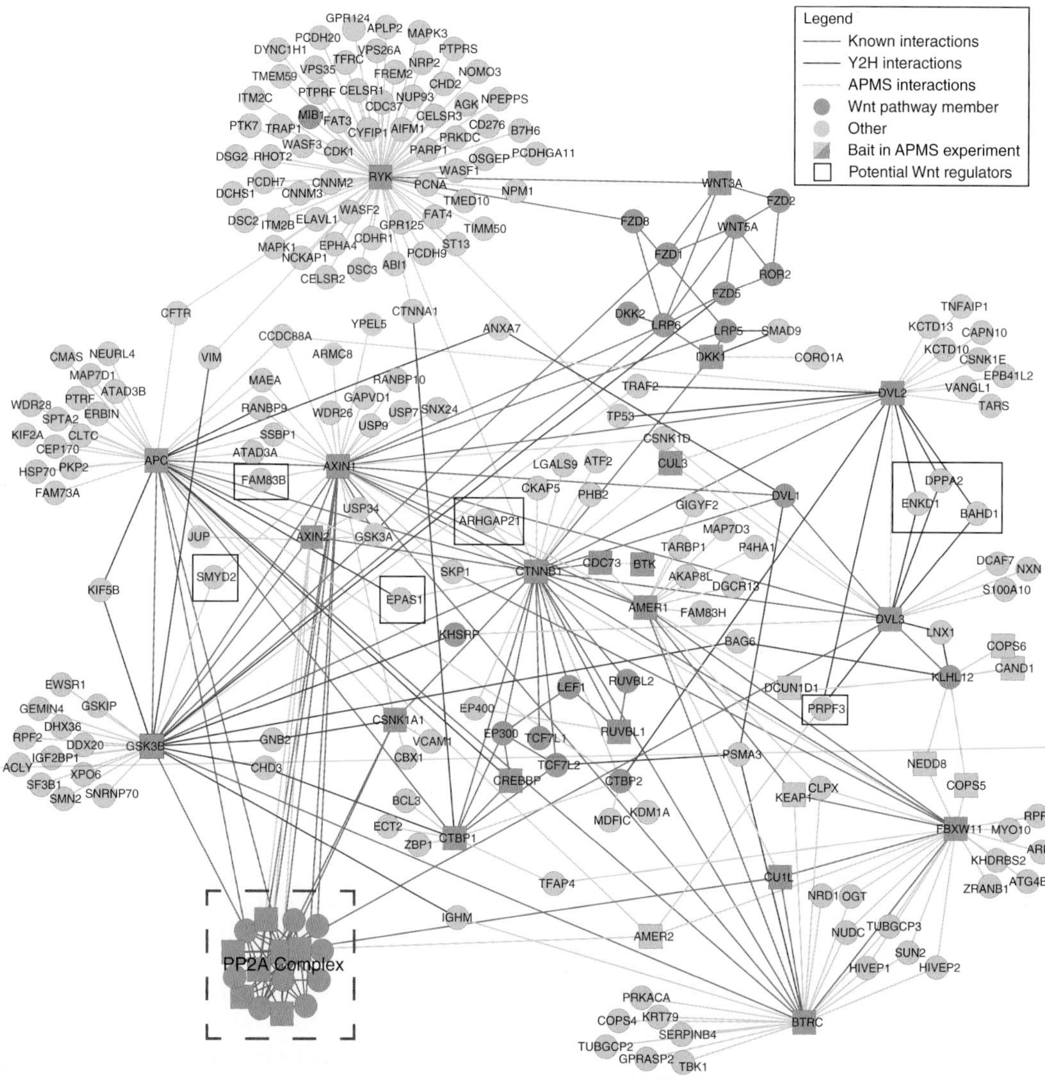

Figure 9.1 Protein–protein interaction network of the WNT/β-catenin pathway. Published APMS experiments and yeast two-hybrid screens were combined with previously known interactions. For purposes of visual clarity, proteins whose only connection to the network was via a single two-hybrid screen interaction were removed from the figure. Protein names correspond to the HUGO nomenclature. *(See insert for color representation of the figure.)*

As illustrated in Figure 9.1, a number of previously unrecognized proteins interact with multiple core components of the WNT pathway. ARHGAP21 is a RHO GTPases activating protein with reported roles in ARP2/3 complex regulation, RHOA activation, and α-catenin recruitment to the membrane that lead to changes in epithelial to mesenchymal transition and cellular migration (Barcellos *et al.*, 2013). By integrating the published APMS studies, we discovered that ARHGAP21 binds CTNNB1, AMER1, and AXIN1 in multiple independent experiments using different affinity purification approaches. Given the established role between WNT signaling and the functional consequences of ARHGAP21 activity, we believe that ARHGAP21 is a promising candidate for further study.

Three other potential WNT pathway interactors, FAM83B, SMYD2, and EPAS1, are of particular interest because of their established connection with human disease. FAM83B interacts with CTNNB1 and AXIN1 and is a mediator of resistance to tyrosine kinase inhibition and oncogenesis, but surprisingly little is known of its molecular function (Cipriano *et al.*, 2012; Grant, 2012). SMYD2 is a described oncogene that is present within AXIN1 and GSK3B protein complexes; it has described lysine methyltransferase activity. Finally, EPAS1 was found to bind CTNNB1 and APC by APMS and yeast two-hybrid, respectively. EPAS1 mRNA expression correlates with a poor prognosis in colon and bladder cancer (Mohammed *et al.*, 2011). Due to their clinical connections, these proteins are interesting topics for future research.

Proteomic discovery of WNT pathway biomarkers

Traditionally, biomarker research has focused on identifying serum proteins indicative of disease. Extensive work has been done to identify biomarkers using an unbiased approach across tumors that have strong ties to WNT signaling. Proteomic analysis of secreted proteins from cancer cells, patient blood sample, and shotgun proteomic profiling of colon cancer tissues resulted in the identification of WNT pathway components and markers for pathway activation. For example, a large-scale study identified over 3500 proteins quantitatively in an attempt to identify proteomic differences between microsatellite-instable tumors and normal colon tissue (Kang *et al.*, 2012). These biomarker studies have yet to yield a robust WNT or cancer biomarker and do not currently represent feasible diagnostic tests.

More recent work has explored using SRM technology to specifically identify WNT pathway components or mutated proteins in a complex sample with the hopes of clinically defining a tumor rather than identifying undiagnosed tumors through serum and/or blood. Early SRM studies in colorectal cell lines and patients samples provided a proof of concept for identifying active WNT signaling and mutations in patient samples (Chen *et al.*, 2010; Halvey *et al.*, 2012). Specifically, APC mutations and levels of CTNNB1 were detected in cells and frozen tissue samples. A promising study used SRM to identify mutated proteins in the serum of cultured cells, but more work is needed to determine the effect of these biomarkers in disease progression (Mathivanan *et al.*, 2012). While these studies are preliminary, the use of SRM as a diagnostic tool is a potential breakthrough in the identification/characterization of WNT pathway activity in patient samples.

Conclusion

Here, we have advocated the continued use of proteomics in WNT signaling by describing MS, highlighting the important discoveries that were made through the use of proteomics and calling attention to the gaps that are in need of proteomic attention. As is common in many scientific fields, we are struggling to keep pace with technological innovation. With unprecedented speed and accuracy, we can now quantitate any protein at will from amazingly complex peptide mixtures. As a result, proteome quantitation, including co-complex protein definition and PTMs, is now affordable, realistic, and achievable. In the coming years, the full impact of this technology will be apparent through increased understanding of WNT signaling, including heretofore unseen context-specific determinants and clinical associations.

Acknowledgments

M.B.M. is supported by grants from the state of North Carolina (University Cancer Research Fund), the National Institutes of Health (New Innovator Award, 1-DP2-OD007149-01), the Sidney Kimmel Cancer Foundation (Scholar Award), and the Greensboro Golfers Against Cancer. M.W. has received support from the National Institutes of Health (T32-CA009156-35).

References

Angers, S., Thorpe, C.J., Biechele, T.L. *et al.* (2006) The KLHL12–Cullin-3 ubiquitin ligase negatively regulates the Wnt–β-catenin pathway by targeting Dishevelled for degradation. *Nature Cell Biology*, **8**, 348–357.

Barcellos, K.S., Bigarella, C.L., Wagner, M.V. *et al.* (2013) ARHGAP21 protein, a new partner of alpha-tubulin involved in cell-cell adhesion formation and essential for epithelial-mesenchymal transition. *The Journal of Biological Chemistry*, **288**, 2179–2189.

Bennett, E.J., Rush, J., Gygi, S.P., and Harper, J.W. (2010) Dynamics of Cullin-RING ubiquitin ligase network revealed by systematic quantitative proteomics. *Cell*, **143**, 951–965.

Berndt, J.D., Aoyagi, A., Yang, P. *et al.* (2011) Mindbomb 1, an E3 ubiquitin ligase, forms a complex with RYK to activate Wnt/β-catenin signaling. *The Journal of Cell Biology*, **194**, 737–750.

Bikkavilli, R.K. and Malbon, C.C. (2010) Dishevelled-KSRP complex regulates Wnt signaling through post-transcriptional stabilization of beta-catenin mRNA. **123**, 1352–1362.

Breitman, M., Zilberberg, A., Caspi, M., and Rosin-Arbesfeld, R. (2008) The armadillo repeat domain of the APC tumor suppressor protein interacts with Striatin family members. *Biochimica et Biophysica Acta*, **1783**, 1792–1802.

Chen, Y., Gruidl, M., Remily-Wood, E. *et al.* (2010) Quantification of beta-catenin signaling components in colon cancer cell lines, tissue sections, and microdissected tumor cells using reaction monitoring mass spectrometry. *Journal of Proteome Research*, **9**, 4215–4227.

Choi, H., Larsen, B., Lin, Z.Y. *et al.* (2011) SAINT: probabilistic scoring of affinity purification–mass spectrometry data. *Nature Methods*, **8**, 70–73.

Cipriano, R., Graham, J., Miskimen, K.L. *et al.* (2012) FAM83B mediates EGFR- and RAS-driven oncogenic transformation. *The Journal of Clinical Investigation*, **122**, 3197–3210.

Dix, M.M., Simon, G.M., and Cravatt, B.F. (2008) Global mapping of the topography and magnitude of proteolytic events in apoptosis. *Cell*, **134**, 679–691.

Dubois, T., Howell, S., Zemlickova, E., and Aitken, A. (2002) Identification of casein kinase Ialpha interacting protein partners. *The FEBS Letters*, **517**, 167–171.

Duncan, J.S., Whittle, M.C., Nakamura, K. *et al.* (2012) Dynamic reprogramming of the kinome in response to targeted MEK inhibition in triple-negative breast cancer. *Cell*, **149**, 307–321.

Ewing, R.M., Chu, P., Elisma, F. *et al.* (2007) Large-scale mapping of human protein-protein interactions by mass spectrometry. *Molecular Systems Biology*, **3**, 89.

Gao, D., Inuzuka, H., Tan, M.-K.M. *et al.* (2011) mTOR drives its own activation via SCF(βTrCP)-dependent degradation of the mTOR inhibitor DEPTOR. *Molecular Cell*, **44**, 290–303.

Gerber, S.A., Rush, J., Stemman, O. *et al.* (2003) Absolute quantification of proteins and phosphoproteins from cell lysates by tandem MS. *Proceedings of the National Academy of Sciences of the United States of America*, **100**, 6940–6945.

Glatter, T., Wepf, A., Aebersold, R., and Gstaiger, M. (2009) An integrated workflow for charting the human interaction proteome: insights into the PP2A system. *Molecular Systems Biology*, **5**, 237.

Grant, S. (2012) FAM83A and FAM83B: candidate oncogenes and TKI resistance mediators. *The Journal of Clinical Investigation*, **122**, 3048–3051.

Guruharsha, K.G., Rual, J.F., Zhai, B. *et al.* (2011) A protein complex network of Drosophila melanogaster. *Cell*, **147**, 690–703.

Halvey, P.J., Zhang, B., Coffey, R.J. *et al.* (2012) Proteomic consequences of a single gene mutation in a colorectal cancer model. *Journal of Proteome Research*, **11**, 1184–1195.

Hilger, M. and Mann, M. (2012) Triple SILAC to determine stimulus specific interactions in the Wnt pathway. *Journal of Proteome Research*, **11**, 982–994.

Jager, S., Cimermancic, P., Gulbahce, N. *et al.* (2012) Global landscape of HIV-human protein complexes. *Nature*, **481**, 365–370.

James, R.G., Biechele, T.L., Conrad, W.H. *et al.* (2009) Bruton's tyrosine kinase revealed as a negative regulator of Wnt-beta-catenin signaling. *Science Signaling*, **2**, ra25.

Kang, U.-B., Yeom, J., Kim, H.-J. *et al.* (2012) Expression profiling of more than 3500 proteins of MSS-type colorectal cancer by stable isotope labeling and mass spectrometry. *Journal of Proteomics*, **75**, 3050–3062.

Kim, J.H., Lee, J.M., Nam, H.J. *et al.* (2007). SUMOylation of pontin chromatin-remodeling complex reveals a signal integration code in prostate cancer cells. *Proceedings of the National Academy Sciences of the United States of America*, **104**, 20793–20798.

Kim, W., Bennett, E.J., Huttlin, E.L. *et al.* (2011) Systematic and quantitative assessment of the ubiquitin-modified proteome. *Molecular Cell*, **44**, 325–340.

Li, V.S.W., Ng, S.S., Boersema, P.J. *et al.* (2012) Wnt signaling through inhibition of β-catenin degradation in an intact Axin1 complex. *Cell*, **149**, 1245–1256.

Liu, C., Xue, H., Lu, Y., and Chi, B. (2012) Stem cell gene Girdin: a potential early liver metastasis predictor of colorectal cancer. *Molecular Biology Reports*, **39**, 8717–8722.

Lui, T.T., Lacroix, C., Ahmed, S.M. *et al.* (2011) The ubiquitin-specific protease USP34 regulates axin stability and Wnt/beta-catenin signaling. *Molecular Cell Biology*, **31**, 2053–2065.

Macheda, M.L., Sun, W.W., Kugathasan, K. *et al.* (2012) The Wnt receptor Ryk plays a role in mammalian planar cell polarity signaling. *The Journal of Biological Chemistry*, **287**, 29312–29323.

MacLean, B., Tomazela, D.M., Shulman, N. *et al.* (2010) Skyline: an open source document editor for creating and analyzing targeted proteomics experiments. *Bioinformatics*, **26**, 966–968.

Major, M.B., Camp, N.D., Berndt, J.D. *et al.* (2007) Wilms tumor suppressor WTX negatively regulates WNT/β-catenin signaling. *Science*, **316**, 1043–1046.

Major, M.B., Roberts, B.S., Berndt, J.D. *et al.* (2008) New regulators of Wnt/beta-catenin signaling revealed by integrative molecular screening. *Science Signaling*, **1**, ra12.

Mathivanan, S., Ji, H., Tauro, B.J. *et al.* (2012) Identifying mutated proteins secreted by colon cancer cell lines using mass spectrometry. *Journal of Proteomics*, **76**, 141–149.

Mohammed, N., Rodriguez, M., Garcia, V. *et al.* (2011) EPAS1 mRNA in plasma from colorectal cancer patients is associated with poor outcome in advanced stages. *Oncology Letters*, **2**, 719–724.

Nomura, D.K., Dix, M.M., and Cravatt, B.F. (2010) Activity-based protein profiling for biochemical pathway discovery in cancer. *Nature Reviews Cancer*, **10**, 630–638.

Olsen, J.V., Blagoev, B., Gnad, F. *et al.* (2006) Global, in vivo, and site-specific phosphorylation dynamics in signaling networks. *Cell*, **127**, 635–648.

Ong, S.E., Blagoev, B., Kratchmarova, I. *et al.* (2002) Stable isotope labeling by amino acids in cell culture, SILAC, as a simple and accurate approach to expression proteomics. *Molecular & Cellular Proteomics*, **1**, 376–386.

Picotti, P., Clement-Ziza, M., Lam, H. *et al.* (2013) A complete mass-spectrometric map of the yeast proteome applied to quantitative trait analysis. *Nature*, **494**, 266–270.

Proffitt, K.D., Madan, B., Ke, Z. *et al.* (2013) Pharmacological inhibition of the Wnt acyltransferase PORCN prevents growth of WNT-driven mammary cancer. *Cancer Research*, **73**, 502–507.

Selbach, M. and Mann, M. (2006) Protein interaction screening by quantitative immunoprecipitation combined with knockdown (QUICK). *Nature Methods*, **3**, 981–983.

Sierra, J., Yoshida, T., Joazeiro, C.A., and Jones, K.A. (2006). The APC tumor suppressor counteracts beta-catenin activation and H3K4 methylation at Wnt target genes. *Genes & Development*, **20**, 586–600.

Siesser, P.F., Motolese, M., Walker, M.P. *et al.* (2012) FAM123A binds to microtubules and inhibits the guanine nucleotide exchange factor ARHGEF2 to decrease actomyosin contractility. *Science Signaling*, **5**, ra64.

Song, J., Hao, Y., Du, Z. *et al.* (2012) Identifying novel protein complexes in cancer cells using epitope-tagging of endogenous human genes and affinity-purification mass-spectrometry. *Journal of Proteome Research*, **11**, 5630–5641.

Sowa, M.E., Bennett, E.J., Gygi, S.P., and Harper, J.W. (2009) Defining the human deubiquitinating enzyme interaction landscape. *Cell*, **138**, 389–403.

Stelzl, U., Worm, U., Lalowski, M. *et al.* (2005) A human protein–protein interaction network: a resource for annotating the proteome. *Cell*, **122**, 957–968.

Sue Ng, S., Mahmoudi, T., Li, V.S.W. *et al.* (2010) MAP3K1 functionally interacts with Axin1 in the canonical Wnt signalling pathway. *Biological Chemistry*, **391**, 171–180.

Takada, R., Satomi, Y., Kurata, T. *et al.* (2006) Monounsaturated fatty acid modification of Wnt protein: its role in Wnt secretion. *Developmental Cell*, **11**, 791–801.

Willert, K., Brown, J.D., Danenberg, E. *et al.* (2003) Wnt proteins are lipid-modified and can act as stem cell growth factors. *Nature*, **423**, 448–452.

Wu, C., Wei, W., Li, C. *et al.* (2012) Delicate analysis of post-translational modifications on Dishevelled 3. *Journal of Proteome Research*, **11**, 3829–3837.

Wu, B., Piloto, S., Zeng, W. *et al.* (2013) Ring Finger Protein 14 is a new regulator of TCF/beta-catenin-mediated transcription and colon cancer cell survival. *EMBO Reports*, **14**, 347–355.

Xia, Z., Guo, M., and Ma, H. (2011) Functional analysis of novel phosphorylation sites of CREB-binding protein using mass spectrometry and mammalian two-hybrid assays. *Proteomics*, **11**, 3444–3451.

Yanfeng, W.A., Berhane, H., Mola, M. *et al.* (2011) Functional dissection of phosphorylation of Disheveled in Drosophila. *Developmental Biology*, **360**, 132–142.

10 New Insights about Wnt/β-Catenin Pathway Mechanisms from Global siRNA Screens

Tenzin Gocha and Ramanuj DasGupta

Department of Biochemistry and Molecular Pharmacology, and the NYU Cancer Institute, New York University Langone Medical Center, New York, NY, USA

Understanding cell signaling pathways in the postgenomic era

One of the fundamental questions during embryonic development is how an individual fertilized egg cell develops into a complex multicellular organism. Research over the past several decades has implicated the critical functions of cell signaling pathways – the mechanism by which cells communicate with each other – in the determination of cell fates that may involve many cell biological processes such as cell proliferation, differentiation, cell shape changes, cell polarity, and death. A major challenge that cells encounter in a developing embryo is that they often simultaneously receive multiple signaling cues, which they have to somehow integrate to promote a unitary developmental outcome or "cell fate." The molecular understanding of the mechanisms by which these parallel signals and their transcriptional outcome are integrated within a cell to generate specific fates remains a major challenge in the field. In the past several years, research in the area of network biology, aided by the rapid growth of high-throughput screening/high-content screening (HTS/HCS), whole-genome sequencing, and proteomic technologies, has attempted to define and better elucidate the mechanisms by which signaling pathways "talk" with each other. This has resulted in an appreciation for the "systems-level" understanding of cell signaling networks (Boxem *et al.*, 2008; Gunsalus *et al.*, 2005; Tewari *et al.*, 2004)– that comprise of multiple nodes (shared cross-regulatory proteins between distinct signal transduction pathways) and edges (direct or indirect interactions between linked nodes) through which information can be integrated and transduced in a regulated fashion within a cell. These observations are therefore beginning to alter the traditional notion of cell signaling pathways from linear cassettes of genes mediating flow of information to complex interconnected gene regulatory networks that need to function in a coordinated manner to dictate specific cell biological outcomes or cell fates.

There is now a rapidly evolving list of genes that were originally identified as part of a given canonical pathway but now are known to regulate multiple signal transduction cascades (De Toni *et al.*, 2006; Hasson *et al.*, 2005; Letamendia, Labbe, and Attisano, 2001; Lum

Wnt Signaling in Development and Disease: Molecular Mechanisms and Biological Functions, First Edition.
Edited by Stefan Hoppler and Randall T. Moon.

et al., 2003; Musgrove, 2004; Von Bubnoff and Cho, 2001)–. Early examples of genes that may function as shared regulators or nodes of cross talk are glycogen synthase kinase 3β (GSK3β) and casein kinase 1α (CK1α), both of which were initially identified as negative regulators in the Wnt/*wingless* (*wg*) pathway but were later shown to function in the Sonic hedgehog (Shh)/*hedgehog* (*hh*) signaling pathway (Lum *et al.*, 2003; Musgrove, 2004; Nusse, 2003). One can argue that such cross-regulation between signaling pathways may have important roles in determining the fate of a cell that is under simultaneous influence of multiple signaling cues. Therefore, the identification of genes that could serve as cross-regulatory nodes and the determination of their mechanisms of action are crucial to understand how the coordinated regulation of cell signaling networks controls many aspects of animal development. Importantly, the nodes of cross talk determined by specific protein–protein interactions or gene regulatory interactions, such as protein–DNA and microRNA (miRNA)–mRNA, could serve as prime therapeutic targets in human disease, such as cancer.

The Wnt/Wg signaling pathway

Regulation of pathway components by members of other signal transduction cascades is especially evident for the evolutionarily conserved Wnt signaling pathway (De Toni *et al.*, 2006; Hasson *et al.*, 2005; Letamendia, Labbe, and Attisano, 2001; Lum *et al.*, 2003; Musgrove, 2004; Von Bubnoff and Cho, 2001)– (Figure 10.1). Wnts belong to a family of conserved signaling molecules that have been shown to regulate many aspects of animal development by controlling basic cell biological processes such as cell proliferation, differentiation, death, polarity, and cell shape changes (Miller *et al.*, 1999; Polakis, 2000; Wodarz and Nusse, 1998)–. Knockout mouse models of the Wnt genes or of those that encode regulators of the Wnt pathway display a variety of birth defects, including those of the central nervous system, axial skeleton, limbs, and occasionally other organs (Ciruna *et al.*, 2006; Grove *et al.*, 1998; Jiang, Bush, and Lidral, 2006; Kokubu *et al.*, 2004; Shu *et al.*, 2002; Staal and Clevers, 2000)–. Aberrant Wnt signaling has also been linked to human disease such as retinal degeneration of the eye (FEVR) (Kaykas *et al.*, 2004; Robitaille *et al.*, 2002) and a variety of cancers including those of the liver, intestine, breast, and skin (Gat *et al.*, 1998; Miyoshi and Hennighausen, 2003; Miyoshi *et al.*, 2002; Polakis, 1997)–. It is therefore not surprising that multiple levels of regulation have to be imposed to control the activity of this important pathway in order to ensure normal development and prevent dysregulated states, such as cancer.

Wnts/*wg* encode secreted glycoproteins that activate receptor-mediated pathways that control expression of numerous transcriptional targets (Moon *et al.*, 2004; Nusse, 1999; Wodarz and Nusse, 1998). The main function of the canonical Wnt pathway is to regulate the stability of the cytosolic pool of a key mediator, β-catenin (β-cat)/*armadillo* (*arm*) (Orsulic and Peifer, 1996; Pai *et al.*, 1997; Wodarz and Nusse, 1998) (Figure 10.1). In the absence of Wnt signaling, β-cat gets phosphorylated by GSK3β within the so-called destruction complex (DC) comprised of the scaffold protein Axin and the tumor suppressor adenomatous polyposis coli (APC) (Ikeda *et al.*, 1998; Kishida *et al.*, 1998; Orford *et al.*, 1997; Polakis, 1997). Phosphorylated β-cat gets polyubiquitinated by beta-transducin repeat containing protein (β-TrCP) (E3 ubiquitin ligase) and is targeted for proteasome-mediated degradation (Hart *et al.*, 1999; Marikawa and Elinson, 1998). Activation of the pathway upon the engagement of the ligand-receptor complex initiates a signal transduction cascade that destabilizes the DC, thereby leading to the cytosolic accumulation of unphosphorylated, active β-cat. Activated β-cat translocates to the nucleus and forms a transcriptional complex with the lymphoid enhancer factor/T-cell factor (LEF/TCF) family of High Mobility Group (HMG)-box transcription factors (Korinek *et al.*, 1998; Van De Wetering *et al.*, 1997). The TCF-/β-cat complex then recruits a host of other transcriptional coactivators, such as B-cell lymphoma 9/*legless* (BCL-9/Lgs) and *pygopus* (PYGO), to induce the expression of downstream target genes (Hoffmans and Basler, 2004; Kramps *et al.*, 2002; Stadeli and Basler, 2005; Townsley, Cliffe, and Bienz, 2004).

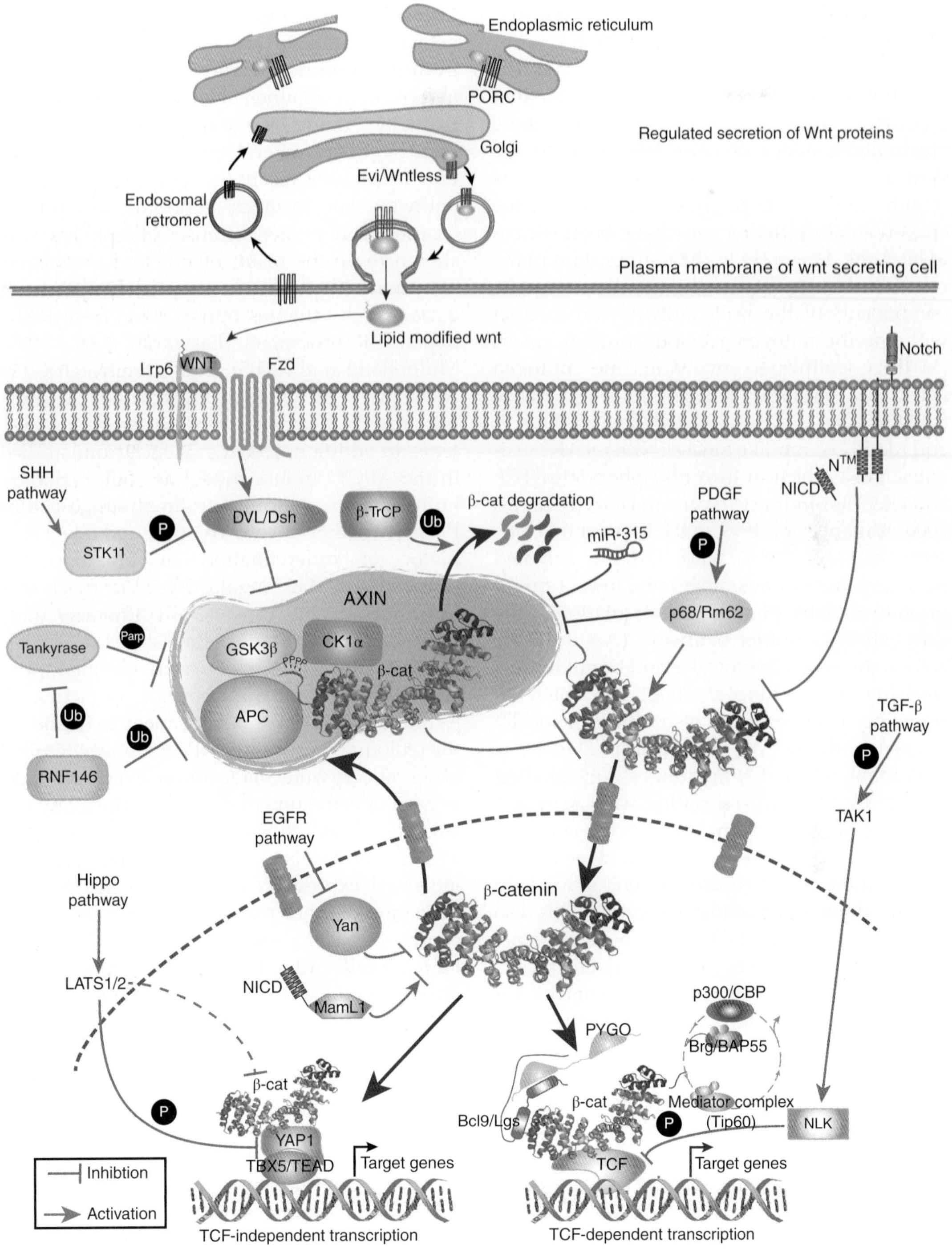

Figure 10.1 Schematic representation of the canonical Wnt pathway and its cross-regulatory interactions with other signaling pathways. Gray arrows represent regulatory interactions that either activate or inhibit the Wnt pathway. Black arrows denote cellular processes. Dashed arrows indicate regulations with indirect mechanism. P, phosphorylation; PARP, poly (ADP-ribose) polymerase; Ub, ubiquitination.

Cross-regulatory interactions between Wnt and other signaling pathways

In addition to the canonical regulators that modulate Wnt/*wg* signaling activity, there are multiple nodes of cross talk between the Wnt/*wg* and other signaling pathways (Figure 10.1). Interestingly, in many cases, the cross-regulatory interactions have been shown to be (Yang, Lin, and Liu, 2006) important in the control of β-cat/TCF activity. For example, components of the Wnt pathway can interact with specific mitogen-activated protein kinase (MAPK) pathways involving the mitogen activated protein kinase kinase kinase (MAPKKK), TGF beta activated kinase 1 (TAK1), and MAPK Nemo-like kinase (NLK). TAK1 activates NLK, which in turn phosphorylates TCF and alters its binding to β-cat and DNA (Behrens, 2000; Ishitani *et al.*, 1999, 2003; Meneghini *et al.*, 1999; Smit *et al.*, 2004). Since TAK1 is activated by TGF-β and various other cytokines, it might provide an entry point for the regulation of the Wnt activity by other pathways (Arsura *et al.*, 2003; Kaminska, Wesolowska, and Danilkiewicz, 2005; Matsumoto-Ida *et al.*, 2006). The Notch (*N*) signaling pathway can also antagonistically interact with the Wg pathway in *Drosophila*. Membrane-tethered *N* modulates Wg signaling activity by physical associating with Arm and prevents it from signaling in the nucleus (Hayward *et al.*, 2005). Consistent with the observations in *Drosophila*, recent studies in embryonic stem cells and colon cancer cells also showed that membrane-tethered N can interact with active β-cat and negatively regulate β-cat's cytosolic accumulation at the posttranslational level (Kim *et al.*, 2012; Kwon *et al.*, 2011). The *mastermind* (*mam*) gene, which was thought to be a specific coactivator of the *N* signaling pathway, has now been shown to genetically interact with components of the Wg pathway, including *arm* and *wg* (Helms *et al.*, 1999; Yedvobnick, Helms, and Barrett, 2001). Recent investigations probing the mechanism of oncogenic properties of β-cat in colon cancer cells identified Mastermind-like 1 (Mam-l1) as a coactivator for β-cat on Wnt target genes, such as cylinD1 and c-myc (Alves-Guerra, Ronchini, and Capobianco, 2007). During epithelial–mesenchymal transition (EMT) in several mammalian cancer cell lines, platelet derived growth factor (PDGF) stimulation of a p68 family RNA helicase results in the displacement of β-cat from the Axin degradation complex in a Wnt-independent manner, thereby activating the pathway (Yang, Lin, and Liu, 2006). Finally, there have been several recent reports of interaction of nuclear receptors with the Wnt/β-cat pathway. For instance, retinoids can inhibit β-cat-dependent gene transcription. This was shown to be the result of physical interaction between retinoic acid receptor (RAR) with β-cat, which activates transcription from RAR-dependent promoters (Easwaran *et al.*, 1999; Mulholland *et al.*, 2005). Interestingly, RAR-γ is also a target of Wnt signaling, thereby suggesting cross talk between the two signaling systems. In addition, recent evidence from studies in the APC^{min} mouse model as well as SW480 human colon cancer cell lines suggest that PPARγ, another nuclear receptor and a key regulator of differentiation of preadipocytes (Kennell and Macdougald, 2005; Prestwich and Macdougald, 2007), physically interacts with the β-cat/TCF4 complex (Jansson *et al.*, 2005). Interestingly, PPARγ levels were elevated, possibly through the sequestration of activated β-cat, both in SW480 colon cancer cells and in the colonic mucosa of APC^{min} mice. Similar observations were made upon overexpression of β-cat or activation of the Wnt pathway by the addition of LiCl to SW480 cells, which resulted in increased levels of PPARγ protein. Moreover, increased expression of β-cat also resulted in activation of a heterologous reporter under the control of PPARγ response element (PPRE) in HEK293 cells. All of these findings allude to the importance of interplay between Wnt/*wg* and other signaling pathways, comprised in a cellular signaling network, that determine and define various developmental and disease states, including cancer (Crosetto, Bienko, and Dikic, 2006; Mcdonald *et al.*, 2006; Van Es, Barker, and Clevers, 2003).

It is thus becoming increasingly clear that the activity of the Wnt pathway, and especially the regulation of β-cat-mediated nuclear signaling, is highly context dependent. Moreover, a comprehensive understanding of the molecular regulation of Wnt/β-cat activity in a given cellular context is still far from being understood. Conventional genetic approaches in model systems, such as *Drosophila melanogaster*, have been enormously successful in identifying core components belonging to the linear cassette that comprises the Wnt/*wg* signaling pathway. However, they have not been as successful in revealing the function of either previously characterized genes or novel factors that may serve as cross-regulatory nodes between Wnt and other signaling pathways. This is likely because mutations in them might lead to early lethality or pleiotropic phenotypes making it impossible to assign them as specific regulators of the Wnt pathway or that their regulatory function can be revealed in only specific genetic backgrounds in a given cell-/context-dependent fashion. Alternatively, functional redundancy and maternal effect (contribution of maternal message that can potentially limit the detection of the zygotic phenotype) may also mask the function of these cross-regulatory factors in the context of development. Therefore, defining a comprehensive network of signaling interactions remains a challenging task.

With the completion of the human genome sequencing project and that of several model organisms, we now have access to an unprecedented resource, which can be tapped to mine useful biological information, and ascribe functions to the novel genes identified from the large-scale genomic studies. One of the most promising functional genomic approaches that has emerged in the past decade is based on the use of double-stranded RNA (dsRNA)/short interfering RNA (siRNA) to knockdown gene function. In several organisms, introduction of a dsRNA/siRNA has proven to be an effective tool to suppress gene expression through a process referred to as RNA interference (RNAi) (Hutvagner *et al.*, 2004; Johannsen *et al.*, 2007; Mello and Conte, 2004; Montgomery, 2006; Sharp, 1999; Tabara, Grishok, and Mello, 1998). The simple transfection of dsRNAs/siRNAs into cells in culture reduces or efficiently eliminates the expression of target genes, thus phenocopying loss-of-function mutations (Clemens *et al.*, 2000). Therefore RNAi-based screens can efficiently link gene sequences to gene function by allowing the query of gene functions in a systematic, rapid, and cost-effective manner.

A major conclusion from the sequencing projects was that while forward genetics had been extremely successful in identifying key genes/core components of many biological processes, including conserved signal transduction cascades, the function(s) of the majority of coding and noncoding genes in the genome remains a mystery. The recent development of high-throughput sequencing technologies and the availability of whole-genome sequences of a variety of living organisms, including that of humans, have led to an enormous push in the quest for a comprehensive understanding of the function of every gene in the genome. Indeed, over the past decade, RNAi-based whole-genome screens have been extensively employed to isolate and determine the function of novel genes in specific cell biological processes and have revolutionized the study of cellular signaling networks. In this review, we discuss the insights we have gained into the molecular regulation of the canonical Wnt signaling pathway from global RNAi screens.

RNAi-based whole-genome screens for modulators of the Wnt/wg pathway

Starting with initial screens in *Drosophila* and subsequently in mammalian cell lines, several research laboratories in the past 7–8 years have completed whole-genome or genome-scale screens in an effort to identify novel modulators of the Wnt/Wg pathway. The first such RNAi screen for the Wnt pathway was designed by Dasgupta *et al.* (2005), as a collaborative effort between the Perrimon and Moon laboratories. The authors developed a high-throughput cell-based assay that took advantage of the known ability of Wnt/β-cat signaling to activate transcription of either a fly cell-optimized (dTF12) or a mammalian cell-optimized (STF16) Wnt-responsive luciferase reporter gene. Specifically, the dTF12 reporter is comprised of 12 multimerized TCF-binding sites cloned upstream of the *Drosophila* heat shock minimal promoter; and STF16 reporter harbors 16 TCF-binding sites upstream of a minimal TATA box from the thymidine kinase (TK) promoter. Both reporters are active in a variety of fly and mammalian cell lines (Dasgupta *et al.*, 2005). Reporters with mutated Tcf-binding sites serve as specificity controls. The cotransfection of a *Renilla luciferase* reporter, *PolIII-RL*, was used as a control for transfection efficiency and cell viability. The ratio of the Wnt pathway-specific firefly luciferase (dTF12/STF16) and the *R. luciferase* activity represented the "normalized" luciferase activity, which was used as a "true" measure of the pathway activity (Dasgupta *et al.*, 2005).

As initial proof of concept and validation of the RNAi approach on a genome scale, a majority of the known core Wnt pathway components were identified in the screen including the ligand-receptor complexes such as *wg*, *arrow* (*arr*)/LDL receptor-related protein-6 (Lrp6) and *frizzled* (*fz*); members of the β-cat/arm degradation complex or genes involved in negative regulation of β-cat activity, such as *axin*, *slmb*/β-TrCP, and *CK1α*; and genes involved in regulating the activity of the β-cat-transcriptional complex in the nucleus, such as *pangolin*/dTCF, *pygo*, and *lgs*/Bcl-9.

The primary screen was followed by a series of secondary screens and follow-up studies that served as "assay filters" to minimize the rate of false-positives, off-target effects (OTEs), including tests for nonspecific effects; and cell-type-specific or reporter-specific factors. In addition, ~50% of the candidate modulators identified in the *Drosophila* screen were also found to have conserved mammalian orthologs, some of which were functionally validated in human HEK293 cells (Dasgupta *et al.*, 2005). Altogether, these observations indicated that similar genome-wide RNAi screens could efficiently identify evolutionarily conserved modulators of the Wnt pathway and also set a paradigm for systematically investigating candidate genes for their function in a variety of cell types and model organisms.

Interestingly, this initial screen for the Wg pathway also identified several genes that had been previously reported in forward genetic screens as genetic interactors of the Wg pathway, including *lilli*, *brahma*, *osa*, *cdc2*, *string* (cdc-25), *N*, *mam*, members of the mediator complex, and specific *brahma*-associated proteins 55 (Collins and Treisman, 2000; Greaves *et al.*, 1999). That said, little was known about the molecular mechanisms by which these proteins could modulate the activity of the Wnt pathway. Detailed mechanistic studies on candidate modulators identified the initial screen are slowly beginning to define the complex regulation of the Wnt pathway, especially in terms of the cross-regulatory interactions between components of the Wnt and other signaling pathways. For instance, novel regulators that were identified in the initial Wg screen have since then been validated in different model systems (Dasgupta *et al.*, 2005). First, the membrane-tethered form of *N* that was initially identified as a negative regulator of Wnt pathway activity in fly cells was also shown to be able to physically interact with Arm protein in *Drosophila* wing imaginal-disc epithelial cells (Hayward *et al.*, 2005). This interaction results in the sequestration of Arm from the nucleus, thereby preventing it from signaling in the nucleus. These studies in the fruit fly have now found interesting and exciting parallels in mammalian stem/progenitor cells, as well as in colon cancer cell lines. In a recent paper, Kwon *et al.* showed that in embryonic stem and colon cancer cells, membrane-tethered N interacts with active β-cat and negatively regulates its accumulation at the posttranslational level (Kwon *et al.*, 2011). Since this cross-regulatory interaction between N and active β-cat was observed only in cells with active Wnt signaling, it has been proposed that this could potentially be a mechanism that cells adopt to titrate level of active β-cat to maintain steady-state proliferation. Second, a report by Yang, Lin, and Liu (2006) described the function of p68 RNA helicase, an ortholog of the *Drosophila* Rm62 gene identified in the Wg screen (He, 2006; Yang, Lin, and Liu, 2006). The authors demonstrated that the p68 RNA helicase could mediate PDGF-induced EMT by displacing Axin from β-cat and promote the nuclear translocation of the latter in human colon cancer cell lines. Third, the DasGupta *et al.* (2005) screen identified Tip60/CG6121 as a positive regulator of Wnt pathway activity. Subsequent studies in human colorectal cancer (CRC) cell lines have shown that the β-cat's C-terminal activation domain associates with TIP60/TRAPPP and a mixed lineage leukemia (MLL1/MLL2) SET1-type chromatin-modifying complex *in vitro* and that this complex promotes H3K4 trimethylation at the c-Myc target gene *in vivo* (Kim *et al.*, 2005; Sierra *et al.*, 2006). Fourth, the transformation specific transcription factor (ETS)-domain transcription factor Yan, a key modulator and repressor of *Drosophila* epidermal growth factor receptor (EGFR) (DER) pathway, was identified as a negative regulator of the Wg pathway in the fly screen. Olson *et al.* (2011) recently demonstrated that cells lacking wild-type *yan* function in the *Drosophila* eye ectopically activate the Wg pathway and that the *yan* loss-of-function phenotype can be rescued by reducing Arm/β-cat levels, thereby validating the inhibitory regulation of Yan on the Wg pathway *in vivo*. Furthermore, biochemical characterization revealed that Yan physically interacts with Arm protein in the nucleus, thereby suggesting a direct inhibitory regulation of Yan on Arm function. These findings have significant implications in the mechanistic understanding of cross-regulatory interactions between the EGFR and Wnt signaling pathways, dysregulation of both of which is closely associated with a variety of human cancers. Finally, a recent genetic screen by Kennell *et al.* to identify novel regulators of the Wg pathway identified the miRNA miR-8 as a negative modulator of Wg activity. The study revealed that miR-8 inhibits the pathway downstream of the Wg signal by repressing the protein levels of TCF and a novel regulator, CG32767 (Kennell *et al.*, 2008). Interestingly, CG32767 was also identified in the initial Wg screen as a strong positive regulator, the RNAi-mediated knockdown of which leads to a significant decrease in the levels of Wg reporter activity (Dasgupta *et al.*, 2005).

While the definitive validation, the physiological relevance, and the function/molecular mechanisms of many of these candidate Wnt pathway modulators remain to be elucidated by biochemical assays and *in vivo* testing in model organisms, such screens offer a good starting point to tackle the complexities of such intricate signaling networks. However, since different cell types express distinct sets of genes, clearly not all modifiers would be found by screening in any given cell type. An interesting case in point is a similar RNAi screen that was performed by Bartscherer *et al.* (2006) in S2R+and Kc167 cells. In this screen, the authors surveyed the effect of knocking down ~2300 putative transmembrane proteins in order to identify genes that may be involved in the secretion of the Wg ligand. Bartscherer *et al.* identified a gene called *evenness interrupted* (*evi*) (also called *wntless* since it was simultaneously identified as a recessive suppressor of a *wg* gain-of-function phenotype in the *Drosophila* eye (Banziger *et al.*, 2006) and was shown to be required for secretion of Wg from the Wg-producing cells). *evi* knockdown in *clone8* (*cl8*) cells did not reduce the activity of the Wg reporter and hence was not identified as a candidate regulator in the initial Wg screen. Therefore, ideally similar screens for cell signaling pathways would benefit from being performed using multiple cell and assay types in parallel to ensure a comprehensive identification of novel candidate regulators while also providing a mechanism to identify cell-type- and context-specific modulators.

Context-dependent modulation of the Wnt/β-cat signaling pathway

From the initial days of RNAi-based HTS, many more whole-genome RNAi screens have been conducted in a variety of mammalian cell culture systems to identify novel, context-dependent modulators of the Wnt signaling pathway. Importantly, many of these screens continue to reveal novel and sometimes surprising new insights into the regulatory interactions that are critical in the control of Wnt pathway activity, especially in the context of disease. More recently, Tang *et al.* (2008) performed genome-wide screen in HeLa cells and identified several novel genes contributing to Wnt/β-cat pathway. In an interesting yet contradicting note, they found that TCF7L2 (TCF4) might have a transcriptional repressor function that restricts colorectal cell growth. In their study, knockdown of TCF4 led to increase in STF16 reporter activity and in cell growth, suggesting a novel tumor suppressor role for TCF4 in some CRCs. The authors demonstrated that the C-terminal tail of TCF4 along with effector(s) such as the corepressor C-terminal binding protein (CtBP) (Arce, Yokoyama, and Waterman, 2006) is required to confer its repressor function. While further work is needed to understand the molecular and cellular contexts that would dictate TCF4's function as an activator or a repressor, an unbiased RNAi screen, such as the one performed by Tang *et al.*, could reveal novel insights into the regulation of this pathway in the context of specific CRC cell lines.

In yet another genome-wide RNAi screen from the Lum laboratory, Jacob *et al.* (2011) revealed a previously unknown association between hh and Wnt signaling pathways. The

β-cat/TCF4 reporter, a large percentage did not affect the reporter readout, suggestive of an alternative mechanism through which β-cat may regulate expression of target genes. Surprisingly, TCF4 was not identified as a hit in the screen with some of the CRC cells. Furthermore, the authors showed that RNAi suppression of TCF4 only modestly reduced proliferation of β-cat-dependent cell lines. These results suggested that β-cat could also function in TCF-independent manner, perhaps by partnering with other transcription factors. Analysis of the top 50 hits revealed a striking enrichment of genes related to YES1-associated protein (YAP1), a key transcriptional regulator of the Hippo signaling pathway (Huang *et al.*, 2005; Zhao *et al.*, 2007). Using extensive biochemical and functional characterization, the authors demonstrated that YAP1 and the transcription factor TBX5 form a complex with β-cat, which results in the activation of antiapoptotic target genes in a TCF4-independent manner (Rosenbluh *et al.*, 2012). This comprehensive screening strategy involving many cancer cell lines not only revealed interesting new insights into β-cat function but also identified yet another node of cross talk between the mammalian Hippo pathway and Wnt signaling in the context of cancer. Interestingly, cross-regulatory interactions between the Hippo and Wnt signaling pathways have been previously reported by a few other studies (Heallen *et al.*, 2011; Hergovich and Hemmings, 2010; Varelas *et al.*, 2010). Varelas *et al.* (2010) demonstrated that TAZ, an integral member of the Hippo pathway, binds Dishevelled (DVL) and negatively regulates the Wnt pathway by modulating DVL phosphorylation in mammalian cells. The authors also showed that the Hippo pathway negatively regulates Wg signaling in *Drosophila* imaginal discs. This is intriguing in the light of previous observation from the Dasgupta *et al.* (2005) screen that identified *Warts* (*Wts*) (LATS1 ortholog), a kinase that phosphorylates TAZ1, as a negative regulator of the Wg reporter.

Integration of RNAi screens with genomic and proteomic approaches

In the future, both whole-genome and functionally targeted RNAi screens using cell-based assays will continue to provide a technology platform for efficient enrichment for potential modulators of Wnt and other signaling pathways in specific cellular contexts. While some individual candidate genes will certainly gain prominence through further experimental analysis of their individual functions, the significance of others may emerge through bioinformatics/computational approaches of integrating these large datasets with those obtained from other similar functional screens. While RNAi screens provide a powerful gene discovery tool for modulators of signaling pathways, they alone reveal little insight into their molecular mechanism(s) of action. A comprehensive and integrative screening strategy was adopted by Major *et al.* (2008) to identify novel modulators of the mammalian Wnt signaling pathway in colon cancer cell lines. They first identified common factors between two different colon cancer cell lines, DLD1 and SW480, that affect the activity of the Wnt pathway. Importantly, the authors focused on genes that share a common transcriptional/gene-signature profile upon their RNAi knockdown, as compared to that of β-cat siRNA. Thereafter, Major *et al.* merged the candidate "hits" onto protein–protein interaction networks (PINs) that they had previously generated (for seven key components of the Wnt pathway) using tandem affinity purification combined with mass spectrometry (TAP-MS). They also expanded their comparison of RNAi-screen-based functional database to an interactome map generated by curating the literature and using the STRING database for PINs using 20 known regulators of the Wnt pathway. This approach allowed them to position siRNA hits into discrete physical complexes of proteins, sometimes with known function, which allowed them to make predictions and test their hypotheses regarding the function of the candidate regulators. Using this methodology, Major *et al.* characterized AGGF1 as a nuclear chromatin-associated protein that participates in β-catenin-mediated transcription in human colon cancer cells. In the future, systematic comparisons and integration of candidate genes (in specific cell-based assays) identified in RNAi screens to known protein–protein interaction databases would be crucial in generating testable hypotheses regarding their function and understanding the "molecular context" of their activity. Mapping the RNAi functional network to that of the PINs could help identify important new regulators that were missed in RNAi screens, thereby generating testable hypotheses regarding their molecular function.

Conclusion

It is now widely recognized that whole-genome RNAi screens using cell-based assays provide a technology platform for efficient enrichment of potential modulators of cell signaling pathways. The early experience with RNAi reagents has led to a better understanding of their specificities and has already resulted in useful recommendations for best usage of the technology. In the future, RNAi-based genome-wide screens will gain tremendously from (i) the use of better siRNA/shRNA reagents with reduced OTEs; (ii) screening multiple cell lines to identify cell-/context-dependent functions of genes that modulate the activity of Wnt and other oncogenic cell signaling pathways; (iii) multiplexing cell-based assays to include "modifier" screens and/or synthetic lethal screens; and (iv) using multiple readouts that can simultaneously monitor activity of signaling pathways along with changes in cell biological outcomes, such as proliferation, differentiation, transformation of cell shape, or death.

While most of the individual screens are powerful by themselves, they cannot complete the big and often complex picture. The use of bioinformatics to integrate and compare the large datasets generated from multiple screens would be essential for extracting meaningful information from genome-scale screens. Additionally, while RNAi screens may be used to assign new function to genes in the context of the Wnt pathway, we also need to simultaneously query the multiprotein complexes in which the core components as well as novel candidate modulators are found within the cell. The application of combinatorial screening approaches will allow researchers in the field to achieve a comprehensive understanding of the molecular mechanisms for candidate genes identified in RNAi screens and also allow them to generate new testable hypotheses about the complex regulatory mechanisms for the Wnt signaling pathway. Finally, the combination of functional genomic screens with small-molecule screens may hold the promise of drug discovery, both for identification of new therapeutic targets and the development of novel small molecules for the treatment of Wnt pathway-related diseases. With this and future knowledge in hand, we expect to see many exciting advances and applications of this powerful screening technology in the near future.

References

Alves-Guerra, M.C., Ronchini, C., and Capobianco, A.J. (2007) Mastermind-like 1 is a specific coactivator of beta-catenin transcription activation and is essential for colon carcinoma cell survival. *Cancer Research*, **67** (18), 8690–8698.

Anton, R., Chatterjee, S.S., Simundza, J., and Dasgupta, R. (2011) A systematic screen for micro-RNAs regulating the canonical Wnt pathway. *PLoS One*, **6** (10): e26257, doi:10.1371/journal.pone.0026257.

Arce, L., Yokoyama, N.N., and Waterman, M.L. (2006) Diversity of LEF/TCF action in development and disease. *Oncogene*, **25** (57), 7492–7504.

Arsura, M., Panta, G.R., Bilyeu, J.D. *et al.* (2003) Transient activation of NF-kappaB through a TAK1/IKK kinase pathway by TGF-beta1 inhibits AP-1/SMAD signaling and apoptosis: implications in liver tumor formation. *Oncogene*, **22** (3), 412–425.

Banziger, C., Soldini, D., Schutt, C. *et al.* (2006) Wntless, a conserved membrane protein dedicated to the secretion of Wnt proteins from signaling cells. *Cell*, **125** (3), 509–522.

Bartscherer, K., Pelte, N., Ingelfinger, D., and Boutros, M. (2006) Secretion of Wnt ligands requires Evi, a conserved transmembrane protein. *Cell*, **125** (3), 523–533.

Behrens, J. (2000) Cross-regulation of the Wnt signalling pathway: a role of MAP kinases. *Journal of Cell Science*, **113** (Pt 6), 911–919.

Biechele, T.L., Kulikauskas, R.M., Toroni, R.A. *et al.* (2012) Wnt/beta-catenin signaling and AXIN1 regulate apoptosis triggered by inhibition of the mutant kinase BRAFV600E in human melanoma. *Science Signaling*, **5** (206), ra3.

Boxem, M., Maliga, Z., Klitgord, N. *et al.* (2008) A protein domain-based interactome network for C. elegans early embryogenesis. *Cell*, **134** (3), 534–545.

Callow, M.G., Tran, H., Phu, L. *et al.* (2011) Ubiquitin ligase RNF146 regulates tankyrase and Axin to promote Wnt signaling. *PLoS One*, **6** (7), e22595.

Chen, B., Dodge, M.E., Tang, W. *et al.* (2009) Small molecule-mediated disruption of Wnt-dependent signaling in tissue regeneration and cancer. *Nature Chemical Biology*, **5** (2), 100–107.

Cheung, H.W., Cowley, G.S., Weir, B.A. *et al.* (2011) Systematic investigation of genetic vulnerabilities across cancer cell lines reveals lineage-specific dependencies in ovarian cancer. *Proceedings of the National Academy of Sciences of the United States of America*, **108** (30), 12372–12377.

Chien, A.J., Moore, E.C., Lonsdorf, A.S. *et al.* (2009) Activated Wnt/beta-catenin signaling in melanoma is associated with decreased proliferation in patient tumors and a murine melanoma model. *Proceedings of the National Academy of Sciences of the United States of America*, **106** (4), 1193–1198.

Ciruna, B., Jenny, A., Lee, D. *et al.* (2006) Planar cell polarity signalling couples cell division and morphogenesis during neurulation. *Nature*, **439** (7073), 220–224.

Clemens, J.C., Worby, C.A., Simonson-Leff, N. *et al.* (2000) Use of double-stranded RNA interference in Drosophila cell lines to dissect signal transduction pathways. *Proceedings of the National Academy of Sciences of the United States of America*, **97** (12), 6499–6503.

Collins, R.T. and Treisman, J.E. (2000) Osa-containing Brahma chromatin remodeling complexes are required for the repression of wingless target genes. *Genes & Development*, **14** (24), 3140–3152.

Crosetto, N., Bienko, M., and Dikic, I. (2006) Ubiquitin hubs in oncogenic networks. *Molecular Cancer Research*, **4** (12), 899–904.

Dasgupta, R., Kaykas, A., Moon, R.T., and Perrimon, N. (2005) Functional genomic analysis of the Wnt-wingless signaling pathway. *Science*, **308** (5723), 826–833.

De Toni, F., Racaud-Sultan, C., Chicanne, G. *et al.* (2006) A crosstalk between the Wnt and the adhesion-dependent signaling pathways governs the chemosensitivity of acute myeloid leukemia. *Oncogene*, **25** (22), 3113–3122.

Easwaran, V., Pishvaian, M., Salimuddin, and Byers, S. (1999) Cross-regulation of beta-catenin-LEF/TCF and retinoid signaling pathways. *Current Biology*, **9** (23), 1415–1418.

Gat, U., Dasgupta, R., Degenstein, L., and Fuchs, E. (1998) De novo hair follicle morphogenesis and hair tumors in mice expressing a truncated beta-catenin in skin. *Cell*, **95** (5), 605–614.

Greaves, S., Sanson, B., White, P., and Vincent, J.P. (1999) A screen for identifying genes interacting with armadillo, the Drosophila homolog of beta-catenin. *Genetics*, **153** (4), 1753–1766.

Grove, E.A., Tole, S., Limon, J. *et al.* (1998) The hem of the embryonic cerebral cortex is defined by the expression of multiple Wnt genes and is compromised in Gli3-deficient mice. *Development (Cambridge, England)*, **125** (12), 2315–2325.

Gunsalus, K.C., Ge, H., Schetter, A.J. *et al.* (2005) Predictive models of molecular machines involved in Caenorhabditis elegans early embryogenesis. *Nature*, **436** (7052), 861–865.

Hart, M., Concordet, J.P., Lassot, I. *et al.* (1999) The F-box protein beta-TrCP associates with phosphorylated beta-catenin and regulates its activity in the cell. *Current Biology*, **9** (4), 207–210.

Hasson, P., Egoz, N., Winkler, C. *et al.* (2005) EGFR signaling attenuates Groucho-dependent repression to antagonize Notch transcriptional output. *Nature Genetics*, **37** (1), 101–105.

Hayward, P., Brennan, K., Sanders, P. *et al.* (2005) Notch modulates Wnt signalling by associating with Armadillo/beta-catenin and regulating its transcriptional activity. *Development (Cambridge, England)*, **132** (8), 1819–1830.

He, X. (2006) Unwinding a path to nuclear beta-catenin. *Cell*, **127** (1), 40–42.

Heallen, T., Zhang, M., Wang, J. *et al.* (2011) Hippo pathway inhibits Wnt signaling to restrain cardiomyocyte proliferation and heart size. *Science*, **332** (6028), 458–461.

Helms, W., Lee, H., Ammerman, M. *et al.* (1999) Engineered truncations in the Drosophila mastermind protein disrupt Notch pathway function. *Developmental Biology*, **215** (2), 358–374.

Hergovich, A. and Hemmings, B.A. (2010) TAZ-mediated crosstalk between Wnt and Hippo signaling. *Developmental Cell*, **18** (4), 508–509.

Hoffmans, R. and Basler, K. (2004) Identification and in vivo role of the Armadillo–Legless interaction. *Development (Cambridge, England)*, **131** (17), 4393–4400.

Huang, J., Wu, S., Barrera, J. *et al.* (2005) The Hippo signaling pathway coordinately regulates cell proliferation and apoptosis by inactivating Yorkie, the Drosophila Homolog of YAP. *Cell*, **122** (3), 421–434.

Huang, S.M., Mishina, Y.M., Liu, S. *et al.* (2009) Tankyrase inhibition stabilizes axin and antagonizes Wnt signalling. *Nature*, **461** (7264), 614–620.

Hutvagner, G., Simard, M.J., Mello, C.C., and Zamore, P.D. (2004) Sequence-specific inhibition of small RNA function. *PLoS Biology*, **2** (4), E98.

Ikeda, S., Kishida, S., Yamamoto, H. *et al.* (1998) Axin, a negative regulator of the Wnt signaling pathway, forms a complex with GSK-3beta and beta-catenin and promotes GSK-3beta-dependent phosphorylation of beta-catenin. *The EMBO Journal*, **17** (5), 1371–1384.

Ishitani, T., Ninomiya-Tsuji, J., Nagai, S. *et al.* (1999) The TAK1-NLK-MAPK-related pathway antagonizes signalling between beta-catenin and transcription factor TCF. *Nature*, **399** (6738), 798–802.

Ishitani, T., Kishida, S., Hyodo-Miura, J. *et al.* (2003) The TAK1-NLK mitogen-activated protein kinase cascade functions in the Wnt-5a/Ca(2+) pathway to antagonize Wnt/beta-catenin signaling. *Molecular and Cellular Biology*, **23** (1), 131–139.

Jacob, L.S., Wu, X.F., Dodge, M.E. *et al.* (2011) Genome-wide RNAi screen reveals disease-associated genes that are common to Hedgehog and Wnt signaling. *Science Signaling*, **4** (157), ra4.

James, R.G., Davidson, K.C., Bosch, K.A. *et al.* (2012) WIKI4, a novel inhibitor of tankyrase and Wnt/ss-catenin signaling. *PLoS One*, **7** (12), e50457.

Jansson, E.A., Are, A., Greicius, G. *et al.* (2005) The Wnt/beta-catenin signaling pathway targets PPARgamma activity in colon cancer cells. *Proceedings of the National Academy of Sciences of the United States of America*, **102** (5), 1460–1465.

Jiang, R., Bush, J.O., and Lidral, A.C. (2006) Development of the upper lip: morphogenetic and molecular mechanisms. *Developmental Dynamics*, **235** (5), 1152–1166.

Johannsen, M., Gneveckow, U., Taymoorian, K. *et al.* (2007) Thermal therapy of prostate cancer using magnetic nanoparticles. *Actas Urologicas Espanolas*, **31** (6), 660–667.

Kageshita, T., Hamby, C.V., Ishihara, T. *et al.* (2001) Loss of beta-catenin expression associated with disease progression in malignant melanoma. *The British Journal of Dermatology*, **145** (2), 210–216.

Kaminska, B., Wesolowska, A., and Danilkiewicz, M. (2005) TGF beta signalling and its role in tumour pathogenesis. *Acta Biochimica Polonica*, **52** (2), 329–337.

Kaykas, A., Yang-Snyder, J., Heroux, M. *et al.* (2004) Mutant Frizzled 4 associated with vitreoretinopathy traps wild-type Frizzled in the endoplasmic reticulum by oligomerization. *Nature Cell Biology*, **6** (1), 52–58.

Kennell, J.A. and Macdougald, O.A. (2005) Wnt signaling inhibits adipogenesis through beta-catenin-dependent and -independent mechanisms. *The Journal of Biological Chemistry*, **280** (25), 24004–24010.

Kennell, J.A., Gerin, I., Macdougald, O.A., and Cadigan, K.M. (2008) The microRNA miR-8 is a conserved negative regulator of Wnt signaling. *Proceedings of the National Academy of Sciences of the United States of America*, **105** (40), 15417–15422.

Kim, J.H., Kim, B., Cai, L. *et al.* (2005) Transcriptional regulation of a metastasis suppressor gene by Tip60 and beta-catenin complexes. *Nature*, **434** (7035), 921–926.

Kim, H.A., Koo, B.K., Cho, J.H. *et al.* (2012) Notch1 counteracts WNT/beta-catenin signaling through chromatin modification in colorectal cancer. *The Journal of Clinical Investigation*, **122** (9), 3248–3259.

Kishida, S., Yamamoto, H., Ikeda, S. *et al.* (1998) Axin, a negative regulator of the wnt signaling pathway, directly interacts with adenomatous polyposis coli and regulates the stabilization of beta-catenin. *The Journal of Biological Chemistry*, **273** (18), 10823–10826.

Kokubu, C., Heinzmann, U., Kokubu, T. *et al.* (2004) Skeletal defects in ringelschwanz mutant mice reveal that Lrp6 is required for proper somitogenesis and osteogenesis. *Development (Cambridge, England)*, **131** (21), 5469–5480.

Korinek, V., Barker, N., Willert, K. *et al.* (1998) Two members of the Tcf family implicated in Wnt/beta-catenin signaling during embryogenesis in the mouse. *Molecular Cell Biology*, **18** (3), 1248–1256.

Kramps, T., Peter, O., Brunner, E. *et al.* (2002) Wnt/wingless signaling requires BCL9/legless-mediated recruitment of pygopus to the nuclear beta-catenin-TCF complex. *Cell*, **109** (1), 47–60.

Kwon, C., Cheng, P., King, I.N. *et al.* (2011) Notch post-translationally regulates beta-catenin protein in stem and progenitor cells. *Nature Cell Biology*, **13** (10), 1244–1251.

Letamendia, A., Labbe, E., and Attisano, L. (2001) Transcriptional regulation by Smads: crosstalk between the TGF-beta and Wnt pathways. *The Journal of Bone and Joint Surgery America*, **83-A Suppl 1** (Pt 1), S31–S39.

Lum, L., Yao, S., Mozer, B. *et al.* (2003) Identification of Hedgehog pathway components by RNAi in Drosophila cultured cells. *Science*, **299** (5615), 2039–2045.

Ly, P., Eskiocak, U., Parker, C.R. *et al.* (2012) RNAi screening of the human colorectal cancer genome identifies multifunctional tumor suppressors regulating epithelial cell invasion. *Cell Research*, **22** (11), 1605–1608.

Major, M.B., Roberts, B.S., Berndt, J.D. *et al.* (2008) New regulators of Wnt/beta-catenin signaling revealed by integrative molecular screening. *Science Signaling*, **1** (45), ra12.

Marikawa, Y. and Elinson, R.P. (1998) Beta-TrCP is a negative regulator of Wnt/beta-catenin signaling pathway and dorsal axis formation in Xenopus embryos. *Mechanisms of Development*, **77** (1), 75–80.

Matsumoto-Ida, M., Takimoto, Y., Aoyama, T. *et al.* (2006) Activation of TGF-beta1-TAK1-p38 MAPK pathway in spared cardiomyocytes is involved in left ventricular remodeling after myocardial infarction in rats. *American Journal of Physiology Heart and Circulatory Physiology*, **290** (2), H709–H715.

Mcdonald, S.A., Preston, S.L., Lovell, M.J. *et al.* (2006) Mechanisms of disease: from stem cells to colorectal cancer. *Nature Clinical Practice Gastroenterology & Hepatology*, **3** (5), 267–274.

Mello, C.C. and Conte, D., Jr. (2004) Revealing the world of RNA interference. *Nature*, **431** (7006), 338–342.

Meneghini, M.D., Ishitani, T., Carter, J.C. *et al.* (1999) MAP kinase and Wnt pathways converge to downregulate an HMG-domain repressor in Caenorhabditis elegans. *Nature*, **399** (6738), 793–797.

Miller, J.R., Hocking, A.M., Brown, J.D., and Moon, R.T. (1999) Mechanism and function of signal transduction by the Wnt/beta-catenin and Wnt/Ca^{2+} pathways. *Oncogene*, **18** (55), 7860–7872.

Miyoshi, K. and Hennighausen, L. (2003) Beta-catenin: a transforming actor on many stages. *Breast Cancer Research*, **5** (2), 63–68.

Miyoshi, K., Rosner, A., Nozawa, M. *et al.* (2002) Activation of different Wnt/beta-catenin signaling components in mammary epithelium induces transdifferentiation and the formation of pilar tumors. *Oncogene*, **21** (36), 5548–5556.

Montgomery, M.K. (2006) RNA interference: unraveling a mystery. *Nature Structural and Molecular Biology*, **13** (12), 1039–1041.

Moon, R.T., Kohn, A.D., De Ferrari, G.V., and Kaykas, A. (2004) WNT and beta-catenin signalling: diseases and therapies. *Nature Reviews Genetics*, **5** (9), 691–701.

Mulholland, D.J., Dedhar, S., Coetzee, G.A., and Nelson, C.C. (2005) Interaction of nuclear receptors with the Wnt/beta-catenin/Tcf signaling axis: Wnt you like to know? *Endocrine Reviews*, **26** (7), 898–915.

Musgrove, E.A. (2004) Wnt signalling via the epidermal growth factor receptor: a role in breast cancer? *Breast Cancer Research*, **6** (2), 65–68.

Nusse, R. (1999) WNT targets. Repression and activation. *Trends in Genetics*, **15** (1), 1–3.

Nusse, R. (2003) Wnts and Hedgehogs: lipid-modified proteins and similarities in signaling mechanisms at the cell surface. *Development (Cambridge, England)*, **130** (22), 5297–5305.

Olson, E.R., Pancratov, R., Chatterjee, S.S. *et al.* (2011) Yan, an ETS-domain transcription factor, negatively modulates the Wingless pathway in the Drosophila eye. *EMBO Report*, **12** (10), 1047–1054.

Orford, K., Crockett, C., Jensen, J.P. *et al.* (1997) Serine phosphorylation-regulated ubiquitination and degradation of beta-catenin. *The Journal of Biological Chemistry*, **272** (40), 24735–24738.

Orsulic, S. and Peifer, M. (1996) An in vivo structure–function study of armadillo, the beta-catenin homologue, reveals both separate and overlapping regions of the protein required for cell adhesion and for wingless signaling. *The Journal of Cell Biology*, **134** (5), 1283–1300.

Pai, L.M., Orsulic, S., Bejsovec, A., and Peifer, M. (1997) Negative regulation of Armadillo, a Wingless effector in Drosophila. *Development*, **124** (11), 2255–2266.

Polakis, P. (1997) The adenomatous polyposis coli (APC) tumor suppressor. *Biochimica Biophysica Acta*, **1332** (3), F127–F147.

Polakis, P. (2000) Wnt signaling and cancer. *Genes & Development*, **14** (15), 1837–1851.

Prestwich, T.C. and Macdougald, O.A. (2007) Wnt/beta-catenin signaling in adipogenesis and metabolism. *Current Opinion in Cell Biology*, **19** (6), 612–617.

Ren, Y., Cheung, H.W., Von Maltzhan, G. *et al.* (2012) Targeted tumor-penetrating siRNA nanocomplexes for credentialing the ovarian cancer oncogene ID4. *Science Translational Medicine*, **4** (147), 147ra112.

Robitaille, J., Macdonald, M.L., Kaykas, A. *et al.* (2002) Mutant frizzled-4 disrupts retinal angiogenesis in familial exudative vitreoretinopathy. *Nature Genetics*, **32** (2), 326–330.

Rosenbluh, J., Nijhawan, D., Cox, A.G. *et al.* (2012) β-Catenin-driven cancers require a YAP1 transcriptional complex for survival and tumorigenesis. *Cell*, **151** (7), 1457–1473.

Sharp, P.A. (1999) RNAi and double-strand RNA. *Genes & Development*, **13** (2), 139–141.

Shu, W., Jiang, Y.Q., Lu, M.M., and Morrisey, E.E. (2002) Wnt7b regulates mesenchymal proliferation and vascular development in the lung. *Development (Cambridge, England)*, **129** (20), 4831–4842.

Sierra, J., Yoshida, T., Joazeiro, C.A., and Jones, K.A. (2006) The APC tumor suppressor counteracts beta-catenin activation and H3K4 methylation at Wnt target genes. *Genes & Development*, **20** (5), 586–600.

Silver, S.J., Hagen, J.W., Okamura, K. *et al.* (2007) Functional screening identifies miR-315 as a potent activator of Wingless signaling. *Proceedings of the National Academy of Sciences of the United States of America*, **104** (46), 18151–18156.

Smit, L., Baas, A., Kuipers, J. *et al.* (2004) Wnt activates the Tak1/Nemo-like kinase pathway. *The Journal of Biological Chemistry*, **279** (17), 17232–17240.

Staal, F.J. and Clevers, H. (2000) Tcf/Lef transcription factors during T-cell development: unique and overlapping functions. *Hematological Journal*, **1** (1), 3–6.

Stadeli, R. and Basler, K. (2005) Dissecting nuclear Wingless signalling: recruitment of the transcriptional co-activator Pygopus by a chain of adaptor proteins. *Mechanisms of Development*, **122** (11), 1171–1182.

Tabara, H., Grishok, A., and Mello, C.C. (1998) RNAi in C. elegans: soaking in the genome sequence. *Science (New York, NY)*, **282** (5388), 430–431.

Tang, W., Dodge, M., Gundapaneni, D. *et al.* (2008) A genome-wide RNAi screen for Wnt/beta-catenin pathway components identifies unexpected roles for TCF transcription factors in cancer. *Proceedings of the National Academy of Sciences of the United States of America*, **105** (28), 9697–9702.

Tap, W.D., Gong, K.W., Dering, J. *et al.* (2010) Pharmacodynamic characterization of the efficacy signals due to selective BRAF inhibition with PLX4032 in malignant melanoma. *Neoplasia*, **12** (8), 637–649.

Tewari, M., Hu, P.J., Ahn, J.S. *et al.* (2004) Systematic interactome mapping and genetic perturbation analysis of a C. elegans TGF-beta signaling network. *Molecular Cell*, **13** (4), 469–482.

Townsley, F.M., Cliffe, A., and Bienz, M. (2004) Pygopus and Legless target Armadillo/beta-catenin to the nucleus to enable its transcriptional co-activator function. *Nature Cell Biology*, **6** (7), 626–633.

Van De Wetering, M., Cavallo, R., Dooijes, D. *et al.* (1997) Armadillo coactivates transcription driven by the product of the Drosophila segment polarity gene dTCF. *Cell*, **88** (6), 789–799.

Van Es, J.H., Barker, N., and Clevers, H. (2003) You Wnt some, you lose some: oncogenes in the Wnt signaling pathway. *Current Opinion in Genetics & Development*, **13** (1), 28–33.

Varelas, X., Miller, B.W., Sopko, R. *et al.* (2010) The Hippo pathway regulates Wnt/beta-catenin signaling. *Developmental Cell*, **18** (4), 579–591.

Von Bubnoff, A. and Cho, K.W. (2001) Intracellular BMP signaling regulation in vertebrates: pathway or network? *Developmental Biology*, **239** (1), 1–14.

Wodarz, A. and Nusse, R. (1998) Mechanisms of Wnt signaling in development. *Annual Review of Cell and Developmental Biology*, **14**, 59–88.

Wood, L.D., Parsons, D.W., Jones, S. *et al.* (2007) The genomic landscapes of human breast and colorectal cancers. *Science*, **318** (5853), 1108–1113.

Yang, L., Lin, C., and Liu, Z.R. (2006) P68 RNA helicase mediates PDGF-induced epithelial mesenchymal transition by displacing axin from beta-catenin. *Cell*, **127** (1), 139–155.

Yedvobnick, B., Helms, W., and Barrett, B. (2001) Identification of chromosomal deficiencies that modify Drosophila mastermind mutant phenotypes. *Genesis*, **30** (4), 250–258.

Yoo, L.I., Chung, D.C., and Yuan, J. (2002) LKB1 – a master tumour suppressor of the small intestine and beyond. *Nature Reviews Cancer*, **2** (7), 529–535.

Zhao, B., Wei, X., Li, W. *et al.* (2007) Inactivation of YAP oncoprotein by the Hippo pathway is involved in cell contact inhibition and tissue growth control. *Genes & Development*, **21** (21), 2747–2761.

11 Mathematical Models of Wnt Signaling Pathways

Michael Kühl[1], Barbara Kracher[2], Alexander Groß[3], and Hans A. Kestler[3]

[1]Institute of Biochemistry and Molecular Biology, Ulm University, Ulm, Germany
[2]Department of Plant-Microbe Interactions, Max Planck Institute for Plant Breeding Research, Cologne, Germany
[3]Research Group Bioinformatics and Systems Biology, Institute of Neural Information Processing, Ulm University, Ulm, Germany

Introduction

Our picture of Wnt signaling has developed from a single linear pathway into a complex Wnt signaling network comprising distinct signaling branches including Wnt/β-catenin, Wnt/rhoA/ROCK, Wnt/rac/JNK, and Wnt/calcium signaling. These branches are highly interconnected and influence the activity of each other (Kestler and Kuhl, 2008). With the increase in complexity, our ability to intuitively understand the behavior of this network is continuously decreasing. This paves the way for the use of mathematical models and computer-based simulations as tools to understand the global behavior of the Wnt signaling network and its interactions with other signaling pathways (Kestler *et al.*, 2008). Here, we first briefly introduce different modeling approaches used to represent intracellular signal transduction by mathematical means and then summarize the currently used models of Wnt signaling. Unexpected findings of these models will be highlighted, and an agenda of future requirements in the field will be developed.

Modeling approaches in signal transduction

In the past, different types of models have been used to model intracellular signal transduction. These can roughly be divided into static and dynamic modeling approaches. The first group simply describes the potential interactions of signaling molecules within a given pathway but does not allow any direct conclusions on the dynamic behavior of the pathway. Nevertheless, these models are of general use as they allow the identification of the so-called hubs, which are signaling components that are involved in many interactions and therefore are of major importance for the pathway. In case of the Wnt/β-catenin pathway, the main components Wnts, Frizzleds, Dishevelled, axin, GSK3β, APC, β-catenin, and TCF/LEF have been identified as hubs (Kestler *et al.*, 2008).

Dynamic modeling approaches in contrast allow statements about the temporal behavior of the network of interest. Again, two main subgroups of model types can be distinguished. Quantitative models rely on concentrations of

Wnt Signaling in Development and Disease: Molecular Mechanisms and Biological Functions, First Edition.
Edited by Stefan Hoppler and Randall T. Moon.

components and changes in concentrations over time, on enzymatic activities, rate constants, etc. They have a high demand on exact kinetic data and usually involve differential equations, either ordinary or partial differential equations (ODE vs. PDE, or stochastic DE). Qualitative models, on the other hand, are much more abstract and rely on more simple entities such as the presence or absence of a certain molecule or simple rules describing the interactions of the components. Boolean networks are a good example for these kinds of modeling approaches.

Wnt signaling has been modeled on different granularities. On the one hand, models have been used to describe gradients of Wnt proteins and their inhibitors such as dkk1 or Sfrp and other growth factors such as BMP4 to describe diverse processes including stemness and differentiation of intestinal stem cells (Murray *et al.*, 2010; Zhang, Lander, and Nie, 2012), differentiation of cardiomyocytes (Gibb, Lavery, and Hoppler, 2013), or hair follicle spacing (Sick *et al.*, 2006). In these models, the intracellular part of Wnt signaling was not directly reflected. Instead, biological effects are explained by extracellular gradients of signaling molecules and their interactions. On the other hand, different quantitative models have been set up to analyze the architecture, behavior, and outcome of Wnt/β-catenin signaling and PCP signaling. Here, we focus on those models that cover the latter aspect.

Wnt/β-catenin signaling: Lessons learned from mathematical models

The first model of Wnt/β-catenin signaling, which was based on ordinary differential equations, was set up by the labs of M. Kirschner and R. Heinrich (Lee *et al.*, 2003). This model focused on the heart of Wnt signaling, the APC/axin/GSK3β-based destruction complex and the subsequent stabilization of cytoplasmic β-catenin. This part of the Wnt/β-catenin pathway could be mirrored in a set of 15 coupled ODEs. Concentrations and kinetic parameters required to run simulations of the model were gained from biochemical measurements in *Xenopus laevis* egg extracts and available literature. A subsequent extended analysis of the mathematical model followed by biochemical experiments revealed that the comparable low concentration of axin is the most critical value for the activation of the pathway. The model was reduced to a set of seven ODEs and eight algebraic equations based on conservation and equilibria conditions (Krüger and Heinrich, 2004). This allowed separating involved reactions by the time scales of their action and the investigation of robustness to structural and kinetic parameters of the Wnt/β-catenin pathway. This model was used for a comparison of effects of truncated APC mutants with differing binding rates on β-catenin (Cho, Baek, and Sung, 2006).

However, the assumption that axin is the rate limiting factor in Wnt/β-catenin signaling was not confirmed by studies comparing the concentration of different signaling mediators in a set of five different cell lines, including HEK293T, MDCK, Caco-2, SW480, and SW480APC cells (Tan *et al.*, 2012). In HEK293T, MDCK, and Caco-2 cells, the axin concentration, for example, was found to be considerably higher than the concentration of APC. This indicates that in these mammalian cells, APC rather than axin might be rate limiting in the assembly of the destruction complex. Using the concentration obtained in this study (Tan *et al.*, 2012), the model by Lee *et al.* revealed significantly different results for the tested mammalian cells in comparison to the *Xenopus* system. This indicates that great care should be taken translating the data obtained from studies based on the *Xenopus* egg extract system into other systems.

An interesting feature of the Wnt/β-catenin pathway becomes apparent when the initial parameters for the Lee–Heinrich model are varied (Wawra, Kuhl, and Kestler, 2007). Although the initial amount of β-catenin can change significantly depending on the input parameters, the fold increase of β-catenin upon stimulation was found to be robustly between 4.5 and 6.0. This finding was later confirmed by others and extended experimentally (Goentoro and Kirschner, 2009; Goentoro *et al.*, 2009). The analyses of the Kirschner lab clearly indicate that the activation of the pathway does not depend on the overall concentration of nuclear β-catenin, but instead uses the fold increase as a readout for target gene activation. This readout resembles Weber's law in physiology (Ferrell, 2009) and can be achieved by a so-called incoherent feedforward loop (Goentoro *et al.*, 2009). In this incoherent type 1 feedforward loop (iFFL-1), a signaling molecule activates a certain target and at the same time is also a repressor of this target (Figure 11.1). Thus, a higher amount of the signaling molecule, for example, β-catenin, can be balanced by the consequential higher activity of the repressor and the strength of the observed pathway response will be robust.

Along the same lines, a recent study comparing the effect of different gene regulatory mechanisms on Wnt target gene expression using an ODE approach found that an incoherent feedforward loop, other than, for example, direct activation or inhibition, can maintain a constant target gene expression level independent of an existing Wnt or APC concentration gradient within a tissue (Benary *et al.*, 2013). In contrast, within the same concentration gradient, a cooperative mode of gene activation together with a TCF feedback loop was shown to result in a clear separation of regions with and without target gene expression, thereby translating the continuous concentration gradient of Wnt or APC into a qualitative transcriptional readout (i.e., whether the target gene is expressed or not).

Target genes of the Wnt pathway include compounds that act as negative feedback components such as axin2 or dkk1 (Aulehla *et al.*, 2003; Niida *et al.*, 2004; Semenov *et al.*, 2001). Meanwhile, several research groups have expanded the initial model of the Wnt/β-catenin pathway by Lee and co-workers to cover this topic (Cho, Baek, and Sung, 2006; Jensen *et al.*, 2010; Kogan *et al.*, 2012; Pedersen, Jensen, and Krishna, 2011; Wawra, Kuhl, and Kestler, 2007). These negative feedback loops can result in an oscillatory behavior of Wnt pathway activity as seen during segmentation of the presomitic mesoderm in vertebrate somitogenesis. This

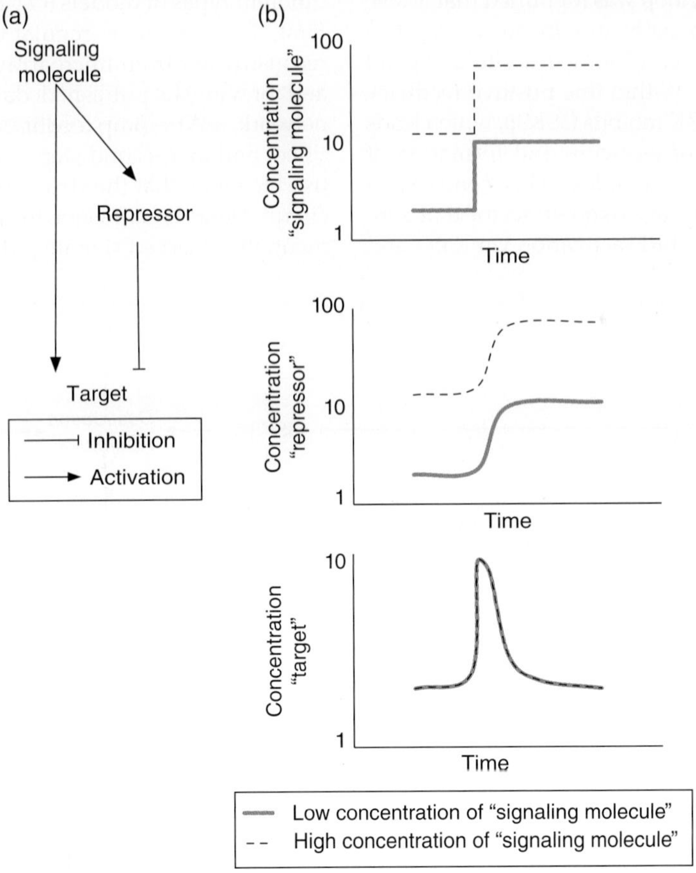

Figure 11.1 Incoherent type 1 feed forward loop. (a) In an iFFL-1, a signaling molecule at the same time activates a certain target and a repressor of this target. (b) In the iFFL-1, a higher concentration of signaling molecule (upper panel) is balanced by a higher concentration of repressor (middle panel). Thus, the response of the target (lower panel) depends on the fold change but not the absolute concentration of signaling molecule.

oscillation is also reflected by fluctuating amounts of axin2 or dkk1 as observed during embryogenesis (Aulehla *et al.*, 2003; Dequeant *et al.*, 2006). Interestingly, dkk1 knockout mice show vertebrae, a derivative of the somites, with reduced size and irregular shape (MacDonald, Adamska, and Meisler, 2004; Mukhopadhyay *et al.*, 2001). Besides Wnt/β-catenin signaling, also Notch and FGF/ERK signaling oscillate during somitogenesis, and this was integrated into different models reflecting the cross talk of Wnt/β-catenin, Notch, and FGF signaling (Agur *et al.*, 2011; Goldbeter and Pourquie, 2008; Kirnasovsky, Kogan, and Agur, 2008; Rodriguez-Gonzalez *et al.*, 2007).

Signaling cross talk also exists, for example, with the ERK pathway and calcium-dependent cell adhesion molecules, the cadherins. In the case of Wnt/β-catenin and ERK signaling, a positive feedback loop was identified that is able to maintain both pathways in an active state even in the absence of an extracellular ligand (Kim *et al.*, 2007). Within this positive feedback loop, activated ERK inhibits GSK3β, which leads to accumulation of β-catenin and formation of the β-catenin/TCF complex. This complex in turn activates a so far unknown factor that subsequently induces ERK activation via Raf-1 and MEK. Through this positive feedback loop, even a small change in the system can be sufficient to induce sustained activity of both pathways and thus carcinogenesis (Kim *et al.*, 2007). This finding demonstrates the power of modeling approaches to uncover an unexpected behavior of a signaling network. Of note, besides the just mentioned positive feedback loop, there is additional cross talk between Wnt and ERK signaling at different layers, and a more recent ODE model of the two pathways includes as many as six coupled feedback loops that were shown to cause diverse dose-response patterns and influence epithelial–mesenchymal transition (EMT) (Shin *et al.*, 2010; Figure 11.2).

Another instance of cross talk occurs between Wnt and Bmp signaling. The interaction of these two pathways in tooth organogenesis was examined in a recent study that combined two different types of models (O'Connell *et al.*, 2012). First, a static gene regulatory network was reconstructed from microarray gene expression and previously published data. Based on this network, a Wnt–Bmp feedback circuit was identified, and in a second step, an ODE model was used to show that the structure of this feedback circuit alone is sufficient to mimic the experimentally observed signaling dynamics.

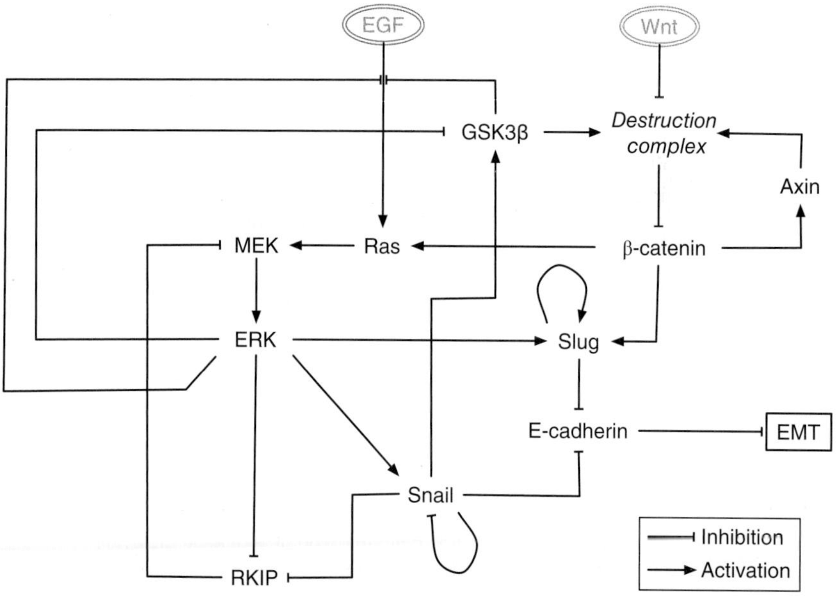

Figure 11.2 Cross talk between Wnt and ERK signaling. Several positive and negative feedback loops exist between and within the Wnt and ERK signaling branches. The figure depicts six feedback loops that were included in a recent computational model of Wnt and ERK signaling and their influence on EMT (Shin *et al.*, 2010).

β-catenin is not only involved in Wnt signaling but has originally been described to be involved in cadherin-mediated cell adhesion suggesting that both processes depend on each other. Indeed, the available amount of cadherins in a cell was shown to influence the expression of β-catenin-dependent target genes (Finnemann *et al.*, 1995). As cell cycle regulators like c-Myc or Cyclin D1 are among the known Wnt/β-catenin target genes, this means that cell adhesion is linked to cell cycle regulation through β-catenin. The cross regulation of these two processes has been described in a mathematical model (van Leeuwen *et al.*, 2007a) which has later been implemented into a multiscale model of colon crypts to describe cell behavior within the crypt and colon cancer (van Leeuwen *et al.*, 2007b, 2009).

PCP signaling: Insights gained from mathematical models

The planar cell polarity pathway organizes the polarity of an epithelial layer of cells perpendicular to the apical/basal axes of the cells. The pathway is also involved in diverse processes such as gastrulation movements, neural tube closure, proper tubulogenesis in the developing kidney, and inner ear and heart development (see Chapter 6 of this book for corresponding references). Whereas Wnt ligands have been shown to be involved in this pathway in vertebrates, no Wnt ligands have been assigned to PCP signaling in Drosophila.

This planar polarity of epithelial cells involves the asymmetric distribution of proteins in (Drosophila) cells such as Frizzled, Flamingo, Van Gogh, Prickle–Spiny legs, Dishevelled, and Diego, for all of which vertebrate homologs exist. Whereas Frizzled and Dishevelled localize to one side of the cell, Van Gogh and Prickle localize to the opposite site (Seifert and Mlodzik, 2007).

An interesting feature of planar cell polarity is the effect of domineering nonautonomy (Axelrod and Tomlin, 2011). In certain mutants that affect only a clone of cells within the epithelial layer, planar cell polarity is not only lost in the affected cells but also in the neighboring cells. Two models have been established to describe this effect. The first hypothesis, called factor X hypothesis, assumes a gradient of a yet unknown factor that reaches through the whole plane of epithelial cells (Axelrod and Tomlin, 2011). The gradient is read out by the cells, resulting in the asymmetric distribution of cells described earlier. The second hypothesis can best be described as a feedback competitive binding mechanism. In this model, it is assumed that the presence of Van Gogh and Prickle prevents the binding and localization of Frizzled and Dishevelled at the same side of the cell and the same is supposed to be true the other way around (Axelrod and Tomlin, 2011). A very sophisticated mathematical model set up by the Axelrod and Tomlin labs supports the latter hypothesis. This model consists of a set of partial differential equations and the predictions generated with the mathematical model were subsequently confirmed experimentally. Of note, the model does not require any factor X to explain domineering nonautonomy suggesting that the factor X hypothesis may be discarded (Amonlirdviman *et al.*, 2005).

Future challenges

What are the future challenges and demands in the field of mathematical models of the Wnt signaling network? The following requirements are clearly listed at the top:

(1) Different hypotheses have been suggested about how the Wnt/β-catenin pathway is activated at the membrane (MacDonald, Adamska, and Meisler, 2004). An initial model of signaling events at the membrane has been implemented (Kogan *et al.*, 2012). Still lacking, however, is a more comprehensive ODE-based model that covers these events including binding of Wnt to LRP and Frizzled receptors, oligomerization of receptors during signalosome formation (MacDonald, Adamska, and Meisler, 2004), and recruitment of the destruction complex to the membrane. A simple ODE model might not suffice to describe these processes as a homogenous distribution of all components in an aqueous, well-mixed solution is normally assumed for these kinds of models, whereas in reality, reactions occur at the two-dimensional membrane. Sophisticated models therefore might include different submodules describing these differences

(compartment models) or will require the use of partial differential equations.

As for the activation at the membrane, there also exist different hypotheses for the exact mechanism on how the Wnt signal triggers β-catenin accumulation (Li *et al.*, 2012; Macdonald, Semenov, and He, 2007). Recent work of the Kirschner lab indicates that partial inhibition of the two β-catenin phosphorylation steps can give rise to the experimentally observed dynamics (Hernandez, Klein, and Kirschner, 2012). Yet, as the applied ODE model concentrated on the core mechanism of β-catenin phosphorylation and degradation, the upstream signal events between membrane and degradation complex still need to be fully resolved.

(2) Different mathematical models of the Wnt/β-catenin pathway so far concentrated on different aspects of the pathway, for example, the destruction complex (Lee *et al.*, 2003), target gene activation, and feedback loops (Jensen *et al.*, 2010; Pedersen, Jensen, and Krishna, 2011; Wawra, Kuhl, and Kestler, 2007) or processes at the membrane (Kogan *et al.*, 2012). Also, the dual role of β-catenin in signal transduction and in cadherin-mediated cell adhesion has been covered (van Leeuwen *et al.*, 2007a). In the future, we will have to develop a unified model that integrates all these different aspects including transcriptional feedback loops.

(3) ODE-based models for other Wnt signaling branches such as Wnt/rhoA/ROCK, Wnt/rac/JNK, or Wnt/calcium signaling are completely lacking so far and desperately need to be set up.

(4) As Wnt signaling branches cannot be seen as isolated entities but work together as a Wnt signaling network, a combination of ODE models integrating these different branches will be a major challenge. The influence of Wnt/calcium signaling on Wnt/β-catenin signaling has recently been investigated in more detail (Nalesso *et al.*, 2011) by showing that a single Wnt ligand (Wnt3a) can activate different Wnt signaling branches in a concentration-dependent manner resulting in a switch-like behavior (Kestler and Kuhl, 2011). These findings need to be extended and confirmed in much greater detail.

(5) Internalization of Wnt receptor complexes is an important biological phenomenon but has not yet been integrated into any mathematical model of Wnt signaling.

(6) These novel Wnt signaling network models will consist of an ever increasing number of differential (ODE, PDE) and algebraic equations combined with time delays but will require a very detailed knowledge about quantitative data describing the individual reactions involved. Thus, there will be the need for more qualitative modeling approaches that do not rely on such detailed information. A recent study (Handorf and Klipp, 2012), for example, modeled the cross talk between Wnt and ERK signaling described earlier using a Boolean approach instead of an ODE model. Interestingly, the authors of this study present a method to infer the Boolean functions from mechanistic information available in public databases in an automated way. This kind of or similar qualitative models will be of particular interest to describe the global behavior of Wnt signaling. In combination with data gained from large-scale screening and sequencing approaches, the use of such models will be a valuable tool to predict the behavior of the network under pathological conditions or during the aging process.

References

Agur, Z., Kirnasovsky, O.U., Vasserman, G. *et al.* (2011) Dickkopf1 regulates fate decision and drives breast cancer stem cells to differentiation: an experimentally supported mathematical model. *PLoS One*, **6**, e24225.

Amonlirdviman, K., Khare, N.A., Tree, D.R. *et al.* (2005) Mathematical modeling of planar cell polarity to understand domineering nonautonomy. *Science*, **307**, 423–426.

Aulehla, A., Wehrle, C., Brand-Saberi, B. *et al.* (2003) Wnt3a plays a major role in the segmentation clock controlling somitogenesis. *Developmental Cell*, **4**, 395–406.

Axelrod, J.D. and Tomlin, C.J. (2011) Modeling the control of planar cell polarity. *Wiley Interdisciplinary Reviews: Systems Biology and Medicine*, **3**, 588–605.

Benary, U., Kofahl, B., Hecht, A., and Wolf, J. (2013) Modeling Wnt/β-catenin target gene expression in APC and Wnt gradients under wild type and mutant conditions. *Frontiers in Physiology*, **4**, 21.

Cho, K.-H., Baek, S., and Sung, M.-H. (2006) Wnt pathway mutations selected by optimal β-catenin signaling for tumorigenesis. *FEBS Letters*, **580**, 3665–3670.

Dequeant, M.L., Glynn, E., Gaudenz, K. *et al.* (2006) A complex oscillating network of signaling genes underlies the mouse segmentation clock. *Science*, **314**, 1595–1598.

Ferrell, J.E., Jr. (2009) Signaling motifs and Weber's law. *Molecular Cell*, **36**, 724–727.

Finnemann, S., Kuhl, M., Otto, G., and Wedlich, D. (1995) Cadherin transfection of Xenopus XTC cells downregulates expression of substrate adhesion molecules. *Molecular and Cellular Biology*, **15**, 5082–5091.

Gibb, N., Lavery, D.L., and Hoppler, S. (2013) sfrp1 promotes cardiomyocyte differentiation in Xenopus via negative-feedback regulation of Wnt signalling. *Development*, **140**, 1537–1549.

Goentoro, L. and Kirschner, M.W. (2009) Evidence that fold-change, and not absolute level, of beta-catenin dictates Wnt signaling. *Molecular Cell*, **36**, 872–884.

Goentoro, L., Shoval, O., Kirschner, M.W., and Alon, U. (2009) The incoherent feedforward loop can provide fold-change detection in gene regulation. *Molecular Cell*, **36**, 894–899.

Goldbeter, A. and Pourquie, O. (2008) Modeling the segmentation clock as a network of coupled oscillations in the Notch, Wnt and FGF signaling pathways. *Journal of Theoretical Biology*, **252**, 574–585.

Handorf, T. and Klipp, E. (2012) Modeling mechanistic biological networks: an advanced Boolean approach. *Bioinformatics*, **28**, 557–563.

Hernandez, A.R., Klein, A.M., and Kirschner, M.W. (2012) Kinetic responses of β-catenin specify the sites of Wnt control. *Science*, **338**, 1337–1340.

Jensen, P.B., Pedersen, L., Krishna, S., and Jensen, M.H. (2010) A Wnt oscillator model for somitogenesis. *Biophysical Journal*, **98**, 943–950.

Kestler, H.A. and Kuhl, M. (2008) From individual Wnt pathways towards a Wnt signalling network. *Philosophical Transactions of the Royal Society of London. Series B, Biological Science*, **363**, 1333–1347.

Kestler, H.A. and Kuhl, M. (2011) Generating a Wnt switch: it's all about the right dosage. *The Journal of Cell Biology*, **193**, 431–433.

Kestler, H.A., Wawra, C., Kracher, B., and Kuhl, M. (2008) Network modeling of signal transduction: establishing the global view. *BioEssays*, **30**, 1110–1125.

Kim, D., Rath, O., Kolch, W., and Cho, K.H. (2007) A hidden oncogenic positive feedback loop caused by crosstalk between Wnt and ERK pathways. *Oncogene*, **26**, 4571–4579.

Kirnasovsky, O.U., Kogan, Y., and Agur, Z. (2008) Analysis of a mathematical model for the molecular mechanism of fate decision in mammary stem cells. *Mathematical Modelling of Natural Phenomena*, **3**, 78–89.

Kogan, Y., Halevi-Tobias, K.E., Hochman, G. *et al.* (2012) A new validated mathematical model of the Wnt signalling pathway predicts effective combinational therapy by sFRP and Dkk. *The Biochemical Journal*, **444**, 115–125.

Krüger, R. and Heinrich, R. (2004) Model reduction and analysis of robustness for the Wnt/β-catenin signal transduction pathway. *Genome Informatics*, **15**, 138–148.

Lee, E., Salic, A., Kruger, R. *et al.* (2003) The roles of APC and Axin derived from experimental and theoretical analysis of the Wnt pathway. *PLoS Biology*, **1**, E10.

Li, V., Ng, S., Boersema, P. *et al.* (2012) Wnt signaling through inhibition of β-catenin degradation in an intact axin1 complex. *Cell*, **149**, 1245–1256.

MacDonald, B.T., Adamska, M., and Meisler, M.H. (2004) Hypomorphic expression of Dkk1 in the doubleridge mouse: dose dependence and compensatory interactions with Lrp6. *Development*, **131**, 2543–2552.

MacDonald, B.T., Semenov, M.V., and He, X. (2007) SnapShot: Wnt/beta-catenin signaling. *Cell*, **131**, 1204.

Mukhopadhyay, M., Shtrom, S., Rodriguez-Esteban, C. *et al.* (2001) Dickkopf1 is required for embryonic head induction and limb morphogenesis in the mouse. *Developmental Cell*, **1**, 423–434.

Murray, P.J., Kang, J.W., Mirams, G.R. *et al.* (2010) Modelling spatially regulated beta-catenin dynamics and invasion in intestinal crypts. *Biophysical Journal*, **99**, 716–725.

Nalesso, G., Sherwood, J., Bertrand, J. *et al.* (2011) WNT-3A modulates articular chondrocyte phenotype by activating both canonical and noncanonical pathways. *The Journal of Cell Biology*, **193**, 551–564.

Niida, A., Hiroko, T., Kasai, M. *et al.* (2004) DKK1, a negative regulator of Wnt signaling, is a target of the β-catenin/TCF pathway. *Oncogene*, **23**, 8520–8526.

O'Connell, D.J., Ho, J.W.K., Mammoto, T. *et al.* (2012) A Wnt-Bmp feedback circuit controls intertissue signaling dynamics in tooth organogenesis. *Science Signaling*, **5**, ra4.

Pedersen, L., Jensen, M.H., and Krishna, S. (2011) Dickkopf1—a new player in modelling the Wnt pathway. *PLoS One*, **6**, e25550.

Rodriguez-Gonzalez, J.G., Santillan, M., Fowler, A.C., and Mackey, M.C. (2007) The segmentation clock in mice: interaction between the Wnt and Notch signalling pathways. *Journal of Theoretical Biology*, **248**, 37–47.

Seifert, J.R.K. and Mlodzik, M. (2007) Frizzled/PCP signalling: a conserved mechanism regulating cell polarity and directed motility. *Nature Review Genetics*, **8**, 126–138.

Semenov, M.V., Tamai, K., Brott, B.K. *et al.* (2001) Head inducer Dickkopf-1 is a ligand for Wnt coreceptor LRP6. *Current Biology*, **11**, 951–961.

Shin, S.-Y., Rath, O., Zebisch, A. *et al.* (2010) Functional roles of multiple feedback loops in extracellular signal-regulated kinase and Wnt signaling pathways that regulate epithelial-mesenchymal transition. *Cancer Research*, **70**, 6715–6724.

Sick, S., Reinker, S., Timmer, J., and Schlake, T. (2006) WNT and DKK determine hair follicle spacing through a reaction-diffusion mechanism. *Science*, **314**, 1447–1450.

Tan, C.W., Gardiner, B.S., Hirokawa, Y. *et al.* (2012) Wnt signalling pathway parameters for mammalian cells. *PLoS One*, **7**, e31882.

van Leeuwen, I.M., Byrne, H.M., Jensen, O.E., and King, J.R. (2007a) Elucidating the interactions between the adhesive and transcriptional functions of beta-catenin in normal and cancerous cells. *Journal of Theoretical Biology*, **247**, 77–102.

van Leeuwen, I.M., Edwards, C.M., Ilyas, M., and Byrne, H.M. (2007b) Towards a multiscale model of colorectal cancer. *World Journal of Gastroenterology*, **13**, 1399–1407.

van Leeuwen, I.M., Mirams, G.R., Walter, A. *et al.* (2009) An integrative computational model for intestinal tissue renewal. *Cell Proliferation*, **42**, 617–636.

Wawra, C., Kuhl, M., and Kestler, H.A. (2007) Extended analyses of the Wnt/beta-catenin pathway: robustness and oscillatory behaviour. *FEBS Letters*, **581**, 4043–4048.

Zhang, L., Lander, A.D., and Nie, Q. (2012) A reaction-diffusion mechanism influences cell lineage progression as a basis for formation, regeneration, and stability of intestinal crypts. *BMC Systems Biology*, **6**, 93.

12 The Wnt's Tale: On the Evolution of a Signaling Pathway

Jenifer C. Croce[1] and Thomas W. Holstein[2]

[1]CNRS, UMR7009, Sorbonne Universités, UPMC Univ Paris 06, Laboratoire de Biologie du Développement de Villefranche-sur-mer, EvoInSiDe Team, Observatoire Océanographique, 06230, Villefranche-sur-mer, France
[2]Department of Molecular Evolution and Genomics, Centre for Organismal Studies (COS), Heidelberg University, Heidelberg, Germany

Introduction

Wnt signaling represents one of the major intercellular signaling pathways during animal development. A multitude of studies have further revealed the involvement of this signal in various human diseases, including several cancers (see Chapters 27 and 28). Wnt signaling encompasses, at the molecular level, at least, three main transduction cascades, all activated by the binding of Wnt ligands onto transmembrane receptors of the Frizzled (Fzd) family. These cascades are the canonical Wnt/β-catenin pathway, characterized by the nuclear localization of β-catenin upon activation (see Chapter 16), and at least two noncanonical signals, the planar cell polarity and the calcium pathways, which do not involve β-catenin (see Chapter 6). Of these three cascades, the most studied to date remains the canonical Wnt pathway, of which evolutionary diversification in metazoans will be discussed in this chapter.

The canonical Wnt pathway is evolutionary old in animals. Genes encoding Wnt ligands and Fzd receptors (the two main activating proteins of Wnt signaling) have been identified in all metazoan phyla ranging from mammals to prebilaterians (i.e., cnidarians and sponges) (Figure 12.1 and Figure 12.2), even though these latter can be traced back to more than 580 million years (Xiao and Laflamme, 2009). In addition, nuclear accumulation of β-catenin (which is the hallmark of activated canonical Wnt signaling) has been reported in animals of all metazoan phyla, including in cnidarians (for details see the succeeding chapters), hence further corroborating the evolutionary conservation of the mode of action of this pathway among multicellular, eukaryotic organisms. However, this classical canonical Wnt signal is metazoan specific. Indeed, although homologs of component proteins of the pathway are present in single-cell eukaryotes, including Fzd-related receptors, no genes encoding Wnt-related ligands have been described yet neither in unicellular eukaryotes nor in fungi (Figure 12.2; Harwood, 2008; King *et al.*, 2008).

At a functional level, a primary role of canonical Wnt signaling, observed in all metazoans, is body axes establishment and patterning. In bilaterians, which possess two body axes determined under a Cartesian coordinate system, growing evidence has established the fundamental role of this canonical pathway in

Wnt Signaling in Development and Disease: Molecular Mechanisms and Biological Functions, First Edition.
Edited by Stefan Hoppler and Randall T. Moon.

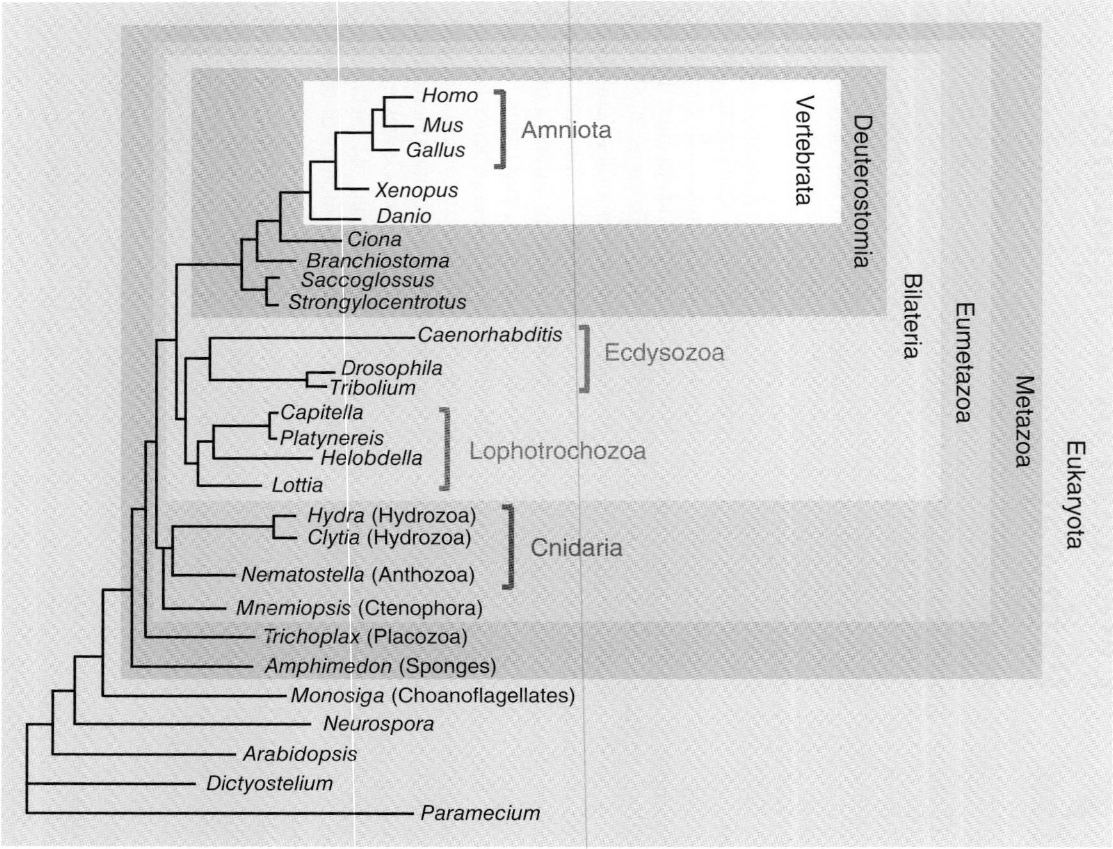

Figure 12.1 Molecular phylogeny of metazoans. *(See insert for color representation of the figure.)* Phylogenetic relationships are based on recent studies from Srivastava *et al.* (2010) and Simakov *et al.* (2013).

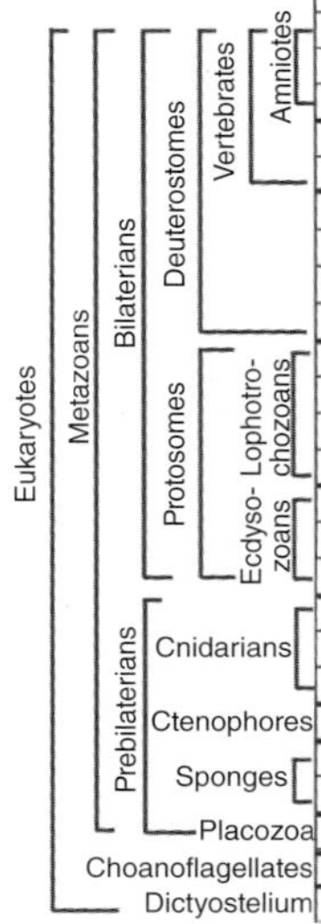

	Wnt1	Wnt2	Wnt3	Wnt4	Wnt5	Wnt6	Wnt7	Wnt8	Wnt9	Wnt10	Wnt11	Wnt16	WntA	Orphans	Total Wnt genes	Fzd 1/2/7	Fzd 3/6	Fzd 4	Fzd 9/10	Fzd 5/8	Orphans	Total Fzd genes
Homo sapiens	1	2	2	1	2	1	2	2	2	2	1	1			19	3	2	1	2	2		10
Mus musculus	1	2	2	1	2	1	2	2	2	2	1	1			19	3	2	1	2	2		10
Gallus gallus	1	2	2	1	2	1	2	2	2	2	2	1			20	3	2	1	2	2		10
Xenopus tropicalis	1	2	3	1	2	1	3	2	3	2	3	1			24	3	2	1	2	2		10
Danio rerio	1	5	2	2	2	2	3	3	2	2	2	1			27	4	2	1	2	3		12
Ciona intestinalis			1		1		1		1	1				5	10		1	1		1	2	5
Branchiostoma floridae	1		1	1	1	1	1	1		1	1				9	1		1	1	2		5
Saccoglossus kowalevskii	1	1	1	1	1		1	1	1		1	1	1		11	1			1	1		3
Strongylocntrotus purpuratus	1		1	1	1	1	1	1	1	1		1	1		11	1		1	1	1		4
Helobdella robusta	1	1		1	2	1	1			1	3	2			13							
Capitella teleta	1	1		1	1	1	1	1	1	1	1	1	1		12							
Platynereis dumerilii	1	1		1	1	1	1	1	1	1	1	1	1		12							
Lottia gigantea	1	1		1	1	1	1		1	1	1	1	1		11							
Tribolium castaneum	1				1	1	1	1	1	1	1		1		9	1		1		1		3
Drosophila melanogaster	1				1	1	1		1	1				1	7	1		2		1		4
Caenorhabditis elegans				1	1				1	1		1			5							
Nematostella vectensis	1	1	1	1	1	1	1	2		1	1	1	1		13	1		2		1		4
Hydra magnipapillata	1	1	1		1		1	1	3		1	1			11	1		1		1		3
Clytia haemispherica			1		1				1					10	13	1		1		1		3
Mnemiopsis leidyi						1			1				1	1	4							
Amphimedon queenslandica											1	1		1	3	1		1				2
Oscarella lobularis														2	2							
Trichoplax adhaerens														3	3						1	1
Mnemiopsis brevicollis															0							
Dictyostelium discoideum															0							16

Figure 12.2 Distribution of *Wnt* and *Fzd* genes in metazoans. *(See insert for color representation of the figure.)* Colors and numbers in the filled boxes indicate the relative abundance of *Wnt* and *Fzd* representatives in each gene subfamily (*white*, no data available; *grey*, zero member; *yellow*, one member; *green*, two members; *blue*, three members; *orange*, four members; *purple*, five members). Data reported here are based on molecular *Wnt* and *Fzd* phylogenies published in Adamska *et al.* (2010), Beermann *et al.* (2011), Cho *et al.* (2010), Croce *et al.* (2006), Freeman *et al.* (2008), Garriock *et al.* (2007), Gloriam, Fredriksson, and Schioth (2007), Hino *et al.* (2003), Janssen *et al.* (2010), King *et al.* (2008), Krishnan *et al.* (2012), Lagerström *et al.* (2006), Lapebie *et al.* (2009), Lengfeld *et al.* (2009), Momose and Houliston (2007), Nordstrom, Fredriksson, and Schioth (2008), Pang *et al.* (2010), Philipp *et al.* (2009), Prabhu and Eichinger (2006), Prud'homme *et al.* (2002), Schubert and Holland (2003), Srivastava *et al.* (2008), and Zhang, Tran, and Wessely (2011).

patterning the anteroposterior (AP) axis (also referred to as the animal–vegetal (AV) axis in some animals), while the dorsoventral (DV) axis is subsequently determined mostly by the action of a TGF-β signaling pathway (i.e., Nodal and/or Bmp) (e.g., De Robertis, 2010; Holstein, Watanabe, and Ozbek, 2011; Niehrs, 2010), although at least in vertebrates the canonical Wnt pathway participates also to that later event (e.g., Hikasa and Sokol, 2013; Niehrs, 2010; Chapter 19). Similarly, in prebilaterians, which exhibit a radial symmetry with a unique primary body axis, Wnt signaling controls the establishment of that axis, and it does so also through interactions with members of the TGF-β pathways (Holstein, 2012; Petersen and Reddien, 2009). Thus, although the transition from radial to bilateral symmetry is not well understood to date, canonical Wnt signaling functions and interactions, at least with the TGF-β pathways, appear similar in all metazoans.

Wnt and *Fzd* genes are deeply embedded in metazoan genomes. They have been reported in a multitude of animals representative of almost all metazoan phyla (Figure 12.2). Several phylogenetic analyses have further been conducted on these *Wnt* and *Fzd* genes, thereby allowing some evolutionary hypotheses regarding these large families of secreted or transmembrane proteins among animals. On the contrary, it is fair to recognize though that the spatiotemporal distribution and the developmental function of these centrally important signaling components, although available for some molecules in some species, remain still largely scarce especially in invertebrates.

Herein, we focused our attention on the complexity and the specific features of canonical Wnt signaling activities and of the *Wnt* and *Fzd* repertoires during early metazoan development. We decided to present this analysis from an atypical perspective, that is, starting from knowledge related to human, mouse, and chick and tracing it back up to prebilaterian animals such as cnidarians and sponges. In this report, we compare the developmental functions of nuclear β-catenin in these several animal phyla in order to extract an essence of the basis of this pathway from an evolutionary point of view. We analyze further the composition of the *Wnt* and *Fzd* inventories in several metazoans to presume the complexity changes of these genes repertoires across animal evolution.

Wnt signaling in amniote vertebrates (human, mouse, and chick)

The development of mammals and birds (i.e., of amniotes) is largely different from the development of most other vertebrates (i.e., amphibians and fishes) and invertebrates. Amniotes evolved reproductive strategies that implied laying fewer eggs and providing them with maternal care. The development of the amnion and of the yolk sac represents thereby a major evolutionary novelty within the vertebrate phyla (Stern and Downs, 2012). In addition, a supplementary important innovation in mammal development is the blastocyst, a thin-walled hollow structure that encloses a pluripotent inner cell mass (embryoblast), from which the embryo will arise, and an outer layer of cells (trophoblast), which will produce supporting tissues for fetal development.

Among amniotes, most contemporary developmental molecular data arise from studies carried out in mouse (i.e., the main mammalian genetic model), whereas knowledge on chick is much reduced and that on human is even further scarce. Nevertheless, concerning canonical Wnt signaling, rodents, birds, and primates appear to be identically represented both in terms of activating molecules and pathway components. For instance, they all share a similar set of *Wnt* and *Fzd* gene subfamilies (Figure 12.2), and they all possess a single β-catenin gene (Chapter 16), hence allowing some translational conclusions.

The developmental function of canonical Wnt signaling in amniotes has extensively been investigated using mouse mutants (e.g., Grigoryan *et al.*, 2008; Wang, Sinha, and Wynshaw-Boris, 2012). Analyses of conditional or permanent mutants for β-catenin or for other components of the pathway have revealed a clear role for the Wnt/β-catenin signal, and for several selective *Wnt* and *Fzd* genes, in the regulation of multiple steps of mammalian embryogenesis, most notably during pre- and postimplantation processes.

Before implantation, the canonical Wnt/β-catenin pathway is critical for the activation and the implantation of the blastocyst into the uterine epithelium (Mohamed *et al.*, 2005; Sonderegger, Pollheimer, and Knofler, 2010). Six of the 19 *Wnt* genes present in the mouse genome are strongly expressed in the blastocyst before implantation (*Wnt1*, *Wnt3a*, *Wnt6*, *Wnt7b*, *Wnt9a*, and *Wnt10b*), and most of the 10 *Fzd* genes are also detectable at that stage by quantitative-PCR (e.g., Kemp *et al.*, 2005, 2007). Additionally, Wnt3a has specifically been involved in the maintenance of the undifferentiated state of the cells of the inner cell mass (e.g., ten Berge *et al.*, 2011; Wang, Sinha, and Wynshaw-Boris, 2012), hence providing some insight into the role of a particular Wnt molecule during this process. In humans, by contrast, no clear functional data are yet available regarding the role of a particular *Wnt* or *Fzd* gene or of canonical Wnt signaling in the implantation of the blastocyst. Nonetheless, mutations in Wnt signaling components, and in human Wnt3a, have been associated with infertility, endometriosis, and/or endometrial cancer, hence supporting a function of this intercellular signal in human blastocyst implantation as well (Sonderegger, Pollheimer, and Knofler, 2010; Sonderegger *et al.*, 2010).

After implantation, the blastocyst undergoes several developmental steps including gastrulation. During this step, in mammals, as in birds, cells ingress through the primitive streak and give rise to the two internal cell fates, the endoderm and the mesoderm. Primitive streak formation is accompanied in mouse by an activation of canonical Wnt signaling activity, and it relies on the action of β-catenin, as well as of the membrane coreceptor Lrp5/6 and of three Wnt ligands Wnt2b, Wnt3, and Wnt8a (Wang, Sinha, and Wynshaw-Boris, 2012). Although no Fzd receptor has yet been functionally implicated in this process, it is noteworthy that *Fzd5*, *Fzd7*, *Fzd8*, and *Fzd10* are expressed in distinct domains of the gastrulating mouse embryo (Kemp *et al.*, 2007), hence providing initial cues for future investigations. Genetic studies performed in mouse and culture cells have further disclosed supplementary information regarding the functions of Wnt3 during mouse gastrulation. For instance, these studies have suggested that in the anterior region of the embryo, Wnt3 activity is most likely suppressed, in a dosage-dependent manner, by the action of Dkk1, an extracellular inhibitor of canonical Wnt signaling (Kimura-Yoshida *et al.*, 2005; Lewis *et al.*, 2008; Mukhopadhyay *et al.*, 2001). They have determined in addition that Wnt3 acts during mesoderm development as part of a Bmp4–Wnt3–Nodal positive feedback loop. In this process, Bmp4 induces *Wnt3* that in turn stimulates *Nodal* expression via an evolutionary conserved proximal epiblast enhancer containing two preserved Tcf consensus motifs (CCTTTGA) (Ben-Haim *et al.*, 2006; Morkel *et al.*, 2003). Finally, Wnt3 has also been shown to be essential in mouse, still through the β-catenin pathway, to set up the molecular segmentation clock that drives, in a periodic manner, segment formation within the presomitic mesoderm, and this is also true in chick (Aulehla *et al.*, 2008; Pourquie, 2011). Thus, these exemplary reports on Wnt3 provide an important set of information on the several functions of that ligand during post-implantation processes, as they outline chiefly the complexity in studying the embryonic role of any given *Wnt* gene in mammals. Each Wnt may indeed be involved in several developmental processes, and this is without taking into account the potential functional redundancy that may exist between Wnts of similar Wnt subfamilies.

In terms of *Wnt* and *Fzd* genes, rodents and primates, hence mammals, appear to share a similar set of gene subfamilies, with further an identical number of genes per subfamily at least when comparing mouse and human (Figure 12.2). This suggests thus that the common ancestor of all mammals may have possessed as well 19 *Wnt* and 10 *Fzd* genes distributed into 12 and 5 gene subfamilies, respectively. By comparison, in birds, such as in chick, an identical collection of *Wnt* and *Fzd* gene subfamilies is observed, with the caveat though of a slightly different number of representatives in particular within some *Wnt* subfamilies (Figure 12.2). This distribution presumes thus that the amniote ancestor also had 12 *Wnt* and 5 *Fzd* gene subfamilies, which however could have encompassed either a similar overall number of genes than that observed in mammals, that is, implying independent acquisition of additional *Wnt* genes in birds through specific duplication events, or potentially a higher number of *Wnt* genes than in mammals, that is, presuming the subsequent loss of these supplementary genes in this group of animals.

Wnt signaling in nonamniote vertebrates (amphibians and fishes)

A major part of the current knowledge on the function of canonical Wnt signaling during vertebrate early embryogenesis, in particular in axis formation, further comes from studies carried out in frogs and more recently in zebrafish (see Chapter 19). In these animals, an initial AV asymmetry is established in the egg by the localization of maternal determinants at the vegetal pole (e.g., vg1, VegT), where components of the canonical Wnt pathway further reside. At sperm entry, a microtubule-dependent rotation of the cortical cytoplasm occurs, which shifts these canonical Wnt components towards the marginal zone of the embryo (e.g., Houliston and Elinson, 1992). There, they contribute in setting up the DV axis during gastrulation, in particular with the formation of the embryonic signaling center (i.e., the Spemann–Mangold organizer in amphibians and the shield in fishes, which correspond to the node organizer in mice) (e.g., Harland and Gerhart, 1997; Schier and Talbot, 2005; Stern, 2006). It is also during gastrulation that the canonical Wnt pathway specifies the AP axis in these animals, through the induction and the maintenance, among others, of the transcription factor T (or Brachyury). However, it should be emphasized that development until gastrulation is governed in frogs and fishes by yolk richness of nonamniote eggs and that overlapping effects of Wnt signaling in early and late embryos should thus be carefully distinguished.

In unfertilized frog and fish eggs, several components of the canonical Wnt pathway are selectively localized at the vegetal pole. These components include more specifically XWnt5a/XWnt11, ZWnt8a, and Dishevelled (Dsh) (Ku and Melton, 1993; Lu, Thisse, and Thisse, 2011; Tao *et al.*, 2005). After sperm entry and by cortical rotation, these molecules are then relocated towards the future dorsal side of the embryo, with the essential outcome of stabilizing β-catenin in the subsequent dorsal/posterior blastomeres and establishing the DV axis (Moon and Kimelman, 1998). Even though the mechanism of β-catenin stabilization remains under discussion (Hikasa and Sokol, 2013), several experimental evidences corroborate these conclusions. For instance, microinjections of *Wnt1* mRNA initially demonstrated in 1989 that a signaling molecule could mimic the transplantation of the dorsal lip of the gastrula (McMahon and Moon, 1989), hence making Wnt signaling the prime candidate for setting up the endogenous axis. Then, β-catenin was shown to be asymmetrically distributed along the future DV axis, as early as at the two-cell stage, in frog embryos (Larabell *et al.*, 1997). Moreover, vesicles associated with Dsh were subsequently described to move along the parallel array of microtubules during cortical rotation and to accumulate in the future dorsal side of the embryo, where Dsh in turns promotes β-catenin stabilization, hence leading to the regulation of *Siamois* and *Goosecoid* expression and the formation of the dorsal lip (Miller *et al.*, 1999). Finally, recent ablation of maternal mRNA encoding *XWnt11* corroborated also the role of this molecule and of canonical Wnt signaling in dorsal development, as did the microinjections of *XWnt11* or *ZWnt8a* mRNA in fertilized frog or fish eggs, respectively, or the analyses of several zebrafish mutants in which reduction of dorsal β-catenin accumulation prompts severe dorsal developmental defects (e.g., Kelly *et al.*, 2000; Lu, Thisse, and Thisse, 2011; Tao *et al.*, 2005). Thus, before gastrulation, the canonical Wnt pathway is clearly essential to trigger DV axis establishment, and it does so further through the inhibition of Bmp signaling on the dorsal side of the embryo. Genes targeted by the canonical Wnt pathway before gastrulation have indeed been identified, and among them, antagonists of the Bmp pathway were found, that is, *Chordin* and *Noggin* (e.g., Hikasa and Sokol, 2013; Kuroda, Wessely, and De Robertis, 2004). Moreover, inhibition of the BMP pathway on the embryonic dorsal side has been shown to be critical for DV axis establishment. Embryos mutant for Chordin exhibit indeed a ventralized phenotype (Schulte-Merker *et al.*, 1997), while microinjections of *Chordin* mRNA reveal this molecule as a potent dorsalizing factor (Sasai *et al.*, 1994).

Later in the development of nonamniote vertebrates, that is, during gastrulation, the relationship between the canonical Wnt pathway and Bmp signaling however greatly differs. The canonical Wnt pathway is no longer involved in the establishment or the maintenance of the pattern along the DV axis, but it rather now controls the shaping of specific organs and/or tissues along the AP axis (i.e., the central nervous system, the mesoderm, and the endoderm), hence perpendicularly to the Bmp pathway (Niehrs, 2010). Several zygotic Wnt ligands (including Wnt3a, Wnt5a, Wnt8, and Wnt11) are expressed in the posterior region of the embryos, while on the opposite anterior region, various Wnt antagonists are present (e.g., Frzb/Sfrp3, Crescent, and Dkk1) (e.g., Hikasa and Sokol, 2013). Moreover, in *amphibians*, a late ectopic overactivation of canonical Wnt signaling through a lithium chloride treatment – a GSK3 inhibitor (Klein and Melton, 1996) – does not affect the formation of specific dorsal or ventral structures but leads to posteriorized embryos (Yamaguchi and Shinagawa, 1989). Similarly, microinjection of *XWnt8* as a plasmid DNA, hence allowing its expression only after the midblastula transition, produces AP defects (Christian and Moon, 1993), as further does depletion of particular zygotic Wnt components in zebrafish mutants (e.g., Erter *et al.*, 2001; Heasman, Kofron, and Wylie, 2000; Lekven *et al.*, 2001; Shimizu *et al.*, 2005). Thus, at gastrulation, the canonical Wnt pathway operates an important molecular switch, shifting from dorsal specification to posterior development, thereby generating the Cartesian coordinate system of positional information required to properly orchestrate specification along the AP and DV axes.

Interestingly, in amphibians and fishes, a remarkable similar set of *Wnt* and *Fzd* gene subfamilies is observed compared to amniotes (Figure 12.2). This suggests thus that the vertebrate ancestor, in all likelihood, encompassed a comparable Wnt/Fzd/β-catenin toolkit consisting of 12 *Wnt* and 5 *Fzd* gene subfamilies. However, it should be outlined that nonamniote vertebrates have an overall higher number of *Wnt* orthologs than amniotes. Assuming that there were two rounds of whole-genome duplication in ancestral vertebrates (Dehal and Boore, 2005), this observation suggests thus that a significant secondary reduction in the complexity of the *Wnt* repertoire in amniotes must have occurred. It should also be stressed that the function of Wnt signaling during early embryogenesis is divergent among vertebrates, most likely due to the formation of distinct modes in development, that is, formation of the blastocyst in amniotes and rapid development of yolk-rich eggs in frogs and fishes. Nonetheless, the canonical Wnt pathway maintained some common roles during early vertebrate embryogenesis, including in establishing an AP body axis and in gastrulation. We presume thus that from an evolutionary point of view, the Wnt/Fzd/β-catenin signaling system present in the ancestral vertebrate also governed AP body axis patterning and regulated gastrulation.

Wnt signaling in nonvertebrate deuterostomes (Urochordata, Cephalochordata, Hemichordata, and Echinodermata)

In all nonvertebrate deuterostome animals, that is, urochordates (e.g., tunicates), cephalochordates (e.g., *Amphioxus*), hemichordates (e.g., acorn worms), and echinoderms (e.g., sea urchins and sea stars), canonical Wnt signaling plays a critical role, such as in vertebrates, in patterning the AP embryonic body axis (also known as the AV axis) and in the formation of a crucial posterior blastoporal signaling center. In urochordates, hemichordates, and echinoderms, β-catenin proteins commonly accumulate selectively in the nuclei of the vegetal blastomeres where they are required for endomesoderm specification, embryo regionalization, and digestive tract development (Darras *et al.*, 2011; Imai *et al.*, 2000; Logan *et al.*, 1999; Miyawaki *et al.*, 2003). In cephalochordates, in contrast, it should be emphasized that β-catenin accumulates initially uniformly in the embryonic nuclei, before disappearing from the vegetal cells, and that this vegetal nonnuclear accumulation of β-catenin seems required for endomesoderm specification and/or gastrulation (Holland *et al.*, 2005). Thus, in cephalochordates, canonical Wnt signaling may have an opposite role during early cleavage than that observed in the other nonvertebrate deuterostome phyla, although that role remains to be fully validated by functional analyses.

In nonvertebrate deuterostomes, many studies have further been conducted to elucidate and characterize the *Wnt* and *Fzd* repertoires present in these animals. In each case, many *Wnt* and *Fzd* genes were identified (Figure 12.2), but information regarding their respective expression profile and individual biological function remains rather disparate among the phyla. For instance, in cephalochordates, the expression profile of most of the nine *Wnt* genes distinguished in the genome of the amphioxus *Branchiostoma floridae* (all except *Wnt10*) is now available (Holland, Holland, and Schubert, 2000; Schubert, Holland, and Holland, 2000a, 2000b; Schubert *et al.*, 2000, 2001), while that of the *Fzd* genes has yet to be determined, as does the developmental function of each of these molecules. By comparison, in hemichordates, the spatiotemporal expression pattern of only two *Wnt* (i.e., *Wnt1* and *Wnt8*) and one *Fzd* (*Fz5/8*) has been established (they are all expressed in the developing brain), and an siRNA knockdown experiment has been conducted, ascertaining the role of Fz5/8 in controlling the AP topology of the proboscis (Pani *et al.*, 2012). Likewise, in urochordates, a single *Wnt* gene, *Wnt5a*, has been extensively studied, revealing its maternal expression with a posterior localization in the early embryos (Sasakura, Ogasawara, and Makabe, 1998), its requirement in the control of the morphogenetic movements of the notochord cells (Niwano *et al.*, 2009; Sasakura and Makabe, 2001), and the role of its zygotic proteins in triggering β-catenin nuclear localization starting at the 64-cell stage (Kawai *et al.*, 2007). However, besides that ligand, only one additional survey has been conducted on an ascidian *Wnt* or *Fzd* gene, which discloses solely the expression profile of *Wnt7a* in the tail neural tube of tail bud embryos (Sasakura and Makabe, 2000). Finally, in echinoderms, and by contrast, multiple analyses have been carried out on *Wnt* and *Fzd* genes. These investigations, mainly performed on sea urchins, have established through PCR and Northern blot analyses the set of ligands and receptors transcribed maternally (Croce *et al.*, 2011; Lhomond *et al.*, 2012; Stamateris, Rafiq, and Ettensohn, 2010). They have determined also the biological functions of three Wnt ligands (Wnt6, Wnt8, Wnt1) and of two Fzd receptors (Fz1/2/7, Fz5/8) (Croce *et al.*, 2006, 2011; Lhomond *et al.*, 2012; Sethi *et al.*, 2012; Wei *et al.*, 2012; Wikramanayake *et al.*, 2004), as well as the identity of a Wnt–Fzd couple mediating β-catenin nuclear localization during early development (Lhomond *et al.*, 2012). Thus, some information is available to date on the spatiotemporal expression profiles and/or the roles of some *Wnt* and *Fzd* genes in nonvertebrate deuterostomes, however that information remains still too heterogeneous to allow any kind of hypotheses regarding the evolutionary diversification of the role of these molecules among these animals.

On the contrary, the distinct phylogenetic analyses conducted between vertebrate and nonvertebrate deuterostome *Wnt* and *Fzd* genes lead to interesting observations. First, it appears that remarkably all nonvertebrate deuterostome genomes surveyed so far encompass about half as many *Wnt* and *Fzd* genes than vertebrates, most likely due to the whole-genome duplication events that occurred in the latter (Dehal and Boore, 2005). Nonetheless, despite this reduction, these analyses establish further that overall all vertebrate *Wnt* and *Fzd* subfamilies are represented in nonvertebrate deuterostomes, with simply only one gene defining each subfamily (except in urochordates where some *Wnt* and *Fzd* genes could not be confidently assigned to any specific subfamily, consistent with the reported fast evolution of the urochordate protein sequences (e.g., Hino *et al.*, 2003; Satou *et al.*, 2003)). Second, the phylogenetic analyses outlined the existence as well, in hemichordates and echinoderms, of a supplementary *Wnt* subfamily, that of *WntA* (Figure 12.2). The presence of that subfamily in these animals but not in chordates suggests that the common ancestor of all deuterostomes may have had actually a complete set of 13 *Wnt* gene subfamilies, rather than 12, which could all have furthermore been represented potentially by a single gene. These hypotheses are further supported by the presence of a *WntA* gene in protostomes and prebilaterians (see the succeeding text), of which each *Wnt* and *Fzd* gene subfamily also encloses a single ortholog per subfamily.

Wnt signaling in nondeuterostome bilaterians (Ecdysozoa and Lophotrochozoa)

Nondeuterostome bilaterians embrace all protostomes, which include two groups of animals, the Ecdysozoa and the Lophotrochozoa. The Ecdysozoa comprise the two major protostome biological models, the fruit fly *Drosophila mela-*

nogaster and the nematode *Caenorhabditis elegans*, while the Lophotrochozoa encompass mainly mollusks and annelids. In protostome animals, nuclear β-catenin, hence the canonical Wnt pathway, has been shown to be critical, as in other metazoans, for the establishment of cell fate identities along the AP axis. This is the case within individual (para)segments such as in *D. melanogaster* (Wieschaus and Riggleman, 1987), in the red flour beetle *Tribolium castaneum* (Bao *et al.*, 2012), in the cricket *Gryllus bimaculatus* (Miyawaki *et al.*, 2004), or in the milkweed bug *Oncopeltus fasciatus* (Angelini and Kaufman, 2005) or in whole developing body plan, alike in the mollusk *Crepidula fornicata* (Henry, Perry, and Martindale, 2010), in the nemertean *Cerebratulus lacteus* (Henry *et al.*, 2008), in the annelid *Platynereis dumerilii* (Schneider and Bowerman, 2007), and in the nematode *C. elegans* (e.g., Thorpe *et al.*, 1997). In addition, in the four latter animals, β-catenin has been shown to be further required for the specification of the posterior endoderm lineage, similar to its role in urochordates, hemichordates, and echinoderms (Henry, Perry, and Martindale, 2010; Henry *et al.*, 2008; Schneider and Bowerman, 2007; Thorpe *et al.*, 1997).

Genes encoding Wnt ligands and Fzd receptors have been reported to date in a number of protostome animals. Expression profiles of *Wnt* genes have been examined in some arthropods (e.g., *D. melanogaster*, *T. castaneum*, *G. bimaculatus*, *Achaearanea tepidariorum* (Janson, Cohen, and Wilder, 2001; Janssen *et al.*, 2010)) as well as in a couple of annelids (e.g., *P. dumerilii*, *Capitella teleta*, and *Helobdella robusta* (Cho *et al.*, 2010; Janssen *et al.*, 2010)), and comparison of these expression patterns in those segmented animals shows that even though arthropods and annelids are evolutionary distant from each other, several of the *Wnt* orthologs display a comparable segmental reiterated stripe pattern. This suggests thus either that the regulatory system leading to segmental expression was independently co-opted by the *Wnt* genes in arthropods and annelids or that it is a primitive character that both phyla inherited from a common ancestor and that was then lost in other Ecdysozoa, for example, in nematodes. In addition, at the level of an individual animal, these analyses further reveal that in most cases each *Wnt* gene displays a distinct expression pattern from its counterparts, although tissue-specific overlaps can be observed. This is the case, for instance, for the nine *Wnt* genes of *T. castaneum* that share partially overlapping expression domains in the head, the segments, the growth posterior zone, and/or the limbs (Bolognesi *et al.*, 2008a). Because of these overlapping expressions, one could presume thus of potential functional redundancies between these *Wnt* genes. However, all-inclusive functional analyses performed in *T. castaneum*, as well as in *D. melanogaster*, indicate that knockdown or overexpression of each of the *Wnt* genes produces developmental defects with no obvious evidence for such redundancies (e.g., Bolognesi *et al.*, 2008b; Janssen *et al.*, 2010). Thus, overall to date, the current knowledge on the expression profile of the activator molecules of Wnt signaling in protostomes is rather restricted to the *Wnt* genes, as it is for their biological functions (except in *T. castaneum* where expression profiles and roles of Fzd receptors have been reported (Beermann *et al.*, 2011)). In addition, it is fair to note that most of the data accessible to date on the protostome *Wnt* genes arise from work carried out on only a couple of selective segmented animals and/or Ecdysozoa, even though the protostome phyla include a much larger animal diversity.

The fruit fly *D. melanogaster* and the nematode *C. elegans* have been for years now the two major protostome models used for biological investigations. For that reason and because of the reduced number of *Wnt* and *Fzd* genes present in these animals (Figure 12.2), it had been suggested that not only ecdysozoans but all protostomes had lost most of the deuterostome *Wnt* genes (e.g., Jockusch and Ober, 2000; Kusserow *et al.*, 2005). However, recent phylogenetic analyses investigating the *Wnt* genes present in the genomes of the annelids *P. dumerilii* and *C. teleta* have shown that rather, at least in these two protostomes, a total of 12 distinct *Wnt* genes could be identified, which further fall into 12 of the 13 deuterostome *Wnt* subfamilies (Cho *et al.*, 2010; Schneider and Bowerman, 2007). Strikingly, the *WntA* subfamily is represented in these animals, such as in hemichordates and echinoderms, whereas that of *Wnt3* is not. Interestingly, the *Wnt3* subfamily appears further to be absent from all protostome animals of which genome has been investigated so far (Figure 12.2). Thus, in sum, these new data on the composition of the *Wnt* and *Fzd* inventories in ecdysozoans and lophotrochozoans corroborate the current idea that the bilaterian ancestor animal most likely encompassed a *Wnt* catalogue of 13 genes (e.g., Holstein, 2012; Kusserow *et al.*, 2005), which underwent then several rounds of gene losses in protostomes, one of which could have occurred in the common ancestor of all protostomes, that of *Wnt3*, while the others would have specifically happened in selective protostome phyla. In addition, it is noteworthy that among lophotrochozoans and ecdysozoans, with the actual datasets available, these losses seem to have occurred predominantly in ecdysozoans, with the highest loss being allocated to date to *C. elegans*, with only five *Wnt* genes present in its genome.

Wnt signaling in prebilaterians (Cnidaria, Ctenophora, Porifera, and Placozoa)

Prebilaterian metazoans comprise four groups of animals: the Cnidaria (corals, jellyfish, and their kin), the Ctenophora (comb jellies), the Porifera (sponges), and the monospecific Placozoa (*Trichoplax adhaerens*). Although the interphylogenetic relationship existing between these groups is still controversial (Schierwater *et al.*, 2009), their analysis remains important for the understanding of the evolution of *Wnt* genes (Holstein, 2012).

All core components of the canonical Wnt pathway, for example, receptor complex and β-catenin destruction complex, have been identified in the genomes of animals from all prebilaterian groups. This has been the case in several cnidarians (i.e., *Nematostella vectensis*, *Hydra magnipapillata*, and *Clytia hemisphaerica*), as well as in the ctenophore *Mnemiopsis leidyi*, the placozoan *T. adhaerens*, and the sponge *Amphimedon queenslandica* (e.g., Adamska *et al.*, 2010; Hobmayer *et al.*, 2000; Holstein, 2012; Kusserow *et al.*, 2005; Pang *et al.*, 2010). In the cnidarians *N. vectensis* and *C. hemisphaerica*, β-catenin has further been shown to accumulate asymmetrically along the primary body axis, that is, the oral–aboral axis, with a selective nuclear localization on the oral side of the embryo where gastrulation will start (Momose and Houliston, 2007; Momose, Derelle, and Houliston, 2008; Wikramanayake *et al.*, 2003). This nuclear localization has moreover experimentally been shown to be essential for the proper establishment of that primary axis and for gastrulation. Privation of β-catenin nuclear localization through the use of morpholino antisense oligonucleotides leads indeed to embryos deprived of oral markers and to absence or delay in gastrulation, while ectopic activation of β-catenin results in the depletion of aboral gene expression (Momose, Derelle, and Houliston, 2008; Wikramanayake *et al.*, 2003) Thus, such as in bilaterians, in cnidarians, at least, the canonical Wnt/β-catenin pathway is similarly required for body axis patterning and gastrulation.

Regarding the *Wnt* and *Fzd* complements, investigations carried out on cnidarians indicate that all bilaterian *Wnt* and *Fzd* gene subfamilies are already represented (Figure 12.2). For instance, the sea anemone *N. vectensis* possesses representatives for the *Wnt* subfamilies *WntA*, *Wnt1*, *Wnt2*, *Wnt3*, *Wnt4*, *Wnt5*, *Wnt6*, *Wnt7*, *Wnt8*, *Wnt10*, *Wnt11*, and *Wnt16* (i.e., all but *Wnt9*) (Cho *et al.*, 2010; Holstein, 2012; Kusserow *et al.*, 2005). Similarly, in *H. magnipapillata*, many *Wnt* genes were found, including though three genes encoding proteins classified as HyWnt9/10-a, HyWnt9/10-b, and HyWnt9/10-c orthologs (Lengfeld *et al.*, 2009), hence suggesting that *Wnt9* representative may have secondarily been lost by some cnidarians, for example, by *N. vectensis*. In agreement, in the jellyfish *C. hemisphaerica*, a *Wnt9* gene is also present, in addition to several other *Wnt* genes, which however for most part appear as orphans as they do not share any strong sequence homology with their bilaterian counterparts (Momose, Derelle, and Houliston, 2008). Likewise, the number of *Fzd* genes present in cnidarians is comparable to that observed in any other nonvertebrate deuterostomes with the same subfamilies being represented for instance in *N. vectensis* compared to protostomes (Figure 12.2). However, at the functional level, the individual role(s) of these Wnt and Fzd molecules, as well as the identity of the potential ligand-receptor pairs they may form, remain however to be largely uncovered. Chiefly, in *C. hemisphaerica*, for instance, CheFz1 and CheWnt3 have been shown to be required for the oral nuclear accumulation of β-catenin, while CheFz3 is necessary on the opposite aboral side to prevent canonical Wnt pathway activation (Momose and Houliston, 2007; Momose, Derelle, and Houliston, 2008).

Outside the Cnidaria, though, information related to *Wnt* and *Fzd* repertoires is rather scarce. *Wnt* and *Fzd* inventories have been reported to date solely in a couple of animals, that is, a ctenophore, a placozoan, and two sponges. From these data, it appears that these repertoires are further much less complex than those found in cnidarians. For instance, in the genome of the ctenophore *M. leidyi*, only four *Wnt* genes (*WntA*, *Wnt6*, *Wnt9*, and one unclassified *Wnt*) have been described (Pang *et al.*, 2010). Likewise, only three *Wnt* genes were reported in the placozoan *T. adhaerens* (Srivastava *et al.*, 2008) and the sponge *A. queenslandica* (Adamska *et al.*, 2010), and mainly two *Wnt* genes were found in the genome of the sponge *Oscarella lobularis* (Lapebie *et al.*, 2009). In addition, most of these *Wnt* genes were difficult to classify among the bilaterian *Wnt* subfamilies, with the exception of the genes found in *M. leidyi* and *A. queenslandica* (Figure 12.2; Adamska *et al.*, 2010). Similarly, regarding the *Fzd* repertoire, fewer *Fzd* genes were reported in prebilaterians other than cnidarians. In the sponge, *A. queenslandica* four *Fzd-related* genes were identified, but only two present some sequence homologies with members of the bilaterian seven-pass transmembrane *Fzd* subfamilies (i.e., *Fzd1/2/7–3/6* and *Fzd 9/10*) (Krishnan *et al.*, 2012). Comparably, in *T. adhaerens*, two *Fzd-related* genes were found, with only one correlating with the bilaterian *Fzd 9/10* subfamily, whereas the other displays no orthologous relationship with any other bilaterian *Fzd* gene (Krishnan *et al.*, 2012). These data indicate thus that the radiation of *Wnt* genes most likely occurred with the emergence of the metazoans and that it was followed then by a diversification of these genes and that of the Fzd receptors during eumetazoan evolution (Figure 12.1) (Holstein, 2012).

Conclusions and outlook

Even though current knowledge on canonical Wnt signaling and its activating components remains rather limited outside the vertebrate phyla, the broad variety of metazoans in which information is currently available allows nevertheless the drawing of some general conclusions regarding the evolutionary diversification of this pathway among animals.

For instance, while maternal activation of canonical Wnt signaling was shown to be driven by maternal Wnt ligands in several organisms, ranging from mammals to cnidarians, it emerges that these ligands are distinct in the various animals and that they do not even belong to a related *Wnt* subfamily, that is, Wnt11 in *Xenopus*, Wnt8 in zebrafish, Wnt6 in sea urchin, and Wnt3 in jellyfish (Lhomond *et al.*, 2012; Lu, Thisse, and Thisse, 2011; Momose, Derelle, and Houliston, 2008; Tao *et al.*, 2005). Thus, since the maternal function of Wnt signaling is conserved among these animals, that is, in primary body axis patterning, these results suggest that a specific adaption of canonical Wnt signaling activation may have pertained to the distinct modes of early embryogenesis of the various species involved. In agreement, rescue experiments performed on *Xenopus laevis* morphants using Wnt ligands originating from the cnidarian *N. vectensis* led to a comparable conclusion. NvWnt5 rescued XWnt11 morphants, while NvWnt11 specifically reconditioned XWnt5a depleted embryos (Rigo-Watermeier *et al.*, 2011).

Canonical Wnt signaling, additionally, is required in all metazoans for the establishment of the primary body axis as stated earlier (i.e., the AP axis in bilaterians and the oral–aboral axis in prebilaterians). Nonetheless, in the urochordate *Ciona intestinalis*, the nematode *C. elegans*, and the annelid *P. dumerilii*, β-catenin has been shown to have also a second broader role during development, that is, in polarizing cell divisions and controlling binary cell fate specification along the AP axis (also called AV axis in ascidians and annelids) (Hudson *et al.*, 2013; Schneider and Bowerman, 2007; Thorpe *et al.*, 1997). In these animals, following each novel equatorial cell divisions, β-catenin is differentially stabilized in the most vegetal daughter cells, where it assumes posterior/vegetal cell fate specification. Thus, a current proposed hypothesis regarding the ancestral mode of action of β-catenin is that it acts as a binary regulator, a mode that still persists in some animals (Martin and Kimelman, 2012; Zhang *et al.*, 2010). This view was further recently reinforced by intriguing experiments where immobilized Wnt proteins on beads were presented to embryonic stem cells in an oriented manner and were found to induce asymmetric mitotic division (Habib *et al.*, 2013).

Finally, while most efforts have been directed to date to depict the evolutionary conservation and most importantly the animal origin of the *Wnt* genes – without any real success so far – much less attention has been given to the *Fzd* genes on that topic, even though suitable conclusions can be reached (Figure 12.2). In vertebrates, up to 10 distinct *Fzd* genes have been identified, *Fzd1* to *Fzd10*, which define five *Fzd* subfamilies (i.e., *Fzd1/2/7*, *Fzd3/6*, *Fzd4*, *Fzd5/8*, and *Fzd9/10*). When traced back to nonvertebrate deuterostomes, that is, in urochordates, cephalochordates, hemichordates, and echinoderms, an average of only four *Fzd* genes is observed; most likely due to the fact that, unlike vertebrates, these animals did not undergo whole-genome duplication events. However, phylogenetic analyses carried out between these genes and those of vertebrates indicate that in nonvertebrate chordate animals (i.e., urochordates and cephalochordates) overall all five vertebrate *Fzd* subfamilies are represented, while in nonchordate deuterostomes (i.e., hemichordates and echinoderms) their *Fzd1/2/7* genes cluster with both the *Fzd1/2/7* and the *Fzd3/6* vertebrate subfamilies, hence suggesting that the deuterostome ancestor only had four *Fzd* genes, that is, a *Fzd1/2/7–3/6*, a *Fzd4*, a *Fzd5/8*, and a *Fzd9/10* ortholog. In addition, protostomes and cnidarians also encompass only three to four *Fzd* genes, and these genes display sequence homologies now with only three vertebrate *Fzd* subfamilies (i.e., *Fzd1/2/7–3/6*, *Fzd4–9/10* and *Fz5/8*). Thus, these observations suggest that specific duplication events of selective *Fzd* genes likely occurred both at the emergence of the deuterostome (*Fzd4*) and at that of the chordates (*Fzd3/6*), before the all set of *Fzd* genes was duplicated through the vertebrate-specific whole-genome duplication events. Interestingly, outside the metazoans, while no *Wnt* genes were observed neither in fungi nor in unicellular eukaryotes, a significant number of Fzd-related receptors have been reported from the cellular slime mold *Dictyostelium discoideum* (e.g., Prabhu and Eichinger, 2006). Here, 16 *Fzd-related* genes encoding proteins with a putative cysteine-rich domain were identified, of which two further exhibit a KTXXXW motif, which is known to be required for signal transduction (Umbhauer *et al.*, 2000). Thus, this clearly indicates that the origin of the Fzd receptor family, unlike that of the Wnt ligands, can be traced back outside the metazoan kingdom and further suggests that the increasing divergence of Fzd receptors was most likely an essential for generating more specific interactions with Wnt ligands, whose complexity (i.e., number of Wnt subfamilies) did not change significantly between humans and cnidarians.

Acknowledgments

The authors thank Herbert Steinbeisser for his critical evaluation of the manuscript, as well as Stefan Hoppler and Randall Moon for the opportunity to contribute to this book. This work was supported by the CNRS and the Université Pierre and Marie Curie (for J.C.C.) and by the DFG (SFB 873 and FOR1036-2) and CellNetworks cluster of excellence at Heidelberg University (for T.H).

References

Adamska, M., Larroux, C., Adamski, M. *et al.* (2010) Structure and expression of conserved Wnt pathway components in the demosponge Amphimedon queenslandica. *Evolution & Development*, **12**, 494–518.

Angelini, D.R. and Kaufman, T.C. (2005) Functional analyses in the milkweed bug Oncopeltus fasciatus (Hemiptera) support a role for Wnt signaling in body segmentation but not appendage development. *Developmental Biology*, **283**, 409–423.

Aulehla, A., Wiegraebe, W., Baubet, V. *et al.* (2008) A beta-catenin gradient links the clock and wavefront systems in mouse embryo segmentation. *Nature Cell Biology*, **10**, 186–193.

Bao, R., Fischer, T., Bolognesi, R. *et al.* (2012) Parallel duplication and partial subfunctionalization of beta-catenin/armadillo during insect evolution. *Molecular Biology & Evolution*, **29**, 647–662.

Beermann, A., Pruhs, R., Lutz, R., and Schroder, R. (2011) A context-dependent combination of Wnt receptors controls axis elongation and leg development in a short germ insect. *Development*, **138**, 2793–2805.

Ben-Haim, N., Lu, C., Guzman-Ayala, M. *et al.* (2006) The nodal precursor acting via activin receptors induces mesoderm by maintaining a source of its convertases and BMP4. *Developmental Cell*, **11**, 313–323.

Bolognesi, R., Beermann, A., Farzana, L. *et al.* (2008a) Tribolium Wnts: evidence for a larger repertoire in insects with overlapping expression patterns that suggest multiple redundant functions in embryogenesis. *Development Genes and Evolution*, **218**, 193–202.

Bolognesi, R., Farzana, L., Fischer, T.D., and Brown, S.J. (2008b) Multiple Wnt genes are required for segmentation in the short-germ embryo of Tribolium castaneum. *Current Biology*, **18**, 1624–1629.

Cho, S.J., Valles, Y., Giani, V.C., Jr. *et al.* (2010) Evolutionary dynamics of the wnt gene family: a lophotrochozoan perspective. *Molecular Biology and Evolution*, **27**, 1645–1658.

Christian, J.L. and Moon, R.T. (1993) Interactions between Xwnt-8 and Spemann organizer signaling pathways generate dorsoventral pattern in the embryonic mesoderm of Xenopus. *Genes & Development*, **7**, 13–28.

Croce, J., Duloquin, L., Lhomond, G. *et al.* (2006) Frizzled5/8 is required in secondary mesenchyme cells to initiate archenteron invagination during sea urchin development. *Development*, **133**, 547–557.

Croce, J.C., Range, R., Wu, S.Y. *et al.* (2011) Wnt 6 activates endoderm in the sea urchin gene regulatory network. *Development*, **138**, 3297–3306.

Darras, S., Gerhart, J., Terasaki, M. *et al.* (2011) {beta}-Catenin specifies the endomesoderm and defines the posterior organizer of the hemichordate Saccoglossus kowalevskii. *Development*, **138**, 959–970.

De Robertis, E.M. (2010) Wnt signaling in axial patterning and regeneration: lessons from planaria. *Science Signalling*, **3**, pe21.

Dehal, P. and Boore, J.L. (2005) Two rounds of whole genome duplication in the ancestral vertebrate. *PLoS Biology*, **3**, e314.

Erter, C.E., Wilm, T.P., Basler, N. *et al.* (2001) Wnt8 is required in lateral mesendodermal precursors for neural posteriorization in vivo. *Development*, **128**, 3571–3583.

Freeman, R.M., Jr., Wu, M., Cordonnier-Pratt, M.M. *et al.* (2008) cDNA sequences for transcription factors and signaling proteins of the hemichordate Saccoglossus kowalevskii: efficacy of the expressed sequence tag (EST) approach for evolutionary and developmental studies of a new organism. *The Biological Bulletin*, **214**, 284–302.

Garriock, R.J., Warkman, A.S., Meadows, S.M. *et al.* (2007) Census of vertebrate Wnt genes: isolation and developmental expression of Xenopus Wnt2, Wnt3, Wnt9a, Wnt9b, Wnt10a, and Wnt16. *Developmental Dynamics*, **236**, 1249–1258.

Gloriam, D.E., Fredriksson, R., and Schioth, H.B. (2007) The G protein-coupled receptor subset of the rat genome. *BMC Genomics*, **8**, 338.

Grigoryan, T., Wend, P., Klaus, A., and Birchmeier, W. (2008) Deciphering the function of canonical Wnt signals in development and disease: conditional loss- and gain-of-function mutations of beta-catenin in mice. *Genes & Development*, **22**, 2308–2341.

Habib, S.J., Chen, B.C., Tsai, F.C. *et al.* (2013) A localized Wnt signal orients asymmetric stem cell division in vitro. *Science*, **339**, 1445–1448.

Harland, R. and Gerhart, J. (1997) Formation and function of Spemann's organizer. *Annual Review of Cell and Developmental Biology*, **13**, 611–667.

Harwood, A.J. (2008) Dictyostelium development: a prototypic Wnt pathway? *Methods in Molecular Biology*, **469**, 21–32.

Heasman, J., Kofron, M., and Wylie, C. (2000) Beta-catenin signaling activity dissected in the early Xenopus embryo: a novel antisense approach. *Developmental Biology*, **222**, 124–134.

Henry, J.Q., Perry, K.J., and Martindale, M.Q. (2010) β-catenin and early development in the gastropod, Crepidula fornicata. *Integrative and Comparative Biology*, **50**, 707–719.

Henry, J.Q., Perry, K.J., Wever, J. *et al.* (2008) Beta-catenin is required for the establishment of vegetal embryonic fates in the nemertean, Cerebratulus lacteus. *Developmental Biology*, **317**, 368–379.

Hikasa, H. and Sokol, S.Y. (2013) Wnt signaling in vertebrate axis specification. *Cold Spring Harbor Perspectives in Biology*, **5**, a007955.

Hino, K., Satou, Y., Yagi, K., and Satoh, N. (2003) A genomewide survey of developmentally relevant genes in Ciona intestinalis. VI. Genes for Wnt, TGFbeta, Hedgehog and JAK/STAT signaling pathways. *Developmental Genes and Evolution*, **213**, 264–272.

Hobmayer, B., Rentzsch, F., Kuhn, K. *et al.* (2000) WNT signalling molecules act in axis formation in the diploblastic metazoan Hydra. *Nature*, **407**, 186–189.

Holland, L.Z., Holland, N.N., and Schubert, M. (2000) Developmental expression of AmphiWnt1, an amphioxus gene in the Wnt1/wingless subfamily. *Developmental Genes and Evolution*, **210**, 522–524.

Holland, L.Z., Panfilio, K.A., Chastain, R., *et al.* (2005) Nuclear beta-catenin promotes non-neural ectoderm and posterior cell fates in amphioxus embryos. *Developmental Dynamics*, **233**, 1430–1443.

Holstein, T.W. (2012) The evolution of the Wnt pathway. *Cold Spring Harbor Perspectives in Biology*, **5**, a007922.

Holstein, T.W., Watanabe, H., and Ozbek, S. (2011) Signaling pathways and axis formation in the lower metazoa. *Current Topics in Developmental Biology*, **97**, 137–177.

Houliston, E. and Elinson, R.P. (1992) Microtubules and cytoplasmic reorganization in the frog egg. *Current Topics in Developmental Biology*, **26**, 53–70.

Hudson, C., Kawai, N., Negishi, T., and Yasuo, H. (2013) β-Catenin-driven binary fate specification segregates germ layers in ascidian embryos. *Current Biology*, **23**, 491–495.

Imai, K., Takada, N., Satoh, N., and Satou, Y. (2000) (beta)-catenin mediates the specification of endoderm cells in ascidian embryos. *Development*, **127**, 3009–3020.

Janson, K., Cohen, E.D., and Wilder, E.L. (2001) Expression of DWnt6, DWnt10, and DFz4 during Drosophila development. *Mechanical Development*, **103**, 117–120.

Janssen, R., Le Gouar, M., Pechmann, M. *et al.* (2010) Conservation, loss, and redeployment of Wnt ligands in protostomes: implications for understanding the evolution of segment formation. *BMC Evolutionary Biology*, **10**, 374.

Jockusch, E.L. and Ober, K.A. (2000) Phylogenetic analysis of the Wnt gene family and discovery of an arthropod Wnt-10 orthologue. *Journal of Experimental Zoology*, **288**, 105–119.

Kawai, N., Iida, Y., Kumano, G., and Nishida, H. (2007) Nuclear accumulation of beta-catenin and transcription of downstream genes are regulated by zygotic Wnt5alpha and maternal Dsh in ascidian embryos. *Developmental Dynamics*, **236**, 1570–1582.

Kelly, C., Chin, A.J., Leatherman, J.L. *et al.* (2000) Maternally controlled (beta)-catenin-mediated signaling is required for organizer formation in the zebrafish. *Development*, **127**, 3899–3911.

Kemp, C., Willems, E., Abdo, S. *et al.* (2005) Expression of all Wnt genes and their secreted antagonists during mouse blastocyst and postimplantation development. *Developmental Dynamics*, **233**, 1064–1075.

Kemp, C.R., Willems, E., Wawrzak, D. *et al.* (2007) Expression of Frizzled5, Frizzled7, and Frizzled10 during early mouse development and interactions with canonical Wnt signaling. *Developmental Dynamics*, **236**, 2011–2019.

Kimura-Yoshida, C., Nakano, H., Okamura, D. *et al.* (2005) Canonical Wnt signaling and its antagonist regulate anterior-posterior axis polarization by guiding cell migration in mouse visceral endoderm. *Developmental Cell*, **9**, 639–650.

King, N., Westbrook, M.J., Young, S.L. *et al.* (2008) The genome of the choanoflagellate Monosiga brevicollis and the origin of metazoans. *Nature*, **451**, 783–788.

Klein, P.S. and Melton, D.A. (1996) A molecular mechanism for the effect of lithium on development. *Proceedings of the National Academy of Sciences of the United States of America*, **93**, 8455–8459.

Krishnan, A., Almen, M.S., Fredriksson, R., and Schioth, H.B. (2012) The origin of GPCRs: identification of mammalian like Rhodopsin, Adhesion, Glutamate and Frizzled GPCRs in fungi. *PLoS One*, **7**, e29817.

Ku, M. and Melton, D.A. (1993) Xwnt-11: a maternally expressed Xenopus Wnt gene. *Development*, **119**, 1161–1173.

Kuroda, H., Wessely, O., and De Robertis, E.M. (2004) Neural induction in Xenopus: requirement for ectodermal and endomesodermal signals via Chordin, Noggin, beta-Catenin, and Cerberus. *PLoS Biology*, **2**, E92.

Kusserow, A., Pang, K., Sturm, C. *et al.* (2005) Unexpected complexity of the Wnt gene family in a sea anemone. *Nature*, **433**, 156–160.

Lagerström, M.C., Hellstrom, A.R., Gloriam, D.E. *et al.* (2006) The G protein-coupled receptor subset of the chicken genome. *PLoS Computational Biology*, **2**, e54.

Lapebie, P., Gazave, E., Ereskovsky, A. *et al.* (2009) WNT/beta-catenin signalling and epithelial patterning in the homoscleromorph sponge Oscarella. *PLoS One*, **4**, e5823.

Larabell, C.A., Torres, M., Rowning, B.A. *et al.* (1997) Establishment of the dorso-ventral axis in Xenopus embryos is presaged by early asymmetries in beta-catenin that are modulated by the Wnt signaling pathway. *The Journal of Cell Biology*, **136**, 1123–1136.

Lekven, A.C., Thorpe, C.J., Waxman, J.S., and Moon, R.T. (2001) Zebrafish wnt8 encodes two wnt8 proteins on a bicistronic transcript and is required for mesoderm and neurectoderm patterning. *Developmental Cell*, **1**, 103–114.

Lengfeld, T., Watanabe, H., Simakov, O. *et al.* (2009) Multiple Wnts are involved in Hydra organizer formation and regeneration. *Developmental Biology*, **330**, 186–199.

Lewis, S.L., Khoo, P.L., De Young, R.A. *et al.* (2008) Dkk1 and Wnt3 interact to control head morphogenesis in the mouse. *Development*, **135**, 1791–1801.

Lhomond, G., Gache, C., McClay, D.R., and Croce, J.C. (2012) Frizzled1/2/7 signaling directs β-catenin nuclearisation and initiates endoderm specification in macromeres during sea urchin embryogenesis. *Development*, **139**, 816–825.

Logan, C.Y., Miller, J.R., Ferkowicz, M.J., and McClay, D.R. (1999) Nuclear beta-catenin is required to specify vegetal cell fates in the sea urchin embryo. *Development*, **126**, 345–357.

Lu, F.I., Thisse, C., and Thisse, B. (2011) Identification and mechanism of regulation of the zebrafish dorsal determinant. *Proceedings of the National Academy of Sciences of the United States of America*, **108**, 15876–15880.

Martin, B.L. and Kimelman, D. (2012) Canonical Wnt signaling dynamically controls multiple stem cell fate decisions during vertebrate body formation. *Developmental Cell*, **22**, 223–232.

McMahon, A.P. and Moon, R.T. (1989) Ectopic expression of the proto-oncogene int-1 in Xenopus embryos leads to duplication of the embryonic axis. *Cell*, **58**, 1075–1084.

Miller, J.R., Rowning, B.A., Larabell, C.A. *et al.* (1999) Establishment of the dorsal–ventral axis in xenopus embryos coincides with the dorsal enrichment of dishevelled that is dependent on cortical rotation. *The Journal of Cell Biology*, **146**, 427–437.

Miyawaki, K., Yamamoto, M., Saito, K. *et al.* (2003) Nuclear localization of beta-catenin in vegetal pole cells during early embryogenesis of the starfish Asterina pectinifera. *Development, Growth & Differentiation*, **45**, 121–128.

Miyawaki, K., Mito, T., Sarashina, I. *et al.* (2004) Involvement of Wingless/Armadillo signaling in the posterior sequential segmentation in the cricket, Gryllus bimaculatus (Orthoptera), as revealed by RNAi analysis. *Mechanical Development*, **121**, 119–130.

Mohamed, O.A., Jonnaert, M., Labelle-Dumais, C. *et al.* (2005) Uterine Wnt/beta-catenin signaling is required for implantation. *Proceedings of the National Academy of Sciences of the United States of America*, **102**, 8579–8584.

Momose, T. and Houliston, E. (2007) Two oppositely localised frizzled RNAs as axis determinants in a cnidarian embryo. *PLoS Biology*, **5**, e70.

Momose, T., Derelle, R., and Houliston, E. (2008) A maternally localised Wnt ligand required for axial patterning in the cnidarian Clytia hemisphaerica. *Development*, **135**, 2105–2113.

Moon, R.T. and Kimelman, D. (1998) From cortical rotation to organizer gene expression: toward a molecular explanation of axis specification in Xenopus. *BioEssays*, **20**, 536–545.

Morkel, M., Huelsken, J., Wakamiya, M. *et al.* (2003) Beta-catenin regulates Cripto- and Wnt3-dependent gene expression programs in mouse axis and mesoderm formation. *Development*, **130**, 6283–6294.

Mukhopadhyay, M., Shtrom, S., Rodriguez-Esteban, C. *et al.* (2001) Dickkopf1 is required for embryonic head induction and limb morphogenesis in the mouse. *Developmental Cell*, **1**, 423–434.

Niehrs, C. (2010) On growth and form: a Cartesian coordinate system of Wnt and BMP signaling specifies bilaterian body axes. *Development*, **137**, 845–857.

Niwano, T., Takatori, N., Kumano, G., and Nishida, H. (2009) Wnt5 is required for notochord cell intercalation in the ascidian Halocynthia roretzi. *Biology of the Cell*, **101**, 645–659.

Nordstrom, K.J., Fredriksson, R., and Schioth, H.B. (2008) The amphioxus (Branchiostoma floridae) genome contains a highly diversified set of G protein-coupled receptors. *BMC Evolutionary Biology*, **8**, 9.

Pang, K., Ryan, J.F., Mullikin, J.C. *et al.* (2010) Genomic insights into Wnt signaling in an early diverging metazoan, the ctenophore Mnemiopsis leidyi. *Evolutionary Developmental Biology*, **1**, 10.

Pani, A.M., Mullarkey, E.E., Aronowicz, J. *et al.* (2012) Ancient deuterostome origins of vertebrate brain signalling centres. *Nature*, **483**, 289–294.

Petersen, C.P. and Reddien, P.W. (2009) Wnt signaling and the polarity of the primary body axis. *Cell*, **139**, 1056–1068.

Philipp, I., Aufschnaiter, R., Ozbek, S. *et al.* (2009) Wnt/beta-catenin and noncanonical Wnt signaling interact in tissue evagination in the simple eumetazoan Hydra. *Proceedings of the National Academy of Sciences of the United States of America*, **106**, 4290–4295.

Pourquie, O. (2011) Vertebrate segmentation: from cyclic gene networks to scoliosis. *Cell*, **145**, 650–663.

Prabhu, Y. and Eichinger, L. (2006) The Dictyostelium repertoire of seven transmembrane domain receptors. *The European Journal of Cell Biology*, **85**, 937–946.

Prud'homme, B., Lartillot, N., Balavoine, G. *et al.* (2002) Phylogenetic analysis of the Wnt gene family. Insights from lophotrochozoan members. *Current Biology*, **12**, 1395.

Rigo-Watermeier, T., Kraft, B., Ritthaler, M. *et al.* (2011) Functional conservation of Nematostella Wnts in canonical and noncanonical Wnt-signaling. *Biology Open*, **1**, 43–51.

Sasai, Y., Lu, B., Steinbeisser, H. *et al.* (1994) Xenopus chordin: a novel dorsalizing factor activated by organizer-specific homeobox genes. *Cell*, **79**, 779–790.

Sasakura, Y. and Makabe, K.W. (2000) Ascidian Wnt-7 gene is expressed exclusively in the tail neural tube of tailbud embryos. *Developmental Genes and Evolution*, **210**, 641–643.

Sasakura, Y. and Makabe, K.W. (2001) Ascidian Wnt-5 gene is involved in the morphogenetic movement of notochord cells. *Development, Growth & Differentiation*, **43**, 573–582.

Sasakura, Y., Ogasawara, M., and Makabe, K.W. (1998) HrWnt-5: a maternally expressed ascidian Wnt gene with posterior localization in early embryos. *The International Journal of Developmental Biology*, **42**, 573–579.

Satou, Y., Imai, K.S., Levine, M. *et al.* (2003) A genome-wide survey of developmentally relevant genes in Ciona intestinalis. I. Genes for bHLH transcription factors. *Developmental Genes and Evolution*, **213**, 213–221.

Schier, A.F. and Talbot, W.S. (2005) Molecular genetics of axis formation in zebrafish. *Annual Review of Genetics*, **39**, 561–613.

Schierwater, B., Eitel, M., Jakob, W. *et al.* (2009) Concatenated analysis sheds light on early metazoan evolution and fuels a modern "urmetazoon" hypothesis. *PLoS Biology*, **7** (1), e20.

Schneider, S.Q. and Bowerman, B. (2007) β-Catenin asymmetries after all animal/vegetal-oriented cell divisions in Platynereis dumerilii embryos mediate binary cell-fate specification. *Developmental Cell*, **13**, 73–86.

Schubert, M. and Holland, L.Z. (2003) The Wnt gene family and the evolutionary conservation of Wnt expression, in *Wnt Signaling in Development* (ed M. Kühl), Landes Biosciences, Georgetown, TX, pp. 210–239.

Schubert, M., Holland, L.Z., and Holland, N.D. (2000a) Characterization of an amphioxus wnt gene, AmphiWnt11, with possible roles in myogenesis and tail outgrowth. *Genesis*, **27**, 1–5.

Schubert, M., Holland, L.Z., and Holland, N.D. (2000b) Characterization of two amphioxus Wnt genes (AmphiWnt4 and AmphiWnt7b) with early expression in the developing central nervous system. *Developmental Dynamics*, **217**, 205–215.

Schubert, M., Holland, L.Z., Panopoulou, G.D. *et al.* (2000) Characterization of amphioxus AmphiWnt8: insights into the evolution of patterning of the embryonic dorsoventral axis. *Evolutionary Development*, **2**, 85–92.

Schubert, M., Holland, L.Z., Stokes, M.D., and Holland, N.D. (2001) Three amphioxus Wnt genes (AmphiWnt3, AmphiWnt5, and AmphiWnt6) associated with the tail bud: the evolution of somitogenesis in chordates. *Developmental Biology*, **240**, 262–273.

Schulte-Merker, S., Lee, K.J., McMahon, A.P., and Hammerschmidt, M. (1997) The zebrafish organizer requires chordino. *Nature*, **387**, 862–863.

Sethi, A.J., Wikramanayake, R.M., Angerer, R.C. *et al.* (2012) Sequential signaling crosstalk regulates endomesoderm segregation in sea urchin embryos. *Science*, **335**, 590–593.

Shimizu, T., Bae, Y.K., Muraoka, O., and Hibi, M. (2005) Interaction of Wnt and caudal-related genes in zebrafish posterior body formation. *Developmental Biology*, **279**, 125–141.

Simakov, O., Marletaz, F., Cho, S.J. *et al.* (2013) Insights into bilaterian evolution from three spiralian genomes. *Nature*, **493**, 526–531.

Sonderegger, S., Haslinger, P., Sabri, A. *et al.* (2010a) Wingless (Wnt)-3A induces trophoblast migration and matrix metalloproteinase-2 secretion through canonical Wnt signaling and protein kinase B/AKT activation. *Endocrinology*, **151**, 211–220.

Sonderegger, S., Pollheimer, J., and Knofler, M. (2010b) Wnt signalling in implantation, decidualisation and placental differentiation—review. *Placenta*, **31**, 839–847.

Srivastava, M., Begovic, E., Chapman, J. *et al.* (2008) The Trichoplax genome and the nature of placozoans. *Nature*, **454**, 955–960.

Srivastava, M., Simakov, O., Chapman, J. *et al.* (2010) The Amphimedon queenslandica genome and the evolution of animal complexity. *Nature*, **466**, 720–726.

Stamateris, R.E., Rafiq, K., and Ettensohn, C.A. (2010) The expression and distribution of Wnt and Wnt receptor mRNAs during early sea urchin development. *Gene Expression Patterns*, **10** (1), 60–64.

Stern, C.D. (2006) Evolution of the mechanisms that establish the embryonic axes. *Current Opinion in Genetics & Development*, **16**, 413–418.

Stern, C.D. and Downs, K.M. (2012) The hypoblast (visceral endoderm): an evo-devo perspective. *Development*, **139**, 1059–1069.

Tao, Q., Yokota, C., Puck, H. *et al.* (2005) Maternal wnt11 activates the canonical wnt signaling pathway required for axis formation in Xenopus embryos. *Cell*, **120**, 857–871.

ten Berge, D., Kurek, D., Blauwkamp, T. *et al.* (2011) Embryonic stem cells require Wnt proteins to prevent differentiation to epiblast stem cells. *Nature Cell Biology*, **13**, 1070–1075.

Thorpe, C.J., Schlesinger, A., Carter, J.C., and Bowerman, B. (1997) Wnt signaling polarizes an early C. elegans blastomere to distinguish endoderm from mesoderm. *Cell*, **90**, 695–705.

Umbhauer, M., Djiane, A., Goisset, C. *et al.* (2000) The C-terminal cytoplasmic Lys-thr-X-X-X-Trp motif in frizzled receptors mediates Wnt/beta-catenin signalling. *The EMBO Journal*, **19**, 4944–4954.

Wang, J., Sinha, T., and Wynshaw-Boris, A. (2012) Wnt signaling in mammalian development: lessons from mouse genetics. *Cold Spring Harbor Perspectives in Biology*, **4**, a007963.

Wei, Z., Range, R., Angerer, R., and Angerer, L. (2012) Axial patterning interactions in the sea urchin embryo: suppression of nodal by Wnt1 signaling. *Development*, **139**, 1662–1669.

Wieschaus, E. and Riggleman, R. (1987) Autonomous requirements for the segment polarity gene armadillo during Drosophila embryogenesis. *Cell*, **49**, 177–184.

Wikramanayake, A.H., Hong, M., Lee, P.N. *et al.* (2003) An ancient role for nuclear beta-catenin in the evolution of axial polarity and germ layer segregation. *Nature*, **426**, 446–450.

Wikramanayake, A.H., Peterson, R., Chen, J. *et al.* (2004) Nuclear beta-catenin-dependent Wnt8 signaling in vegetal cells of the early sea urchin embryo regulates gastrulation and differentiation of endoderm and mesodermal cell lineages. *Genesis*, **39**, 194–205.

Xiao, S. and Laflamme, M. (2009) On the eve of animal radiation: phylogeny, ecology and evolution of the Ediacara biota. *Trends in Ecology & Evolution*, **24**, 31–40.

Yamaguchi, Y. and Shinagawa, A. (1989) Marked alteration at midblastula transition in the effect of lithium on formation of the larval body pattern of Xenopus laevis. *Development, Growth & Differentiation*, **31**, 531–541.

Zhang, B., Tran, U., and Wessely, O. (2011) Expression of Wnt signaling components during Xenopus pronephros development. *PLoS One*, **6**, e26533.

Zhang, J., Woodhead, G.J., Swaminathan, S.K. *et al.* (2010) Cortical neural precursors inhibit their own differentiation via N-cadherin maintenance of beta-catenin signaling. *Developmental Cell*, **18**, 472–479.

Part 2

Selected Key Molecules in Wnt Signaling

The Wnt signaling mechanisms, whether they are seen as linear pathways or as complex networks, are carried out by individual protein molecules that physically interact with each other, thereby influencing the activity of other molecules and being influenced in turn by other molecules. Detailed study of individual Wnt signaling components reveals important aspects of Wnt signaling mechanisms, such as the importance of multiprotein complexes to assemble and disassemble in what we now know as the Wnt/β-catenin pathway (e.g., β-catenin-degradation complex from studies of Axin and APC; see Chapter 3). Detailed studies of the relationship between protein structure and protein function lead to the discovery of functionally important subdomains. Recognizing such a functional domain can sometimes give great insight into a suggested function for the particular molecule, for instance, the seven-transmembrane domain at the time singled out the frizzled proteins as prime candidates for the then long-sought-after Wnt signal receptor. Sharing of a particular domain can also be suggestive, such as the identification of a Wnt-signal-binding cysteine-rich domain in not only the frizzled receptors but also in secreted frizzled-related protein (sFRP) molecules, which suggested they function as extracellular Wnt inhibitors (see Chapter 13). However, the functional significance for Wnt signaling of the many structural features that frizzled receptors share with G-protein-coupled receptors remained for many years controversial and have only recently become clearer (see Chapter 14).

By their very nature or for technical reasons, some proteins have been much more difficult to study in the past. The Wnt signal itself after hard and arduous work is only more recently starting to reveal more details about its pivotal molecular and cellular function (see Chapter 1). The dishevelled protein (see Chapter 15) and β-catenin (see Chapter 16) both function at important but different crossroads in the Wnt signaling network. The clearly defined functional domain structure of dishevelled protein promises to hold the key for understanding the choice of different Wnt signaling pathway mechanisms (see Chapters 7 and 15). Contrary to that, some of the many biological functions of the β-catenin protein are structurally difficult to separate, possibly because the same large central armadillo repeat domain mediates binding to different

Wnt Signaling in Development and Disease: Molecular Mechanisms and Biological Functions, First Edition.
Edited by Stefan Hoppler and Randall T. Moon.

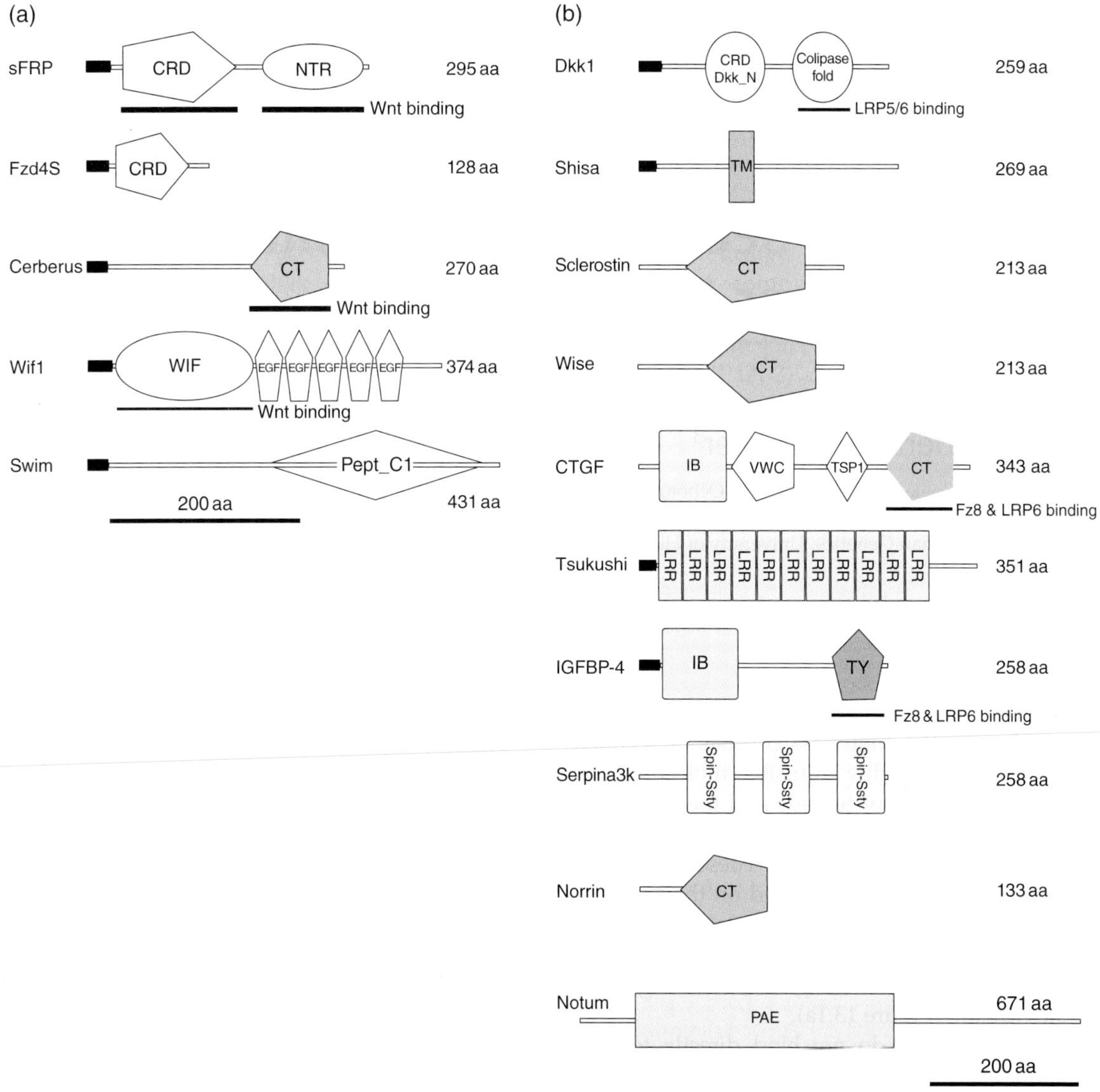

Figure 13.1 Schematic summary of the structure of the domains that compose secreted Wnt modulators subdivided in Wnt specific and Wnt pathway specific. The first class binds directly to Wnt ligands influencing pathway activation and signaling range. In contrast, the second class of modulators interacts with proteins of the Wnt receptor complex such as Fzd and Lrp interfering with the formation of active Wnt complex formation (except Notum). Thereby different domains are involved in protein interactions. In the graphic, individual domains of modulators are underlined to emphasize the respective interaction domains of Wnt, Fzd, or Lrp binding as far as known. The protein structure of secreted Wnt modulators has been determined using the SMART domain architecture tool. Data bank entries are listed for proteins described here. (a) Wnt-specific modulators: *Homo sapiens* sFRP2 (NCBI Reference Sequence: NP_003004.1), *Xenopus laevis* Fzd4S, *X. laevis* Cer (accession number GenBank BAC54274.1), *X. laevis* Wif1 (GenBank AAI69623.1), and *Drosophila melanogaster* Swim (GenBank: AAF46818.2). (b) Wnt pathway-specific modulators: *X. laevis* Dkk1 (NCBI reference sequence NP_001079061.1), *X. laevis* Shisa (GenBank: AAI60672.1), *H. sapiens* (GenBank AAK16158.1), *X. laevis* Wise (GenBank AAQ57627.1), *X. laevis* CTGF (NCBI reference NP 001081697.1), *X. laevis* TSK (NCBI reference: NP-001088996.1), *H. sapiens* IGFBP-4 (UniProtKB/Swiss-Prot P22692.2), *H. sapiens* Serpina3k (NCBI reference NP-001006684.1), *H. sapiens* Norrin (NCBI reference NP-000257.1), and *D. melanogaster* Notum (GenBank AAF49550.3). *Abbreviations*: CRD, cysteine-rich domain; CT, C-terminal cysteine knot-like domain; EGF, epidermal growth factor-like domain; IGFBP, insulin-like growth factor-binding protein homologs; LRR, leucine-rich repeats; NTR, Netrin-like domain; PAE, Pectin acylesterase domain; Pept-C1, Papain family cysteine protease; Tsp1, thrombospondin type 1 repeats; TY, thyroglobulin type I repeats; Spin-Ssty, Spindin domain; VWC, von Willebrand factor (vWF) type C domain. The scale bar is identical for A and B with the exception of Notum.

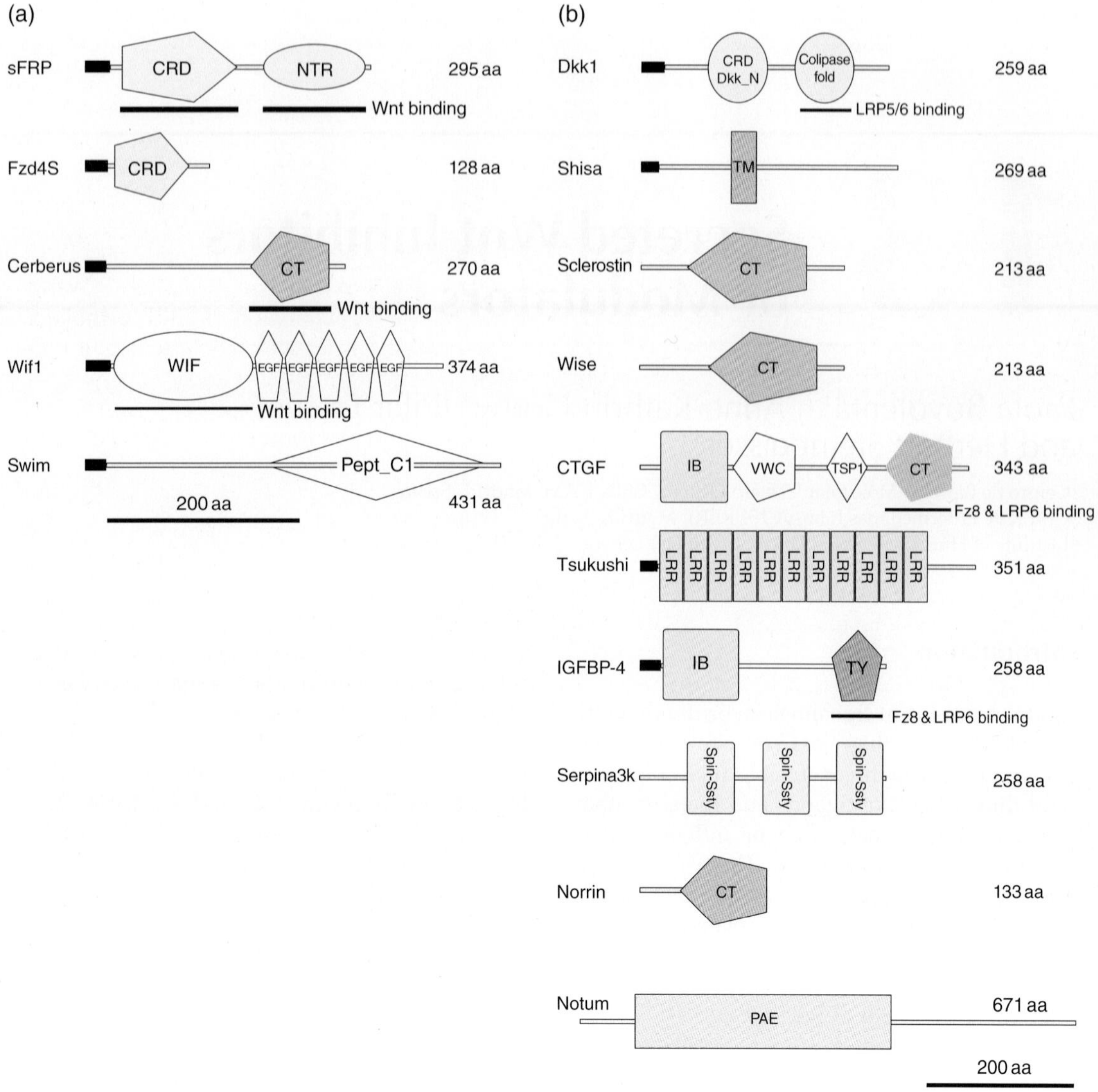

Figure 13.1 Schematic summary of the structure of the domains that compose secreted Wnt modulators subdivided in Wnt specific and Wnt pathway specific. The first class binds directly to Wnt ligands influencing pathway activation and signaling range. In contrast, the second class of modulators interacts with proteins of the Wnt receptor complex such as Fzd and Lrp interfering with the formation of active Wnt complex formation (except Notum). Thereby different domains are involved in protein interactions. In the graphic, individual domains of modulators are underlined to emphasize the respective interaction domains of Wnt, Fzd, or Lrp binding as far as known. The protein structure of secreted Wnt modulators has been determined using the SMART domain architecture tool. Data bank entries are listed for proteins described here. (a) Wnt-specific modulators: *Homo sapiens* sFRP2 (NCBI Reference Sequence: NP_003004.1), *Xenopus laevis* Fzd4S, *X. laevis* Cer (accession number GenBank BAC54274.1), *X. laevis* Wif1 (GenBank AAI69623.1), and *Drosophila melanogaster* Swim (GenBank: AAF46818.2). (b) Wnt pathway-specific modulators: *X. laevis* Dkk1 (NCBI reference sequence NP_001079061.1), *X. laevis* Shisa (GenBank: AAI60672.1), *H. sapiens* (GenBank AAK16158.1), *X. laevis* Wise (GenBank AAQ57627.1), *X. laevis* CTGF (NCBI reference NP-001081697.1), *X. laevis* TSK (NCBI reference: NP-001088996.1), *H. sapiens* IGFBP-4 (UniProtKB/Swiss-Prot P22692.2), *H. sapiens* Serpina3k (NCBI reference NP-001006684.1), *H. sapiens* Norrin (NCBI reference NP-000257.1), and *D. melanogaster* Notum (GenBank AAF49550.3). *Abbreviations*: CRD, cysteine-rich domain; CT, C-terminal cysteine knot-like domain; EGF, epidermal growth factor-like domain; IGFBP, insulin-like growth factor-binding protein homologs; LRR, leucine-rich repeats; NTR, Netrin-like domain; PAE, Pectin acylesterase domain; Pept-C1, Papain family cysteine protease; Tsp1, thrombospondin type 1 repeats; TY, thyroglobulin type I repeats; Spin-Ssty, Spindin domain; VWC, von Willebrand factor (vWF) type C domain. The scale bar is identical for A and B with the exception of Notum.

13 Secreted Wnt Inhibitors or Modulators

Paola Bovolenta[1,2], Anne-Kathrin Gorny[3], Pilar Esteve[1,2], and Herbert Steinbeisser[3]

[1]Centro de Biología Molecular "Severo Ochoa," CSIC-UAM, Madrid, Spain
[2]CIBER de Enfermedades Raras (CIBERER), Madrid, Spain
[3]Institute of Human Genetics, University of Heidelberg, Heidelberg, Germany

Introduction

Modulators of Wnt signaling can be divided into two classes. The first class consists of secreted and in general highly diffusible molecules that bind directly to Wnt ligands and thereby either prevent pathway activation or influence their activity and signaling range. These factors fall into four distinct groups based on their protein structure: secreted Frizzled-related proteins (sFRPs), Cerberus (Cer), Wnt-inhibitory factor 1 (WIF-1), and secreted wingless interacting molecules (Swim) (Figure 13.1a).

Molecules that do not bind directly to Wnt ligands but essentially interfere with the formation of active ligand/receptor/coreceptor complexes compose the second class. Their mechanism of interference is however quite variable as detailed in the following text. These modulators can be grouped in 10 subclasses: Dickkopf (Dkk), Wnt modulator in surface ectoderm (Wise), IGF-binding protein-4 (IGFBP-4), Tsukushi (TSK), Notum, Serpina3k, connective tissue growth factor (CTGF), Sclerostin/SOST, Shisa, and Norrin (Figure 13.1b).

The majority of secreted modulators contain a domain with conserved cysteine sequences, with the exception of Swim, TSK, and Serpina3k. The cysteine-rich domains (CRD) of WIF, sFRPs, DKK, and Shisa do not bear a phylogenetic relationship, whereas Cer, Wise, CTGF, Norrin, and Sclerostin/SOST share a common cysteine knot motif, which is also found in secreted growth factors such as TGF-β, PDGF, NGF and GNDF, and BMP antagonists of the differential screening-selected gene aberrative in neuroblastoma (Dan) family (Vitt, Hsu, and Hsueh, 2001). This C-terminal knot domain mediates modulators' interaction with Frizzled (Fzd) receptors and Lrp coreceptors.

In this chapter, we discuss the evidence supporting the role and mechanism of action of each one of these modulators and summarize the main pathological conditions that have been so far associated with their dysfunction.

Wnt Signaling in Development and Disease: Molecular Mechanisms and Biological Functions, First Edition.
Edited by Stefan Hoppler and Randall T. Moon.

functional protein partners in different cellular compartments and for different molecular functions, which may reflect a conserved fundamental functional interdependency relevant for the very evolution of multicellular animals (see Chapter 16). Wnt signaling mechanisms and protein components are on the whole remarkably conserved throughout evolution (see Chapter 12). However, there is a clear radiation of TCF/LEF genes and protein isoforms with the evolution of vertebrates, which provides much larger versatility of protein function (see Chapter 17) that may have enabled novel functional roles for Wnt signaling (see Part 3).

Since multiprotein complexes and regulation of protein–protein interactions are so fundamental to Wnt signaling mechanisms, the study of each molecule in isolation ultimately will become limiting. Structural analysis of not only individual proteins but also of proteins interacting and ultimately of multiprotein complexes (see Chapter 18) provides a much more integrated understanding of how individual components discussed here in Part 2 function in the molecular mechanisms introduced in Part 1 to carry out biological functions (Part 3) and possibly contributing to human disease (Part 4).

Part 2

Selected Key Molecules in Wnt Signaling

The Wnt signaling mechanisms, whether they are seen as linear pathways or as complex networks, are carried out by individual protein molecules that physically interact with each other, thereby influencing the activity of other molecules and being influenced in turn by other molecules. Detailed study of individual Wnt signaling components reveals important aspects of Wnt signaling mechanisms, such as the importance of multiprotein complexes to assemble and disassemble in what we now know as the Wnt/β-catenin pathway (e.g., β-catenin-degradation complex from studies of Axin and APC; see Chapter 3). Detailed studies of the relationship between protein structure and protein function lead to the discovery of functionally important subdomains. Recognizing such a functional domain can sometimes give great insight into a suggested function for the particular molecule, for instance, the seven-transmembrane domain at the time singled out the frizzled proteins as prime candidates for the then long-sought-after Wnt signal receptor. Sharing of a particular domain can also be suggestive, such as the identification of a Wnt-signal-binding cysteine-rich domain in not only the frizzled receptors but also in secreted frizzled-related protein (sFRP) molecules, which suggested they function as extracellular Wnt inhibitors (see Chapter 13). However, the functional significance for Wnt signaling of the many structural features that frizzled receptors share with G-protein-coupled receptors remained for many years controversial and have only recently become clearer (see Chapter 14).

By their very nature or for technical reasons, some proteins have been much more difficult to study in the past. The Wnt signal itself after hard and arduous work is only more recently starting to reveal more details about its pivotal molecular and cellular function (see Chapter 1). The dishevelled protein (see Chapter 15) and β-catenin (see Chapter 16) both function at important but different crossroads in the Wnt signaling network. The clearly defined functional domain structure of dishevelled protein promises to hold the key for understanding the choice of different Wnt signaling pathway mechanisms (see Chapters 7 and 15). Contrary to that, some of the many biological functions of the β-catenin protein are structurally difficult to separate, possibly because the same large central armadillo repeat domain mediates binding to different

Wnt Signaling in Development and Disease: Molecular Mechanisms and Biological Functions, First Edition.
Edited by Stefan Hoppler and Randall T. Moon.

Modulators that interact with Wnt proteins

Secreted Frizzled-related proteins

sFRPs are the largest family of extracellular Wnt modulators. sFRPs, as such, have not been found in protozoan lineages and seem to have appeared early in metazoan evolution (Leclere and Rentzsch, 2012). sFRP homologs have been found in cnidarians (Guder *et al.*, 2006), planarians (Gurley *et al.*, 2010), *Caenorhabditis elegans* (Harterink *et al.*, 2011), and vertebrates, but, notably, not in *Drosophila*, suggesting that sFRP genes have probably been lost in the genome of this species (Mii and Taira, 2009).

In mammals, there are five family members (sFRP1–sFRP5), subdivided into two subfamilies (sFRP1, sFRP2, sFRP5 and sFRP3, sFRP4) on the basis of their structure similarities and possible independent origin (Leclere and Rentzsch, 2012). A third subgroup, composed by Sizzled and Crescent, has been identified in *Xenopus*, zebrafish, and chick but not in mammals and invertebrates, suggesting that this subclass is likely a vertebrate invention then lost in mammals (Kuraku and Kuratani, 2011).

sFRPs are modular proteins that fold into two independent domains (Figure 13.1a). The N-terminal contains a secretion signal peptide followed by a CRD, characterized by the presence of 10 cysteine residues at conserved positions, which form a pattern of five disulfide bridges. This domain shares similarities with the extracellular CRD of Fzd and Ror receptors and is responsible for the name of the family. Sizzled and Crescent contain an additional cysteine residue that can form an interdomain disulfide bridge (Chong *et al.*, 2002). The C-terminal portion of sFRPs is characterized by segments of hydrophobic residues that confer heparin binding properties (Uren *et al.*, 2000) and by six cysteine residues that form three disulfide bridges. These features, which define the Netrin (NTR) module (NTR related), have been identified in other secreted but unrelated proteins, including the axon guidance molecule Netrin1, tissue inhibitors of metalloproteinases (TIMPS), type 1 procollagen C-proteinase enhancer proteins (PCOLCEs), and the complement proteins C3, C4, and C5 (Banyai and Patthy, 1999). The NTR module of sFRP1, sFRP2, and sFRP5 presents a pattern of spacing related to that of Netrin1, whereas sFRP3 and sFRP4 display a different spacing and pattern of disulfide bonds (Chong *et al.*, 2002).

The founding family member, Frzb/sFRP3, was identified in *Xenopus* as a soluble factor, expressed in the head organizer and capable of antagonizing Wnt/canonical signaling during the establishment of the embryonic anterior–posterior (A–P) axis (Leyns *et al.*, 1997; Wang *et al.*, 1997). The expression of sFRP orthologs in anterior structures and their protective role against the activation of canonical Wnt signaling in these regions is conserved throughout evolution and may represent the most ancestral function of these proteins (Harterink *et al.*, 2011). sFRPs were also independently identified with the name of secreted apoptosis-related proteins (Sarps) in mammalian cell cultures as proteins implicated in the modulation of apoptotic cell death (Melkonyan *et al.*, 1997). This function has been thereafter observed in different contexts (Bovolenta *et al.*, 2008).

There are multiple mechanisms by which sFRPs appear to modulate Wnt activity. Initial studies demonstrated that sFRPs bind Wnt ligands through the CRD, thus preventing receptor activation (Bafico *et al.*, 1999; Lin *et al.*, 1997). However, truncated forms of sFRP1 lacking the CRD retain strong interaction with Wingless and Wnt8 and mimic the function of the entire protein in both cell culture and fish embryos (Lopez-Rios *et al.*, 2008; Uren *et al.*, 2000; Xavier *et al.*, 2014). These data suggest that sFRPs have multiple Wnt binding sites (Figure 13.1a), which may be crucial to establish specific sFRP–Wnt pairs with different affinities (Wawrzak *et al.*, 2007). Furthermore, sFRPs cannot be considered as simple Wnt scavengers because there is increasing evidence that they also have important functions in the activation of the pathway. Their expression in fact seems crucial to promote Wnt diffusion in the extracellular space, enabling long-range signaling at least for the canonical branch of the pathway (Esteve *et al.*, 2011b; Mii and Taira, 2009). Likely due to the reported affinity among CRD, sFRPs can also bind to different Fzd receptors, including Fzd2, Fzd5, and Fzd8 (Dufourcq *et al.*, 2008; Kawano *et al.*, 2009; Rodriguez *et al.*, 2005). In retinal neurons, this binding results in the activation of intracellular mediators implicated in noncanonical signaling

(Rodriguez *et al.*, 2005). Thus, activation or inhibition of the intricate Wnt signaling cascade may strongly depend on context-dependent specific interactions among different Wnt ligands, sFRPs, and Fzd receptors (Bovolenta *et al.*, 2008; Xavier *et al.*, 2014).

sFRP1, sFRP2, and sFRP5 seem to play redundant roles during mammalian development, possibly because of their sequence similarities and their partially overlapping expression patterns in some tissues. For example, proper development of the eye, trunk, gut epithelium, and male gonads depends on both sFRP1 and sFRP2 (Bovolenta *et al.*, 2008), whereas A–P axis formation is ensured by the cooperative action of the three genes (Satoh *et al.*, 2008). However, several unique functions have also been reported. For example, genetic inactivation of sFRP1 accelerates mouse chondrocyte differentiation and endochondral ossification (Gaur *et al.*, 2006) and causes trabecular bone formation in aged animals (Bodine *et al.*, 2004, 2005). The homeostasis of lymphocytes and leukocytes is also altered in *Sfrp1*$^{-/-}$ mice (Renstrom *et al.*, 2009). Moreover, sFRP1 is strongly expressed in the developing cortical neuroepithelium, controlling its proliferation and differentiation (P.E. and P.B., unpublished) and its ortholog in *C. elegans* is involved in the regulation of the anterior migration of neuroblasts (Harterink *et al.*, 2011).

Notably, sFRP function has been also associated to Wnt-unrelated pathways. In fact, sFRP1 binds to and inhibits the activity of RANKL, a member of the TNF family involved in osteoclast formation (Hausler *et al.*, 2004). sFRP2 interacts with an integrin–fibronectin complex modulating apoptosis (Lee *et al.*, 2004), and Sfrp3 binds to EGF, blocking its function in different cellular contexts (Scardigli *et al.*, 2008). Binding and modulation of metalloproteases is also emerging as an additional important function of sFRP family members, relating their function to the modulation of other signaling cascades. Indeed, Sizzled binds to tolloid, a metalloproteinase directly involved in the control of BMP signaling (Lee *et al.*, 2006; Muraoka *et al.*, 2006), whereas sFRP1 regulates Notch signaling activation by acting as a negative modulator of ADAM10, a sheddase crucial to Notch receptor cleavage and thus to pathway activation (Esteve *et al.*, 2011a).

Frizzled-4 secreted splice variant

Frizzled-4 splice variant (Fzd4s) encodes a putative secreted polypeptide that derives from the alternative splicing of the transmembrane Fzd4 receptor gene. This polypeptide contains the N-terminal part of the CRD extracellular domain of the Fzd4 but lacks the C-terminal NTR domain, which is otherwise typical of sFRPs. *Fzd4s* mRNA is expressed in the human fetal lung and kidney, adult heart, and lung (Sagara *et al.*, 2001) and has been detected at all stages of *Xenopus* embryogenesis (Swain *et al.*, 2005). The activity of this splice variant has been determined using a *Xenopus* axis duplication assay, which provides an easy readout of Wnt/β-catenin signaling activation.

When coinjected with Wnt3 or Wnt8 into the ventral *Xenopus* blastomeres, *Fzd4s* mRNA could enhance or inhibit axis duplication, suggesting that this variant may act as a biphasic modulator of Wnt-signaling (Sagara *et al.*, 2001; Swain *et al.*, 2005; Gorny *et al.*, 2013). The physiological relevance of this molecule is however still unclear, because the existence of its mRNA has not been correlated with the demonstration of a corresponding protein in *Xenopus* nor in human tissues. This leaves open the possibility that this variant may represent a non-coding RNA of yet unknown function.

Cerberus and Coco

Cer is the founding member of the Dan family of secreted proteins, characterized by the presence of a cysteine knot functional motif in their tertiary structure (Figure 13.1a). There are several additional members in the family, such as Dan, protein related to DAN and Cerberus (PRDC), dante, caronte, gremlin/drm, Coco, sclerostin/SOST, and Cerberus-like (Cer-l), which is expressed only during mouse embryogenesis (Shawlot, Deng, and Behringer, 1998); however, only Cer and Coco have been functionally related to Wnt signaling, whereas the other proteins have diverse functions. Cer was discovered in a differential screening designed to isolate dorsal specific cDNAs in *Xenopus*. Accordingly, Cer is expressed in the *Xenopus* dorsovegetal organizer that induces anterior structures, and its overexpression induces the formation of ectopic heads

(Bouwmeester *et al.*, 1996). Initial studies proposed that this effect could be explained by the multifunctional antagonistic interaction of Cer with Wnt, BMP, and Nodal (a member of the TGF-β family) ligands (Piccolo *et al.*, 1999), since the respective signaling pathways need to be simultaneously repressed to allow anterior neural tissue formation (Glinka *et al.*, 1997; Silva *et al.*, 2003). A similar function and mechanism of action has also been proposed for Coco – a Spanish nickname for head – also isolated and characterized in *Xenopus* (Bell *et al.*, 2003).

A conserved role for Cer and Cer-l in the specification of anterior structures and Wnt signaling inhibition is however questioned by the observation that inactivation of the respective genes has no impact in mouse head formation (Belo *et al.*, 2000; Borges, Marques, and Belo, 2001; Shawlot *et al.*, 2000). Furthermore, Cer-l does not antagonize Wnt signaling (Belo *et al.*, 2000), and studies on the establishment of vertebrate left–right asymmetry favor the view that Cer and Cer-l are specific modulators of Nodal and BMP signaling, similarly to what was reported for other DAN family members (Kattamuri *et al.*, 2012).

Wnt inhibitory factor 1 (WIF)

WIF-1 was initially isolated in vertebrates as a secreted protein that physically interacts with Wnt ligands, thereby inhibiting canonical signaling (Figure 13.2; Hsieh *et al.*, 1999a, b). WIF-1 is highly conserved from *Drosophila* to humans and expressed in a variety of developing and adult tissues, particularly in the brain, retina, lung, and cartilage (Hsieh *et al.*, 1999a; Kansara *et al.*, 2009). WIF-1 contains a C-terminal hydrophilic tail followed by five EGF-like repeats that strengthen the interaction with both the ligand and heparan sulfate proteoglycans known to contribute to Wnt diffusion (Avanesov *et al.*, 2012; Malinauskas *et al.*, 2011). The N-terminal

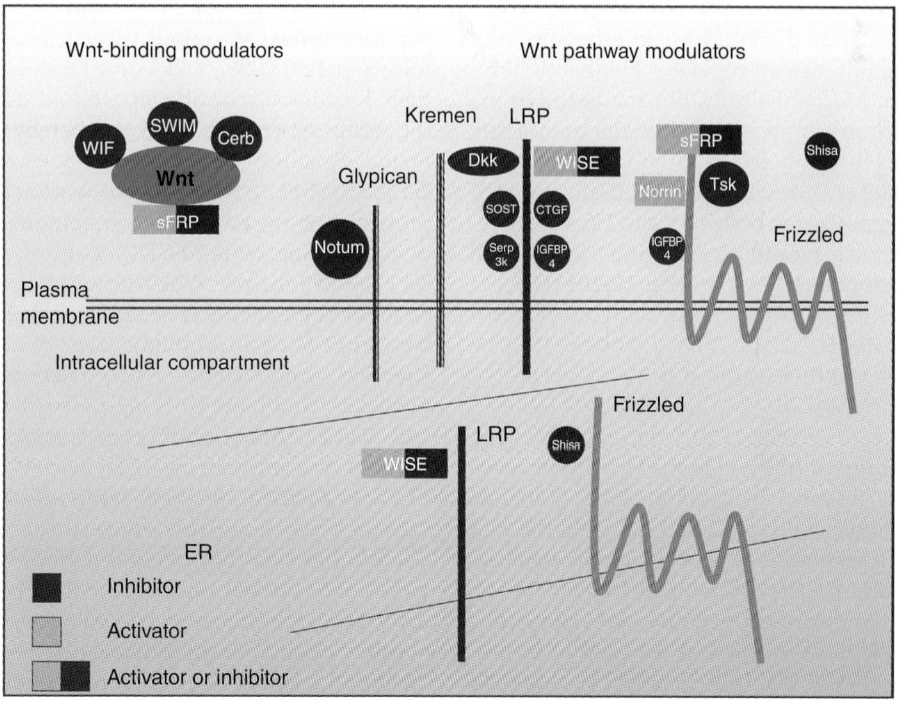

Figure 13.2 Schematic summary of the mode of action of secreted Wnt signaling modulators. Wnt binding modulators such as WIF, SWIM, or Cer bind to Wnt ligands and antagonize their activity. sFRPs can function as activators or inhibitors. Wnt pathway modulators interact with Fzd receptors, the coreceptors LRP and Kremen, or components of the extracellular matrix such as glypican. They can either activate the Wnt cascade (Norrin, sFRP, WISE) or inhibit it (Tsk, IGFBP-4, CTGF, SOST, Serp3, Dkk, Notum). Shisa and WISE can also interact with Fzd receptors and LRP in the ER and thereby regulate the amount of receptors available at the cell membrane.

region of the protein is instead composed of a characteristic WIF domain, which is also found in the extracellular portion of Ryk, a Wnt receptor. In both proteins the WIF domain is responsible for Wnt binding (Macheda *et al.*, 2012; Malinauskas *et al.*, 2011). When coinjected with Wnt ligands into the ventral blastomeres of the *Xenopus* embryo, WIF-1 counteracts Wnt-induced secondary axis formation (Hsieh *et al.*, 1999a; Hunter *et al.*, 2004). Conversely, WIF-1 knockdown in zebrafish embryos results in increased Wnt canonical signaling, causing shortening of the anteroposterior axis as well as somite and swim bladder defects. In contrast, developmental events mediated by the β-catenin-independent Wnt pathway appear normal, suggesting that WIF-1 might be specific for the canonical branch, despite its interaction with noncanonical ligands in coimmunoprecipitation assays (Yin, Korzh, and Gong, 2012). In mammals, other Wnt antagonists may compensate WIF-1 function, because *Wif-1* null mice develop and reproduce normally with no obvious defect during adulthood (Kansara *et al.*, 2009).

Notably, the *Drosophila* homolog of WIF-1, known as Shifted (Shf), has no effect on Wnt signaling but instead regulates Hedgehog (Hh) diffusion. Mechanistically, Shf binds to Hh via the WIF domain and reinforces the interaction between Hh and heparan sulfate proteoglycans (Gorfinkiel *et al.*, 2005). Together, these observations suggest that in both Wnt and Hh signaling, WIF-1 acts by modulating the interaction with ECM components (Avanesov *et al.*, 2012).

Secreted wingless interacting molecule (Swim)

The most recent addition to the list of Wnt modulators is Swim, which favors Wnt diffusion. This protein has been isolated in *Drosophila* as a putative member of the lipocalin family of extracellular transporter proteins (Mulligan *et al.*, 2012). Owing to their palmitic and palmitoleic modifications, Wnt proteins are highly hydrophobic and poorly diffusible molecules that tend to stick to the plasma membrane of the producing cells (Mii and Taira, 2009; Takada *et al.*, 2006; Willert *et al.*, 2003). Functional studies suggest that palmitate-dependent interaction between Swim and lipid-modified wingless maintains the solubility of the ligand, promoting its diffusion and thus its signaling range (Mulligan *et al.*, 2012), as observed for sFRPs in specific contexts (Esteve *et al.*, 2011b; Mii and Taira, 2009). Other lipocalin family members have been copurified with mammalian Wnts (Mulligan *et al.*, 2012), suggesting that they may have a similar function.

Modulators that interact with Wnt pathway components

Dickkopf

The founding member of the Dickkopf protein family, Dkk1, was identified in an expression screen for head inducers in *Xenopus* embryos. Indeed, Dkk1-mediated repression of canonical Wnt signaling is a prerequisite for head induction during early *Xenopus* development (Glinka *et al.*, 1998). Dkks are evolutionary conserved between cnidarians and humans and share two conserved CRD at the N- and C-terminus. The N-terminal CRD (Dkk-N) is unique to Dkks, whereas the pattern of the 10 conserved cysteine residues in the C-terminus is related to colipase folds (Niehrs, 2006). Dkk1 and Dkk2 bind with their functional critical colipase fold domain to the Wnt coreceptors Lrp5/6, thereby antagonizing signaling (Figure 13.2; Mao *et al.*, 2001; Semenov *et al.*, 2001). This mode of action was proposed on the basis of *in vitro* binding assays, demonstrating that Dkk1 simply prevents the formation of Wnt-1-induced Fzd/Lrp6 complexes (Semenov *et al.*, 2001). Other studies have implicated an additional class of molecules Kremen1 and Kremen2, two transmembrane molecules that bind with high affinity to Dkk1 and Dkk2. Upon binding to Kremen, Dkk1 induces the formation of a ternary Dkk1/LRP5/6/Kremen complex, which is then rapidly internalized. This mechanism assures LRP5/6 removal from the cell surface, preventing Wnt/β-catenin signaling activation (Mao *et al.*, 2002). However, not all Dkk proteins act as negative modulators of canonical Wnt signaling. The most distantly related member of the family Dkk3 does not interact with Lrp6 and seems to have no Wnt signaling inhibitory activity (Mao *et al.*, 2001). There is also evidence that Dkk2 promotes pathway activation (Wu *et al.*, 2000), and Wnt-independent functions for Dkk1 and Dkk2 have been reported (Niehrs, 2006).

Shisa

Shisa family members are characterized by the presence of an N-terminal CRD and a proline-rich region at the C-terminus (Figure 13.1b). There are nine Shisa family members in vertebrates as well as a distantly related homolog that possesses an N-terminal domain with six conserved cysteines (Pei and Grishin, 2012). The founding member, Shisa1, was identified in a screen for novel genes involved in *Xenopus* head formation, and, accordingly, it promotes the formation of anterior embryonic structures (Yamamoto *et al.*, 2005). Shisa transcripts localize to the Spemann organizer and its derivatives, the prospective head mesoderm/endomesoderm and the anterior neuroectoderm. In a paracrine signaling assay, Shisa1 did not prevent Wnt ligand/receptor interaction, but overexpression of Shisa1 in Fzd8- and Lrp6-expressing cells suppressed Wnt signaling, indicating a cell-autonomous effect. Localization studies indicated that Shisa retains Fzd8 in the endoplasmic reticulum (ER), preventing posttranslational maturation of the receptor and thus its localization at the plasma membrane. Notably, this novel and interesting mechanism by which a modulator controls signaling activation applies also to the FGF receptor (Yamamoto *et al.*, 2005). How Shisa1 retains Fzd and FGF receptors in the ER remains unknown. Shisa2, a *Xenopus* paralog, inhibits Wnt and FGF signaling during somitogenesis with a similar mechanism (Nagano *et al.*, 2006).

Sclerostin/SOST

Sclerostin is structurally closely related to WISE and both proteins contain a cysteine knot domain (Figure 13.1b). Loss-of-function mutations in the Sclerostin gene, *SOST*, are responsible for autosomal recessive sclerosteosis and van Buchem diseases, which are characterized by a substantial increase of bone mass. Both diseases share similarities with high bone mass (HBM) diseases caused by hyperactive Lrp5, which lead to the suggestion that SOST could antagonize Lrp function (Brunkow *et al.*, 2001; Semenov, Tamai, and He, 2005). Indeed, Sclerostin can bind to both Lrp5 and Lrp6 through the first two extracellular YWTD-EGF repeats of these coreceptors. In contrast, Sclerostin does not interfere with Wnt binding to the same repeats, indicating that it interferes with the formation of the Fzd/Lrp complexes rather than acting on the ligand (Li *et al.*, 2005; Semenov, Tamai, and He, 2005). Notably, mutations in the *Lrp5* gene responsible for the HBM disease are single amino acid substitutions in the putative Sclerostin binding site (YWTD β-propeller), which would result in impaired coreceptor regulation by SOST.

Wnt modulator in surface ectoderm (Wise)

The secreted protein Wise, also known as USAG1/Ectodin/Sostdc1, was first identified in a screen for factors that could alter A–P patterning of neuralized *Xenopus* ectoderm (Itasaki *et al.*, 2003). The rat homolog uterine sensitization-associated gene 1 (USAG1) was isolated independently in a screen for genes involved in embryo implantation to the endometrium (Simmons and Kennedy, 2002). Wise shares 38% amino acid identity with Sclerostin and also contains a cysteine knot (Figure 13.1b; Vitt, Hsu, and Hsueh, 2001). Experiments in *Xenopus* revealed that secreted Wise acts as both an activator and an inhibitor of Wnt signaling (Figure 13.2), depending on the cellular context (Itasaki *et al.*, 2003). Similarly to Sclerostin, Wise inhibits Wnt signaling by binding to the YWTD ß-propeller of Lrp5/6 (Itasaki *et al.*, 2003; Li *et al.*, 2005). However, differently from Sclerostin, Wise likely competes with Wnt ligands for binding to LRP6 (Itasaki *et al.*, 2003). In addition to its secreted form, Wise has been also found in an ER-retained fraction. Resembling the mode of action of Shisa, this ER-localized form of Wise can inhibit Wnt signaling by trapping Lrp6 in the ER and thereby reducing the amount of Lrp6 available at the cell surface. As a result, cells lose their responsiveness to extracellular Wnt signals (Guidato and Itasaki, 2007).

Connective tissue growth factor (CTGF)

CTGF belongs to the vertebrate-specific CCN family (named after three prototypical members CTGF, Cyr61, and Nov) of cysteine knot motif containing proteins (Mercurio *et al.*, 2004; Figure 13.1b). CTGF was initially characterized for its ability to bind to integrins and to modulate TGF-ß and BMP signaling. Recent studies have however demonstrated that CTGF interacts with the low-density lipoprotein receptor-related protein (Lrp)/α_2-macroglobulin receptor and possibly also with Lrp6 an additional member of the LDL receptor family (Abreu *et al.*, 2002; Lau and Lam, 1999; Segarini *et al.*, 2001). Coimmunoprecipitation experiments showed that the CT domain of CTGF mediated the interaction with Lrp6 and Fzd8 (Mercurio *et al.*, 2004). CTGF binds with high affinities the first two EGF-like repeats of Lrp6, which also represent the characterized Wnt binding site. In *Xenopus* embryos, CTGF inhibits Wnt/β-catenin signaling triggered by Wnt8 and also interferes with noncanonical signaling, thereby interfering with morphogenetic cell movements. The mechanism of inhibition is still unclear but likely involves competition with the ligand on Lrp6. Alternatively, CTGF could induce a conformational change in Lrp6 that would alter the interaction with other pathway components (Mercurio *et al.*, 2004). Other Wnt modulators might control CTGF activity because Wif-1 seems to inhibit CTGF function in primary murine chondrocytes and Serpina3k and Dkk1 attenuated CTGF overexpression in the retina of diabetic rats (Surmann-Schmitt *et al.*, 2012; Zhang, Zhou, and Ma, 2010).

Tsukushi (TSK)

TSK belongs to a class of the small leucine-rich proteoglycan (SLRP) family, which comprises five classes based on structural and functional properties (Ohta *et al.*, 2004; Schaefer and Iozzo, 2008). These classes are defined by the presence of four N-terminal cysteine residues. There are two TSK isoforms (A and B) that arise from the alternative splicing of the mRNA (Ohta *et al.*, 2004). The two isoforms differ in their C-terminal region and are both characterized by 11 homologous leucine-rich repeats (LRRs) flanked by the N-terminal cysteine-rich region (Schaefer and Iozzo, 2008; Figure 13.1b). TSK was first identified as an inhibitor of BMP signaling because it forms a complex with BMP and chordin, thereby acting as a dorsalizing factor in the induction of the Hensen's node in chicken embryos (Ohta *et al.*, 2004). Subsequent studies have shown that TSK interacts also with Fzd4-CRD but not with Wnt2b, Lrp5, and/or Fzd5-CRD. In chick eye development, both TSK B and TSK A bind to Fzd4. Further studies with TSK B have shown that the molecules compete with Wnt2b for receptor occupancy, thereby inhibiting Wnt signaling activation in the peripheral retina (Figure 13.2). Deletion studies have shown that at least 4 of 11 LRRs of TSK B are necessary to completely abrogate Wnt2b activity (Ohta *et al.*, 2011).

GF-binding protein 4 (IGFBP-4)

The mammalian IGFBP family comprises six members, 1–6. These proteins are composed of three distinct domains: a conserved N-terminal domain, a variable mid region, and a conserved C-terminal domain characterized by six cysteine residues (Figure 13.1b). The amino acid sequences encompassing the last five cysteines share 37% similarity with the thyroglobulin type I domain. IGFBPs belong to the insulin growth factor (IGF) system. These proteins bind to IGFs and have been mostly characterized as carrier and modulators of IGF (Hwa, Oh, and Rosenfeld, 1999). In addition to IGFs, IGFBPs interact with integrins and TGF-ß as well as with intracellular proteins. However, a recent study demonstrated that IGFBP-4 inhibits canonical Wnt signaling during cardiogenesis in an IGF-independent and non-cell-autonomous manner. Indeed, morpholino-mediated knockdown of IGFBP-4 in *Xenopus* embryos caused cardiac defects, which could be rescued by the coexpression of an IGF-binding-defective IGFBP-4 mutant or a dominant-negative form of Lrp6 (Zhu *et al.*, 2008). Mechanistically, the C-terminal thyroglobulin type I domain of IGFBP-4 specifically binds to the extracellular domain of Fzd8 and Lrp6 (Zhu *et al.*, 2008; Figure 13.2). Similar interactions have been also observed with IGFBP-1, IGFBP-2, and IGFBP-6, supporting the role of IGFBPs in the regulation of the Wnt pathway (Zhu *et al.*, 2008).

Serpina3k

Serpina3k belongs to the serine protease inhibitors, a functionally diverse superfamily of proteins that inhibit serine proteinases of the chymotrypsin family, and to cysteine proteinases, albeit with less efficiency. Members of the family serve also as hormone transporters, as corticosteroid-binding globulins and regulators of blood pressure (Silverman *et al.*, 2001). Serpina3k was first discovered as an inhibitor of tissue kallikrein, a subgroup of serine proteases and thereafter related to Wnt/β-catenin signaling in diabetic retinopathy. Serpina3k blocks β-catenin accumulation and Lrp6 phosphorylation in ARPE19 (human retinal pigment epithelial) cells (Zhang *et al.*, 2010), by binding to the extracellular domain of Lrp6 with high affinity (Figure 13.2). In contrast, Serpina3k does not interact with Fzd4, a known angiogenesis modulator. Binding of Serpina3k to Lrp6 prevents dimerization of Wnt1-induced Fzd8/Lrp6 complexes, thereby inhibiting Wnt signaling at the receptor level (Zhang *et al.*, 2010).

Notum

Notum was codiscovered in *Drosophila* as a Wnt activity repressor that mechanistically acts as a negative regulator of Wnt diffusion (Gerlitz and Basler, 2002; Giraldez, Copley, and Cohen, 2002). As other morphogens, Wnt diffusion and interaction with their receptors involve GPI-linked heparan sulfate proteoglycans (glypicans) (Kreuger *et al.*, 2004). Notum encodes a secreted enzyme with α/β hydrolase activity that cleaves glypican anchors, reducing the amount of glypicans at the cell surface. As a consequence, Wnt diffusion and signaling range is limited. Consistently, loss of Notum function in *Drosophila* extends the wingless gradient and increases its signaling range. Notum has been found in many metazoans; its function seems to be conserved in evolution (Flowers, Topczewska, and Topczewski, 2012) and rather ancestral as demonstrated by regeneration studies in Planaria. This species can regenerate both head and tails after amputation. Wnt signaling promotes tail regeneration and prevents head regeneration in posterior wounds. Notably, Notum expression is upregulated in anterior wounds enabling head regeneration (Petersen and Reddien, 2011). In all species analyzed so far, Notum appears to be a target of Wnt signaling, indicating that Wnt ligands modulate the formation of their own gradients with a feedback regulator mechanism (Flowers, Topczewska, and Topczewski, 2012; Giraldez, Copley, and Cohen, 2002; Petersen and Reddien, 2011). A controversial issue is whether Notum activity is limited to the Wnt pathway or may also influence other glypican-dependent signaling (Flowers, Topczewska, and Topczewski, 2012; Giraldez, Copley, and Cohen, 2002; Traister, Shi, and Filmus, 2008).

Norrin

Norrin is a secreted protein containing a terminal cysteine knot present, as mentioned earlier, in other Wnt modulators and growth factors (Clevers, 2004; Meitinger *et al.*, 1993). Norrin occupies a special place among Wnt signaling regulators because it is a well-recognized specific ligand of the Fzd4 receptor. Norrin bears no sequence similarity with Wnt ligands, but its binding site on Fzd4 largely overlaps with that of Wnt8, suggesting that, when present, Norrin might compete with Wnts for receptor binding. Notably, this interaction is rather specific because Norrin does not interact with any other Fzd receptor or with the CRD of sFRPs (Smallwood *et al.*, 2007). High-affinity binding of Norrin to Fzd4 activates the canonical pathway in an Lrp5-/Lrp6-dependent manner (Xu *et al.*, 2004; Ye *et al.*, 2009). Norrin/Fzd4 signaling is crucial for vasculature formation as part of an angiogenic pathway that includes also Lrp5 and Tspan12 (Junge *et al.*, 2009; Xu *et al.*, 2004; Ye *et al.*, 2009). Mutations in *Norrin* or *Fzd4* genes cause different ocular pathologies that are phenotypically reproduced in $Fzd4^{-/-}$ mice. These pathologies are characterized by abnormal growth and development of the retinal blood vessels, and the integrity of blood–brain barrier in the retina, cerebellum, and olfactory bulb is altered in both *Fzd4* and *Norrin* adult knockout mice (Wang *et al.*, 2012).

Pathological implications of Wnt modulators' malfunction

Wnt signaling controls a large number of developmental and homeostatic events. Alteration of its functions has thus many pathological consequences, a number of which are also related to abnormal expression or function of the antagonists and modulators described in this chapter. In general, abnormal activity of Wnt modulators, including sFRPs, WIF-1, and Cer, is linked to the alteration of their epigenetic regulation, for example, epigenetic silencing by hypermethylation, or to the presence of single-nucleotide polymorphisms (SNPs) that appear to confer susceptibility to some diseases. No pathogenic mutations have been so far reported in most of the corresponding genes with the exception of Norrin and SOST.

Mutations in the Norrin gene (*NDP*) cause Norrie disease, familial exudative vitreoretinopathy (FEVR), retinopathy of prematurity (ROP), and Coats disease, all hereditary retinopathies resulting from abnormal blood supply to the eye (Black *et al.*, 1999; Xu *et al.*, 2004). Mutations in the *SOST* gene instead cause sclerosteosis, a disease characterized by massive skeletal overgrowth (Balemans *et al.*, 2001; Brunkow *et al.*, 2001). Canonical Wnt signaling plays a critical role in bone metabolism, and Wnt modulators other than SOST are involved in bone disorders, including age-related osteoporosis (Bodine *et al.*, 2004). For example, different polymorphisms in the 3′ untranslated region of *sFRP1* have been associated with variations in bone mineral density in women (Ohnaka *et al.*, 2009; Sims *et al.*, 2008), and SNPs in *sFRP3* confer risk of developing osteolysis and heterotopic ossification (Gordon *et al.*, 2007). Antibody-based therapies that target SOST and DKK1 have been successful in clinical trials for the treatment of osteoporosis.

Aberrant Wnt/β-catenin activation, either due to effectors' activation or due to the loss of tumor suppressors, has been widely associated to carcinogenesis. The suppression or downregulation of *sFRPs*, *WIF-1*, and *DKK3* expression is thus considered a key event in tumor formation as the hypermethylation, and silencing of their respective genes seems to be one of the earliest events in the development of basically all solid tumors and hematological malignancies (Cain and Manilay, 2013; Esteve and Bovolenta 2010; Veeck and Dahl, 2012). This hypermethylation is rather predictive, and the status of *sFRP1* and *DKK3* promoter methylation has been proposed as a useful epigenetic biomarker for the detection of the progression and prognosis of the disease.

Alterations in sFRP1 expression have been observed also in a number of inflammatory autoimmune diseases. For example, high sFRP1 levels have been reported in the synovial fluid of patients affected by rheumatoid arthritis (Lee *et al.*, 2012), in the orbital adipose tissue of individuals with Graves' ophthalmopathy (Kumar *et al.*, 2005), or in cases of lung emphysemas (Foronjy *et al.*, 2010). Similarly, the aqueous fluid of glaucoma patients is enriched in sFRP1, which is considered among the factors responsible for the elevated intraocular pressure that characterizes the disease (Wang *et al.*, 2008). Notably, there are also reports indicating that different sFRPs are elevated in the retina of patients affected by retinitis pigmentosa (Jones and Jomary, 2002) that invariably leads to photoreceptors' degeneration. Overexpression of another sFRP family member, sFRP4, has been reported in individuals affected by type 2 diabetes, and this alteration seems linked to the reduction in insulin secretion (Mahdi *et al.*, 2012). In contrast, low levels of Sfrp5 have been found in obese individuals, although the significance of this decrease is still unclear (Ouchi *et al.*, 2010).

Besides these reports in humans, the use of animal models points to the potential implication of Wnt modulators in several additional pathological conditions. For example, genetic alterations in the levels of sFRP1 or sFRP2 affect heart repair after myocardial infarction in mouse models with a mechanism that implicates Sfrp2-mediated regulation of tolloid metalloproteinases (Alfaro *et al.*, 2010; Barandon *et al.*, 2003; Kobayashi *et al.*, 2009; Mirotsou *et al.*, 2007). Recent studies have however questioned that Sfrp2 can modulate tolloid activity (Bijakowski *et al.*, 2012). Furthermore, in a murine model of periodontitis, suppression of the elevated Sfrp1 levels alleviates the inflammation and destruction of periodontal tissue that characterize this disease (Li and Amar, 2007).

Conclusions

In recent years, our knowledge on the regulation of the Wnt signaling pathway has increased dramatically. Novel regulatory mechanisms have been described, and notably, they regulate the pathway not only in the extracellular space but also inside the cells. It has also become clear that Wnt modulators cannot be simply divided into activators and inhibitors. Several biphasic regulators have been identified, which act on the Wnt pathway in a context-dependent manner. Furthermore, molecules known for their regulatory activity on other signaling pathways or physiological processes have been shown to have important roles in Wnt signaling modulation. These new discoveries broadened our insight of the mechanics of Wnt signaling and will hopefully help to unravel the additional aspects of its function in development and diseases. In this respect, recent observations have implicated Wnt modulators in stem cell biology. For example, sFRP2 can induce neural identity in embryonic stem cells (Aubert *et al.*, 2002), whereas reduction of Dkk1 or sFRP3 promotes hippocampal neurogenesis, counteracting in the first case age-related cognitive decline (Seib *et al.*, 2013) or accelerating the development of new neurons in the second case (Jang *et al.*, 2013). Together, these findings point to the potential value of these molecules also in regenerative medicine.

References

Abreu, J.G., Ketpura, N.I., Reversade, B., and De Robertis, E.M. (2002) Connective-tissue growth factor (CTGF) modulates cell signalling by BMP and TGF-beta. *Nature Cell Biology*, **4** (8), 599–604.

Alfaro, M.P., Vincent, A., Saraswati, S. *et al.* (2010) sFRP2 suppression of bone morphogenic protein (BMP) and Wnt signaling mediates mesenchymal stem cell (MSC) self-renewal promoting engraftment and myocardial repair. *The Journal of Biological Chemistry*, **285** (46), 35645–35653.

Aubert, J., Dunstan, H., Chambers, I., and Smith, I. (2002) Functional gene screening in embryonic stem cells implicates Wnt antagonism in neural differentiation. *Nature Biotechnology*, **20** (12), 1240–1245.

Avanesov, A., Honeyager, S.M., Malicki, J., and Blair, S.S. (2012) The role of glypicans in Wnt inhibitory factor-1 activity and the structural basis of Wif1's effects on Wnt and Hedgehog signaling. *PLoS Genetics*, **8** (2), e1002503.

Bafico, A., Gazit, A., Pramila, T. *et al.* (1999) Interaction of frizzled related protein (FRP) with Wnt ligands and the frizzled receptor suggests alternative mechanisms for FRP inhibition of Wnt signaling. *The Journal of Biological Chemistry*, **274** (23), 16180–16187.

Balemans, W., Ebeling, J., Patel, A. *et al.* (2001) Increased bone density in sclerosteosis is due to the deficiency of a novel secreted protein (SOST). *Human Molecular Genetics*, **10** (5), 537–543.

Banyai, L. and Patthy, L. (1999) The NTR module: domains of netrins, secreted frizzled related proteins, and type I procollagen C-proteinase enhancer protein are homologous with tissue inhibitors of metalloproteases. *Protein Science*, **8** (8), 1636–1642.

Barandon, L., Couffinhal, T., Ezan, J. *et al.* (2003) Reduction of infarct size and prevention of cardiac rupture in transgenic mice overexpressing FrzA. *Circulation*, **108** (18), 2282–2289.

Bell, E., Munoz-Sanjuan, I., Altmann, C.R. *et al.* (2003) Cell fate specification and competence by Coco, a maternal BMP, TGFbeta and Wnt inhibitor. *Development*, **130** (7), 1381–1389.

Belo, J.A., Bachiller, D., Agius, E. *et al.* (2000) Cerberus-like is a secreted BMP and nodal antagonist not essential for mouse development. *Genesis*, **26** (4), 265–270.

Bijakowski, C., Vadon-Le Goff, S., Delolme, F. *et al.* (2012) Sizzled is unique among secreted Frizzled-related proteins for its ability to specifically inhibit bone morphogenetic protein-1 (BMP-1)/tolloid-like proteinases. *The Journal of Biological Chemistry*, **287** (40), 33581–33593.

Black, G.C., Perveen, R., Bonshek, R. *et al.* (1999) Coats' disease of the retina (unilateral retinal telangiectasis) caused by somatic mutation in the NDP gene: a role for norrin in retinal angiogenesis. *Human Molecular Genetics*, **8** (11), 2031–2035.

Bodine, P.V., Zhao, W., Kharode, Y.P. *et al.* (2004) The Wnt antagonist secreted frizzled-related protein-1 is a negative regulator of trabecular bone formation in adult mice. *Molecular Endocrinology*, **18** (5), 1222–1237.

Bodine, P.V., Billiard, J., Moran, R.A. *et al.* (2005) The Wnt antagonist secreted frizzled-related protein-1 controls osteoblast and osteocyte apoptosis. *Journal of Cellular Biochemistry*, **96** (6), 1212–1230.

Borges, A.C., Marques, S., and Belo, J.A. (2001) The BMP antagonists cerberus-like and noggin do not interact during mouse forebrain development. *The International Journal of Developmental Biology*, **45** (2), 441–444.

Bouwmeester, T., Kim, S., Sasai, Y. *et al.* (1996) Cerberus is a head-inducing secreted factor expressed in the anterior endoderm of Spemann's organizer. *Nature*, **382** (6592), 595–601.

Bovolenta, P., Esteve, P., Ruiz, J.M. *et al.* (2008) Beyond Wnt inhibition: new functions of secreted Frizzled-related proteins in development and disease. *Journal of Cell Science*, **121** (Pt 6), 737–746.

Brunkow, M.E., Gardner, J.C., Ness, J. *et al.* (2001) Bone dysplasia sclerosteosis results from loss of the SOST gene product, a novel cystine knot-containing protein. *American Journal of Human Genetics*, **68** (3), 577–589.

Cain, C.J. and Manilay, J.O. (2013) Hematopoietic stem cell fate decisions are regulated by Wnt antagonists: comparisons and current controversies. *Experimental Hematology*, **41** (1), 3–16.

Chong, J.M., Uren, A., Rubin, J.S., and Speicher, D.W. (2002) Disulfide bond assignments of secreted Frizzled-related protein-1 provide insights about Frizzled homology and netrin modules. *The Journal of Biological Chemistry*, **277** (7), 5134–5144.

Clevers, H. (2004) Wnt signaling: Ig-norrin the dogma. *Current Biology*, **14** (11), R436–R437.

Dufourcq, P., Leroux, T., Ezan, J. *et al.* (2008) Regulation of endothelial cell cytoskeletal reorganization by a secreted frizzled-related protein-1 and frizzled 4- and frizzled 7-dependent pathway: role in neovessel formation. *The American Journal of Pathology*, **172** (1), 37–49.

Esteve, P. and Bovolenta, P. (2010) The advantages and disadvantages of sfrp1 and sfrp2 expression in pathological events. *The Tohoku Journal of Experimental Medicine*, **221** (1), 11–17.

Esteve, P., Sandonis, A., Cardozo, M. *et al.* (2011a) SFRPs act as negative modulators of ADAM10 to regulate retinal neurogenesis. *Nature Neuroscience*, **14** (5), 562–569.

Esteve, P., Sandonis, A., Ibanez, C. *et al.* (2011b) Secreted frizzled-related proteins are required for Wnt/beta-catenin signalling activation in the vertebrate optic cup. *Development*, **138** (19), 4179–4184.

Flowers, G.P., Topczewska, J.M., and Topczewski, J. (2012) A zebrafish Notum homolog specifically blocks the Wnt/beta-catenin signaling pathway. *Development*, **139** (13), 2416–2425.

Foronjy, R., Imai, K., Shiomi, T. *et al.* (2010) The divergent roles of secreted frizzled related protein-1 (SFRP1) in lung morphogenesis and emphysema. *The American Journal of Pathology*, **177** (2), 598–607.

Gaur, T., Rich, L., Lengner, C.J. *et al.* (2006) Secreted frizzled related protein 1 regulates Wnt signaling for BMP2 induced chondrocyte differentiation. *Journal of Cellular Physiology*, **208** (1), 87–96.

Gerlitz, O. and Basler, K. (2002) Wingful, an extracellular feedback inhibitor of Wingless. *Genes & Development*, **16** (9), 1055–1059.

Giraldez, A.J., Copley, R.R., and Cohen, S.M. (2002) HSPG modification by the secreted enzyme Notum shapes the Wingless morphogen gradient. *Developmental Cell*, **2** (5), 667–676.

Glinka, A., Wu, W., Onichtchouk, D. *et al.* (1997) Head induction by simultaneous repression of Bmp and Wnt signalling in Xenopus. *Nature*, **389** (6650), 517–519.

Glinka, A., Wu, W., Delius, H. *et al.* (1998) Dickkopf-1 is a member of a new family of secreted proteins and functions in head induction. *Nature*, **391** (6665), 357–362.

Gordon, A., Southam, L., Loughlin, J. *et al.* (2007) Variation in the secreted frizzled-related protein-3 gene and risk of osteolysis and heterotopic ossification after total hip arthroplasty. *Journal of Orthopaedic Research*, **25** (12), 1665–1670.

Gorfinkiel, N., Sierra, J., Callejo, A. *et al.* (2005) The Drosophila ortholog of the human Wnt inhibitor factor Shifted controls the diffusion of lipid-modified Hedgehog. *Developmental Cell*, **8** (2), 241–253.

Gorny, A.-K., Kaufmann, L.T., Swain, R.K., Steinbeisser, H. (2013) A splice variant of the Xenopus frizzled-4 receptor is a biphasic modulator of Wnt signalling. *Cell Communication and Signaling*, **11**, 89.

Guder, C., Pinho, S., Nacak, T.G. *et al.* (2006) An ancient Wnt-Dickkopf antagonism in Hydra. *Development*, **133** (5), 901–911.

Guidato, S. and Itasaki, N. (2007) Wise retained in the endoplasmic reticulum inhibits Wnt signaling by reducing cell surface LRP6. *Developmental Biology*, **310** (2), 250–263.

Gurley, K.A., Elliott, S.A., Simakov, O. *et al.* (2010) Expression of secreted Wnt pathway components reveals unexpected complexity of the planarian amputation response. *Developmental Biology*, **347** (1), 24–39.

Harterink, M., Kim, D.H., Middelkoop, T.C. *et al.* (2011) Neuroblast migration along the anteroposterior axis of C. elegans is controlled by opposing gradients of Wnts and a secreted Frizzled-related protein. *Development*, **138** (14), 2915–2924.

Hausler, K.D., Horwood, N.J., Chuman, Y. *et al.* (2004) Secreted frizzled-related protein-1 inhibits RANKL-dependent osteoclast formation. *Journal of Bone and Mineral Research*, **19** (11), 1873–1881.

Hsieh, J.C., Kodjabachian, L., Rebbert, M.L. *et al.* (1999a) A new secreted protein that binds to Wnt proteins and inhibits their activities. *Nature*, **398** (6726), 431–436.

Hsieh, J.C., Rattner, A., Smallwood, P.M., and Nathans, J. (1999b) Biochemical characterization of Wnt-frizzled interactions using a soluble, biologically active vertebrate Wnt protein. *Proceedings of the National Academy of Sciences of the United States of America*, **96** (7), 3546–3551.

Hunter, D.D., Zhang, M., Ferguson, J.W. *et al.* (2004) The extracellular matrix component WIF-1 is expressed during, and can modulate, retinal development. *Molecular and Cellular Neurosciences*, **27** (4), 477–488.

Hwa, V., Oh, Y., and Rosenfeld, R.G. (1999) The insulin-like growth factor-binding protein (IGFBP) superfamily. *Endocrine Reviews*, **20** (6), 761–787.

Itasaki, N., Jones, C.M., Mercurio, S. *et al.* (2003) Wise, a context-dependent activator and inhibitor of Wnt signalling. *Development*, **130** (18), 4295–4305.

Jang, M.H., Bonaguidi, M.A., Kitabatake, Y. *et al.* (2013) Secreted frizzled-related protein 3 regulates activity-dependent adult hippocampal neurogenesis. *Cell Stem Cell*, **12** (2), 215–223.

Jones, S.E. and Jomary, C. (2002) Secreted Frizzled-related proteins: searching for relationships and patterns. *BioEssays*, **24** (9), 811–820.

Junge, H.J., Yang, S., Burton, J.B. *et al.* (2009) TSPAN12 regulates retinal vascular development by promoting Norrin- but not Wnt-induced FZD4/beta-catenin signaling. *Cell*, **139** (2), 299–311.

Kansara, M., Tsang, M., Kodjabachian, L. *et al.* (2009) Wnt inhibitory factor 1 is epigenetically silenced in human osteosarcoma, and targeted disruption accelerates osteosarcomagenesis in mice. *The Journal of Clinical Investigation*, **119** (4), 837–851.

Kattamuri, C., Luedeke, D.M., Nolan, K. *et al.* (2012) Members of the DAN family are BMP antagonists that form highly stable noncovalent dimers. *Journal of Molecular Biology*, **424** (5), 313–327.

Kawano, Y., Diez, S., Uysal-Onganer, E.S. *et al.* (2009) Secreted Frizzled-related protein-1 is a negative regulator of androgen receptor activity in prostate cancer. *British Journal of Cancer*, **100** (7), 1165–1174.

Kobayashi, K., Luo, M., Zhang, Y. *et al.* (2009) Secreted Frizzled-related protein 2 is a procollagen C proteinase enhancer with a role in fibrosis associated with myocardial infarction. *Nature Cell Biology*, **11** (1), 46–55.

Kreuger, J., Perez, L., Giraldez, A.J., and Cohen, S.M. (2004) Opposing activities of Dally-like glypican at high and low levels of Wingless morphogen activity. *Developmental Cell*, **7** (4), 503–512.

Kumar, S., Leontovich, A., Coenen, M.J., and Bahn, R.S. (2005) Gene expression profiling of orbital adipose tissue from patients with Graves' ophthalmopathy: a potential role for secreted frizzled-related protein-1 in orbital adipogenesis. *The Journal of Clinical Endocrinology and Metabolism*, **90** (8), 4730–4735.

Kuraku, S. and Kuratani, S. (2011) Genome-wide detection of gene extinction in early mammalian evolution. *Genome Biology and Evolution*, **3**, 1449–1462.

Lau, L.F. and Lam, S.C. (1999) The CCN family of angiogenic regulators: the integrin connection. *Experimental Cell Research*, **248** (1), 44–57.

Leclere, L. and Rentzsch, F. (2012) Repeated evolution of identical domain architecture in metazoan netrin domain-containing proteins. *Genome Biology and Evolution*, **4** (9), 883–899.

Lee, J.L., Lin, C.T., Chueh, L.L., and Chang, C.J. (2004) Autocrine/paracrine secreted Frizzled-related protein 2 induces cellular resistance to apoptosis: a possible mechanism of mammary tumorigenesis. *The Journal of Biological Chemistry*, **279** (15), 14602–14609.

Lee, H.X., Ambrosio, A.L., Reversade, B., and De Robertis, E.M. (2006) Embryonic dorsal-ventral signaling: secreted frizzled-related proteins as inhibitors of tolloid proteinases. *Cell*, **124** (1), 147–159.

Lee, Y.S., Lee, K.A., Yoon, H.B. *et al.* (2012) The Wnt inhibitor secreted Frizzled-related protein 1 (sFRP1) promotes human Th17 differentiation. *European Journal of Immunology*, **42** (10), 2564–2573.

Leyns, L., Bouwmeester, T., Kim, S.H. *et al.* (1997) Frzb-1 is a secreted antagonist of Wnt signaling expressed in the Spemann organizer. *Cell*, **88** (6), 747–756.

Li, C.H. and Amar, S. (2007) Inhibition of SFRP1 reduces severity of periodontitis. *Journal of Dental Research*, **86** (9), 873–877.

Li, X., Zhang, Y., Kang, H. *et al.* (2005) Sclerostin binds to LRP5/6 and antagonizes canonical Wnt signaling. *The Journal of Biological Chemistry*, **280** (20), 19883–19887.

Lin, K., Wang, S., Julius, M.A. *et al.* (1997) The cysteine-rich frizzled domain of Frzb-1 is required and sufficient for modulation of Wnt signaling. *Proceedings of the National Academy of Sciences of the United States of America*, **94** (21), 11196–11200.

Lopez-Rios, J., Esteve, P., Ruiz, J.M., and Bovolenta, P. (2008) The Netrin-related domain of Sfrp1 interacts with Wnt ligands and antagonizes their activity in the anterior neural plate. *Neural Development*, **3**, 19.

Macheda, M.L., Sun, W.W., Kugathasan, K. *et al.* (2012) The Wnt receptor Ryk plays a role in mammalian planar cell polarity signaling. *The Journal of Biological Chemistry*, **287** (35), 29312–29323.

Mahdi, T., Hanzelmann, S., Salehi, A. *et al.* (2012) Secreted frizzled-related protein 4 reduces insulin secretion and is overexpressed in type 2 diabetes. *Cell Metabolism*, **16** (5), 625–633.

Malinauskas, T., Aricescu, A.R., Lu, W. *et al.* (2011) Modular mechanism of Wnt signaling inhibition by Wnt inhibitory factor 1. *Nature Structural & Molecular Biology*, **18** (8), 886–893.

Mao, B., Wu, W., Li, Y. *et al.* (2001) LDL-receptor-related protein 6 is a receptor for Dickkopf proteins. *Nature*, **411** (6835), 321–325.

Mao, B., Wu, W., Davidson, G. *et al.* (2002) Kremen proteins are Dickkopf receptors that regulate Wnt/beta-catenin signalling. *Nature*, **417** (6889), 664–667.

Meitinger, T., Meindl, A., Bork, P. *et al.* (1993) Molecular modelling of the Norrie disease protein predicts a cystine knot growth factor tertiary structure. *Nature Genetics*, **5** (4), 376–380.

Melkonyan, H.S., Chang, W.C., Shapiro, J.P. *et al.* (1997) SARPs: a family of secreted apoptosis-related proteins. *Proceedings of the National Academy of Sciences of the United States of America*, **94** (25), 13636–13641.

Mercurio, S., Latinkic, B., Itasaki, N. *et al.* (2004) Connective-tissue growth factor modulates WNT signalling and interacts with the WNT receptor complex. *Development*, **131** (9), 2137–2147.

Mii, Y. and Taira, M. (2009) Secreted Frizzled-related proteins enhance the diffusion of Wnt ligands and expand their signalling range. *Development*, **136** (24), 4083–4088.

Mirotsou, M., Zhang, Z., Deb, A. *et al.* (2007) Secreted frizzled related protein 2 (Sfrp2) is the key Akt-mesenchymal stem cell-released paracrine factor mediating myocardial survival and repair. *Proceedings of the National Academy of Sciences of the United States of America*, **104** (5), 1643–1648.

Mulligan, K.A., Fuerer, C., Ching, W. *et al.* (2012) Secreted Wingless-interacting molecule (Swim) promotes long-range signaling by maintaining Wingless solubility. *Proceedings of the National Academy of Sciences of the United States of America*, **109** (2), 370–377.

Muraoka, O., Shimizu, T., Yabe, T. *et al.* (2006) Sizzled controls dorso-ventral polarity by repressing cleavage of the Chordin protein. *Nature Cell Biology*, **8** (4), 329–338.

Nagano, T., Takehara, S., Takahashi, M. *et al.* (2006) Shisa2 promotes the maturation of somitic precursors and transition to the segmental fate in Xenopus embryos. *Development*, **133** (23), 4643–4654.

Niehrs, C. (2006) Function and biological roles of the Dickkopf family of Wnt modulators. *Oncogene*, **25** (57), 7469–7481.

Ohnaka, K., Yamamoto, K., Nakamura, K. *et al.* (2009) Association of single nucleotide polymorphisms in secreted frizzled-related protein 1 gene with bone mineral density in Japanese women. *Geriatrics & Gerontology International*, **9** (3), 304–309.

Ohta, K., Lupo, G., Kuriyama, S. *et al.* (2004) Tsukushi functions as an organizer inducer by inhibition of BMP activity in cooperation with chordin. *Developmental Cell*, **7** (3), 347–358.

Ohta, K., Ito, A., Kuriyama, S. *et al.* (2011) Tsukushi functions as a Wnt signaling inhibitor by competing with Wnt2b for binding to transmembrane protein Frizzled4. *Proceedings of the National Academy of Sciences of the United States of America*, **108** (36), 14962–14967.

Ouchi, N., Higuchi, A., Ohashi, K. *et al.* (2010) Sfrp5 is an anti-inflammatory adipokine that modulates metabolic dysfunction in obesity. *Science*, **329** (5990), 454–457.

Pei, J. and Grishin, N.V. (2012) Unexpected diversity in Shisa-like proteins suggests the importance of their roles as transmembrane adaptors. *Cellular Signalling*, **24** (3), 758–769.

Petersen, C.P. and Reddien, P.W. (2011) Polarized notum activation at wounds inhibits Wnt function to promote planarian head regeneration. *Science*, **332** (6031), 852–855.

Piccolo, S., Agius, E., Leyns, L. *et al.* (1999) The head inducer Cerberus is a multifunctional antagonist of Nodal, BMP and Wnt signals. *Nature*, **397** (6721), 707–710.

Renstrom, J., Istvanffy, R., Gauthier, K. *et al.* (2009) Secreted frizzled-related protein 1 extrinsically regulates cycling activity and maintenance of hematopoietic stem cells. *Cell Stem Cell*, **5** (2), 157–167.

Rodriguez, J., Esteve, P., Weinl, C. *et al.* (2005) SFRP1 regulates the growth of retinal ganglion cell axons through the Fz2 receptor. *Nature Neuroscience*, **8** (10), 1301–1309.

Sagara, N., Kirikoshi, H., Terasaki, H. *et al.* (2001) FZD4S, a splicing variant of frizzled-4, encodes a soluble-type positive regulator of the WNT signaling pathway. *Biochemical and Biophysical Research Communications*, **282** (3), 750–756.

Satoh, W., Matsuyama, M., Takemura, H. *et al.* (2008) Sfrp1, Sfrp2, and Sfrp5 regulate the Wnt/beta-catenin and the planar cell polarity pathways during early trunk formation in mouse. *Genesis*, **46** (2), 92–103.

Scardigli, R., Gargioli, C., Tosoni, D. *et al.* (2008) Binding of sFRP-3 to EGF in the extra-cellular space affects proliferation, differentiation and morphogenetic events regulated by the two molecules. *PLoS One*, **3** (6), e2471.

Schaefer, L. and Iozzo, R.V. (2008) Biological functions of the small leucine-rich proteoglycans: from genetics to signal transduction. *The Journal of Biological Chemistry*, **283** (31), 21305–21309.

Segarini, P.R., Nesbitt, J.E., Li, D. *et al.* (2001) The low density lipoprotein receptor-related protein/alpha2-macroglobulin receptor is a receptor for connective tissue growth factor. *The Journal of Biological Chemistry*, **276** (44), 40659–40667.

Seib, D.R., Corsini, N.S., Ellwanger, K. *et al.* (2013) Loss of dickkopf-1 restores neurogenesis in old age and counteracts cognitive decline. *Cell Stem Cell*, **12** (2), 204–214.

Semenov, M., Tamai, K., and He, X. (2005) SOST is a ligand for LRP5/LRP6 and a Wnt signaling inhibitor. *The Journal of Biological Chemistry*, **280** (29), 26770–26775.

Semenov, M., Tamai, K., Brott, B.K. *et al.* (2001) Head inducer Dickkopf-1 is a ligand for Wnt coreceptor LRP6. *Current Biology*, **11** (12), 951–961.

Shawlot, W., Deng, J.M., and Behringer, R.R. (1998) Expression of the mouse cerberus-related gene, Cerr1, suggests a role in anterior neural induction and somitogenesis. *Proceedings of the National Academy of Sciences of the United States of America*, **95** (11), 6198–6203.

Shawlot, W., Deng, J., Wakamiya, M., and Behringer, R.R. (2000) The cerberus-related gene, Cerr1, is not essential for mouse head formation. *Genesis*, **26** (4), 253–258.

Silva, A.C., Filipe, M., Kuerner, K.M. *et al.* (2003) Endogenous Cerberus activity is required for anterior head specification in Xenopus. *Development*, **130** (20), 4943–4953.

Silverman, G.A., Bird, P.I., Carrell, R.W. *et al.* (2001) The serpins are an expanding superfamily of structurally similar but functionally diverse proteins. Evolution, mechanism of inhibition, novel functions, and a revised nomenclature. *The Journal of Biological Chemistry*, **276** (36), 33293–33296.

Simmons, D.G. and Kennedy, T.G. (2002) Uterine sensitization-associated gene-1: a novel gene induced within the rat endometrium at the time of uterine receptivity/sensitization for the decidual cell reaction. *Biology of Reproduction*, **67** (5), 1638–1645.

Sims, A.M., Shephard, N., Carter, K. *et al.* (2008) Genetic analyses in a sample of individuals with high or low BMD shows association with multiple Wnt pathway genes. *Journal of Bone and Mineral Research*, **23** (4), 499–506.

Smallwood, P.M., Williams, J., Xu, Q. *et al.* (2007) Mutational analysis of Norrin-Frizzled4 recognition. *The Journal of Biological Chemistry*, **282** (6), 4057–4068.

Surmann-Schmitt, C., Sasaki, T., Hattori, T. *et al.* (2012) The Wnt antagonist Wif-1 interacts with CTGF and inhibits CTGF activity. *Journal of Cellular Physiology*, **227** (5), 2207–2216.

Swain, R.K., Katoh, M., Medina, A., and Steinbeisser, H. (2005) Xenopus frizzled-4S, a splicing variant of Xfz4 is a context-dependent activator and inhibitor of Wnt/beta-catenin signaling. *Cell Communication and Signaling*, **3**, 12.

Takada, R., Satomi, Y., Kurata, T. *et al.* (2006) Monounsaturated fatty acid modification of Wnt protein: its role in Wnt secretion. *Developmental Cell*, **11** (6), 791–801.

Traister, A., Shi, W., and Filmus, J. (2008) Mammalian Notum induces the release of glypicans and other GPI-anchored proteins from the cell surface. *The Biochemical Journal*, **410** (3), 503–511.

Uren, A., Reichsman, F., Anest, V. *et al.* (2000) Secreted frizzled-related protein-1 binds directly to Wingless and is a biphasic modulator of Wnt signaling. *The Journal of Biological Chemistry*, **275** (6), 4374–4382.

Veeck, J. and Dahl, E. (2012) Targeting the Wnt pathway in cancer: the emerging role of Dickkopf-3. *Biochimica et Biophysica Acta*, **1825** (1), 18–28.

Vitt, U.A., Hsu, S.Y., and Hsueh, A.J. (2001) Evolution and classification of cystine knot-containing hormones and related extracellular signaling molecules. *Molecular Endocrinology*, **15** (5), 681–694.

Wang, S., Krinks, M., Lin, K. *et al.* (1997) Frzb, a secreted protein expressed in the Spemann organizer, binds and inhibits Wnt8. *Cell*, **88** (6), 757–766.

Wang, W.H., McNatt, L.G., Pang, I.H. *et al.* (2008) Increased expression of the WNT antagonist sFRP-1 in glaucoma elevates intraocular pressure. *The Journal of Clinical Investigation*, **118** (3), 1056–1064.

Wang, Y., Rattner, A., Zhou, Y. *et al.* (2012) Norrin/Frizzled4 signaling in retinal vascular development and blood brain barrier plasticity. *Cell*, **151** (6), 1332–1344.

Wawrzak, D., Metioui, M., Willems, E. *et al.* (2007) Wnt3a binds to several sFRPs in the nanomolar range. *Biochemical and Biophysical Research Communications*, **357** (4), 1119–1123.

Willert, K., Brown, J.D., Danenberg, E. *et al.* (2003) Wnt proteins are lipid-modified and can act as stem cell growth factors. *Nature*, **423** (6938), 448–452.

Wu, W., Glinka, A., Delius, H., and Niehrs, C. (2000) Mutual antagonism between dickkopf1 and dickkopf2 regulates Wnt/beta-catenin signalling. *Current Biology*, **10** (24), 1611–1614.

Xavier, C.P., Melikowa, M., Chuman, Y. *et al.* (2014) Secreted Frizzled-related protein potentiation versus inhibition of Wnt3a/β-catenin signaling. *Cell Signal.*, **26**, 94–101.

Xu, Q., Wang, Y., Dabdoub, A. *et al.* (2004) Vascular development in the retina and inner ear: control by Norrin and Frizzled-4, a high-affinity ligand-receptor pair. *Cell*, **116** (6), 883–895.

Yamamoto, A., Nagano, T., Takehara, S. *et al.* (2005) Shisa promotes head formation through the inhibition of receptor protein maturation for the caudalizing factors, Wnt and FGF. *Cell*, **120** (2), 223–235.

Ye, X., Wang, Y., Cahill, H. *et al.* (2009) Norrin, frizzled-4, and Lrp5 signaling in endothelial cells controls a genetic program for retinal vascularization. *Cell*, **139** (2), 285–298.

Yin, A., Korzh, V., and Gong, Z. (2012) Perturbation of zebrafish swimbladder development by enhancing Wnt signaling in Wif1 morphants. *Biochimica et Biophysica Acta*, **1823** (2), 236–244.

Zhang, B., Zhou, K.K., and Ma, J.X. (2010) Inhibition of connective tissue growth factor overexpression in diabetic retinopathy by SERPINA3K via blocking the WNT/beta-catenin pathway. *Diabetes*, **59** (7), 1809–1816.

Zhang, B., Abreu, J.G., Zhou, K. *et al.* (2010) Blocking the Wnt pathway, a unifying mechanism for an angiogenic inhibitor in the serine proteinase inhibitor family. *Proceedings of the National Academy of Sciences of the United States of America*, **107** (15), 6900–6905.

Zhu, W., Shiojima, I., Ito, Y. *et al.* (2008) IGFBP-4 is an inhibitor of canonical Wnt signalling required for cardiogenesis. *Nature*, **454** (7202), 345–349.

14 Frizzleds as G Protein-Coupled Receptors

Gunnar Schulte

Department of Physiology & Pharmacology, Section for Receptor Biology & Signaling, Karolinska Institutet, Stockholm, Sweden

Introduction

The 10 mammalian Frizzleds (FZD_{1-10}) are – together with Smoothened – classified as G protein-coupled receptors (GPCRs) in the GPCR database of the International Union of Basic and Clinical Pharmacology (IUPHAR). FZDs mediate cellular effects of the WNT family of lipoglycoproteins, whereas Smoothened relays hedgehog signaling. Based on the seven transmembrane-spanning (7TM) architecture of the FZDs and both structural and functional peculiarities, these receptors were grouped separately in the Class Frizzled of the GPCR superfamily (Foord *et al.*, 2005; Schulte, 2010). In this chapter, I will focus on the molecular pharmacology of FZDs regarding ligand-receptor interaction and receptor function and hallmarks of their GPCR nature and G protein-dependent and G protein-independent pathways. The aim is to summarize our current understanding of the mechanisms of FZD signaling through heterotrimeric G proteins and to strengthen the idea that Class Frizzled receptors indeed are GPCRs.

Heterotrimeric G proteins are composed of a GTP-binding α-subunit and βγ-subunits and are grouped according to the α-subunit's characteristics into four subgroups: Gαi/o, Gαs, Gαq/11, and Gα12/13 proteins, with several representatives in each subgroup (Gilman, 1987). Gαi/o proteins (with the exception of Gαz) are ADP ribosylated and thereby inhibited by the bacterial toxin from *Bordetella pertussis* (pertussis toxin, PTX), whereas Gαs proteins are affected by a toxin from *Vibrio cholerae* (cholera toxin, CTX). Investigation of Gαq/11 and Gα12/13 is more difficult since no pharmacological tools for their inhibition are readily available. Downstream of the different G protein subfamilies, distinct signaling pathways are typically activated (Dorsam and Gutkind, 2007). For example, Gαi/o and Gαs decrease or increase the production of cAMP, respectively. Gαq/11 mainly couples to phospholipase C resulting in IP_3 and diacylglycerol formation, and Gα12/13 is implicated in the regulation of the cytoskeleton through communication with small GTPases. GPCRs typically act as guanine nucleotide exchange factors at heterotrimeric G proteins, and agonist binding to the receptor promotes exchange of GDP in the inactive α-subunit to GTP, which results in the dissociation of the trimer and signal activation (Oldham and Hamm, 2008).

Wnt Signaling in Development and Disease: Molecular Mechanisms and Biological Functions, First Edition.
Edited by Stefan Hoppler and Randall T. Moon.

From the moment the first *frizzled* gene was cloned from *Drosophila melanogaster* and the 7TM architecture of the resulting protein became obvious, the idea was born that the gene product could not only be seen as a 7TM receptor but that it in fact might be a GPCR (Vinson, Conover, and Adler, 1989). Sequence similarities initially suggested close relationship to secretin receptors (Barnes, Duckworth, and Beeley, 1998). Architectural resemblance, however, does not necessarily predict functional similarity, raising the question about whether FZDs indeed signal through and directly associate with heterotrimeric G proteins.

In the late 1990s, the first evidence based on functional approaches employing different means of inhibiting heterotrimeric G proteins supported the GPCR nature of FZDs (Liu *et al.*, 1999b; Sheldahl *et al.*, 1999; Slusarski, Corces, and Moon, 1997; Slusarski *et al.*, 1997). However, the direct coupling of FZDs to heterotrimeric G proteins remains an issue of discussion despite more recent and direct proof of WNT-induced and FZD-mediated activation of heterotrimeric G proteins and their involvement in both classical WNT signaling and separate pathways (Malbon, 2011). Remarkably, underlying mechanisms of receptor–G protein selectivity, coupling mode, and stoichiometry still remain obscure.

Frizzled structure and molecular pharmacology

The far N-terminus of the FZD sequence presents a signal sequence important for proper membrane embedding during protein synthesis at the rough endoplasmic reticulum. This short stretch of amino acids is followed by a highly conserved cysteine-rich domain (CRD) comprising 10 cysteines engaging in five disulfide bonds. The structure of the CRD has been solved, and structure analysis also supports the functional data indicating that the CRD of FZDs indeed binds ligands of the WNT family of lipoglycoproteins (Dann *et al.*, 2001; Janda *et al.*, 2012). Despite the functional and structural evidence that the FZD-CRD interacts with WNT (Bhanot *et al.*, 1996; Janda *et al.*, 2012; Povelones and Nusse, 2005; Wang *et al.*, 1996), additional data suggest that the FZD-CRD might not be required for transmission of WNT signals (Chen *et al.*, 2004), a conundrum, which still awaits clarification. A possible explanation could be that the CRD/WNT interaction precedes the formation of a high-affinity interaction with other regions of the receptor, which then leads to conformational changes and finally receptor activation. In this case, it would be disputable if the CRD could be seen as the orthosteric site of WNT binding to FZDs.

Comparable to all 7TM molecules, the FZDs consist of seven transmembrane domains, three extracellular and three intracellular loops and a C-terminal region of various lengths (Schulte, 2010). Even though sequence homology among the Class FZD is high, four clusters of FZDs have emerged from sequence comparisons grouping $FZD_{1,2,7}$, $FZD_{3,6}$, $FZD_{4,9,10}$, and $FZD_{5,8}$ together.

On the intracellular face of the receptor, there are three loops (i1, i2, i3) and a C-terminal domain of varying length, which have important functional implications for association of FZD with proteins such as the phosphoprotein Dishevelled (DVL) and heterotrimeric G proteins. DVL interacts with FZDs primarily through the conserved KTxxxW sequence in the C-terminus close to the seventh helix of the receptor core (Umbhauer *et al.*, 2000) (Chapter 2). In addition, the flanking regions of i3 are crucial for DVL interaction and its FZD-induced membrane recruitment (Tauriello *et al.*, 2012). These interaction sites between the phosphoprotein DVL and FZDs are functionally important since they overlap substantially with domains involved in G protein coupling described in conventional GPCRs (Rasmussen *et al.*, 2011; Wess, 1998).

Frizzled dimerization

Even though monomeric GPCRs are functional units (Whorton *et al.*, 2007), homo- and hetero-oligomerization has functional relevance for many GPCRs (Milligan, 2009). FZDs are also known to form receptor dimers even though the functional importance is still unclear. In the case of FZD_4, dimerization appears to be crucial for receptor maturation and cell surface exposure (Kaykas *et al.*, 2004). The interaction between FZDs could be mediated either by CRD–CRD interaction or by receptor core interaction mediated by transmembrane helices (Dann *et al.*, 2001; Parker *et al.*, 2011). So far, it is unclear if FZD dimerization is required for WNT-induced signaling or if it is important for signal specification, for example, toward heterotrimeric G proteins. Xenopus FZD_3 apparently dimerizes and mediates thereby activation of the WNT/β-catenin pathway, whereas monomeric FZD_7 neither dimerizes nor activates the WNT/β-catenin pathway. On the other hand, forced dimerization of FZD_7 contributes to receptor-mediated WNT/β-catenin signaling (Carron *et al.*, 2003).

Activating ligand of Frizzleds

All 10 FZDs but not Smoothened are thought to be activated by WNTs (Willert and Nusse, 2012). In mammals, 19 different WNTs are known, but only few are available in purified and active form. So far, the interaction profile between the 10 mammalian FZDs and the 19 WNTs is basically unknown. Extensive mapping of binding affinities between *D. melanogaster* WNTs and FZDs, however, suggests a distinct binding profile and selectivity also for the mammalian WNT/FZD system (Wu and Nusse, 2002). Due to the nature of the WNTs as lipophilic glycoproteins, it is inherently difficult to establish classical ligand binding assays providing pharmacological binding data. Several labs have on the other hand used isolated CRDs for interaction studies suggesting that FZD-CRD/WNT binding has a dissociation constant in the lower nanomolar range (Hsieh *et al.*, 1999; Rulifson, Wu, and Nusse, 2000). As pointed out earlier (Schulte and Bryja, 2007), a sensitive binding assay could be a powerful tool to prove the GPCR nature of FZDs, since the agonist affinity shifts with the activation-dependent G protein association and dissociation.

In addition to WNTs, other proteins are also known to bind and activate FZDs. The most prominent and a structurally unrelated example is Norrin, which selectively interacts with and signals through FZD_4 (Smallwood *et al.*, 2007; Xu *et al.*, 2004). Mutations of Norrin are associated with Norrin disease and familial exudative vitreoretinopathy (FEVR) both of which are retinal vascular diseases based on dysfunctional Norrin/FZD_4/LRP6 communication (Ye, Wang, and Nathans, 2010). Secreted Frizzled-related proteins (SFRPs) act by sequestering WNTs to prevent their interaction with FZDs. Also, those soluble FZD-CRD homologs can interact with FZDs probably through CRD–CRD dimerization (Bovolenta *et al.*, 2008; Dann *et al.*, 2001). The functional relevance of this interaction remains still obscure, but it could possibly lead to receptor activation (Uren *et al.*, 2000).

To date, small-molecule compounds that could serve as agonists, antagonist, or allosteric modulators of FZDs have not been described. However, short WNT-derived peptides were reported to mimic (FOXY-5) and have antagonistic effects (BOX5, UM206) with a certain degree of receptor selectivity (Jenei *et al.*, 2009; Laeremans *et al.*, 2011; Säfholm *et al.*, 2006). In addition to their potential therapeutic value, these and similar peptides may serve very useful to dissect underlying mechanisms of FZD–ligand interaction and FZD–effector coupling.

Ligand-receptor selectivity and possibilities for ligand trafficking

Conceptionally, it is feasible to imagine that the complexity of the WNT/FZD signaling system has developed to allow fine-tuning of signal transduction pathway selection through certain WNT/FZD combinations. Thus, distinct WNT/FZD pairings should show high affinity to each other in order to define selective signaling trafficking. Mapping WNT/FZD selectivity and relating complex affinity to signaling outcome and efficacy are some of the challenges in the WNT field. Problems arise from (i) difficulties to obtain WNTs in purified and active form, (ii) the WNT's intrinsic high affinity for membranes and extracellular matrix components, (iii) the lack of a suitable cell system allowing the analysis of individual FZD isoforms, and (iv) the lack of pharmacologically active FZD-targeting small-molecule compounds.

A recent study from my laboratory employing a microglia-like cell line that predominantly, but not solely, expresses FZD_5 showed that different commercially available WNTs have very distinct signaling profiles (Kilander, Halleskog, and Schulte, 2011). In order to visualize the signaling profiles of WNT3A, WNT4, WNT5A, WNT5B, WNT7A, and WNT9B with regard to the stimulation of LRP6 phosphorylation,

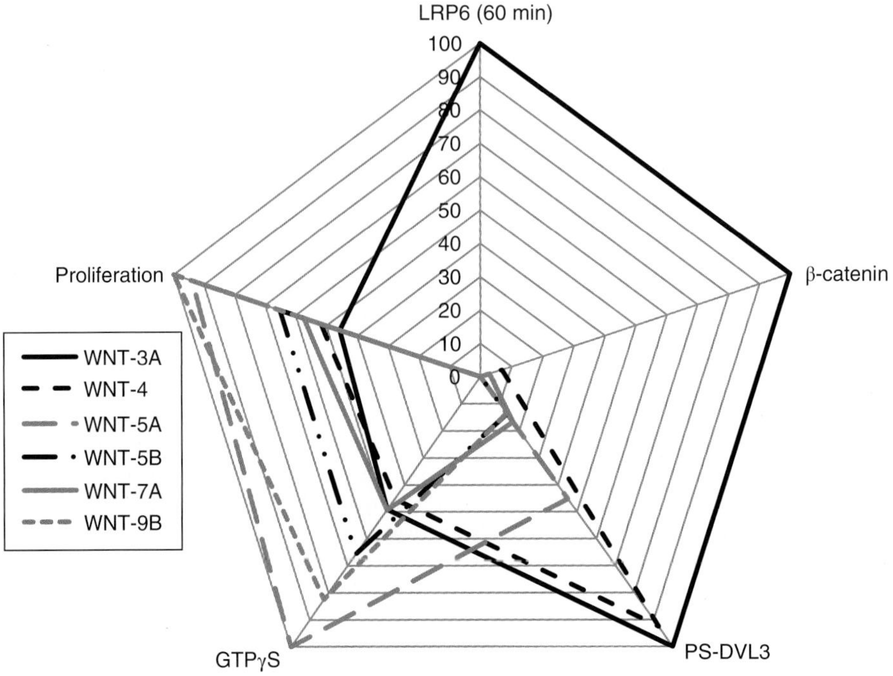

Figure 14.1 Signaling profiles of various WNTs in the microglia-like cell line N13. WNTs were tested for their capacity to induce LRP6 phosphorylation, β-catenin stabilization, G protein activation (GTPγS binding), and cellular proliferation in N13 cells (Kilander, Halleskog, and Schulte, 2011b). The mean values from published data for WNT3A, WNT4, WNT5A, WNT5B, WNT7A, and WNT9B were summarized in the presented radar plot. Maximal response in each individual assay was normalized to 100%. Center point represents 0% activation. Differences in size and distribution of the areas encircled by the connecting lines in the radar plot visualize differential efficacy of WNTs in N13 cells, which predominantly express FZD_5. Note that only WNT3A activated the WNT/β-catenin pathway represented by LRP6 phosphorylation and β-catenin stabilization. All WNTs activate heterotrimeric G proteins (GTPγS) and shift DVL in an immunoblot assay (phosphorylated and shifted DVL: PS-DVL) with varying efficacies.

β-catenin stabilization, formation of the phosphorylated and shifted form of DVL, activation of heterotrimeric G proteins, and cell proliferation, published data were summarized in a radar plot (Figure 14.1). Note the various sizes and shapes of the areas encompassed by the connecting lines in the radar graph for the individual WNTs. These data (i) argue for a general capacity of WNTs to activate heterotrimeric G proteins and (ii) clearly pinpoint functional selectivity of WNTs in cells with a given FZD expression pattern. Another example for the WNT's functional selectivity is the ability of WNT5A, generally associated with β-catenin-independent signaling, to activate the WNT/β-catenin pathway when FZD_4 is highly expressed (Mikels and Nusse, 2006).

Differences in efficacy could primarily be dictated by varying ligand-receptor affinity. In other words, functional selectivity of WNTs could be achieved through stabilization of pathway-specific FZD conformations. Recruitment of coreceptors appears to play an important role in pathway selectivity as, for example, the WNT/β-catenin signaling activated by WNT3A depends on concurrent activation of LRP5/6.

The role of coreceptors for FZD signaling and specification

For WNT/FZD communication, several single transmembrane-spanning receptors were described as coreceptors of WNTs (see also

Chapters 2 and 7). Most importantly, LRP5/6 is intrinsically connected to the WNT/β-catenin pathway. WNT-induced recruitment of LRP5/6 to FZD initiates the formation of the so-called signalosomes comprised of the WNT/FZD/LRP5/6 complex as well as DVL and axin (Bilic *et al.*, 2007). In addition to LRP5/6, other single transmembrane receptors related to receptor tyrosine kinases, such as ROR1/2, RYK, and PTK7, are suggested as coreceptors even though they also might act independent of FZDs (Hendrickx and Leyns, 2008). With the exception of LRP6, these receptors have, however, not been related to heterotrimeric G protein signaling. Surprisingly, it was shown recently that functional interaction of LRP6 and βγ is at the center point of cross talk between the Gαo/i/q-coupled GPCRs and the stabilization of β-catenin (Jernigan *et al.*, 2010).

Doubtful liaison (or just a bit different?): FZDs and heterotrimeric G proteins

The WNT/FZD signaling network steadily grows in complexity and comprises the WNT/β-catenin, WNT/RAC and WNT/RHO, WNT/Ca^{2+}, WNT/cAMP, and WNT/ROR pathway as well as planar cell polarity signaling (Schulte, 2010) as the main routes for intracellular communication. Most of these pathways involve DVL and have also been associated with heterotrimeric G proteins, an interaction, which is mechanistically not fully understood yet. The mechanistic difficulty compared to conventional GPCR–G protein coupling is that FZDs accommodate interaction to at least two signaling compounds: Direct interaction with DVL (see Chapter 15) and heterotrimeric G proteins seems to be required to accomplish activation of the WNT/FZD signaling network. Experimental evidence clearly supports the FZD-mediated activation of heterotrimeric G proteins in overexpressing cell systems, tissues, primary cells, and cell lines with physiological receptor–G protein stoichiometry (Halleskog *et al.*, 2012; Katanaev and Buestorf, 2009; Kilander, Halleskog, and Schulte, 2011; Kilander *et al.*, 2011; Koval and Katanaev, 2011). However, most of these studies have not addressed the role and relative location of DVL systematically. Symptomatic of the general difficulties to dissect the FZD-induced G protein-dependent signaling pathways, heterotrimeric G proteins have been localized both upstream and downstream of DVL without being able to pinpoint exact mechanistic details (Bikkavilli, Feigin, and Malbon, 2008; Katanaev *et al.*, 2005). Indeed, DVL is known to interact with free βγ-subunits resulting in DVL membrane recruitment and degradation (Angers *et al.*, 2006; Jung *et al.*, 2009). It remains unclear, however, how DVL–βγ interaction translates into WNT/FZD-mediated signal transduction. Thus, signal integration through FZD, DVL, and heterotrimeric G proteins is still poorly understood and requires further investigations.

FZD–G protein signaling

G protein-dependent signaling to calcium

The first functional indications for the involvement of heterotrimeric G proteins in WNT/FZD communication were obtained in *Xenopus laevis* and *Danio rerio* embryos (Sheldahl *et al.*, 1999; Slusarski, Corces, and Moon, 1997; Slusarski *et al.*, 1997). Those studies argue for the involvement of PTX-sensitive Gαi/o proteins signaling through the release of βγ-subunits, the activation of phospholipase C, and the release of calcium from intracellular stores to activation of calcium-dependent protein kinases. This pathway resembles what is known for classical GPCRs coupling to Gαi/o protein (Dorsam and Gutkind, 2007) and was early on coined the WNT/Ca^{2+} pathway (Kühl *et al.*, 2000). Several other studies have indeed found a WNT-induced and PTX-sensitive induction of $[Ca^{2+}]_i$ in various cell systems ranging from diverse cancer cells to primary microglia (Dejmek *et al.*, 2006; Halleskog *et al.*, 2012; Liu *et al.*, 1999b). Additionally, another G protein-dependent route from FZD to calcium was identified in pluripotent F9 teratocarcinoma cells; this pathway was mediated by PTX-sensitive G proteins and was sensitive to inhibition of cGMP-selective phosphodiesterases (Ahumada *et al.*, 2002). Since calcium is a central mediator of cellular signaling with a plethora of calcium-dependent downstream targets, it is not surprising that calcium-dependent protein kinase (PKC), calcium-/calmodulin-dependent protein kinase II (CamKII), extracellular signal-regulated kinases1/2 (ERK1/2), and nuclear factor of activated T cells (NFAT) were reported as downstream targets of WNT-induced and G protein-dependent signaling (Dejmek *et al.*, 2006; Halleskog and Schulte, 2013; Halleskog *et al.*, 2012; Kohn and Moon, 2005; Sheldahl *et al.*, 2003).

Regulation of adenylyl cyclase

Since the early days of receptor pharmacology, it was recognized that the second messenger cyclic AMP (cAMP) is a central signaling molecule that is rapidly produced from ATP by adenylyl cyclases (Sutherland and Robison, 1966). The early evidence of FZD-mediated activation of PTX-sensitive Gαi/o proteins suggested a direct communication from FZDs to decrease production of cAMP through inhibition of adenylyl cyclases (Birnbaumer, Abramowitz, and Brown, 1990). However, it took over 10 years that activation of this pathway could be measured upon WNT stimulation in mammalian cells. WNT5A stimulation decreased the forskolin-induced levels of $[cAMP]_i$ in mouse primary microglia in a dose-dependent manner (Halleskog *et al.*, 2012). On the other hand, any FZD-mediated activation of stimulatory, CTX-sensitive Gαs proteins would be expected to elevate $[cAMP]_i$ and thereby activate cAMP-dependent protein kinase (PKA). Indeed, WNT5A is capable of inducing cAMP production through FZD_3 and eliciting cAMP/PKA-dependent antimigratory effects in breast cancer cells (Hansen *et al.*, 2009). The cAMP/PKA pathway seems also to mediate antiapoptotic effects in dermal fibroblasts and to induce myogenic effects in mammalian development (Chen, Ginty, and Fan, 2005; Torii *et al.*, 2008).

Heterotrimeric G proteins, WNT/β-catenin pathway, and WNT/RAC signaling

Historically, the first signaling pathways that were discovered to be activated by WNTs were WNT/β-catenin signaling and pathways regulating convergent extension movements and planar cell polarity, such as the WNT/RHO or WNT/RAC pathways (Klaus and Birchmeier, 2008; Veeman, Axelrod, and Moon, 2003). Later, the β-catenin-independent WNT/Ca^{2+} pathway was identified as the first WNT signaling branch being dependent on heterotrimeric G proteins. This signaling branch appeared crucial for the WNT-mediated regulation of convergence extension movements in vertebrates (Moon *et al.*, 1993). During recent years, interesting molecular details of the importance of heterotrimeric G proteins in the pathways that were originally perceived as being G protein-independent were revealed. Thus, the original picture of a clear distinction between G protein-dependent and G protein-independent signaling has changed dramatically. As summarized in Table 14.1, several components of the WNT/β-catenin pathway, for example, the phosphoprotein DVL – important

Table 14.1 WNT pathway compounds interacting with heterotrimeric G proteins

WNT pathway component	Interaction with (by immunoprecipitation)	Functional relevance	References
FZD	GDP-Gαo/I	Signaling	Katanaev and Buestorf (2009), Liu, Rubin, and Kimmel (2005)
LRP5/6	βγ	Signaling	Jernigan *et al.* (2010)
DVL	βγ	Membrane recruitment, DVL degradation	Angers *et al.* (2006), Egger-Adam and Katanaev (2009), Jernigan *et al.* (2010), Jung *et al.* (2009)
DVL	Gαo	Signaling	Liu, Rubin, and Kimmel (2005)
Axin	GTP-Gαo	Membrane recruitment	Egger-Adam and Katanaev (2009), Jernigan *et al.* (2010)
Axin	Gα12	Signaling	Stemmle, Fields, and Casey (2006)
GSK3	βγ	Membrane recruitment	Jernigan *et al.* (2010)

for both β-catenin-dependent and β-catenin-independent signaling – associate with, are spatially restricted, or are actively modulated by heterotrimeric G proteins.

The involvement of heterotrimeric G proteins in the WNT/β-catenin pathway remains still a matter of debate (Malbon, 2011), even though the following experimental evidence strongly argues for an involvement of heterotrimeric G proteins in both cellular and *in vivo* assays. Epistatic analysis of Gαo function in *D. melanogaster* indicated that the heterotrimeric G protein is directly downstream of the receptor and upstream of DVL regulating both WNT/β-catenin and PCP signaling (Katanaev *et al.*, 2005). Additionally, experiments in mammalian cells underline the importance of Gαi/o and Gαq proteins for WNT-induced communication to β-catenin in loss-of-function approaches (Liu *et al.*, 2001). Further, coimmunoprecipitation experiments support the dynamic, agonist-dependent interaction between FZD, Gαi/o, and DVL with functional relevance for axin/GSK3 dissociation and thereby cytosolic β-catenin stabilization (Liu, Rubin, and Kimmel, 2005). Interestingly, recent findings obtained in mouse primary microglia indicate that WNT3A-induced LRP6 phosphorylation, DVL electrophoretic mobility changes, β-catenin stabilization, and ERK1/2 phosphorylation are sensitive to Gαi/o blockade through PTX. On the other hand, the βγ-inhibitor M119 only inhibited the WNT3A-induced signal to ERK1/2, arguing that the pathway bifurcates downstream of the heterotrimeric G proteins into an α- and βγ-subunit-mediated branch (Halleskog and Schulte, 2013).

Also, it was shown in *D. melanogaster* that Gαo has dual function in the WNT/β-catenin pathway: first, Gαo binds axin, which is followed by axin recruitment to the membrane and consequently the release of the inhibitory input of axin on β-catenin. Furthermore, the βγ-subunits released from Gαo serve to constrain DVL to the membrane where it can further interact with axin in an inhibitory manner (Egger-Adam and Katanaev, 2009). In addition to PTX-sensitive Gαi/o proteins, also PTX-insensitive Gαq/11 proteins participate in WNT/β-catenin signaling through the WNT3A-evoked and PLC-mediated formation of inositol pentakisphosphate (IP_5) (Gao and Wang, 2007).

Despite recent progress, the underlying mechanisms of WNT-induced receptor–G protein coupling, the factors that determine the FZD–G protein selectivity, signaling integration, and structure of receptor–G protein complexes, remain obscure. Strikingly, several non-FZD GPCRs were recently shown to activate WNT/β-catenin signaling in different cells, which adds a level of complexity (Shevtsov, Haq, and Force, 2006). Activation of prostaglandin PGE_2 and lysophosphatidic acid receptors leads to β-catenin stabilization through diverse mechanisms: Gαs-mediated axin recruitment and Gαq/11- or Gα12/13-dependent phosphatidylinositol-3′-kinase- or PKC-dependent mechanisms contribute to β-catenin stabilization. In addition, the parathyroid hormone receptor is able to interact with DVL directly through a KSxxxW in its C-terminus (Romero *et al.*, 2010; Shevtsov, Haq, and Force, 2006).

Biological relevance of WNT-induced G protein signaling

Evidence that the Class Frizzled receptors function as *bona fide* GPCRs emerges and it is supported by direct biochemical and pharmacological data. In parallel, it becomes obvious that FZD–G protein signaling has also highly relevant physiological and pathophysiological implications in organisms, such as *X. laevis* or *D. melanogaster*, as well as in mammals. The FZD–G protein liaison is not only crucial in models for embryonic development such as *D. melanogaster*, where it is involved, for example, in wing development and asymmetric cell division (Katanaev and Tomlinson, 2006; Katanaev *et al.*, 2005), but turns out to be important also during mammalian development as well as in the adult.

WNTs are strongly expressed in the central nervous system even though the subcellular distribution and functional implications might not always be clear. In addition, they have been implicated in many different neurological diseases (De Ferrari and Moon, 2006; Inestrosa and Arenas, 2010) (Chapters 29–31). Therefore, it was important to realize that purified WNT3A exerts a strong activation of PTX-sensitive heterotrimeric G proteins in membranes prepared from the brains from 5-day-old rats (Koval and Katanaev, 2011). Due to the complexity of the membrane preparation, it is however not possible to draw conclusions on either cell type or activated FZD in these experiments. One important cell type in the CNS, the microglia, reacts to WNT stimulation with the activation of heterotrimeric G proteins (Halleskog and Schulte, 2013; Halleskog *et al.*, 2012; Kilander, Halleskog, and Schulte, 2011). WNT5A, for example, activates a Gαi/o, phospholipase C, calcium-dependent protein kinase, and extracellular signal-regulated kinase 1/2 signaling axis to modulate microglia activity. Among the WNT-induced and G protein-dependent changes in microglia are activation of ERK1/2, cyclooxygenase 2 and metalloprotease expression, invasion, and proliferation – cornerstones of the microglia's proinflammatory transformation (Halleskog and Schulte, 2013; Halleskog *et al.*, 2012). Thus, with regard to the important function of microglia to maintain tissue homeostasis and to react to brain injury, WNTs present a novel, previously unappreciated modulator of the CNS immune response (Hanisch and Kettenmann, 2007).

Further, the regulation of stem cells is an important target of WNT signaling (Kühl and Kühl, 2012). In fact, several instrumental findings on the FZD–G protein connection were obtained using pluripotent F9 teratocarcinoma cells (Bikkavilli, Feigin, and Malbon, 2008; Liu *et al.*, 1999a, b, 2001), a well-established cellular stem cell model (Alonso *et al.*, 1991) arguing for a general role of this pathway in stem cell differentiation. This assumption is further supported by data reporting a Gαq/11-dependent PKC-mediated pathway regulating WNT-induced bone formation both *in vitro* and *in vivo* (Tu *et al.*, 2007).

The versatility of WNT–G protein signaling becomes especially obvious regarding the WNT–Gαs–cAMP pathway, which was shown to exert antimigratory effects in breast cancer cells, antiapoptotic effects in fibroblasts, and myogenic input. Surely, these phenomena are of physiological and putative therapeutic interest (Chen, Ginty, and Fan, 2005; Hansen *et al.*, 2009; Torii *et al.*, 2008).

In summary, the FZD–G protein connection is not only physiologically relevant and important; it has also been implicated as a novel possibility in the quest for new drugs targeting FZDs (Koval and Katanaev, 2012). Several well-established high-throughput-compatible platforms could complement the currently available, biased, and very limited array of screening methods, for example, the TOPflash technology. Facing the recent data (as presented in Figure 14.1) on the wide ability of WNTs to mediate G protein signaling rather than WNT/β-catenin signaling, alternative screening methods might be considered: measurement of WNT-induced activation of heterotrimeric G proteins by GTP-binding assays and measurement of cAMP production and calcium fluxes could provide inexpensive and effective tools for discovery of FZD-targeting compounds.

Conclusions and future challenges

Even though our understanding of FZDs as GPCRs has improved, many questions remain to be answered. Detailed understanding of the functional selectivity of WNTs and the FZD–G protein interaction will provide a better understanding of WNT signaling and might offer novel targets for therapy in human. As depicted in Figure 14.2, several (hypothetic) WNT

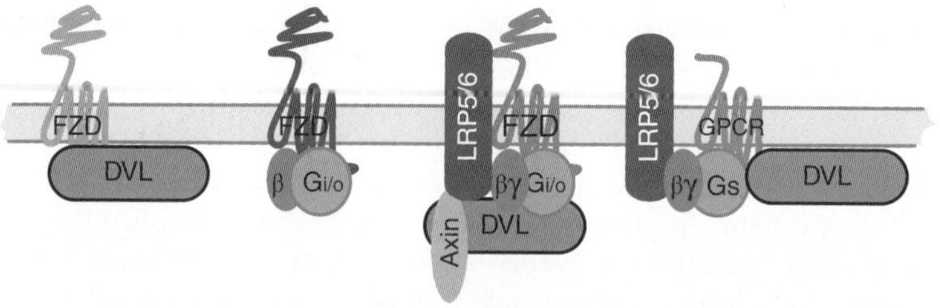

Figure 14.2 Schematic presentation of putative Frizzled complexes in association with single transmembrane-spanning receptors, DVL, and heterotrimeric G proteins. Complex composition determined by the agonists or expression levels of the involved components might be important to drive signaling selectivity. Presented complexes are hypothetic and schematic.

receptor complexes are supported by current data. However, structural details of FZD–G protein receptor complexes remain to be clarified to understand signaling specification, integration, and also cross talk with other cellular communication systems.

Acknowledgments

Work in the Schulte laboratory is supported by Karolinska Institutet, the Swedish Research Council (K2008-68P-20810-01-4, K2008-333 68X-20805-01-4, K2012-67X-20805-05-3), the Swedish Cancer Society (CAN 2008/539, 2011/690), and the Knut and Alice Wallenberg Foundation (KAW2008.0149).

References

Ahumada, A., Slusarski, D.C., Liu, X. *et al.* (2002) Signaling of rat Frizzled-2 through phosphodiesterase and cyclic GMP. *Science*, **298** (5600), 2006–2010.

Alonso, A., Breuer, B., Steuer, B., and Fischer, J. (1991) The F9-EC cell line as a model for the analysis of differentiation. *The International Journal of Developmental Biology*, **35** (4), 389–397.

Angers, S., Thorpe, C.J., Biechele, T.L. *et al.* (2006) The KLHL12-Cullin-3 ubiquitin ligase negatively regulates the Wnt-beta-catenin pathway by targeting Dishevelled for degradation. *Nature Cell Biology*, **8** (4), 348–357.

Barnes, M.R., Duckworth, D.M., and Beeley, L.J. (1998) Frizzled proteins constitute a novel family of G protein-coupled receptors, most closely related to the secretin family. *Trends in Pharmacological Sciences*, **19** (10), 399–400.

Bhanot, P., Brink, M., Samos, C.H. *et al.* (1996) A new member of the frizzled family from Drosophila functions as a Wingless receptor. *Nature*, **382** (6588), 225–230.

Bikkavilli, R.K., Feigin, M.E., and Malbon, C.C. (2008) G alpha o mediates WNT-JNK signaling through dishevelled 1 and 3, RhoA family members, and MEKK 1 and 4 in mammalian cells. *Journal of Cell Science*, **121** (PT 2), 234–245.

Bilic, J., Huang, Y.L., Davidson, G. *et al.* (2007) Wnt induces LRP6 signalosomes and promotes dishevelled-dependent LRP6 phosphorylation. *Science*, **316** (5831), 1619–1622.

Birnbaumer, L., Abramowitz, J., and Brown, A.M. (1990) Receptor-effector coupling by G proteins. *Biochimica et Biophysica Acta*, **1031** (2), 163–224.

Bovolenta, P., Esteve, P., Ruiz, J.M. *et al.* (2008) Beyond Wnt inhibition: new functions of secreted Frizzled-related proteins in development and disease. *Journal of Cell Science*, **121** (PT 6), 737–746.

Carron, C., Pascal, A., Djiane, A. *et al.* (2003) Frizzled receptor dimerization is sufficient to activate the Wnt/beta-catenin pathway. *Journal of Cell Science*, **116** (PT 12), 2541–2550.

Chen, A.E., Ginty, D.D., and Fan, C.M. (2005) Protein kinase A signalling via CREB controls myogenesis induced by Wnt proteins. *Nature*, **433** (7023), 317–322.

Chen, C.M., Strapps, W., Tomlinson, A., and Struhl, G. (2004) Evidence that the cysteine-rich domain of Drosophila Frizzled family receptors is dispensable for transducing Wingless. *Proceedings of the National Academy of Sciences of the United States of America*, **101** (45), 15961–15966.

Dann, C.E., Hsieh, J.C., Rattner, A. *et al.* (2001) Insights into Wnt binding and signalling from the structures of two Frizzled cysteine-rich domains. *Nature*, **412** (6842), 86–90.

De Ferrari, G.V. and Moon, R.T. (2006) The ups and downs of Wnt signaling in prevalent neurological disorders. *Oncogene*, **25** (57), 7545–7553.

Dejmek, J., Säfholm, A., Kamp Nielsen, C. *et al.* (2006) Wnt-5a/Ca^{2+}-induced NFAT activity is counteracted by Wnt-5a/Yes-Cdc42-casein kinase 1alpha signaling in human mammary epithelial cells. *Molecular and Cell Biology*, **26** (16), 6024–6036.

Dorsam, R.T. and Gutkind, J.S. (2007) G-protein-coupled receptors and cancer. *Nature Reviews Cancer*, **7** (2), 79–94.

Egger-Adam, D. and Katanaev, V.L. (2009) The trimeric G protein Go inflicts a double impact on axin in the Wnt/frizzled signaling pathway. *Developmental Dynamics*, **239** (1), 168–183.

Foord, S.M., Bonner, T.I., Neubig, R.R. *et al.* (2005) International Union of Pharmacology. XLVI. G protein-coupled receptor list. *Pharmacological Reviews*, **57** (2), 279–288.

Gao, Y. and Wang, H.Y. (2007) Inositol pentakisphosphate mediates Wnt/beta-catenin signaling. *The Journal of Biological Chemistry*, **282** (36), 26490–26502.

Gilman, A.G. (1987) G proteins: transducers of receptor-generated signals. *Annual Review of Biochemistry*, **56**, 615–649.

Halleskog, C. and Schulte, G. (2013) Pertussis toxin-sensitive heterotrimeric G(αi/o) proteins mediate WNT/β-catenin and WNT/ERK1/2 signaling in mouse primary microglia stimulated with purified WNT-3A. *Cell Signaling*, **25** (4), 822–828.

Halleskog, C., Dijksterhuis, J.P., Kilander, M.B. *et al.* (2012) Heterotrimeric G protein-dependent WNT-5A signaling to ERK1/2 mediates distinct aspects of microglia proinflammatory transformation. *Journal of Neuroinflammation*, **9** (1), 111.

Hanisch, U.K. and Kettenmann, H. (2007) Microglia: active sensor and versatile effector cells in the normal and pathologic brain. *Nature Neuroscience*, **10** (11), 1387–1394.

Hansen, C., Howlin, J., Tengholm, A. *et al.* (2009) Wnt-5a-induced phosphorylation of DARPP-32 inhibits breast cancer cell migration in a CREB-dependent manner. *The Journal of Biological Chemistry*, **284** (40), 27533–27543.

Hendrickx, M. and Leyns, L. (2008) Non-conventional Frizzled ligands and Wnt receptors. *Development, Growth & Differentiation*, **50** (4), 229–243.

Hsieh, J.C., Rattner, A., Smallwood, P.M., and Nathans, J. (1999) Biochemical characterization of Wnt-Frizzled interactions using a soluble, biologically active vertebrate Wnt protein. *Proceedings of the National Academy of Sciences of the United States of America*, **96** (7), 3546–3551.

Inestrosa, N.C. and Arenas, E. (2010) Emerging roles of Wnts in the adult nervous system. *Nature Reviews Neuroscience*, **11** (2), 77–86.

Janda, C.Y., Waghray, D., Levin, A.M. *et al.* (2012) Structural basis of Wnt recognition by Frizzled. *Science*, **337** (6090), 59–64.

Jenei, V., Sherwood, V., Howlin, J. *et al.* (2009) A t-butyloxycarbonyl-modified Wnt5a-derived hexapeptide functions as a potent antagonist of Wnt5a-dependent melanoma cell invasion. *Proceedings of the National Academy of Sciences of the United States of America*, **106** (46), 19473–19478.

Jernigan, K.K., Cselenyi, C.S., Thorne, C.A. *et al.* (2010) Gbetagamma activates GSK3 to promote LRP6-mediated beta-catenin transcriptional activity. *Science Signaling*, **3** (121), ra37.

Jung, H., Kim, H.J., Lee, S.K. *et al.* (2009) Negative feedback regulation of Wnt signaling by Gbetagamma-mediated reduction of Dishevelled. *Experimental & Molecular Medicine*, **41** (10), 695–706.

Katanaev, V.L. and Tomlinson, A. (2006) Dual roles for the trimeric G protein Go in asymmetric cell division in Drosophila. *Proceedings of the National Academy of Sciences of the United States of America*, **103** (17), 6524–6529.

Katanaev, V.L. and Buestorf, S. (2009) Frizzled Proteins are bona fide G protein-coupled receptors. *Nature Precedings*, http://hdl.handle.net/10101/npre.2009.2765.1 (accessed October 24, 2013).

Katanaev, V.L., Ponzielli, R., Sémériva, M., and Tomlinson, A. (2005) Trimeric G protein-dependent frizzled signaling in Drosophila. *Cell*, **120** (1), 111–122.

Kaykas, A., Yang-Snyder, J., Héroux, M. *et al.* (2004) Mutant Frizzled 4 associated with vitreoretinopathy traps wild-type Frizzled in the endoplasmic reticulum by oligomerization. *Nature Cell Biology*, **6** (1), 52–58.

Kilander, M., Halleskog, C., and Schulte, G. (2011) Purified WNTs differentially activate beta-catenin-dependent and -independent pathways in mouse microglia-like cells. *Acta Physiologica*, **203** (3), 363–372.

Kilander, M., Dijksterhuis, J., Ganji, R.S. *et al.* (2011) WNT-5A stimulates the GDP/GTP exchange at pertussis toxin-sensitive heterotrimeric G proteins. *Cell Signaling*, **23** (3), 550–554.

Klaus, A. and Birchmeier, W. (2008) Wnt signalling and its impact on development and cancer. *Nature Reviews Cancer*, **8** (5), 387–398.

Kohn, A.D. and Moon, R.T. (2005) Wnt and calcium signaling: beta-catenin-independent pathways. *Cell Calcium*, **38** (3–4), 439–446.

Koval, A. and Katanaev, V.L. (2011) Wnt3a stimulation elicits G-protein-coupled receptor properties of mammalian Frizzled proteins. *The Biochemical Journal*, **433** (3), 435–440.

Koval, A. and Katanaev, V.L. (2012) Platforms for high-throughput screening of Wnt/Frizzled antagonists. *Drug Discovery Today*, **17**, 1316–1319.

Kühl, S.J. and Kühl, M. (2012) On the role of Wnt/β-catenin signaling in stem cells. *Biochimica et Biophysica Acta*, **1830**, 2297–2306.

Kühl, M., Sheldahl, L.C., Park, M. *et al.* (2000) The Wnt/Ca^{2+} pathway: a new vertebrate Wnt signaling pathway takes shape. *Trends in Genetics*, **16** (7), 279–283.

Laeremans, H., Hackeng, T.M., van Zandvoort, M.A. *et al.* (2011) Blocking of frizzled signaling with a homologous peptide fragment of wnt3a/wnt5a reduces infarct expansion and prevents the development of heart failure after myocardial infarction. *Circulation*, **124** (15), 1626–1635.

Liu, X., Rubin, J.S., and Kimmel, A.R. (2005) Rapid, Wnt-induced changes in GSK3beta associations that regulate beta-catenin stabilization are mediated by Galpha proteins. *Current Biology*, **15** (22), 1989–1997.

Liu, T., Liu, X., Wang, H. *et al.* (1999a) Activation of rat frizzled-1 promotes Wnt signaling and differentiation of mouse F9 teratocarcinoma cells via pathways that require Galpha(q) and Galpha(o) function. *The Journal of Biological Chemistry*, **274** (47), 33539–33544.

Liu, X., Liu, T., Slusarski, D.C. *et al.* (1999b) Activation of a frizzled 2/beta adrenergic receptor chimera promotes Wnt signaling and differentiation of mouse F9 teratocarcinoma cells via Galphao and Galphat. *Proceedings of the National Academy of Sciences of the United States of America*, **96** (25), 14383–14388.

Liu, T., DeCostanzo, A.J., Liu, X. *et al.* (2001) G protein signaling from activated rat frizzled-1 to the beta-catenin-Lef-Tcf pathway. *Science*, **292** (5522), 1718–1722.

Malbon, C.C. (2011) Wnt signalling: the case of the 'missing' G-protein. *Biochemical Journal*, **433** (3), e3–e5.

Mikels, A.J. and Nusse, R. (2006) Purified Wnt5a protein activates or inhibits beta-catenin-TCF signaling depending on receptor context. *PLoS Biology*, **4** (4), e115.

Milligan, G. (2009) G protein-coupled receptor hetero-dimerization: contribution to pharmacology and function. *British Journal of Pharmacology*, **158** (1), 5–14.

Moon, R.T., Campbell, R.M., Christian, J.L. *et al.* (1993) Xwnt-5A: a maternal Wnt that affects morphogenetic movements after overexpression in embryos of Xenopus laevis. *Development*, **119** (1), 97–111.

Oldham, W.M. and Hamm, H.E. (2008) Heterotrimeric G protein activation by G-protein-coupled receptors. *Nature Reviews Molecular Cell Biology*, **9** (1), 60–71.

Parker, M.S., Park, E.A., Sallee, F.R., and Parker, S.L. (2011) Two intracellular helices of G-protein coupling receptors could generally support oligomerization and coupling with transducers. *Amino Acids*, **40** (2), 261–268.

Povelones, M. and Nusse, R. (2005) The role of the cysteine-rich domain of Frizzled in Wingless-Armadillo signaling. *The EMBO Journal*, **24** (19), 3493–3503.

Rasmussen, S.G., DeVree, B.T., Zou, Y. *et al.* (2011) Crystal structure of the β2 adrenergic receptor-Gs protein complex. *Nature*, **477** (7366), 549–555.

Romero, G., Sneddon, W.B., Yang, Y. *et al.* (2010) Parathyroid hormone receptor directly interacts with dishevelled to regulate {beta}-catenin signaling and osteoclastogenesis. *The Journal of Biological Chemistry*, **285** (19), 14756–14763.

Rulifson, E.J., Wu, C.H., and Nusse, R. (2000) Pathway specificity by the bifunctional receptor frizzled is determined by affinity for wingless. *Molecular Cell*, **6** (1), 117–126.

Säfholm, A., Leandersson, K., Dejmek, J. *et al.* (2006) A formylated hexapeptide ligand mimics the ability of Wnt-5a to impair migration of human breast epithelial cells. *The Journal of Biological Chemistry*, **281** (5), 2740–2749.

Schulte, G. (2010) International union of basic and clinical pharmacology. LXXX. The class frizzled receptors. *Pharmacological Reviews*, **62** (4), 632–667.

Schulte, G. and Bryja, V. (2007) The Frizzled family of unconventional G-protein-coupled receptors. *Trends in Pharmacological Sciences*, **28** (10), 518–525.

Sheldahl, L.C., Park, M., Malbon, C.C., and Moon, R.T. (1999) Protein kinase C is differentially stimulated by Wnt and Frizzled homologs in a G-protein-dependent manner. *Current Biology*, **9** (13), 695–698.

Sheldahl, L.C., Slusarski, D.C., Pandur, P. *et al.* (2003) Dishevelled activates Ca2+ flux, PKC, and CamKII in vertebrate embryos. *The Journal of Cell Biology*, **161** (4), 769–777.

Shevtsov, S.P., Haq, S., and Force, T. (2006) Activation of beta-catenin signaling pathways by classical G-protein-coupled receptors: mechanisms and consequences in cycling and non-cycling cells. *Cell Cycle*, **5** (20), 2295–2300.

Slusarski, D.C., Corces, V.G., and Moon, R.T. (1997) Interaction of Wnt and a Frizzled homologue triggers G-protein-linked phosphatidylinositol signalling. *Nature*, **390** (6658), 410–413.

Slusarski, D.C., Yang-Snyder, J., Busa, W.B., and Moon, R.T. (1997) Modulation of embryonic intracellular Ca2+ signaling by Wnt-5A. *Developmental Biology*, **182** (1), 114–120.

Smallwood, P.M., Williams, J., Xu, Q. *et al.* (2007) Mutational analysis of Norrin-Frizzled4 recognition. *The Journal of Biological Chemistry*, **282** (6), 4057–4068.

Stemmle, L.N., Fields, T.A., and Casey, P.J. (2006) The regulator of G protein signaling domain of axin selectively interacts with Galpha12 but not Galpha13. *Molecular Pharmacology*, **70** (4), 1461–1468.

Sutherland, E.W. and Robison, G.A. (1966) The role of cyclic-3′,5′-AMP in responses to catecholamines and other hormones. *Pharmacological Reviews*, **18** (1), 145–161.

Tauriello, D.V., Jordens, I., Kirchner, K. *et al.* (2012) Wnt/β-catenin signaling requires interaction of the Dishevelled DEP domain and C terminus with a discontinuous motif in Frizzled. *Proceedings of the National Academy of Sciences of the United States of America*, **109** (14), 812–820.

Torii, K., Nishizawa, K., Kawasaki, A. *et al.* (2008) Anti-apoptotic action of Wnt5a in dermal fibroblasts is mediated by the PKA signaling pathways. *Cell Signaling*, **20** (7), 1256–1266.

Tu, X., Joeng, K.S., Nakayama, K.I. *et al.* (2007) Noncanonical Wnt signaling through G protein-linked PKCdelta activation promotes bone formation. *Developmental Cell*, **12** (1), 113–127.

Umbhauer, M., Djiane, A., Goisset, C. *et al.* (2000) The C-terminal cytoplasmic Lys-thr-X-X-X-Trp motif in frizzled receptors mediates Wnt/beta-catenin signalling. *The EMBO Journal*, **19** (18), 4944–4954.

Uren, A., Reichsman, F., Anest, V. *et al.* (2000) Secreted frizzled-related protein-1 binds directly to Wingless and is a biphasic modulator of Wnt signaling. *The Journal of Biological Chemistry*, **275** (6), 4374–4382.

Veeman, M.T., Axelrod, J.D., and Moon, R.T. (2003) A second canon. Functions and mechanisms of beta-catenin-independent Wnt signaling. *Developmental Cell*, **5** (3), 367–377.

Vinson, C.R., Conover, S., and Adler, P.N. (1989) A Drosophila tissue polarity locus encodes a protein containing seven potential transmembrane domains. *Nature*, **338** (6212), 263–264.

Wang, Y., Macke, J.P., Abella, B.S. *et al.* (1996) A large family of putative transmembrane receptors homologous to the product of the Drosophila tissue polarity gene frizzled. *The Journal of Biological Chemistry*, **271** (8), 4468–4476.

Wess, J. (1998) Molecular basis of receptor/G-protein-coupling selectivity. *Pharmacological Therapy*, **80** (3), 231–264.

Whorton, M.R., Bokoch, M.P., Rasmussen, S.G. *et al.* (2007) A monomeric G protein-coupled receptor isolated in a high-density lipoprotein particle efficiently activates its G protein. *Proceedings of the National Academy of Sciences of the United States of America*, **104** (18), 7682–7687.

Willert, K. and Nusse, R. (2012) Wnt proteins. *Cold Spring Harbor Perspectives in Biology*, **4**, a007864.

Wu, C.H. and Nusse, R. (2002) Ligand receptor interactions in the Wnt signaling pathway in Drosophila. *The Journal of Biological Chemistry*, **277** (44), 41762–41769.

Xu, Q., Wang, Y., Dabdoub, A. *et al.* (2004) Vascular development in the retina and inner ear: control by Norrin and Frizzled-4, a high-affinity ligand-receptor pair. *Cell*, **116** (6), 883–895.

Ye, X., Wang, Y., and Nathans, J. (2010) The Norrin/Frizzled4 signaling pathway in retinal vascular development and disease. *Trends in Molecular Medicine*, **16** (9), 417–425.

15 Dishevelled at the Crossroads of Pathways

Vítězslav Bryja and Ondřej Bernatík

Faculty of Science, Institute of Experimental Biology, Masaryk University & Institute of Biophysics, Academy of Sciences of the Czech Republic, Brno, Czech Republic

Introduction

Dishevelled (Dvl) is a cytoplasmic phosphoprotein required for signal transduction in the Wnt/β-catenin pathway and in β-catenin-independent, collectively called the noncanonical, Wnt pathways. Dvl was named after the phenotype of the *Drosophila* mutant with defects in planar cell polarity (PCP) manifested by disoriented body and wing hairs (Fahmy and Fahmy, 1959). Later on, Dvl was identified to function genetically downstream of Frizzled receptors in this process (Krasnow, Wong, and Adler, 1995). Moreover, approximately at the same time, it was shown that Dvl acts as a component of Wnt/β-catenin pathway influencing segment polarity of *Drosophila* larvae (Klingensmith, Nusse, and Perrimon, 1994; Krasnow, Wong, and Adler, 1995; Noordermeer *et al.*, 1994; Theisen *et al.*, 1994). Already these early genetic experiments in flies demonstrated that Dvl has distinct molecular functions and plays the dual role in both PCP and β-catenin-mediated signaling.

There is only one Dvl gene in *Drosophila*, in contrast to three genes (Dvl1, Dvl2, and Dvl3) in mammalian genomes (Klingensmith *et al.*, 1996; Sussman *et al.*, 1994; Tsang *et al.*, 1996). Dvl protein sequence similarity is around 50% between *Drosophila* and mouse Dvls, consistent with the conserved function between species. The Dvl proteins encoded by the three mouse genes are around 60% identical. Dvl proteins have mostly identical molecular activities as demonstrated by *Dvl* transgene rescue experiments in knockout mice (Etheridge *et al.*, 2008). The phenotypes of individual Dvl gene knockouts however differ due to differential gene expression patterns. *Dvl1*$^{-/-}$ mice were viable with no obvious developmental malformations but showed social deficits due to impaired synaptic assembly, neurotransmitter release, and dendritic branching (Ahmad-Annuar *et al.*, 2006; Lijam *et al.*, 1997; Rosso *et al.*, 2005) (see also Chapter 29). Phenotypes of *Dvl2*$^{-/-}$ and *Dvl3*$^{-/-}$ mouse were much more severe and included perinatal lethality, abnormal cardiac morphogenesis due to neural crest defects, and skeletal malformations as a result of improper somite formation (Etheridge *et al.*, 2008; Hamblet *et al.*, 2002; Kioussi *et al.*, 2002). The phenotypes typically associated with the PCP defects such as craniorachischisis and defects in the organ of Corti were observed frequently in compound mutants of two Dvl isoforms (the most severe in *Dvl2*$^{-/-}$ and *Dvl3*$^{-/-}$ mice) (Etheridge *et al.*, 2008). Interestingly, double Dvl mutants lacked phenotypes typically

Wnt Signaling in Development and Disease: Molecular Mechanisms and Biological Functions, First Edition.
Edited by Stefan Hoppler and Randall T. Moon.

associated with the Wnt/β-catenin pathway defects and did not show global changes in the TCF-/LEF-driven transcription (Etheridge *et al.*, 2008), which suggests that even low levels of the remaining single Dvl isoform are capable of mediating physiological Wnt/β-catenin but not Wnt/PCP signaling during development.

Structure of Dvl

Dvl is a modular protein that contains three well-defined structural domains linked by the intrinsically disordered regions, which are predicted to have no secondary structure. N-terminal DIX (Dvl, Axin; amino acids 1–82, hDvl3) domain is connected with the central PDZ (Postsynaptic density, Discs large, Zonula occludens; aa 249–321, hDvl3) domain by the serine (Ser)/threonine (Thr)-rich unstructured region with the conserved stretch of basic amino acids (aa 175–248, hDvl3). PDZ domain and the C-terminal DEP domain (Dvl, Egl-10, Pleckstrin; aa 422–496, hDvl3) are linked by the region enriched in proline (Pro-rich region). The C-terminus is formed by relatively large but poorly conserved disordered region (Figure 15.1). Rescue analysis using

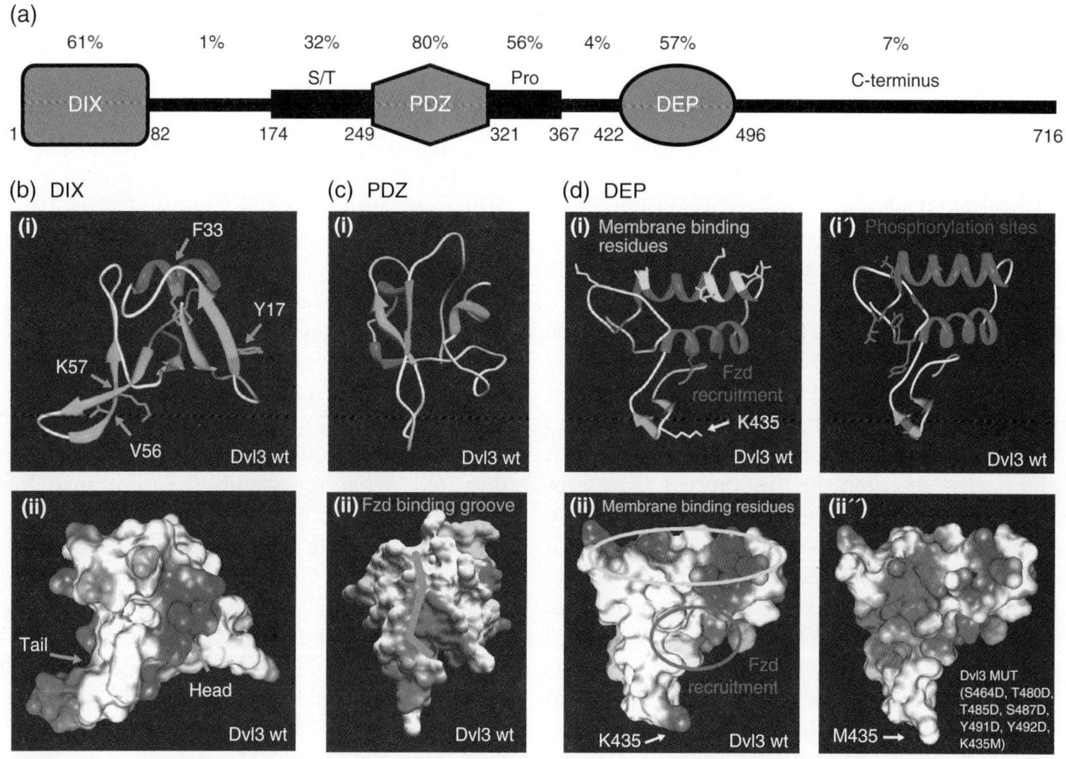

Figure 15.1 Schematic representation of Dvl structure. (a) Amino acid (aa) sequence identity (%) based on alignment of *Drosophila* Dsh, and mouse (m) and human (h) Dvl isoforms 1, 2, and 3. Position numbering of aa bordering individual regions is based on hDvl3. (b–d) 3D models of individual domains (hDvl3) are provided either as ribbon models (i) or as surface models (ii). Alpha-helices are shown in *magenta*, and beta-sheets in *cyan*. Electrostatic surface charge is depicted as negative in *red* and positive in *blue*. (b) DIX domain. (i) Residues mutated in Dvl polymerization mutants corresponding to F43S, V67A/K68A, and Y27D in mDvl2 are depicted in orange (Schwarz-Romond *et al.*, 2007). (ii) Regions of DIX domain required for head-to-tail polymerization are marked by *arrows*. (c) PDZ domain. (i) α-helix and β-sheet forming the Fzd binding groove are indicated in *green* (Wong *et al.*, 2003). (ii) Fzd binding groove is indicated by a *green line*. (d) DEP domain. (i, ii) Residues and surface required for membrane binding are indicated in *yellow* (Simons *et al.*, 2009) and residues required for Fzd binding are shown in *dark blue* (Tauriello *et al.*, 2012). Position of K435 corresponding to K417 mutated in *Drosophila* Dsh[1] mutation (Axelrod *et al.*, 1998) is indicated in *white*. (i′) Residues (S464, T480, T485, S487, Y491, Y492) phosphorylated following Fzd overexpression (Yanfeng *et al.*, 2011) are indicated in *red*. (ii′) Phosphomimicking mutations of these residues change electrostatic surface potential and disrupt the positive charge of the membrane-binding region. Please note also the change caused by the mutation K435M (Dsh[1], *white*). *(See insert for color representation of the figure.)*

Dvl deletion mutants in flies suggested that DIX and PDZ domains are necessary for Wnt/β-catenin pathway, whereas PDZ and DEP domains are required for Wnt/PCP pathway (Axelrod *et al.*, 1998; Boutros *et al.*, 1998).

DIX domain

Mutational studies of Dvl2 DIX domain and modeling based on the crystal structure of the Axin DIX domain (Schwarz-Romond *et al.*, 2007) have convincingly demonstrated that the key feature of DIX domain is its ability to form head-to-tail polymers. DIX domain mutants that lack the ability to polymerize were unable to form Dvl polymers visible as cytoplasmatic Dvl punctae (when overexpressed) and are inactive in the Wnt/β-catenin pathway (Schwarz-Romond *et al.*, 2007). It has been proposed that DIX domain-mediated Dvl polymers are a physical platform for the complexes of Dvl and other signaling components, called signalosomes, required for Lrp6 phosphorylation (Bilic *et al.*, 2007). The requirement of DIX domain for the phosphorylation of Lrp6 and its capacity to dimerize seems to be the molecular reason for the absolute requirement of DIX domain in the Wnt/β-catenin signaling.

PDZ domain

PDZ domain is commonly found in various proteins, where it serves as a universal interaction and docking platform. Dvl PDZ domain is inherently more flexible than canonical PDZ domains (Zhang *et al.*, 2009). PDZ domain seems to functions as an important scaffolding region of Dvl protein, and it was shown to be required for the interaction with many Dvl binding partners (for review see Gao and Chen, 2010). Despite the fact that the PDZ domain of Dvl has been implicated in many protein–protein interactions, not much is known about the molecular details and the significance of the individual interactions. Notable exception is the NMR characterization of the interaction between the PDZ domain (mDvl1) and the internal PDZ-binding motif of FZD7 (Wong *et al.*, 2003). The same binding cleft is bound by natural Wnt pathway inhibitors such as proteins from the Dact family (homologs of fly Dapper/Frodo) (Wong *et al.*, 2003) and serves also as a target for the development of novel drugs targeting FZD–Dvl interaction (Shan *et al.*, 2012).

DEP domain

DEP domain of Dvl is responsible for the recruitment of Dvl to the membrane and for signal transduction in the Wnt/PCP pathway (Axelrod *et al.*, 1998; Boutros *et al.*, 1998). The crystal structure of the DEP domain of Dvl is known (Wong *et al.*, 2000). A positively charged region of Dvl DEP domain is essential for the binding to the negatively charged lipids in the plasma membrane, and mutation of the positively charged residues abolished recruitment of Dvl to the plasma membrane and caused PCP phenotypes in *Drosophila* (Simons *et al.*, 2009). It was proposed that ion pumps (such as Nhe2, Na^+/H^+ exchanger) in the plasma membrane control Dvl membrane recruitment by regulating local pH (Simons *et al.*, 2009). It should be however noted that Fzd-induced phosphorylation within the DEP domain (Yanfeng *et al.*, 2011), which also changes the charge of the surface-interacting region from positive to negative, may have similar consequences (see model in Figure 15.1). Furthermore, it was recently proposed that the DEP domain of Dvl is the second interaction interface (in addition to the PDZ domain) mediating interaction with FZD (Tauriello *et al.*, 2012). Last but not least, currently, there are only two known point mutations which completely disrupt PCP signaling. Both mutations in the fully conserved Lys (Dsh^1 mutant K417M *Drosophila* Dvl) and Tyr (Y473A in *Drosophila*) are in the DEP domain. The analysis of 3D structure reveals that they are located close to each other and do not overlap with the membrane or Fzd interaction interface, which implicates existence of the third functionally important binding interface within DEP domain (Figure 15.1).

Dvl-associated kinases and consequences of Dvl phosphorylation

Dvl is a dynamically phosphorylated protein. Precise mapping and functional analysis of Dvl phosphorylation sites is complicated by the fact that ~15–20% (depends on the isoform) of Dvl is composed by Ser, Thr, or tyrosine (Tyr). Treatment with Wnt ligands, irrespective of their ability to activate the β-catenin-dependent or β-catenin-independent pathways, triggers phosphorylation-dependent mobility shift of endogenous Dvl on SDS-PAGE (Bryja *et al.*, 2007; Gonzalez-Sancho *et al.*, 2004). The shifted form of endogenous Dvl, which is induced by Wnts, is referred to as phosphorylated and shifted Dvl (PS-Dvl). The term PS-Dvl is primarily used to distinguish the Wnt-induced modification of Dvl from other constitutive phosphorylations, which are constitutive and/or Wnt pathway independent (Bernatik *et al.*, 2011).

Among the several kinases that phosphorylate Dvl proteins, the best described and the first one to mention is casein kinase (CK) 1 δ/ε. CK1δ and CK1ε are redundant in most of their functions in Dvl biology, and in further text, we will thus refer only to CK1ε. CK1ε is both required and sufficient for PS-Dvl formation. Following Wnt stimulation, CK1ε gets activated by a poorly understood mechanism, which requires dephosphorylation of the inhibitory residues in its C-terminus (Swiatek *et al.*, 2004). Of note, C-terminus is well conserved in CK1δ and CK1ε but is missing in other CK1 isoforms. CK1ε-mediated phosphorylation of Dvl subsequently triggers interaction of Dvl with endogenous Axin1 and activates downstream Wnt/β-catenin signaling (Bernatik *et al.*, 2011; Kishida *et al.*, 2001; Peters *et al.*, 1999). Apart from these clearly positive effects on Wnt/β-catenin signaling, CK1ε phosphorylation dissolves (and CK1 inhibition promotes formation of) DIX domain-dependent Dvl polymers, which are required for the phosphorylation of Lrp6 and downstream signaling (Bilic *et al.*, 2007; Bryja *et al.*, 2007; Cong, Schweizer, and Varmus, 2004) (see also Chapter 2). These opposing functions of CK1ε require distinct domains of Dvl – activation requires phosphorylation in the PDZ domain (and likely also elsewhere), whereas inhibitory activity requires the Dvl C-terminus (Bernatik *et al.*, 2011). CK1ε thus seems to act as the component of the intrinsic negative feedback loop, which by numerous sequential phosphorylation (up to 40 phosphorylated residues, Bryja and Bernatík, unpublished) activates and subsequently inactivates Dvl. The function of CK1ε is probably even more complex, and the role of CK1-driven Dvl phosphorylation in noncanonical Wnt signaling pathways is only starting to emerge (Bryja *et al.*, 2007, 2008; Cong, Schweizer, and Varmus, 2004; Klein *et al.*, 2006; Strutt, Price, and Strutt, 2006; Witte *et al.*, 2010).

The second important kinase is casein kinase 2 (CK2), which was identified as the first Dvl-associated kinase during the search for the kinase activity coimmunoprecipitating with Dvl (Lee, Ishimoto, and Yanagawa, 1999; Willert *et al.*, 1997). CK2 is not structurally related to CK1 and functions as a tetramer with two catalytic α (or α′) and two regulatory β subunits. When compared to dynamic activation of CK1ε, CK2 behaves with respect to Dvl as a rather constitutive kinase, which is not regulated by Wnt ligands although CK2 activity is required for the PS-Dvl activation and downstream signaling in both Wnt/β-catenin and Wnt/PCP pathway (Bernatik *et al.*, 2011; Bryja *et al.*, 2008).

The third important kinase is tyrosine kinase Abl, identified in the screen for molecules required for membrane localization of Dvl. Abl phosphorylates Dvl on several Tyr residues in the DEP domain, and it was shown that both mutation of one of these sites (Y473F) and Abl depletion caused PCP phenotypes but not Wnt/β-catenin phenotypes in *Drosophila* (Singh *et al.*, 2010). Any information about the function of Abl-driven phosphorylation of Dvl in vertebrates is missing.

Several other kinases were shown to bind and phosphorylate Dvl. However, according to our current knowledge, their role is restricted to specific model systems or limited to one report only. These kinases include (i) PAR-1 (in Xenopus, homologous to MAP/microtubule affinity-regulating kinase [MARK] family of kinases in mammals), which was shown to activate canonical Wnt signaling as well as gastrulation movements, a process controlled by noncanonical Wnt signaling (Kusakabe and Nishida, 2004; Ossipova *et al.*, 2005; Sun *et al.*, 2001); (ii) metastasis-associated kinase 1 (MAK1), homologous to the mammalian HUNK, a member of sucrose nonfermenting-related kinase 1 (SNF1) family (Kibardin, Ossipova, and Sokol, 2006); (iii) Polo-like kinase 1 (Plk1), a mitosis-associated kinase, which via Dvl phosphorylation at T206 affects division

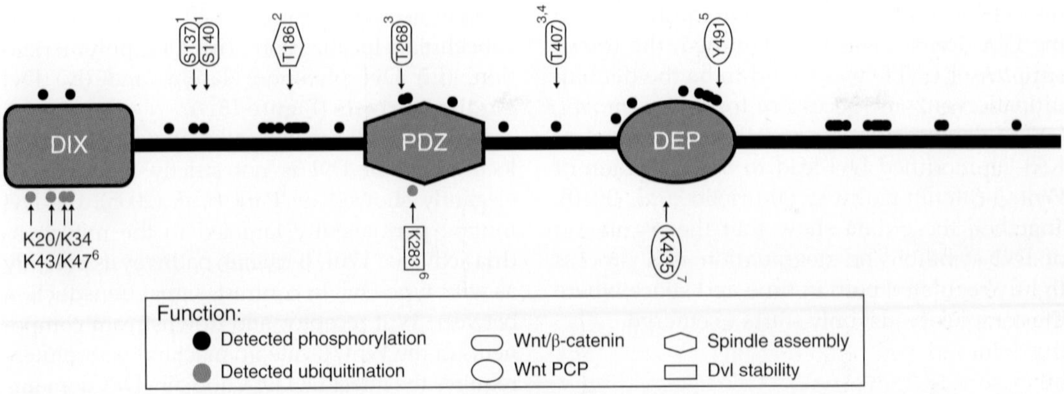

Figure 15.2 Posttranslational modifications of Dvl. Summary of Dvl posttranslational modifications mapped onto hDvl3. Position of detected phosphorylation (•) and ubiquitination (•) sites is shown. Functionally relevant modifications are specified in detail. The figure is based on the following studies: [1]Klimowski *et al.* (2006); [2]Kikuchi *et al.* (2010); [3]Cong, Schweizer, and Varmus (2004); [4]Yokoyama *et al.* (2012); [5]Singh *et al.*, (2010); [6]Tauriello *et al.* (2010); [7]Axelrod *et al.* (1998).

plane during mitosis (Kikuchi *et al.*, 2010); and (iv) PKCδ regulating noncanonical Wnt signaling (Chen *et al.*, 2003; Kinoshita *et al.*, 2003). The comprehensive list of identified phosphorylation sites and their known function is provided in Figure 15.2.

Dvl stability and ubiquitination

Dvl is a key protein, which controls direction and amplitude of downstream Wnt pathway activity. As mentioned previously, Dvl exists in several so-called Dvl pools, which differ in their phosphorylation state and subcellular localization (cytoplasm, cell membrane, ciliary compartment). Levels of Dvl in individual pools are tightly regulated. Several proteins, which differ both in the route used for Dvl degradation and in the ability to target specific Dvl pools, were shown to trigger degradation of Dvl. Dvl is degraded both via proteasome and lysosome/autophagy pathways. The lysosomal pathway, which degrades Dvl mainly in starved cells, is dependent on the Von Hippel–Lindau protein (Gao *et al.*, 2010) and was shown to be promoted by the Wnt inhibitor Dapper (Zhang *et al.*, 2006).

Proteosomal degradation of Dvl is unspecific – triggered by an E3 ubiquitin (Ub) ligase NEDL1, associated with the amyotrophic lateral sclerosis (Miyazaki *et al.*, 2004), or by KLHL12, an adaptor protein linking Dvl to Cullin 3 E3 Ub ligase (Angers *et al.*, 2006). Complete degradation of Dvl then leads both to inhibition of Wnt/β-catenin pathway activity and to disruption of convergent extension controlled by noncanonical Wnt signaling. Alternatively, in order to control Dvl function more precisely, there are mechanisms that control degradation of specific pools of Dvl. The examples include (i) inversin (Invs), a protein mutated in the cystic renal disease nephronopthisis, which enhances degradation of cytoplasmic but not membrane-bound Dvl (Simons *et al.*, 2005); (ii) myristoylated Naked2, which on the contrary promotes degradation of Dvl bound to the cell membrane (Hu *et al.*, 2010); (iii) Rpgrip1l, which is essential for stabilization of Dvl at the base of the cilium (Mahuzier *et al.*, 2012); and, last but not least, (iv) Itch, an HECT-domain containing Ub ligase that specifically promotes degradation of phosphorylated and activated PS-Dvl (Wei *et al.*, 2012). A unique function was also proposed for nucleoredoxin (NXN), which prevents KLHL12-mediated degradation of the inactive Dvl pool (Funato *et al.*, 2010).

Ubiquitination does not only target proteins for degradation, but it also has a potential to regulate the signaling abilities of modified proteins by differential linkages (e.g., via K63) of Ub side chains (Chen and Sun, 2009). It was confirmed that Dvl can be modified by K48 (at K285)- and K63 (in DIX domain)-linked Ub chains, but the sites of modification were different (see Figure 15.2; Kim *et al.*, 2011; Tauriello *et al.*, 2010). Although the E3 ligase responsible for the K63-linked ubiquitination in the DIX domain was not identified, the tumor suppressor CYLD was found to be the deubiquitination enzyme necessary for their removal. CYLD downregulation and increased level of K63-Ub-modified Dvl lead to the activation of Wnt/β-catenin pathway (Tauriello *et al.*, 2010). Together, these data show that the regulation of Dvl stability and degradation is a process tightly regulated both in time and space where the complexity just only starts to emerge.

Dvl in the Wnt/β-catenin and Wnt/PCP pathway: What makes a difference?

Dvl serves as a branching point between Wnt/β-catenin and Wnt/PCP signaling. A lot of attention is paid on uncovering the mechanisms, which control Dvl function and define the direction of downstream signaling. The answer is not completely solved yet, but it is obvious that the pathway selection is made based on the combination of several mutually interconnected factors, which involve (i) Dvl subcellular localization, (ii) Dvl polymerization, (iii) Dvl phosphorylation, and (iv) Dvl binding partners (Figure 15.3).

In the Wnt/β-catenin pathway, cell membrane localization of Dvl is not strictly required. As elegantly showed by Park *et al.* (2005), the Dvl mutant permanently targeted to the mitochondria activated Wnt/β-catenin pathway as potently as wild-type Dvl. In contrast, signal transduction between Wnt receptor and downstream components of the Wnt/β-catenin machinery absolutely requires the intact Dvl DIX domain. DIX domains mediate Dvl polymerization, which is crucial for the activating phosphorylation of Lrp6 (Bilic *et al.*, 2007; Schwarz-Romond *et al.*, 2007). The current view is that following ligand binding, Dvl gets activated by a poorly understood mechanism involving CK1ε-controlled phosphorylation, which subsequently triggers production of phosphatidylinositol 4,5-bisphosphates (Pan *et al.*, 2008; Tanneberger *et al.*, 2011) and formation of signalosomes composed of polymerized Dvl, Axin, and other Wnt pathway components (Bilic *et al.*, 2007).

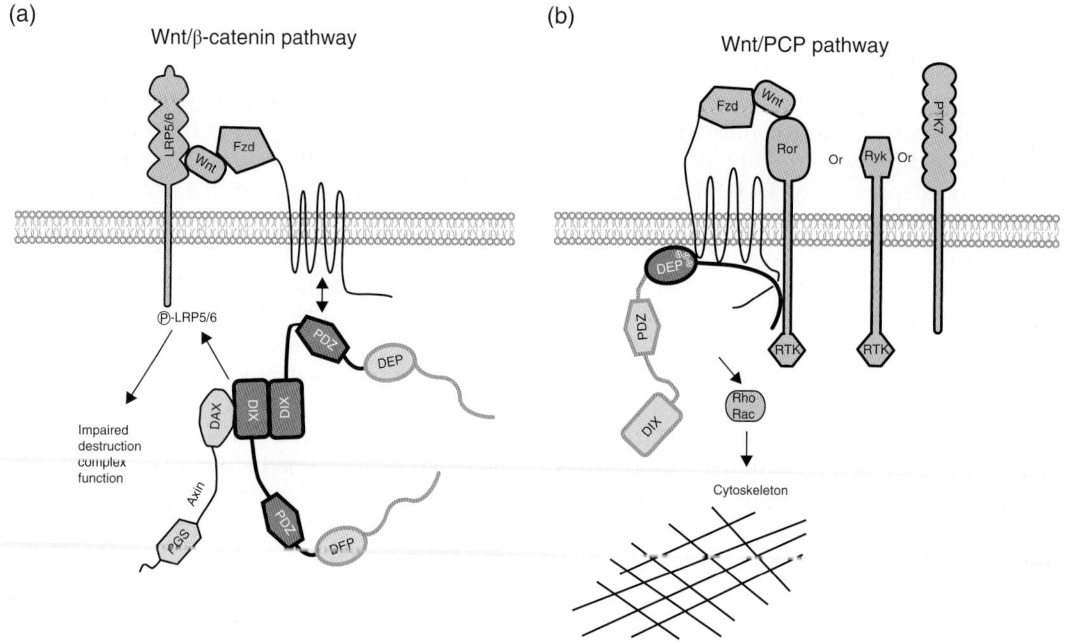

Figure 15.3 Dvl in Wnt pathways. (a) Schematic representation of Dvl function in Wnt/β-catenin pathway. Dvl is required for Lrp6 phosphorylation and subsequent inhibition of the destruction complex function. Polymerization via DIX domains but not membrane localization is required for this function. (b) Schematic representation of Dvl function in Wnt/PCP pathway. Dvl is recruited to the membrane (DEP domain dependently), where it interacts with Fzd and other coreceptors (Ror1/2, Ryk, PTK7) and activates small GTPases Rho and Rac, which subsequently control reorganization of cytoskeleton. C-terminus of Dvl is required for binding to Ror2.

This complex is required for the activation of Lrp6, recruitment of Axin, and inhibition of the destruction complex. It should be however noted that alternative explanation exists: Polymerization-deficient Dvl mutants (see Figure 15.1), which represent key evidence for the concept of signalosomes, are not only unable to interact with each other but also fail to interact with axin1 DIX domain (Fiedler *et al.*, 2011). This opens the possibility that Dvl can rip out Axin1 directly from the β-catenin destruction complex via interaction of Axin and Dvl DIX domains. Future work is required to clarify whether it is the DIX-mediated Dvl–Dvl or DIX-mediated Dvl–Axin interaction which is the key for activation of downstream signaling.

In the Wnt/PCP pathway, DEP domain and membrane localization of Dvl is the absolute prerequisite for the signaling activity. So far, all the Wnt/PCP-deficient mutants of Dvl also failed to be recruited to the membrane by Fzds (Axelrod *et al.*, 1998; Simons *et al.*, 2009; Singh *et al.*, 2010) and even wild-type Dvl constitutively trafficked to other compartment such as the outer mitochondrial membrane (Park *et al.*, 2005) was inactive in the Wnt/PCP pathway. Various mechanisms, which control membrane localization of Dvl and subsequently favor PCP pathway over Wnt/β-catenin, were described. They include regulation of local pH (Simons *et al.*, 2009), PCP-specific Dvl phosphorylation by Abl (Singh *et al.*, 2010), or the targeted degradation of cytoplasmic Dvl by Invs (Simons *et al.*, 2005). Lrp6, a coreceptor dedicated to the Wnt/β-catenin, is not positively involved in the PCP pathway (albeit is involved as a negative regulator (Bryja *et al.*, 2009)). Other receptors, mainly Ror1/2, Ryk, and PTK, cooperate with Dvl in a noncanonical pathway. Importantly, and in contrast to the Wnt/β-catenin pathway, where Dvl functions upstream of LRP5/6, it is currently not clear whether Dvl acts downstream or upstream of these receptors. There is evidence for both scenarios – the support for Dvl acting downstream comes from the phenotype of Ror1/2$^{-/-}$ cells, which show low levels of PS-Dvl, and phenocopy Wnt5a$^{-/-}$ cells (Ho *et al.*, 2012) and from the description of the Wnt5a-induced Dvl polymers, which are dependent on Ror2/Fzd (Nishita *et al.*, 2010). On the other hand, it was suggested that activation of Ror2 (phosphorylation dependent shift) requires Dvl and axin (Grumolato *et al.*, 2010). This scenario is supported by the observations that Ror2 only binds to the activated PS-Dvl, which is phosphorylated by CK1 (Witte *et al.*, 2010). The other unresolved question is the role of Dvl DIX domain in noncanonical signaling. Genetic evidence suggested that DIX domain is clearly dispensable for PCP signaling in *Drosophila* (Axelrod *et al.*, 1998; Boutros *et al.*, 1998) and mouse (ΔDIX–Dvl2 transgene can fully rescue PCP phenotypes in Dvl2$^{-/-}$ mouse) (Wang *et al.*, 2006). In contrast, it was shown that Wnt5A induces DIX-dependent polymerization of Dvl, which is required for AP1-activation (Nishita *et al.*, 2010), and Rac1, a major downstream effector, binds to the DIX domain or to its close proximity (Cajanek *et al.*, 2013). It should be however noted that polymerization-dependent mutants of Dvl2 activate Rac1 normally when overexpressed (Cajanek *et al.*, 2013).

Remaining secrets of Dvl

Dvls are intensively studied proteins, which – despite the efforts of numerous labs – still resist our understanding and hide many of their secrets. The biggest challenge for the near future will be to describe the sequence of events following interaction of the Wnt ligands with their receptors and relevance for downstream pathway selection and activation. The lines of research, which will at the endogenous level define the dynamics of Dvl post-translational modifications, Dvl interaction partners, Dvl quantity, and Dvl subcellular localization following activation of discrete Wnt pathways, will answer these questions. It is probable that detailed mechanistic understanding of Dvl biology will shed light not only on the mechanism discriminating between β-catenin-mediated and PCP downstream signaling but also onto the details of various noncanonical Wnt pathways (see Chapter 6), which diversified during the course of evolution and differ in humans substantially from the core PCP pathway postulated in fly.

It is also becoming obvious that Dvl controls processes, which are currently not directly linked with Wnt/β-catenin and Wnt/PCP pathway. Accumulating body of evidence suggests a role of Dvl in the biology of cilia. Dvl was shown in numerous contexts to interact with centriolar proteins and to be required for the positioning of the basal body and the proper function of cilia (Hashimoto *et al.*, 2010; Park *et al.*, 2008). This function of Dvl is evolutionary conserved from planarians (Almuedo-Castillo, Salo, and Adell, 2011). Recently, Dvl was shown to be involved not only in the biogenesis of cilia but also in the process of cilia disassembly (Lee *et al.*, 2012). The last exciting avenue is the unexpected role of Dvl in the regulation of cell cycle. Very recent evidence suggests that Dvl2 can regulate spindle orientation in mitosis via its interaction with Plk1 (Kikuchi *et al.*, 2010) and cytokinesis by the regulation of midbody dynamics (Fumoto *et al.*, 2012). All these diverse aspects of Dvl biology are tightly controlled and provide unique outcome in each cell and each developmental process. The mechanism of such error-resistant integration of all Dvl functions remains the biggest challenge for future research.

References

Ahmad-Annuar, A., Ciani, L., Simeonidis, I. *et al.* (2006) Signaling across the synapse: a role for Wnt and Dishevelled in presynaptic assembly and neurotransmitter release. *The Journal of Cell Biology*, **174** (1), 127–139.

Almuedo-Castillo, M., Salo, E., and Adell, T. (2011) Dishevelled is essential for neural connectivity and planar cell polarity in planarians. *Proceedings of the National Academy of Sciences of the United States of America*, **108** (7), 2813–2818.

Angers, S., Thorpe, C.J., Biechele, T.L. *et al.* (2006) The KLHL12-Cullin-3 ubiquitin ligase negatively regulates the Wnt-beta-catenin pathway by targeting Dishevelled for degradation. *Nature Cell Biology*, **8** (4), 348–357.

Axelrod, J.D., Miller, J.R., Shulman, J.M. *et al.* (1998) Differential recruitment of Dishevelled provides signaling specificity in the planar cell polarity and Wingless signaling pathways. *Genes & Development*, **12** (16), 2610–2622.

Bernatik, O., Ganji, R.S., Dijksterhuis, J.P. *et al.* (2011) Sequential activation and inactivation of Dishevelled in the Wnt/beta-catenin pathway by casein kinases. *The Journal of Biological Chemistry*, **286** (12), 10396–10410.

Bilic, J., Huang, Y.L., Davidson, G. *et al.* (2007) Wnt induces LRP6 signalosomes and promotes dishevelled-dependent LRP6 phosphorylation. *Science*, **316** (5831), 1619–1622.

Boutros, M., Paricio, N., Strutt, D.I., and Mlodzik, M. (1998) Dishevelled activates JNK and discriminates between JNK pathways in planar polarity and wingless signaling. *Cell*, **94** (1), 109–118.

Bryja, V., Schulte, G., Rawal, N. *et al.* (2007) Wnt-5a induces Dishevelled phosphorylation and dopaminergic differentiation via a CK1-dependent mechanism. *Journal of Cell Science*, **120** (Pt 4), 586–595.

Bryja, V., Schambony, A., Cajanek, L. *et al.* (2008) Beta-arrestin and casein kinase 1/2 define distinct branches of non-canonical WNT signalling pathways. *EMBO Reports*, **9** (12), 1244–1250.

Bryja, V., Andersson, E.R., Schambony, A. *et al.* (2009) The extracellular domain of Lrp5/6 inhibits noncanonical Wnt signaling in vivo. *Molecular Biology of the Cell*, **20** (3), 924–936.

Cajanek, L., Ganji, R.S., Henriques-Oliveira, C. *et al.* (2013) Tiam1 regulates the Wnt/Dvl/Rac1 signaling pathway and the differentiation of midbrain dopaminergic neurons. *Molecular and Cellular Biology*, **33** (1), 59–70.

Chen, Z.J. and Sun, L.J. (2009) Nonproteolytic functions of ubiquitin in cell signaling. *Molecular Cell*, **33** (3), 275–286.

Chen, W., ten Berge, D., Brown, J. *et al.* (2003) Dishevelled 2 recruits beta-arrestin 2 to mediate Wnt5A-stimulated endocytosis of Frizzled 4. *Science*, **301** (5638), 1391–1394.

Cong, F., Schweizer, L., and Varmus, H. (2004) Casein kinase Iepsilon modulates the signaling specificities of dishevelled. *Molecular and Cellular Biology*, **24** (5), 2000–2011.

Etheridge, S.L., Ray, S., Li, S. *et al.* (2008) Murine dishevelled 3 functions in redundant pathways with dishevelled 1 and 2 in normal cardiac outflow tract, cochlea, and neural tube development. *PLoS Genetics*, **4** (11), e1000259.

Fahmy, O.G. and Fahmy, M.J. (1959) Differential gene response to mutagens in Drosophila melanogaster. *Genetics*, **44** (6), 1149–1171.

Fiedler, M., Mendoza-Topaz, C., Rutherford, T.J. *et al.* (2011) Dishevelled interacts with the DIX domain polymerization interface of Axin to interfere with its function in down regulating beta-catenin. *Proceedings of the National Academy of Sciences of the United States of America*, **108** (5), 1937–1942.

Fumoto, K., Kikuchi, K., Gon, H., and Kikuchi, A. (2012) Wnt5a signaling controls cytokinesis by correctly positioning ESCRT-III at the midbody. *The Journal of Cell Science*, **125** (Pt 20), 4822–4832.

Funato, Y., Terabayashi, T., Sakamoto, R. *et al.* (2010) Nucleoredoxin sustains Wnt/beta-catenin signaling by retaining a pool of inactive dishevelled protein. *Current Biology*, **20** (21), 1945–1952.

Gao, C. and Chen, Y.G. (2010) Dishevelled: the hub of Wnt signaling. *Cell Signaling*, **22** (5), 717–727.

Gao, C., Cao, W., Bao, L. *et al.* (2010) Autophagy negatively regulates Wnt signalling by promoting Dishevelled degradation. *Nature Cell Biology*, **12** (8), 781–790.

Gonzalez-Sancho, J.M., Brennan, K.R., Castelo-Soccio, L.A., and Brown, A.M. (2004) Wnt proteins induce dishevelled phosphorylation via an LRP5/6- independent mechanism, irrespective of their ability to stabilize beta-catenin. *Molecular and Cellular Biology*, **24** (11), 4757–4768.

Grumolato, L., Liu, G., Mong, P. *et al.* (2010) Canonical and noncanonical Wnts use a common mechanism to activate completely unrelated coreceptors. *Genes & Development*, **24** (22), 2517–2530.

Hamblet, N.S., Lijam, N., Ruiz-Lozano, P. *et al.* (2002) Dishevelled 2 is essential for cardiac outflow tract development, somite segmentation and neural tube closure. *Development*, **129** (24), 5827–5838.

Hashimoto, M., Shinohara, K., Wang, J. *et al.* (2010) Planar polarization of node cells determines the rotational axis of node cilia. *Nature Cell Biology*, **12** (2), 170–176.

Ho, H.Y., Susman, M.W., Bikoff, J.B. *et al.* (2012) Wnt5a-Ror-Dishevelled signaling constitutes a core developmental pathway that controls tissue morphogenesis. *Proceedings of the National Academy of Sciences of the United States of America*, **109** (11), 4044–4051.

Hu, T., Li, C., Cao, Z. *et al.* (2010) Myristoylated Naked2 antagonizes Wnt-beta-catenin activity by degrading Dishevelled-1 at the plasma membrane. *The Journal of Biological Chemistry*, **285** (18), 13561–13568.

Kibardin, A., Ossipova, O., and Sokol, S.Y. (2006) Metastasis-associated kinase modulates Wnt signaling to regulate brain patterning and morphogenesis. *Development*, **133** (15), 2845–2854.

Kikuchi, K., Niikura, Y., Kitagawa, K., and Kikuchi, A. (2010) Dishevelled, a Wnt signalling component, is involved in mitotic progression in cooperation with Plk1. *The EMBO Journal*, **29** (20), 3470–3483.

Kim, W., Bennett, E.J., Huttlin, E.L. *et al.* (2011) Systematic and quantitative assessment of the ubiquitin-modified proteome. *Molecular Cell*, **44** (2), 325–340.

Kinoshita, N., Iioka, H., Miyakoshi, A., and Ueno, N. (2003) PKC delta is essential for Dishevelled function in a noncanonical Wnt pathway that regulates Xenopus convergent extension movements. *Genes & Development*, **17** (13), 1663–1676.

Kioussi, C., Briata, P., Baek, S.H. *et al.* (2002) Identification of a Wnt/Dvl/beta-Catenin → Pitx2 pathway mediating cell-type-specific proliferation during development. *Cell*, **111** (5), 673–685.

Kishida, M., Hino, S., Michiue, T. *et al.* (2001) Synergistic activation of the Wnt signaling pathway by Dvl and casein kinase Iepsilon. *The Journal of Biological Chemistry*, **276** (35), 33147–33155.

Klein, T.J., Jenny, A., Djiane, A., and Mlodzik, M. (2006) CKIepsilon/discs overgrown promotes both Wnt-Fz/beta-catenin and Fz/PCP signaling in Drosophila. *Current Biology*, **16** (13), 1337–1343.

Klimowski, L.K., Garcia, B.A., Shabanowitz, J. *et al.* (2006) Site-specific casein kinase 1epsilon-dependent phosphorylation of Dishevelled modulates beta-catenin signaling. *The FEBS Journal*, **273** (20), 4594–4602.

Klingensmith, J., Nusse, R., and Perrimon, N. (1994) The Drosophila segment polarity gene dishevelled encodes a novel protein required for response to the wingless signal. *Genes & Development*, **8** (1), 118–130.

Klingensmith, J., Yang, Y., Axelrod, J.D. *et al.* (1996) Conservation of dishevelled structure and function between flies and mice: isolation and characterization of Dvl2. *Mechanisms of Development*, **58** (1–2), 15–26.

Krasnow, R.E., Wong, L.L., and Adler, P.N. (1995) Dishevelled is a component of the frizzled signaling pathway in Drosophila. *Development*, **121** (12), 4095–4102.

Kusakabe, M. and Nishida, E. (2004) The polarity-inducing kinase Par-1 controls Xenopus gastrulation in cooperation with 14-3-3 and aPKC. *The EMBO Journal*, **23** (21), 4190–4201.

Lee, J.S., Ishimoto, A., and Yanagawa, S. (1999) Characterization of mouse dishevelled (Dvl) proteins in Wnt/Wingless signaling pathway. *The Journal of Biological Chemistry*, **274** (30), 21464–21470.

Lee, K.H., Johmura, Y., Yu, L.R. *et al.* (2012) Identification of a novel Wnt5a-CK1varepsilon-Dvl2-Plk1-mediated primary cilia disassembly pathway. *The EMBO Journal*, **31** (14), 3104–3117.

Lijam, N., Paylor, R., McDonald, M.P. *et al.* (1997) Social interaction and sensorimotor gating abnormalities in mice lacking Dvl1. *Cell*, **90** (5), 895–905.

Mahuzier, A., Gaude, H.M., Grampa, V. *et al.* (2012) Dishevelled stabilization by the ciliopathy protein Rpgrip1l is essential for planar cell polarity. *The Journal of Cell Biology*, **198** (5), 927–940.

Miyazaki, K., Fujita, T., Ozaki, T. *et al.* (2004) NEDL1, a novel ubiquitin-protein isopeptide ligase for dishevelled-1, targets mutant superoxide dismutase-1. *The Journal of Biological Chemistry*, **279** (12), 11327–11335.

Nishita, M., Itsukushima, S., Nomachi, A. *et al.* (2010) Ror2/Frizzled complex mediates Wnt5a-induced AP-1 activation by regulating Dishevelled polymerization. *Molecular and Cellular Biology*, **30** (14), 3610–3619.

Noordermeer, J., Klingensmith, J., Perrimon, N., and Nusse, R. (1994) dishevelled and armadillo act in the wingless signalling pathway in Drosophila. *Nature*, **367** (6458), 80–83.

Ossipova, O., Dhawan, S., Sokol, S., and Green, J.B. (2005) Distinct PAR-1 proteins function in different branches of Wnt signaling during vertebrate development. *Developmental Cell*, **8** (6), 829–841.

Pan, W., Choi, S.C., Wang, H. *et al.* (2008) Wnt3a-mediated formation of phosphatidylinositol 4,5-bisphosphate regulates LRP6 phosphorylation. *Science*, **321** (5894), 1350–1353.

Park, T.J., Gray, R.S., Sato, A. *et al.* (2005) Subcellular localization and signaling properties of dishevelled in developing vertebrate embryos. *Current Biology*, **15** (11), 1039–1044.

Park, T.J., Mitchell, B.J., Abitua, P.B. *et al.* (2008) Dishevelled controls apical docking and planar polarization of basal bodies in ciliated epithelial cells. *Nature Genetics*, **40** (7), 871–879.

Peters, J.M., McKay, R.M., McKay, J.P., and Graff, J.M. (1999) Casein kinase I transduces Wnt signals. *Nature*, **401** (6751), 345–350.

Rosso, S.B., Sussman, D., Wynshaw-Boris, A., and Salinas, P.C. (2005) Wnt signaling through Dishevelled, Rac and JNK regulates dendritic development. *Nature Neuroscience*, **8** (1), 34–42.

Schwarz-Romond, T., Fiedler, M., Shibata, N. *et al.* (2007) The DIX domain of Dishevelled confers Wnt signaling by dynamic polymerization. *Nature Structural and Molecular Biology*, **14** (6), 484–492.

Shan, J., Zhang, X., Bao, J. *et al.* (2012) Synthesis of potent dishevelled PDZ domain inhibitors guided by virtual screening and NMR studies. *Chemical Biology & Drug Design*, **79** (4), 376–383.

Simons, M., Gloy, J., Ganner, A. *et al.* (2005) Inversin, the gene product mutated in nephronophthisis type II, functions as a molecular switch between Wnt signaling pathways. *Nature Genetics*, **37** (5), 537–543.

Simons, M., Gault, W.J., Gotthardt, D. *et al.* (2009) Electrochemical cues regulate assembly of the Frizzled/Dishevelled complex at the plasma membrane during planar epithelial polarization. *Nature Cell Biology*, **11** (3), 286–294.

Singh, J., Yanfeng, W.A., Grumolato, L. *et al.* (2010) Abelson family kinases regulate Frizzled planar cell polarity signaling via Dsh phosphorylation. *Genes & Development*, **24** (19), 2157–2168.

Strutt, H., Price, M.A., and Strutt, D. (2006) Planar polarity is positively regulated by casein kinase Iepsilon in Drosophila. *Current Biology*, **16** (13), 1329–1336.

Sun, T.Q., Lu, B., Feng, J.J. *et al.* (2001) PAR-1 is a Dishevelled-associated kinase and a positive regulator of Wnt signalling. *Nature Cell Biology*, **3** (7), 628–636.

Sussman, D.J., Klingensmith, J., Salinas, P. *et al.* (1994) Isolation and characterization of a mouse homolog of the Drosophila segment polarity gene dishevelled. *Developmental Biology*, **166** (1), 73–86.

Swiatek, W., Tsai, I.C., Klimowski, L. *et al.* (2004) Regulation of casein kinase I epsilon activity by Wnt signaling. *The Journal of Biological Chemistry*, **279** (13), 13011–13017.

Tanneberger, K., Pfister, A.S., Brauburger, K. *et al.* (2011) Amer1/WTX couples Wnt-induced formation of PtdIns(4,5)P2 to LRP6 phosphorylation. *The EMBO Journal*, **30** (8), 1433–1443.

Tauriello, D.V., Haegebarth, A., Kuper, I. *et al.* (2010) Loss of the tumor suppressor CYLD enhances Wnt/beta-catenin signaling through K63-linked ubiquitination of Dvl. *Molecular Cell*, **37** (5), 607–619.

Tauriello, D.V., Jordens, I., Kirchner, K. *et al.* (2012) Wnt/beta-catenin signaling requires interaction of the Dishevelled DEP domain and C terminus with a discontinuous motif in Frizzled. *Proceedings of the National Academy of Sciences of the United States of America*, **109** (14), E812–E820.

Theisen, H., Purcell, J., Bennett, M. *et al.* (1994) dishevelled is required during wingless signaling to establish both cell polarity and cell identity. *Development*, **120** (2), 347–360.

Tsang, M., Lijam, N., Yang, Y. *et al.* (1996) Isolation and characterization of mouse dishevelled-3. *Developmental Dynamics*, **207** (3), 253–262.

Wang, J., Hamblet, N.S., Mark, S. *et al.* (2006) Dishevelled genes mediate a conserved mammalian PCP pathway to regulate convergent extension during neurulation. *Development*, **133** (9), 1767–1778.

Wei, W., Li, M., Wang, J. *et al.* (2012) The E3 ubiquitin ligase ITCH negatively regulates canonical Wnt signaling by targeting dishevelled protein. *Molecular and Cellular Biology*, **32** (19), 3903–3912.

Willert, K., Brink, M., Wodarz, A. *et al.* (1997) Casein kinase 2 associates with and phosphorylates dishevelled. *The EMBO Journal*, **16** (11), 3089–3096.

Witte, F., Bernatik, O., Kirchner, K. *et al.* (2010) Negative regulation of Wnt signaling mediated by CK1-phosphorylated Dishevelled via Ror2. *The FASEB Journal*, **24** (7), 2417–2426.

Wong, H.C., Mao, J., Nguyen, J.T. *et al.* (2000) Structural basis of the recognition of the dishevelled DEP domain in the Wnt signaling pathway. *Nature Structural Biology*, **7** (12), 1178–1184.

Wong, H.C., Bourdelas, A., Krauss, A. *et al.* (2003) Direct binding of the PDZ domain of Dishevelled to a conserved internal sequence in the C-terminal region of Frizzled. *Molecular Cell*, **12** (5), 1251–1260.

Yanfeng, W.A., Berhane, H., Mola, M. *et al.* (2011) Functional dissection of phosphorylation of Disheveled in Drosophila. *Developmental Biology*, **360** (1), 132–142.

Yokoyama, N., Markova, N.G., Wang, H.Y., and Malbon, C.C. (2012) Assembly of Dishevelled 3-based supermolecular complexes via phosphorylation and Axin. *Journal of Molecular Signaling*, **7** (1), 8.

Zhang, L., Gao, X., Wen, J. *et al.* (2006) Dapper 1 antagonizes Wnt signaling by promoting dishevelled degradation. *The Journal of Biological Chemistry*, **281** (13), 8607–8612.

Zhang, Y., Appleton, B.A., Wiesmann, C. *et al.* (2009) Inhibition of Wnt signaling by Dishevelled PDZ peptides. *Nature Chemical Biology*, **5** (4), 217–219.

16 β-Catenin: a Key Player in Both Cell Adhesion and Wnt Signaling

Jonathan Pettitt
Institute of Medical Sciences, University of Aberdeen, Aberdeen, Scotland

Introduction

β-catenin was discovered independently by researchers investigating the molecular basis of cadherin cell adhesion in vertebrate epithelial cells (Ozawa, Baribault, and Kemler, 1989) and through genetic screens in *Drosophila* for mutants with defective embryonic segmentation (Peifer and Wieschaus, 1990), which was subsequently discovered to be a Wnt-dependent signaling event. Multiple studies revealed that both cell adhesion and cell signaling functions of β-catenin are conserved between *Drosophila* and vertebrates, leading to the concept of β-catenin as a dual-function molecule that mediates distinct adhesive and signaling activities, often in the same cell. The physiological consequences of integrating these two functions in one protein are still not fully understood, but β-catenin clearly has the potential to functionally coordinate Wnt signaling and cadherin-dependent adhesion events, which may be relevant to common developmental processes such as epithelial–mesenchymal transitions (EMTs) and the pathogenic transitions leading to cancers.

Cell adhesion function of β-catenin

One of the major roles for β-catenin is at the cell–cell adhesion junctions between epithelial cells. These junctions are required not just to allow individual cells to stick to one another to form an epithelial sheet but also to regulate and coordinate the behavior of individual epithelial cells during tissue development and morphogenesis (Takeichi, 2011). The cadherin–catenin complex forms one of several discrete adhesion junctions that are conserved throughout the animal kingdom (Oda and Takeichi, 2011). Cadherins are single-pass transmembrane proteins that form extracellular adhesion interfaces between adjacent cells (Pokutta and Weis, 2007). The search for components of cadherin-based adhesion junctions led to the identification of a group of proteins termed catenins (Ozawa, Baribault, and Kemler, 1989), the naming of which was derived from the Latin *catena*, meaning chain, referring to their apparent function in linking the transmembrane cadherin adhesion molecules to the actin cytoskeleton. Two of these, α- and β-catenin, are conserved in all metazoans (Oda and Takeichi, 2011).

Wnt Signaling in Development and Disease: Molecular Mechanisms and Biological Functions, First Edition.
Edited by Stefan Hoppler and Randall T. Moon.

Molecular characterization of the cadherin–catenin complex in multiple organisms has shown that the complex evolved prior to the metazoan radiation (Hiroki, 2012), and credible homologs of both α- and β-catenin have been identified in the nonmetazoan *Dictyostelium* (Dickinson, Nelson, and Weis, 2011; Grimson *et al.*, 2000), although cadherins appear to be absent. The complex, which is very well understood in terms of molecular interactions and structure, forms around the extracellular adhesion interface produced by the single-pass transmembrane classical cadherins (Brasch *et al.*, 2012). The cytoplasmic tails of classical cadherins contain a distinct motif that binds to β-catenin, and α-catenin subsequently binds to β-catenin; it is not itself able to directly bind cadherin (Pokutta *et al.*, 2008). The β-catenin-dependent recruitment of α-catenin is essential for cadherin-dependent adhesion and leads to localized changes in actin filament organization necessary for strong cell adhesion in both vertebrates and invertebrates (Pokutta *et al.*, 2008).

Wnt signaling function of β-catenin

Several lines of evidence place β-catenin in the Wnt signaling pathway. Firstly, screens for embryonic lethal mutations led to the identification of Armadillo (Arm), the *Drosophila* homolog of β-catenin (Nüsslein-Volhard and Wieschaus, 1980). Subsequent genetic analysis of a suite of mutations conferring related phenotypes led to the delineation of the basic Wnt signaling pathway in *Drosophila* and the placement of β-catenin at the regulatory center of the canonical Wnt signaling machinery. Secondly, studies in *Xenopus* showed that ectopic β-catenin expression caused axis duplication, a phenotype ultimately caused by ectopic canonical Wnt signaling activity (Funayama *et al.*, 1995). Finally, human β-catenin mutations that affect its regulation have been implicated in several types of cancer (Polakis, 2012). Taken together, these studies show that as in its function in cadherin adhesion, the Wnt signaling function of β-catenin is similarly conserved across the animal phyla, and both loss- and gain-of-function studies in multiple organisms and cultured cells demonstrate that it is an essential component of canonical Wnt signaling.

β-catenin is the major protein target of Wnt signaling, which acts to regulate its cytoplasmic and ultimately nuclear levels. In the absence of a Wnt signal, most cytoplasmic β-catenin is targeted for degradation by the proteasome, through the activity of a destruction complex, which results in the N-terminal phosphorylation of β-catenin (Chapter 3). β-catenin localized to cadherin adhesion junctions is protected from this because it is unable to interact with the destruction complex. In the presence of Wnt ligand, the levels of cytoplasmic β-catenin accumulate, allowing it to enter the nucleus and bind to TCF and activate transcription of Wnt target genes (Chapter 4).

Recent studies suggest that N-terminally phosphorylated β-catenin also accumulates at centrosomes, consistent with a recently identified third, less well-understood function for β-catenin in centrosome separation and possibly spindle orientation during mitosis (Bahmanyar *et al.*, 2008; Chilov *et al.*, 2011; Hadjihannas, Brückner, and Behrens, 2010; Huang, Senga, and Hamaguchi, 2007).

Relating structure to function

A key feature of β-catenin is the central, rigid superhelix formed by 12 imperfectly repeated motifs, termed Arm repeats after the eponymous *Drosophila* β-catenin homolog (Huber, Nelson, and Weis, 1997; Xing *et al.*, 2008; Figure 16.1). These repeats are not confined to β-catenin and are also found in another family of cadherin binding cytoplasmic proteins, the p120 catenins, as well as other more distantly related proteins (Tewari *et al.*, 2010). Each Arm repeat consists of three alpha-helices and displays repeat-specific primary sequence conservation indicative of functional specialization between repeats that has been maintained across large phylogenetic distances. The superhelix of β catenin forms an extended, positively charged groove that forms the binding interface between β-catenin and the cadherin cytoplasmic domain (Huber and Weis, 2001); however, this region is also involved in β-catenin's interaction with TCF and APC, and it is likely that interactions occurring along this binding interface are mutually exclusive (Sampietro *et al.*, 2006). This might suggest that the dual functions of β-catenin are interdependent; however, in both mice and *Drosophila*, β-catenin's two functions arise from distinct regions of the molecule (Orsulic and Peifer, 1996;

Valenta *et al.*, 2011). The molecular basis of this behavior is largely the effect of the less well-structured N- and C-terminal domains that flank the central Arm repeat domain (Gottardi and Peifer, 2008).

Choosing where to go and what to do

The key to the ability of β-catenin to mediate more than one function lies in the regulation of its subcellular localization. It can localize and function at both the cell–cell junctions and the nucleus (after accumulating in the cytoplasm) (Figure 16.1). This raises several intriguing and long-standing questions. What are the determinants that regulate the location and function of a multifunction protein? Do the different pools of β-catenin interact, and if they do, what are the functional consequences? Many of β-catenin's binding partners interact with the same site (s), so there is clearly potential for competitive interactions, and thus, association with the different functional compartments must be subject to regulation.

A key consideration when understanding the cell biology of β-catenin is that perturbation of β-catenin function could potentially affect cell adhesion, Wnt signaling, or both processes. In the absence of Wnt signaling, cytoplasmic levels of β-catenin are likely to be low since it would be targeted for destruction (Chapter 3). Thus, β-catenin released from cadherin junctions is not thought to be capable of activating Wnt target genes. However, in principle, if it could be protected from the vicissitudes of the destruction complex, release of β-catenin from cadherin junctions could enter the nucleus and lead to activation of Wnt target genes independent of the presence of Wnt signaling.

Besides the bulk disassembly of the cadherin–catenin complex, there are posttranslational modifications that affect the ability of β-catenin to bind to its cadherin and α-catenin binding partners. There are two key conserved tyrosine residues, at positions 142 and 654 (mammalian protein), which when phosphorylated reduce the binding of β-catenin to α-catenin and cadherin, respectively (Brembeck *et al.*, 2004; Bustos *et al.*, 2006; Huber and Weis, 2001; Piedra *et al.*, 2003; Roura *et al.*, 1999). The physiological relevance of the Y142 phosphorylation is not clear, however, and there is no strong evidence to link this modification to a specific β-catenin-dependent process in mammals. Moreover, Y142 modification has no obvious effect of β-catenin function

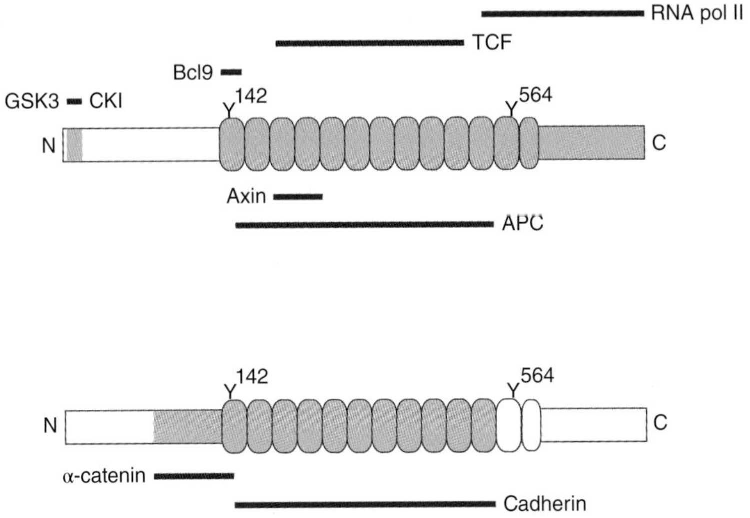

Figure 16.1 Major binding partners of β-catenin. Schematic representation of β-catenin structure, with the individual Arm repeats and the C-terminal helix (C) represented as rounded rectangles. Shading represents the regions of β-catenin that engage in binding partners involved in Wnt signaling (upper figure) and cadherin-mediated adhesion (lower figure). The horizontal lines represent known, experimentally determined binding interfaces with specific interaction partners. The two tyrosine residues that potentially regulate β-catenin function are indicated.

in *Drosophila* (Hoffmans and Basler, 2007). Nevertheless, Y142 and its surrounding region are highly conserved in all animals from sponges to vertebrates, suggesting that this residue plays an important role in β-catenin function.

In contrast, the evidence for a role for phosphorylation of Y654 is clearer. Conditional mouse mutants that express β-catenin harboring the phosphor-mimicking version of this residue (Y5654E) die as embryos with phenotypes consistent with constitutively activated canonical Wnt signaling (van Veelen *et al.*, 2011). However, despite this, cadherin-based cell adhesion does not seem to be affected, indicating that *in vivo* the mutation does not appreciably deplete β-catenin levels at the junction. Thus, phosphorylation of this residue is not sufficient to drive a global change in the localization of β-catenin from the junctions to the cytoplasm/nucleus. The effect of this mutation on Wnt signaling is thus not clear, but might reflect the limited release of small amounts of β-catenin from the junctions, coupled with an effect on the conformation of β-catenin increasing the affinity for transcriptional coactivators (Piedra *et al.*, 2001; Xing *et al.*, 2008).

Epithelial cells that lose the ability to express their main cadherin, E-cadherin, undergo a process termed EMT, whereby the cell undergoes profound changes in its cellular phenotype and leaves the epithelial sheet (Nieto, 2011). At present, beyond such bulk disassembly of cadherin junctions, there are no known single modifications that lead to recruitment of β-catenin from cell junctions to the cytoplasm/nucleus. However, Gottardi and Gumbiner have reported a conformational change that favors Wnt signaling over cadherin binding (Gottardi and Gumbiner, 2004). They found that β-catenin in cells exposed to a Wnt signal underwent a conformational shift such that the C-terminus folds back to interact with the Arm repeat region, effectively blocking the interaction with cadherin, but leaving the binding of TCF unaffected. It is not clear whether this conformation of β-catenin is also protected from the activity of the destruction complex, but this work shows that conformational changes that affect the protein–protein interactions experienced by β-catenin can shift the balance between the subcellular locations to which it is destined.

As well as interacting with the cadherin–catenin complex, TCF and its associated coactivators, and the destruction complex, β-catenin also interacts with a conserved pair of proteins, BCL9/Legless and Pygopus (Brembeck *et al.*, 2004; Kramps *et al.*, 2002). In *Drosophila*, both proteins are essential for Wnt signaling, whereas in mammals, which have two Bcl9/Legless paralogs and two Pygopus homologs, the roles of these proteins in Wnt signaling are not essential for the canonical Wnt signaling pathway in general, and they appear to be restricted to the regulation of Wnt signaling in specific tissues and developmental events (Brack *et al.*, 2009; Deka *et al.*, 2010; Gu *et al.*, 2009; Kennedy *et al.*, 2010).

Bcl9/Legless binds to the first Arm repeat of β-catenin, and it acts as a linker between β-catenin and Pygopus (Kramps *et al.*, 2002; Sampietro *et al.*, 2006). The role of Pygopus based on experiments in *Drosophila* is as a transcriptional activator in association with β-catenin (Hoffmans and Basler, 2004; Städeli and Basler, 2005). However, the precise function of Pygopus is still unclear, and there is evidence that it can positively regulate Wnt target genes independent of its interaction with β-catenin, apparently antagonizing the activity of the Groucho transcriptional repressor (Mieszczanek, de La Roche, and Bienz, 2008). In addition, Pygopus possesses a plant homology domain (PHD) through which it is able to interact with methylated histones (Bienz, 2006; Fiedler *et al.*, 2008). There exist differences between this aspect of Pygopus function when the *Drosophila* and vertebrate homologs are compared, which make it hard to determine the conserved functions of this protein in Wnt signaling.

Subfunctionalization of β-catenins

Studies in both *Drosophila* and mice have shown that the cell adhesion and Wnt signaling functions of β-catenin can be separately mutated, showing that they are mediated by distinct motifs (Orsulic and Peifer, 1996; Valenta *et al.*, 2011). Strikingly, there are a number of evolutionary lineages where these functions have become completely partitioned into separate proteins through the processes of gene duplication of the ancestral β-catenin and subfunctionalization of the resulting paralogs (Schneider, Finnerty, and Martindale, 2003).

In vertebrates, a gene duplication of the ancestral β-catenin gene resulted in the current vertebrate β-catenin, which has likely retained the ancestral β-catenin functions, and the novel plakoglobin, which has undergone changes that have reduced its ability to interact with the Wnt signaling pathway and allowed it to interact with a specialized class of cadherins, the desmogleins (Butz *et al.*, 1992). This may have facilitated the formation of a derived, vertebrate-specific adhesion junction known as the desmosome.

A later gene duplication also occurred in the lineage leading to zebrafish and pufferfish species (Zhang *et al.*, 2012); however, although there is some functional diversification between the two paralogs, both appear to have retained the ability to function in Wnt signaling, and there is no evidence that either have lost the ability to participate in cadherin-mediated cell adhesion.

Subfunctionalization of β-catenins has also been reported for the planarian *Schmidtea mediterranea* (Chai *et al.*, 2010), where there is evidence for partitioning of the cell adhesion and Wnt signaling functions into two distinct β-catenins, giving rise to monofunctional β-catenins that are capable of participating only in either Wnt signaling or cell adhesion, at least in terms of their interactions with the various β-catenin binding partners (see also Chapter 26).

However, the most dramatic example of β-catenin subfunctionalization has occurred in the lineage leading to the nematode *Caenorhabditis elegans*. This organism has four β-catenin homologs, HMP-2, BAR-1, WRM-1, and SYS-1, which are the products of serial gene duplication events (Korswagen, Herman, and Clevers, 2000; Liu *et al.*, 2008; Natarajan, Witwer, and Eisenmann, 2001). Comparison between nematode genomes reveals that the first gene duplication clearly took place after the radiation of the major nematode clades (Pettitt *et al.*, manuscript in preparation), giving rise to the ancestral *hmp-2* and *bar-1* genes. Later gene duplications, likely involving the ancestral *bar-1* gene, generated *wrm-1* and *sys-1* in the *Caenorhabditis* lineage (Liu *et al.*, 2008). All four of the *C. elegans* β-catenins are highly derived in terms of primary sequence; however, WRM-1 and SYS-1 are extreme cases, with SYS-1 only being recognized as a β-catenin after it was implicated in Wnt signaling through functional studies and its tertiary structure determined (Kidd *et al.*, 2005; Liu *et al.*, 2008).

C. elegans HMP-2 appears to have lost the ability to be regulated by Wnt signaling, and its normal function appears to be confined to the cadherin–catenin complex (Costa *et al.*, 1998; Korswagen, Herman, and Clevers, 2000). In contrast, BAR-1 is not able to interact with either cadherin or α-catenin, but has retained the ancestral β-catenin role as an essential regulatory component in the canonical Wnt signaling pathway (Korswagen, Herman, and Clevers, 2000; Natarajan, Witwer, and Eisenmann, 2001). Thus, it is clear that the two β-catenin homologs have undergone subfunctionalization, partitioning the cell adhesion and Wnt signaling functions of the ancestral β-catenin into two separate proteins.

The situation for WRM-1 and SYS-1 is less straightforward since these molecules appear to have evolved highly derived functions, participating in a noncanonical Wnt signaling pathway, which seems likely to have arisen after the evolution of the major nematode lineages. This pathway, termed the Wnt/β-catenin asymmetry pathway (WβA), is used to direct asymmetric cell fate determination in many, and possibly all, anterior–posterior cell divisions (Mizumoto and Sawa, 2007; Phillips and Kimble, 2009). As in the canonical Wnt signaling pathway, the WβA pathway leads to the activation of target genes through the modulation of TCF transcriptional activity. However, unlike this pathway, the WβA pathway employs two β-catenins, and their roles are clearly distinct. The function of WRM-1 is to mediate the phosphorylation and nuclear export of TCF, whereas SYS-1 acts as a transcriptional coactivator of TCF. The current model for this unusual dual β-catenin requirement is that high levels of TCF repress Wnt target gene expression, even in the presence of the SYS-1 transcriptional activator; thus, WRM-1 is required to reduce the levels of this repressive form of TCF and allow the SYS-1 activated version of TCF to predominate and transcriptionally activate Wnt target genes (Mizumoto and Sawa, 2007; Phillips and Kimble, 2009). While it is clear that this highly derived pathway is not present in other organisms (or even in other nematodes), it is possible that some aspects of the pathway might operate more generally in canonical Wnt signaling (Chapter 17).

Perspective

The evolution of β-catenin as an essential component in two key processes within animal cells, cell adhesion and cell signaling, clearly occurred prior to the radiation of the major animal groups. With only a few notable exceptions, these two functions have been maintained in the diverse lineages leading to the major animal phyla. The conservation of a single, dual-function protein in both processes throughout animal evolution has led to speculations that there has been selection for a functional link between the two processes. However, despite a wealth of proposed mechanisms, to date, there is no strong evidence for any physiologically relevant process that would indicate that the junctional and signaling pools of β-catenin functionally interact. The picture that emerges from several decades of studying β-catenin function is that its two functions are spatially and functionally partitioned within the cell. This has important implications not only for understanding the biology of β-catenin but also for developing therapeutic interventions that specifically target β-catenin's Wnt signaling function.

References

Bahmanyar, S., Kaplan, D.D., Deluca, J.G. *et al.* (2008) beta-Catenin is a Nek2 substrate involved in centrosome separation. *Genes & Development*, **22**, 91–105.

Bienz, M. (2006) The PHD finger, a nuclear protein-interaction domain. *Trends in Biochemical Science*, **31**, 35–40.

Brack, A.S., Murphy-Seiler, F., Hanifi, J. *et al.* (2009) BCL9 is an essential component of canonical Wnt signaling that mediates the differentiation of myogenic progenitors during muscle regeneration. *Developmental Biology*, **335**, 1–51.

Brasch, J., Harrison, O.J., Honig, B., and Shapiro, L. (2012) Thinking outside the cell: how cadherins drive adhesion. *Trends in Cell Biology*, **22**, 299–310.

Brembeck, F.H., Schwarz-Romond, T., Bakkers, J. *et al.* (2004) Essential role of BCL9-2 in the switch between beta-catenin's adhesive and transcriptional functions. *Genes & Development*, **18**, 2225–2230.

Bustos, V.H., Ferrarese, A., Venerando, A. *et al.* (2006) The first armadillo repeat is involved in the recognition and regulation of beta-catenin phosphorylation by protein kinase CK1. *Proceedings of the National Academy of Sciences of the United States of America*, **103**, 19725–19730.

Butz, S., Stappert, J., Weissig, H., and Kemler, R. (1992) Plakoglobin and beta-catenin: distinct but closely related. *Science*, **257**, 1142–1144.

Chai, G., Ma, C., Bao, K. *et al.* (2010) Complete functional segregation of planarian-catenin-1 and -2 in mediating Wnt signaling and cell adhesion. *The Journal of Biological Chemistry*, **285**, 24120–24130.

Chilov, D., Sinjushina, N., Rita, H. *et al.* (2011) Phosphorylated β-catenin localizes to centrosomes of neuronal progenitors and is required for cell polarity and neurogenesis in developing midbrain. *Developmental Biology*, **357**, 259–268.

Costa, M., Raich, W., Agbunag, C. *et al.* (1998) A putative catenin-cadherin system mediates morphogenesis of the *Caenorhabditis elegans* embryo. *The Journal of Cell Biology*, **141**, 297–308.

Deka, J., Wiedemann, N., Anderle, P. *et al.* (2010) Bcl9/Bcl9l are critical for Wnt-mediated regulation of stem cell traits in colon epithelium and adenocarcinomas. *Cancer Research*, **70**, 6619–6628.

Dickinson, D.J., Nelson, W.J., and Weis, W.I. (2011) A polarized epithelium organized by beta- and alpha-catenin predates cadherin and metazoan origins. *Science*, **331**, 1336–1339.

Fiedler, M., Sánchez-Barrena, M.J., Nekrasov, M. *et al.* (2008) Decoding of methylated histone H3 tail by the Pygo-BCL9 Wnt signaling complex. *Molecular Cell*, **30**, 507–518.

Funayama, N., Fagotto, F., McCrea, P., and Gumbiner, B.M. (1995) Embryonic axis induction by the armadillo repeat domain of beta-catenin: evidence for intracellular signaling. *The Journal of Cell Biology*, **128**, 959–968.

Gottardi, C.J. and Gumbiner, B.M. (2004) Distinct molecular forms of beta-catenin are targeted to adhesive or transcriptional complexes. *The Journal of Cell Biology*, **167**, 339–349.

Gottardi, C.J. and Peifer, M. (2008) Terminal regions of beta-catenin come into view. *Structure*, **16**, 336–338.

Grimson, M.J., Coates, J.C., Reynolds, J.P. *et al.* (2000) Adherens junctions and beta-catenin-mediated cell signalling in a non-metazoan organism. *Nature*, **408**, 727–731.

Gu, B., Sun, P., Yuan, Y. *et al.* (2009) Pygo2 expands mammary progenitor cells by facilitating histone H3 K4 methylation. *The Journal of Cell Biology*, **185**, 811–826.

Hadjihannas, M.V., Brückner, M., and Behrens, J. (2010) Conductin/axin2 and Wnt signalling regulates centrosome cohesion. *EMBO Reports*, **11**, 317–324.

Hiroki, O. (2012) Evolution of the cadherin-catenin complex. *Subcellular Biochemistry*, **60**, 9–35.

Hoffmans, R. and Basler, K. (2004) Identification and in vivo role of the Armadillo-Legless interaction. *Development*, **131**, 4393–4400.

Hoffmans, R. and Basler, K. (2007) BCL9-2 binds Arm/beta-catenin in a Tyr142-independent manner and requires Pygopus for its function in Wg/Wnt signaling. *Mechanisms of Development*, **124**, 59–67.

Huang, P., Senga, T., and Hamaguchi, M. (2007) A novel role of phospho-beta-catenin in microtubule regrowth at centrosome. *Oncogene*, **26**, 4357–4371.

Huber, A.H. and Weis, W.I. (2001) The structure of the beta-catenin/E-cadherin complex and the molecular basis of diverse ligand recognition by beta-catenin. *Cell*, **105**, 391–402.

Huber, A.H., Nelson, W.J., and Weis, W.I. (1997) Three-dimensional structure of the armadillo repeat region of beta-catenin. *Cell*, **90**, 871–882.

Kennedy, M.W., Cha, S.-W., Tadjuidje, E. *et al.* (2010) A co-dependent requirement of xBcl9 and Pygopus for embryonic body axis development in Xenopus. *Developmental Dynamics*, **239**, 271–283.

Kidd, A.R., Miskowski, J.A., Siegfried, K.R. *et al.* (2005) A beta-catenin identified by functional rather than sequence criteria and its role in Wnt/MAPK signaling. *Cell*, **121**, 761–772.

Korswagen, H.C., Herman, M.A., and Clevers, H.C. (2000) Distinct beta-catenins mediate adhesion and signalling functions in *C. elegans*. *Nature*, **406**, 527–532.

Kramps, T., Peter, O., Brunner, E. *et al.* (2002) Wnt/wingless signaling requires BCL9/legless-mediated recruitment of pygopus to the nuclear beta-catenin-TCF complex. *Cell*, **109**, 47–60.

Liu, J., Phillips, B.T., Amaya, M.F. *et al.* (2008) The *C. elegans* SYS-1 protein is a bona fide beta-catenin. *Developmental Cell*, **14**, 751–761.

Mieszczanek, J., de La Roche, M., and Bienz, M. (2008) A role of Pygopus as an anti-repressor in facilitating Wnt-dependent transcription. *Proceedings of the National Academy of Sciences of the United States of America*, **105**, 19324–19329.

Mizumoto, K. and Sawa, H. (2007) Two betas or not two betas: regulation of asymmetric division by beta-catenin. *Trends in Cell Biology*, **17**, 465–473.

Natarajan, L., Witwer, N.E., and Eisenmann, D.M. (2001) The divergent *Caenorhabditis elegans* beta-catenin proteins BAR-1, WRM-1 and HMP-2 make distinct protein interactions but retain functional redundancy in vivo. *Genetics*, **159**, 159–172.

Nieto, M.A. (2011) The ins and outs of the epithelial to mesenchymal transition in health and disease. *Annual Reviews of Cell and Developmental Biology*, **27**, 347–376.

Nüsslein-Volhard, C. and Wieschaus, E. (1980) Mutations affecting segment number and polarity in *Drosophila*. *Nature*, **287**, 795–801.

Oda, H. and Takeichi, M. (2011) Evolution: structural and functional diversity of cadherin at the adherens junction. *The Journal of Cell Biology*, **193**, 1137–1146.

Orsulic, S. and Peifer, M. (1996) An in vivo structure-function study of armadillo, the beta-catenin homolog, reveals both separate and overlapping regions of the protein required for cell adhesion and for wingless signaling. *The Journal of Cell Biology*, **134**, 1283–1300.

Ozawa, M., Baribault, H., and Kemler, R. (1989) The cytoplasmic domain of the cell adhesion molecule uvomorulin associates with three independent proteins structurally related in different species. *The EMBO Journal*, **8**, 1711–1717.

Peifer, M. and Wieschaus, E. (1990) The segment polarity gene armadillo encodes a functionally modular protein that is the *Drosophila* homolog of human plakoglobin. *Cell*, **63**, 1167–1176.

Phillips, B.T. and Kimble, J. (2009) A new look at TCF and beta-catenin through the lens of a divergent *C. elegans* Wnt pathway. *Developmental Cell*, **17**, 27–34.

Piedra, J., Martinez, D., Castano, J. *et al.* (2001) Regulation of beta-catenin structure and activity by tyrosine phosphorylation. *The Journal of Biological Chemistry*, **276**, 20436–20443.

Piedra, J., Miravet, S., Castaño, J. *et al.* (2003) p120 Catenin-associated Fer and Fyn tyrosine kinases regulate beta-catenin Tyr-142 phosphorylation and beta-catenin-alpha-catenin Interaction. *Molecular and Cellular Biology*, **23**, 2287–2297.

Pokutta, S. and Weis, W.I. (2007) Structure and mechanism of cadherins and catenins in cell-cell contacts. *Annual Reviews of Cell and Developmental Biology*, **23**, 237–261.

Pokutta, S., Drees, F., Yamada, S. *et al.* (2008) Biochemical and structural analysis of alpha-catenin in cell-cell contacts. *Biochemical Society Transactions*, **36**, 141–147.

Polakis, P. (2012) Wnt signaling in cancer. *Cold Spring Harbor Perspectives in Biology*, **4**, a008052.

Roura, S., Miravet, S., Piedra, J. *et al.* (1999) Regulation of E-cadherin/Catenin association by tyrosine phosphorylation. *The Journal of Biological Chemistry*, **274**, 36734–36740.

Sampietro, J., Dahlberg, C.L., Cho, U.S. *et al.* (2006) Crystal structure of a beta-catenin/BCL9/Tcf4 complex. *Molecular Cell*, **24**, 293–300.

Schneider, S.Q., Finnerty, J.R., and Martindale, M.Q. (2003) Protein evolution: structure–function relationships of the oncogene beta-catenin in the evolution of multicellular animals. *Journal of Experimental Zoology Part B: Molecular and Developmental Evolution*, **295**, 25–44.

Städeli, R. and Basler, K. (2005) Dissecting nuclear Wingless signalling: recruitment of the transcriptional co-activator Pygopus by a chain of adaptor proteins. *Mechanisms of Development*, **122**, 1171–1182.

Takeichi, M. (2011) Self-organization of animal tissues: cadherin-mediated processes. *Developmental Cell*, **21**, 24–26.

Tewari, R., Bailes, E., Bunting, K.A., and Coates, J.C. (2010) Armadillo-repeat protein functions: questions for little creatures. *Trends in Cell Biology*, **20**, 470–481.

Valenta, T., Gay, M., Steiner, S. *et al.* (2011) Probing transcription-specific outputs of β-catenin in vivo. *Genes & Development*, **25**, 2631–2643.

van Veelen, W., Le, N.H., Helvensteijn, W. *et al.* (2011) β-catenin tyrosine 654 phosphorylation increases Wnt signalling and intestinal tumorigenesis. *Gut*, **60**, 1204–1212.

Xing, Y., Takemaru, K., Liu, J. *et al.* (2008) Crystal structure of a full-length β-catenin. *Structure*, **16**, 478–487.

Zhang, M., Zhang, J., Lin, S.-C., and Meng, A. (2012) β-Catenin 1 and β-catenin 2 play similar and distinct roles in left-right asymmetric development of zebrafish embryos. *Development*, **139**, 2009–2019.

17 Evolutionary Diversification of Vertebrate TCF/LEF Structure, Function, and Regulation

Stefan Hoppler[1] and Marian L. Waterman[2]

[1]Institute of Medical Sciences, University of Aberdeen, Aberdeen, Scotland
[2]Department of Microbiology and Molecular Genetics, University of California, Irvine, CA, USA

Genome duplications enrich the vertebrate complement of Wnt signaling components

Two rounds of genome duplications occurred early in the evolution of vertebrates (Dehal and Boore, 2005). The most well-known consequence are the four Hox gene clusters in vertebrates (Duboule, 2007), but genome duplication has also affected other important gene families (e.g., *GATA* genes; Gillis *et al.*, 2009). As a general rule, most genome duplications tend to be lost (Van de Peer, Maere, and Meyer, 2009), and indeed, the complete complement of Wnt signaling components in vertebrates is not four times larger than in invertebrates (see Chapter 12). The most notable exception, and the most relevant to this chapter, is the TCF/LEF family of DNA-binding factors. TCF/LEFs are the major mediators of Wnt/β-catenin-dependent regulation of gene expression in the nucleus (see Chapter 4). Multiple copies of TCF/LEF genes have been retained from genome duplications since vertebrates commonly have four TCF/LEF family genes (four paralogs; see Figure 17.1), *TCF7* (also known as *Tcf1*), *LEF1* (*TCF7L3*), *TCF7l1* (*Tcf3*), and *TCF7l2* (*Tcf4*), while invertebrates typically only have one TCF/LEF gene.

TCF/LEF gene duplications allowed the different vertebrate TCF genes to assume more specialized (subfunctionalization) and novel (neofunctionalization) structures, functions, and regulatory mechanisms (Klingel *et al.*, 2012), a degree of diversification that seems necessary for mediating the complex functional roles of Wnt/β-catenin signaling in vertebrates. One might tilt the argument on its head to suggest that it is only because the different vertebrate TCF genes have assumed specialized and novel biological functions that they were retained during subsequent vertebrate evolutionary radiation. This argument further suggests that the major innovations in vertebrate Wnt signaling therefore pertain particularly to nuclear Wnt signaling mechanisms and the transcriptional response to Wnt signaling. These vertebrate innovations in TCF/LEF protein structure, function, and regulation need to be understood to appreciate the innovation of added complexity of Wnt signaling mechanisms and functions in vertebrates.

Wnt Signaling in Development and Disease: Molecular Mechanisms and Biological Functions, First Edition.
Edited by Stefan Hoppler and Randall T. Moon.

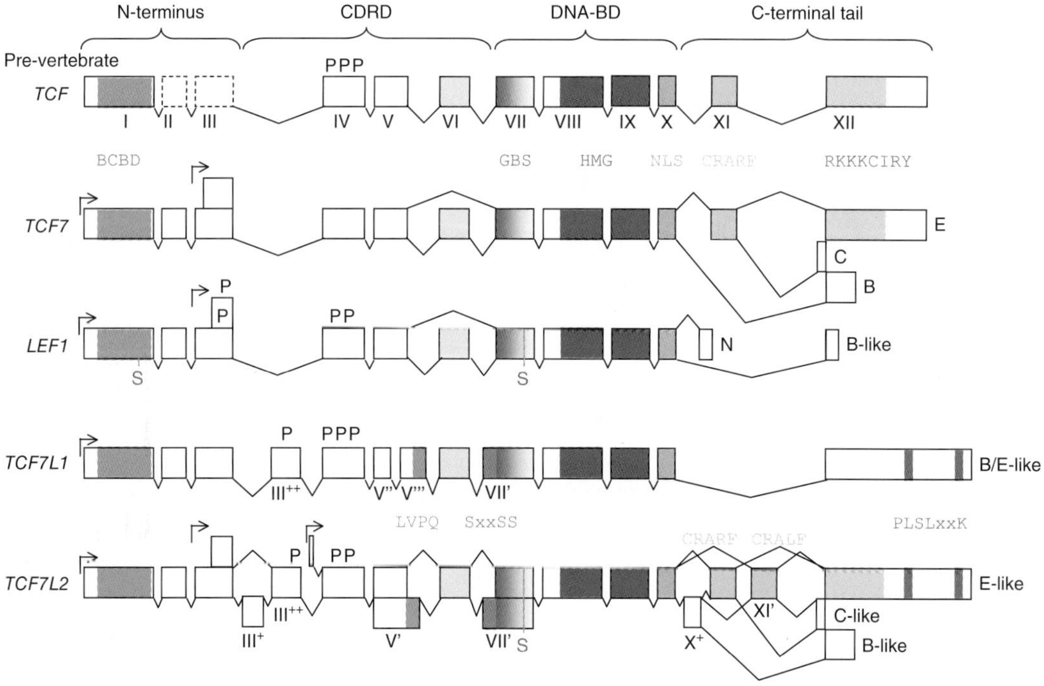

Figure 17.1 Conservation and innovation in exon structure and functional domains of TCF proteins. The exon structures of vertebrate TCF/LEF genes are variations on a conserved exon structure inherited from a single prevertebrate ancestral TCF gene. Invertebrate TCF/LEF proteins are often encoded by 11 conserved exons; however, the split of the original ninth exon meant that the shared ancestor of all vertebrate TCF/LEF proteins must have been encoded by 12 exons (here in roman numbers, to differentiate them from the Arabic numbers used in gene-specific literature). Experimentally identified functional domains of TCF proteins generally correspond to different exons or groups of exons, such as the BCBD; the groucho binding sequence (GBS)/TLE transcriptional corepressor binding sequence; the HMG-fold DNA-binding domain, a basic domain that assists in DNA binding and functions as an NLS; and additional domains at the C-terminus of certain TCF proteins, such as a short polypeptide sequence CRARF (or CRALF), which together with another polypeptide sequence (RKKKCIRY) forms an auxiliary DNA-binding domain called C-clamp and a proposed CtBP transcriptional corepressor binding domain (PLSLxxK). Vertebrate genomes generally have four TCF/LEF genes, *TCF7* (also called *TCF1*), *LEF1* (*TCF7L3*), *TCF7L1* (also called *TCF3*), and *TCF7L2* (also referred to as *TCF4*), which (apart from *TCF7L1*) encode several protein isoforms each, which are generated through alternative transcription and translation start sites and alternative splicing. Four major domains in TCF/LEF proteins can thus be identified: an N-terminus (containing the BCBD), a CDRD situated between the N-terminus and the core DNA-binding domain, and a C-terminal tail. The protein-coding exon structure of the major TCF7 transcript is most similar to the prevertebrate ancestral TCF/LEF gene (apart from the otherwise conserved phosphorylation sites; see following text). Alternative use of different transcription and translation start sites modify the N-terminus, but differences in splicing primarily affect the CDRD and the structure of the C-terminus, while the DNA-binding domain remains relatively unaltered. Some vertebrate TCF/LEF genes encode one (*TCF7L1*) or few (*LEF1*) isoforms, while others (*TCF7* and *TCF7L1*) encode many. Through differences in their domain structures, vertebrate TCF/LEF proteins can become specialists at mediating transcriptional activation (LEF1 and several TCF7 isoforms) or transcriptional repression (TCF7L1 and several TCF7L2 isoforms). Further evolved differences in the protein sequence within the CDRD release TCF7 proteins from being regulated by phosphorylation that regulates DNA binding and nuclear export of other TCF/LEF proteins (for additional complexity in exon structure of mammalian TCF/LEF genes and in particular *TCF7L2*, consult Mao and Byers, 2011; Weise *et al.*, 2010). Length of illustrated exons is proportional to relevant peptide sequences in *TCF7*. *(See insert for color representation of the figure.)*

Shared features of invertebrate and vertebrate TCF structure and function

Vertebrate and invertebrate TCF/LEF genes share inherited structures and functions. TCF/LEF proteins are DNA-binding proteins that provide a platform on target genes containing a Wnt response element (WRE) for assembly of multiprotein complexes that include non-DNA-binding transcriptional coregulators (see Chapter 4). DNA-associated TCF/LEF proteins mediate transcriptional activation when they are bound by nuclear β-catenin, which links to transcriptional coactivators (see Chapter 16), or they mediate transcriptional repression when bound by transcriptional corepressors (transcriptional switch, Figure 17.2a and b; see also Chapter 4).

Bipartite *cis*-regulatory DNA sequences (i.e., WREs) have recently been described for invertebrate (Chang *et al.*, 2008) and vertebrate (Atcha *et al.*, 2007) TCF/LEF proteins consisting of a core site (e.g., Van de Wetering *et al.*, 1997) comprising the nucleotide sequence CTTTGWW (where W stands for A or T) and after a gap a second less well-defined, GC-rich helper site (Atcha *et al.*, 2007). The interaction with this bipartite DNA recognition sequence is reflected in the conserved domain structure of TCF/LEF proteins, as are the binding domains for β-catenin and transcriptional corepressors.

The prevertebrate ancestral TCF/LEF gene must have contained 12 protein-coding exons (I–XII) to produce a conserved TCF/LEF protein structure, which we describe as four major functional parts (see Figure 17.1):

(1) The N-terminal well-defined β-catenin binding domain (BCBD) is the site through which Wnt signaling mainly regulates TCF activity and is encoded by the first protein-coding exon I followed by two less well-conserved exons II + III.
(2) The context-dependent regulatory domain (CDRD) is less conserved and appears structurally flexible but functions to integrate additional levels of regulation and is encoded by protein-coding exons IV–VI.
(3) The core of TCF/LEF proteins contains a well-defined DNA-binding domain (high-mobility group (HMG) fold), preceded by a less well-defined binding sequence for the groucho/TLE transcriptional corepressor (GBS) and followed by a basic nuclear localization signal (NLS), encoded by protein-coding exons VII–X.
(4) The C-terminal tail containing an auxiliary DNA-binding domain (C-clamp) encoded by exons XI and XII.

Wnt signaling directs nuclear localization of β-catenin, which via its binding to the N-terminal BCBD domain of TCF/LEF proteins is recruited to Wnt regulatory elements (WREs) for transcription activation (Behrens *et al.*, 1996;

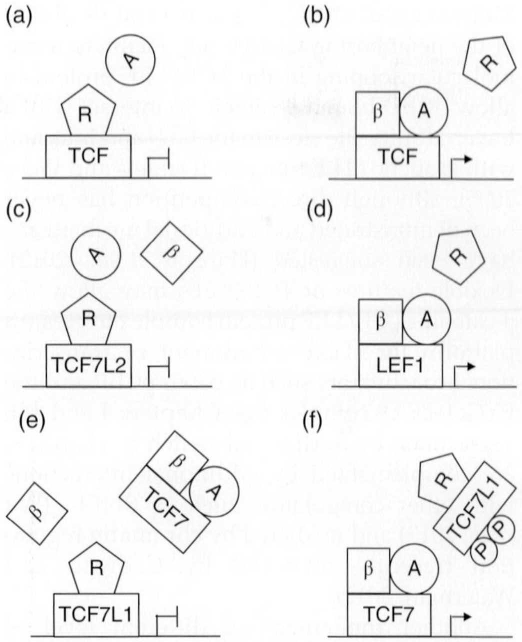

Figure 17.2 β-catenin-dependent transcriptional switch and TCF exchange mechanisms regulate vertebrate Wnt target gene expression. TCF/LEF proteins provide a platform bound to specific DNA sequences for the assembly of multiprotein complexes to regulate target gene transcription. Wnt signaling controls the nuclear concentration of β-catenin to affect a so-called transcriptional switch in both vertebrates and invertebrates (a, b). With low nuclear β-catenin levels, DNA-bound TCF proteins associate with transcriptional corepressors ("R") to repress target gene expression (a), but when Wnt signaling causes higher nuclear levels of β-catenin, which binds to these TCF proteins and recruits transcriptional coactivators ("A"), target gene expression is increased (b) (for more detail, see Chapter 4). The heterogeneity of TCF/LEF proteins encoded by vertebrate TCF/LEF genes (Figure 17.1) allows for the so-called TCF exchange mechanisms to further influence Wnt target gene expression, at a range of intermediate levels of nuclear β-catenin (c–f). Differential gene expression of TCF/LEF proteins that are either adept at repression (such as certain isoforms of TCF7L2) (c) or better at activation (such as LEF1) (d) will result in repression or expression of Wnt target genes, respectively, even at similar intermediate levels of nuclear β-catenin (e.g., Eichhoff *et al.*, 2011). Phosphorylation of some vertebrate TCF/LEF proteins regulates the strength of their DNA association and can therefore affect a TCF exchange and to regulate Wnt target gene expression even if TCF/LEF repressor and activator proteins are coexpressed in the same cell (e, f). TCF7L1 proteins mediate strong transcriptional repression at intermediate levels of nuclear β-catenin even in the presence of TCF7 isoforms that could mediate transcriptional activation (e); however, phosphorylation of TCF7L1 reduces its affinity to DNA and allows the TCF7 isoform to replace them in order to activate target gene expression, since TCF7 proteins are refractory to this regulation by phosphorylation (f) (Hikasa *et al.*, 2010).

Molenaar *et al.*, 1996). The structural flexibility of the neighboring CDRD may facilitate intramolecular looping in the TCF/LEF protein to allow BCBD-bound β-catenin to interact with a lower-affinity site close to the GBS and interfere with groucho/TLE function (Daniels and Weis, 2005), although direct competition has never been demonstrated and additional mechanisms have been suggested (Hanson *et al.*, 2012). Flexible features of TCF/LEFs may allow the β-catenin–TCF/LEF protein complex to create a platform for stable recruitment of transcriptional coactivators such as p300/pCBP and the PYGO–BCL9 complex (see Chapters 4 and 16). These main transcriptional switch mechanisms are complemented by additional interactions with other coregulators such as RNF14 (Wu *et al.*, 2013) and mediated by chromatin regulation (recently reviewed by Cadigan and Waterman, 2012).

Another fundamentally different level of regulation has recently come to prominence (reviewed by Sokol, 2011). TCF/LEF proteins are regulated by protein kinases, such as HIPK2 (Hikasa *et al.*, 2010), TNIK (Mahmoudi *et al.*, 2009), and NLK (Ishitani *et al.*, 2003; Ota *et al.*, 2012; Shetty *et al.*, 2005). They target at least partially overlapping phosphorylation sites in the CDRD (Figure 17.1). Phosphorylation can regulate TCF/LEF protein activity by promoting their dissociation from DNA (Hikasa *et al.*, 2010; Ishitani *et al.*, 2003) and their nuclear export (in *Caenorhabditis elegans*; Shetty *et al.*, 2005) but in some cases also by promoting their activating function (Mahmoudi *et al.*, 2009; Ota *et al.*, 2012). β-catenin is often required for regulation of TCF/LEF proteins by phosphorylation, which suggests both that these kinases form part of a Wnt signaling pathway and that they specifically regulate TCF/LEF transcriptional activators. Wnt signaling may therefore control TCF/LEF-mediated transcription not only by regulating nuclear localization of β-catenin but also nuclear localization and DNA-binding abilities of TCF/LEF proteins. The Wnt signaling pathway may therefore establish an optimal ratio between nuclear β-catenin and available TCF/LEF proteins that are competent for transcriptional activation (Goentoro and Kirschner, 2009; Phillips and Kimble, 2009; reviewed by Sokol, 2011). That is, a scarcity of TCF/LEF proteins would prevent nuclear β-catenin from reaching WREs to activate target gene transcription, while a surfeit of TCF/LEF proteins could form competing complexes with transcriptional corepressors to repress target gene transcription.

Invertebrate TCF/LEF proteins deviate little from this general structure and have few alternative isoforms (TCF/LEF homogeneity), whereas in vertebrates, evolutionary radiation created a multitude of different TCF/LEF isoforms (TCF heterogeneity). This major innovation profoundly enriches the functional complexity of Wnt signaling in vertebrates. In order to understand this complexity of Wnt signaling in vertebrates, the structural mechanisms, functional consequences, and molecular regulation of this TCF heterogeneity needs to be analyzed in detail.

Novel and specialized TCF structures and functions created by differential transcriptional regulation

Differential expression of vertebrate TCF genes

Duplicated genes tend to be retained in evolution when they assume novel or more specialized functions. They become nonredundant and essential (Dehal and Boore, 2005; Klingel *et al.*, 2012; Van de Peer, Maere, and Meyer, 2009). One way in which duplicated genes are retained is through evolution of unique expression patterns. In fact, differential regulation of gene expression of the four TCF/LEF paralog genes in vertebrates may have been the initial pressure to retain four gene copies during vertebrate evolution. Vertebrate TCF/LEF genes are differentially expressed in both exclusive and overlapping patterns in different tissues and at different stages of development. Genetic analysis of vertebrate TCF/LEF genes reflects these patterns of expression in that knockout phenotypes reveal only partial redundancy and each family member exhibits unique phenotypes in specific tissues (e.g., Galceran *et al.*, 1999; Gregorieff, Grosschedl, and Clevers, 2004).

Since each vertebrate TCF/LEF locus produces proteins that carry out divergent molecular activities (see following text), differential TCF/LEF gene expression strongly influences the tissue-specific response to Wnt signaling. The regulation of Wnt target genes in vertebrates is therefore not only determined by the ratio of nuclear β-catenin to TCF/LEF proteins (see preceding text and Phillips and Kimble, 2009) but also by the specific TCF/LEF gene that is expressed in each tissue at each developmental stage (Figure 17.2c and d). In the hair follicle, for instance, the expression of *TCF7L1* and *TCF7L2* in the stem cells is replaced by *LEF1* as the cells exit the stem cell niche of the bulge and migrate to the appropriate destinations as they differentiate (Merrill *et al.*, 2001; Nguyen *et al.*, 2009). During early Xenopus development, a dramatic change in the response to Wnt signaling around the start of zygotic gene expression is linked to an increase in *TCF7* and *LEF1* gene expression (Liu *et al.*, 2005; Molenaar *et al.*, 1998; Roël *et al.*, 2002, 2003). The relevance of this vertebrate-specific mechanism has also recently become evident in cancer biology. An exchange of gene expression from *LEF1* to *TCF7L2* causes melanoma cell-type switching (see Chapter 28 and Eichhoff *et al.*, 2011) because LEF1 proteins activate, while TCF7L2 proteins tend to repress, Wnt target gene expression (see following text). In another example, expression of the *TCF7L1* gene, which encodes an especially efficient repressor (see following text), is actively inhibited so as not to interfere with other TCF family members that promote high Wnt target gene expression in colorectal cancer cells (Atlasi *et al.*, 2013).

Differential expression of the four vertebrate TCF/LEF genes allows them to be associated with the development of specific tissues, organs, and also diseases; and it provides additional mechanisms for regulating context-specific responses to Wnt signaling.

TCF protein isoforms with long and short N-termini from alternative transcription start sites within TCF genes

Vertebrates produce TCF/LEF proteins that lack the BCBD. This removes an important lever for the transcriptional switch from repressor to activator. These proteins therefore tend to function as constitutive repressors that are mostly refractory to Wnt regulation. These alternative transcriptional isoforms are created from alternative promoters in downstream introns and rely on newly evolved translation start sites in downstream exons to produce these shorter TCF/LEF protein isoforms (Figure 17.1).

These short N-terminal isoforms have also been described as ΔN isoforms (referring to their lacking the conserved N-terminus of the full-length protein isoforms) or also dominant-negative proteins (since they function to displace TCF/β-catenin complexes from WREs and interfere with transcriptional responses to activated Wnt signaling). The last two terms derive from artificially truncated constructs (ΔN) that were experimentally deployed to inhibit the Wnt signaling response (dominant-negative activity). Such constructs were instrumental in demonstrating that TCF/LEF proteins are downstream mediators of Wnt/β-catenin signaling and that the BCBD is a functional domain (Behrens *et al.*, 1996; Molenaar *et al.*, 1996). It was subsequently discovered that vertebrate evolution had beaten the research scientists to this trick of producing constitutive TCF/LEF repressors. Endogenous short N-terminal gene products have now been described from the *TCF7* (Atcha *et al.*, 2007; Duboule, 2007; Van de Wetering *et al.*, 1996), *LEF1* (Hovanes *et al.*, 2001; e.g., Van de Wetering *et al.*, 1996), and *TCF7L2* locus (Duval *et al.*, 2000; Vacik, Stubbs, and Lemke, 2011), but not from the *TCF7L1* locus (at least so far).

This unique vertebrate aspect of Wnt signaling appears to have important relevance for cancer and possibly for type 2 diabetes. Short N-terminal TCF7 repressors are expressed in healthy colorectal tissue to function as tumor suppressors, but are not expressed in colorectal cancer cells (Najdi *et al.*, 2009). Experimental reexpression of short N-terminal TCF7 or TCF7L2 isoforms in colorectal cancer cells can stall tumor growth (van de Wetering *et al.*, 2002), and expression of a short N-terminal truncated LEF1 protein can slow proliferation of colon cancer cells (Yokoyama *et al.*, 2010). Changes in TCF/LEF expression during the transformation from normal gut epithelium to colorectal cancer tumor are complex and appear to involve several TCF/LEF genes, but an observed tendency towards expression of full-length versions that are expected to promote

Wnt target gene expression may prove to be highly relevant (Najdi, Holcombe, and Waterman, 2011). The *TCF7L2* locus is particularly associated with polymorphisms and mutations that are linked to diseases such as cancer (e.g., Bass *et al.*, 2011) and type 2 diabetes (e.g., Grant *et al.*, 2006). The strongest known genetic association with diabetes maps to a genomic region that may be involved in regulating promoters producing short N-terminal TCF7L2 isoforms (Duval *et al.*, 2000; Vacik, Stubbs, and Lemke, 2011), although other effects on alternative splicing (see following text) cannot currently be ruled out (e.g., Prokunina-Olsson *et al.*, 2009).

It is difficult to detect the relative expression of mRNAs that encode the shorter N-terminally truncated isoforms, compared to mRNAs that encode the longer, full-length TCF/LEF isoforms in microarray and RNA-seq analyses. It is therefore very likely that the relevance of this vertebrate innovation for normal development and disease is currently underestimated. Caution is therefore advised when interpreting previous studies aimed at dissecting TCF/LEF gene function through mouse knockout experiments that targeted the shared portion (usually the core DNA-binding domain) of the long and the short N-terminal isoforms, since they would interfere with both opposing functions of short and long N-terminal TCF/LEF isoforms. On the other hand, this insight also provides an explanation for the confusing observation of why the mouse *TCF7* knockout leads to adenomas, yet *TCF7* knockdown in colorectal cancer cells slows growth (Tang *et al.*, 2008). That is, a *TCF7* knockout in normal intestinal epithelia may interfere with a tumor suppressor function by removing a constitutively repressing short N isoform, while *TCF7* knockout in colon cancer cells removes a long N isoform that can interact with β-catenin to activate gene transcription and promote proliferation.

The transcriptional regulatory mechanisms to produce the short and long N-terminal TCF/LEF isoforms are currently being investigated (recently reviewed by Cadigan and Waterman, 2012), but positive autoregulation emerges as a likely trend (e.g., Hatzis *et al.*, 2008; Hovanes *et al.*, 2001; Roose *et al.*, 1999). Interestingly, while such positive feedback mechanisms likely represent a conserved feature of Wnt/β-catenin signaling with invertebrates, vertebrates appear to have developed additional mechanisms to deliberately avoid feedback on vertebrate-specific TCF/LEF transcriptional repressor isoforms. For instance, the promoter to produce the short N-terminal LEF1 isoform is silenced in colorectal cancer cells and is specifically prevented from being autoregulated by Wnt signaling (Yokoyama *et al.*, 2010). Also as we will discuss later, the vertebrate TCF7L1 protein excels as transcriptional repressor and may therefore be encoded from the only vertebrate TCF/LEF gene that does not appear to be regulated by Wnt/β-catenin signaling. Such positive autoregulation would represent an interesting example of a coherent type of the feedforward loop regulatory network motif that is widely distributed in biological networks (Davidson, 2010; Mangan and Alon, 2003).

Vertebrates have therefore evolved novel promoters and transcriptional start sites to express TCF/LEF proteins that function as constitutive repressors, which are refractory to regulation by upstream Wnt signaling mechanisms.

Novel and specialized TCF proteins from altered and alternative splicing

There is little variation in the exon structure of the N-terminus (i.e., in long N-terminus transcripts; see preceding text) or in the core DNA-binding domain (except in mammalian isoforms that lack it entirely; Kennell *et al.*, 2003). However, the vertebrate TCF/LEF gene family has evolved distinct differences in exon organization and exon structures in the CDRD and the C-terminus (Figure 17.1), which leads to heterogeneity of coding potential for protein structure and function. In addition to differences between gene loci, alternative splicing of gene products from one gene produces additional TCF/LEF isoforms and therefore additional complexity (Figure 17.1). It is interesting to note that the extent of alternative splicing and the degree of similarity to invertebrate orthologs differs among the members of the vertebrate TCF/LEF gene family. *LEF1* and the *TCF7L1* loci encode proteins that have evolved significantly compared to the invertebrate homolog, yet these genes use little (*LEF1*) or no (*TCF7L1*) alternative splicing (Figure 17.1). In contrast,

TCF7 and *TCF7L2* encode isoforms that are structurally similar to invertebrate homolog, but these genes also enable extensive alternative splicing to produce many different isoforms, some of which resemble LEF1 and TCF7L1 proteins in structure and therefore probably also in molecular function.

Variety in the context-dependent regulatory domain

Altered and alternative splicing in the CDRD generally maintains the open reading frame with downstream exons to encode proteins with DNA-binding activity. The heterogeneity of vertebrate TCF/LEF isoforms in the CDRD is therefore primarily about including or excluding exons. An additional obligatory exon (III++) is encoded in *TCF7L1* (exon 4) and *TCF7L2* (exon 5), which has recently been shown to be the target of phosphorylation by TNIK (Mahmoudi *et al.*, 2009), suggesting regulation by TNIK is confined to vertebrate TCF7L1 and TCF7L2 proteins. Additionally, this exon is preceded in some TCF7L2 transcripts by an alternative exon (III+), which was linked to reduced β-catenin-mediated transcriptional activation (Weise *et al.*, 2010) and increased repression (Struewing *et al.*, 2010) and may be preferentially expressed in stem cells (discussed in Mao and Byers, 2011). Exon VI has been identified as the binding domain for the LIM-domain protein HIC5, which enhances repressor activities and reduces activator function of TCF/LEF proteins (Ghogomu *et al.*, 2006; Li *et al.*, 2011). In *TCF7* and *LEF1* loci, alternative splicing creates isoforms with and without exon VI, which suggests that those TCF7 and LEF1 isoforms lacking exon VI possess enhanced capacity as transcriptional activators.

While the homologous exon VI is always included in TCF7L1 and TCF7L2 gene products, altered splicing (in *TCF7L1*) and alternative splicing (in *TCF7L2*) using different splice donor sites in the preceding exon V and different acceptor sites in the following exon VII insert short additional peptide sequences: LVPQ before the exon VI sequence and SxxSS thereafter (Gradl, König, and Wedlich, 2002; see Figure 17.1; Pukrop *et al.*, 2001). Remarkably, these short sequences create TCF7L1 and SxxSS-containing TCF7L2 isoforms that function as strong transcriptional repressors (Liu *et al.*, 2005). This subtle difference seems to be relevant for cancer biology since the SxxSS-containing TCF7L2 isoform in liver cancer cells suppresses proliferation, while alternatively spliced TCF7L2 isoforms lacking SxxSS promote growth (Koga *et al.*, 2012).

Currently, we have insufficient understanding of the detailed structural consequences and molecular mechanisms through which the altered and alternative CDRD in vertebrate TCF/LEF proteins affects their activity. One strong possibility is that these modifications influence the TCF/LEF transcriptional switch from DNA-bound TCF/LEF repressor to TCF/LEF activator. While the short N-terminal isoforms (discussed earlier) are generally not regulated by Wnt/β-catenin signaling (since they lack the BCBD), altered CDRD sequences may more subtly influence protein structure (and assembly of protein complexes) to make it easier or more difficult to switch into transcription activating complexes with β-catenin and associated cofactors. Even though all long N-terminal, full-length vertebrate TCF/LEF proteins can mediate both transcriptional activation (Hsu, Galceran, and Grosschedl, 1998; Korinek *et al.*, 1997; Molenaar *et al.*, 1996; Van de Wetering *et al.*, 1997) and transcriptional repression (Brantjes *et al.*, 2001) in *in vitro* settings, loss-of-function (and rescue) experiments *in vivo* reveal stark differences in the talents and requirements of particular TCF/LEFs (and their isoforms) for activation or repression (Kim *et al.*, 2000; Korinek *et al.*, 1998; Liu *et al.*, 2005; Nguyen *et al.*, 2009; Reya *et al.*, 2000; Standley *et al.*, 2006; Tang *et al.*, 2008; Yi *et al.*, 2011). Defining the molecular mechanisms that form the basis for these different regulatory potentials is an ongoing challenge.

Ultimately, the heterogeneity in CDRD determines the threshold for which different vertebrate TCF/LEF proteins enable nuclear β-catenin to induce a transcriptional switch from repression to activation. It can therefore generally be observed that *LEF1* mostly encodes transcriptional activator proteins (e.g., Reya *et al.*, 2000) and *TCF7L1* repressors (e.g., Kim *et al.*, 2000), while *TCF7* and *TCF7L2* genes encode different isoforms depending on the sequence of their particular CDRD (although heterogeneity in the C-terminus is also important; see following text). LEF1 proteins may be so much optimized for transcriptional activation that even short N-terminal LEF1 isoforms (which lack the BCBD) have been shown to bind β-catenin and activate transcription (Hoeppner *et al.*, 2011), presumably by interacting with low-affinity binding sites that have been observed *in vitro* (Daniels and Weis, 2005). In fact, in certain vertebrate tissues, the TCF7L1 transcriptional repressor has assumed biological functions that are no longer regulated by upstream Wnt signaling (Gribble *et al.*, 2009), and TCF7 and LEF1 transcriptional activators have assumed functions that are independent of β-catenin function (Grumolato *et al.*, 2013; Jeannet *et al.*, 2008).

The outcome of Wnt/β-catenin signaling in vertebrates is therefore not only determined by the level of nuclear β-catenin or even the optimal ratio between β-catenin and TCF/LEF proteins (Phillips and Kimble, 2009; see preceding text) but by the composition of the expressed TCF/LEF protein pool in specific cells and at particular developmental stages. The expression of TCF/LEF isoforms that are more or less adept at transcriptional repression or activation leads to a context-dependent response to a given Wnt/β-catenin signaling activity. Thus, in vertebrates, the exact time and place where a particular TCF/LEF protein is active is especially important. As has been emphasized earlier in this chapter, this complex pattern is controlled not only by regulating transcription (Figure 17.2c and d) but also by alternative splicing (particularly of the CDRD-encoding exons; see preceding text) and by phosphorylation (recently reviewed by Sokol, 2011).

Phosphorylation of target residues for TCF regulation has been mapped to residues in the CDRD, especially in the conserved exon IV (Figure 17.1). Regulation by phosphorylation is not vertebrate specific (see preceding text and Chapter 4), and differences in patterns of phosphorylation between different vertebrate TCF/LEF proteins are not created via altered or the alternative splicing. Instead, vertebrate TCF7 has evolved to evade this regulation since the relevant targeted residues are absent (Figure 17.1). A TCF exchange can therefore be regulated in vertebrates in tissues where both potent TCF/LEF repressor proteins (e.g., TCF7L1) and TCF7-encoded adept activator proteins are expressed. For example, HIPK2 regulation can induce a TCF exchange by causing dissociation of TCF7L1 proteins from their DNA target sites to be then replaced by TCF7 proteins (Hikasa and Sokol, 2011; Hikasa *et al.*, 2010). A TCF exchange from TCF7L1 to TCF7 protein has also been recently discovered in embryonic stem cells (Yi *et al.*, 2011), which may possibly also be regulated by HIPK2.

Vertebrate TCF/LEF genes have evolved heterogeneity in the CDRD that therefore primarily influences quantitative differences in determining the particular threshold for the transcriptional switch from repressor to activator. LEF1 tends to be particularly good as an activator and TCF7L1 as a repressor, while the transcriptional switch threshold in TCF7 and TCF7L2 isoforms may be determined by alternative splicing. In vertebrates, the tissue-specific transcriptional response to Wnt/β-catenin signaling is therefore also regulated by TCF exchange mechanisms, where altering TCF/LEF transcription, alternative splicing, or posttranslational regulation of TCF/LEF protein association with DNA control which vertebrate TCF/LEF proteins mediate Wnt/β-catenin signaling regulation at the promoter of Wnt target genes. However, the molecular functions of TCF7- and TCF7L2-encoded proteins at the promoters of Wnt target genes are also fundamentally influenced by their C-terminal structure.

Variety in the C-terminal tail

Another hallmark of vertebrate TCF/LEF genes is a pattern of altered and alternative splicing that creates different C-termini. This heterogeneity is mostly brought about by different pairing of splice donor with splice acceptor sites in the final exon, a pattern of splicing that creates different translational reading frames in the final, conserved exon XII. The C-terminus of both invertebrate and some vertebrate TCF proteins encodes an auxiliary DNA-binding domain called the C-clamp (Atcha *et al.*, 2007), which functions to recognize a GC-rich sequence referred to as the helper site (e.g., RCCG) in the conserved elongated WRE (see preceding text). The cysteine-rich C-clamp domain is encoded by the conserved exons XI and XII. The sequence of the Exon XI encodes a CRARF amino acid motif, recognized early on as highly conserved (Van de Wetering *et al.*, 1997), while exon XI encodes the rest of the C-clamp domain (including a RKKKCIRY peptide). Importantly, the C-clamp appears in the ancestral invertebrate TCF/LEF genes, as well as in vertebrate TCF7 and TCF7L2. In fact, the C-clamp is as conserved as the HMG DNA-binding domain and exhibits greater amino acid identity among all TCF/LEFs than the BCBD (Cadigan and Waterman, 2012; Van de Wetering *et al.*, 1997). The heterogeneity of vertebrate TCF/LEF C-termini therefore mainly affects whether a full C-clamp is encoded, whether a disrupted C-clamp with only the CRARF sequence is present, or whether there is no C-clamp domain at all.

C-terminal alternative splicing patterns were first described in detail for TCF7 (Van de Wetering *et al.*, 1996; see Figure 17.1). In this locus, the conserved splicing pattern that includes exon XI and XII retains the conserved open reading frame as in invertebrate TCFs to encode the "E-tail," a C-terminus with a complete C-clamp domain. However, in vertebrates, direct splicing of exon X (encoding the basic NLS) to exon XII can occur, which not only omits the CRARF-encoding exon XI but also changes the reading frame in exon XII to lead to a translational stop codon and therefore a TCF7 protein isoform with a short C-terminus lacking any C-clamp sequences, described as a "B-tail." There is also an alternative splice acceptor site immediately preceding conserved exon XII sequences that splices the CRARF-encoding exon XI to a stop codon to encode another short C-terminal TCF7 protein isoform with a disrupted C-clamp ("C-tail").

This pattern of alternative splicing in TCF7 provides a useful framework for describing altered and alternative structures in other vertebrate TCF/LEF genes (Figure 17.1). While the *TCF7L2* locus also encodes isoforms with a great variety of C-termini even exceeding that of TCF7, the *TCF7L1* and *LEF1* loci encode a limited isoform repertoire of C-termini. The splicing patterns of *TCF7L2* share many features with *TCF7* in that splicing incorporates the conserved exon XII to create E-like, C-like, and B-like tails (Figure 17.1). In addition, *TCF7L2* splicing is even more complicated by having two versions of the conserved exon XI (exon XI encoding the conserved CRARF and exon XI′ the related CRALF peptide sequence) and an optional additional exon preceding either of them (recently reviewed in more detail by Mao and Byers, 2011; Weise *et al.*, 2010). The *LEF1* locus encodes isoforms with a B-like tail and a unique N-tail (Atcha *et al.*, 2003). *TCF7L1* also splices directly from exon X to conserved exon XII in a B-like manner but regains the E-like reading frame. However, the conserved RKKKCIRY sequence is no longer encoded by the TCF7L1 equivalent of exon XII (Brannon *et al.*, 1999; Molenaar *et al.*, 1996), creating a chimeric version of a B- and E-like tail that is highly modified. Although produced by different splicing paths, both LEF1 isoforms and TCF7L1 end up encoding C-terminal tails lacking any C-clamp sequences.

The relevance of the C-clamp was recently demonstrated for gene regulation in human cancer cells (Atcha *et al.*, 2007; Hoverter *et al.*, 2012) and for regulating specific Wnt target genes (Weise *et al.*, 2010; Wöhrle, Wallmen, and Hecht, 2007). Genome-wide chromatin immunoprecipitation studies have revealed an enrichment for extended WRE sequences (core plus helper site; see preceding text) among TCF7L2 binding sequences (Hatzis *et al.*, 2008). This is currently a highly active area of research, but the evolved heterogeneity at the C-terminus of vertebrate TCF/LEF genes could either influence primarily qualitative differences in determining which DNA sequences are bound or could influence overall binding affinity, since two DNA-binding domains provide more contact with nucleic acid. Presumably, C-clamp TCF/LEF proteins preferentially bind elongated WRE sequences (core plus helper site; see preceding text), while those with disrupted C-termini or isoforms missing the C-clamp domain altogether bind shorter WRE sequences only containing the core site.

TCF/LEF isoforms missing the C-clamp should not necessarily be considered as weaker, handicapped forms, but may reflect how evolutionary opportunities have opened up cooperative networking with other transcription-regulating mechanisms. In other words, LEF1 and TCF7L1 may be less potent at binding promoters with elongated WREs, but they might also be able to contribute to gene regulation as part of complex *cis*-regulatory elements occupied by other transcription factors. While there are several examples of such interactions, much more detailed investigations are required to define a coherent pattern in the evolution of vertebrate regulatory mechanisms that have evolved to take advantage of C-clamp-less TCF/LEF proteins.

The C-terminus of TCF7L1 and the E-like tail of TCF7L2 also contain conserved peptide sequences (PLSLxxK) that have been described as binding sites for the transcriptional coregulator CtBP (Brannon *et al.*, 1999; Valenta, Lukas, and Korinek, 2003). CtBP is clearly involved in regulating Wnt target gene regulation in both vertebrates and invertebrates (see Chapter 4 and Fang *et al.*, 2006; Hamada and Bienz, 2004; Valenta *et al.*, 2006), but invertebrate TCF proteins do not contain these peptide sequences, and experimentally mutating these sequences did not alter the molecular activity of TCF7L1 when tested in Xenopus development (Liu *et al.*, 2005).

Conclusions

The structures and functions of vertebrate TCF/LEF proteins, and the described regulatory mechanisms they make possible, continue to be studied for an understanding of their fundamental actions as Wnt mediators, as well as other mechanisms that we have not been able to consider in this review, such as TCF/LEF protein SUMOylation and acetylation and regulatory protein interactions with other cotranscription factors (see Chapter 4 and Cadigan, 2012). However, we particularly need to investigate the molecular, cellular, and developmental mechanisms that regulate alternative splicing of TCF/LEF transcripts (particularly *TCF7* and *TCF7L2*) in a tissue- and stage-specific manner to better appreciate the impact that these vertebrate innovations bring to Wnt/β-catenin signaling in development and disease (recently reviewed by Mao and Byers, 2011).

It is now undisputed that vertebrate innovations in TCF/LEF structure, function, and regulation are relevant for human health and well-being. It is not just colorectal cancer but also other cancers that are affected by altered TCF/LEF gene and isoform expression (recently reviewed by Najdi, Holcombe, and Waterman, 2011). Intronic sequences surrounding CDRD-encoding exons in the *TCF7L2* locus are particularly important to study, as there are human polymorphisms in this region clearly associated with type 2 diabetes (e.g., Grant *et al.*, 2006) and cancer (e.g., Bass *et al.*, 2011). By uncovering the biological effects of these polymorphisms and the relative contributions of TCF/LEF isoforms to normal developmental processes versus their contributions to disease, we will gain not only new insights into the etiology and possible treatment of these diseases but will also deepen our understanding of the associated normal physiology. Further insight into the innovations within the Wnt/β-catenin signaling network that have their origin in modifications of the evolutionary ancient and successful design of TCF/LEF genes will help us understand how Wnt signaling has kept pace with the increase in complexity and lifespan in vertebrates and particularly in mammals.

References

Atcha, F.A., Munguia, J.E., Li, T.W.H. *et al.* (2003) A new beta-catenin-dependent activation domain in T cell factor. *The Journal of Biological Chemistry*, **278**, 16169–16175.

Atcha, F.A., Syed, A., Wu, B. *et al.* (2007) A unique DNA binding domain converts T-cell factors into strong Wnt effectors. *Molecular and Cellular Biology*, **27**, 8352–8363.

Atlasi, Y., Noori, R., Gaspar, C. *et al.* (2013) Wnt signaling regulates the lineage differentiation potential of mouse embryonic stem cells through Tcf3 down-regulation. *PLoS Genetics*, **9**, e1003424.

Bass, A.J., Lawrence, M.S., Brace, L.E. *et al.* (2011) Genomic sequencing of colorectal adenocarcinomas identifies a recurrent VTI1A-TCF7L2 fusion. *Nature Genetics*, **43**, 964–968.

Behrens, J., von Kries, J.P., Kuhl, M. *et al.* (1996) Functional interaction of beta-catenin with the transcription factor LEF-1. *Nature*, **382**, 638–642.

Brannon, M., Brown, J.D., Bates, R. *et al.* (1999) XCtBP is a XTcf-3 co-repressor with roles throughout Xenopus development. *Development*, **126**, 3159–3170.

Brantjes, H., Roose, J., Van de Wetering, M., and Clevers, H. (2001) All Tcf HMG box transcription factors interact with Groucho-related co-repressors. *Nucleic Acids Research*, **29**, 1410–1419.

Cadigan, K.M. (2012) TCFs and Wnt/β-catenin signaling: more than one way to throw the switch. *Current Topics in Developmental Biology*, **98**, 1–34.

Cadigan, K.M. and Waterman, M.L. (2012) TCF/LEFs and Wnt signaling in the nucleus. *Cold Spring Harbor Perspectives in Biology*, **4**, a007906.

Chang, M.V., Chang, J.L., Gangopadhyay, A. *et al.* (2008) Activation of wingless targets requires bipartite recognition of DNA by TCF. *Current Biology*, **18**, 1877–1881.

Daniels, D.L. and Weis, W.I. (2005) Beta-catenin directly displaces Groucho/TLE repressors from Tcf/Lef in Wnt-mediated transcription activation. *Nature Structural & Molecular Biology*, **12**, 364–371.

Davidson, E.H. (2010).Emerging properties of animal gene regulatory networks. *Nature*, **468**, 911–920.

Dehal, P. and Boore, J.L. (2005) Two rounds of whole genome duplication in the ancestral vertebrate. *PLoS Biology*, **3**, e314.

Duboule, D. (2007) The rise and fall of Hox gene clusters. *Development*, **134**, 2549–2560.

Duval, A., Rolland, S., Tubacher, E. *et al.* (2000) The human T-cell transcription factor-4 gene: structure, extensive characterization of alternative splicings, and mutational analysis in colorectal cancer cell lines. *Cancer Research*, **60**, 3872–3879.

Eichhoff, O.M., Weeraratna, A., Zipser, M.C. *et al.* (2011) Differential LEF1 and TCF4 expression is involved in melanoma cell phenotype switching. *Pigment Cell & Melanoma Research*, **24**, 631–642.

Fang, M., Li, J., Blauwkamp, T. *et al.* (2006) C-terminal-binding protein directly activates and represses Wnt transcriptional targets in Drosophila. *The EMBO Journal*, **25**, 2735–2745.

Galceran, J., Fariñas, I., Depew, M.J. *et al.* (1999) Wnt3a$^{-/-}$-like phenotype and limb deficiency in Lef1(−/−)Tcf1(−/−) mice. *Genes & Development*, **13**, 709–717.

Ghogomu, S.M., Van Venrooy, S., Ritthaler, M. *et al.* (2006) HIC-5 is a novel repressor of lymphoid enhancer factor/T-cell factor-driven transcription. *The Journal of Biological Chemistry*, **281**, 1755–1764.

Gillis, W.Q., St John, J., Bowerman, B., and Schneider, S.Q. (2009) Whole genome duplications and expansion of the vertebrate GATA transcription factor gene family. *BMC Evolutionary Biology*, **9**, 207.

Goentoro, L. and Kirschner, M.W. (2009) Evidence that fold-change, and not absolute level, of beta-catenin dictates Wnt signaling. *Molecular Cell*, **36**, 872–884.

Gradl, D., König, A., and Wedlich, D. (2002) Functional diversity of Xenopus lymphoid enhancer factor/T-cell factor transcription factors relies on combinations of activating and repressing elements. *The Journal of Biological Chemistry*, **277**, 14159–14171.

Grant, S.F.A., Thorleifsson, G., Reynisdottir, I. *et al.* (2006) Variant of transcription factor 7-like 2 (TCF7L2) gene confers risk of type 2 diabetes. *Nature Genetics*, **38**, 320–323.

Gregorieff, A., Grosschedl, R., and Clevers, H. (2004) Hindgut defects and transformation of the gastro-intestinal tract in Tcf4(−/−)/Tcf1(−/−) embryos. *The EMBO Journal*, **23**, 1825–1833.

Gribble, S.L., Kim, H.-S., Bonner, J. *et al.* (2009) Tcf3 inhibits spinal cord neurogenesis by regulating sox4a expression. *Development*, **136**, 781–789.

Grumolato, L., Liu, G., Haremaki, T. *et al.* (2013) β-catenin-independent activation of TCF1/LEF1 in human hematopoietic tumor cells through interaction with ATF2 transcription factors. *PLoS Genetics*, **9**, e1003603.

Hamada, F. and Bienz, M. (2004) The APC tumor suppressor binds to C-terminal binding protein to divert nuclear beta-catenin from TCF. *Developmental Cell*, **7**, 677–685.

Hanson, A.J., Wallace, H.A., Freeman, T.J. *et al.* (2012) XIAP monoubiquitylates Groucho/TLE to promote canonical Wnt signaling. *Molecular Cell*, **45**, 619–628.

Hatzis, P., van der Flier, L.G., van Driel, M.A. *et al.* (2008) Genome-wide pattern of TCF7L2/TCF4 chromatin occupancy in colorectal cancer cells. *Molecular and Cellular Biology*, **28**, 2732–2744.

Hikasa, H. and Sokol, S. Y. (2011) Phosphorylation of TCF proteins by homeodomain-interacting protein kinase 2. *The Journal of Biological Chemistry*, **286**, 12093–12100.

Hikasa, H., Ezan, J., Itoh, K. *et al.* (2010) Regulation of TCF3 by Wnt-dependent phosphorylation during vertebrate axis specification. *Developmental Cell*, **19**, 521–532.

Hoeppner, L.H., Secreto, F.J., Razidlo, D.F. *et al.* (2011) Lef1DeltaN binds beta-catenin and increases osteoblast activity and trabecular bone mass. *The Journal of Biological Chemistry*, **286**, 10950–10959.

Hovanes, K., Li, T.W., Munguia, J.E. *et al.* (2001) Beta-catenin-sensitive isoforms of lymphoid enhancer factor-1 are selectively expressed in colon cancer. *Nature Genetics*, **28**, 53–57.

Hoverter, N.P., Ting, J.-H., Sundaresh, S. *et al.* (2012) A WNT/p21 circuit directed by the C-clamp, a sequence-specific DNA binding domain in TCFs. *Molecular and Cellular Biology*, **32**, 3648–3662.

Hsu, S.C., Galceran, J., and Grosschedl, R. (1998) Modulation of transcriptional regulation by LEF-1 in response to Wnt-1 signaling and association with beta-catenin. *Molecular and Cellular Biology*, **18**, 4807–4818.

Ishitani, T., Kishida, S., Hyodo-Miura, J. *et al.* (2003) The TAK1-NLK mitogen-activated protein kinase cascade functions in the Wnt-5a/Ca(2+) pathway to antagonize Wnt/beta-catenin signaling. *Molecular and Cellular Biology*, **23**, 131–139.

Jeannet, G., Scheller, M., Scarpellino, L. *et al.* (2008) Long-term, multilineage hematopoiesis occurs in the combined absence of beta-catenin and gamma-catenin. *Blood*, **111**, 142–149.

Kennell, J.A., O'Leary, E.E., Gummow, B.M. *et al.* (2003) T-cell factor 4N (TCF-4N), a novel isoform of mouse TCF-4, synergizes with beta-catenin to coactivate C/EBPalpha and steroidogenic factor 1 transcription factors. *Molecular and Cellular Biology*, **23**, 5366–5375.

Kim, C.H., Oda, T., Itoh, M. *et al.* (2000) Repressor activity of Headless/Tcf3 is essential for vertebrate head formation. *Nature*, **407**, 913–916.

Klingel, S., Morath, I., Strietz, J. *et al.* (2012) Subfunctionalization and neofunctionalization of vertebrate Lef/Tcf transcription factors. *Developmental Biology*, **368**, 44–53.

Koga, H., Tsedensodnom, O., Tomimaru, Y. *et al.* (2012) Loss of the SxxSS motif in a human T-cell factor-4 isoform confers hypoxia resistance to liver cancer: an oncogenic switch in Wnt signaling. *PLoS One*, **7**, e39981.

Korinek, V., Barker, N., Morin, P.J. *et al.* (1997) Constitutive transcriptional activation by a beta-catenin-Tcf complex in $APC^{-/-}$ colon carcinoma. *Science*, **275**, 1784–1787.

Korinek, V., Barker, N., Moerer, P. *et al.* (1998) Depletion of epithelial stem-cell compartments in the small intestine of mice lacking Tcf-4. *Nature Genetics*, **19**, 379–383.

Li, X., Martinez-Ferrer, M., Botta, V. *et al.* (2011) Epithelial Hic-5/ARA55 expression contributes to prostate tumorigenesis and castrate responsiveness. *Oncogene*, **30**, 167–177.

Liu, F., Van Den Broek, O., Destrée, O., and Hoppler, S. (2005) Distinct roles for Xenopus Tcf/Lef genes in mediating specific responses to Wnt/beta-catenin signalling in mesoderm development. *Development*, **132**, 5375–5385.

Mahmoudi, T., Li, V.S.W., Ng, S.S. *et al.* (2009) The kinase TNIK is an essential activator of Wnt target genes. *The EMBO Journal*, **28**, 3329–3340.

Mangan, S. and Alon, U. (2003) Structure and function of the feed-forward loop network motif. *Proceedings of the National Academy of Sciences of the United States of America*, **100**, 11980–11985.

Mao, C.D. and Byers, S.W. (2011) Cell-context dependent TCF/LEF expression and function: alternative tales of repression, de-repression and activation potentials. *Critical Reviews in Eukaryotic Gene Expression*, **21**, 207–236.

Merrill, B.J., Gat, U., Dasgupta, R., and Fuchs, E. (2001) Tcf3 and Lef1 regulate lineage differentiation of multipotent stem cells in skin. *Genes & Development*, **15**, 1688–1705.

Molenaar, M., Van de Wetering, M., Oosterwegel, M. *et al.* (1996) XTcf-3 transcription factor mediates beta-catenin-induced axis formation in Xenopus embryos. *Cell*, **86**, 391–399.

Molenaar, M., Roose, J., Peterson, J. *et al.* (1998) Differential expression of the HMG box transcription factors XTcf-3 and XLef-1 during early xenopus development. *Mechanisms of Development*, **75**, 151–154.

Najdi, R., Holcombe, R.F., and Waterman, M.L. (2011) Wnt signaling and colon carcinogenesis: beyond APC. *Journal of Carcinogenesis*, **10**, 5.

Najdi, R., Syed, A., Arce, L. *et al.* (2009) A Wnt kinase network alters nuclear localization of TCF-1 in colon cancer. *Oncogene*, **28**, 4133–4146.

Nguyen, H., Merrill, B.J., Polak, L. *et al.* (2009) Tcf3 and Tcf4 are essential for long-term homeostasis of skin epithelia. *Nature Genetics*, **41**, 1068–1075.

Ota, S., Ishitani, S., Shimizu, N. *et al.* (2012) NLK positively regulates Wnt/β-catenin signalling by phosphorylating LEF1 in neural progenitor cells. *The EMBO Journal*, **31**, 1904–1915.

Phillips, B.T. and Kimble, J. (2009) A new look at TCF and beta-catenin through the lens of a divergent C. elegans Wnt pathway. *Developmental Cell*, **17**, 27–31.

Prokunina-Olsson, L., Welch, C., Hansson, O. *et al.* (2009) Tissue-specific alternative splicing of TCF7L2. *Human Molecular Genetics*, **18**, 3795–3804.

Pukrop, T., Gradl, D., Henningfeld, K.A. *et al.* (2001) Identification of two regulatory elements within the high mobility group box transcription factor XTCF-4. *The Journal of Biological Chemistry*, **276**, 8968–8978.

Reya, T., O'Riordan, M., Okamura, R. *et al.* (2000) Wnt signaling regulates B lymphocyte proliferation through a LEF-1 dependent mechanism. *Immunity*, **13**, 15–24.

Roël, G., Hamilton, F.S., Gent, Y. *et al.* (2002) Lef-1 and Tcf-3 transcription factors mediate tissue-specific Wnt signaling during Xenopus development. *Current Biology*, **12**, 1941–1945.

Roël, G., Van Den Broek, O., Spieker, N. *et al.* (2003) Tcf-1 expression during Xenopus development. *Gene Expression Patterns*, **3**, 123–126.

Roose, J., Huls, G., van Beest, M. *et al.* (1999) Synergy between tumor suppressor APC and the beta-catenin-Tcf4 target Tcf1. *Science*, **285**, 1923–1926.

Shetty, P., Lo, M.-C., Robertson, S.M., and Lin, R. (2005) C. elegans TCF protein, POP-1, converts from repressor to activator as a result of Wnt-induced lowering of nuclear levels. *Developmental Biology*, **285**, 584–592.

Sokol, S.Y. (2011) Wnt signaling through T-cell factor phosphorylation. *Cell Research*, **21**, 1002–1012.

Standley, H.J., Destrée, O., Kofron, M. *et al.* (2006) Maternal XTcf1 and XTcf4 have distinct roles in regulating Wnt target genes. *Developmental Biology*, **289**, 318–328.

Struewing, I., Boyechko, T., Barnett, C. *et al.* (2010) The balance of TCF7L2 variants with differential activities in Wnt-signaling is regulated by lithium in a

GSK3beta-independent manner. *Biochemical and Biophysical Research Communications*, **399**, 245–250.

Tang, W., Dodge, M., Gundapaneni, D. *et al.* (2008) A genome-wide RNAi screen for Wnt/beta-catenin pathway components identifies unexpected roles for TCF transcription factors in cancer. *Proceedings of the National Academy of Sciences of the United States of America*, **105**, 9697–9702.

Vacik, T., Stubbs, J.L., and Lemke, G. (2011) A novel mechanism for the transcriptional regulation of Wnt signaling in development. *Genes & Development*, **25**, 1783–1795.

Valenta, T., Lukas, J., and Korinek, V. (2003) HMG box transcription factor TCF-4's interaction with CtBP1 controls the expression of the Wnt target Axin2/Conduction in human embryonic kidney cells. *Nucleic Acids Research*, **31**, 2369–2380.

Valenta, T., Lukas, J., Doubravska, L. *et al.* (2006) HIC1 attenuates Wnt signaling by recruitment of TCF-4 and beta-catenin to the nuclear bodies. *The EMBO Journal*, **25**, 2326–2337.

Van de Peer, Y., Maere, S., and Meyer, A. (2009) The evolutionary significance of ancient genome duplications. *Nature Review Genetics*, **10**, 725–732.

Van de Wetering, M., Castrop, J., Korinek, V., and Clevers, H. (1996) Extensive alternative splicing and dual promoter usage generate Tcf-1 protein isoforms with differential transcription control properties. *Molecular and Cellular Biology*, **16**, 745–752.

Van de Wetering, M., Cavallo, R., Dooijes, D. *et al.* (1997) Armadillo coactivates transcription driven by the product of the Drosophila segment polarity gene dTCF. *Cell*, **88**, 789–799.

Van de Wetering, M., Sancho, E., Verweij, C. *et al.* (2002) The beta-catenin/TCF-4 complex imposes a crypt progenitor phenotype on colorectal cancer cells. *Cell*, **111**, 241–250.

Weise, A., Bruser, K., Elfert, S. *et al.* (2010) Alternative splicing of Tcf7l2 transcripts generates protein variants with differential promoter-binding and transcriptional activation properties at Wnt/beta-catenin targets. *Nucleic Acids Research*, **38**, 1964–1981.

Wöhrle, S., Wallmen, B., and Hecht, A. (2007) Differential control of Wnt target genes involves epigenetic mechanisms and selective promoter occupancy by T-cell factors. *Molecular and Cellular Biology*, **27**, 8164–8177.

Wu, B., Piloto, S., Zeng, W. *et al.* (2013) Ring Finger Protein 14 is a new regulator of TCF/β-catenin-mediated transcription and colon cancer cell survival. *EMBO Reports*, **14**, 347–355.

Yi, F., Pereira, L., Hoffman, J.A. *et al.* (2011) Opposing effects of Tcf3 and Tcf1 control Wnt stimulation of embryonic stem cell self-renewal. *Nature Cell Biology*, **13**, 762–770.

Yokoyama, N.N., Pate, K.T., Sprowl, S., and Waterman, M.L. (2010) A role for YY1 in repression of dominant negative LEF-1 expression in colon cancer. *Nucleic Acids Research*, **38**, 6375–6388.

18 Insights from Structural Analysis of Protein–Protein Interactions by Wnt Pathway Components and Functional Multiprotein Complex Formation

Zhihong Cheng and Wenqing Xu

Department of Biological Structure, University of Washington, Seattle, WA, USA

Structure of Wnt and a Wnt–Frizzled complex

Wnt proteins are cysteine-rich glycoproteins highly conserved through the animal kingdom to human (see Chapter 12), with 22 conserved cysteine residues and ~350 residues in length. There are 19 Wnt genes in human. Wnts activate at least three signaling pathways (see Chapter 6), including the canonical Wnt/β-catenin, Wnt/planar cell polarity (PCP), and Wnt/Ca^{2+} pathways, by binding to N-terminal cysteine-rich domain (CRD) of their cognate receptor Frizzled (Fzd). The crystal structure of CRD of mFzd8 was determined a decade ago (Dann *et al.*, 2001), whereas the first image of Wnt structure was not available until 2012, largely due to its unique lipid modifications – the addition of palmitoleic acid to a conserved Ser residue (Ser187 in XWnt8 and Ser209 in Wnt3a) (Takada *et al.*, 2006). Janda *et al.* (2012) got around the Wnt solubility problem for structural analysis by working with the Wnt–Fzd complexes. After screening various Wnts for better expression and more stable complex with various Fzd-CRD pairs, 30 years after the first Wnt genes (*Int-1*, *Wingless*) were discovered, the Garcia group finally determined the crystal structure of XWnt8 in complex with the mFzd8-CRD (Figure 18.1a).

In the complex structure, the ~120-residue mFzd8-CRD adopts almost identical conformation as the unliganded CRD, including 10 conserved cysteines forming five disulfide bonds. In contrast, the overall shape of XWnt8, with two subdomains joint together by a conserved interface, was totally surprising. The XWnt8 N-terminal α-helical domain (NTD) is composed of seven helices with a lipid-modified thumb as an insertion between two helices. The C-terminal cysteine-rich domain (CTD) contains basically a helix and two β-strands forming the index finger. Both thumb and index finger are composed of β-sheets stabilized by an extensive disulfide bond network (Figure 18.1a).

Wnt utilizes this unusual pair of extended β-hairpin fingers to grasp the opposite sides of the Fzd-CRD domain. At site 1 (the tip of the thumb), the interface is mainly mediated by the palmitoleic moiety (PAM) of the modified Ser187 and the deep hydrophobic groove formed by helix B and D of mFzd8-CRD. The residues of Fzd8-CRD in the interface are highly conserved, so is the modified Ser187 of Wnt8,

Wnt Signaling in Development and Disease: Molecular Mechanisms and Biological Functions, First Edition.
Edited by Stefan Hoppler and Randall T. Moon.

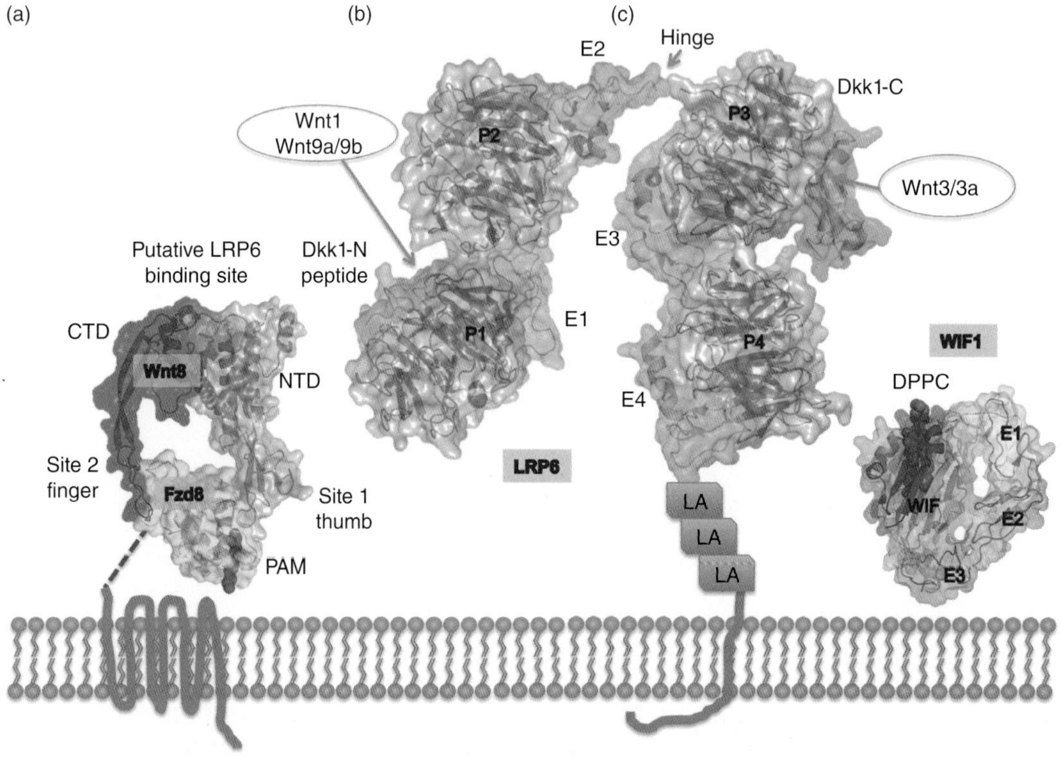

Figure 18.1 Wnt recognition by Fzd and LRP6 at the cell membrane, as well as its inhibition by Dkk1 and WIF1. Wnt8, Fzd8, LRP6, Dkk1, and WIF1 are all shown in cartoon illustration with semitransparent space-filling models. (**a**) Structure of a Wnt–Fzd complex. XWnt8 interacts with Fzd8-CRD through thumb- and fingerlike structures from its NTD and CTD domain, respectively. A PAM (shown as purple spheres) of modified Ser187 of XWnt8 mediated the site 1 interaction. A putative LRP6-binding surface is shown in pink on the XWnt8 surface. (**b**) Structure of LRP6 and its complex with Dkk1. The P1E1P2E2 and P3E3P4E4 domains of LRP6 form two relatively rigid structural blocks jointed by a short hinge, which restrains the relative orientation of these two blocks. The Dkk1 N-terminal peptide sits on the top surface of β-propeller 1 (P1). The Dkk1 C-terminal domain interacts with the top surface of β-propeller 3 (P3). EGF-like domains after each β-propeller are shown in *cyan*. The binding sites of Dkk1 N-terminal peptide and C-terminal domain partially overlap with those of Wnts (in circles with *arrows*) on LRP6. (**c**) Structure of a lipid-bound near-full-length WIF1(WIF-1ΔC). WIF1 contains a WIF domain, five EGF domains, and a hydrophilic C-terminal tail. The cell membrane lipid 1,2-dipalmitoylphosphatidylcholine (DPPC), bound to the WIF1 domain, is shown as purple spheres. WIF1 may inhibit Wnt activities by binding to the palmitoleic group of Wnt proteins in a similar manner. The WIF1 EGF3 (E3) domain folds back to contact the WIF domain, whereas the EGF4 and EGF5 domains have flexible orientations related to the rest of molecule and are not shown. *(See insert for color representation of the figure.)*

suggesting that other Wnt–Fzd pairs may share the same binding mode with this unusual lipid-recognition feature. The residues at the interface around the base of the thumb are less conservative, providing possibly interaction domains for ligand-receptor specificity. At site 2 (the tip of the index finger), the loop region of the Wnt8 C-terminal cysteine-rich cytokine-like domain contacts the helix C/D loop region and the C-terminus of Fzd8-CRD. Fzd8-CRD residues at site 2 are highly conserved but with some substitutions between several Fzd proteins, which may provide Wnt-subtype binding preferences. A binding assay with engineered mini-Wnt proteins suggests that the site 2 may make key contributions to the overall binding affinity as well as specificity for different Fzd receptors (Janda *et al.*, 2012).

Based on the modular structure of XWnt8 in the Wnt8/Fzd8 complex, it was proposed that Wnt proteins are evolved from the fusion of two distinct types of domains (Bazan, Janda, and Garcia, 2012). While the XWnt8 NTD has structural similarities with the saposin-like domain, which is a multipurpose lipid-interacting and carrier helical fold, the XWnt8 CTD contains half of a cysteine knot structure that is found in many receptor-binding growth factors including PDGF, interleukin, and Noggin.

Structures of LRP5/6 and LRP6-containing complexes

Low-density lipoprotein receptor-related proteins 5 and 6 (LRP5/6) function as Wnt coreceptor for mediating canonical Wnt/β-catenin signaling, in which the formation of the Wnt-mediated Fzd–Wnt–LRP5/6 trimeric complex is a crucial step (see Chapter 2). How Wnt interacts with LRP5/6 remains poorly understood. Based on the XWnt8 structure and the conservation pattern of its surface residues, a continuous patch on the XWnt8 surface opposite from the Fzd-CRD binding region was proposed as a potential LRP5/6 binding site for a ternary complex formation (Janda *et al.*, 2012). Future structural studies of the Wnt–LRP5/6 and Fzd–Wnt–LRP5/6 complexes will provide much-awaited insight into how Wnt initiates the canonical Wnt signaling pathway.

LRP5/6 is a type I transmembrane protein with an N-terminal ectodomain (ECD) containing four YWTD β-propeller–EGF-like domains, three LDLR type A (LA) domains, a single transmembrane helix, and a cytoplasmic domain. Extensive genetic and mutagenesis studies revealed that Wnt ligands, as well as the negative regulators Dickkopf (Dkk) and sclerostin (SOST), bind to the four propeller–EGF pairs of LRP5/6 ECD. Both Dkk and SOST play important roles in embryonic (including skeletal) development and adult tissue homeostasis (see Chapters 13 and 25). It is believed that Dkk and SOST regulate skeletal development by acting as inhibitors of canonical Wnt/β-catenin signaling in osteoblast cells, and they inhibit Wnt signaling by directly binding to LRP5/6 and preventing Wnt protein binding and signaling complex formation. Modulation of the interactions between LRP5/6 and Dkk/SOST would be an attractive therapeutic goal for treating multiple diseases, including osteoporosis.

Five groups have recently reported the crystal structures of the first two β-propeller–EGF-like domain of LRP6 (P1E1P2E2) (Cheng *et al.*, 2011; Holdsworth *et al.*, 2012), LRP6 P1E1 in complex with an antibody and in complexes with short peptides of Dkk1 or SOST (Bourhis *et al.*, 2011), the third and fourth β-propeller–EGF-like domain of LRP6 (P3E3P4E4) (Chen *et al.*, 2011; Cheng *et al.*, 2011), and LRP6 P3E3P4E4 in complex with Dkk1 C-terminal cysteine-rich domain (Dkk1c) (Ahn *et al.*, 2011; Cheng *et al.*, 2011). These crystal structures showed that these four β-propeller structures are similar to one another and that in all these propeller–EGF pairs, the propellers are structurally coupled with the following EGF-like domains (Figure 18.1b). Furthermore, the relative orientation between the first and second propeller–EGF pairs, as well as the third and fourth pairs, is also restrained due to the interactions between neighboring β-propellers and EGF-like domains. Negative staining EM studies of LRP6 ECD without three LA repeats revealed two rigid LRP6 structural blocks (P1E1P2E2 and P3E3P4E4, respectively) hinged by a short restrained linker (Cheng *et al.*, 2011). In the presence of three LA repeat, LRP6 ECD adopts a skewed horseshoe structure (Chen *et al.*, 2011; see Figure 18.1b).

There are four Dkks in human. Dkk1, Dkk2, and Dkk4 (but not Dkk3) can bind to LRP5/6 coreceptor and inhibit Wnt signaling. The Dkk proteins are composed of two conserved CRD, with five disulfide bonds in each domain. It has been shown that the C-terminal domain of Dkk1 alone is sufficient for inhibition of Wnt signaling in many scenarios (Brott and Sokol, 2002; Li *et al.*, 2002). The NMR structure of the C-terminal CRD of mouse Dkk2 (Dkk2C) showed that Dkk2C is a flat six β-sheet architecture stabilized by five disulfide bonds (Chen *et al.*, 2008). The six β-sheet structure is divided into two subdomains connected by a long flexible loop region that is involved in LRP5/6 binding.

The structures of LRP6 P3E3P4E4 in complex with DKK1c (Ahn *et al.*, 2011; Cheng *et al.*, 2011) revealed that Dkk1c interacts with the top surface of the third LRP6 propeller (Figure 18.1b). In both crystal structures, two LRP6 P3E3P4E4 molecules sandwich one Dkk1c molecule. One of the interfaces involves the long Dkk1 loop between two subdomains of DKK1c, as well as the short loop region in the first subdomain of DKK1c. The biological relevance of the second interface between DKK1c and the second copy of LRP6 P3 propeller needs further investigation. In the meantime, identification of a LRP6 E1 binding motif, which is also found in the Dkk1 N-terminus, and biochemical analysis of the interaction suggested a bipartite binding model in which full-length Dkk1 may use its N- and C-termini for interacting with the LRP6 propellers 1 and 3, respectively (Ahn *et al.*, 2011; Bourhis *et al.*, 2011; Figure 18.1b). Structure of LRP6 ECD in complex with a full-length Dkk will be needed to resolve the complete picture of LRP–Dkk interaction.

SOST contains a cysteine knot motif and belongs to a member of DAN/Cerberus protein family. NMR structures of SOST from human and mouse (Veverka *et al.*, 2009; Weidauer *et al.*, 2009) revealed the structure of the cysteine knot motif, which is composed of two twisted fingers stabilized by four disulfide bonds. There is a long flexible loop (loop 2) between two fingers and also flexible N- and C-terminal arms. Structural modeling and docking combined with mutagenesis also identified a potential heparan sulfate binding site, which may play a role in controlling SOST's inhibition activity (Veverka *et al.*, 2009). More recently, biochemical and NMR analysis illustrated that SOST binds to LRP6 via its central cysteine knot, in particular the NXI motif in SOST loop 2, and does not involve the amino- and carboxyl-terminal flexible arm regions (Holdsworth *et al.*, 2012).

Other complexes regulating the formation of the Fzd–Wnt–LRP signaling complex: WIF1, sFRP, and the LGR–RSPO complexes

Due to its important role in development and adult life, Wnt signaling is subject to extensive regulation inside the cell as well as outside the cell. The latter is achieved by fine-tuning of the Wnt ligands production, secretion, and their recognition by their cognate receptors (see Chapter 1). The lipidation of Wnt ligands is essential for its function and performed by Porcupine (Porc), a member of the membrane-bound O-acyltransferase family (MBOAT). The Wnt secretion process is also regulated by an important multipass transmembrane protein Wntless (Wls), the so-called Wnt cargo receptor. Serine O-esterification of Wnt is required for binding to Wls and its sorting function. Wls contains a lipocalin-like domain in its N-terminal region, which may bind to Wnt lipid moiety. The retromer complex, required for recycling of Wls from endosomes to the Golgi, is also critical for Wnt secretion. The structural basis for Wnt posttranslational modification, secretion, and paracrine gradient formation remains poorly understood.

In addition to its secretion, release, and diffusion, Wnt signals are tightly controlled by a number of secreted regulatory proteins, which regulate the interaction between Wnt and its receptor and coreceptor(s) (see Chapter 13). Among them, R-spondins (RSPO) are potent Wnt signaling activators and function as stem cell growth factors (see Chapters 2 and 8). Mechanistically, RSPO mediates the interaction between the extracellular domains of LGR4/5/6 and transmembrane ubiquitin ligase E3s (ZNRF3 and RNF43), which promote the turnover of Fzd and LRP6. Formation of the LGR–RSPO–ZNRF3 trimeric complex leads to the clearance of ZNRF3 and thus enhances Wnt-dependent signaling by stabilizing Fzd and LRP6 (Hao *et al.*, 2012; Koo *et al.*, 2012). The structural basis of function and regulation of LGR, RSPO, and ZNRF3 in Wnt signaling remains enigmatic. Among the secreted Wnt inhibitors, other than Dkk and SOST, there are also Wnt inhibitory factor (WIF) and secreted Frizzled-related protein (sFRP), with which we have gain substantial insights.

WIF1 serves as a negative regulator of Wnt signaling by binding directly to Wnts and preventing Wnt/receptor signaling complex formation (see Chapter 13). The mature WIF1 protein contains a WIF domain about 140 residues, followed by five EGF-like domains and a hydrophilic C-terminus. WIF1 binds to Wnt proteins mainly through its WIF domain, which is sufficient for Wnt inhibition. The NMR structure of the WIF domain of WIF1 (Liepinsh *et al.*, 2006) revealed that the WIF domain structure has an immunoglobulin-like fold of an eight-stranded β-sandwich. Authors identified an unexpected, tight association between refolded WIF1 protein and the alkyl chain of the detergent Brij-35 used in the refolding protocol, suggesting that WIF domain serves as a recognition motif for lipid-modified Wnt proteins. Recently, the crystal structure of WIF1 containing both WIF and EGF-like domains revealed a tightly bound lipid in the core of the WIF domain with the head group exposing to the surface (Malinauskas *et al.*, 2011; Figure 18.1c). Mutagenesis and biochemical studies identified a Wnt-binding surface on WIF1 domain proximal to the lipid head group. The EGF-like domains I–V fold back around WIF domain with EGF III and may also be involved in Wnt binding. This work suggested a mechanism in which the WIF domain and EGF-like domains act synergistically to restrain Wnt molecules in an HSPG-dependent manner.

Similar to WIF1, sFRP can inhibit or modulate Wnt signaling pathway by binding to Wnt proteins or dimerizing with the CRD of Fzd (see Chapter 13). In human, there are five sFRP homologs, each containing N-terminal CRD and C-terminal Netrin (NTR) domain. The function of the C-terminal NTR domain is unclear. The N-terminal CRDs of sFRPs are homologous to the extracellular CRD of Fzd proteins. Wnt proteins bind to the CRD of sFRP or Wnt receptor Fzd. The overall fold of CRD of sFRP-3 and Fzd8 are very similar, and it is plausible that the CRD/Wnt binding mode is shared between sFRP and Fzd (Dann *et al.*, 2001; Janda *et al.*, 2012; see Figure 18.1a).

Interactions of Axin and Dishevelled with the cytoplasmic side of Wnt receptor and coreceptors

Upon the formation of the Wnt-mediated Fzd–Wnt–LRP5/6 complex, the Wnt protein promotes the formation of a large protein assembly (the "signalosome") to transduce the signal across membrane (Bilic *et al.*, 2007; see Chapter 2). The exact composition and the molecular mechanism of the signalosome remain unclear. Nevertheless, it is known that phosphorylation of the LRP5/6 intracellular domain and subsequent Axin binding are crucial steps of transmembrane Wnt signaling (Davidson *et al.*, 2005; Zeng *et al.*, 2005) and the dynamic properties of the DIX domain-dependent puncta structures play an important role in this process. A number of important protein–protein interactions related to this process, including DIX domain-dependent oligomerization and the Fzd–Dvl interaction, have been structurally characterized.

Dvl contains three conserved domains: DIX, PDZ, and DEP (see Chapter 15). The DIX domain of Dvl mediates dynamic polymerization, which is essential for the signaling activity of Dvl. Axin also contains a DIX domain (referred to as DAX) in its C-terminus, which also mediates dynamic, reversible polymerization and is required for canonical Wnt signaling. Both Dvl and Axin interact with numerous partners, and Dvl DIX domain can interact with the Axin DAX domain to modulate Wnt signaling. DIX–DAX domain-dependent multiple-protein complexes are visible in cells as puncta. Crystal structures of DIX and DAX domains and biochemical analysis showed that heterotypic DIX–DAX interactions involve the same DAX surface residues as those mediating homotypic DAX–DAX polymerization, and so the two interactions are mutually exclusive. Thus, it was proposed that while Axin DAX-mediated homotypic polymerization promotes the phosphorylation and subsequent degradation of β-catenin in the Axin-scaffold destruction complex, canonical Wnt signals promote the heterotypic DIX–DAX interactions which turn off the activity of the destruction complex (Fiedler *et al.*, 2011). It should be noted that DIX domains of other proteins (e.g., Ccd1) may also regulate Wnt signaling in a manner similar with the Dvl DIX domain (Liu *et al.*, 2011).

During canonical Wnt signaling, the PDZ domain of Dvl interacts with the C-terminal domain of Fzd. NMR analysis demonstrated the direct interaction between the internal KTXXXW motif in Fzd C-terminal domain and the conventional binding site on Dvl PDZ domain, which typically interacts with the C-terminal motif of other binding partners (such as the Dvl inhibitor Dapper (Dpr)) (Wong *et al.*, 2003). This interaction provides a critical link between Wnt-mediated Fzd–LRP6 colocalization and the controlled Dvl–Axin interactions. Using structure-based approaches (virtual screening followed by biochemical and functional characterization), small-molecule inhibitors of Dvl PDZ domain have been developed to disrupt the Fzd–Dvl interaction, which have been shown to effectively inhibit the Wnt signaling (Grandy *et al.*, 2009).

In certain experimental settings, different Wnt ligands were found to activate canonical or noncanonical Wnt pathways. Dvl is required for both canonical and noncanonical Wnt/PCP signaling. Understanding how Dvl activates these distinct pathways in specific manner is of critical importance (Chapter 15). For canonical Wnt signaling, the DIX and PDZ domains of Dvl are required by mediating the formation of a supercomplex (signalosome) that consists of Wnt, Fzd, LRP6, Dvl, and Axin (Chapter 2). In this case, Axin is recruited to the membrane through the interaction between phosphorylated LRP5/6 cytoplasmic domain and Axin (the structural basis of this interaction remains unclear), and Dvl is recruited via its weak/dynamic interactions with both Fzd and Axin, as well as the intracellular surface of the plasma membrane. In contrast, Axin and LRP5/6 are not directly involved in Wnt/PCP signaling; the association of Dvl with plasma membrane thus relies critically on the DEP domain. The DEP domain-mediated plasma membrane association appears to involve multiple distinct, perhaps synergistic, molecular interactions. The NMR structure of Dvl1 DEP domain revealed a positively charged surface that is likely to interact with the negatively charged plasma membrane and contributes to its membrane association (Wong *et al.*, 2000). Importantly, the Dvl DEP and C-terminal domains were found to interact with the intracellular loop 3 and the C-terminal tail of Fzd (Tauriello *et al.*, 2012). In addition to the Fzd–Dvl interactions, the μ2 subunit of the clathrin adaptor protein complex AP2 may contribute to Dvl membrane association by simultaneously interacting with the DEP domain and the YHEL motif of Dvl C-terminal domain (Yu *et al.*, 2010). All these and other studies have provided important insights regarding how Dvl interacts with membrane and suggested a model in which Dvl is regulated by both intramolecular and intermolecular interactions (see Figure 15.1 and Figure 15.3). More work will be needed to understand how all these interactions elaborate and generate specific downstream signals.

Structural basis of β-catenin turnover and the β-catenin destruction complex

β-catenin is the central effector protein in canonical Wnt signaling. In the absence of a Wnt signal, free cytosolic β-catenin is captured and earmarked for degradation by the β-catenin destruction complex. In the presence of a Wnt signal, the destruction complex is inhibited, which leads to the accumulation of β-catenin in the cytosol and nucleus (see Chapter 3) and the transcriptional activation of Wnt responsive genes (see Chapter 4). Phosphorylation of β-catenin in the cytosolic destruction complex is thus one of the most crucial regulatory steps in Wnt signaling. Essential components of the β-catenin destruction complex include the scaffold Axin, the substrate β-catenin, CK1 and GSK3 – the two kinases responsible for catalyzing β-catenin phosphorylation – and the tumor suppressor APC. There are also other proteins that are found in the complex and play important roles in the destruction complex function, including tumor suppressors PP2A and WTX (Amer1). It is clear that the large β-catenin destruction complex is tightly regulated by Wnt signals through changes in the localization, composition, and likely conformations of the complex. Although structural studies of individual protein–protein interactions in the destruction complex have shed light on its molecular mechanisms (see following text), two fundamental questions remain unclear: how does Axin organize the functional assembly of the destruction complex, and how APC functions in the complex?

Axin is the scaffold of the destruction complex. Loss-of-function mutations of Axin are associated with cancers, and Axin has been thus proposed as a tumor suppressor. From N- to C-terminus, Axin contains binding sites for tankyrase (TNKS), APC, GSK-3, β-catenin, CK1, PP2A, Dvl, and LRP6. Interestingly, except for the RGS domain and the C-terminal DAX domain, Axin is predicted to be largely an intrinsically disordered protein (Cortese, Uversky, and Dunker, 2008). Many of these interactions are mediated by Axin regions that may be flexible till forming a complex with the partner. The DAX domain promotes Axin oligomerization, which may enhance the potential weak interactions among different Axin regions or between Axin and Axin partners. Indeed, many Axin binding partners can interact with more than one region of Axin. Importantly, a recent report demonstrated that Wnt signaling is governed by phosphorylation regulation of the Axin scaffolding function.

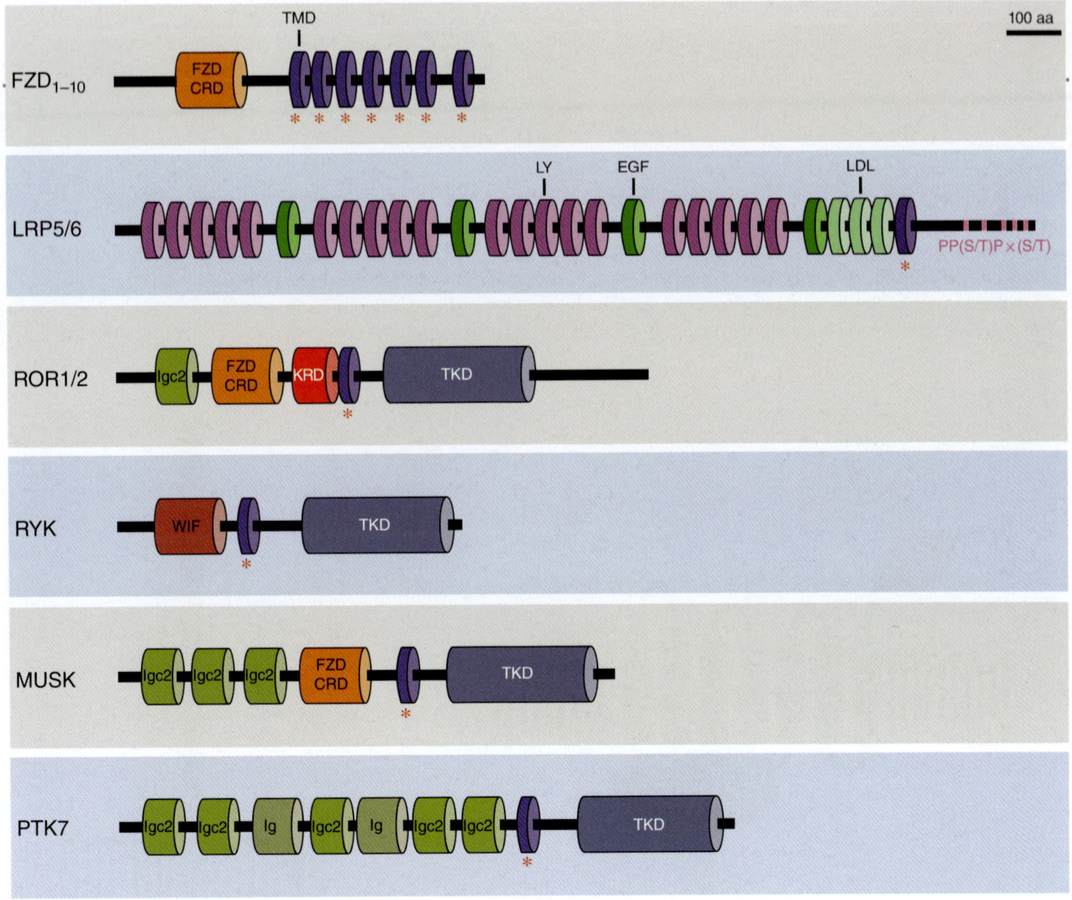

Figure 2.1 Wnt receptors and coreceptors. Domain structure of Wnt receptors and coreceptors according to SMART database (http://smart.embl-heidelberg.de). ECDs are shown to the left, intracellular to the right, and position of transmembrane domains indicated by a red asterisk. Domain abbreviations: FZD-CRD, Frizzled cysteine-rich domain; TMD, transmembrane domain; LY, low-density lipoprotein receptor YWTD domain; EGF, epidermal growth factor-like domain; LDL, low-density lipoprotein receptor domain class A; KRD, Kringle domain; TKD, tyrosine kinase catalytic domain; WIF, Wnt-inhibitory factor-1-like domain; IgC2, immunoglobulin C-2 type; Ig, immunoglobulin domain. The five PPSPxS motifs on the ICD of LRP5/6 are indicated.

Wnt Signaling in Development and Disease: Molecular Mechanisms and Biological Functions, First Edition.
Edited by Stefan Hoppler and Randall T. Moon.

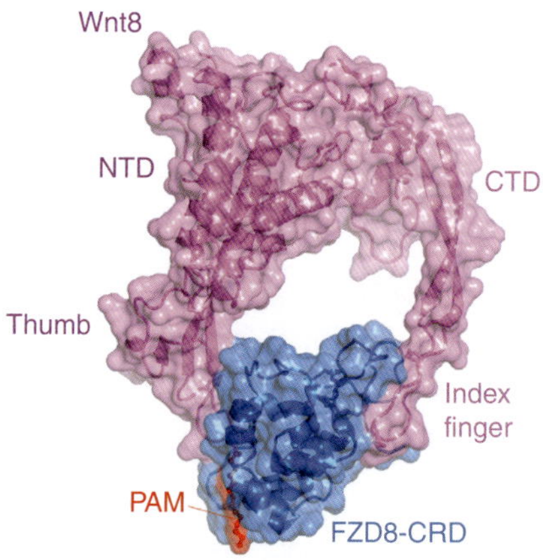

Figure 2.2 Structure of Wnt8 in complex with the extracellular CRD of Frizzled 8. Surface representation model developed from crystal structure data of *Xenopus* Wnt8 in complex with a CRD of mouse Frizzled 8 (FZD8-CRD). The palmitoleic acid (PAM) group is shown in red, extending from the tip of the Wnt thumb and passing through a hydrophobic groove in the FZD-CRD. Wnt8 is shown in pink with its N-terminal domain (NTD) and C-terminal domain (CTD), from which the thumb and index finger regions protrude. The figure was kindly supplied by, and with permission to publish from, Chris Garcia and Claudia Janda, Howard Hughes Medical Institute, Stanford University School of Medicine, Stanford.

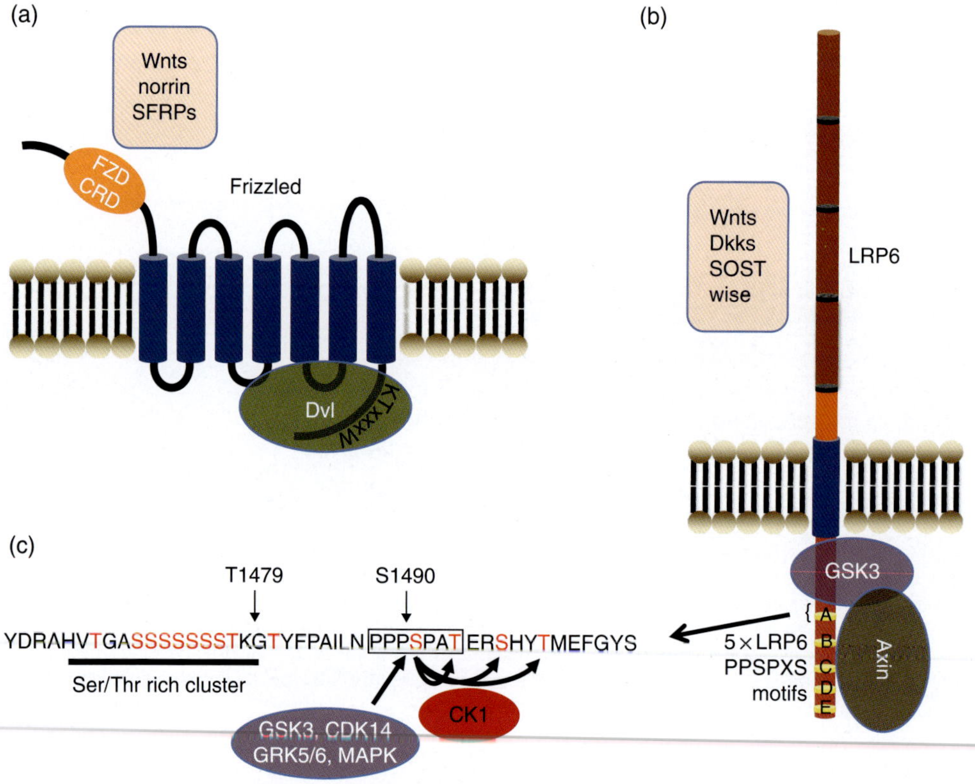

Figure 2.3 FZD and LRP6 and their interaction partners. (a) Extracellularly, Wnt, Norrin, and sFRP proteins bind the FZD-CRD of FZD receptors, whereas intracellularly, and in a Wnt-dependent manner, Dvl interacts with a KTxxxW motif on the C-Terminal domain and two additional motifs within the third intracellular loop of FZD. (b) Wnts, Dkks (Dkk1, 2, and 4), Wise, and SOST bind to the extracellular EGF/YWTD repeats of LRP6, whereas Axin/GSK3 interact with phosphorylated motifs on the ICD of LRP6. (c) Short region of LRP6 ICD spanning the Ser/Thr-rich cluster (important for GSK3 interaction) and PPSPxS motif A (boxed). The dual nature of PPSPxS site phosphorylation is illustrated, where both PPSP site kinases (GSK3, CDK14, GRK5/6, MAPK) and CK1 contribute to phosphorylation of the motifs to create Axin binding sites. Within PPSPxS motif A, the priming of CK1 site phosphorylation by prior PPSP site phosphorylation is also illustrated. Phosphospecific antibodies recognizing S1490 and T1479 are commonly used to detect PPSP and CK1 site phosphorylation of LRP6, respectively.

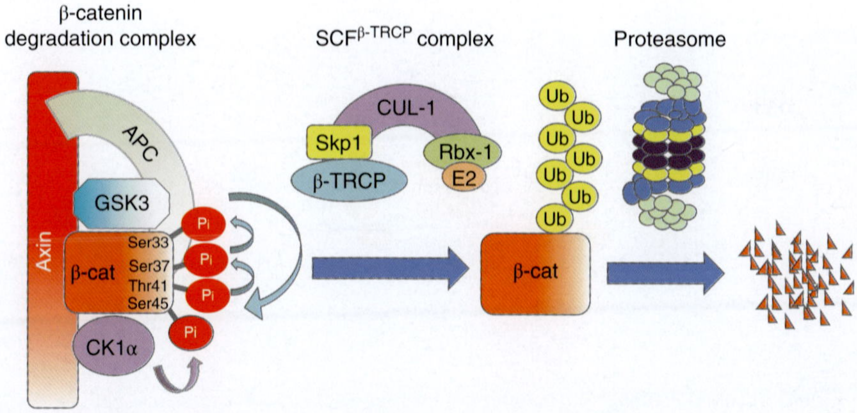

Figure 3.1 The core β-catenin destruction complex. The core β-catenin destruction complex consists of Axin, APC, GSK3, CK1α, and β-catenin. In the absence of a Wnt signal, these proteins form a complex that promotes β-TrCP-mediated ubiquitination of β-catenin, which is ultimately degraded by the proteasome. SCF$^{β\text{-TrCP}}$ consists of the F-box protein β-TrCP that binds phosphorylated β-catenin and the core catalytic complex containing Cul1, Skp1, and the Rbx1 RING finger protein. Binding of the E2 ubiquitin-conjugating enzyme to the SCF complex occurs via interaction with the Rbx1 protein.

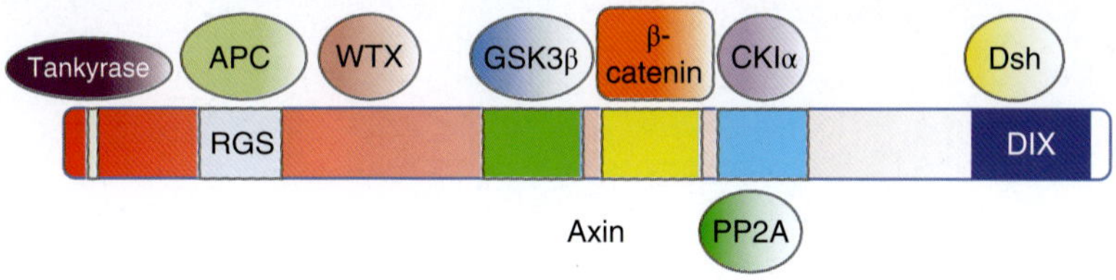

Figure 3.2 Axin is a multidomain scaffold protein that nucleates assembly of the β-catenin destruction complex. Axin contains binding sites for the core components of the destruction complex as well as Tankyrase, PP2A, WTX, and Dsh. The capacity of Axin to bind with high affinity to other components of the destruction complex and its limiting concentration make it a critical regulator of Wnt/β-catenin signaling.

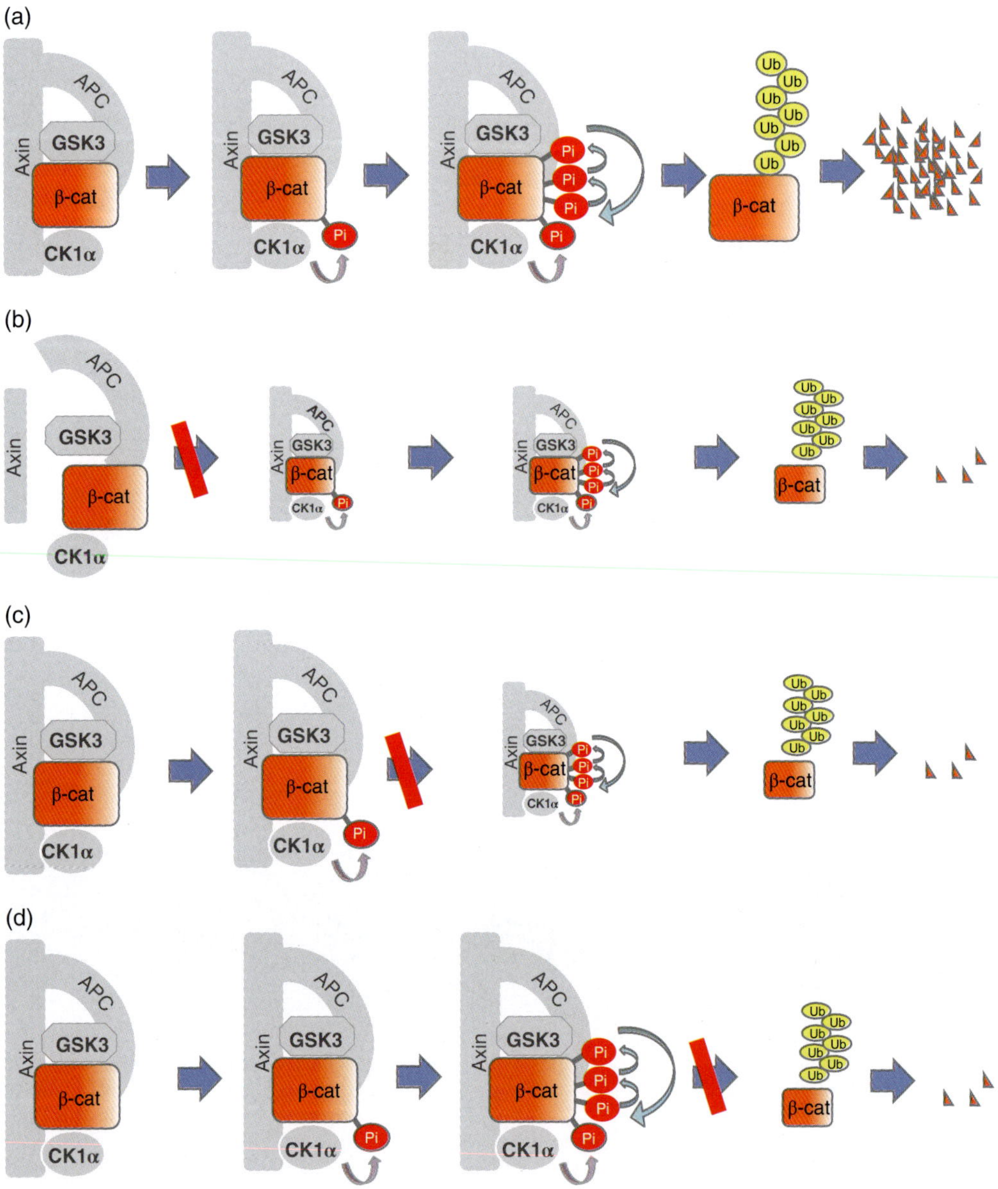

Figure 3.3 Proposed models for inhibition of the β-catenin degradation complex upon Wnt pathway activation. (a) In the absence of a Wnt signal, β-catenin is assembled into a complex where it is phosphorylated by CK1α and GSK3 and targeted for ubiquitin-mediated proteasomal degradation. (b) Wnt signaling inhibits β-catenin turnover by blocking its phosphorylation by CK1α and GSK3 (possibly due to disruption of complex assembly via subunit dissociation and/or Axin degradation). (c) Wnt signaling inhibits β-catenin phosphorylation by GSK3 only. (d) Wnt signaling does not inhibit β-catenin phosphorylation, but rather its ubiquitination. In b–d, blocking the degradation of β-catenin ultimately results in its increased cytoplasmic concentration, thereby promoting its entry into the nucleus and initiation of a Wnt-specific transcriptional program (see Chapter 4).

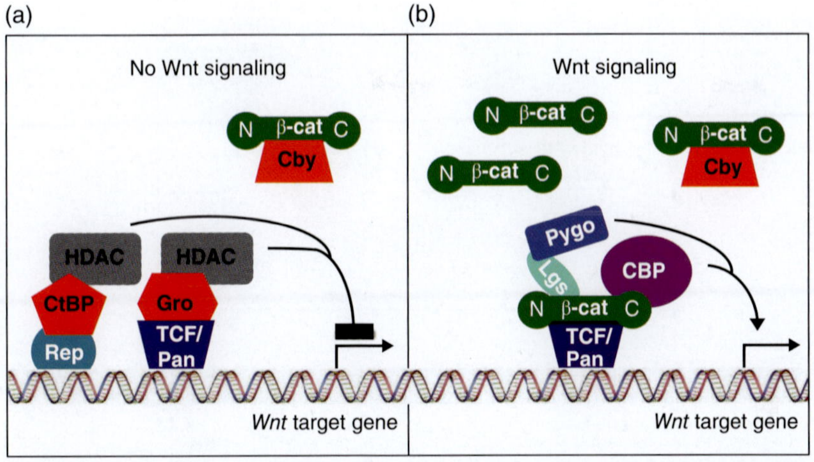

Figure 4.1 Summary of the TCF transcriptional switch in *Drosophila* cells. Depiction of a Wnt target gene and surrounding nucleoplasm in the absence or presence of Wnt/β-catenin signaling. (a) Under conditions of low nuclear β-catenin (β-cat), TCF/Pan is complexed with corepressors such as Gro, which recruits HDACs to inhibit transcription. Other factors act in concert with TCF/Pan to silence Wnt targets, such as CtBP, which is recruited to WREs by other, as yet unidentified transcriptional repressors (Rep). In addition, β-catenin binding proteins such as Cby can associate with β-catenin and block its interaction with TCF/Pan. (b) Wnt signaling causes higher levels of nuclear β-catenin, which overcomes the binding proteins and TCF corepressors and binds to TCF/Pan on target gene chromatin. β-catenin recruits a variety of coactivators, such as Lgs and Pygo through Arm repeats in the N-terminal half of β-catenin and CBP through its C-terminal transactivation domain, leading to activation of transcription.

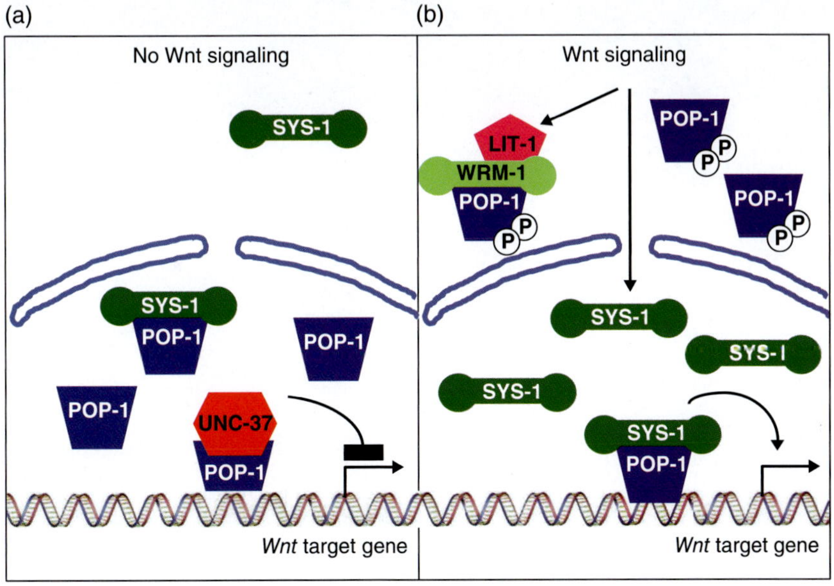

Figure 4.2 Summary of the TCF transcriptional switch in the Wnt/β-catenin asymmetry pathway in *C. elegans*. (a) In unstimulated cells, there are low levels of the worm β-catenin Sys-1 and high levels of POP-1 in the nucleus. The Gro homolog Unc-37 contributes to repression (Calvo *et al.*, 2001). (b) Wnt signaling increases the nuclear concentration of Sys-1 and lowers the level of POP-1 through its phosphorylation via the Lit-1 kinase, which promotes nuclear efflux. This shifts the equilibrium on WRE chromatin from POP-1 (or POP-1-Unc-37) to POP-1-Sys-1 heterodimers, resulting in transcriptional activation.

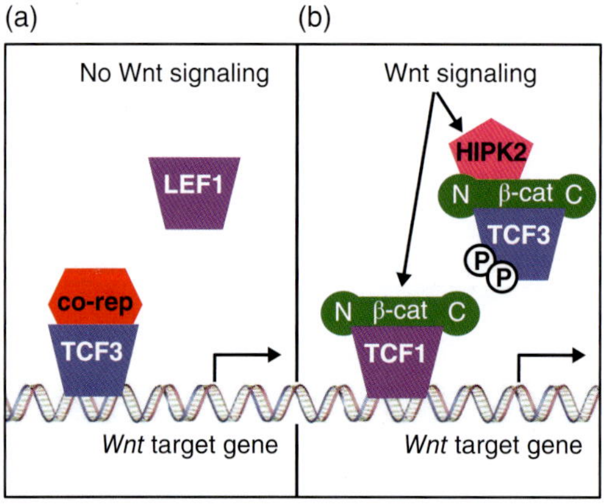

Figure 4.3 Summary of the TCF exchange between TCF3 and TCF1 on Wnt targets in *Xenopus* embryos. (a) In cells with low nuclear β-catenin, TCF3 represses Wnt target gene transcription. (b) Wnt signaling activates HIPK2, which acts with β-catenin to phosphorylate TCF3, removing it from target gene chromatin, where it is replaced by TCF1, which activates target gene transcription (Hikasa and Sokol, 2011; Hikasa *et al.*, 2010).

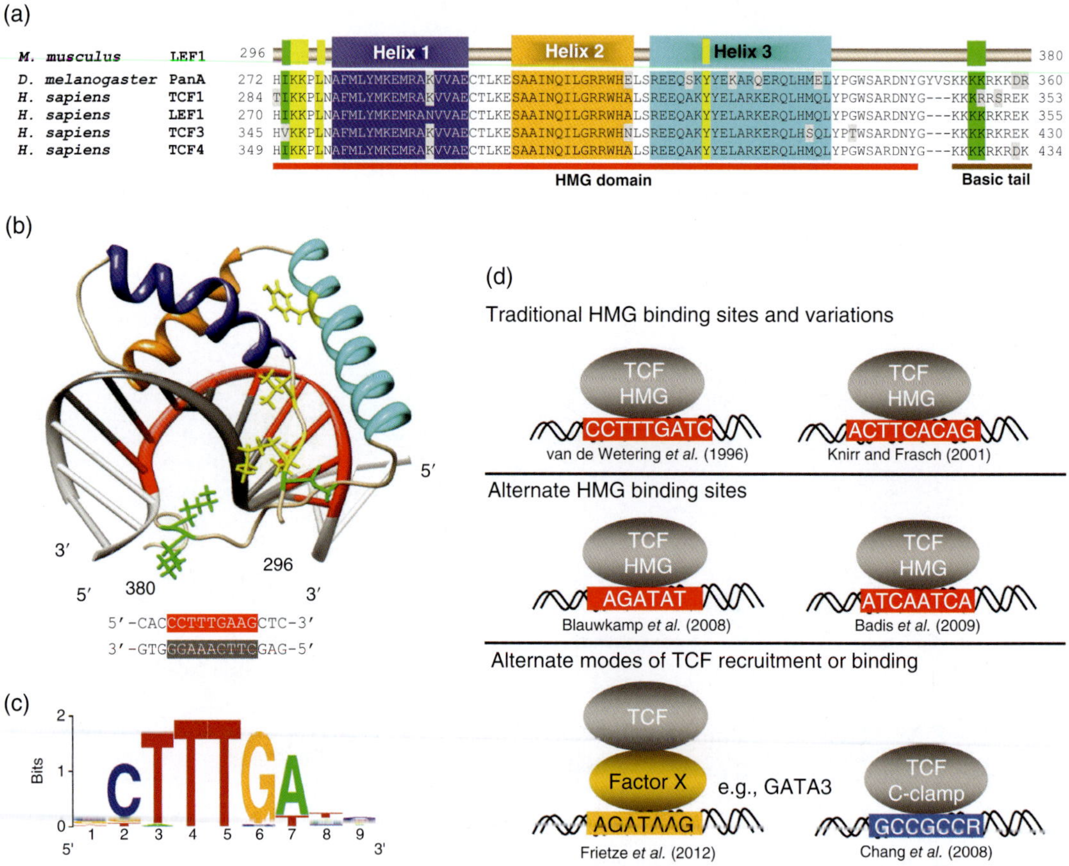

Figure 5.1 (a) Sequence alignment of the HMG domain (*red bar*) and basic tail (*brown bar*) of the *Drosophila* TCF/Pan (isoform A; NP_726522), human TCF1E (EAW62279.1), human LEF1 (NP_001124185), human TCF3 (NP_112573.1), and human TCF4E (CAB97213.1). Above them is a cartoon showing the corresponding region of mouse LEF1 (mLEF1; NP_034833.2) for which the solution structure is available (Love *et al.*, 1995), with the three α-helices in *blue*, *orange*, and *cyan*. Residues not conserved with mLEF1in the sequence alignment are shown in gray. Residues in yellow are essential for DNA binding and the ones in green are required for optimal binding (Giese, Amsterdam, and Grosschedl, 1991). (b) Ribbon diagram based on the solution structure of mLef1 (modified from Love *et al.*, 1995; PDB id 2LEF). Shown is the sequence of the double-stranded DNA motif used to determine the structure in complex with mLEF1, with nine base pairs of the consensus shown in *red* and *dark gray*. (c) Sequence logo of TCF functional binding sites derived from 32 WREs of various species, highlighting the CTTTGA consensus. (d) Models representing the diversity in HMG binding sites or alternative mechanisms of TCF recruitment to WREs. See text for further explanation.

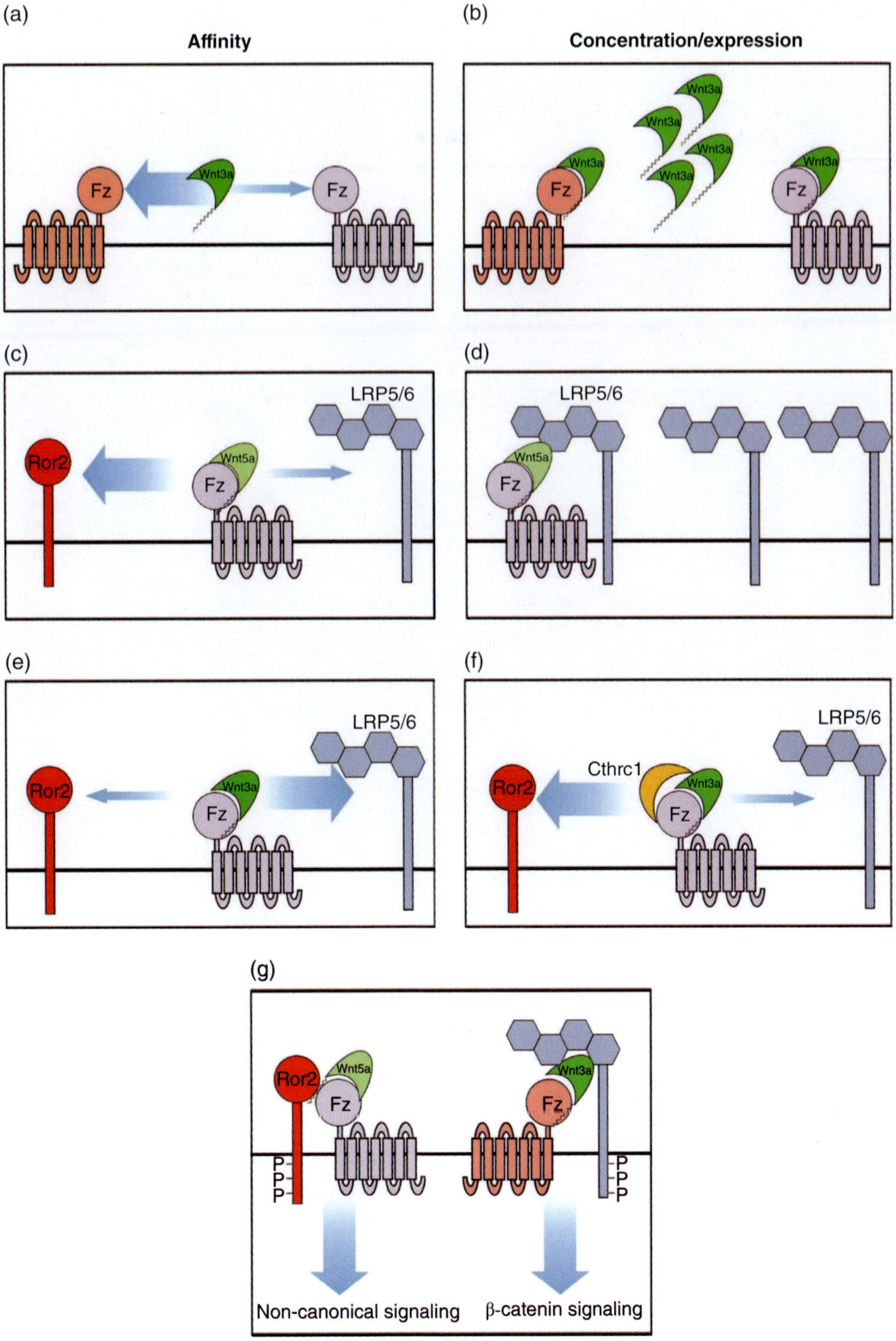

Figure 7.1 Multiple layers of Wnt pathway specificity. Wnt pathway specificity is dependent on affinities between the Wnt signaling components (left panels, i.e., a, c, and e) and the expression levels or concentrations of these components (right panels, i.e., b, d, and f). Wnt proteins differentially associate with Fz receptor family members depending on Wnt–Fz affinities (a) and Fz expression levels and Wnt concentrations (b). Furthermore, Wnt proteins bind coreceptor molecules with different affinity. For instance, Wnt5a preferentially interacts with Ror2 (c), unless LRP5/6 levels greatly exceed Ror2 levels (d). Conversely, Wnt3a has greater affinity for LRP5/6 than for Ror2 (e) but can interact with Ror2 in the presence of additional factors, such as Cthrc1 (f). In general (g), Wnt–Fz–LRP5/6 complexes transduce β-catenin signaling, whereas complexes between Wnt, Fz, and Ror2 mediate noncanonical signaling pathways.

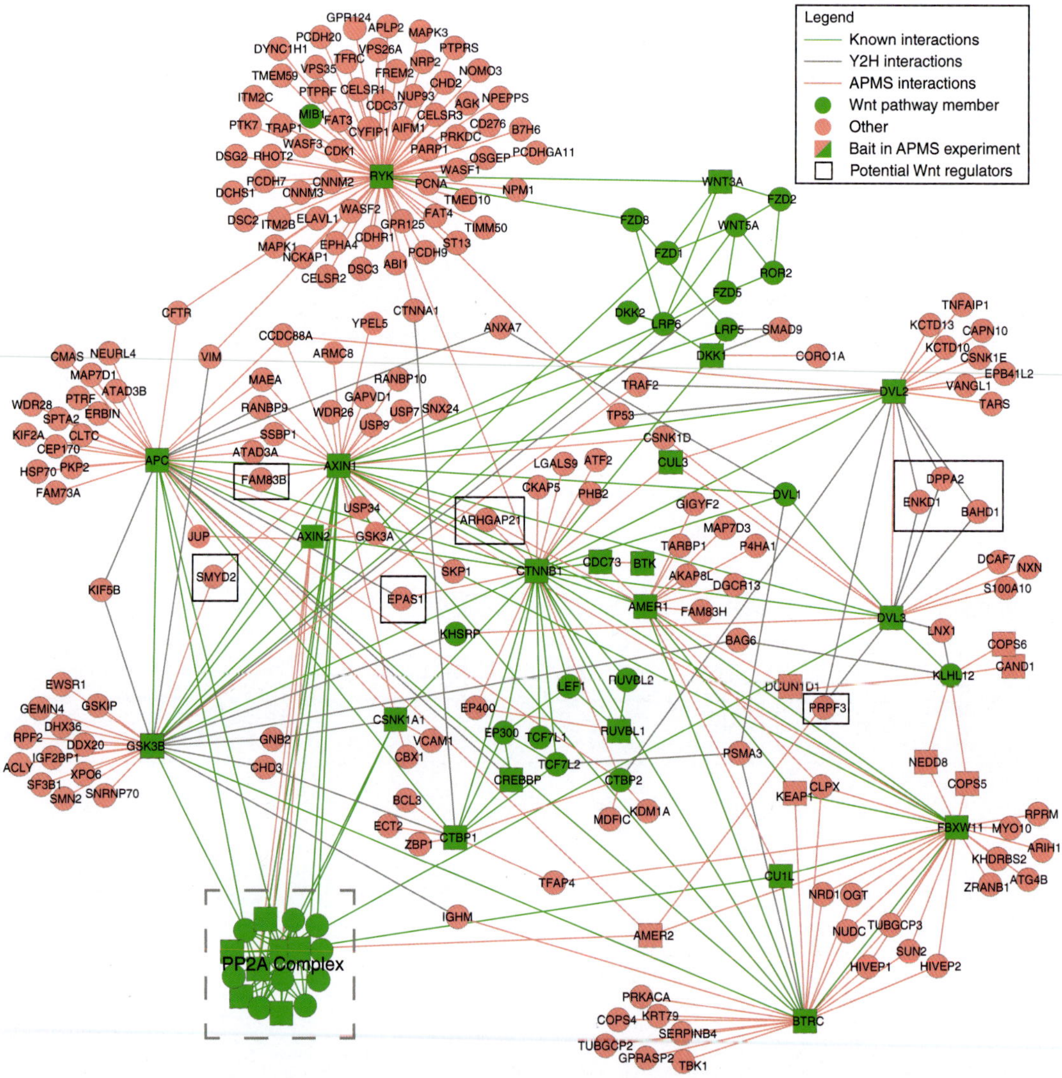

Figure 9.1 Protein–protein interaction network of the WNT/β-catenin pathway. Published APMS experiments and yeast two-hybrid screens were combined with previously known interactions. For purposes of visual clarity, proteins whose only connection to the network was via a single two-hybrid screen interaction were removed from the figure. Protein names correspond to the HUGO nomenclature.

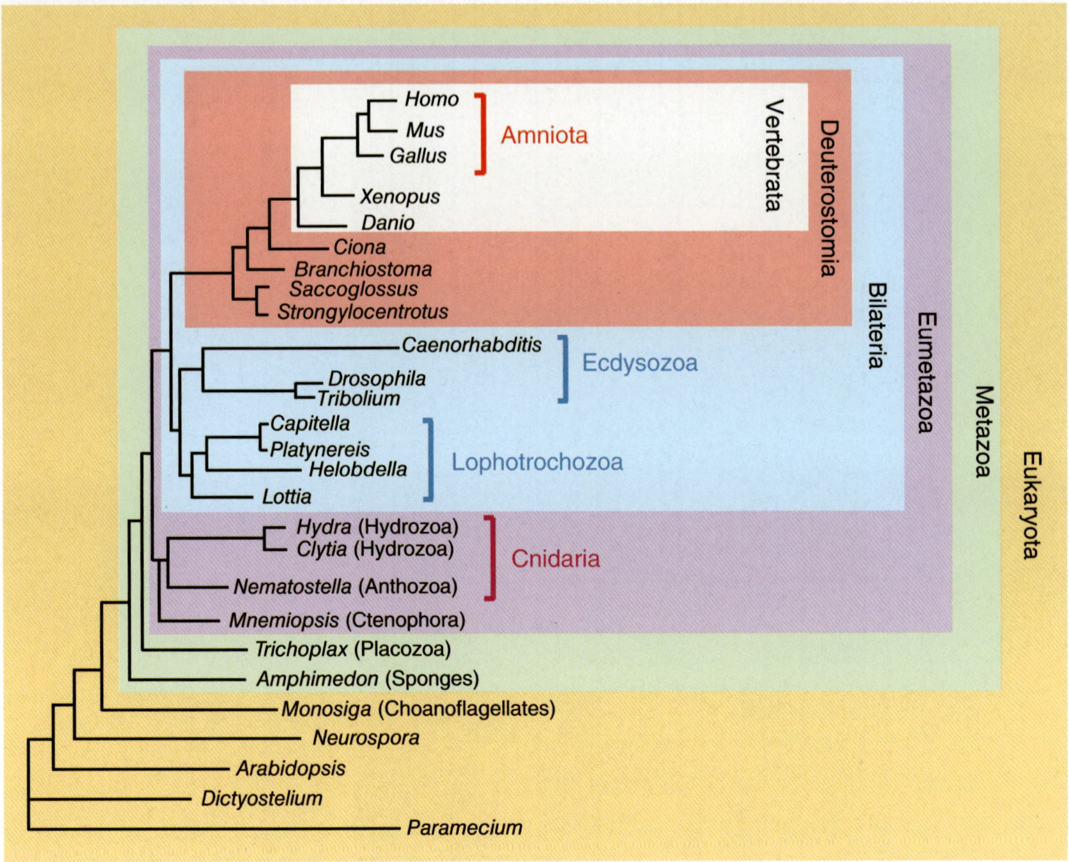

Figure 12.1 Molecular phylogeny of metazoans. Phylogenetic relationships are based on recent studies from Srivastava *et al.* (2010) and Simakov *et al.* (2013).

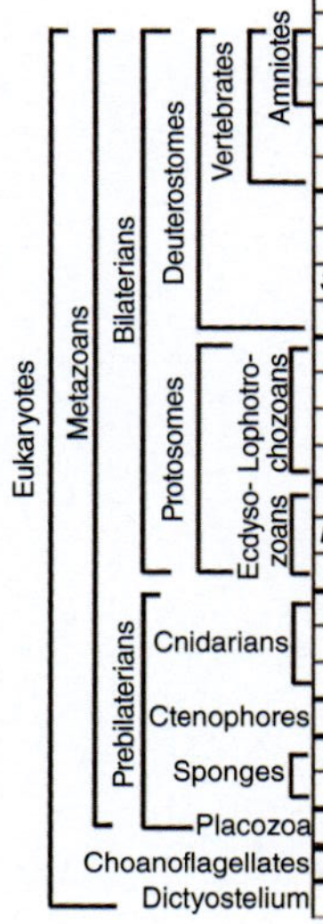

	Wnt1	Wnt2	Wnt3	Wnt4	Wnt5	Wnt6	Wnt7	Wnt8	Wnt9	Wnt10	Wnt11	Wnt16	WntA	Orphans	Total Wnt genes	Fzd 1/2/7	Fzd 3/6	Fzd 4	Fzd 9/10	Fzd 5/8	Orphans	Total Fzd genes
Homo sapiens	1	2	2	1	2	1	2	2	2	2	1	1			19	3	2	1	2	2		10
Mus musculus	1	2	2	1	2	1	2	2	2	2	1	1			19	3	2	1	2	2		10
Gallus gallus	1	2	2	1	2	1	2	2	2	2	2	1			20	3	2	1	2	2		10
Xenopus tropicalis	1	2	3	1	2	1	3	2	3	2	3	1			24	3	2	1	2	2		10
Danio rerio	1	5	2	2	2	2	3	3	2	2	2	1			27	4	2	1	2	3		12
Ciona intestinalis			1		1		1		1	1				5	10		1	1		1	2	5
Branchiostoma floridae	1		1	1	1	1	1	1		1	1				9	1		1	1	2		5
Saccoglossus kowalevskii	1	1	1	1	1		1	1	1		1	1	1		11	1			1	1		3
Strongylocentrotus purpuratus	1		1	1	1	1	1	1	1	1		1	1		11	1		1	1	1		4
Helobdella robusta	1	1		1	2	1	1			1	3	2			13							
Capitella teleta	1	1		1	1	1	1	1	1	1	1	1	1		12							
Platynereis dumerilii	1	1		1	1	1	1	1	1	1	1	1	1		12							
Lottia gigantea	1	1		1	1	1	1		1	1	1	1	1		11							
Tribolium castaneum	1				1	1	1	1	1	1	1		1		9	1		1		1		3
Drosophila melanogaster	1				1	1	1		1	1				1	7	1		2		1		4
Caenorhabditis elegans				1	1				1	1		1			5							
Nematostella vectensis	1	1	1	1	1	1	1	2		1	1	1	1		13	1		2		1		4
Hydra magnipapillata	1	1	1		1		1	1	3		1	1			11	1		1		1		3
Clytia haemispherica			1		1				1					10	13	1		1		1		3
Mnemiopsis leidyi						1			1				1	1	4							
Amphimedon queenslandica											1	1		1	3	1		1				2
Oscarella lobularis														2	2							
Trichoplax adhaerens														3	3						1	1
Mnemiopsis brevicollis															0							
Dictyostelium discoideum															0							16

Figure 12.2 Distribution of *Wnt* and *Fzd* genes in metazoans. Colors and numbers in the filled boxes indicate the relative abundance of *Wnt* and *Fzd* representatives in each gene subfamily (*white*, no data available; *grey*, zero member; *yellow*, one member; *green*, two members; *blue*, three members; *orange*, four members; *purple*, five members). Data reported here are based on molecular *Wnt* and *Fzd* phylogenies published in Adamska *et al.* (2010), Beermann *et al.* (2011), Cho *et al.* (2010), Croce *et al.* (2006), Freeman *et al.* (2008), Garriock *et al.* (2007), Gloriam, Fredriksson, and Schioth (2007), Hino *et al.* (2003), Janssen *et al.* (2010), King *et al.* (2008), Krishnan *et al.* (2012), Lagerström *et al.* (2006), Lapebie *et al.* (2009), Lengfeld *et al.* (2009), Momose and Houliston (2007), Nordstrom, Fredriksson, and Schioth (2008), Pang *et al.* (2010), Philipp *et al.* (2009), Prabhu and Eichinger (2006), Prud'homme *et al.* (2002), Schubert and Holland (2003), Srivastava *et al.* (2008); and Zhang, Tran, and Wessely (2011).

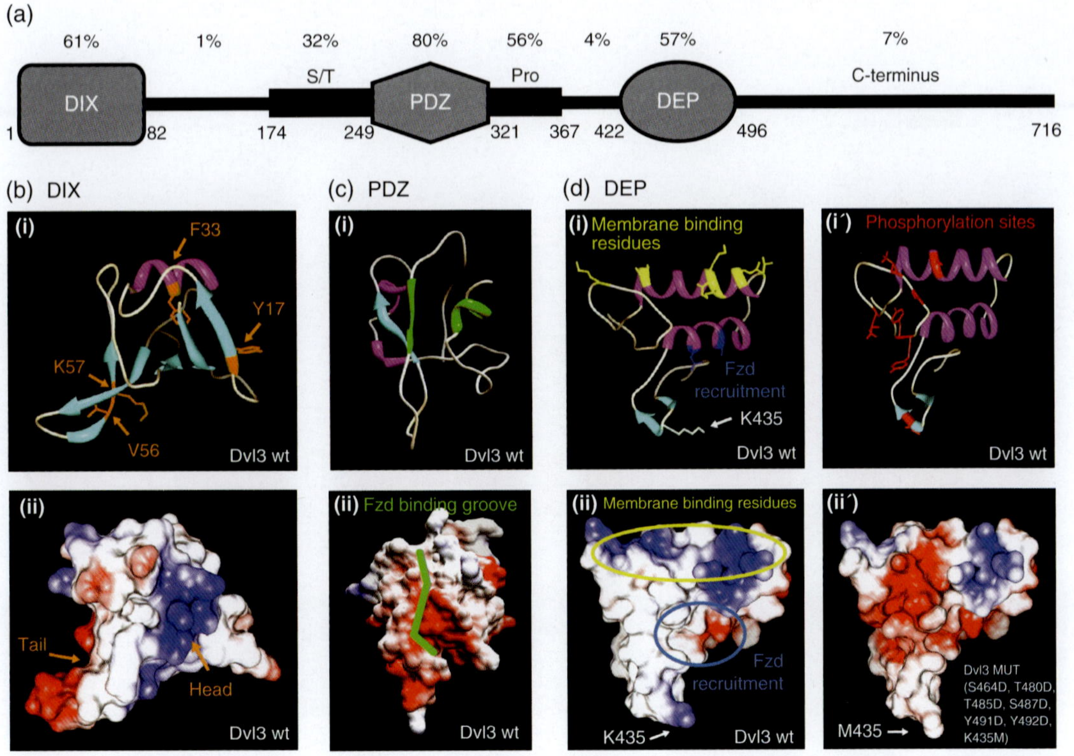

Figure 15.1 Schematic representation of Dvl structure. (a) Amino acid (aa) sequence identity (%) based on alignment of *Drosophila* Dsh, and mouse (m) and human (h) Dvl isoforms 1, 2, and 3. Position numbering of aa bordering individual regions is based on hDvl3. (b–d) 3D models of individual domains (hDvl3) are provided either as ribbon models (i) or as surface models (ii). Alpha-helices are shown in *magenta*, and beta-sheets in *cyan*. Electrostatic surface charge is depicted as negative in *red* and positive in *blue*. (b) DIX domain. (i) Residues mutated in Dvl polymerization mutants corresponding to F43S, V67A/K68A, and Y27D in mDvl2 are depicted in orange (Schwarz-Romond *et al.*, 2007). (ii) Regions of DIX domain required for head-to-tail polymerization are marked by *arrows*. (c) PDZ domain. (i) α-helix and β-sheet forming the Fzd binding groove are indicated in *green* (Wong *et al.*, 2003). (ii) Fzd binding groove is indicated by a *green line*. (d) DEP domain. (i, ii) Residues and surface required for membrane binding are indicated in *yellow* (Simons *et al.*, 2009) and residues required for Fzd binding are shown in *dark blue* (Tauriello *et al.*, 2012). Position of K435 corresponding to K417 mutated in *Drosophila* Dsh[1] mutation (Axelrod *et al.*, 1998) is indicated in *white*. (i′) Residues (S464, T480, T485, S487, Y491, Y492) phosphorylated following Fzd overexpression (Yanfeng *et al.*, 2011) are indicated in *red*. (ii′) Phosphomimicking mutations of these residues change electrostatic surface potential and disrupt the positive charge of the membrane-binding region. Please note also the change caused by the mutation K435M (Dsh[1], *white*).

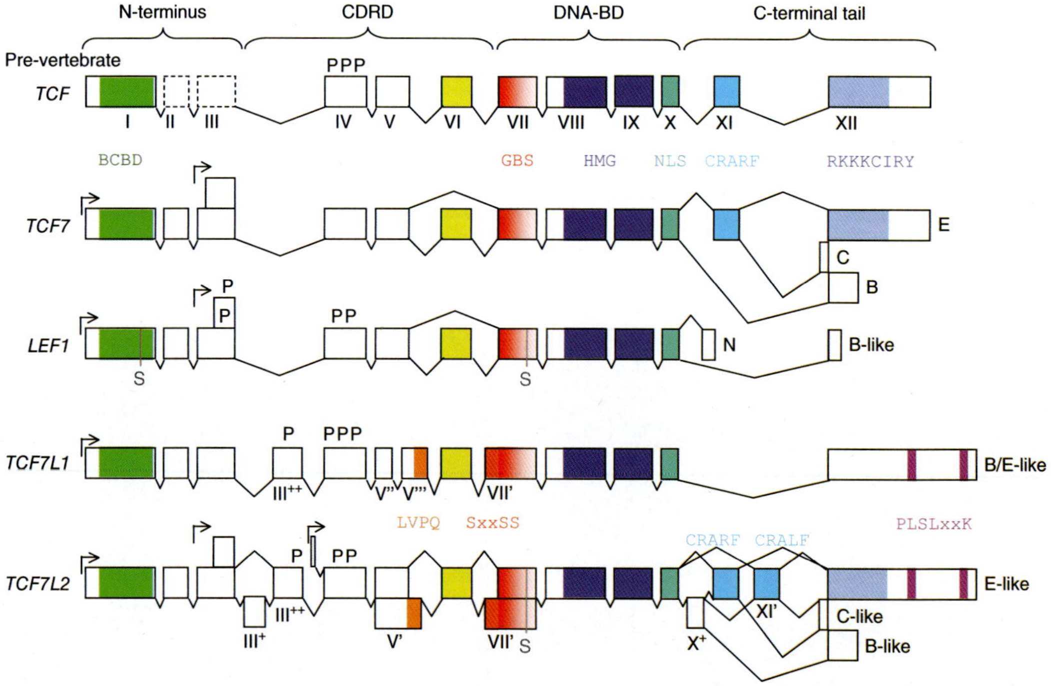

Figure 17.1 Conservation and innovation in exon structure and functional domains of TCF proteins. The exon structures of vertebrate TCF/LEF genes are variations on a conserved exon structure inherited from a single prevertebrate ancestral TCF gene. Invertebrate TCF/LEF proteins are often encoded by 11 conserved exons; however, the split of the original ninth exon meant that the shared ancestor of all vertebrate TCF/LEF proteins must have been encoded by 12 exons (here in roman numbers, to differentiate them from the Arabic numbers used in gene-specific literature). Experimentally identified functional domains of TCF proteins generally correspond to different exons or groups of exons, such as the BCBD; the groucho binding sequence (GBS)/TLE transcriptional corepressor binding sequence; the HMG-fold DNA-binding domain, a basic domain that assists in DNA binding and functions as an NLS; and additional domains at the C-terminus of certain TCF proteins, such as a short polypeptide sequence CRARF (or CRALF), which together with another polypeptide sequence (RKKKCIRY) forms an auxiliary DNA-binding domain called C-clamp and a proposed CtBP transcriptional corepressor binding domain (PLSLxxK). Vertebrate genomes generally have four TCF/LEF genes, *TCF7* (also called *TCF1*), *LEF1* (*TCF7L3*), *TCF7L1* (also called *TCF3*), and *TCF7L2* (also referred to as *TCF4*), which (apart from *TCF7L1*) encode several protein isoforms each, which are generated through alternative transcription and translation start sites and alternative splicing. Four major domains in TCF/LEF proteins can thus be identified: an N-terminus (containing the BCBD), a CDRD situated between the N-terminus and the core DNA-binding domain, and a C-terminal tail. The protein-coding exon structure of the major TCF7 transcript is most similar to the prevertebrate ancestral TCF/LEF gene (apart from the otherwise conserved phosphorylation sites; see following text). Alternative use of different transcription and translation start sites modify the N-terminus, but differences in splicing primarily affect the CDRD and the structure of the C-terminus, while the DNA-binding domain remains relatively unaltered. Some vertebrate TCF/LEF genes encode one (*TCF7L1*) or few (*LEF1*) isoforms, while others (*TCF7* and *TCF7L1*) encode many. Through differences in their domain structures, vertebrate TCF/LEF proteins can become specialists at mediating transcriptional activation (LEF1 and several TCF7 isoforms) or transcriptional repression (TCF7L1 and several TCF7L2 isoforms). Further evolved differences in the protein sequence within the CDRD release TCF7 proteins from being regulated by phosphorylation that regulates DNA binding and nuclear export of other TCF/LEF proteins (for additional complexity in exon structure of mammalian TCF/LEF genes and in particular *TCF7L2*, consult Mao and Byers, 2011; Weise *et al.*, 2010). Length of illustrated exons is proportional to relevant peptide sequences in *TCF7*.

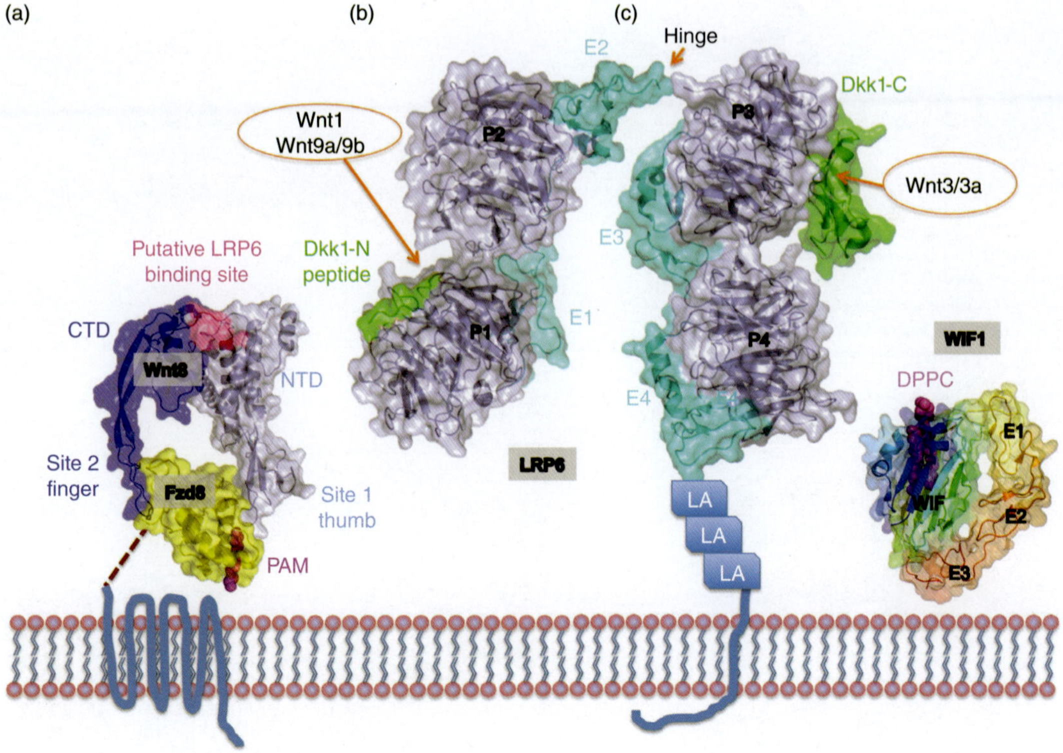

Figure 18.1 Wnt recognition by Fzd and LRP6 at the cell membrane, as well as its inhibition by Dkk1 and WIF1. Wnt8, Fzd8, LRP6, Dkk1, and WIF1 are all shown in cartoon illustration with semitransparent space-filling models. (a) Structure of a Wnt–Fzd complex. XWnt8 interacts with Fzd8-CRD through thumb- and fingerlike structures from its NTD and CTD domain, respectively. A PAM (shown as purple spheres) of modified Ser187 of XWnt8 mediated the site 1 interaction. A putative LRP6-binding surface is shown in pink on the XWnt8 surface. (b) Structure of LRP6 and its complex with Dkk1. The P1E1P2E2 and P3E3P4E4 domains of LRP6 form two relatively rigid structural blocks jointed by a short hinge, which restrains the relative orientation of these two blocks. The Dkk1 N-terminal peptide sits on the top surface of β-propeller 1 (P1). The Dkk1 C-terminal domain interacts with the top surface of β-propeller 3 (P3). EGF-like domains after each β-propeller are shown in *cyan*. The binding sites of Dkk1 N-terminal peptide and C-terminal domain partially overlap with those of Wnts (in circles with *arrows*) on LRP6. (c) Structure of a lipid-bound near-full-length WIF1(WIF-1ΔC). WIF1 contains a WIF domain, five EGF domains, and a hydrophilic C-terminal tail. The cell membrane lipid 1,2-dipalmitoylphosphatidylcholine (DPPC), bound to the WIF1 domain, is shown as purple spheres. WIF1 may inhibit Wnt activities by binding to the palmitoleic group of Wnt proteins in a similar manner. The WIF1 EGF3 (E3) domain folds back to contact the WIF domain, whereas the EGF4 and EGF5 domains have flexible orientations related to the rest of molecule and are not shown.

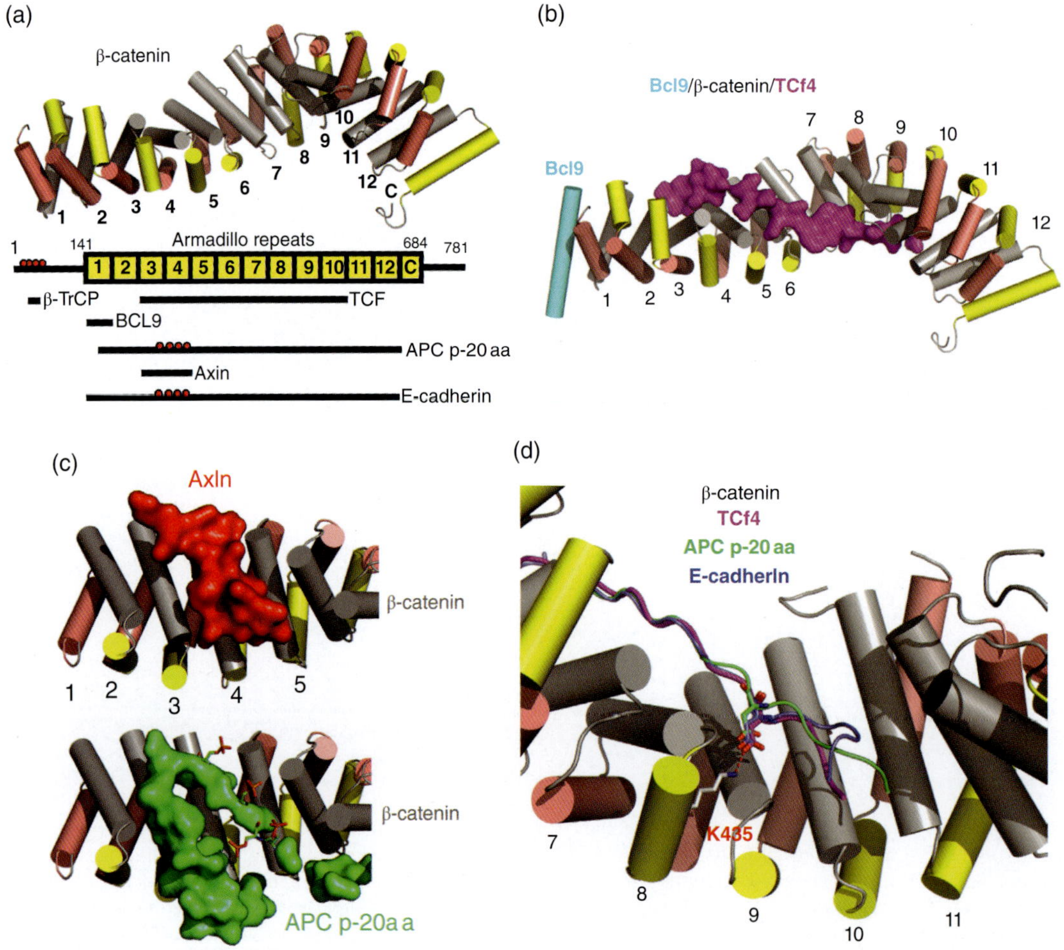

Figure 18.2 Structures of β-catenin and β-catenin-containing complexes. (a) Structures of β-catenin, which contains 12 armadillo repeats, an ordered C-terminal helix, and flexible N- and C-terminal tails. The groove in the armadillo repeat region serves as a common binding site for several β-catenin binding partners, including Tcf, APC, and cadherin. *Black bars* indicate regions of β-catenin bound by the different key partners involved in the Wnt signaling, whereas *red dots* indicate critical phosphorylation sites. APC p-20 aa represents the phosphorylated form of the APC 20 aa repeat region. (b) Structure of β-catenin complexes in the β-catenin destruction complex. Crystal structures of the β-catenin armadillo repeat domain in complex with the β-catenin binding domain of Axin (in *red*) and the phosphorylated third 20 aa repeat of APC (in *green*) are shown in the same orientation. The positions of the four phosphorylated residues in APC 20 aa repeat 3 are shown in *red* sticks. (c) Structure of a β-catenin transcriptional complex. The β-catenin armadillo repeat domain is shown in complex with Tcf4 (*purple*) and BCL9 (*cyan*). (d) The critical lysine residue (K435) of β-catenin involved in recognition of multiple partners.

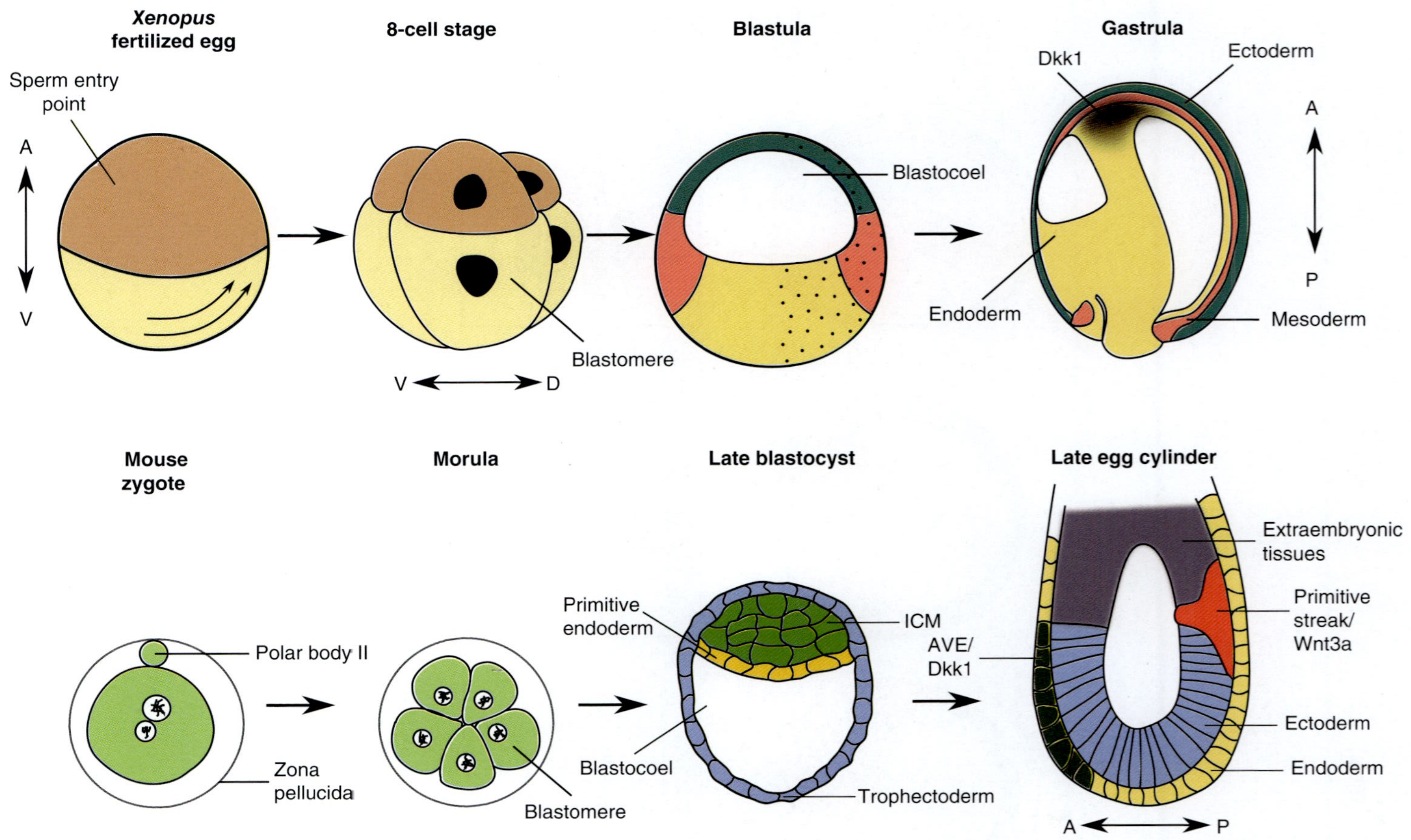

Figure 19.1 Schematic diagram of early mouse and *Xenopus* development. (a) Early *Xenopus* development from fertilized egg to gastrula. Note nuclear localization of β-catenin in the blastula at the dorsal side of the embryo (indicated with *black dots*, see text for more details) and expression of the Wnt inhibitor Dkk1 in anterior cells in the gastrula (indicated in *black*, see text for more details). (b) Early mouse development from zygote to late egg cylinder stage. Note at the late egg cylinder stage localized expression of *Dkk1* in the AVE (anterior visceral endoderm), which will become the anterior of the embryo and of Wnt3a in the primitive streak. ICM, inner cell mass.

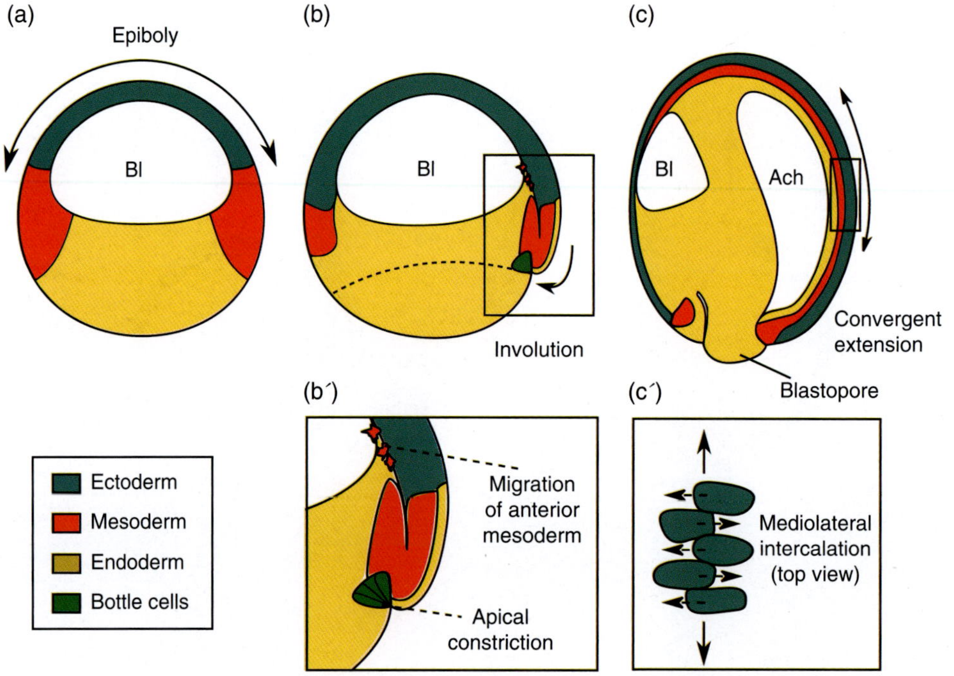

Figure 19.2 Morphogenetic processes during *Xenopus* gastrulation. (a) Late blastula, (b) early and (c) late gastrula in *Xenopus* embryos in the same dorsoventral and anteroposterior orientation as in Figure 19.1. Prospective ectoderm (*dark green*), mesoderm (*red*), and endoderm (*amber*) as indicated. Note in (b′) formation of bottle cells (*light green*) as the mesoderm (*red*) involutes and in (c′) the mediolateral intercalation characteristic of conversion extension movements (for detail, see text).

(a)

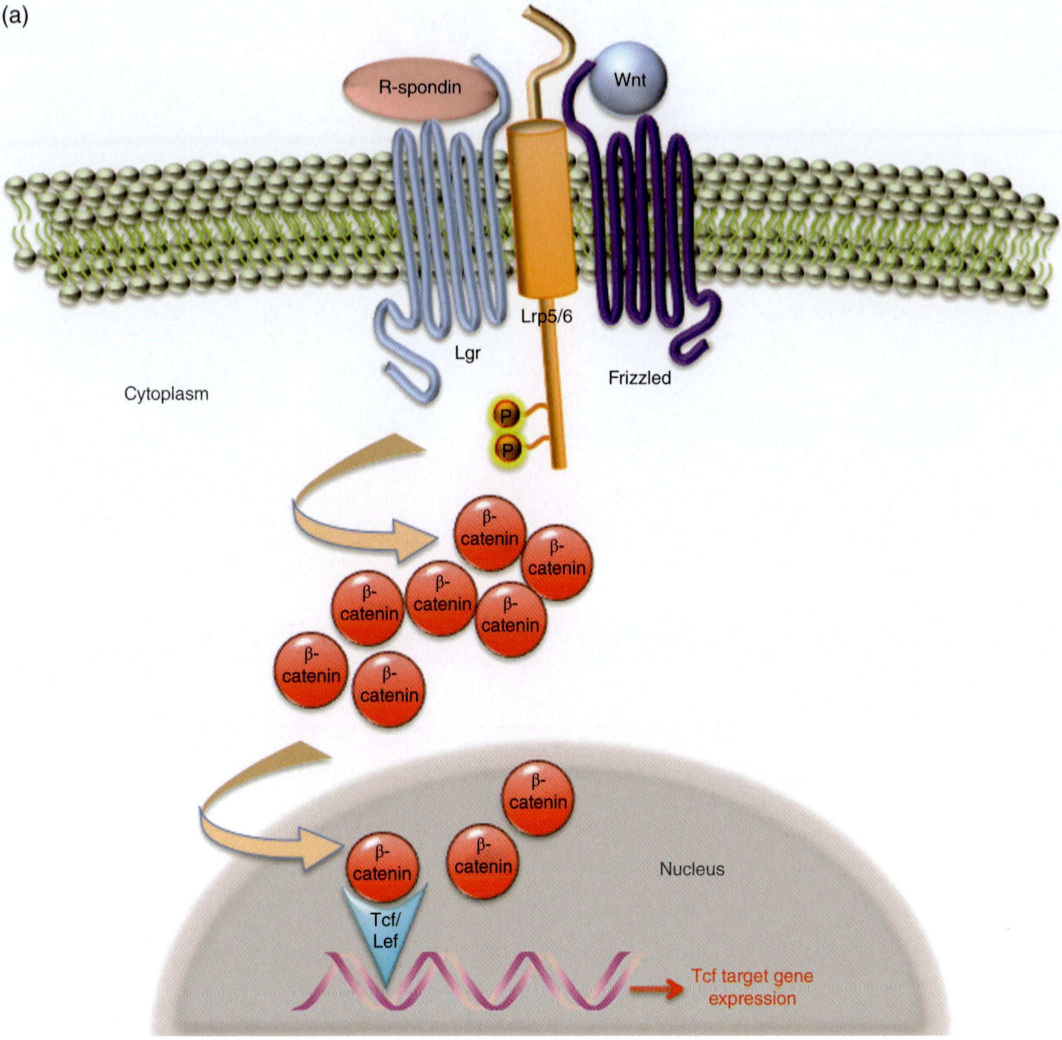

Figure 24.1 Overview of Wnt signaling pathways. (a) Upon binding of Wnt to Fz in a receptor complex that contains LRP and R-spondin bound to LGR, β-catenin accumulates in the cytoplasm and translocates to the nucleus. Once in the nucleus, together with TCF (TCF/lymphoid-enhancer binding protein factor), β-catenin forms a transcriptionally active complex to activate the expression of target genes.

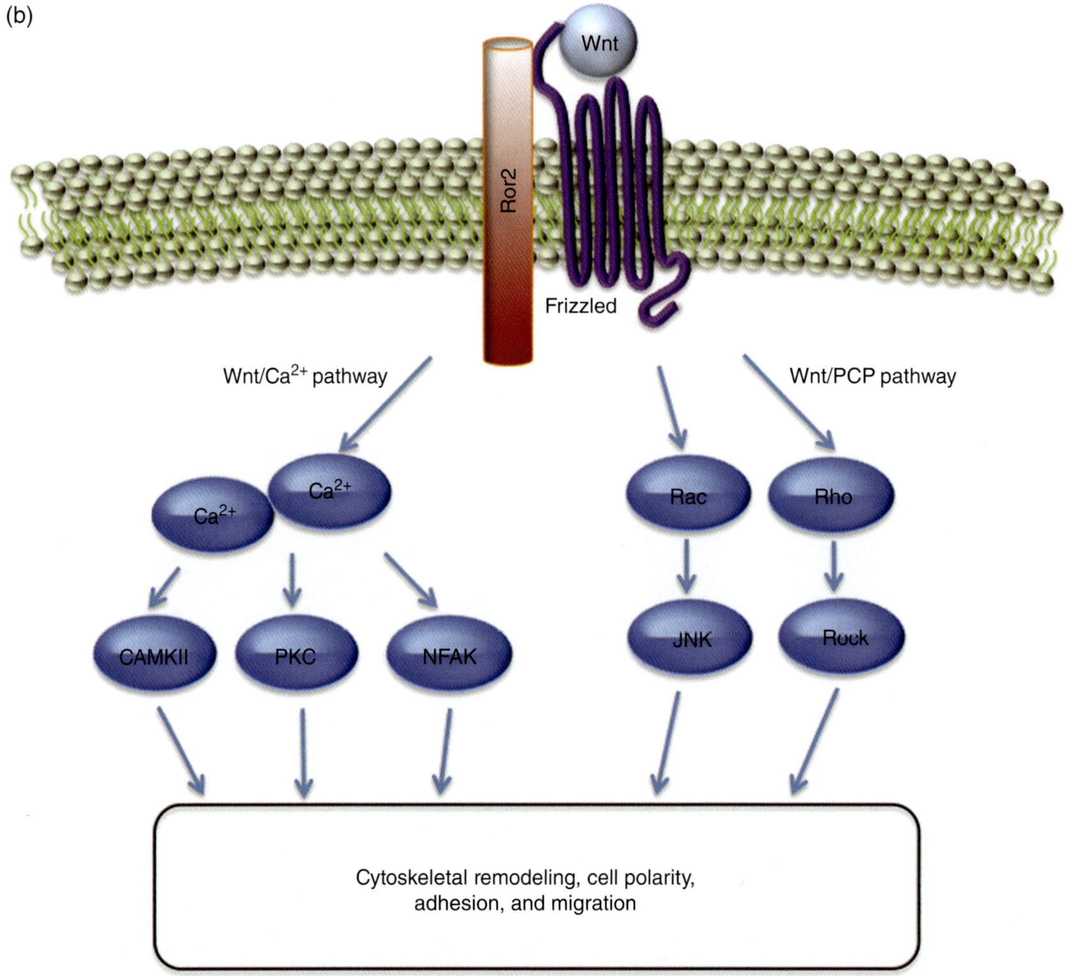

Figure 24.1 (continued) (b) In the β-catenin-independent arm of Wnt signaling, the Wnt signal is relayed into the cell via calcium (Ca^{2+}) and the subsequent activation of CAMKII (calmodulin-dependent protein kinase II), PKC (protein kinase C), NFAK (nuclear factor of activated T-cells) Ca^{2+}-responsive proteins (referred to as Wnt/Ca^{2+} pathway) or Rac/Rho GTPases and JNK/ROCK (referred to as Wnt/PCP pathway).

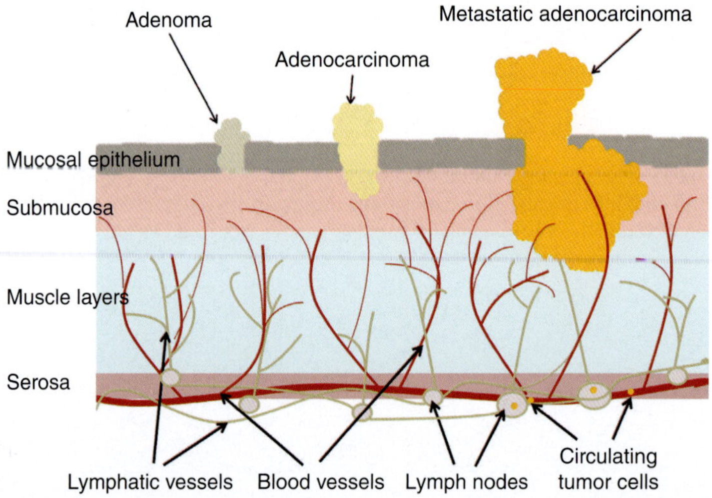

Figure 24.2 Progression in CRC. Initial mutations most frequently involving the APC genes result in benign adenoma growths. Accumulated mutations in other genes over time enable the tumors to become invasive and metastatic through the hematogenous and the lymphatic routes.

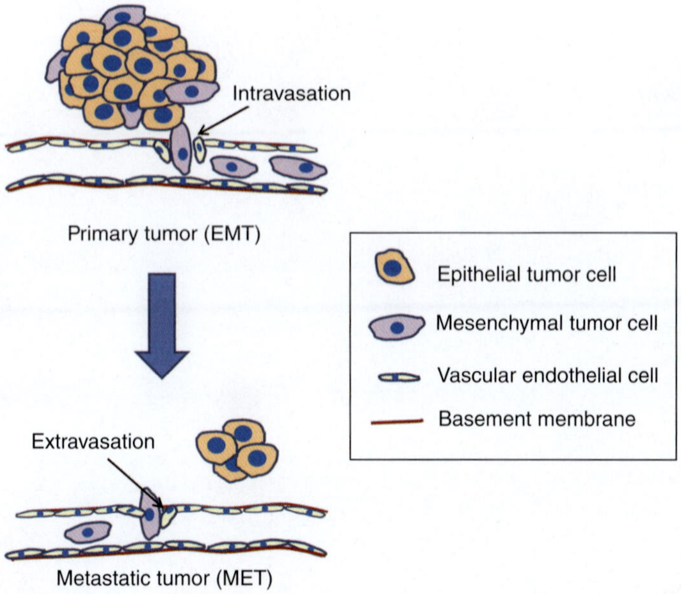

Figure 24.3 Tumor EMT and MET in metastasis. Selective epithelial tumor cells at the tumor periphery acquire a mesenchymal morphology (EMT) and the ability to degrade and invade the basement membrane and enter the circulation. At the metastatic site, the tumor cells undergo the reverse process (MET) to establish micrometastases.

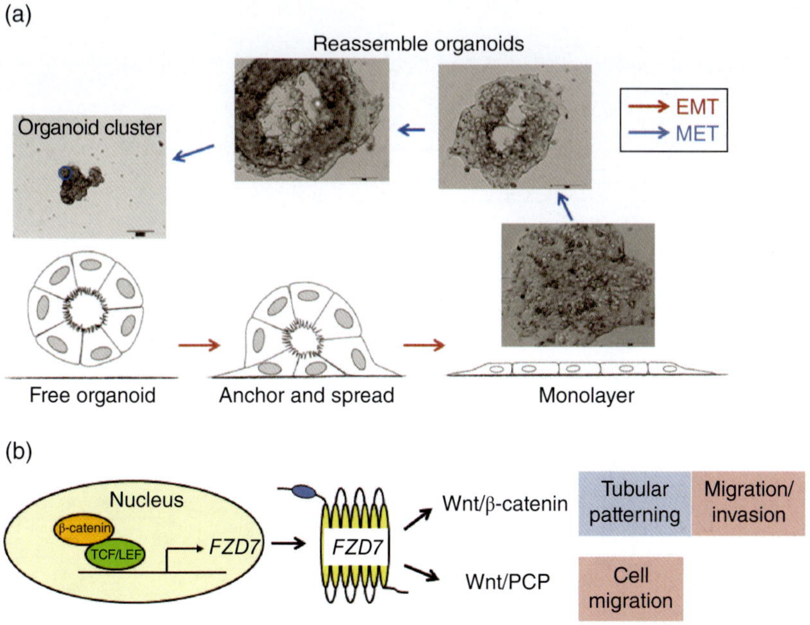

Figure 24.4 Wnt signaling in CRC morphogenesis. (a) The LIM1863-*Mph* model recapitulates many features of CRC tumor morphogenesis. The cells spontaneously undergo transitions between monolayer and organoid states in tissue culture medium. The phenotype change from organoid to monolayer is similar to EMT (*red arrows*), while the reverse is similar to MET (*blue arrows*) observed in carcinoma tissues (shown are DIC images, *blue circle* indicates one organoid in a cluster of organoids, scale bar 100 μM). (b) Experimental evidence indicates that the Wnt receptor FZD7, which is itself a TCF/β-catenin target gene, dictates patterning of the organoids via TCF/β-catenin signaling and monolayer cell migration via Wnt/PCP signaling.

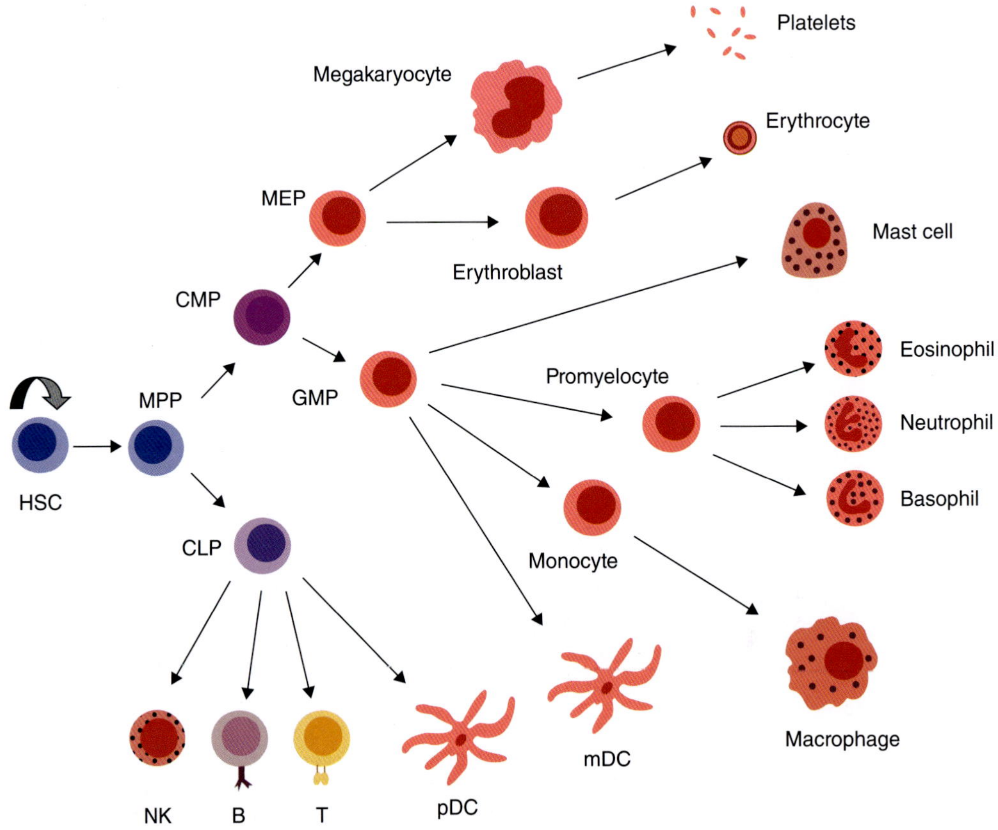

Figure 25.1 Schematic overview of hematopoiesis. HSCs are responsible for blood cell production throughout the lifetime of an individual. Pluripotent HSC can give rise to several different hematopoietic lineages while retaining the capacity for self-renewal. Lineage-committed progenitor cells produce progeny destined to differentiate into red cells, granulocytes, lymphoid cells, and platelets.

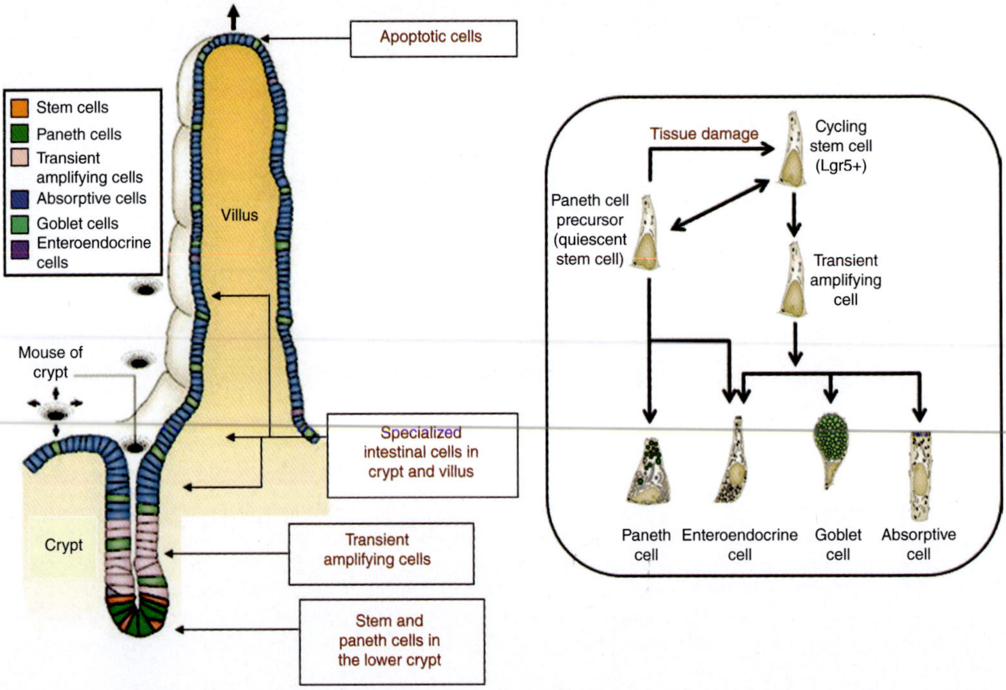

Figure 25.2 Intestinal stem cells in the crypt. The cell population that renews the epithelium of the small intestine occurs in the intestinal crypts, deep invaginations in the lining of the gut. The crypt contains stem cells and other cell types, giving rise to TA cells and differentiating in several other cell types. For details, see main text.

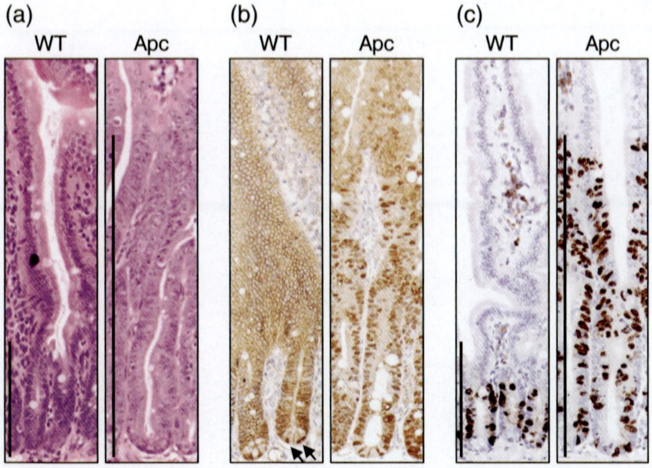

Figure 27.1 Crypt progenitor phenotype following *Apc* loss. (a) Hematoxylin and eosin staining of normal (WT) and *VilCre-ERT Apc*$^{580S/580S}$ (*Apc*) small intestines 4 days following Cre induction with tamoxifen. The expanded *Apc*-deficient crypt is indicated with a black vertical line. (b) Immunohistochemical staining for β-catenin in WT and *Apc*-deficient small intestines. Arrows indicate cells with nuclear localized β-catenin in WT crypt. β-catenin is localized to the nucleus in all *Apc*-deficient cells. (c) Immunohistochemical staining for BrdU in WT and *Apc*-deficient small intestines. The expanded proliferative zone in *Apc*-deficient small intestine is indicated with a black vertical line.

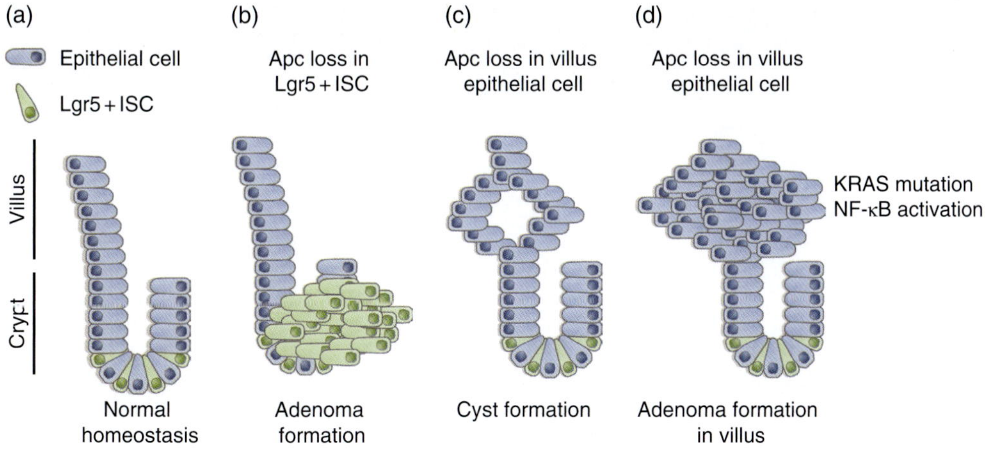

Figure 27.2 Cell of origin of CRC. (a) Graphical representation of a normal intestinal crypt. Lgr5+ ISCs are depicted in green residing at the crypt base; all other intestinal epithelial cells are depicted in blue. (b) *Apc* loss in Lgr5+ ISCs leads to rapid adenoma formation. (c) *Apc* loss in villus epithelial cells gives rise to formation of cyst-like structures that take much longer to progress to adenomas. (d) If *Apc* loss in villus epithelial cells is combined with an activating mutation in *KRAS* or activation of NF-κB signaling, adenoma formation rapidly occurs.

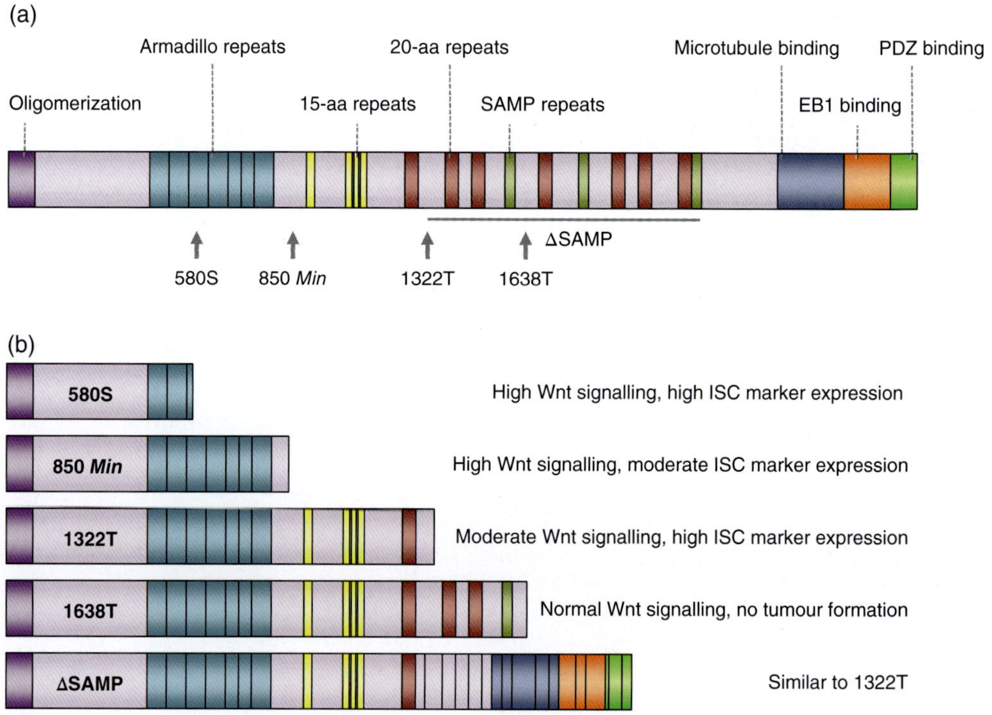

Figure 27.3 Mutant alleles used to study *Apc* function *in vivo*. (a) Graphical representation of the Apc protein structure with key domains highlighted. 15-aa and 20-aa repeats are important for β-catenin binding, and SAMP repeats mediate binding to Axin. The positions of mutations creating five mutant *Apc* alleles are shown below the structure. (b) Graphical representations of the truncated protein products produced from these mutations and their effects on Wnt signaling and ISC marker expression.

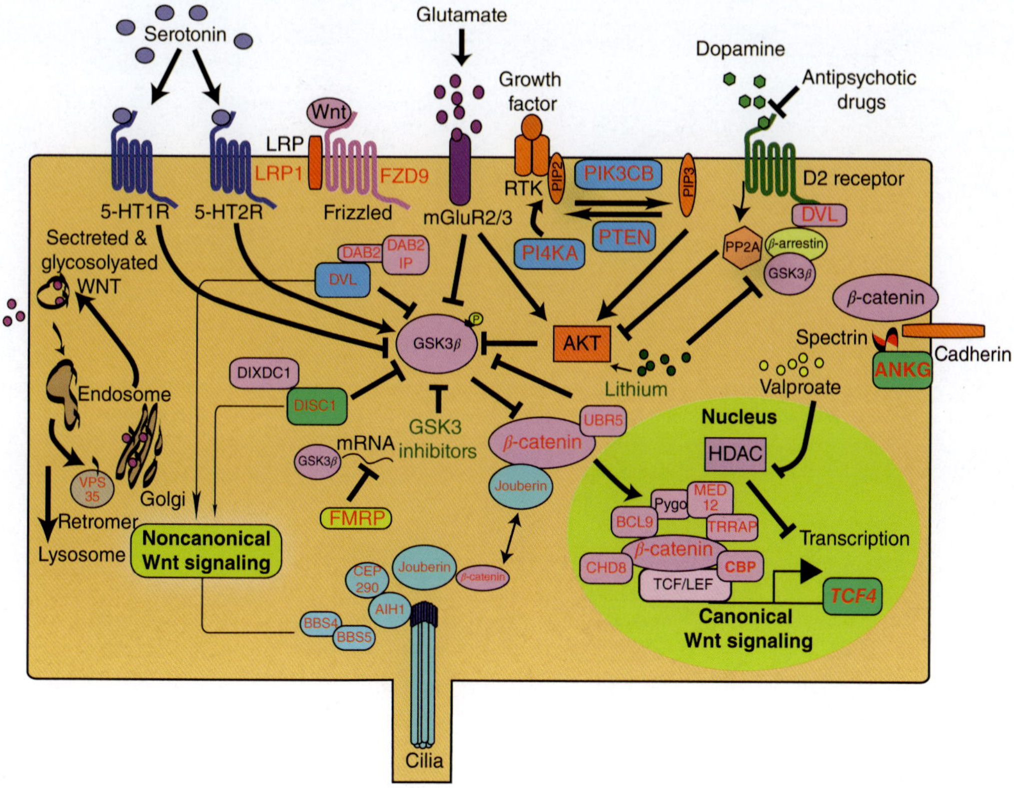

Figure 29.1 Overview of Wnt signaling pathway components implicated in neuropsychiatric disorders by human genetics, neuropharmacology, and functional genomic studies.

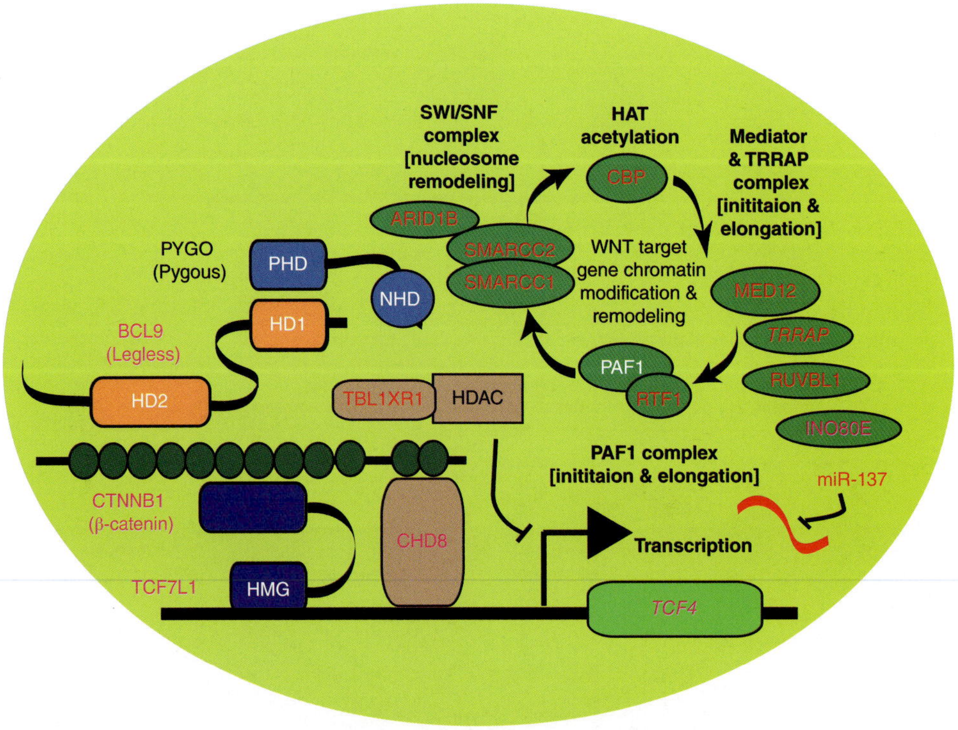

Figure 30.1 Wnt signaling pathway components involved in chromatin remodeling and transcription that have been implicated in neuropsychiatric disorders by human genetics, neuropharmacology, and functional genomic studies.

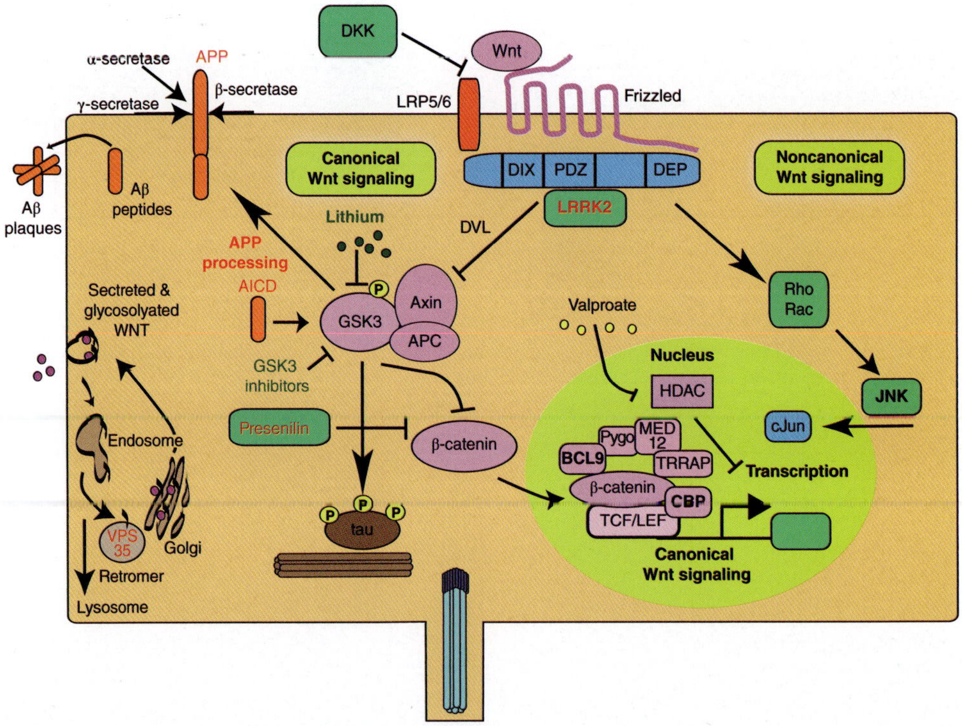

Figure 31.1 Wnt signaling pathway components implicated in dementias by human genetics and functional genomic studies.

Axin phosphorylated in a region C-terminal to its β-catenin binding domain (Axin-CBD) keeps Axin in an activated ("open") state, which promotes the β-catenin–Axin interaction. Formation of the Wnt-induced signalosome promotes Axin dephosphorylation and inactivates Axin through an intramolecular interaction between the Axin-CBD region and C-terminal DAX domain (in a "closed" state), which inhibits the β-catenin–Axin binding, thereby inhibiting β-catenin phosphorylation (Kim *et al.*, 2013). This attractive model puts the regulation of Axin phosphorylation and conformation in the central spot of canonical Wnt signaling.

APC is the tumor suppressor whose mutations are associated with the vast majority of colorectal cancers (see Chapter 27). *APC* encodes a large ~310 kDa protein, which contains multiple functional domains. The central region of APC, which is necessary and sufficient for canonical Wnt signaling, contains multiple β-catenin binding domains, including four 15 amino acid (15 aa) repeats and seven 20 amino acid (20 aa) repeats. Three SAMP repeats, interspersed among the 20 aa repeats, mediate APC's interaction with Axin. Crystal structures of an Axin RGS domain in complex with a SAMP motif from APC showed that the SAMP motif forms a short helix and docks on a conserved groove on RGS domain surface, which is distinct from the G protein interface of classical RGS proteins (Spink, Polakis, and Weis, 2000). How APC plays a crucial role in β-catenin turnover has been the focus of intense research. Although a number of models have been proposed, which have been the topic of several recent reviews (Kimelman and Xu, 2006; Stamos and Weis, 2012), it remains a critical issue that requires further investigation. While it is possible that several nonexclusive models are right and APC is a multifaceted protein playing roles in multiple steps of β-catenin turnover, here, we will only mention one of the models derived from structural studies.

The β-catenin destruction complex can be considered as a megasize complex kinase for β-catenin. Crystal structures of β-catenin proteins revealed that the β-catenin central domain contains 12 armadillo repeats. Each repeat consists of 3 helices, and the 12 armadillo repeats form a twisted superhelix of helices, which is relatively rigid and with a positively charged groove that is the binding site for many β-catenin partners, including Tcf, cadherin, and APC (Huber, Nelson, and Weis, 1997; Xing *et al.*, 2008; Figure 18.2a). Complex crystal structure showed that the β-catenin binding domain of Axin forms a helix that occupies the groove formed by the third and fourth armadillo repeats of β-catenin (Ha *et al.*, 2004; Xing *et al.*, 2003). In comparison, crystal structures of β-catenin in complex with phosphorylated APC 20 aa repeats illustrated that the Axin and the phosphorylated segments of APC 20 aa repeats overlap with each other (Ha *et al.*, 2004; Xing *et al.*, 2004; Figure 18.2b). Consistent with these structures, biochemical analysis showed that APC competes with Axin for β-catenin binding in a phosphorylation-dependent manner. Since APC's loss-of-function phenotypes can be rescued by Axin overexpression (Hart *et al.*, 1998), we proposed that one of APC's functions may be to enhance Axin's efficiency by removing the phosphorylated β-catenin from the Axin binding site (Ha *et al.*, 2004; Xing *et al.*, 2003).

This model is consistent with the other structural and biochemical analysis. Close to the β-catenin binding site is Axin's GSK3 binding site. By bringing together β-catenin and GSK3, the Axin scaffold, when present at equal molar ratio, can dramatically enhance phosphorylation of β-catenin (Dajani *et al.*, 2003). Crystal structure of the GSK3β–Axin complex showed that the GSK3-binding domain of Axin also forms a helix and dock at a site on the GSK3 C-terminal domain surface that has a distance from the GSK3 active site and does not directly affect GSK3 kinase activity (Dajani *et al.*, 2003), suggesting that Axin's enhancement effect is predominantly derived from the scaffolding effect. However, since the Axin level is much lower than these of β-catenin and APC in the cell and each Axin molecule has only one β-catenin binding site, there is a need for a mechanism to remove the phosphorylated β-catenin product from the active site. Phosphorylated APC is perfectly suited for such a role.

In comparison to the clear snapshots of the β-catenin–Axin, β-catenin/APC, and GSK3β–Axin complexes, we still only have a rough map of the binding domains for some of other key partners, such as LRP6, CK1, and PP2A. Nevertheless, crystal structure of a β-TrCP/β-catenin

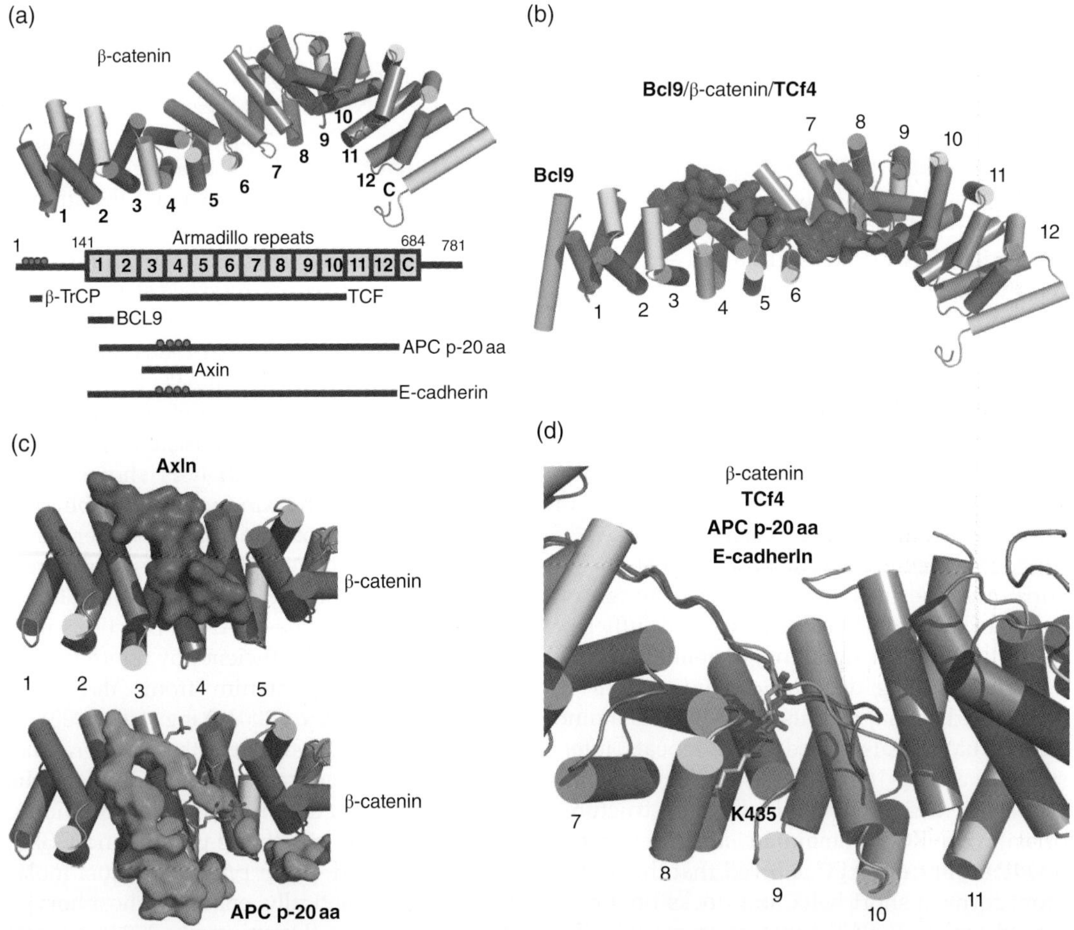

Figure 18.2 Structures of β-catenin and β-catenin-containing complexes. (a) Structures of β-catenin, which contains 12 armadillo repeats, an ordered C-terminal helix, and flexible N- and C-terminal tails. The groove in the armadillo repeat region serves as a common binding site for several β-catenin binding partners, including Tcf, APC, and cadherin. *Black bars* indicate regions of β-catenin bound by the different key partners involved in the Wnt signaling, whereas *red dots* indicate critical phosphorylation sites. APC p-20 aa represents the phosphorylated form of the APC 20 aa repeat region. (b) Structure of β-catenin complexes in the β-catenin destruction complex. Crystal structures of the β-catenin armadillo repeat domain in complex with the β-catenin binding domain of Axin (in *red*) and the phosphorylated third 20 aa repeat of APC (in *green*) are shown in the same orientation. The positions of the four phosphorylated residues in APC 20 aa repeat 3 are shown in *red* sticks. (c) Structure of a β-catenin transcriptional complex. The β-catenin armadillo repeat domain is shown in complex with Tcf4 (*purple*) and BCL9 (*cyan*). (d) The critical lysine residue (K435) of β-catenin involved in recognition of multiple partners. *(See insert for color representation of the figure.)*

complex have revealed how the β-TrCP E3 ubiquitin ligase recognizes the $D_pSG\Phi X_pS$ (Φ: hydrophobic residue) destruction motif formed on β-catenin upon its phosphorylation at S33 and S37 (Wu *et al.*, 2003). It remains enigmatic if and how proteins such as APC and WTX play a role in regulating β-catenin ubiquitination and degradation after its phosphorylation in the destruction complex. Given the dynamic nature and structural complexity of the destruction complex, a combination of structural, biophysical, and biochemical approaches will be needed to dissect its complex molecular mechanism, which remains a major challenge.

As the essential scaffold of the destruction complex, Axin has become an important target for developing Wnt pathway inhibitors. In fact, a number of cell-based screening of Wnt signaling pathway inhibitors have led to the discovery of compounds that inhibit Wnt signaling by inhibiting Axin degradation. Axin turnover is controlled by polyubiquitination, which requires a premodification of Axin by TNKS. TNKS is a member of poly(ADP-ribose) polymerase (PARP) family, which adds poly(ADP-ribose) chains to Axin and earmarks it for E3 ubiquitin ligase (Huang *et al.*, 2009). TNKS contains a large N-terminal ankyrin repeat domain, responsible for Axin binding, and a C-terminal catalytic PARP domain, which is the target of known Wnt pathway inhibitors such as XAV939 and IWR-1 (see Chapter 32). Since TNKS is also involved in many other functions in the cell, it is desirable to develop inhibitors that can disrupt the recruitment of Axin by TNKS ankyrin repeat domain. Crystal structure of an Axin–TNKS complex demonstrated that the N-terminal domain of Axin (residues 1–80 in Axin1) contains two discrete TNKS-binding segments, each of which binds to one copy of a TNKS homodimer (Morrone *et al.*, 2012). The Axin–TNKS interface involves the recognition of two critical glycine residues via a Tyr–Tyr gate on the TNKS surface. With the dual TNKS-binding segments, Axin stabilizes TNKS dimerization. Comparison with other TNKS-substrate peptide complex crystal structures suggests that the TNKS–Axin protein–protein interface may not be the ultimate solution for developing Wnt-specific inhibitors, due to the similar binding sites of different TNKS partners on TNKS surface (Guettler *et al.*, 2011). One difference for Axin–TNKS interaction is that Axin contains two TNKS-binding segments and both are required for Axin turnover. It remains unclear if it will be possible to develop Wnt-specific inhibitors by targeting at TNKS oligomerization states.

Nuclear protein complexes crucial for Wnt signaling

In the presence of a Wnt signal, β-catenin accumulates in the cytosol and migrates into the nucleus, where it forms complexes with the Tcf/LEF family of DNA-binding proteins and activates transcription of Wnt target genes (see Chapters 4, 16, and 17). Since most cancer-associated mutations, in particular those found in colon cancers, are found in genes encoding components of the β-catenin destruction complex, targets downstream of the destruction complex are of particular interests for developing novel anticancer drugs. These targets include proteins/enzymes specifically required for β-catenin nuclear import/localization and targets in the nuclear transcriptional complex. While many β-catenin binding proteins, including Tcf, APC, and Axin, are involved in β-catenin retention in either cytosol or nucleus, no protein has been shown to be sufficient to export β-catenin from the nucleus. Thus, the β-catenin/Tcf-based Wnt transactivation complex is considered as a key drug target.

Crystal structure of the β-catenin/Tcf complex showed that the N-terminal ~50 residues of Tcf3/4 bind to the positively charged groove of β-catenin armadillo domain (Graham *et al.*, 2000, 2001; Poy *et al.*, 2001; Figure 18.2c). Structure-oriented mutagenesis revealed the core β-catenin binding segment of Tcf and a crucial salt bridge formed between β-catenin K435 and Tcf4 D16 (referred to as the charged button). However, the perspective of using this hot spot for drug discovery is challenged by structural observations that the Tcf4 core binding site on β-catenin surface, in particular K435, is the shared binding site for Tcf, cadherins, APC, and other β-catenin partners (Xu and Kimelman, 2007; Figure 18.2d). While disrupting the β-catenin/Tcf complex may be beneficial for cancer treatment, disrupting the β-catenin/cadherin interface may promote tumorigenesis and metastasis (see Chapter 27). Therefore, transcriptional coactivators interacting with the β-catenin N- and C-terminal transactivation domains are of particular interest for developing specific Wnt pathway inhibitors.

One protein of particular interest is BCL9, which interacts with β-catenin's N-terminal transactivation domain. BCL9 is required for Wnt/β-catenin signaling, and its upregulation is associated with cancers (Kramps *et al.*, 2002). Two conserved domains of BCL9, HD1 and HD2, are responsible for interaction with Pygopus (Pygo) and β-catenin, respectively. Crystal structure of a β-catenin/Tcf/BCL9-HD2 complex demonstrated that BCL9-HD2 forms a helix, which docks on the surface formed by β-catenin's first armadillo repeat (Sampietro *et al.*, 2006; Figure 18.2c). It should be noted that the BCL9 binding site is distinct from these for all other β-catenin binding partners and thus is a potential site for developing specific Wnt pathway inhibitors. A screen of compounds that disrupt the β-catenin/BCL9 interaction led to the discovery of carnosic acid, which attenuates β-catenin transcriptional activity in colorectal cancer cells most likely through its interaction with the BCL9-HD binding site on the β-catenin surface (de la Roche *et al.*, 2012). More recently, a stapled peptide (SAH-BCL9) based on the BCL9-HD's β-catenin binding helix was shown to dissociate the β-catenin/BCL9 complex, selectively inhibits Wnt-induced transcription, and exhibited mechanism-based antitumor activity. Importantly, SAH-BCL9 suppresses tumor growth, angiogenesis, invasion, and metastasis in mouse xenograft models of colorectal carcinoma and INA-6 multiple myeloma (Takada *et al.*, 2012). All these findings provided a proof of principle that the β-catenin/BCL9 interface is a valid target for developing specific Wnt pathway inhibitors that may be useful for cancer treatment.

The main BCL9 function is the recruitment of Pygo, another transcriptional coactivator, to the β-catenin/Tcf complex (Kramps *et al.*, 2002). The PHD domain of Pygo interacts with the HD1 domain of BCL9. This interaction not only recruits Pygo to the Tcf/LEF-binding promoters but also boosts the recognition of methylated histone H3 tail by the PHD domain (Fiedler *et al.*, 2008). Crystal structure of a BCL9-HD1/Pygo-PHD/H3 peptide ternary complex demonstrated how BCL9-HD1 modulates the PHD–H3 peptide interaction by buttressing the PHD A1 cavity from the back (Fiedler *et al.*, 2008). These binding interfaces between Pygo and its partners are also potential targets for small-molecule Wnt pathway inhibitors (see Chapter 32).

There are a number of β-catenin partners, including CBP/p300, MED12, PAF1, and Brg1, which interact with the C-terminal transactivation domain (Cadigan and Waterman, 2012). The interactions between these transcriptional coactivators and β-catenin remain to be structurally characterized. Some of these interfaces are potential drug targets. Crystal structure of full-length β-catenin revealed an additional C-terminal helix packing at the end of β-catenin armadillo domain, which is likely involved in interactions with some of these C-terminal transcriptional coactivators (Xing *et al.*, 2008; Figure 18.2a). In addition, CDK8 has been identified as a colon cancer oncoprotein that regulates β-catenin transcriptional activity, either directly as a mediator component or via suppression of E2F1 (Firestein *et al.*, 2008; Morris *et al.*, 2008). CDK8 may be a key enzymatic drug target downstream of β-catenin.

Last but not least, it should be mentioned that β-catenin can interact with DNA-binding proteins other than Tcf/LEF family members, including FOXO, SOX17, HIF1, and some of the nuclear receptor (e.g., androgen receptor). It remains unclear how they cross talk with the Wnt pathway and if some of the Wnt signaling phenotypes are derived from these complexes. How β-catenin interacts with these DNA-binding proteins remains to be structurally characterized. Crystal structures of these β-catenin complexes will allow people to design specific mutations on these proteins and examine the specific contributions of these interactions in Wnt signaling.

Due to space limit, many structural studies on protein–protein interactions regulating Wnt signaling cannot be included in this chapter. Nevertheless, it is clear that structural studies have provided many important insights into Wnt signaling mechanism. Many more complexes remain to be structurally characterized, which will be critical for both understanding the signaling mechanism and drug discovery.

References

Ahn, V.E., Chu, M.L., Choi, H.J. *et al.* (2011) Structural basis of Wnt signaling inhibition by Dickkopf binding to LRP5/6." *Developmental Cell*, **21** (5), 862–873.

Bazan, J.F., Janda, C.Y., and Garcia, K.C. (2012) "Structural architecture and functional evolution of Wnts." *Developmental Cell*, **23** (2), 227–232.

Bilic, J., Huang, Y.L., Davidson, G. *et al.* (2007) "Wnt induces LRP6 signalosomes and promotes dishevelled-dependent LRP6 phosphorylation." *Science*, **316** (5831), 1619–1622.

Bourhis, E., Wang, W., Tam, C. *et al.* (2011) "Wnt antagonists bind through a short peptide to the first beta-propeller domain of LRP5/6." *Structure*, **19** (10), 1433–1442.

Brott, B.K. and Sokol, S.Y. (2002) "Regulation of Wnt/LRP signaling by distinct domains of Dickkopf proteins." *Molecular and Cellular Biology*, **22** (17), 6100–6110.

Cadigan, K.M. and Waterman, M.L. (2012) "TCF/LEFs and Wnt signaling in the nucleus." *Cold Spring Harbor Perspectives in Biology*, **4** (11), a007906.

Chen, L., Wang, K., Shao, Y. *et al.* (2008) "Structural insight into the mechanisms of Wnt signaling antagonism by Dkk." *The Journal of Biological Chemistry*, **283** (34), 23364–23370.

Chen, S., Bubeck, D., MacDonald, B.T. *et al.* (2011) "Structural and functional studies of LRP6 ectodomain reveal a platform for Wnt signaling." *Developmental Cell*, **21** (5), 848–861.

Cheng, Z., Biechele, T., Wei, Z. *et al.* (2011) "Crystal structures of the extracellular domain of LRP6 and its complex with DKK1." *Nature Structural & Molecular Biology*, **18** (11), 1204–1210.

Cortese, M.S., Uversky, V.N., and Dunker, A.K. (2008) "Intrinsic disorder in scaffold proteins: getting more from less." *Progress in Biophysics and Molecular Biology*, **98** (1), 85–106.

Dajani, R., Fraser, E., Roe, S.M. *et al.* (2003) "Structural basis for recruitment of glycogen synthase kinase 3beta to the axin-APC scaffold complex." *The EMBO Journal*, **22** (3), 494–501.

Dann, C.E., Hsieh, J.C., Rattner, A. *et al.* (2001) "Insights into Wnt binding and signalling from the structures of two Frizzled cysteine-rich domains." *Nature*, **412** (6842), 86–90.

Davidson, G., Wu, W., Shen, J. *et al.* (2005) "Casein kinase 1 gamma couples Wnt receptor activation to cytoplasmic signal transduction." *Nature*, **438** (7069), 867–872.

de la Roche, M., Rutherford, T.J., Gupta, D. *et al.* (2012) "An intrinsically labile alpha-helix abutting the BCL9-binding site of beta-catenin is required for its inhibition by carnosic acid." *Nature Communications*, **3**, 680.

Fiedler, M., Sanchez-Barrena, M.J., Nekrasov, M. *et al.* (2008) Decoding of methylated histone H3 tail by the Pygo-BCL9 Wnt signaling complex. *Molecular Cell*, **30** (4), 507–518.

Fiedler, M., Mendoza-Topaz, C., Rutherford, T.J. *et al.* (2011) "Dishevelled interacts with the DIX domain polymerization interface of Axin to interfere with its function in down-regulating beta-catenin." *Proceedings of the National Academy of Sciences of the United States of America*, **108** (5), 1937–1942.

Firestein, R., Bass, A.J., Kim, S.Y. *et al.* (2008) "CDK8 is a colorectal cancer oncogene that regulates beta-catenin activity." *Nature*, **455** (7212), 547–551.

Graham, T.A., Weaver, C., Mao, F. *et al.* (2000) "Crystal structure of a beta-catenin/Tcf complex." *Cell*, **103** (6), 885–896.

Graham, T.A., Ferkey, D.M., Mao, F. *et al.* (2001) Tcf4 can specifically recognize beta-catenin using alternative conformations. *Nature Structural Biology*, **8** (12), 1048–1052.

Grandy, D., Shan, J., Zhang, X. *et al.* (2009) "Discovery and characterization of a small molecule inhibitor of the PDZ domain of dishevelled." *The Journal of Biological Chemistry*, **284** (24), 16256–16263.

Guettler, S., LaRose, J., Petsalaki, E. *et al.* (2011) "Structural basis and sequence rules for substrate recognition by Tankyrase explain the basis for cherubism disease." *Cell*, **147** (6), 1340–1354.

Ha, N.C., Tonozuka, T., Stamos, J.L. *et al.* (2004) "Mechanism of phosphorylation-dependent binding of APC to beta-catenin and its role in beta-catenin degradation." *Molecular Cell*, **15** (4), 511–521.

Hao, H.X., Xie, Y., Zhang, Y. *et al.* (2012) "ZNRF3 promotes Wnt receptor turnover in an R-spondin-sensitive manner." *Nature*, **485** (7397), 195–200.

Hart, M.J., de los Santos, R., Albert, I.N. *et al.* (1998) "Downregulation of beta-catenin by human Axin and its association with the APC tumor suppressor, beta-catenin and GSK3 beta." *Current Biology*, **8** (10), 573– 581.

Holdsworth, G., Slocombe, P., Doyle, C. *et al.* (2012) "Characterization of the interaction of sclerostin with the low density lipoprotein receptor-related protein (LRP) family of Wnt co-receptors." *The Journal of Biological Chemistry*, **287** (32), 26464 – 26477.

Huang, S.M., Mishina, Y.M., Liu, S. *et al.* (2009) "Tankyrase inhibition stabilizes axin and antagonizes Wnt signalling." *Nature*, **461** (7264), 614–620.

Huber, A.H., Nelson, W.J., and Weis, W.I. (1997) "Three-dimensional structure of the armadillo repeat region of beta-catenin." *Cell*, **90** (5), 871–882.

Janda, C.Y., Waghray, D., Levin, A.M. *et al.* (2012) "Structural basis of Wnt recognition by Frizzled." *Science*, **337** (6090), 59–64.

Kim, S.E., Huang, H., Zhao, M. *et al.* (2013) "Wnt stabilization of beta-catenin reveals principles for morphogen receptor-scaffold assemblies." *Science*, **340** (6134), 867–870.

Kimelman, D. and Xu, W. (2006) "β-Catenin destruction complex: insights and questions from a structural perspective." *Oncogene*, **25** (57), 7482–7491.

Koo, B.K., Spit, M., Jordens, I. *et al.* (2012) "Tumour suppressor RNF43 is a stem-cell E3 ligase that induces endocytosis of Wnt receptors." *Nature*, **488** (7413), 665–669.

Kramps, T., Peter, O., Brunner, E. *et al.* (2002) "Wnt/wingless signaling requires BCL9/legless-mediated recruitment of pygopus to the nuclear beta-catenin-TCF complex." *Cell*, **109** (1), 47–60.

Li, L., Mao, J., Sun, L. *et al.* (2002) "Second cysteine-rich domain of Dickkopf-2 activates canonical Wnt signaling pathway via LRP-6 independently of dishevelled." *The Journal of Biological Chemistry*, **277** (8), 5977–5981.

Liepinsh, E., Banyai, L., Patthy, L., and Otting, G. (2006) "NMR structure of the WIF domain of the human Wnt-inhibitory factor-1." *The Journal of Molecular Biology*, **357** (3), 942–950.

Liu, Y.T., Dan, Q.J., Wang, J., *et al.* (2011) "Molecular basis of Wnt activation via the DIX-domain protein Ccd1." *The Journal of Biological Chemistry*, **286** (10), 8597–8608.

Malinauskas, T., Aricescu, A.R., Lu, W. *et al.* (2011) "Modular mechanism of Wnt signaling inhibition by Wnt inhibitory factor 1." *Nature Structural & Molecular Biology*, **18** (8), 886–893.

Morris, E.J., Ji, J.Y., Yang, F. *et al.* (2008) "E2F1 represses beta-catenin transcription and is antagonized by both pRB and CDK8." *Nature*, **455** (7212), 552–556.

Morrone, S., Cheng, Z., Moon, R.T. *et al.* (2012) "Crystal structure of a Tankyrase-Axin complex and its implications for Axin turnover and Tankyrase substrate recruitment." *Proceedings of the National Academy of Sciences of the United States of America*, **109** (5), 1500–1505.

Poy, F., Lepourcelet, M., Shivdasani, R.A., and Eck, M.J. (2001) "Structure of a human Tcf4-beta-catenin complex." *Nature Structural Biology*, **8** (12), 1053–1057.

Sampietro, J., Dahlberg, C.L., Cho, U.S. *et al.* (2006) "Crystal structure of a beta-catenin/BCL9/Tcf4 complex." *Molecular Cell*, **24** (2), 293–300.

Spink, K.E., Polakis, P., and Weis, W.I. (2000) "Structural basis of the Axin-adenomatous polyposis coli interaction." *The EMBO Journal*, **19** (10), 2270–2279.

Stamos, J.L. and Weis, W.I. (2012) "The beta-catenin destruction complex." *Cold Spring Harbor Perspectives Biology*, **5** (1), a007898.

Takada, R., Satomi, Y., Kurata, T. *et al.* (2006) Monounsaturated fatty acid modification of Wnt protein: its role in Wnt secretion. *Developmental Cell*, **11** (6), 791–801.

Takada, K., Zhu, D., Bird, G.H. *et al.* (2012) "Targeted disruption of the BCL9/beta-catenin complex inhibits oncogenic Wnt signaling." *Science Translational Medicine*, **4** (148), 148ra117.

Tauriello, D.V., Jordens, I., Kirchner, K. *et al.* (2012) "Wnt/beta-catenin signaling requires interaction of the Dishevelled DEP domain and C terminus with a discontinuous motif in Frizzled." *Proceedings of the National Academy of Sciences of the United States of America*, **109** (14), E812–E820.

Veverka, V., Henry, A.J., Slocombe, P.M. *et al.* (2009) "Characterization of the structural features and interactions of sclerostin: molecular insight into a key regulator of Wnt-mediated bone formation." *The Journal of Biological Chemistry*, **284** (16), 10890–10900.

Weidauer, S.E., Schmieder, P., Beerbaum, M. *et al.* (2009) "NMR structure of the Wnt modulator protein Sclerostin." *Biochemical and Biophysical Research Communications*, **380** (1), 160–165.

Wong, H.C., Mao, J., Nguyen, J.T. *et al.* (2000) Structural basis of the recognition of the dishevelled DEP domain in the Wnt signaling pathway. *Nature Structural Biology*, **7** (12), 1178–1184.

Wong, H.C., Bourdelas, A., Krauss, A. *et al.* (2003) "Direct binding of the PDZ domain of Dishevelled to a conserved internal sequence in the C-terminal region of Frizzled." *Molecular Cell*, **12** (5), 1251–1260.

Wu, G., Xu, G., Schulman, B.A. *et al.* (2003) "Structure of a beta-TrCP1-Skp1-beta-catenin complex: destruction motif binding and lysine specificity of the SCF(beta-TrCP1) ubiquitin ligase." *Molecular Cell*, **11** (6), 1445–1456.

Xing, Y., Clements, W.K., Kimelman, D., and Xu, W. (2003) "Crystal structure of a beta-catenin/axin complex suggests a mechanism for the beta-catenin destruction complex." *Genes & Development*, **17** (22), 2753–2764.

Xing, Y., Clements, W.K., Le Trong, I. *et al.* (2004) "Crystal structure of a beta-catenin/APC complex reveals a critical role for APC phosphorylation in APC function." *Molecular Cell*, **15** (4), 523–533.

Xing, Y., Takemaru, K., Liu, J. *et al.* (2008) "Crystal structure of a full-length beta-catenin." *Structure*, **16** (3), 478–487.

Xu, W. and Kimelman, D. (2007) "Mechanistic insights from structural studies of beta-catenin and its binding partners." *Journal of Cell Science*, **120** (Pt 19), 3337–3344.

Yu, A., Xing, Y., Harrison, S.C., and Kirchhausen, T. (2010) "Structural analysis of the interaction between Dishevelled2 and clathrin AP-2 adaptor, a critical step in noncanonical Wnt signaling." *Structure*, **18** (10), 1311–1320.

Zeng, X., Tamai, K., Doble, B. *et al.* (2005) "A dual kinase mechanism for Wnt co-receptor phosphorylation and activation." *Nature*, **438** (7069), 873–877.

Part 3

Wnt Signaling in Embryonic Development and Adult Tissue Homeostasis

The Wnt signaling mechanisms are central to regulating embryonic development in a remarkably diverse and large number of functional roles. Studies of early animal embryogenesis were not only often initially aimed at understanding Wnt signaling mechanisms (see Part 1) and the molecular function of individual protein components (see Part 2) but also provided insight into the functional roles of diverse Wnt signaling pathways for animal development and in extension for human embryology.

The Wnt/β-catenin pathway was found to have a prominent role in controlling cell fate decisions during embryonic development, which reflects its molecular ability to regulate gene expression. Wnt/β-catenin signaling is also often associated with regulating the balance between cell proliferation and differentiation. β-catenin-independent Wnt signaling mechanisms have important roles in regulating morphogenesis in embryos, such as gastrulation movements and cell migration and particularly the coordinated polarity of cells in tissues. But regulation of cell biology in tissues is not the exclusive domain of β-catenin-independent Wnt signaling mechanisms, since the Wnt/β-catenin pathway has an evolutionary conserved and possibly fundamentally important link to regulation of epithelia formation (see Chapters 16 and 24). These functional roles of Wnt signaling mechanisms are not only evident during early embryogenesis (see Chapter 19) but also prove to be equally important during organogenesis (e.g., see Chapters 22 and 23), which may have more direct relevance to human conditions and birth defects. However, in terms of complexity of gene expression and quickly changing functional roles for Wnt signaling, the patterning of the developing central nervous system and associated tissues appears to be quite unique (see Chapter 21). Wnt function in neural tissue is of course also probably linked to psychiatric disease and dementia (see Chapters 29–31).

Embryonic stem cells do not only represent potentially powerful future therapeutic agents for regenerative medicine, but studying the role of Wnt signaling in embryonic stem cells also provides important insight into aspects of mammalian early development and tissue

Wnt Signaling in Development and Disease: Molecular Mechanisms and Biological Functions, First Edition.
Edited by Stefan Hoppler and Randall T. Moon.

differentiation that can be difficult to study in the intact mammalian embryo (see Chapter 20).

Similar mechanisms as initially discovered in the embryo are also found to operate in the adult for the maintenance of healthy tissues. Wnt signaling function is also prominently associated with regulation of proliferation versus differentiation of tissue stem cells for normal tissue homeostasis in adults (see Chapter 25) but particularly for repair of damaged tissue after acute tissue injury (see Chapter 26). It is precisely these roles in adult tissue homeostasis that have been suggested often to be related to chronic human disease (Part 4).

19 Wnt Signaling in Early Vertebrate Development: From Fertilization to Gastrulation

Eliza Zylkiewicz[1], Sergei Y. Sokol[1] and Stefan Hoppler[2]

[1]Department of Developmental and Regenerative Biology, Icahn School of Medicine at Mount Sinai, New York, NY, USA

[2]Institute of Medical Sciences, University of Aberdeen, Aberdeen, Scotland

Introduction

During vertebrate development, the fertilized egg develops embryonic tissues that are patterned along the dorsal–ventral, the anterior–posterior, and the left–right body axes. In the model vertebrate *Xenopus laevis*, the radial symmetry of the egg with only an animal–vegetal axis is broken by the sperm entry during fertilization, to establish the dorsoventral body axis. These events also define the site for the formation of the organizer, a special dorsal signaling center that subsequently confers positional information to all embryonic tissues (Harland and Gerhart, 1997). The Spemann's organizer in *Xenopus* corresponds to the shield in zebrafish (Schier and Talbot, 1998), Hensen's node in avian embryos (Viebahn, 2001), and the node in mammals (Beddington and Robertson, 1999). The organizer forms by the late blastula stages, at which time the conventionally described three germ layers become apparent by the expression of characteristic molecular markers. Extensive cell movements during gastrulation result in the involution of mesoderm and endoderm and the spreading of ectoderm to cover the whole embryo from the outside. The locations of the future anterior (head) and posterior (tail) become defined. This completes the formation of the basic body plan, in which embryonic tissues become positioned in the approximate locations that they occupy later in the larva.

Traditionally, the Wnt/β-catenin pathway is thought to specify cell fate through β-catenin-mediated target gene activation, whereas the noncanonical Wnt branches are associated with various morphogenetic events (Logan and Nusse, 2004; Veeman, Axelrod, and Moon, 2003). However, this distinction has been blurred by reports implicating canonical receptors and β-catenin in morphogenesis (Colosimo and Tolwinski, 2006; Hill *et al.*, 2006; Tahinci *et al.*, 2007) and those demonstrating a role for noncanonical Wnt signaling in cell fate determination (Ossipova and Sokol, 2011; Tao *et al.*, 2005). Together, the combination of multiple canonical and noncanonical signals in the early vertebrate embryo leads to the establishment of the vertebrate body plan by the end of gastrulation.

Wnt Signaling in Development and Disease: Molecular Mechanisms and Biological Functions, First Edition.
Edited by Stefan Hoppler and Randall T. Moon.

Induction of embryonic axis by maternal Wnt/β-catenin signaling

Wnt signaling in early vertebrate embryogenesis is best understood in *Xenopus* and zebrafish embryos. These eggs and early embryos have an intrinsic so-called animal–vegetal polarity, which via gastrulation movements translates approximately into the inside to outside axis; with vegetal pole derived tissue mostly ending up in the endoderm and animal pole tissue in the neuroectoderm. Signaling interactions between animal and vegetal tissues establish mesoderm between them (Kimelman and Griffin, 2000). This basic arrangement is fundamentally modified in mammalian embryos, which must prioritize development of extraembryonic and embryonic tissues (Downs, 2009). Axis induction is the first priority in *Xenopus* and zebrafish. The radial symmetry of the egg is broken by sperm entry at fertilization, which rearranges maternal Wnt signaling components in the early embryo (Figure 19.1).

Nuclear β-catenin is the early hallmark of the establishment of dorsal signaling centers in vertebrate embryos (Larabell *et al.*, 1997; Schneider *et al.*, 1996; Schohl and Fagotto, 2002) whereas other cells, for instance, at the prospective ventral side, only have cytoplasmic and cell-membrane-associated β-catenin. β-catenin is not only specifically localized to nuclei in dorsal cells, but overexpression of β-catenin is also sufficient to induce ectopic dorsal development (Funayama *et al.*, 1995; Kelly, Erezyilmaz, and Moon, 1995) or rescue dorsal development (Guger and Gumbiner, 1995). Most importantly, endogenous β-catenin is required for embryonic axis development (Bellipanni *et al.*, 2006; Heasman *et al.*, 1994; Huelsken *et al.*, 2000). Nuclear β-catenin functions with TCF transcription factors (see Chapters 4 and 17) to regulate transcription of some of the first genes in the embryo (Blythe *et al.*, 2010; Molenaar *et al.*, 1996; Standley *et al.*, 2006; Yang *et al.*, 2002). They include homeodomain transcription factor genes such as *siamois* (Brannon *et al.*, 1997), *twin* (Laurent *et al.*, 1997), and *nieuwkoid/dharma/bozozok* (Ryu *et al.*, 2001), which regulate gene expression important for induction of the dorsal signaling center (Fan and Sokol, 1997) and *nodal* genes, which function in mesoderm induction and patterning (Ben-Haim *et al.*, 2006; Schohl and Fagotto, 2003).

But how does β-catenin get localized to nuclei of dorsal cells? Classic embryology experiments demonstrated that sperm entry at fertilization establishes a microtubule network associated with a visible rotation of the cortical cytoplasm (Chan and Etkin, 2001; Elinson and Rowning, 1988) along which an initially vegetally localized dorsal-inducing determinant (Fujisue, Kobayakawa, and Yamana, 1993; Mizuno *et al.*, 1999) is transported to the future dorsal side (Sakai, 1996; Tran *et al.*, 2012). The hunt for the endogenous dorsal determinant has been on ever since and also for the maternal Wnt signaling components regulating differential nuclear localization of β-catenin in the early embryo.

The fact that dominant-negative Wnt8 (Hoppler, Brown, and Moon, 1996), dominant-negative Dishevelled (Dvl) (Sokol, 1996), or precocious Dickkopf-1 expression (Brott and Sokol, 2002; Glinka *et al.*, 1998) did not interfere with induction of the dorsal signaling center initially suggested that β-catenin is localized by Wnt signaling components that act downstream of Dvl (Harland and Gerhart, 1997). Indeed, a GSK3 binding and inhibiting protein FRAT was found to be required for *Xenopus* dorsal development (also known as GBP, Yost *et al.*, 1998), but not for mammalian development (van Amerongen *et al.*, 2005). However, the experimental expression of functional dominant-negative Wnt and Dvl proteins may come too late to influence endogenous maternal Wnt signaling mechanisms involved in regulating differential nuclear localization of β-catenin in the early embryo. Certainly, overexpression of upstream Wnt signaling components is capable of ectopically inducing or rescuing a dorsal signaling center, including Wnt ligands (e.g., Wnt1, McMahon and Moon, 1989; Wnt8, e.g., Sokol *et al.*, 1991) and Dvl (Sokol *et al.*, 1995). Furthermore, not only β-catenin but also Dvl proteins are enriched on the prospective dorsal side in a cortical rotation-dependent process (Miller *et al.*, 1999; Rowning *et al.*, 1997).

Supporting evidence for the involvement of upstream Wnt signaling components comes from antisense oligonucleotide-mediated depletion of maternal gene products for the Wnt receptors Frizzled7 (Sumanas *et al.*, 2000) and LRP6 (Kofron *et al.*, 2007) as well as Wnt11 (Tao *et al.*, 2005). Wnt11 mRNA is also localized to the

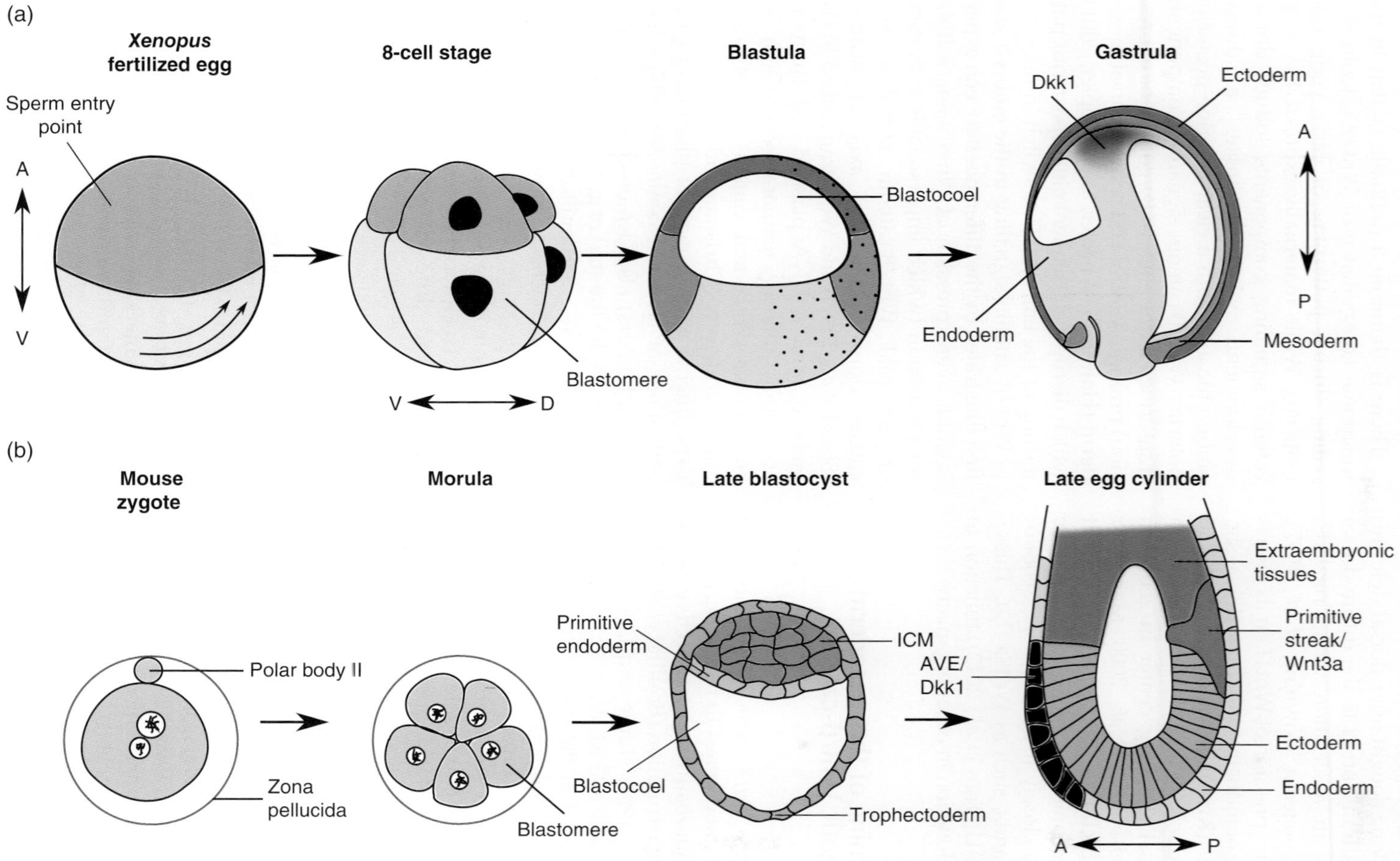

Figure 19.1 Schematic diagram of early mouse and *Xenopus* development. (a) Early *Xenopus* development from fertilized egg to gastrula. Note nuclear localization of β-catenin in the blastula at the dorsal side of the embryo (indicated with *black dots*, see text for more details) and expression of the Wnt inhibitor Dkk1 in anterior cells in the gastrula (indicated in *black*, see text for more details). (b) Early mouse development from zygote to late egg cylinder stage. Note at the late egg cylinder stage localized expression of *Dkk1* in the AVE (anterior visceral endoderm), which will become the anterior of the embryo and of Wnt3a in the primitive streak. ICM, inner cell mass. *(See insert for color representation of the figure.)*

vegetal pole in the egg and early embryo (Ku and Melton, 1993), suggesting that Wnt11 may represent the long sought after dorsal determinant, which is transported to the future dorsal side (possibly in association with Frizzled7, Dvl, and/or β-catenin, Rowning *et al.*, 1997). However, overexpression of Wnt11 on its own is barely capable of inducing an ectopic dorsal signaling center (Ku and Melton, 1993), and genetically reduced *Wnt11* does not lead to a defective dorsal signaling center in zebrafish and mice (Heisenberg *et al.*, 2000; Majumdar *et al.*, 2003; Ulrich *et al.*, 2003), suggesting that additional mechanisms are required to regulate Wnt11 activity closely (e.g., Cha *et al.*, 2008) or other Wnt ligands, such as Wnt8a (Lu, Thisse, and Thisse, 2011; Tran *et al.*, 2012), function to carry out this function in other organisms.

Axial patterning of the mesoderm through zygotic Wnt/β-catenin signaling

Wnt8a and other *Wnt8* paralogs are among the earliest Wnt ligand genes to be transcribed in vertebrate embryos in the prospective mesoderm during gastrulation (Bouillet *et al.*, 1996; Christian *et al.*, 1991; Hume and Dodd, 1993; Kelly, Erezyilmaz, and Moon, 1995). If experimentally activated at these embryonic stages (Christian and Moon, 1993; Hoppler and Moon, 1998), Wnt8 causes ventralization in the mesoderm and subsequent posteriorization of the dorsal embryonic axis; and if its activity is inhibited (Hoppler, Brown, and Moon, 1996; Lekven *et al.*, 2001) an extension of dorsal mesoderm. This Wnt activity in the ventral mesoderm is mediated by the Wnt/β-catenin pathway (Hamilton, Wheeler, and Hoppler, 2001). This is reflected after onset of transcription in the embryo in the stabilization of β-catenin in the ventral mesoderm (Schohl and Fagotto, 2002), which is dependent on Wnt8a function (Hikasa *et al.*, 2010). During the tissue rearrangements of ensuing gastrulation (see succeeding text), the dorsoventral patterning of the mesoderm has fundamental consequences for induction of neural tissue in the dorsal ectoderm and subsequent anterior–posterior patterning of the induced neural plate, which is also prominently and directly mediated by Wnt/β-catenin signaling (see Chapter 21).

There is therefore a dramatic switch in the functional consequence of Wnt/β-catenin signaling during patterning of early vertebrate embryos. While maternally expressed Wnt/β-catenin signaling components induce dorsal development, which promotes a dorsal embryonic axis with a clear anterior–posterior polarity, Wnt/β-catenin signaling components transcribed in the embryo (zygotic gene expression) promote ventral and restrict dorsal mesoderm (Hamilton, Wheeler, and Hoppler, 2001), which disrupts subsequent anteroposterior patterning of the embryo. This switch in response to Wnt/β-catenin signaling in the embryo was first illustrated by the difference between experimental overexpression of Wnt8 using mRNA microinjection (which impersonates activation of maternal Wnt/β-catenin signaling and causes dorsalization and axis duplication (Sokol *et al.*, 1991)) and Wnt8 transcribed in the embryo from a DNA plasmid (which more accurately reflects temporal expression of endogenous *Wnt8* and causes ventralization and axial defects (Christian and Moon, 1993)). Experiments with sequential pharmacological activation of Wnt/β-catenin signaling (i.e., GSK3 inhibitors) also demonstrate this switch associated with the start of gene transcription in the embryo (e.g., Hamilton, Wheeler, and Hoppler, 2001; Stachel, Grunwald, and Myers, 1993). At this stage, the targets of maternal Wnt/β-catenin signaling, such as *siamois* and *nodal* genes (see preceding text), are no longer activated; and different genes are activated instead, such as genes encoding the homeodomain transcription factors of the *vent* (Hikasa *et al.*, 2010; Hoppler and Moon, 1998) and *caudal* families (Haremaki *et al.*, 2003); and *HoxD1* (In der Rieden, Vilaspasa, and Durston, 2010).

The four vertebrate homologs of the TCF/LEF transcription factor family (Chapter 17) are differentially expressed in early embryonic development and are instrumental in mediating this remarkable shift in response to Wnt signaling (Hamilton, Wheeler, and Hoppler, 2001; Liu *et al.*, 2005). While TCF3 mediates transcriptional repression particularly in early stages, Lef1 and Tcf1 function as transcriptional activators, particularly during mesoderm patterning (Liu *et al.*, 2005; Standley *et al.*, 2006).

This shift in TCF/LEF function is not only influenced by strong TCF1 and LEF1 expression in the embryo (Roel *et al.*, 2002) but also further supported by HIPK2-mediated replacement of TCF3 repressor by TCF1 activator on *cis*-regulatory sequences of Wnt target genes (Hikasa *et al.*, 2010; see also Chapter 17). This striking example of context-dependent activation of Wnt target genes in early vertebrate embryos is therefore clearly linked to specific activities of tissue-specific TCF proteins (Hoppler and Kavanagh, 2007), their distinct responses to posttranslational phosphorylation (Sokol, 2011), but probably also directly influenced by cross talk with other developmentally important cell signaling pathways (e.g., BMP signaling (Itasaki and Hoppler, 2010)).

Wnt/β-catenin signaling in early mammalian development

Placental mammals pursue a different reproductive strategy from frogs and fish, which requires a fundamentally different design of early embryogenesis (e.g., Duboule, 1994; Irie and Kuratani, 2011). Axial patterning is only a relatively late priority in mammalian embryogenesis after development of extraembryonic tissues necessary for interaction with the mother's uterus and long after maternal mRNAs are degraded and embryonic transcription activated (Hamatani *et al.*, 2006). Consistent with these different priorities, conventional Wnt signaling reporters or immunocytochemistry for nonphosphorylated active β-catenin do not indicate active Wnt signaling at early preimplantation embryonic stages (Mohamed, Clarke, and Dufort, 2004), which is also consistent with an absence of any early phenotype in β-catenin (Huelsken *et al.*, 2000) and LRP5/6 mutants (Kelly, Pinson, and Skarnes, 2004) or even in β-catenin gain of function experiments (Kemler *et al.*, 2004). However, an Axin2-based β-catenin reporter is active in the mouse blastocyst, indicating that Wnt signaling might play a role in early mouse embryos (ten Berge *et al.*, 2011). Wnt/β-catenin signaling is clearly required for axial patterning in mammals (Huelsken *et al.*, 2000) when they finally get to it after implantation. Wnt/β-catenin signaling in embryonic cells induces via nodal signaling a distal visceral endoderm cell population in the extraembryonic endoderm (Brennan *et al.*, 2001; Morkel *et al.*, 2003), which then migrates away from the source of Wnt signaling in embryonic cells (Kimura-Yoshida *et al.*, 2005). This cell population becomes the anterior visceral endoderm, which by expressing Dkk1 establishes a Wnt signaling free side in the embryo opposite from where Wnt ligand expression becomes restricted and the primitive streak subsequently initiated (reviewed by Yamaguchi, 2008). This "no Wnt versus Wnt" axis becomes the anteroposterior axis (see Chapter 21). Conventional Wnt signaling reporters detect active Wnt signaling where the primitive streak is about to form and also during gastrulation as the mesoderm is induced (Mohamed, Clarke, and Dufort, 2004). The expression of *Wnt8* ligand genes in the prospective mesoderm is conserved in amniotes (Bouillet *et al.*, 1996; Hume and Dodd, 1993) but joined by *Wnt3* and *Wnt2b* (reviewed by Yamaguchi, 2008). *Wnt3* has been genetically shown to be required for mesoderm induction (Barrow *et al.*, 2007; Liu *et al.*, 1999) and so have genes encoding Wnt signaling components, such as Lrp5/6 (Kelly, Pinson, and Skarnes, 2004) and β-catenin (Huelsken *et al.*, 2000), yet *Wnt8a* and *Wnt2b* are not absolutely required (reviewed by Yamaguchi, 2008) and may therefore have redundant functions. The gastrula is referred to as a relatively short embryonic stage, but gastrulation-like processes continue in an anterior to posterior progression, eventually ending up in the tail bud. In mammals, as *Wnt3* expression ceases in the primitive streak (Liu *et al.*, 1999), it is replaced by *Wnt3a* at later stages (Nakaya *et al.*, 2005), which is required for the maintenance of the primitive streak (Takada *et al.*, 1994).

In mammals, *Wnt3* and *Wnt3a* regulate a gene regulatory network (GRN) associated with gastrulation and mesoderm induction that may be evolutionary conserved (Loose and Patient, 2004). As in *Xenopus* and zebrafish (see preceding text), Wnt/β-catenin signaling during mammalian gastrulation regulates *nodal* gene expression (Ben-Haim *et al.*, 2006; Nakaya *et al.*, 2005) and is required for maintenance of *Fgf8* expression (Aulehla *et al.*, 2003), with all three signaling pathways implicated in regulating *brachyury* expression and mesoderm induction (Galceran, Hsu, and Grosschedl,

2001; Yamaguchi *et al.*, 1999b). BMP signaling in turn regulates *Wnt8a* expression during *Xenopus* development (Hoppler and Moon, 1998) and *Wnt3* expression in the mouse embryo (Ben-Haim *et al.*, 2006).

Wnt signaling in the control of morphogenesis

The basic vertebrate body plan is assembled during gastrulation by highly coordinated movements of groups of cells. These morphogenetic rearrangements are mediated by dynamic changes in cell shape and polarity in all germ layers (Keller, 2002; Solnica-Krezel and Sepich, 2012). Radial intercalation, apical constriction, and different cellular events associated with gastrulation have been studied most extensively in zebrafish and *Xenopus* embryos (Figure 19.2). Radial intercalation, *or epiboly*, precedes gastrulation in *Xenopus* and leads to thinning of ectoderm from several to two-cell layers to allow its spreading over the surface of the whole embryo by the time gastrulation is concluded. In *Xenopus* embryos, gastrulation begins with the formation of the dorsal blastopore lip, which is marked by *apical constriction*, a reduction in apical cell surface. This process results in the formation of *bottle cells* and allows tissue invagination at the dorsal marginal zone (Hardin and Keller, 1988; Keller, 2002). In contrast to *Xenopus* embryos, in which mesendodermal cells move together as tissue, the same process consists of individual cell movements in many other species, including mammals (Ferrer-Vaquer, Viotti, and Hadjantonakis, 2010; Lawson, Meneses, and Pedersen, 1991). Epiblast cells undergo the so-called epithelial–mesenchymal transition (EMT), in which they become delaminated from the epithelial sheet and actively migrate through the primitive streak (Arnold and Robertson, 2009; Nowotschin and Hadjantonakis, 2010). In *Xenopus*, individual cell motility is thought to be limited to anterior mesoderm cells that

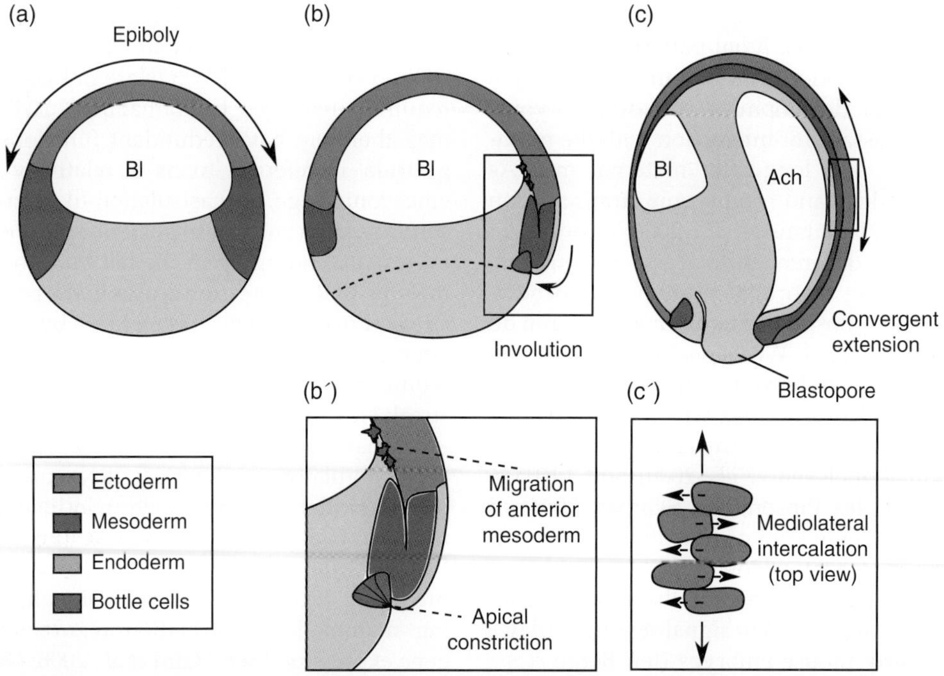

Figure 19.2 Morphogenetic processes during *Xenopus* gastrulation. (a) Late blastula, (b) early and (c) late gastrula in *Xenopus* embryos in the same dorsoventral and anteroposterior orientation as in Figure 19.1. Prospective ectoderm (*dark green*), mesoderm (*red*), and endoderm (*amber*) as indicated. Note in (b′) formation of bottle cells (*light green*) as the mesoderm (*red*) involutes and in (c′) the mediolateral intercalation characteristic of conversion extension movements (for detail, see text). *(See insert for color representation of the figure.)*

actively migrate on the blastocoel roof towards the future anterior pole of the embryo. Later on, mesoderm and endoderm progenitor cells undergo shape changes and form dynamic protrusions, which are used to attach to and crawl between neighboring cells (Elul and Keller, 2000; Keller, 2002; Keller, Shih, and Sater, 1992; Keller *et al.*, 1989). These cell intercalations along the mediolateral plane generate pushing forces leading to the extension of the anterior–posterior embryonic axis (Keller *et al.*, 1985; Wallingford, Fraser, and Harland, 2002). The neuroectoderm overlying the mesoderm undergoes similar elongation (Keller *et al.*, 1985; Wallingford, Fraser, and Harland, 2002). These processes are often referred to as convergent extension movements and are conserved from ascidians to mammals (Glickman *et al.*, 2003; Munro and Odell, 2002; Sausedo and Schoenwolf, 1994; Solnica-Krezel and Sepich, 2012). Whereas convergent extension during body elongation has been commonly accepted to result from mediolateral cell intercalations, additional mechanisms have been demonstrated. Oriented cell divisions were proposed to play an important role in this process in zebrafish (Gong, Mo, and Fraser, 2004; Segalen *et al.*, 2010), but the timing and the degree of the involvement of this process in convergent extension is still unclear (Quesada-Hernandez *et al.*, 2010). In chick embryos, massive cellular flows (also known as polonaise movements) are thought to contribute to formation of the primitive streak (Chuai and Weijer, 2009). Regardless of the underlying mechanism, by the end of gastrulation in all model organisms, the germ layers are placed in their final locations in the embryo, with mesendoderm inside and ectoderm covering the whole embryo surface along the defined anterior–posterior axis.

The Wnt pathways significantly contribute to the molecular control of morphogenetic processes during gastrulation. Although multiple Wnt proteins are expressed in lower vertebrates before gastrulation (Christian and Moon, 1993; Cui *et al.*, 1995; Tada and Smith, 2000), their role in radial cell intercalation during epiboly has not been investigated (Figure 19.2). At the onset of gastrulation, Wnt signaling has been proposed to play a role in apical constriction together with the epithelial polarity regulator Lgl to allow bottle cell formation (Choi and Sokol, 2009; Dollar *et al.*, 2005). Furthermore, tissue involution across the blastopore has been shown to require Dvl and its interacting protein Daam1 (Ewald *et al.*, 2004; Habas, Kato, and He, 2001; Sokol, 1996). Currently, there is no direct evidence for Wnt involvement in anterior mesoderm migration, although it is highly anticipated, based on the known role of Wnt5a and Ror2 signaling in mammalian cell migration *in vitro* (Nishita *et al.*, 2010). In addition, the *silberblick/Wnt11* mutation in zebrafish revealed a defect in prechordal mesoderm, consistent with impaired prechordal/anterior mesoderm migration (Heisenberg *et al.*, 2000), and this phenotype was proposed to be due to impaired cadherin-based cell adhesion (Ulrich *et al.*, 2005).

Considerable evidence has accumulated for the involvement of the noncanonical Wnt signaling branch in regulating convergent extension movements. In *Xenopus* and zebrafish, modulating Wnt5a and Wnt11 levels disrupted convergent extension (Heisenberg *et al.*, 2000; Kilian *et al.*, 2003; Moon *et al.*, 1993; Schambony and Wedlich, 2007; Sokol, 1996; Tada and Smith, 2000). Whereas some of these studies have used dominant-negative approaches (Moon *et al.*, 1993; Sokol, 1996; Tada and Smith, 2000), others described the effects of relevant genetic mutations (Heisenberg *et al.*, 2000; Kilian *et al.*, 2003), demonstrating comparable results. Similarly, in chick embryos, cell migration from the primitive streak has been reported to be controlled by Wnt5a and Wnt3, although it is less clear whether this process is equivalent to convergent extension in *Xenopus* (Sweetman *et al.*, 2008). In mouse embryos, lack of a functional copy of the Wnt5a gene resulted in a shortened anterior–posterior axis, consistent with defects in convergent extension (Yamaguchi *et al.*, 1999a). By contrast, the removal of the *Wnt 11* gene from early mouse embryos revealed a function of this protein in kidney branching rather than body axis elongation (Majumdar *et al.*, 2003), indicating possible functional redundancy between mammalian Wnt genes in early morphogenetic events.

The molecules that function downstream of noncanonical Wnt ligand-receptor complexes in convergent extension are thought to involve Dvl, small GTPases of the Rho family, and the Dvl binding partner Daam1 (Habas, Kato, and He, 2001; Schlessinger, Hall, and Tolwinski, 2009; Sokol, 1996; Wallingford, Fraser, and

Harland, 2002), which together provide a direct link to regulation of cell morphology via the actomyosin cytoskeleton (Schlessinger, Hall, and Tolwinski, 2009). In addition to changes in actin dynamics, membrane trafficking is another important mechanism regulating cellular shape during gastrulation downstream of Wnt signaling (Kikuchi and Yamamoto, 2007; see also Chapter 8). Despite intense research aimed to identify downstream effectors of Wnt signaling, molecular links between specific Wnt components and these general regulators of cell shape and behavior largely remain to be determined.

Although morphogenetic processes are usually linked to noncanonical Wnt signaling, there is considerable cross talk between canonical and noncanonical Wnt pathways. Many known Wnt/β-catenin components modulate noncanonical Wnt signaling, and vice versa, noncanonical Wnt components (such as PCP proteins, see succeeding text) interfere with β-catenin-dependent gene expression (Tao *et al.*, 2005; Veeman *et al.*, 2003). The observed competition between Wnt signaling branches is presumably due to availability of receptors or pathway intermediates (Axelrod *et al.*, 1998; Gordon and Nusse, 2006; Niehrs, 2012). In some contexts, Wnt5a has been shown to antagonize canonical Wnt proteins (Torres *et al.*, 1996; Westfall *et al.*, 2003). Surprisingly, Wnt1 and Wnt8/3a but not Wnt5a activate Rho signaling during gastrulation (Habas, Kato, and He, 2001). In addition, depletion of the canonical receptor LRP6 results in gastrulation defects in *Xenopus* embryos (Tahinci *et al.*, 2007). Thus, cross talk between Wnt pathway branches takes place at multiple levels of regulation, and this emerging complexity needs to be investigated in future studies.

Relationship between Wnt signaling and core planar cell polarity (PCP) proteins

Planar cell polarity (PCP) genes were first defined in *Drosophila* embryos as required for cell polarization in the plane of epithelial tissues (Lawrence and Shelton, 1975; Singh and Mlodzik, 2012; Vinson and Adler, 1987). The PCP pathway consists of the core components Frizzled, Dishevelled, Prickle, Flamingo/Celsr, and Strabismus/Van Gogh (or Vangl in vertebrates) (Chae *et al.*, 1999; Feiguin *et al.*, 2001; Wolff and Rubin, 1998), which form distinct molecular complexes at different sides of polarized epithelial cells. Specifically, in Drosophila, Frizzled colocalizes with Dvl on one side, whereas Vang and Prickle are distributed to the other side of the cell (Singh and Mlodzik, 2012; Wang *et al.*, 2006). This localization appears to be conserved in vertebrates (Antic *et al.*, 2010; Curtin *et al.*, 2003; Guo, Hawkins, and Nathans, 2004; Wang, Guo, and Nathans, 2006; Wang *et al.*, 2005, 2006). This mutually exclusive subcellular localization of core PCP proteins has been attributed to their functional interactions reinforcing the initial localization bias through a positive feedback loop (Jenny *et al.*, 2003; Lawrence, Casal, and Struhl, 2004; Singh and Mlodzik, 2012; Tree *et al.*, 2002).

Since both Frizzled and Dvl have been identified as key components of Wnt signaling, the idea that PCP should be controlled by Wnt ligands seems very attractive, despite lack of evidence from the fly model. Indeed, consistent with noncanonical Wnt signaling regulating PCP, vertebrate Wnt5a protein has been shown to control PCP in the mammalian inner ear (Qian *et al.*, 2007). This hypothesis can be extended to early morphogenetic processes, since they are affected by both Wnt signaling and PCP components, even though evidence for classical PCP at these developmental stages is minimal.

Interfering with different core PCP components (Darken *et al.*, 2002; Park and Moon, 2002; Takeuchi *et al.*, 2003) invariably leads to morphogenesis defects in *Xenopus* embryos, similar to those observed after compromising the function of Wnt11 or Dvl (Hamblet *et al.*, 2002; Sokol, 1996; Tada and Smith, 2000; Wallingford *et al.*, 2000). Genetic loss-of-function studies and morpholino-mediated depletion of Vang and Pk in mouse, zebrafish, and *Xenopus* embryos resulted in shortened body axis which was linked to disrupted cell polarity (Carreira-Barbosa *et al.*, 2003; Darken *et al.*, 2002; Goto *et al.*, 2005; Jessen *et al.*, 2002; Kibar *et al.*, 2001; Park and Moon, 2002; Takeuchi *et al.*, 2003; Tao *et al.*, 2009; Veeman *et al.*, 2003). Also, the PCP protein Flamingo/Celsr has been reported to affect epiboly in zebrafish embryos (Carreira-Barbosa *et al.*, 2009) and to antagonize Wnt/β-catenin signaling in

Xenopus (Morgan, El-Kadi, and Theokli, 2003). These observations strongly support the view that the cascade of PCP protein interactions underlying collective cell movements during gastrulation is closely linked to Wnt signaling.

The molecular connection of Wnt ligands to PCP signaling is likely due to the known function of Frizzled proteins as Wnt receptors, but the downstream events are far from clear. Wnt proteins are expected to modify specific PCP components in a spatially-restricted manner, for example, by targeting them for degradation (Narimatsu *et al.*, 2009), thereby modulating their polarized distribution within each cell. Additional studies are warranted to identify downstream cellular targets of noncanonical Wnt signaling and explain the role of PCP proteins during early morphogenetic processes.

Conclusions

Wnt signaling retains conserved roles in anteroposterior body axis specification and early morphogenetic movements in all vertebrate model organisms. However, spatial and temporal variations in the developmental programs of different vertebrate embryos lead to diversification of specific roles played by Wnt pathways. While in *Xenopus* and zebrafish the function of an early Wnt/β-catenin signaling center is essential, in mammalian embryos, the activity of the Wnt/β-catenin pathway becomes prominent only after implantation. Similarly, conserved noncanonical Wnt signaling mechanisms direct embryonic cells to undergo morphogenetic movements to create the vertebrate body plan; however, the particular movements regulated by this pathway are specific for distinct vertebrate models.

References

Antic, D., Stubbs, J.L., Suyama, K. *et al.* (2010) Planar cell polarity enables posterior localization of nodal cilia and left-right axis determination during mouse and Xenopus embryogenesis. *PLoS One*, **5**, e8999.

Arnold, S.J. and Robertson, E.J. (2009) Making a commitment: cell lineage allocation and axis patterning in the early mouse embryo. *Nature Reviews Molecular Cell Biology*, **10**, 91–103.

Aulehla, A., Wehrle, C., Brand-Saberi, B. *et al.* (2003) Wnt3a plays a major role in the segmentation clock controlling somitogenesis. *Developmental Cell*, **4**, 395–406.

Axelrod, J.D., Miller, J.R., Shulman, J.M. *et al.* (1998) Differential recruitment of Dishevelled provides signaling specificity in the planar cell polarity and Wingless signaling pathways. *Genes & Development*, **12**, 2610–2622.

Barrow, J.R., Howell, W.D., Rule, M. *et al.* (2007) Wnt3 signaling in the epiblast is required for proper orientation of the anteroposterior axis. *Developmental Biology*, **312**, 312–320.

Beddington, R.S. and Robertson, E.J. (1999) Axis development and early asymmetry in mammals. *Cell*, **96**, 195–209.

Bellipanni, G., Varga, M., Maegawa, S. *et al.* (2006) Essential and opposing roles of zebrafish beta-catenins in the formation of dorsal axial structures and neurectoderm. *Development*, **133**, 1299–1309.

Ben-Haim, N., Lu, C., Guzman-Ayala, M. *et al.* (2006) The nodal precursor acting via activin receptors induces mesoderm by maintaining a source of its convertases and BMP4. *Developmental Cell*, **11**, 313–323.

Blythe, S.A., Cha, S.W., Tadjuidje, E. *et al.* (2010) β-Catenin primes organizer gene expression by recruiting a histone H3 arginine 8 methyltransferase, Prmt2. *Developmental Cell*, **19**, 220–231.

Bouillet, P., Oulad-Abdelghani, M., Ward, S.J. *et al.* (1996) A new mouse member of the Wnt gene family, mWnt-8, is expressed during early embryogenesis and is ectopically induced by retinoic acid. *Mechanisms of Development*, **58**, 141–152.

Brannon, M., Gomperts, M., Sumoy, L. *et al.* (1997) A beta-catenin/XTcf-3 complex binds to the siamois promoter to regulate dorsal axis specification in Xenopus. *Genes & Development*, **11**, 2359–2370.

Brennan, J., Lu, C.C., Norris, D.P. *et al.* (2001) Nodal signalling in the epiblast patterns the early mouse embryo. *Nature*, **411**, 965–969.

Brott, B.K. and Sokol, S.Y. (2002) Regulation of Wnt/LRP signaling by distinct domains of Dickkopf proteins. *Molecular and Cellular Biology*, **22**, 6100–6110.

Carreira-Barbosa, F., Concha, M.L., Takeuchi, M. *et al.* (2003) Prickle 1 regulates cell movements during gastrulation and neuronal migration in zebrafish. *Development*, **130**, 4037–4046.

Carreira-Barbosa, F., Kajita, M., Morel, V. *et al.* (2009) Flamingo regulates epiboly and convergence/extension movements through cell cohesive and signalling functions during zebrafish gastrulation. *Development*, **136**, 383–392.

Cha, S.W., Tadjuidje, E., Tao, Q. *et al.* (2008) Wnt5a and Wnt11 interact in a maternal Dkk1-regulated fashion to activate both canonical and non-canonical signaling in Xenopus axis formation. *Development*, **135**, 3719–3729.

Chae, J., Kim, M.J., Goo, J.H. *et al.* (1999) The Drosophila tissue polarity gene starry night encodes a member of the protocadherin family. *Development*, **126**, 5421–5429.

Chan, A.P. and Etkin, L.D. (2001) Patterning and lineage specification in the amphibian embryo. *Current Topics in Developmental Biology*, **51**, 1–67.

Choi, S.C. and Sokol, S.Y. (2009) The involvement of lethal giant larvae and Wnt signaling in bottle cell formation in Xenopus embryos. *Developmental Biology*, **336**, 68–75.

Christian, J.L. and Moon, R.T. (1993) Interactions between Xwnt-8 and Spemann organizer signaling pathways generate dorsoventral pattern in the embryonic mesoderm of Xenopus. *Genes & Development*, **7**, 13–28.

Christian, J.L., McMahon, J.A., McMahon, A.P., and Moon, R.T. (1991) Xwnt-8, a Xenopus Wnt-1/int-1-related gene responsive to mesoderm-inducing growth factors, may play a role in ventral mesodermal patterning during embryogenesis. *Development*, **111**, 1045–1055.

Chuai, M. and Weijer, C.J. (2009) Who moves whom during primitive streak formation in the chick embryo. *HFSP Journal*, **3**, 71–76.

Colosimo, P.F. and Tolwinski, N.S. (2006) Wnt, Hedgehog and junctional Armadillo/beta-catenin establish planar polarity in the Drosophila embryo. *PLoS One*, **1**, e9.

Cui, Y., Brown, J.D., Moon, R.T., and Christian, J.L. (1995) Xwnt-8b: a maternally expressed Xenopus Wnt gene with a potential role in establishing the dorsoventral axis. *Development*, **121**, 2177–2186.

Curtin, J.A., Quint, E., Tsipouri, V. *et al.* (2003) Mutation of Celsr1 disrupts planar polarity of inner ear hair cells and causes severe neural tube defects in the mouse. *Current Biology*, **13**, 1129–1133.

Darken, R.S., Scola, A.M., Rakeman, A.S. *et al.* (2002) The planar polarity gene strabismus regulates convergent extension movements in Xenopus. *The EMBO Journal*, **21**, 976–985.

Dollar, G.L., Weber, U., Mlodzik, M., and Sokol, S.Y. (2005) Regulation of Lethal giant larvae by dishevelled. *Nature*, **437**, 1376–1380.

Downs, K.M. (2009) The enigmatic primitive streak: prevailing notions and challenges concerning the body axis of mammals. *BioEssays*, **31**, 892–902.

Duboule, D. (1994) Temporal colinearity and the phylotypic progression: a basis for the stability of a vertebrate Bauplan and the evolution of morphologies through heterochrony. *Development Supplement*, **1994**, 135–142.

Elinson, R.P. and Rowning, B. (1988) A transient array of parallel microtubules in frog eggs: potential tracks for a cytoplasmic rotation that specifies the dorso-ventral axis. *Developmental Biology*, **128**, 185–197.

Elul, T. and Keller, R. (2000) Monopolar protrusive activity: a new morphogenic cell behavior in the neural plate dependent on vertical interactions with the mesoderm in Xenopus. *Developmental Biology*, **224**, 3–19.

Ewald, A.J., Peyrot, S.M., Tyszka, J.M. *et al.* (2004) Regional requirements for Dishevelled signaling during Xenopus gastrulation: separable effects on blastopore closure, mesendoderm internalization and archenteron formation. *Development*, **131**, 6195–6209.

Fan, M.J. and Sokol, S.Y. (1997) A role for Siamois in Spemann organizer formation. *Development*, **124**, 2581–2589.

Feiguin, F., Hannus, M., Mlodzik, M., and Eaton, S. (2001) The ankyrin repeat protein Diego mediates Frizzled-dependent planar polarization. *Developmental Cell*, **1**, 93–101.

Ferrer-Vaquer, A., Viotti, M., and Hadjantonakis, A.K. (2010) Transitions between epithelial and mesenchymal states and the morphogenesis of the early mouse embryo. *Cell Adhesion & Migration*, **4**, 447–457.

Fujisue, M., Kobayakawa, Y., and Yamana, K. (1993) Occurrence of dorsal axis-inducing activity around the vegetal pole of an uncleaved Xenopus egg and displacement to the equatorial region by cortical rotation. *Development*, **118**, 163–170.

Funayama, N., Fagotto, F., McCrea, P., and Gumbiner, B.M. (1995) Embryonic axis induction by the armadillo repeat domain of beta-catenin: evidence for intracellular signaling. *The Journal of Cell Biology*, **128**, 959–968.

Galceran, J., Hsu, S.C., and Grosschedl, R. (2001) Rescue of a Wnt mutation by an activated form of LEF-1: regulation of maintenance but not initiation of Brachyury expression. *Proceedings of the National Academy of Sciences of the United States of America*, **98**, 8668–8673.

Glickman, N.S., Kimmel, C.B., Jones, M.A., and Adams, R.J. (2003) Shaping the zebrafish notochord. *Development*, **130**, 873–887.

Glinka, A., Wu, W., Delius, H. *et al.* (1998) Dickkopf-1 is a member of a new family of secreted proteins and functions in head induction. *Nature*, **391**, 357–362.

Gong, Y., Mo, C., and Fraser, S.E. (2004) Planar cell polarity signalling controls cell division orientation during zebrafish gastrulation. *Nature*, **430**, 689–693.

Gordon, M.D. and Nusse, R. (2006) Wnt signaling: multiple pathways, multiple receptors, and multiple transcription factors. *The Journal of Biological Chemistry*, **281**, 22429–22433.

Goto, T., Davidson, L., Asashima, M., and Keller, R. (2005) Planar cell polarity genes regulate polarized extracellular matrix deposition during frog gastrulation. *Current Biology*, **15**, 787–793.

Guger, K.A. and Gumbiner, B.M. (1995) β-Catenin has Wnt-like activity and mimics the Nieuwkoop signaling center in Xenopus dorsal-ventral patterning. *Developmental Biology*, **172**, 115–125.

Guo, N., Hawkins, C., and Nathans, J. (2004) Frizzled6 controls hair patterning in mice. *Proceedings of the National Academy of Sciences of the United States of America*, **101**, 9277–9281.

Habas, R., Kato, Y., and He, X. (2001) Wnt/Frizzled activation of Rho regulates vertebrate gastrulation and requires a novel Formin homology protein Daam1. *Cell*, **107**, 843–854.

Hamatani, T., Ko, M., Yamada, M. *et al.* (2006) Global gene expression profiling of preimplantation embryos. *Human Cell*, **19**, 98–117.

Hamblet, N.S., Lijam, N., Ruiz-Lozano, P. *et al.* (2002) Dishevelled 2 is essential for cardiac outflow tract development, somite segmentation and neural tube closure. *Development*, **129**, 5827–5838.

Hamilton, F.S., Wheeler, G.N., and Hoppler, S. (2001) Difference in XTcf-3 dependency accounts for change in response to beta-catenin-mediated Wnt signalling in Xenopus blastula. *Development*, **128**, 2063–2073.

Hardin, J. and Keller, R. (1988) The behaviour and function of bottle cells during gastrulation of Xenopus laevis. *Development*, **103**, 211–230.

Haremaki, T., Tanaka, Y., Hongo, I. *et al.* (2003) Integration of multiple signal transducing pathways on Fgf response elements of the Xenopus caudal homologue Xcad3. *Development*, **130**, 4907–4917.

Harland, R. and Gerhart, J. (1997) Formation and function of Spemann's organizer. *Annual Review of Cell and Developmental Biology*, **13**, 611–667.

Heasman, J., Crawford, A., Goldstone, K. *et al.* (1994) Overexpression of cadherins and underexpression of beta-catenin inhibit dorsal mesoderm induction in early Xenopus embryos. *Cell*, **79**, 791–803.

Heisenberg, C.P., Tada, M., Rauch, G.J. *et al.* (2000) Silberblick/Wnt11 mediates convergent extension movements during zebrafish gastrulation. *Nature*, **405**, 76–81.

Hikasa, H., Ezan, J., Itoh, K. *et al.* (2010) Regulation of TCF3 by Wnt-dependent phosphorylation during vertebrate axis specification. *Developmental Cell*, **19**, 521–532.

Hill, T.P., Taketo, M.M., Birchmeier, W., and Hartmann, C. (2006) Multiple roles of mesenchymal beta-catenin during murine limb patterning. *Development*, **133**, 1219–1229.

Hoppler, S. and Moon, R.T. (1998) BMP-2/-4 and Wnt-8 cooperatively pattern the Xenopus mesoderm. *Mechanisms of Development*, **71**, 119–129.

Hoppler, S. and Kavanagh, C.L. (2007) Wnt signalling: variety at the core. *Journal of Cell Science*, **120**, 385–393.

Hoppler, S., Brown, J.D., and Moon, R.T. (1996) Expression of a dominant-negative Wnt blocks induction of MyoD in Xenopus embryos. *Genes & Development*, **10**, 2805–2817.

Huelsken, J., Vogel, R., Brinkmann, V. *et al.* (2000) Requirement for beta-catenin in anterior-posterior axis formation in mice. *The Journal of Cell Biology*, **148**, 567–578.

Hume, C.R. and Dodd, J. (1993) Cwnt-8C: a novel Wnt gene with a potential role in primitive streak formation and hindbrain organization. *Development*, **119**, 1147–1160.

In der Rieden, P.M., Vilaspasa, F.L., and Durston, A.J. (2010) Xwnt8 directly initiates expression of labial Hox genes. *Developmental Dynamics*, **239**, 126–139.

Irie, N. and Kuratani, S. (2011) Comparative transcriptome analysis reveals vertebrate phylotypic period during organogenesis. *Nature Communications*, **2**, 248.

Itasaki, N. and Hoppler, S. (2010) Crosstalk between Wnt and bone morphogenic protein signaling: a turbulent relationship. *Developmental Dynamics*, **239**, 16–33.

Jenny, A., Darken, R.S., Wilson, P.A., and Mlodzik, M. (2003) Prickle and Strabismus form a functional complex to generate a correct axis during planar cell polarity signaling. *The EMBO Journal*, **22**, 4409–4420.

Jessen, J.R., Topczewski, J., Bingham, S. *et al.* (2002) Zebrafish trilobite identifies new roles for Strabismus in gastrulation and neuronal movements. *Nature Cell Biology*, **4**, 610–615.

Keller, R. (2002) Shaping the vertebrate body plan by polarized embryonic cell movements. *Science*, **298**, 1950–1954.

Keller, R., Shih, J., and Sater, A. (1992) The cellular basis of the convergence and extension of the Xenopus neural plate. *Developmental Dynamics*, **193**, 199–217.

Keller, R.E., Danilchik, M., Gimlich, R., and Shih, J. (1985) The function and mechanism of convergent extension during gastrulation of Xenopus laevis. *Journal of Embryology and Experimental Morphology*, **89** (Suppl), 185–209.

Keller, R., Cooper, M.S., Danilchik, M. *et al.* (1989) Cell intercalation during notochord development in Xenopus laevis. *Journal of Experimental Zoology*, **251**, 134–154.

Kelly, G.M., Erezyilmaz, D.F., and Moon, R.T. (1995) Induction of a secondary embryonic axis in zebrafish occurs following the overexpression of beta-catenin. *Mechanisms of Development*, **53**, 261–273.

Kelly, O.G., Pinson, K.I., and Skarnes, W.C. (2004) The Wnt co-receptors Lrp5 and Lrp6 are essential for gastrulation in mice. *Development*, **131**, 2803–2815.

Kemler, R., Hierholzer, A., Kanzler, B. *et al.* (2004) Stabilization of beta-catenin in the mouse zygote leads to premature epithelial-mesenchymal transition in the epiblast. *Development*, **131**, 5817–5824.

Kibar, Z., Vogan, K.J., Groulx, N. *et al.* (2001) Ltap, a mammalian homolog of Drosophila Strabismus/Van Gogh, is altered in the mouse neural tube mutant Loop-tail. *Nature Genetics*, **28**, 251–255.

Kikuchi, A. and Yamamoto, H. (2007) Regulation of Wnt signalling by receptor-mediated endocytosis. *Journal of Biochemistry*, **141**, 443–451.

Kilian, B., Mansukoski, H., Barbosa, F.C. *et al.* (2003) The role of Ppt/Wnt5 in regulating cell shape and movement during zebrafish gastrulation. *Mechanisms of Development*, **120**, 467–476.

Kimelman, D. and Griffin, K.J. (2000) Vertebrate mesendoderm induction and patterning. *Current Opinion in Genetics & Development*, **10**, 350–356.

Kimura-Yoshida, C., Nakano, H., Okamura, D. *et al.* (2005) Canonical Wnt signaling and its antagonist regulate anterior-posterior axis polarization by guiding cell migration in mouse visceral endoderm. *Developmental Cell*, **9**, 639–650.

Kofron, M., Birsoy, B., Houston, D. *et al.* (2007) Wnt11/beta-catenin signaling in both oocytes and early embryos acts through LRP6-mediated regulation of axin. *Development*, **134**, 503–513.

Ku, M. and Melton, D.A. (1993) Xwnt-11: a maternally expressed Xenopus wnt gene. *Development*, **119**, 1161–1173.

Larabell, C.A., Torres, M., Rowning, B.A. *et al.* (1997) Establishment of the dorso-ventral axis in Xenopus embryos is presaged by early asymmetries in beta-catenin that are modulated by the Wnt signaling pathway. *The Journal of Cell Biology*, **136**, 1123–1136.

Laurent, M.N., Blitz, I.L., Hashimoto, C. *et al.* (1997) The Xenopus homeobox gene twin mediates Wnt induction of goosecoid in establishment of Spemann's organizer. *Development*, **124**, 4905–4916.

Lawrence, P.A. and Shelton, P.M. (1975) The determination of polarity in the developing insect retina. *Journal of Embryology & Experimental Morphology*, **33**, 471–486.

Lawrence, P.A., Casal, J., and Struhl, G. (2004) Cell interactions and planar polarity in the abdominal epidermis of Drosophila. *Development*, **131**, 4651–4664.

Lawson, K.A., Meneses, J.J., and Pedersen, R.A. (1991) Clonal analysis of epiblast fate during germ layer formation in the mouse embryo. *Development*, **113**, 891–911.

Lekven, A.C., Thorpe, C.J., Waxman, J.S., and Moon, R.T. (2001) Zebrafish wnt8 encodes two wnt8 proteins on a bicistronic transcript and is required for mesoderm and neurectoderm patterning. *Developmental Cell*, **1**, 103–114.

Liu, P., Wakamiya, M., Shea, M.J. *et al.* (1999) Requirement for Wnt3 in vertebrate axis formation. *Nature Genetics*, **22**, 361–365.

Liu, F., van den Broek, O., Destree, O., and Hoppler, S. (2005) Distinct roles for Xenopus Tcff/Lef genes in mediating specific responses to Wnt/beta-catenin signalling in mesoderm development. *Development*, **132**, 5375–5385.

Logan, C.Y. and Nusse, R. (2004) The Wnt signaling pathway in development and disease. *Annual Review in Cell and Developmental Biology*, **20**, 781–810.

Loose, M. and Patient, R. (2004) A genetic regulatory network for Xenopus mesendoderm formation. *Developmental Biology*, **271**, 467–478.

Lu, F.I., Thisse, C., and Thisse, B. (2011) Identification and mechanism of regulation of the zebrafish dorsal determinant. *Proceedings of the National Academy of Sciences of the United States of America*, **108**, 15876–15880.

Majumdar, A., Vainio, S., Kispert, A. *et al.* (2003) Wnt11 and Ret/Gdnf pathways cooperate in regulating ureteric branching during metanephric kidney development. *Development*, **130**, 3175–3185.

McMahon, A.P. and Moon, R.T. (1989) Ectopic expression of the proto-oncogene int-1 in Xenopus embryos leads to duplication of the embryonic axis. *Cell*, **58**, 1075–1084.

Miller, J.R., Rowning, B.A., Larabell, C.A. *et al.* (1999) Establishment of the dorsal-ventral axis in Xenopus embryos coincides with the dorsal enrichment of dishevelled that is dependent on cortical rotation. *The Journal of Cell Biology*, **146**, 427–437.

Mizuno, T., Yamaha, E., Kuroiwa, A., and Takeda, H. (1999) Removal of vegetal yolk causes dorsal deficencies and impairs dorsal-inducing ability of the yolk cell in zebrafish. *Mechanisms of Development*, **81**, 51–63.

Mohamed, O.A., Clarke, H.J., and Dufort, D. (2004) Beta-catenin signaling marks the prospective site of primitive streak formation in the mouse embryo. *Developmental Dynamics*, **231**, 416–424.

Molenaar, M., van de Wetering, M., Oosterwegel, M. *et al.* (1996) XTcf-3 transcription factor mediates beta-catenin-induced axis formation in Xenopus embryos. *Cell*, **86**, 391–399.

Moon, R.T., Campbell, R.M., Christian, J.L. *et al.* (1993) Xwnt-5A: a maternal Wnt that affects morphogenetic movements after overexpression in embryos of Xenopus laevis. *Development*, **119**, 97–111.

Morgan, R., El-Kadi, A.M., and Theokli, C. (2003) Flamingo, a cadherin-type receptor involved in the Drosophila planar polarity pathway, can block signaling via the canonical wnt pathway in Xenopus laevis. *The International Journal of Developmental Biology*, **47**, 245–252.

Morkel, M., Huelsken, J., Wakamiya, M. *et al.* (2003) Beta-catenin regulates Cripto- and Wnt3-dependent gene expression programs in mouse axis and mesoderm formation. *Development*, **130**, 6283–6294.

Munro, E.M. and Odell, G.M. (2002) Polarized basolateral cell motility underlies invagination and convergent extension of the ascidian notochord. *Development*, **129**, 13–24.

Nakaya, M.A., Biris, K., Tsukiyama, T. *et al.* (2005) Wnt3a links left-right determination with segmentation and anteroposterior axis elongation. *Development*, **132**, 5425–5436.

Narimatsu, M., Bose, R., Pye, M. *et al.* (2009) Regulation of planar cell polarity by Smurf ubiquitin ligases. *Cell*, **137**, 295–307.

Niehrs, C. (2012) The complex world of WNT receptor signalling. *Nature Reviews Molecular Cell Biology*, **13**, 767–779.

Nishita, M., Itsukushima, S., Nomachi, A. *et al.* (2010) Ror2/Frizzled complex mediates Wnt5a-induced AP-1 activation by regulating Dishevelled polymerization. *Molecular and Cellular Biology*, **30**, 3610–3619.

Nowotschin, S. and Hadjantonakis, A.K. (2010) Cellular dynamics in the early mouse embryo: from axis formation to gastrulation. *Current Opinion in Genetics & Development*, **20**, 420–427.

Ossipova, O. and Sokol, S.Y. (2011) Neural crest specification by noncanonical Wnt signaling and PAR-1. *Development*, **138**, 5441–5450.

Park, M. and Moon, R.T. (2002) The planar cell-polarity gene stbm regulates cell behaviour and cell fate in vertebrate embryos. *Nature Cell Biology*, **4**, 20–25.

Qian, D., Jones, C., Rzadzinska, A. *et al.* (2007) Wnt5a functions in planar cell polarity regulation in mice. *Developmental Biology*, **306**, 121–133.

Quesada-Hernandez, E., Caneparo, L., Schneider, S. *et al.* (2010) Stereotypical cell division orientation controls neural rod midline formation in zebrafish. *Current Biology*, **20**, 1966–1972.

Roel, G., Hamilton, F.S., Gent, Y. *et al.* (2002) Lef-1 and Tcf-3 transcription factors mediate tissue-specific Wnt signaling during Xenopus development. *Current Biology*, **12**, 1941–1945.

Rowning, B.A., Wells, J., Wu, M. *et al.* (1997) Microtubule-mediated transport of organelles and localization of beta-catenin to the future dorsal side of Xenopus eggs. *Proceedings of the National Academy of Sciences of the United States of America*, **94**, 1224–1229.

Ryu, S.L., Fujii, R., Yamanaka, Y. *et al.* (2001) Regulation of dharma/bozozok by the Wnt pathway. *Developmental Biology*, **231**, 397–409.

Sakai, M. (1996) The vegetal determinants required for the Spemann organizer move equatorially during the first cell cycle. *Development*, **122**, 2207–2214.

Sausedo, R.A. and Schoenwolf, G.C. (1994) Quantitative analyses of cell behaviors underlying notochord formation and extension in mouse embryos. *Anatomical Record*, **239**, 103–112.

Schambony, A. and Wedlich, D. (2007) Wnt-5A/Ror2 regulate expression of XPAPC through an alternative noncanonical signaling pathway. *Developmental Cell*, **12**, 779–792.

Schier, A.F. and Talbot, W.S. (1998) The zebrafish organizer. *Current Opinion in Genetics & Development*, **8**, 464–471.

Schlessinger, K., Hall, A., and Tolwinski, N. (2009) Wnt signaling pathways meet Rho GTPases. *Genes & Development*, **23**, 265–277.

Schneider, S., Steinbeisser, H., Warga, R.M., and Hausen, P. (1996) Beta-catenin translocation into nuclei demarcates the dorsalizing centers in frog and fish embryos. *Mechanisms of Development*, **57**, 191–198.

Schohl, A. and Fagotto, F. (2002) Beta-catenin, MAPK and Smad signaling during early Xenopus development. *Development*, **129**, 37–52.

Schohl, A. and Fagotto, F. (2003) A role for maternal beta-catenin in early mesoderm induction in Xenopus. *The EMBO Journal*, **22**, 3303–3313.

Segalen, M., Johnston, C.A., Martin, C.A. *et al.* (2010) The Fz-Dsh planar cell polarity pathway induces oriented cell division via Mud/NuMA in Drosophila and zebrafish. *Developmental Cell*, **19**, 740–752.

Singh, J. and Mlodzik, M. (2012) Planar cell polarity signaling: coordination of cellular orientation across tissues. *Wiley Interdisciplinary Reviews Developmental Biology*, **1**, 479–499.

Sokol, S.Y. (1996) Analysis of Dishevelled signalling pathways during Xenopus development. *Current Biology*, **6**, 1456–1467.

Sokol, S.Y. (2011) Maintaining embryonic stem cell pluripotency with Wnt signaling. *Development*, **138**, 4341–4350.

Sokol, S., Christian, J.L., Moon, R.T., and Melton, D.A. (1991) Injected Wnt RNA induces a complete body axis in Xenopus embryos. *Cell*, **67**, 741–752.

Sokol, S.Y., Klingensmith, J., Perrimon, N., and Itoh, K. (1995) Dorsalizing and neuralizing properties of Xdsh, a maternally expressed Xenopus homolog of dishevelled. *Development*, **121**, 1637–1647.

Solnica-Krezel, L. and Sepich, D.S. (2012) Gastrulation: making and shaping germ layers. *Annual Reviews of Cell & Developmental Biology*, **28**, 687–717.

Stachel, S.E., Grunwald, D.J., and Myers, P.Z. (1993) Lithium perturbation and goosecoid expression identify a dorsal specification pathway in the pregastrula zebrafish. *Development*, **117**, 1261–1274.

Standley, H.J., Destree, O., Kofron, M. *et al.* (2006) Maternal XTcf1 and XTcf4 have distinct roles in regulating Wnt target genes. *Developmental Biology*, **289**, 318–328.

Sumanas, S., Strege, P., Heasman, J., and Ekker, S.C. (2000) The putative wnt receptor Xenopus frizzled-7 functions upstream of beta-catenin in vertebrate dorsoventral mesoderm patterning. *Development*, **127**, 1981–1990.

Sweetman, D., Wagstaff, L., Cooper, O. *et al.* (2008) The migration of paraxial and lateral plate mesoderm cells emerging from the late primitive streak is controlled by different Wnt signals. *BMC Developmental Biology*, **8**, 63.

Tada, M. and Smith, J.C. (2000) Xwnt11 is a target of Xenopus Brachyury: regulation of gastrulation movements via Dishevelled, but not through the canonical Wnt pathway. *Development*, **127**, 2227–2238.

Tahinci, E., Thorne, C.A., Franklin, J.L. *et al.* (2007) Lrp6 is required for convergent extension during Xenopus gastrulation. *Development*, **134**, 4095–4106.

Takada, S., Stark, K.L., Shea, M.J. *et al.* (1994) Wnt-3a regulates somite and tailbud formation in the mouse embryo. *Genes & Development*, **8**, 174–189.

Takeuchi, M., Nakabayashi, J., Sakaguchi, T. *et al.* (2003) The prickle-related gene in vertebrates is essential for gastrulation cell movements. *Current Biology*, **13**, 674–679.

Tao, Q., Yokota, C., Puck, H. *et al.* (2005) Maternal wnt11 activates the canonical wnt signaling pathway required for axis formation in Xenopus embryos. *Cell*, **120**, 857–871.

Tao, H., Suzuki, M., Kiyonari, H. *et al.* (2009) Mouse prickle1, the homolog of a PCP gene, is essential for epiblast apical-basal polarity. *Proceedings of the National Academy of Sciences of the United States of America*, **106**, 14426–14431.

ten Berge, D., Kurek, D., Blauwkamp, T. *et al.* (2011) Embryonic stem cells require Wnt proteins to prevent differentiation to epiblast stem cells. *Nature Cell Biology*, **13**, 1070–1075.

Torres, M.A., Yang-Snyder, J.A., Purcell, S.M. *et al.* (1996) Activities of the Wnt-1 class of secreted signaling factors are antagonized by the Wnt-5A class and by a dominant negative cadherin in early Xenopus development. *The Journal of Cell Biology*, **133**, 1123–1137.

Tran, L.D., Hino, H., Quach, H. *et al.* (2012) Dynamic microtubules at the vegetal cortex predict the embryonic axis in zebrafish. *Development*, **139**, 3644–3652.

Tree, D.R., Shulman, J.M., Rousset, R. *et al.* (2002) Prickle mediates feedback amplification to generate asymmetric planar cell polarity signaling. *Cell*, **109**, 371–381.

Ulrich, F., Concha, M.L., Heid, P.J. *et al.* (2003) Slb/Wnt11 controls hypoblast cell migration and morphogenesis at the onset of zebrafish gastrulation. *Development*, **130**, 5375–5384.

Ulrich, F., Krieg, M., Schotz, E.M. *et al.* (2005) Wnt11 functions in gastrulation by controlling cell cohesion through Rab5c and E-cadherin. *Developmental Cell*, **9**, 555–564.

van Amerongen, R., Nawijn, M., Franca-Koh, J. *et al.* (2005) Frat is dispensable for canonical Wnt signaling in mammals. *Genes & Development*, **19**, 425–430.

Veeman, M.T., Axelrod, J.D., and Moon, R.T. (2003) A second canon. Functions and mechanisms of beta-catenin-independent Wnt signaling. *Developmental Cell*, **5**, 367–377.

Veeman, M.T., Slusarski, D.C., Kaykas, A. *et al.* (2003) Zebrafish prickle, a modulator of noncanonical Wnt/Fz signaling, regulates gastrulation movements. *Current Biology*, **13**, 680–685.

Viebahn, C. (2001) Hensen's node. *Genesis*, **29**, 96–103.

Vinson, C.R. and Adler, P.N. (1987) Directional non-cell autonomy and the transmission of polarity information by the frizzled gene of Drosophila. *Nature*, **329**, 549–551.

Wallingford, J.B., Fraser, S.E., and Harland, R.M. (2002) Convergent extension: the molecular control of polarized cell movement during embryonic development. *Developmental Cell*, **2**, 695–706.

Wallingford, J.B., Rowning, B.A., Vogeli, K.M. *et al.* (2000) Dishevelled controls cell polarity during Xenopus gastrulation. *Nature*, **405**, 81–85.

Wang, Y., Guo, N., and Nathans, J. (2006a) The role of Frizzled3 and Frizzled6 in neural tube closure and in the planar polarity of inner-ear sensory hair cells. *The Journal of Neuroscience*, **26**, 2147–2156.

Wang, J., Mark, S., Zhang, X. *et al.* (2005) Regulation of polarized extension and planar cell polarity in the cochlea by the vertebrate PCP pathway. *Nature Genetics*, **37**, 980–985.

Wang, J., Hamblet, N.S., Mark., S. *et al.* (2006b) Dishevelled genes mediate a conserved mammalian PCP pathway to regulate convergent extension during neurulation. *Development*, **133**, 1767–1778.

Westfall, T.A., Brimeyer, R., Twedt, J. *et al.* (2003) Wnt-5/pipetail functions in vertebrate axis formation as a negative regulator of Wnt/beta-catenin activity. *The Journal of Cell Biology*, **162**, 889–898.

Wolff, T. and Rubin, G.M. (1998) Strabismus, a novel gene that regulates tissue polarity and cell fate decisions in Drosophila. *Development*, **125**, 1149–1159.

Yamaguchi, T.P. (2008) Genetics of Wnt signaling during early mammalian development. *Methods of Molecular Biology*, **468**, 287–305.

Yamaguchi, T.P., Bradley, A., McMahon, A.P., and Jones, S. (1999a) A Wnt5a pathway underlies outgrowth of multiple structures in the vertebrate embryo. *Development*, **126**, 1211–1223.

Yamaguchi, T.P., Takada, S., Yoshikawa, Y. *et al.* (1999b) T (Brachyury) is a direct target of Wnt3a during paraxial mesoderm specification. *Genes & Development*, **13**, 3185–3190.

Yang, J., Tan, C., Darken, R.S. *et al.* (2002) Beta-catenin/Tcf-regulated transcription prior to the midblastula transition. *Development*, **129**, 5743–5752.

Yost, C., Farr, G.H., 3rd, Pierce, S.B. *et al.* (1998) GBP, an inhibitor of GSK-3, is implicated in Xenopus development and oncogenesis. *Cell*, **93**, 1031–1041.

20 Wnt/β-Catenin Signaling in Embryonic Stem Cells: Insights into Early Mammalian Development

Kathryn C. Davidson

Department of Pharmacology, Institute for Stem Cell and Regenerative Medicine, Howard Hughes Medical Institute, University of Washington School of Medicine, Seattle, WA, USA
Centre for Eye Research Australia, University of Melbourne, and Royal Victorian Eye and Ear Hospital, East Melbourne, Victoria, Australia

Introduction to pluripotent stem cells

Stem cells are essentially primordial cells capable of both self-renewal and subsequent differentiation into other cell types within an organism. Traditionally, stem cells have been defined according to the range and extent to which they can differentiate, as well as their source. Pluripotent stem cells have the potential to differentiate into any of the three primary germ lineages (ectoderm, endoderm, and mesoderm) but not necessarily extraembryonic tissues. In contrast, multipotent stem cells are capable of differentiating into a limited number of more restricted cell fates.

Embryonic stem cells (ESCs) are pluripotent stem cells derived from the inner cell mass of preimplantation stage mammalian embryos. In blastocyst-stage embryos, the inner cell mass is the small cluster of cells that gives rise to the embryo proper. Although the inner cell mass exists transiently in development, an established ESC line can be expanded indefinitely *in vitro* in an undifferentiated state without undergoing any restriction in pluripotency. ESCs are also genetically stable, with high telomerase activity and a normal diploid karyotype.

Differentiation potential of stem cells becomes gradually more restricted as development proceeds; however, this restriction is reversible. Somatic cells may be reprogrammed to become induced pluripotent stem cells (iPSCs) via ectopic expression of several key transcription factors (Takahashi *et al.*, 2007; Yu *et al.*, 2007). In this chapter, iPSCs and ESCs are referred to collectively as pluripotent stem cells. However, iPSCs and ESCs do not always behave equivalently across experimental contexts. More research is required to fully understand the nuanced differences between ESCs and iPSCs.

Because of their unique features, pluripotent stem cells are considered valuable tools for scientific research and regenerative medicine. ESCs represent the earliest developmental cell type available for *in vitro* studies, providing a stable experimental system in which to query early developmental processes and to test for embryonic drug toxicity. The ability of ESCs and iPSCs to differentiate into certain cell types that cannot be isolated and/or propagated efficiently as primary cells allows for development and scalability of cell-based assays that would otherwise be unfeasible. In addition, pluripotent stem cells may provide an unlimited source

Wnt Signaling in Development and Disease: Molecular Mechanisms and Biological Functions, First Edition.
Edited by Stefan Hoppler and Randall T. Moon.

of material for cell replacement therapies in the future. Finally, understanding the mechanisms that control stem cell pluripotency, self-renewal, and differentiation is fundamentally important to the fields of cellular and developmental biology, as acknowledged by the award of the 2012 Nobel Prize for Medicine or Physiology to Gurdon and Yamanaka for their discoveries in cell fate plasticity and induced pluripotency.

Pluripotent stem cells are characterized by the expression of cell surface markers and enzymatic and molecular markers as well as their ability to self-renew and differentiate into all three germ lineages (reviewed in Davidson, Dottori, and Pebay, 2008; Merrill, 2012; Nichols and Smith, 2012; Zhao *et al.*, 2012, and summarized in Table 20.1). No single marker or assay on its own informs whether a cell is pluripotent; rather, a panel of markers in combination with assays for both self-renewal and differentiation are required to determine the effects of experimental manipulations on pluripotency. In addition to expressing known stem cell markers, undifferentiated pluripotent stem cells also do not express markers associated with differentiation.

Pluripotent stem cells have complex culture requirements and are typically cocultured on a mouse embryonic fibroblast feeder layer in medium that is optimized for that species and/or pluripotent state. Mouse pluripotent stem cells are conventionally propagated in medium containing leukemia inhibitory factor (LIF) in combination with either bone morphogenetic protein 4 (BMP4), serum, or a combination of small-molecule inhibitors known as 2i (Ying *et al.*, 2003, 2008) (see succeeding text), while human pluripotent stem cells are propagated in medium with fibroblast growth factor 2 (FGF2) and knockout serum replacement (Amit *et al.*, 2000) and require endogenous Activin/Smad2/3 signaling (James *et al.*, 2005; Vallier, Alexander, and Pedersen, 2005). Assays for self-renewal vary between laboratories but routinely involve culturing stem cells in conditions permissive for self-renewal and determining if marker expression and undifferentiated colony morphology is maintained over several passages.

Table 20.1 Characteristic markers of human and mouse ESCs

Cell surface markers	Human ESC	Mouse ESC
SSEA-1	–	+
SSEA-3	+	–
TRA-1-60	+	NR
TRA-1-81	+	NR
GCTM2	+	NR
TG343	+	NR
TG30 (CD9)	+	+
CD133 (Prominin, AC133)	+	+
Podocalyxin	+	±
gp130	±	+
Enzymatic markers		
Alkaline phosphatase	+	+
Telomerase	+	+
Transcription factors		
Oct4	+	+
Nanog	+	+
Sox2	+	+
Klf4	±	+
Rex1 (Zpf42)	±	+
FoxD3	±	+
Stat3	±	+
Other markers		
Fgf4	–	+
Cripto (TDGF1)	+	+
GDF3	+	+
Nodal	+	+
Lefty	+	+
Stella (Dppa3)	±	+

+, expressed; –, not expressed; ±, expression varies between lines and sometimes is not detected; NR, nonreactive species.

Multilineage differentiation is often assessed by transcriptional, histological, and/or immunological analysis of cells differentiated in one of several *in vitro* or *in vivo* assays of pluripotency. Spontaneous differentiation occurs *in vitro* when adherent stem cells are cultured at high density without passage for several weeks or when cells are grown in suspension as embryoid bodies. For specific cell types of interest, established directed differentiation protocols may also be tested. The *in vivo* assay for spontaneous differentiation involves injection of pluripotent stem cells into testis, muscle, or renal capsule of immunocompromised mice for teratoma formation. Normal pluripotent stem cells readily give rise to cell types with markers representative of the three germ lineages and do not persist in significant numbers as undifferentiated cells in these assays.

Naïve versus prime state pluripotency

Although human and mouse ESCs (hESCs and mESCs) share many key features associated with pluripotency, the species differ in a number of characteristics, particularly the signal transduction pathways that influence self-renewal and differentiation. Upon closer examination, these differences may be explained because hESCs more precisely resemble epiblast stem cells from the mouse (mEpiSC), which correspond to a slightly later developmental stage than inner cell mass cells (Brons *et al.*, 2007; Tesar *et al.*, 2007). Both mouse and human ESCs are pluripotent; however, recent evidence suggests that they may actually represent different distinct states of pluripotency: naïve and primed states. The distinguishing features of naïve and primed pluripotent states are summarized in Table 20.2 and reviewed in Buecker and Geijsen (2010), Nichols and Smith (2009, 2012), and Silva *et al.* (2009).

Mouse and human iPSCs generally reflect the pluripotent state (naïve or primed) of their respective ESC counterparts (i.e., mouse iPSC resembles naïve mESC, and human iPSC resembles primed hESC). Human and mouse cells can convert between these states of pluripotency if given the right intrinsic (ectopic expression of transcription factors) and/or extrinsic factors (media/growth factors) (Buecker *et al.*, 2010; Guo and Smith, 2010; Guo *et al.*, 2009; Hanna *et al.*, 2010; Li *et al.*, 2009; Zhou *et al.*, 2010). However, very few naïve human pluripotent stem cell lines have been described, and these remain poorly characterized and unstable in the absence of forced transgene expression (Buecker *et al.*, 2010; Hanna *et al.*, 2010). The methods that convert mEpiSCs to naïve state do not translate directly to hESCs (Zhou *et al.*, 2010). However, it's unclear if this instability is due in part to existing hESCs being exposed to selective extrinsic signals that are optimal for primed pluripotent cells rather than for naïve cells at the time they are isolated from the embryo and stabilized in culture. If this is the case, then naïve hESCs may need to be isolated from embryos into culture conditions optimal for the stabilization of human naïve pluripotency instead of trying to "reprogram" primed hESCs back to a naïve state.

Table 20.2 Properties of naïve versus primed pluripotent mammalian stem cells

Naïve pluripotent state cells (mouse ESCs)	Primed pluripotent state cells (human ESCs, mouse EpiSCs)
Inner cell mass like	Epiblast like
Compact dome colony morphology	Flattened colony morphology
Express Oct4, Nanog, Sox2, Klf2, Klf4, Rex1, Fgf4	Express, Oct4, Nanog, Sox2, Fgf5
LIF-dependent	FGF2- and Activin/TGF-β-dependent
High clonogenicity	Low clonogenicity
XaXa (no X inactivation)	XaXi (X inactivated)
Amenable to gene targeting	Difficult to genetically manipulate
High chimeric contribution in rodents	Low chimeric contribution in rodents
Response to Wnt activation: *self-renewal*	Response to Wnt activation: *differentiation*
Response to Wnt inhibition: *convert to primed pluripotent state*	Response to Wnt inhibition: *self-renewal*

Wnt/β-catenin signaling in embryonic stem cells

One feature that differs significantly between naïve and primed pluripotent stem cells is their response to Wnt/β-catenin signaling. Historically, the literature regarding the effects of Wnt signaling in mouse and human ESCs has been wrought with controversy due to widely conflicting results. However, by reviewing the literature with an understanding of naïve and primed pluripotent states, one is better equipped to draw meaningful conclusions from various studies with results that may otherwise appear contradictory.

Wnt/β-catenin signaling positively regulates self-renewal of naïve ESCs

Many studies report that Wnt/β-catenin signaling promotes self-renewal of mESCs *in vitro*, although most of these earlier studies were carried out in classical mESC conditions containing serum and LIF rather than in conditions specifically optimized for naïve ESC maintenance and thus may include a mixture of naïve and primed state ESCs (Hao *et al.*, 2006; Kielman *et al.*, 2002; Miyabayashi *et al.*, 2007; Ogawa *et al.*, 2006; Sato *et al.*, 2004; Singla *et al.*, 2006; Takao, Yokota, and Koide, 2007; ten Berge *et al.*, 2011; Wagner *et al.*, 2010). Activating Wnt/β-catenin signaling using Wnt-conditioned medium, recombinant Wnt, GSK3 inhibitors, or coculture with Wnt-expressing feeder cells promotes self-renewal of mESCs in medium without LIF or with low/suboptimal concentrations of LIF (Hao *et al.*, 2006; Ogawa *et al.*, 2006; Sato *et al.*, 2004). Similarly, forced expression of a mutant stable form of β-catenin that constitutively activates the Wnt pathway also maintains mESC self-renewal in medium with reduced LIF or on STO fibroblasts, both conditions that would otherwise be nonsupportive (Ogawa *et al.*, 2006; Takao, Yokota, and Koide, 2007).

Further evidence for the role of Wnt/β-catenin signaling as a positive regulator of naïve pluripotency comes from an examination of the culture conditions that have been established empirically for propagation of "ground-state" naïve mESCs. Self-renewal of mESCs in the naïve pluripotent state is achieved by using chemically defined medium (N2B27 medium) containing LIF and two inhibitors (2i): PD0325901 (MEK inhibitor) and CHIR99021 (GSK3 inhibitor and Wnt/β-catenin pathway agonist) (Ying *et al.*, 2008). Neither inhibitor on its own is sufficient for long-term maintenance of naïve ESCs in the absence of LIF; however, combining any two of the three components (LIF and 2i) is permissive for naïve mESC maintenance, though at slightly reduced efficiency (Wray, Kalkan, and Smith, 2010; Ying *et al.*, 2008). GSK3 inhibitor-mediated enhancement of self-renewal requires β-catenin (Wray *et al.*, 2011), but interestingly, the optimal concentration of GSK3 inhibitor for mESC culture only partially inhibits GSK3 (Ying *et al.*, 2008). Partial GSK3 inhibition is important because higher concentrations of five structurally distinct GSK3 inhibitors, as well as CHIR99021, fail to maintain mESCs (Wray *et al.*, 2011).

Conversely, the inhibition of endogenous Wnt signaling drives mESCs (naïve) to convert to mEpiSC-like cells (primed) (ten Berge *et al.*, 2011). Upon transfer to mEpiSC culture conditions (i.e., FGF2 and Activin), mESCs readily convert to mEpiSC-like cells (Guo *et al.*, 2009). Exogenous Wnt3a blocks this transition from naïve to primed pluripotent state (ten Berge *et al.*, 2011). Thus, Wnt/β-catenin signaling not only reinforces self-renewal of naïve pluripotent cells but also inhibits transition to a primed pluripotent state.

Although Wnt signaling positively regulates the naïve pluripotent state, β-catenin is not required *per se* for maintenance of naïve mESCs *in vitro* if the right extrinsic factors are present, as demonstrated by the maintenance of β-catenin-null mESC lines in media containing LIF + PD0325901 (MEK inhibitor) (Lyashenko *et al.*, 2011; Wray *et al.*, 2011). β-catenin-null mESCs express similar levels of naïve state markers, Klf2 and Kfl4, as control mESCs and do not upregulate Fgf5 (primed state marker). However, the loss of junctional β-catenin causes adhesion defects in knockout mESCs, which diminishes their ability to form compact colonies with the morphology associated with naïve mESCs. So although β-catenin-knockout mESCs can survive in naïve ESC conditions, they form fewer colonies with reduced alkaline phosphatase staining compared to control mESCs, suggesting that they may be disadvantaged (Lyashenko *et al.*, 2011; Wray *et al.*, 2011).

In development, β-catenin is dispensable for inner cell mass formation, providing further evidence that β-catenin is not absolutely required for naïve pluripotency. β-catenin-knockout embryos form blastocysts but fail at gastrulation to induce primitive streak and subsequently lack anterior and mesodermal structures (Haegel *et al.*, 1995; Huelsken *et al.*, 2000). Collectively, these studies show that Wnt/β-catenin signaling positively reinforces, but is not required for, naïve pluripotency *in vitro* and *in vivo*.

Wnt/β-catenin signaling promotes differentiation of primed ESCs

Primed pluripotent cells respond quite differently to Wnt/β-catenin pathway activation in comparison to naïve pluripotent cells. Instead of promoting self-renewal, Wnt/β-catenin signaling drives the differentiation of hESCs and mEpiSCs towards mesodermal and endodermal lineages (Blauwkamp *et al.*, 2012; Bone *et al.*, 2011; Davidson *et al.*, 2012; Greber *et al.*, 2010; Nakanishi *et al.*, 2009), which is consistent with the requirement for Wnt/β-catenin in the epiblast at the equivalent developmental stage to induce primitive streak formation (Huelsken *et al.*, 2000; Kelly, Pinson, and Skarnes, 2004; Liu *et al.*, 1999). Unlike in naïve mESCs, the inhibition of Wnt/β-catenin signaling in primed hESCs is compatible with self-renewal (Davidson *et al.*, 2012; James *et al.*, 2012). Thus, the effects of activating or inhibiting Wnt/β-catenin signaling in pluripotent stem cells are context dependent, depending on the state of pluripotency.

Endogenous Wnt signaling activity also differs between naïve and primed pluripotent states. Naïve mESCs exhibit endogenous β-catenin/TCF reporter activity, as do inner cell mass cells from preimplantation blastocyst-stage mouse embryos (ten Berge *et al.*, 2011). In contrast, hESCs and the epiblast region of E5.5 implanted mouse embryos have low/no endogenous reporter activity (Davidson *et al.*, 2012; ten Berge *et al.*, 2011). In one study, detectable levels of endogenous β-catenin/TCF reporter activity were identified in a distinct subpopulation within a clonally established culture of self-renewing hESCs (Blauwkamp *et al.*, 2012). However, the high levels of primitive streak-associated markers expressed by the reporter-positive subpopulation in this study, combined with a failure to generate neural ectoderm lineage, suggest that these cells represent a subset of early mesendoderm progenitors.

Maintenance of ESCs requires a delicate balance of proliferation, survival, and positive regulation of the core pluripotency gene network plus negative regulation of differentiation genes. Wnt/β-catenin signaling in hESCs has other short-term effects, such as increased proliferation (Dravid *et al.*, 2005), increased clonal survival (Blauwkamp *et al.*, 2012; Hasegawa *et al.*, 2006), or increased transient expression of pluripotency markers (Ullmann *et al.*, 2008). In these cases, Wnt pathway activation ultimately leads to differentiation after several passages, but these transient effects may explain some of the conflicting results reported from short-term experiments with Wnts in hESCs. To further complicate matters, functional effects of Wnt ligand and GSK3 inhibitor in hESCs are augmented by other factors present in conventional hESC medium containing knockout serum replacement (Bone *et al.*, 2011; Blauwkamp *et al.*, 2012). When experiments are conducted using chemically defined medium, then a clear role for Wnt/β-catenin signaling as a driver of differentiation in primed hESCs emerges (Blauwkamp *et al.*, 2012; Bone *et al.*, 2011).

Once naïve and primed stem cells exit the self-renewing conditions artificially created *in vitro*, Wnt/β-catenin signaling acts as a key regulator of differentiation. Genetic loss-of-function experiments demonstrate a role for Wnt/β-catenin in multiple lineage differentiation of mESCs, similar to the observed role in development (see Chapter 19). β-catenin-null mESCs fail to differentiate towards mesodermal, endodermal, and neuronal lineages (Lyashenko *et al.*, 2011). Mesoderm differentiation requires β-catenin/TCF-mediated target gene activation (Arnold *et al.*, 2000; Yamaguchi *et al.*, 1999), but endoderm and neuronal differentiation only require β-catenin-mediated cell adhesion function rather than transcriptional activation (Lyashenko *et al.*, 2011), implying a β-catenin-dependent/TCF-independent mechanism. This is demonstrated by rescue of endoderm and neuroepithelial differentiation in β-catenin-knockout mESCs by expressing a variant of β-catenin that lacks the transactivation domain but restores adhesion function (Lyashenko *et al.*, 2011) (see also Chapter 16). One caveat to interpreting experiments with β-catenin variants is that the effects of Wnt/β-catenin signaling in ESCs may be regulated strongly by abrogating levels of Tcf7l1-mediated transcriptional repression, rather than by classical β-catenin/TCF transcriptional activation, adding an additional level of complexity to the system. β-catenin can bind Tcf7l1 and alleviate gene repression without requirement for its transactivation domain (Wray *et al.*, 2011) (see succeeding text and Chapter 17). In the early mouse embryo, β-catenin is not required in a cell-autonomous manner for the formation of anterior or mesodermal structures because β-catenin-null mESCs contribute to anterior and mesodermal structures (head folds and paraxial mesoderm) when injected into wild-type blastocysts (Huelsken *et al.*, 2000).

Genetic gain-of-function studies further highlight the importance of regulation of Wnt/β-catenin signaling levels in the differentiation of ESCs. In particular, constitutively high levels of Wnt/β-catenin signaling resulting from mutant APC or double knockout of GSK3-α and GSK3-β in mESCs lead to neural differentiation defects both *in vitro* and *in vivo* because signaling must be inhibited to initiate neural differentiation of mESCs (Doble *et al.*, 2007; Kelly *et al.*, 2011; Kielman *et al.*, 2002). Indeed, in the converse experiment, the inhibition of endogenous Wnt/β-catenin signaling using antagonists on mESCs enhances neural differentiation (Aubert *et al.*, 2002; Watanabe *et al.*, 2005).

Wnt/β-catenin signaling promotes acquisition of pluripotency

In addition to regulating naïve and primed pluripotent states, Wnt/β-catenin signaling also promotes the acquisition of pluripotency during reprogramming of mouse and human somatic cells to iPSCs (Li *et al.*, 2009, 2011; Lluis *et al.*, 2008; Marson *et al.*, 2008a). Adding exogenous Wnt3a to mouse embryonic fibroblasts enhances the number of colonies about 20-fold when cells are induced to express the Yamanaka reprogramming factors without c-Myc (i.e., Oct4, Sox2, and Klf4) (Marson *et al.*, 2008a). Although c-Myc is reportedly a Wnt/β-catenin target gene in other cellular contexts (He *et al.*, 1998), no upregulation of c-Myc expression was apparent in this case, so the mechanism is unclear. In human fibroblasts, the inhibition of GSK3 promoted the reprogramming of fibroblasts with the Yamanaka factors in the absence of c-Myc (Li *et al.*, 2009). The same study also demonstrated a role for β-catenin in promoting iPSC formation even without ectopic expression of Sox2, thus with only Oct4 and Klf4 in the transcription factor cocktail (Li *et al.*, 2009). In addition, it has been shown that Wnt3a treatment promotes the reprogramming of mouse somatic cells into pluripotent stem cells through hybrid fusion and that specific β-catenin levels are necessary for this process to occur (Lluis *et al.*, 2008). Taken together, these reports suggest that Wnt/β-catenin signaling has a positive effect on the reprogramming process.

Most mouse strains and other rodents, including rats, are nonpermissive for ESC derivation using conventional serum and mouse embryonic fibroblast feeder conditions (Blair, Wray, and Smith, 2011; Gardner and Brook, 1997). This phenomenon, known as recalcitrance, can be overcome by utilizing naïve cell culture conditions (LIF + 2i) during derivation, which includes a GSK3 inhibitor, CHIR99021 (Buehr *et al.*, 2008; Li *et al.*, 2008; Nichols *et al.*, 2009; Ying *et al.*, 2008). Wnt3a can also facilitate derivation of ESCs from recalcitrant mouse strains in the absence of CHI99021 (ten Berge *et al.*, 2011). These results suggest that Wnt/β-catenin signaling promotes the stabilization of naïve pluripotency.

Considering these results collectively, the Wnt/β-catenin pathway appears to be a *master regulator of naïve pluripotency*, with the ability to promote the acquisition and stabilization of naïve pluripotency and to inhibit transition to a primed pluripotent state (Figure 20.1). In this pluripotency model, we have included human-induced pluripotent stem cells (hiPSCs) under the naïve pluripotent cell heading with a question mark because it is not known whether human cells transiently pass through a naïve state during the reprogramming process before stabilizing in a primed pluripotent state. However, it is entirely possible that human cells appear to reprogram directly to a primed state because standard culture methods utilize extrinsic conditions that are highly selective for this outcome and incompatible with the stabilization of a naïve state *in vitro*.

A core pluripotency transcriptional network

The core transcriptional network that regulates pluripotency, consisting of OCT4 (gene name *Pou5f1*), SOX2, and NANOG, is conserved in both naïve and primed ESCs (reviewed in Ng and Surani, 2011; Young, 2011). Each factor is genetically required for self-renewal and pluripotency of inner cell mass cells as they transition to the epiblast stage *in vivo* (Avilion *et al.*, 2003; Mitsui *et al.*, 2003; Nichols *et al.*, 1998). OCT4 and SOX2 are required for self-renewal of ESCs (Niwa, Miyazaki, and Smith, 2000), while NANOG is dispensable, although NANOG-deficient mESCs exhibit a higher propensity to differentiate compared to wild-type cells (Chambers *et al.*, 2007).

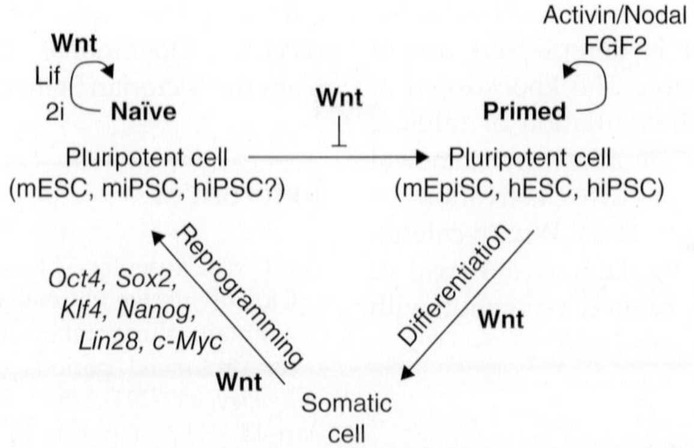

Figure 20.1 Pluripotency model. Wnt positively regulates the acquisition of induced pluripotency, reinforces self-renewal of naïve pluripotent cells, and blocks conversion to the primed pluripotent state. Wnt promotes the differentiation of primed pluripotent cells. Intrinsic and extrinsic factors reported to regulate various states of pluripotency are also included (2i: two inhibitors, i.e., MEK inhibitor (PD0325901) and GSK3 inhibitor (CHIR99021)). It is currently unclear whether hiPSCs initially pass through a naïve state before settling in as primed pluripotent cells (hence the question mark).

Genome-wide chromatin immunoprecipitation (ChIP) experiments indicate that OCT4, SOX2, and NANOG transcription factor proteins co-occupy an overlapping set of target genes in mESCs and hESCs (Boyer *et al.*, 2005; Chen *et al.*, 2008; Cole *et al.*, 2008; Kim *et al.*, 2008; Loh *et al.*, 2006; Marson *et al.*, 2008b). Together, these core factors form a complex autoregulatory network whereby they positively regulate each others' promoters at steady-state levels but repress one or several core promoters at elevated levels (Boer *et al.*, 2007; Masui *et al.*, 2007; Pan and Thomson, 2007). Thus, OCT4, SOX2, and NANOG form core components of a feedforward network of transcription factors that drive self-renewal and pluripotency in stem cells.

TCF7L1 repression of core pluripotency gene network

The mechanism behind Wnt/β-catenin-mediated enhancement of self-renewal in mESCs remains somewhat controversial but clearly involves TCF7L1. The most abundant Tcf/Lef member expressed in ESCs, TCF7L1, co-occupies much of the genome with the core pluripotency transcription factors, OCT4, SOX2, and NANOG (Cole *et al.*, 2008; Marson *et al.*, 2008b; Pereira, Yi, and Merrill, 2006; Tam *et al.*, 2008). Initially, it was speculated that TCF7L1 recruited β-catenin to pluripotency network genes as a transcriptional activator, which provided a potential mechanism for positive regulation of self-renewal by Wnt/β-catenin pathway activation. However, genetic deletion of *Tcf7l1* leads to increased expression of stem cell network genes, while overexpression of *Tcf7l1* inhibited stem cell network genes, inconsistent with this hypothesis. On the contrary, now, evidence suggests that TCF7L1, unlike conventional TCF/LEF proteins, acts as a repressor and is thus a negative regulator of core pluripotency genes (Pereira, Yi, and Merrill, 2006; Sokol, 2011; Yi *et al.*, 2011). Indeed, genetic deletion of *Tcf7l1* leads to enhanced self-renewal and suppressed differentiation of mESCs (Guo *et al.*, 2011; Pereira, Yi, and Merrill, 2006) and replaces the requirement for GSK3 inhibition for mESC self-renewal (Yi *et al.*, 2011), confirming in fact that the inhibition of TCF7L1 repressor function is the downstream effect for Wnt/β-catenin signaling-mediated self-renewal of mESCs. The updated model suggests that TCF7L1 inhibits ESC self-renewal network through its repressor activity. Wnt pathway activation leads to β-catenin binding to TCF7L1 and suppression of TCF7L1 repression. Hence, Wnt/β-catenin pathway promotes self-renewal of mESC via inhibiting TCF7L1 repression (see Chapter 17).

Recently, the orphan nuclear receptor, ESRRB, was identified as the downstream effector of GSK3 inhibitor-mediated ESC self-renewal

(Martello *et al.*, 2012). *Esrrb* is a direct functional target of TCF7L1 that is derepressed downstream of GSK3 inhibition. The knockdown of *Esrrb* leads to rapid differentiation of mESCs, while *Esrrb* overexpression allows self-renewal independent of Wnt pathway activation or LIF (Martello *et al.*, 2012). Thus, Wnt/β-catenin appears to positively regulate self-renewal of naïve mESCs through β-catenin interaction with TCF7L1, which derepresses *Esrrb*.

Summary

The effects of Wnt/β-catenin signaling in pluripotent stem cells are complex and vary between naïve and primed pluripotent states. However, decades of research suggest the Wnt/β-catenin pathway is a master regulator of naïve pluripotency: promoting the acquisition of naïve pluripotency during reprogramming, driving self-renewal of naïve stem cells, and preventing naïve stem cells from transitioning to a primed state. For years, researchers have looked for consensus and overlap between human and mouse ESCs to identify conserved features of pluripotency, but primed state hESCs respond differently from naïve state mESCs in terms of Wnt/β-catenin signaling. In primed state pluripotent stem cells, Wnt/β-catenin signaling becomes a driver of differentiation. With this understanding in mind, future research can now focus on how Wnt signaling achieves such different outcomes in naïve versus primed pluripotent state cells that share so many cellular features. A recent report by Gafni *et al.* (2013) confirms a conserved role for Wnt/β-catenin signaling in naïve state hESCs.

Acknowledgments

I wish to thank the members, past and present, of the Moon lab for their enthusiasm, insight, and dominating drive to understand all things Wnt. I am especially grateful to Randy Moon for providing the environment and encouragement for such pursuits. The author's research is supported by the National Institutes of Health (NIH) Grants 1P01-GM081619 and NIH U01 HL100395 and the National Stem Cell Foundation of Australia. The Centre for Eye Research Australia receives Operational Infrastructure Support from the Victorian State Government.

References

Amit, M., Carpenter, M.K., Inokuma, M.S. *et al.* (2000) Clonally derived human embryonic stem cell lines maintain pluripotency and proliferative potential for prolonged periods of culture. *Developmental Biology*, **227**, 271–278.

Arnold, S.J., Stappert, J., Bauer, A. *et al.* (2000) Brachyury is a target gene of the Wnt/beta-catenin signaling pathway. *Mechanisms of Development*, **91**, 249–258.

Aubert, J., Dunstan, H., Chambers, I., and Smith, A. (2002) Functional gene screening in embryonic stem cells implicates Wnt antagonism in neural differentiation. *Nature Biotechnology*, **20**, 1240–1245.

Avilion, A.A., Nicolis, S.K., Pevny, L.H. *et al.* (2003) Multipotent cell lineages in early mouse development depend on SOX2 function. *Genes & Development*, **17**, 126–140.

Blair, K., Wray, J., and Smith, A. (2011) The liberation of embryonic stem cells. *PLoS Genetics*, **7**, e1002019.

Blauwkamp, T.A., Nigam, S., Ardehali, R. *et al.* (2012) Endogenous Wnt signalling in human embryonic stem cells generates an equilibrium of distinct lineage-specified progenitors. *Nature Communications*, **3**, 1070.

Boer, B., Kopp, J., Mallanna, S. *et al.* (2007) Elevating the levels of Sox2 in embryonal carcinoma cells and embryonic stem cells inhibits the expression of Sox2:Oct-3/4 target genes. *Nucleic Acids Research*, **35**, 1773–1786.

Bone, H.K., Nelson, A.S., Goldring, C.E. *et al.* (2011) A novel chemically directed route for the generation of definitive endoderm from human embryonic stem cells based on inhibition of GSK-3. *Journal of Cell Science*, **124**, 1992–2000.

Boyer, L.A., Lee, T.I., Cole, M.F. *et al.* (2005) Core transcriptional regulatory circuitry in human embryonic stem cells. *Cell*, **122**, 947–956.

Brons, I.G., Smithers, L.E., Trotter, M.W. *et al.* (2007) Derivation of pluripotent epiblast stem cells from mammalian embryos. *Nature*, **448**, 191–195.

Buecker, C. and Geijsen, N. (2010) Different flavors of pluripotency, molecular mechanisms, and practical implications. *Cell Stem Cell*, **7**, 559–564.

Buecker, C., Chen, H.H., Polo, J.M. *et al.* (2010) A murine ESC-like state facilitates transgenesis and homologous recombination in human pluripotent stem cells. *Cell Stem Cell*, **6**, 535–546.

Buehr, M., Meek, S., Blair, K. *et al.* (2008) Capture of authentic embryonic stem cells from rat blastocysts. *Cell*, **135**, 1287–1298.

Chambers, I., Silva, J., Colby, D. *et al.* (2007) Nanog safeguards pluripotency and mediates germline development. *Nature*, **450**, 1230–1234.

Chen, X., Xu, H., Yuan, P. *et al.* (2008) Integration of external signaling pathways with the core transcriptional network in embryonic stem cells. *Cell*, **133**, 1106–1117.

Cole, M.F., Johnstone, S.E., Newman, J.J. *et al.* (2008) Tcf3 is an integral component of the core regulatory circuitry of embryonic stem cells. *Genes & Development*, **22**, 746–755.

Davidson, K.C., Dottori, M., and Pebay, A. (2008) Human embryonic stem cells: key characteristics and main applications in disease research, in *Stem Cell Applications in Disease* (ed. M.L. Sorensen), Nova Science Publishers, Hauppauge.

Davidson, K.C., Adams, A.M., Goodson, J.M. *et al.* (2012) Wnt/beta-catenin signaling promotes differentiation, not self-renewal, of human embryonic stem cells and is repressed by Oct4. *Proceedings of the National Academy of Sciences of the United States of America*, **109**, 4485–4490.

Doble, B.W., Patel, S., Wood, G.A. *et al.* (2007) Functional redundancy of GSK-3alpha and GSK-3beta in Wnt/beta-catenin signaling shown by using an allelic series of embryonic stem cell lines. *Development Cell*, **12**, 957–971.

Dravid, G., Ye, Z., Hammond, H. *et al.* (2005) Defining the role of Wnt/beta-catenin signaling in the survival, proliferation, and self-renewal of human embryonic stem cells. *Stem Cells*, **23**, 1489–1501.

Gafni, O., Weinberger, L., Mansour, L. *et al.* (2013) Derivation of novel human ground state naive pluripotent stem cells. *Nature*, **504** (7479), 282–286.

Gardner, R.L. and Brook, F.A. (1997) Reflections on the biology of embryonic stem (ES) cells. *The International Journal of Developmental Biology*, **41**, 235–243.

Greber, B., Wu, G., Bernemann, C. *et al.* (2010) Conserved and divergent roles of FGF signaling in mouse epiblast stem cells and human embryonic stem cells. *Cell Stem Cell*, **6**, 215–226.

Guo, G. and Smith, A. (2010) A genome-wide screen in EpiSCs identifies Nr5a nuclear receptors as potent inducers of ground state pluripotency. *Development*, **137**, 3185–3192.

Guo, G., Yang, J., Nichols, J. *et al.* (2009) Klf4 reverts developmentally programmed restriction of ground state pluripotency. *Development*, **136**, 1063–1069.

Guo, G., Huang, Y., Humphreys, P. *et al.* (2011) A PiggyBac-based recessive screening method to identify pluripotency regulators. *PLoS One*, **6**, e18189.

Haegel, H., Larue, L., Ohsugi, M. *et al.* (1995) Lack of beta-catenin affects mouse development at gastrulation. *Development*, **121**, 3529–3537.

Hanna, J., Cheng, A.W., Saha, K. *et al.* (2010) Human embryonic stem cells with biological and epigenetic characteristics similar to those of mouse ESCs. *Proceedings of the National Academy of Sciences of the United States of America*, **107**, 9222–9227.

Hao, J., Li, T.G., Qi, X. *et al.* (2006) WNT/beta-catenin pathway up-regulates Stat3 and converges on LIF to prevent differentiation of mouse embryonic stem cells. *Developmental Biology*, **290**, 81–91.

Hasegawa, K., Fujioka, T., Nakamura, Y. *et al.* (2006) A method for the selection of human embryonic stem cell sublines with high replating efficiency after single-cell dissociation. *Stem Cells*, **24**, 2649–2660.

He, T.C., Sparks, A.B., Rago, C. *et al.* (1998) Identification of c-MYC as a target of the APC pathway. *Science*, **281**, 1509–1512.

Huelsken, J., Vogel, R., Brinkmann, V. *et al.* (2000) Requirement for beta-catenin in anterior-posterior axis formation in mice. *The Journal of Cell Biology*, **148**, 567–578.

James, D., Levine, A.J., Besser, D., and Hemmati-Brivanlou, A. (2005) TGF{beta}/activin/nodal signaling is necessary for the maintenance of pluripotency in human embryonic stem cells. *Development*, **132**, 1273–1282.

James, R.G., Davidson, K.C., Bosch, K.A. *et al.* (2012) WIKI4, a novel inhibitor of tankyrase and Wnt/ss-catenin signaling. *PLoS One*, **7**, e50457.

Kelly, O.G., Pinson, K.I., and Skarnes, W.C. (2004) The Wnt co-receptors Lrp5 and Lrp6 are essential for gastrulation in mice. *Development*, **131**, 2803–2815.

Kelly, K.F., Ng, D.Y., Jayakumaran, G. *et al.* (2011) β-Catenin enhances Oct-4 activity and reinforces pluripotency through a TCF-independent mechanism. *Cell Stem Cell*, **8**, 214–227.

Kielman, M.F., Rindapaa, M., Gaspar, C. *et al.* (2002) Apc modulates embryonic stem-cell differentiation by controlling the dosage of beta-catenin signaling. *Nature Genetics*, **32**, 594–605.

Kim, J., Chu, J., Shen, X. *et al.* (2008) An extended transcriptional network for pluripotency of embryonic stem cells. *Cell*, **132**, 1049–1061.

Li, P., Tong, C., Mehrian-Shai, R. *et al.* (2008) Germline competent embryonic stem cells derived from rat blastocysts. *Cell*, **135**, 1299–1310.

Li, W., Zhou, H., Abujarour, R. *et al.* (2009) Generation of human-induced pluripotent stem cells in the absence of exogenous Sox2. *Stem Cells*, **27**, 2992–3000.

Li, Y., Zhang, Q., Yin, X. *et al.* (2011) Generation of iPSCs from mouse fibroblasts with a single gene, Oct4, and small molecules. *Cell Research*, **21**, 196–204.

Liu, P., Wakamiya, M., Shea, M.J. *et al.* (1999) Requirement for Wnt3 in vertebrate axis formation. *Nature Genetics*, **22**, 361–365.

Lluis, F., Pedone, E., Pepe, S., and Cosma, M.P. (2008) Periodic activation of Wnt/beta-catenin signaling enhances somatic cell reprogramming mediated by cell fusion. *Cell Stem Cell*, **3**, 493–507.

Loh, Y.H., Wu, Q., Chew, J.L. *et al.* (2006) The Oct4 and Nanog transcription network regulates pluripotency in mouse embryonic stem cells. *Nature Genetics*, **38**, 431–440.

Lyashenko, N., Winter, M., Migliorini, D. *et al.* (2011) Differential requirement for the dual functions of beta-catenin in embryonic stem cell self-renewal and germ layer formation. *Nature Cell Biology*, **13**, 753–761.

Marson, A., Foreman, R., Chevalier, B. *et al.* (2008a) Wnt signaling promotes reprogramming of somatic cells to pluripotency. *Cell Stem Cell*, **3**, 132–135.

Marson, A., Levine, S.S., Cole, M.F. *et al.* (2008b) Connecting microRNA genes to the core transcriptional regulatory circuitry of embryonic stem cells. *Cell*, **134**, 521–533.

Martello, G., Sugimoto, T., Diamanti, E. *et al.* (2012) Esrrb is a pivotal target of the Gsk3/Tcf3 axis regulating embryonic stem cell self-renewal. *Cell Stem Cell*, **11**, 491–504.

Masui, S., Nakatake, Y., Toyooka, Y. *et al.* (2007) Pluripotency governed by Sox2 via regulation of Oct3/4 expression in mouse embryonic stem cells. *Nature Cell Biology*, **9**, 625–635.

Merrill, B.J. (2012) Wnt pathway regulation of embryonic stem cell self-renewal. *Cold Spring Harbor Perspectives in Biology*, **4**, a007971.

Mitsui, K., Tokuzawa, Y., Itoh, H. *et al.* (2003) The homeoprotein Nanog is required for maintenance of pluripotency in mouse epiblast and ES cells. *Cell*, **113**, 631–642.

Miyabayashi, T., Teo, J.L., Yamamoto, M. *et al.* (2007) Wnt/beta-catenin/CBP signaling maintains long-term murine embryonic stem cell pluripotency. *Proceedings of the National Academy of Sciences of the United States of America*, **104**, 5668–5673.

Nakanishi, M., Kurisaki, A., Hayashi, Y. *et al.* (2009) Directed induction of anterior and posterior primitive streak by Wnt from embryonic stem cells cultured in a chemically defined serum-free medium. *The FASEB Journal*, **23**, 114–122.

Ng, H.H. and Surani, M.A. (2011) The transcriptional and signalling networks of pluripotency. *Nature Cell Biology*, **13**, 490–496.

Nichols, J. (2012) Pluripotency in the embryo and in culture. *Cold Spring Harbor Perspectives in Biology*, **4**, a008128.

Nichols, J. and Smith, A. (2009) Naive and primed pluripotent states. *Cell Stem Cell*, **4**, 487–492.

Nichols, J., Zevnik, B., Anastassiadis, K. *et al.* (1998) Formation of pluripotent stem cells in the mammalian embryo depends on the POU transcription factor Oct4. *Cell*, **95**, 379–391.

Nichols, J., Jones, K., Phillips, J.M. *et al.* (2009) Validated germline-competent embryonic stem cell lines from nonobese diabetic mice. *Nature Medicine*, **15**, 814–818.

Niwa, H., Miyazaki, J., and Smith, A.G. (2000) Quantitative expression of Oct-3/4 defines differentiation, dedifferentiation or self-renewal of ES cells. *Nature Genetics*, **24**, 372–376.

Ogawa, K., Nishinakamura, R., Iwamatsu, Y. *et al.* (2006) Synergistic action of Wnt and LIF in maintaining pluripotency of mouse ES cells. *Biochemical and Biophysical Research Communications*, **343**, 159–166.

Pan, G. and Thomson, J.A. (2007) Nanog and transcriptional networks in embryonic stem cell pluripotency. *Cell Research*, **17**, 42–49.

Pereira, L., Yi, F., and Merrill, B.J. (2006) Repression of Nanog gene transcription by Tcf3 limits embryonic stem cell self-renewal. *Molecular and Cellular Biology*, **26**, 7479–7491.

Sato, N., Meijer, L., Skaltsounis, L. *et al.* (2004) Maintenance of pluripotency in human and mouse embryonic stem cells through activation of Wnt signaling by a pharmacological GSK-3-specific inhibitor. *Nature Medicine*, **10**, 55–63.

Silva, J., Nichols, J., Theunissen, T.W. *et al.* (2009) Nanog is the gateway to the pluripotent ground state. *Cell*, **138**, 722–737.

Singla, D.K., Schneider, D.J., LeWinter, M.M., and Sobel, B.E. (2006) wnt3a but not wnt11 supports self-renewal of embryonic stem cells. *Biochemical and Biophysical Research Communications*, **345**, 789–795.

Sokol, S.Y. (2011) Maintaining embryonic stem cell pluripotency with Wnt signaling. *Development*, **138**, 4341–4350.

Takahashi, K., Tanabe, K., Ohnuki, M. *et al.* (2007) Induction of pluripotent stem cells from adult human fibroblasts by defined factors. *Cell*, **131**, 861–872.

Takao, Y., Yokota, T., and Koide, H. (2007) Beta-catenin up-regulates Nanog expression through interaction with Oct-3/4 in embryonic stem cells. *Biochemical and Biophysical Research Communications*, **353**, 699–705.

Tam, W.L., Lim, C.Y., Han, J. *et al.* (2008) T-cell factor 3 regulates embryonic stem cell pluripotency and self-renewal by the transcriptional control of multiple lineage pathways. *Stem Cells*, **26**, 2019–2031.

ten Berge, D., Kurek, D., Blauwkamp, T. *et al.* (2011) Embryonic stem cells require Wnt proteins to prevent differentiation to epiblast stem cells. *Nature Cell Biology*, **13**, 1070–1075.

Tesar, P.J., Chenoweth, J.G., Brook, F.A. *et al.* (2007) New cell lines from mouse epiblast share defining features with human embryonic stem cells. *Nature*, **448**, 196–199.

Ullmann, U., Gilles, C., De Rycke, M. *et al.* (2008) GSK-3-specific inhibitor-supplemented hESC medium prevents the epithelial-mesenchymal transition process and the up-regulation of matrix metalloproteinases in hESCs cultured in feeder-free conditions. *Molecular Human Reproduction*, **14**, 169–179.

Vallier, L., Alexander, M., and Pedersen, R.A. (2005) Activin/Nodal and FGF pathways cooperate to maintain pluripotency of human embryonic stem cells. *Journal of Cell Science*, **118**, 4495–4509.

Wagner, R.T., Xu, X., Yi, F. *et al.* (2010) Canonical Wnt/beta-catenin regulation of liver receptor homolog-1 mediates pluripotency gene expression. *Stem Cells*, **28**, 1794–1804.

Watanabe, K., Kamiya, D., Nishiyama, A. *et al.* (2005) Directed differentiation of telencephalic precursors from embryonic stem cells. *Nature Neuroscience*, **8**, 288–296.

Wray, J., Kalkan, T., and Smith, A.G. (2010) The ground state of pluripotency. *Biochemical Society Transactions*, **38**, 1027–1032.

Wray, J., Kalkan, T., Gomez-Lopez, S. *et al.* (2011) Inhibition of glycogen synthase kinase-3 alleviates Tcf3 repression of the pluripotency network and increases embryonic stem cell resistance to differentiation. *Nature Cell Biology*, **13**, 838–845.

Yamaguchi, T.P., Takada, S., Yoshikawa, Y. *et al.* (1999) T (Brachyury) is a direct target of Wnt3a during paraxial mesoderm specification. *Genes & Development*, **13**, 3185–3190.

Yi, F., Pereira, L., Hoffman, J.A. *et al.* (2011) Opposing effects of Tcf3 and Tcf1 control Wnt stimulation of embryonic stem cell self-renewal. *Nature Cell Biology*, **13**, 762–770.

Ying, Q.L., Nichols, J., Chambers, I., and Smith, A. (2003) BMP induction of Id proteins suppresses differentiation and sustains embryonic stem cell self-renewal in collaboration with STAT3. *Cell*, **115**, 281–292.

Ying, Q.L., Wray, J., Nichols, J. *et al.* (2008) The ground state of embryonic stem cell self-renewal. *Nature*, **453**, 519–523.

Young, R.A. (2011) Control of the embryonic stem cell state. *Cell*, **144**, 940–954.

Yu, J., Vodyanik, M.A., Smuga-Otto, K. *et al.* (2007) Induced pluripotent stem cell lines derived from human somatic cells. *Science*, **318**, 1917–1920.

Zhao, W., Ji, X., Zhang, F. *et al.* (2012) Embryonic stem cell markers. *Molecules*, **17**, 6196–6236.

Zhou, H., Li, W., Zhu, S. *et al.* (2010) Conversion of mouse epiblast stem cells to an earlier pluripotency state by small molecules. *The Journal of Biological Chemistry*, **285**, 29676–29680.

21 Wnt Signaling in Neural Development

Richard I. Dorsky

Department of Neurobiology and Anatomy, University of Utah, Salt Lake City, UT, USA

Introduction

The importance of Wnt signaling in nervous system formation has been recognized since the initial characterization of a vertebrate *Wnt* homolog. The *Wnt1* gene (initially called *int-1*) was found to be expressed throughout the dorsal midline of the embryonic central nervous system (CNS) (Gavin, McMahon, and McMahon, 1990), suggesting a role in early development. Following this observation, loss-of-function analysis showed that *Wnt1* mutants lack a midbrain and cerebellum (McMahon and Bradley, 1990). Intriguingly, many of the critical functions of Wnt signaling in the developing nervous system were revealed in this initial phenotype, as noted by the authors. There was a positional component to the defects, with dorsal tissues primarily affected. Large, anatomically distinct structures containing diverse cell types failed to be properly specified, and cell populations failed to grow appropriately. Finally, some areas expressing *Wnt1* were unaffected, suggesting redundancy with other genes. Subsequent work has validated and extended all of these insightful observations to provide a clearer picture of Wnt function during nervous system development.

Analysis of the expression and function of *Wnt* pathway components has continued to support a critical role in the formation of neuronal and glial cell populations throughout the CNS and peripheral nervous system (PNS). While the roles of Wnt signaling touch almost every aspect of neural development, their conservation in various vertebrate and invertebrate species has revealed a few common themes. First, Wnt signaling plays reiterated roles in nervous system formation, acting on the same cell populations multiple times. Thus, a single lineage can respond to Wnt signals differently depending on the developmental or molecular context. Second, the most highly conserved functions for Wnt signaling in the vertebrate nervous system are CNS patterning, neural crest induction, neuronal differentiation, and the maintenance of specific areas of continued neurogenesis throughout life. Finally, there is an emerging picture of functions for Wnt signaling in neuronal connectivity, although the specific roles may vary in different species. This chapter will cover the roles of Wnt signaling in developmental sequence, from the earliest steps of embryogenesis to the terminal differentiation of functional neurons. Due to the

Wnt Signaling in Development and Disease: Molecular Mechanisms and Biological Functions, First Edition.
Edited by Stefan Hoppler and Randall T. Moon.

overwhelming number of individual studies published in this field, some examples will inevitably be missed in an attempt to focus on general concepts and summarize larger themes.

CNS induction and early axial patterning

The CNS arises from ectoderm in all triploblastic animals, and differentiates from nonneural ectoderm as a result of extracellular signals, through a process called neural induction. While there is little evidence of a direct role for the Wnt pathway in neural induction, it is quite clear that in all vertebrates this event depends on a tissue itself induced by Wnt signals – the dorsal organizer or node. As described in Chapter 19, activation of maternally expressed Wnt components during early embryogenesis leads to specification of the dorsal organizer, through an adjacent endodermal signaling center (Joubin and Stern, 2001). The dorsal organizer expresses many important signaling molecules, including several factors that act on the neighboring ectoderm to promote neural gene expression (Harland and Gerhart, 1997). Because embryos lacking maternal Wnt signaling fail to make a dorsal organizer and often lack neuroectoderm, it can therefore be argued that this represents the earliest (albeit indirect) role for the pathway in CNS formation.

The first direct function of Wnt signaling on CNS development occurs shortly after neural induction, as the presumptive neural plate extends during gastrulation (Figure 21.1a). Embryological assays in *Xenopus* have shown that the "default" fate of induced neural ectoderm has anterior character, expressing markers of forebrain (Sharpe, Lawrence, and Arias, 2001). Wnt/β-catenin signals, primarily arising from the nonaxial mesoderm, convert CNS tissue to more posterior fates in a dose-dependent manner (Domingos *et al.*, 2001; Erter *et al.*, 2001). In addition, multiple Wnt pathway antagonists are specifically expressed in the anterior neuroectoderm, creating a gradient of Wnt/β-catenin activity that patterns the anterior–posterior axis of the brain and spinal cord. Using a variety of approaches, many studies have demonstrated that excess Wnt signaling leads to a "posteriorized" CNS lacking forebrain and midbrain structures, while the loss of Wnt signaling leads to an "anteriorized" CNS with an expanded forebrain (Dorsky *et al.*, 2003;

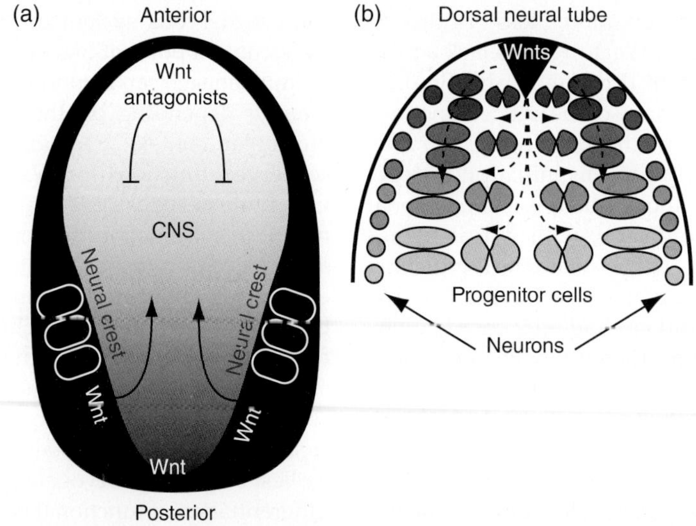

Figure 21.1 (a) During neural plate formation, Wnt signals from the paraxial mesoderm and nonneural ectoderm induce posterior CNS fates as well as neural crest at the neural plate border. Wnt antagonists expressed in the anterior embryo counteract the posteriorizing Wnt signal and allow anterior CNS structures to be formed. (b) In the developing neural tube, Wnt signals from the roof plate act as mitogens to promote progenitor proliferation, as well as patterning signals to induce dorsal progenitor fates.

Lagutin *et al.*, 2003; Nordstrom, Jessell, and Edlund, 2002). This patterning is further refined by Wnt signaling as the forebrain is subdivided into presumptive eye and brain regions. Specification of the eye field requires local inhibition of Wnt/β-catenin activity, as well as activation of Wnt/PCP signaling (Cavodeassi *et al.*, 2005).

In contrast, initial dorsal/ventral CNS patterning is not thought to be directly regulated by Wnt signaling. Midline-expressed morphogens induce medial fates within the neural plate (Dessaud, McMahon, and Briscoe, 2008), which eventually occupy the basal (ventral) plate of the neural tube. These are antagonized by signals arising from adjacent nonneural ectoderm, which induce lateral fates that eventually occupy the alar (dorsal) plate of the neural tube (Zhuang and Sockanathan, 2006). However, later in development, Wnt signals arising from the roof plate and floor plate refine dorsal/ventral fate specification during the process of neurogenesis, as will be discussed later.

The Wnt/PCP pathway also plays an important role in early CNS development, controlling the anterior migration of neural plate cells during convergent extension in all vertebrates examined. While there is not strong evidence for this pathway directly functioning in cell fate specification or tissue patterning, the correct positioning of neural tissues is required for their subsequent receipt of other environmental signals. For example, due to convergent extension defects in zebrafish embryos lacking *wnt11* function, the ventral forebrain fails to be specified properly (Heisenberg *et al.*, 2000). Furthermore, Wnt/PCP signaling regulates the orientation of neural progenitor cells within the neural tube, affecting apical/basal polarity and organizing cell division during neurogenesis (Ciruna *et al.*, 2006).

Establishment of CNS organizers

The first observable expression of Wnt ligands within the CNS occurs in organizer regions, which often are found at boundaries between major subdivisions of the brain. Like the mesodermal dorsal organizer, these tissues secrete morphogens that act to pattern surrounding tissues. In addition, organizer cells often remain undifferentiated as development proceeds, serving as a progenitor source through embryogenesis and even beyond. Many boundaries are maintained following development as morphologically distinct regions and contain specific neuronal populations as well as harboring defined axon tracts.

The best understood Wnt-dependent CNS organizer is the midbrain/hindbrain boundary (MHB or isthmic organizer), located at the junction of the posterior mesencephalon and first hindbrain rhombomere. Several Wnt ligands are expressed at the MHB including Wnt1, important not only for its status as the first vertebrate Wnt examined but also due to the knockout phenotype. As discussed previously, the analysis of *Wnt1* mutants revealed a functional requirement in formation of the midbrain and cerebellum by regulating the specification of these tissues (McMahon and Bradley, 1990). In addition, the MHB and its associated source of Wnt/β-catenin signaling play an important role in patterning surrounding tissues through downstream targets including transcription factors such as Engrailed or Gbx2 (Danielian and McMahon, 1996; Millet *et al.*, 1999), other secreted signals such as FGF8 (Lee *et al.*, 1997; Liu and Joyner, 2001), and molecules directly functioning in cell morphogenesis and adhesion (E-cadherin and alpha N-catenin; Shimamura *et al.*, 1994).

Other important boundaries in the developing brain have been identified as sites of Wnt expression and function, although the roles of Wnt signaling are not as well characterized as in the MHB. Several Wnts, as well as the transcriptional mediator Tcf7l2 (formerly named Tcf4), are expressed in the dorsal diencephalon, which resides at the forebrain/midbrain boundary (Cho and Dressler, 1998; Hollyday, McMahon, and McMahon, 1995; Salinas and Nusse, 1992). This tissue lies adjacent to other Wnt-dependent structures such as the dorsal telencephalon, dorsal thalamus, and epithalamus (Carl *et al.*, 2007; Theil *et al.*, 2002). In addition, Wnt/β-catenin signals from the dorsal diencephalon function to induce an adjacent organizer tissue, the zona limitans intrathalamica (Braun *et al.*, 2003; Peukert *et al.*, 2011). The developing hindbrain rhombomeres also represent segmental patterning units for central and PNS structures. Wnt ligand expression is present at interrhombomere boundaries

(Amoyel *et al.*, 2005; Riley *et al.*, 2004), although the role of the pathway in these regions remains uncertain.

The roof plate of the neural tube serves as a signaling center for dorsal/ventral patterning along the entire body axis, and as mentioned previously, this structure expresses several Wnt ligands (Chizhikov and Millen, 2005). Wnt signaling plays an important role in specifying particular dorsal neuronal fates following initial tissue patterning and as a mitogen for dorsal progenitor cells in the spinal cord (Figure 21.1b; Megason and McMahon, 2002; Muroyama *et al.*, 2002). In the brain, Wnt signaling from the roof plate functions to promote the growth of specific dorsal structures such as the optic tectum and the cerebral cortex (Nyholm *et al.*, 2007; Zhou *et al.*, 2006). The most dorsal neural structure arising from the telencephalon in mammals is the hippocampus, which depends on Wnt signaling from a specialized roof plate region, the cortical hem, for its induction. Mutation or inhibition of Wnt pathway components such as Wnt3a, Lrp6, and Lef1 disrupts formation of a complete hippocampus, often with loss of the most dorsally derived region, the dentate gyrus, due to failures in the specification and proliferation of neural progenitors of this tissue (Galceran *et al.*, 2000; Lee *et al.*, 2000; Zhou, Zhao, and Pleasure, 2004).

Regulation of CNS neurogenesis

Once the CNS is patterned and progenitor cells acquire regional identity, they undergo the process of neurogenesis. The two main events occurring during this period are the proliferative expansion and eventual cell cycle exit of neural progenitors and the progressive expression of a neuronal gene program. Evidence from many studies indicates that Wnt/β-catenin signaling plays important roles in both events, but the precise functions have been difficult to ascertain due to the variety of experimental systems and approaches used. It is clear that Wnt signaling can promote cell division, as demonstrated by a study showing that activated β-catenin expressed in neural progenitor cells under control of the *Nestin* promoter caused overproliferation and a resulting increase in cortical size (Chenn and Walsh, 2002). Combined with other studies showing that Wnt signaling acts as a mitogen in the CNS (Dickinson, Krumlauf, and McMahon, 1994; Megason and McMahon, 2002), this work supports a general role for the pathway in expanding the neural progenitor population.

With regard to neuronal differentiation, the evidence is more complex. The activation of β-catenin-mediated transcription can delay neurogenesis (Imura *et al.*, 2010; Machon *et al.*, 2007), and the inhibition of β-catenin function causes premature differentiation of neural progenitors (Machon *et al.*, 2007; Woodhead *et al.*, 2006). However, equally convincing evidence has shown that in some situations Wnt activity can promote neuronal differentiation. Pathway activation can induce the expression of SoxB1 neural competence factors (Lee *et al.*, 2006; Van Raay *et al.*, 2005), as well as bHLH proneural factors (Hirabayashi *et al.*, 2004), both hallmarks of neural fate specification. Likewise, the loss of Wnt signaling can limit the number of neurons produced during development (Galceran *et al.*, 2000; Zhou, Zhao, and Pleasure, 2004; Zhou *et al.*, 2006). Together, these studies lead to an apparent conundrum, in which Wnt signaling can either positively or negatively influence neurogenesis. However, diverse experimental approaches can affect the strength and duration of Wnt signals, as well as the precise cell types responding. When considering these data as a whole, several general themes appear. First, transient activation of Wnt signaling tends to increase the number of neurons, while constitutive activation tends to block neurogenesis. Second, Wnt/β-catenin activity is most often observed during neural progenitor stages, when proneural factors are expressed, and can promote this identity. Third, the inhibition of Wnt signaling can block neural progenitor formation in immature cells but conversely can promote differentiation once neural progenitors are specified.

The sum of these observations suggests that Wnt activity generally promotes a neural progenitor fate, which is compatible with proliferative expansion and ultimately contributes mature neurons to the CNS (Figure 21.2a). However, it also appears that Wnt signaling must be inhibited for progenitors to fully differentiate. Thus, the Wnt pathway may provide a permissive signal for the maintenance of neural

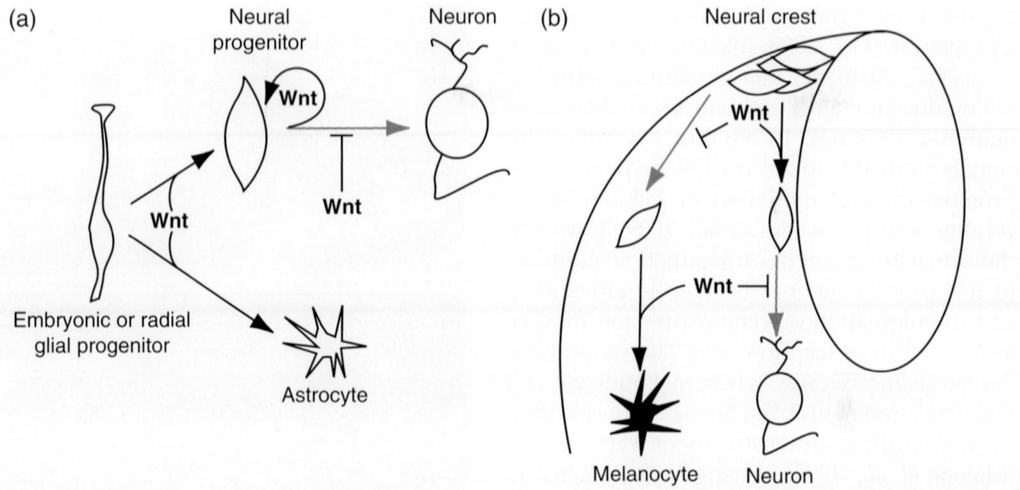

Figure 21.2 (a) In embryonic and postembryonic neurogenesis, Wnt/β-catenin signaling drives the transition of radial glia into neural progenitors and astrocytes. The pathway also promotes the maintenance and proliferation of neural progenitors but inhibits their terminal differentiation into neurons. (b) During neural crest specification, Wnt/β-catenin signaling plays a dual role. In premigratory neural crest, Wnt signals promote sensory neurogenesis and inhibit melanogenesis. Conversely, in migratory neural crest, Wnt signals promote melanocyte formation and inhibit neurogenesis.

progenitors, which can then be expanded, specified into individual fates, and driven to differentiate as needed.

Cell-type specificity in the CNS

Both descriptive and functional analyses show that the requirements for Wnt/β-catenin signaling in neurogenesis seem to be restricted to specific CNS regions. Transcriptional activity downstream of β-catenin, as visualized with reporter transgenes, is not universal throughout the developing CNS (Dorsky, Sheldahl, and Moon, 2002; Maretto *et al.*, 2003; Moro *et al.*, 2012), and neurogenesis can proceed normally in many regions despite the widespread loss of Wnt pathway function. Yet clear requirements have been reported in the development of individual CNS populations. Wnt signaling promotes the initiation of retinal neurogenesis in *Xenopus* (Van Raay *et al.*, 2005), and in other species, the maintenance of retinal progenitors at the peripheral margin of the growing retina is regulated by the pathway (Kubo, Takeichi, and Nakagawa, 2003; Liu *et al.*, 2007). Wnt activity is required for neurogenesis throughout the cerebral cortex of mammals, where it functions to regulate neuronal production by intermediate cortical progenitors (Munji *et al.*, 2011; Mutch *et al.*, 2010). Specific neuronal populations in the midbrain and hypothalamus also require Wnt signaling for their generation, where pathway activity is necessary to initiate neuronal gene expression and differentiation (Lee *et al.*, 2006; Prakash *et al.*, 2006; Wang *et al.*, 2012). One interesting feature that many of these cell populations have in common is a persistent, long-lasting progenitor pool that continues to make neurons throughout, and even beyond, embryogenesis. In a sense, Wnt signaling could be considered to function in the continuous growth, rather than the establishment, of CNS neuron populations.

Beyond acting to promote the expression of general neural competence and proneural factors, there is little evidence that Wnt signaling functions to specify individual neuronal subtypes. Depending on the tissue analyzed, several different types of neurons can arise from Wnt-dependent progenitors, and target gene analysis has failed to produce data supporting particular lineages. In cases where restricted lineages arise from progenitors in a Wnt-dependent manner, it is not clear that pathway activity acts instructively to determine cell fate. For example, dopaminergic neurons arising from the ventral vertebrate midbrain require Wnt signaling for their development (Andersson *et al.*, 2008; Joksimovic *et al.*, 2009; Tang *et al.*, 2010), yet manipulation of Wnt signaling does not shift progenitors to alternative neuronal identities. In another example, Wnt signals from the spinal cord floor plate act to promote the differentiation of median motor column neurons (Agalliu *et al.*, 2009); however, the downstream signal transduction pathway in this process has not been fully elucidated, and the regulation of gene expression may be indirect. An exception can be found in *Caenorhabditis elegans*, where asymmetric cell fates in paired neuroblast lineages are specified by transcription downstream of Wnt signals (Maloof *et al.*, 1999; Whangbo and Kenyon, 1999). However, this example may represent a unique binary fate choice that does not occur in vertebrates.

Wnt signaling can also influence the production of specific glial lineages during CNS development. Several studies have shown that oligodendrocyte precursor formation is negatively regulated by Wnt/β-catenin activity (Langseth *et al.*, 2010; Shimizu *et al.*, 2005), which may partly reflect the promotion of neurogenesis from multipotent progenitors, although the pathway later plays an important positive role in myelination by these same cells (Fancy *et al.*, 2009; Tawk *et al.*, 2011). Other evidence suggests that Wnt signaling regulates the differentiation of radial glia into astrocytes in the mammalian cerebellum and hypothalamus (Wang *et al.*, 2011, 2012). Interestingly, quiescent radial glia appear to have low levels of endogenous Wnt activity (Meyers *et al.*, 2012; Wang *et al.*, 2012), consistent with a role for Wnt signaling in their differentiation into other neuronal or glial fates (Figure 21.2a).

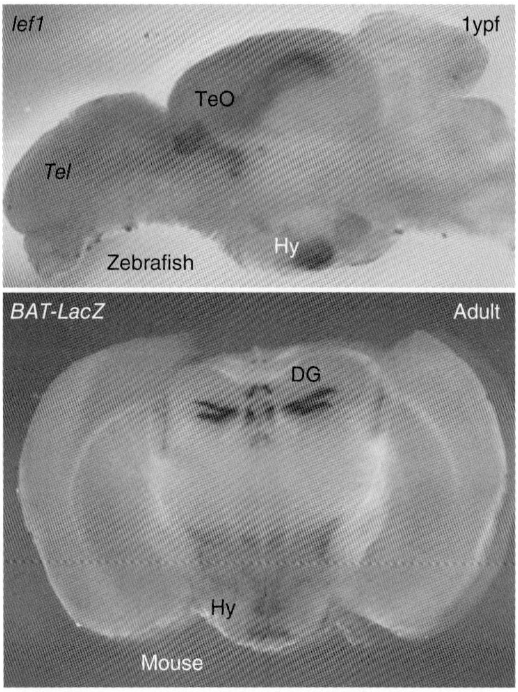

Figure 21.3 Throughout the adult vertebrate CNS, Wnt/β-catenin activity is found in areas of continued neurogenesis. In a midsagittal section of an adult zebrafish brain (top), the expression of the Wnt mediator and feedback target *lef1* is observed in the optic tectum (TeO) and hypothalamus (Hy), as well as other locations. In a coronal section of an adult mouse brain (bottom), the expression of the Wnt reporter *BAT-LacZ* is observed in the dentate gyrus (DG) and hypothalamus (Hy) as well as other locations.

Maintenance of continuous CNS neurogenesis

Several studies have established a strong association between Wnt signaling and areas of the CNS harboring progenitors beyond embryogenesis and even into adult stages. Specialized regions of the vertebrate brain contain neural stem cells that contribute to behavioral plasticity and response to environmental stimuli. In mammals, the two best-characterized postembryonic neurogenic niches are the dentate gyrus of the hippocampus and telencephalic lateral ventricles. In both of these areas, stem cells constantly produce neural progenitors through asymmetric divisions, recapitulating the process of embryonic neurogenesis. While the identity and specific behavior of the two stem cell populations is different, they both require extracellular signals from the surrounding niche to regulate self-renewal and neurogenic differentiation. The dentate gyrus has been shown to maintain a high level of Wnt activity (Figure 21.3), which is necessary for the formation of new granule cell neurons (Lie *et al.*, 2005), partly by regulating the expression of the proneural factor NeuroD. Less is known about the molecular pathway downstream of Wnt signaling in the telencephalic subventricular zone, but Wnt activity has been shown to promote symmetric division of neural stem cells (Piccin and Morshead, 2011), as well as the proliferation of neural progenitors (Adachi *et al.*, 2007).

In nonmammalian vertebrates, postembryonic neurogenesis is more widespread, extending to multiple regions of the brain (Figure 21.3). Wnt signaling has been shown to function in the neuronal differentiation of stem cells residing in the ventricular region of the adult zebrafish hypothalamus (Wang *et al.*, 2012). Interestingly, this brain region has recently been shown to harbor neural stem cells in mammals as well (Lee *et al.*, 2012; Li, Tang, and Cai, 2012), suggesting that the role of Wnt signaling in regulating adult hypothalamic neurogenesis may be evolutionarily conserved. In addition, other brain regions such as the cerebellum and midbrain tectum contain identified stem cell populations in zebrafish (Ito *et al.*, 2010; Kaslin *et al.*, 2009), and while the role of Wnt signaling in these areas has not been functionally tested, the expression of Wnt reporters and pathway components strongly suggests a function in stem cell regulation (Wang *et al.*, 2012).

Formation of neural crest: A source of the PNS

The vertebrate PNS arises from the neural crest and epidermal placodes, specialized tissues that are not part of the neuroectoderm. Neural crest is induced at the boundary of neural and nonneural ectoderm along the entire anterior–posterior body axis, while placodes form at discrete locations in the head corresponding to future sensory organs and cranial ganglia. Following the initial specification of the neural plate, Wnt signaling plays a direct role in neural crest induction (Figure 21.1a), mediated by signals from the underlying paraxial mesoderm and surrounding ectoderm (Chang and Hemmati-Brivanlou, 1998; Garcia-Castro, Marcelle, and Bronner-Fraser, 2002; Lewis *et al.*, 2004). Specific transcription factors that mark premigratory neural crest and regulate its differentiation have been shown to be direct targets of Wnt signaling during the process of neural crest specification (Berndt and Halloran, 2006; Dorsky, Raible, and Moon, 2000; Piloto and Schilling, 2010). In addition, the positioning and specification of individual neurogenic epidermal placodes is Wnt dependent (Lassiter *et al.*, 2007; Litsiou, Hanson, and Streit, 2005).

Control of neural crest migration

The formation of specific PNS populations from neural crest also requires proper cell migration, both to control the location and number of peripheral ganglia and to expose precursor cells to the correct terminal differentiation signals and trophic factors. Neural crest migration is dependent on Wnt/PCP signaling, which regulates cell polarity and motility (De Calisto *et al.*, 2005; Matthews *et al.*, 2008). Wnt signals expressed along the migratory pathway control the directionality of cell protrusions and contact-mediated inhibition of locomotion (Carmona-Fontaine *et al.*, 2008). In trunk neural crest, Wnt/PCP signaling through the MuSK tyrosine kinase receptor organizes the segmental migration of cells through the developing somites (Banerjee *et al.*, 2011). Together, these processes allow neural crest cells to migrate in a segmentally restricted, directional fashion towards their ultimate targets.

Regulation of PNS neurogenesis

As they undergo an epithelial–mesenchymal transition and begin migration, neural crest cells become specified towards a number of diverse lineages including neurons, glia, pigment cells, smooth muscle, and skeleton (Dorsky, Moon, and Raible, 2000). Wnt signaling has been implicated in the regulation of multiple neural crest-derived cell types, but different requirements for the pathway occur at different steps of the differentiation process. The activation of Wnt/β-catenin signaling in neural crest progenitors can promote a program of melanogenesis, including the expression of *Mitf*, a master regulator of melanocyte formation, at the expense of neurogenesis (Dorsky, Moon, and Raible, 1998; Dorsky, Raible, and Moon, 2000; Takeda *et al.*, 2000). However, these findings are counterbalanced by evidence that Wnt signaling is also necessary and sufficient to promote sensory neuron formation from neural crest *in vivo* and *in vitro* (Hari *et al.*, 2002; Lee *et al.*, 2004). Further studies have attempted to reconcile these findings by showing that the activation of Wnt signaling in the premigratory neural crest of the mouse inhibits melanocyte formation, while the activation in migratory neural crest increases the number of melanocytes (Hari *et al.*, 2012). These results suggest that the regulation of sensory neurogenesis by Wnt signaling is context dependent and occurs during an early time window of neural crest fate specification (Figure 21.2b).

Once PNS fates have been specified, Wnt activsity plays a role in promoting the proliferative progenitor state. The activation of β-catenin signaling through Fzd3 maintains a population of dividing sympathetic neuroblasts (Armstrong *et al.*, 2011), and the inhibition of this pathway results in a decrease in the size of sympathetic ganglia. While other PNS populations have not been examined at this level, these data suggest that the continued production of neurons from PNS progenitor populations may generally require Wnt signaling.

Neural connectivity: Pathfinding and synaptogenesis

After neurons are specified and begin to differentiate, they establish connectivity through axon pathfinding and synapse formation. Both of these important processes are regulated by Wnt signaling in organisms from flies to humans (Figure 21.4). Generally, such developmental events require local reorganization of the cytoskeleton, membrane, and related structures and occur rapidly, suggesting that their regulation requires local signaling as opposed to nuclear transcription.

Genetic evidence first demonstrated that Wnt signaling regulated axon guidance in *Drosophila*, acting through the tyrosine kinase receptor Ryk (Yoshikawa *et al.*, 2003). This Wnt/Ryk signaling pathway and its function in axon guidance are conserved in vertebrates, although in vertebrates Ryk acts as a coreceptor with Frizzled (Lu *et al.*, 2004; Schmitt *et al.*, 2006). The analysis of guidance phenotypes in multiple species supports the idea that Wnts can act as either attractive or repulsive cues, functioning in a gradient to establish the polarity of major axon tracts. The downstream signal transduction pathway requires Dishevelled and likely involves the modulation of planar cell polarity, suggesting a direct effect on the cytoskeleton during axon outgrowth (Shafer *et al.*, 2011; Shimizu, Sato, and Tabata, 2011). Interestingly, similar mechanisms to those used for axon guidance may be employed in the orientation of polarity in neurons of *C. elegans* (Hilliard and Bargmann, 2006) and migration of branchiomotor neuron soma in the vertebrate hindbrain (Jessen *et al.*, 2002; Vivancos *et al.*, 2009).

The formation of both central and peripheral synapses has been shown to be Wnt dependent in multiple model organisms, although the precise function of the pathway appears to differ between systems and cell types. In *Drosophila*, Wingless acts as a required signal

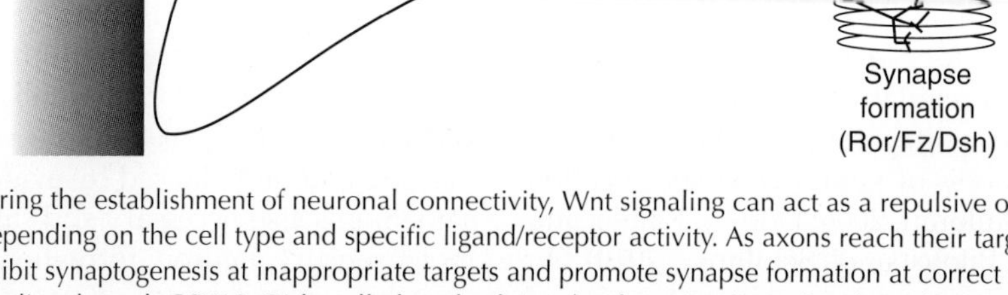

Figure 21.4 During the establishment of neuronal connectivity, Wnt signaling can act as a repulsive or attractive guidance cue, depending on the cell type and specific ligand/receptor activity. As axons reach their targets, Wnt signaling can inhibit synaptogenesis at inappropriate targets and promote synapse formation at correct targets. Downstream signaling through GSK3β, Dishevelled, and calcium has been implicated in these functions, while β-catenin-mediated transcription has not been shown to play an important role.

for target-dependent synaptogenesis at the neuromuscular junction, acting bidirectionally to regulate presynaptic and postsynaptic development (Packard *et al.*, 2002), whereas Wnt4 functions to inhibit synapse formation in specific muscles, thus generating target specificity (Inaki *et al.*, 2007). In *C. elegans*, Wnt signaling inhibits synaptogenesis onto target muscles from particular domains of motor neuron axons (Klassen and Shen, 2007). This general role in neuromuscular synapse formation also extends to vertebrates, in which Wnt signaling through MuSK has been shown to regulate receptor clustering in muscle and provide a feedback signal to guide the approaching growth cone (Jing *et al.*, 2009). Finally, Wnt activity regulates synaptogenesis in the vertebrate CNS, where Wnt7a promotes presynaptic development of cerebellar granule cell neurons as well as hippocampal neurons (Farias *et al.*, 2007; Hall, Lucas, and Salinas, 2000), and also plays a role in postsynaptic development within dendritic spines (Ciani *et al.*, 2011). Most of these functions require Dishevelled signaling, and many require GSK3β inhibition, but none so far appear to require the transcriptional function of β-catenin.

The function of Wnt signaling in organizing synaptic machinery also implicates the pathway in the important process of plasticity. Both presynaptic and postsynaptic sites have an extraordinary capacity to regulate cellular components and tune activity to external stimuli, providing both homeostatic regulation and adaptability. The expression of Wnts and other pathway components has been shown to be activity dependent (Ataman *et al.*, 2008; Chen, Park, and Tang, 2006; Wayman *et al.*, 2006), and in *C. elegans*, activity-dependent Wnt signaling through the receptor tyrosine kinase ROR regulates the translocation of synaptic receptors and subsequent changes in synaptic strength (Jensen *et al.*, 2012). Thus, Wnt pathway activity may be a general mechanism to allow presynaptic and postsynaptic sites to communicate and modify their structure beyond development.

Conclusions

Despite its ubiquity throughout neural development, several fundamental principles that distinguish the Wnt pathway from other secreted signals emerge from the analysis of both descriptive and functional experiments: (i) Generally, signaling does not appear to mediate different effects at different distances from the source, acting more like a niche signal than a classical morphogen. (ii) Wnt/β-catenin activity can increase neural progenitor proliferation, but this effect is most apparent in distinct populations that persist beyond embryogenesis and may reflect a unique status of these cells. (iii) The Wnt pathway does not usually act instructively to promote particular cell fates during neurogenesis, but rather functions as a permissive signal that activates general mediators of differentiation. (iv) Signaling in postmitotic neurons, particularly during the establishment of connectivity, tends to act locally on the cytoskeleton and associated membrane structures, rather than through nuclear transcription. While there is no single unifying model for the multifaceted functions of Wnt activity, it is fascinating that one family of signals mediates the patterning, growth and maintenance, and wiring of the nervous system. From a therapeutic point of view, these functions make Wnt signaling an attractive target for modulating both neural regeneration and plasticity.

References

Adachi, K., Mirzadeh, Z., Sakaguchi, M. *et al.* (2007) Beta-catenin signaling promotes proliferation of progenitor cells in the adult mouse subventricular zone. *Stem Cells*, **25**, 2827–2836.

Agalliu, D., Takada, S., Agalliu, I. *et al.* (2009) Motor neurons with axial muscle projections specified by Wnt4/5 signaling. *Neuron*, **61**, 708–720.

Amoyel, M., Cheng, Y.C., Jiang, Y.J., and Wilkinson, D.G. (2005) Wnt1 regulates neurogenesis and mediates lateral inhibition of boundary cell specification in the zebrafish hindbrain. *Development*, **132**, 775–785.

Andersson, E.R., Prakash, N., Cajanek, L. *et al.* (2008) Wnt5a regulates ventral midbrain morphogenesis and the development of A9-A10 dopaminergic cells in vivo. *PLoS One*, **3**, e3517.

Armstrong, A., Ryu, Y.K., Chieco, D., and Kuruvilla, R. (2011) Frizzled3 is required for neurogenesis and target innervation during sympathetic nervous system development. *The Journal of Neuroscience*, **31**, 2371–2381.

Ataman, B., Ashley, J., Gorczyca, M. *et al.* (2008) Rapid activity-dependent modifications in synaptic structure and function require bidirectional Wnt signaling. *Neuron*, **57**, 705–718.

Banerjee, S., Gordon, L., Donn, T.M. *et al.* (2011) A novel role for MuSK and non-canonical Wnt signaling during segmental neural crest cell migration. *Development*, **138**, 3287–3296.

Berndt, J.D. and Halloran, M.C. (2006) Semaphorin 3d promotes cell proliferation and neural crest cell development downstream of TCF in the zebrafish hindbrain. *Development*, **133**, 3983–3992.

Braun, M.M., Etheridge, A., Bernard, A. *et al.* (2003) Wnt signaling is required at distinct stages of development for the induction of the posterior forebrain. *Development*, **130**, 5579–5587.

Carl, M., Bianco, I.H., Bajoghli, B. *et al.* (2007) Wnt/Axin1/beta-catenin signaling regulates asymmetric nodal activation, elaboration, and concordance of CNS asymmetries. *Neuron*, **55**, 393–405.

Carmona-Fontaine, C., Matthews, H.K., Kuriyama, S. *et al.* (2008) Contact inhibition of locomotion in vivo controls neural crest directional migration. *Nature*, **456**, 957–961.

Cavodeassi, F., Carreira-Barbosa, F., Young, R.M. *et al.* (2005) Early stages of zebrafish eye formation require the coordinated activity of Wnt11, Fz5, and the Wnt/beta-catenin pathway. *Neuron*, **47**, 43–56.

Chang, C.B. and Hemmati-Brivanlou, A. (1998) Neural crest induction by Xwnt7B in Xenopus. *Developmental Biology*, **194**, 129–134.

Chen, J., Park, C.S., and Tang, S.J. (2006) Activity-dependent synaptic Wnt release regulates hippocampal long term potentiation. *The Journal of Biological Chemistry*, **281**, 11910–11916.

Chenn, A. and Walsh, C.A. (2002) Regulation of cerebral cortical size by control of cell cycle exit in neural precursors. *Science*, **297**, 365–369.

Chizhikov, V.V. and Millen, K.J. (2005) Roof plate-dependent patterning of the vertebrate dorsal central nervous system. *Developmental Biology*, **277**, 287–295.

Cho, E.A. and Dressler, G.R. (1998) TCF-4 binds beta-catenin and is expressed in distinct regions of the embryonic brain and limbs. *Mechanisms of Development*, **77**, 9–18.

Ciani, L., Boyle, K.A., Dickins, E. *et al.* (2011) Wnt7a signaling promotes dendritic spine growth and synaptic strength through Ca(2)(+)/Calmodulin-dependent protein kinase II. *Proceedings of the National Academy of Sciences of the United States of America*, **108**, 10732–10737.

Ciruna, B., Jenny, A., Lee, D. *et al.* (2006) Planar cell polarity signalling couples cell division and morphogenesis during neurulation. *Nature*, **439**, 220–224.

Danielian, P.S. and McMahon, A.P. (1996) Engrailed-1 as a target of the Wnt-1 signalling pathway in vertebrate midbrain development. *Nature*, **383**, 332–334.

De Calisto, J., Araya, C., Marchant, L. *et al.* (2005) Essential role of non-canonical Wnt signalling in neural crest migration. *Development*, **132**, 2587–2597.

Dessaud, E., McMahon, A.P., and Briscoe, J. (2008) Pattern formation in the vertebrate neural tube: a sonic hedgehog morphogen-regulated transcriptional network. *Development*, **135**, 2489–2503.

Dickinson, M.E., Krumlauf, R., and McMahon, A.P. (1994) Evidence for a mitogenic effect of Wnt-1 in the developing mammalian central nervous system. *Development*, **120**, 1453–1471.

Domingos, P.M., Itasaki, N., Jones, C.M. *et al.* (2001) The Wnt/beta-catenin pathway posteriorizes neural tissue in Xenopus by an indirect mechanism requiring FGF signalling. *Developmental Biology*, **239**, 148–160.

Dorsky, R.I., Moon, R.T., and Raible, D.W. (1998) Control of neural crest cell fate by the Wnt signalling pathway. *Nature*, **396**, 370–373.

Dorsky, R.I., Moon, R.T., and Raible, D.W. (2000) Environmental signals and cell fate specification in premigratory neural crest. *BioEssays*, **22**, 708–716.

Dorsky, R.I., Raible, D.W., and Moon, R.T. (2000) Direct regulation of nacre, a zebrafish MITF homolog required for pigment cell formation, by the Wnt pathway. *Genes & Development*, **14**, 158–162.

Dorsky, R.I., Sheldahl, L.C., and Moon, R.T. (2002) A transgenic lef1/beta-catenin-dependent reporter is expressed in spatially restricted domains throughout zebrafish development. *Developmental Biology*, **241**, 229–237.

Dorsky, R.I., Itoh, M., Moon, R.T., and Chitnis, A. (2003) Two tcf3 genes cooperate to pattern the zebrafish brain. *Development*, **130**, 1937–1947.

Erter, C.E., Wilm, T.P., Basler, N. *et al.* (2001) Wnt8 is required in lateral mesendodermal precursors for neural posteriorization in vivo. *Development*, **128**, 3571–3583.

Fancy, S.P., Baranzini, S.E., Zhao, C. *et al.* (2009) Dysregulation of the Wnt pathway inhibits timely myelination and remyelination in the mammalian CNS. *Genes & Development*, **23**, 1571–1585.

Farias, G.G., Valles, A.S., Colombres, M. *et al.* (2007) Wnt 7a induces presynaptic colocalization of alpha 7-nicotinic acetylcholine receptors and adenomatous polyposis coli in hippocampal neurons. *The Journal of Neuroscience*, **27**, 5313–5325.

Galceran, J., Miyashita-Lin, E.M., Devaney, E. *et al.* (2000) Hippocampus development and generation of dentate gyrus granule cells is regulated by LEF1. *Development*, **127**, 469–482.

Garcia-Castro, M.I., Marcelle, C., and Bronner-Fraser, M. (2002) Ectodermal Wnt function as a neural crest inducer. *Science*, **297**, 848–851.

Gavin, B.J., McMahon, J.A., and McMahon, A.P. (1990) Expression of multiple novel Wnt-1/int-1-related genes during fetal and adult mouse development. *Genes & Development*, **4**, 2319–2332.

Hall, A.C., Lucas, F.R., and Salinas, P.C. (2000) Axonal remodeling and synaptic differentiation in the cerebellum is regulated by WNT-7a signaling. *Cell*, **100**, 525–535.

Hari, L., Brault, V., Kleber, M. *et al.* (2002) Lineage-specific requirements of beta-catenin in neural crest development. *The Journal of Cell Biology*, **159**, 867–880.

Hari, L., Miescher, I., Shakhova, O. *et al.* (2012) Temporal control of neural crest lineage generation by Wnt/beta-catenin signaling. *Development*, **139**, 2107–2117.

Harland, R. and Gerhart, J. (1997) Formation and function of Spemann's organizer. *Annual Review of Cell and Developmental Biology*, **13**, 611–667.

Heisenberg, C.P., Tada, M., Rauch, G.J. *et al.* (2000) Silberblick/Wnt11 mediates convergent extension movements during zebrafish gastrulation. *Nature*, **405**, 76–81.

Hilliard, M.A. and Bargmann, C.I. (2006) Wnt signals and frizzled activity orient anterior-posterior axon outgrowth in C. elegans. *Developmental Cell*, **10**, 379–390.

Hirabayashi, Y., Itoh, Y., Tabata, H. *et al.* (2004) The Wnt/beta-catenin pathway directs neuronal differentiation of cortical neural precursor cells. *Development*, **131**, 2791–2801.

Hollyday, M., McMahon, J.A., and McMahon, A.P. (1995) Wnt expression patterns in chick embryo nervous system. *Mechanisms of Development*, **52**, 9–25.

Imura, T., Wang, X., Noda, T. *et al.* (2010) Adenomatous polyposis coli is essential for both neuronal differentiation and maintenance of adult neural stem cells in subventricular zone and hippocampus. *Stem Cells*, **28**, 2053–2064.

Inaki, M., Yoshikawa, S., Thomas, J.B. *et al.* (2007) Wnt4 is a local repulsive cue that determines synaptic target specificity. *Current Biology*, **17**, 1574–1579.

Ito, Y., Tanaka, H., Okamoto, H., and Ohshima, T. (2010) Characterization of neural stem cells and their progeny in the adult zebrafish optic tectum. *Developmental Biology*, **342**, 26–38.

Jensen, M., Hoerndli, F.J., Brockie, P.J. *et al.* (2012) Wnt signaling regulates acetylcholine receptor translocation and synaptic plasticity in the adult nervous system. *Cell*, **149**, 173–187.

Jessen, J.R., Topczewski, J., Bingham, S. *et al.* (2002) Zebrafish trilobite identifies new roles for Strabismus in gastrulation and neuronal movements. *Nature Cell Biology*, **4**, 610–615.

Jing, L., Lefebvre, J.L., Gordon, L.R., and Granato, M. (2009) Wnt signals organize synaptic prepattern and axon guidance through the zebrafish unplugged/MuSK receptor. *Neuron*, **61**, 721–733.

Joksimovic, M., Yun, B.A., Kittappa, R. *et al.* (2009) Wnt antagonism of Shh facilitates midbrain floor plate neurogenesis. *Nature Neuroscience*, **12**, 125–131.

Joubin, K. and Stern, C.D. (2001) Formation and maintenance of the organizer among the vertebrates. *The International Journal of Developmental Biology*, **45**, 165–175.

Kaslin, J., Ganz, J., Geffarth, M. *et al.* (2009) Stem cells in the adult zebrafish cerebellum: initiation and maintenance of a novel stem cell niche. *The Journal of Neuroscience*, **29**, 6142–6153.

Klassen, M.P. and Shen, K. (2007) Wnt signaling positions neuromuscular connectivity by inhibiting synapse formation in C. elegans. *Cell*, **130**, 704–716.

Kubo, F., Takeichi, M., and Nakagawa, S. (2003) Wnt2b controls retinal cell differentiation at the ciliary marginal zone. *Development*, **130**, 587–598.

Lagutin, O.V., Zhu, C.C., Kobayashi, D. *et al.* (2003) Six3 repression of Wnt signaling in the anterior neuroectoderm is essential for vertebrate forebrain development. *Genes & Development*, **17**, 368–379.

Langseth, A.J., Munji, R.N., Choe, Y. *et al.* (2010) Wnts influence the timing and efficiency of oligodendrocyte precursor cell generation in the telencephalon. *The Journal of Neuroscience*, **30**, 13367–13372.

Lassiter, R.N., Dude, C.M., Reynolds, S.B. *et al.* (2007) Canonical Wnt signaling is required for ophthalmic trigeminal placode cell fate determination and maintenance. *Developmental Biology*, **308**, 392–406.

Lee, S.M., Danielian, P.S., Fritzsch, B., and McMahon, A.P. (1997) Evidence that FGF8 signalling from the midbrain-hindbrain junction regulates growth and polarity in the developing midbrain. *Development*, **124**, 959–969.

Lee, S.M., Tole, S., Grove, E., and McMahon, A.P. (2000) A local Wnt-3a signal is required for development of the mammalian hippocampus. *Development*, **127**, 457–467.

Lee, H.Y., Kleber, M., Hari, L. *et al.* (2004) Instructive role of Wnt/beta-catenin in sensory fate specification in neural crest stem cells. *Science*, **303**, 1020–1023.

Lee, J.E., Wu, S.F., Goering, L.M., and Dorsky, R.I. (2006) Canonical Wnt signaling through Lef1 is required for hypothalamic neurogenesis. *Development*, **133**, 4451–4461.

Lee, D.A., Bedont, J.L., Pak, T. *et al.* (2012) Tanycytes of the hypothalamic median eminence form a diet-responsive neurogenic niche. *Nature Neuroscience*, **15**, 700–702.

Lewis, J.L., Bonner, J., Modrell, M. *et al.* (2004) Reiterated Wnt signaling during zebrafish neural crest development. *Development*, **131**, 1299–1308.

Li, J., Tang, Y., and Cai, D. (2012) IKKbeta/NF-kappaB disrupts adult hypothalamic neural stem cells to mediate a neurodegenerative mechanism of

dietary obesity and pre-diabetes. *Nature Cell Biology*, **14**, 999–1012.
Lie, D.C., Colamarino, S.A., Song, H.J. *et al.* (2005) Wnt signalling regulates adult hippocampal neurogenesis. *Nature*, **437**, 1370–1375.
Litsiou, A., Hanson, S., and Streit, A. (2005) A balance of FGF, BMP and WNT signalling positions the future placode territory in the head. *Development*, **132**, 4051–4062.
Liu, A. and Joyner, A.L. (2001) EN and GBX2 play essential roles downstream of FGF8 in patterning the mouse mid/hindbrain region. *Development*, **128**, 181–191.
Liu, H., Xu, S., Wang, Y. *et al.* (2007) Ciliary margin transdifferentiation from neural retina is controlled by canonical Wnt signaling. *Developmental Biology*, **308**, 54–67.
Lu, W., Yamamoto, V., Ortega, B., and Baltimore, D. (2004) Mammalian Ryk is a Wnt coreceptor required for stimulation of neurite outgrowth. *Cell*, **119**, 97–108.
Machon, O., Backman, M., Machonova, O. *et al.* (2007) A dynamic gradient of Wnt signaling controls initiation of neurogenesis in the mammalian cortex and cellular specification in the hippocampus. *Developmental Biology*, **311**, 223–237.
Maloof, J.N., Whangbo, J., Harris, J.M. *et al.* (1999) A Wnt signaling pathway controls hox gene expression and neuroblast migration in C. elegans. *Development*, **126**, 37–49.
Maretto, S., Cordenonsi, M., Dupont, S. *et al.* (2003) Mapping Wnt/beta-catenin signaling during mouse development and in colorectal tumors. *Proceedings of the National Academy of Sciences of the United States of America*, **100**, 3299–3304.
Matthews, H.K., Broders-Bondon, F., Thiery, J.P., and Mayor, R. (2008) Wnt11r is required for cranial neural crest migration. *Developmental Dynamics*, **237**, 3404–3409.
McMahon, A.P. and Bradley, A. (1990) The Wnt-1 (int-1) proto-oncogene is required for development of a large region of the mouse brain. *Cell*, **62**, 1073–1085.
Megason, S.G. and McMahon, A.P. (2002) A mitogen gradient of dorsal midline Wnts organizes growth in the CNS. *Development*, **129**, 2087–2098.
Meyers, J.R., Hu, L., Moses, A. *et al.* (2012) ss-catenin/Wnt signaling controls progenitor fate in the developing and regenerating zebrafish retina. *Neural Development*, **7**, 30.
Millet, S., Campbell, K., Epstein, D.J. *et al.* (1999) A role for Gbx2 in repression of Otx2 and positioning the mid/hindbrain organizer. *Nature*, **401**, 161–164.
Moro, E., Ozhan-Kizil, G., Mongera, A. *et al.* (2012) In vivo Wnt signaling tracing through a transgenic biosensor fish reveals novel activity domains. *Developmental Biology*, **366**, 327–340.
Munji, R.N., Choe, Y., Li, G. *et al.* (2011) Wnt signaling regulates neuronal differentiation of cortical intermediate progenitors. *The Journal of Neuroscience*, **31**, 1676–1687.
Muroyama, Y., Fujihara, M., Ikeya, M. *et al.* (2002) Wnt signaling plays an essential role in neuronal specification of the dorsal spinal cord. *Genes & Development*, **16**, 548–553.
Mutch, C.A., Schulte, J.D., Olson, E., and Chenn, A. (2010) Beta-catenin signaling negatively regulates intermediate progenitor population numbers in the developing cortex. *PLoS One*, **5**, e12376.
Nordstrom, U., Jessell, T.M., and Edlund, T. (2002) Progressive induction of caudal neural character by graded Wnt signaling. *Nature Neuroscience*, **5**, 525–532.
Nyholm, M.K., Wu, S.F., Dorsky, R.I., and Grinblat, Y. (2007) The zebrafish zic2a-zic5 gene pair acts downstream of canonical Wnt signaling to control cell proliferation in the developing tectum. *Development*, **134**, 735–746.
Packard, M., Koo, E.S., Gorczyca, M. *et al.* (2002) The Drosophila Wnt, wingless, provides an essential signal for pre- and postsynaptic differentiation. *Cell*, **111**, 319–330.
Peukert, D., Weber, S., Lumsden, A., and Scholpp, S. (2011). Lhx2 and Lhx9 determine neuronal differentiation and compartition in the caudal forebrain by regulating Wnt signaling. *PLoS Biology*, **9**, e1001218.
Piccin, D. and Morshead, C.M. (2011) Wnt signaling regulates symmetry of division of neural stem cells in the adult brain and in response to injury. *Stem Cells*, **29**, 528–538.
Piloto, S. and Schilling, T.F. (2010) Ovo1 links Wnt signaling with N-cadherin localization during neural crest migration. *Development*, **137**, 1981–1990.
Prakash, N., Brodski, C., Naserke, T. *et al.* (2006) A Wnt1-regulated genetic network controls the identity and fate of midbrain-dopaminergic progenitors in vivo. *Development*, **133**, 89–98.
Riley, B.B., Chiang, M.Y., Storch, E.M. *et al.* (2004) Rhombomere boundaries are Wnt signaling centers that regulate metameric patterning in the zebrafish hindbrain. *Developmental Dynamics*, **231**, 278–291.
Salinas, P.C. and Nusse, R. (1992) Regional expression of the Wnt-3 gene in the developing mouse forebrain in relationship to diencephalic neuromeres. *Mechanisms of Development*, **39**, 151–160.
Schmitt, A.M., Shi, J., Wolf, A.M. *et al.* (2006) Wnt-Ryk signalling mediates medial-lateral retinotectal topographic mapping. *Nature*, **439**, 31–37.
Shafer, B., Onishi, K., Lo, C. *et al.* (2011) Vangl2 promotes Wnt/planar cell polarity-like signaling by antagonizing Dvl1-mediated feedback inhibition in growth cone guidance. *Developmental Cell*, **20**, 177–191.

Sharpe, C., Lawrence, N., and Arias, A.M. (2001) Wnt signalling: a theme with nuclear variations. *BioEssays*, **23**, 311–318.

Shimamura, K., Hirano, S., McMahon, A.P., and Takeichi, M. (1994) Wnt-1-dependent regulation of local E-cadherin and alpha N-catenin expression in the embryonic mouse brain. *Development*, **120**, 2225–2234.

Shimizu, K., Sato, M., and Tabata, T. (2011) The Wnt5/planar cell polarity pathway regulates axonal development of the Drosophila mushroom body neuron. *The Journal of Neuroscience*, **31**, 4944–4954.

Shimizu, T., Kagawa, T., Wada, T. *et al.* (2005) Wnt signaling controls the timing of oligodendrocyte development in the spinal cord. *Developmental Biology*, **282**, 397–410.

Takeda, K., Yasumoto, K., Takada, R. *et al.* (2000) Induction of melanocyte-specific microphthalmia-associated transcription factor by Wnt-3a. *The Journal of Biological Chemistry*, **275**, 14013–14016.

Tang, M., Villaescusa, J.C., Luo, S.X. *et al.* (2010) Interactions of Wnt/beta-catenin signaling and sonic hedgehog regulate the neurogenesis of ventral midbrain dopamine neurons. *The Journal of Neuroscience*, **30**, 9280–9291.

Tawk, M., Makoukji, J., Belle, M. *et al.* (2011) Wnt/beta-catenin signaling is an essential and direct driver of myelin gene expression and myelinogenesis. *The Journal of Neuroscience*, **31**, 3729–3742.

Theil, T., Aydin, S., Koch, S. *et al.* (2002) Wnt and Bmp signalling cooperatively regulate graded Emx2 expression in the dorsal telencephalon. *Development*, **129**, 3045–3054.

Van Raay, T.J., Moore, K.B., Iordanova, I. *et al.* (2005) Frizzled 5 signaling governs the neural potential of progenitors in the developing Xenopus retina. *Neuron*, **46**, 23–36.

Vivancos, V., Chen, P., Spassky, N. *et al.* (2009) Wnt activity guides facial branchiomotor neuron migration, and involves the PCP pathway and JNK and ROCK kinases. *Neural Development*, **4**, 7.

Wang, X., Imura, T., Sofroniew, M.V., and Fushiki, S. (2011) Loss of adenomatous polyposis coli in Bergmann glia disrupts their unique architecture and leads to cell nonautonomous neurodegeneration of cerebellar Purkinje neurons. *Glia*, **59**, 857–868.

Wang, X., Kopinke, D., Lin, J. *et al.* (2012) Wnt signaling regulates postembryonic hypothalamic progenitor differentiation. *Developmental Cell*, **23**, 624–636.

Wayman, G.A., Impey, S., Marks, D. *et al.* (2006) Activity-dependent dendritic arborization mediated by CaM-kinase I activation and enhanced CREB-dependent transcription of Wnt-2. *Neuron*, **50**, 897–909.

Whangbo, J. and Kenyon, C. (1999) A Wnt signaling system that specifies two patterns of cell migration in C. elegans. *Molecular Cell*, **4**, 851–858.

Woodhead, G.J., Mutch, C.A., Olson, E.C., and Chenn, A. (2006) Cell-autonomous beta-catenin signaling regulates cortical precursor proliferation. *The Journal of Neuroscience*, **26**, 12620–12630.

Yoshikawa, S., McKinnon, R.D., Kokel, M., and Thomas, J.B. (2003) Wnt-mediated axon guidance via the Drosophila Derailed receptor. *Nature*, **422**, 583–588.

Zhou, C.J., Zhao, C., and Pleasure, S.J. (2004) Wnt signaling mutants have decreased dentate granule cell production and radial glial scaffolding abnormalities. *The Journal of Neuroscience*, **24**, 121–126.

Zhou, C.J., Borello, U., Rubenstein, J.L., and Pleasure, S.J. (2006) Neuronal production and precursor proliferation defects in the neocortex of mice with loss of function in the canonical Wnt signaling pathway. *Neuroscience*, **142**, 1119–1131.

Zhuang, B. and Sockanathan, S. (2006) Dorsal-ventral patterning: a view from the top. *Current Opinion in Neurobiology*, **16**, 20–24.

22 Wnt Signaling in Heart Organogenesis

Stefan Hoppler[1], Silvia Mazzotta[1], and Michael Kühl[2]

[1]Institute of Medical Sciences, University of Aberdeen, Aberdeen, Scotland
[2]Institute of Biochemistry and Molecular Biology, Ulm University, Ulm, Germany

Introduction

A relatively simple treatment with canonical Wnt agonists followed by experimental inhibition of Wnt signaling can coax human embryonic stem cells (ESCs) to differentiate into beating heart muscle cells or cardiomyocytes (Lian *et al.*, 2012a, b). This remarkable finding supports a wealth of other evidence from many research fields that regulation by Wnt signaling is absolutely paramount for heart development.

Understanding heart development is of fundamental importance, because the heart organ is the first functional organ in mammalian embryos and heart function is unremittingly required throughout subsequent life. There is also direct practical relevance, since heart defects are a major health issue, with injury particularly to the adult heart muscle showing limited regeneration potential.

Canonical Wnt/β-catenin signaling has multiphasic roles in heart muscle differentiation

The heart muscle is central to the functioning heart. In vertebrate embryos, the heart muscle tissue derives from the anterior mesoderm in close association with the underlying endoderm. A prominent role for Wnt/β-catenin signaling involves restricting heart muscle differentiation, which was inferred from experiments over a decade ago. Experimental overexpression of known inhibitors of Wnt/β-catenin signaling (such as Dkk1) promoted heart muscle differentiation and experimental overexpression of Wnt ligands known to activate Wnt/β-catenin signaling repressed heart muscle differentiation in chick and Xenopus embryos (Marvin *et al.*, 2001; Schneider and Mercola, 2001; Tzahor and Lassar, 2001). Experiments using cultured stem cells confirmed that activated Wnt/β-catenin signaling is able to reduce subsequent cardiomyocyte differentiation (Naito *et al.*, 2006; Qyang *et al.*, 2007; Ueno *et al.*, 2007) and inhibited Wnt/β-catenin signaling to promote it (Lanier *et al.*, 2012; Paige *et al.*, 2012; Singh *et al.*, 2007).

However, these cultured stem cell experiments also highlighted positive effects of canonical Wnt signaling on subsequent cardiomyocyte differentiation (Naito *et al.*, 2006; Nakamura *et al.*, 2003; Ueno *et al.*, 2007). An early function for Wnt/β-catenin signaling relates to its requirement for mesoderm induction from which cardiac tissue subsequently differentiates

Wnt Signaling in Development and Disease: Molecular Mechanisms and Biological Functions, First Edition.
Edited by Stefan Hoppler and Randall T. Moon.

(Lindsley *et al.*, 2006). Additionally, there is an important role for expanding cardiac precursor populations prior to differentiation (Kwon *et al.*, 2007).

These findings and others therefore suggest a generic model where Wnt/β-catenin signaling is initially required for mesoderm induction and also for subsequent expansion of progenitor populations, while restricting specification of cardiac precursors, and subsequent differentiation into functional cardiomyocytes (Figure 22.1). Such a model would obviously be consistent with the simple human ESC differentiation protocol described in the preceding text (Lian *et al.*, 2012a, b).

There are apparent differences in different experimental systems. Particularly the early positive effect appears more evident in stem cell experiments but is often less prominent in experiments with embryonic model systems. Wnt/β-catenin signaling is clearly required for mesoderm induction (see Chapter 19), but experimentally elevated signaling often does not further enhance subsequent cardiac specification (e.g., Samuel and Latinkic, 2009). A possible explanation for this observation might be the fact that sufficient suitable mesoderm for subsequent cardiac development is usually already available in embryos. On the other hand, the restriction of cell fate specification of cardiogenic progenitors and of cardiomyocyte differentiation has been studied in more detail in embryonic model systems (Afouda *et al.*, 2008; Gibb, Lavery, and Hoppler, 2013; Martin, Afouda, and Hoppler, 2010; Pandur *et al.*, 2002).

Canonical Wnt/β-catenin signaling regulates first and second heart field development

There also appear to be differences in the regulation between different cell lineages that can normally give rise to functional cardiomyocytes in the embryo: the so-called first heart field (FHF) and the second heart field (SHF) (reviewed by Kelly, 2012). These two cardiac progenitor fields emerge close to each other in the embryo, but the main difference is that cardiomyocytes first differentiate from the FHF and only later from the SHF (e.g., Meilhac *et al.*, 2004). The requirement for Wnt/β-catenin signaling for expansion of progenitor populations appears therefore more evident in the SHF lineage and its derivatives (Klaus *et al.*, 2007), which include the right ventricle and outflow and inflow tracts, and less so in the FHF lineage (Kwon *et al.*, 2007), which forms the muscle tissue of the linear heart tube

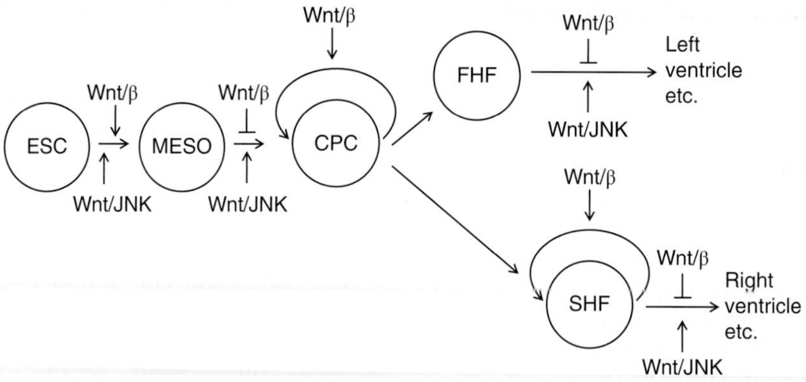

Figure 22.1 Summary of Wnt signaling function in cardiomyocyte differentiation. Canonical Wnt/β-catenin signaling (Wnt/β) and noncanonical Wnt/JNK signaling (WNT/JNK) regulate heart muscle or cardiomyocyte differentiation from pluripotent ESCs in vertebrate embryos and in cultured stem cell experiments. Both Wnt signaling pathways promote mesoderm induction (MESO) from pluripotent ESCs, WNT/JNK slightly before Wnt/β signaling, but they oppose each other in the regulation of common cardiovascular progenitor (CPC) development; while Wnt/β signaling champions the hematopoietic lineage at the expense of CPC development, Wnt/JNK signaling instead promotes CPC differentiation at the expense of hematopoietic development. Wnt/β signaling later promotes proliferation of CPCs and of SHF progenitors, which derive from CPCs, and restricts differentiation of functional cardiomyocyte from either the FHF or the SHF, while Wnt/JNK signaling promotes cardiomyocyte differentiation.

that subsequently gives rise primarily to the left ventricle. However, a common cardiac precursor population earlier in development has been proposed that gives rise not only to the FHF and the SHF but also other cardiovascular cell types (e.g., Moretti *et al.*, 2006), which may also be regulated by Wnt/β-catenin signaling (Kwon *et al.*, 2007, 2009; Qyang *et al.*, 2007) (reviewed by Gessert and Kühl, 2010). The exact role of Wnt/β-catenin signaling in this early population of cardiac progenitors awaits further characterization.

Wnt signals activating canonical Wnt/β-catenin signaling in heart development

Despite these apparent differences between experimental systems and different progenitor lineages, a consensus about the role of the Wnt/β-catenin pathway in heart muscle development is slowly emerging along the model proposed earlier. However, it is currently far less clear which Wnt signaling components are deployed to activate and restrain the Wnt/β-catenin pathway in heart development. Wnt3a is the favorite candidate for promoting mesoderm induction via Wnt/β-catenin signaling, particularly in mammalian development (see Chapter 19). While Wnt signals expressed in dorsal tissue are sufficient for inhibiting cardiac specification in experiments (Tzahor and Lassar, 2001), a requirement for normal heart development could not yet be formally demonstrated, possibly due to functional redundancy. Xenopus Wnt6 is not only capable of inhibiting heart development in experiments, but endogenous Wnt6 function is also required during Xenopus heart organogenesis to prevent development of an excessively large heart (Lavery *et al.*, 2008). Wnt6 function in regulating cardiomyocyte differentiation also appears to be conserved in adult mammalian cardiac progenitor cells (Verma *et al.*, 2012). In mouse embryos, canonical Wnt2 signaling is required for proliferation in the posterior part of the SHF (Tian *et al.*, 2010), which gives rise to the atria and portions of the atrioventricular canal. In ES cell experiments, however, Wnt2 is also required for efficient cardiomyocyte differentiation (Wang *et al.*, 2006) but surprisingly promotes cardiomyocyte differentiation via a noncanonical Wnt pathway (Onizuka *et al.*, 2012). A similar role may be carried out by the closely related molecule Wnt2b (reviewed by Gessert and Kühl, 2010). These findings require the notion that individual Wnt ligands can activate different Wnt signaling pathways in different cellular contexts. The role of Wnt2 and Wnt2b in noncanonical Wnt signaling is discussed in the succeeding text.

Extracellular Wnt inhibitors regulate canonical Wnt/β-catenin signaling in heart development

Extracellular Wnt inhibitors such as Dkk1 were instrumental for initially demonstrating a requirement for Wnt/β-catenin signaling in heart development (see the preceding text). Endogenous dickkopf genes are indeed expressed in the embryo close to the developing heart (Monaghan *et al.*, 1999), but a Dkk1 knockout showed no heart defects (Mukhopadhyay *et al.*, 2001) and even double Dkk1 and Dkk2 knockout mice embryos only have late heart defects (Phillips *et al.*, 2011). There is also still discussion about whether Dkk1 expression is induced in the early cardiovascular lineage (Bondue *et al.*, 2008; David *et al.*, 2008; Lindsley *et al.*, 2008) and whether it regulates other signaling pathways in heart development (Korol, Gupta, and Mercola, 2008). Thus, while a role for endogenous Dkk1 in regulating Wnt/β-catenin signaling during early heart development cannot be ruled out, a requirement remains to be demonstrated.

Expression studies in embryos and stem cell experiments highlighted a possible role for another molecular family of extracellular Wnt signaling inhibitors. Increased expression of sFRP genes is associated with differentiation into a cardiac lineage, and experimental addition of sFRP1 together with other factors promoted cardiomyocyte differentiation (Chen *et al.*, 2008). Endogenous Xenopus sFRP1 is expressed in cardiac mesoderm (Xu, D'Amore, and Sokol, 1998), and experimental overexpression can increase the size of the developing heart muscle, and importantly endogenous sFRP1 is also required for regulating the size of the developing heart muscle (Gibb, Lavery, and Hoppler, 2013) by inhibiting Wnt6 signaling, which otherwise functions to restrict heart muscle development during organogenesis (Lavery *et al.*, 2008). Similar roles for sFRP genes have recently been suggested in healing and tissue homeostasis following myocardial infarction (e.g., Alfaro *et al.*, 2008; Barandon *et al.*, 2003) (reviewed by Bergmann, 2010).

Gene targets regulated canonical Wnt/β-catenin signaling in heart development

Because of the multiphasic role of Wnt/β-catenin, the genes regulated by Wnt/β-catenin signaling in heart development are likely to be different depending on stage and cell lineage. Brachyury is regulated by Wnt signaling in mesoderm induction (see Chapter 19). Consistent with the inhibitory role on cardiac development, Wnt/β-catenin signaling in Xenopus functions to reduce sustained gata6 expression (Afouda *et al.*, 2008; Martin, Afouda, and Hoppler, 2010), yet gata6 is a positive Wnt target in the mouse (Tian *et al.*, 2010). Islet1 expression is closely linked to expanding cardiogenic precursors (for review, see Pandur *et al.*, 2013), particularly in the SHF (Kwon *et al.*, 2009), and is a direct target of Wnt/β-catenin signaling (Lin *et al.*, 2007; Qyang *et al.*, 2007). Clearly, a much more comprehensive understanding is required of how Wnt/β-catenin signaling controls the gene regulatory networks leading to heart development and cardiomyocyte differentiation.

However, because there is Wnt feedback regulation, a false impression of opposing phases of Wnt signaling function could appear. Wnt/β-catenin signaling induces gene targets such as Axin2 (e.g., Lustig *et al.*, 2002) and Dkk1 (e.g., Chamorro *et al.*, 2005) that regulate Wnt signaling in feedback regulatory loops. Experimental activation (or inhibition) of Wnt signaling at an earlier developmental stage may therefore influence expression of endogenous Wnt signaling components at a later stage. Due to the inherent time delay of these processes, it may be altered Wnt signaling function at the later stage that primarily regulates subsequent heart development, rather than the initial experimental manipulation.

Controversy about importance of canonical Wnt/β-catenin signaling in the cardiogenic endoderm

The cardiomyocytes of the heart normally develop in an intact embryo next to other embryonic tissues. These tissues interact in order to integrate organ development in the embryo. The embryonic endoderm in particular is known to promote heart development, not only in the embryo (e.g., Sater and Jacobson, 1990) but also if cocultured with ESCs (Brown *et al.*, 2010; Mummery *et al.*, 2003; Uosaki *et al.*, 2012). Wnt signaling can influence cardiac mesoderm development indirectly through regulation of adjacent, cardiogenesis-promoting endodermal tissue, which expresses the transcription factors Hex and Sox17 (Foley and Mercola, 2005; Liu *et al.*, 2007). These findings could suggest that the role of Wnt/β-catenin signaling in restricting specification of cardiac precursors is indirect via inhibiting cardiogenesis-promoting endodermal tissue. Such a mechanism would be consistent with the phenotype of an endoderm-specific mouse β-catenin knockout, which shows excessive and probably precocious cardiac tissue differentiation (Lickert, 2005). On the other hand, cardiac induction by anterior endoderm in Xenopus experiments was shown to be independent of Wnt/β-catenin signaling (Samuel and Latinkic, 2009). There may well be parallel and redundant pathways in the endoderm and the mesoderm, which develop from the same precursors adjacent to each other in the embryo. The absence of early heart defects in Sox17 knockout mice who lack definitive gut endoderm (Kanai-Azuma *et al.*, 2002) but an absolute requirement for cardiogenesis in some ES cell experiments (Liu *et al.*, 2007) suggests there may also be differences between experimental systems or stages of development.

Noncanonical Wnt/JNK signaling supports cardiac specification, morphogenesis, and terminal differentiation

Noncanonical Wnt signaling is clearly a strong stimulator of cardiac specification and subsequent cardiomyocyte differentiation (e.g., Eisenberg and Eisenberg, 1999; He *et al.*, 2011; Pandur *et al.*, 2002; Terami *et al.*, 2004). It has been shown that this effect is mediated by the Wnt/JNK signaling pathway (Pandur *et al.*, 2002). While these positive effects of experimental Wnt/JNK signaling were evident, there has previously been some controversy about whether endogenous noncanonical Wnt signaling was actually required for normal cardiogenesis. In Xenopus, a clear requirement for Wnt11b/JNK signaling had been demonstrated (Afouda and Hoppler, 2011; Afouda *et al.*, 2008; Pandur *et al.*, 2002), but mouse knockout of the mammalian Wnt11 gene initially revealed no early heart phenotype (Majumdar, 2003). Stem cell experiments on the other hand suggested that Wnt5a may function in cardiomyocyte differentiation via noncanonical Wnt signaling (Chen *et al.*, 2008). It has recently been revealed that there is some redundancy in mammals between Wnt5a and Wnt11 since Wnt5a/Wnt11 double knockout embryos do have clear heart defects (Cohen *et al.*, 2012). Furthermore, Wnt2 – and probably Wnt2b – functions in heart development via activation of noncanonical signaling (Onizuka *et al.*, 2012); and Wnt2/2b and Wnt11 are coexpressed in the developing cardiac tissue of mouse embryos (Kispert *et al.*, 1996; Monkley *et al.*, 1996) but not in *Xenopus* (Ku and Melton, 1993; Landesman and Sokol, 1997). This redundancy in mammals makes studying the requirement for Wnt ligands that activate noncanonical Wnt signaling difficult. On the other hand, Xenopus embryos provide an opportunity, since they have two Wnt11 genes, to dissect several functions of noncanonical Wnt11 signaling in vertebrate heart development. One Xenopus wnt11 gene (*wnt11b*) is involved in specification of cardiac progenitor cells (Afouda *et al.*, 2008; Pandur *et al.*, 2002), whereas another wnt11 gene (wnt11a or XWnt11-R) regulates cell adhesion and morphogenesis (Garriock *et al.*, 2005), probably via transcriptional regulation of the cell adhesion protein Alcam/DM-GRASP/CD166 (Choudhry and Trede, 2013; Gessert *et al.*, 2008).

Controversy about role for noncanonical Wnt/JNK signaling beyond inhibiting canonical Wnt/β-catenin signaling in heart development

At least some of the positive roles of noncanonical Wnt signaling on cardiac specification and subsequent cardiomyocyte differentiation coincide with when canonical Wnt signaling restricts heart development (Figure 22.1). Interaction between different Wnt signaling pathways is known in other tissues (e.g., Maye *et al.*, 2004; see Chapter 6), which leads to the suggestion that at least part of the function of noncanonical Wnt signaling in heart development is via inhibition of canonical Wnt signaling and that therefore the heart promoting effects of noncanonical Wnt signaling simply reflect interference with the function of canonical Wnt/β-catenin signaling in restricting cardiac specification and subsequent cardiomyocyte differentiation. Inhibition by Wnt/JNK signaling of components of the canonical Wnt signaling pathway has been demonstrated in heart development (Abdul-Ghani *et al.*, 2011) and requires to be investigated in more detail. However, the roles of noncanonical Wnt signaling in heart morphogenesis may go beyond inhibition of Wnt/β-catenin signaling (Garriock *et al.*, 2005; Nagy *et al.*, 2010). Furthermore, Wnt/β-catenin and Wnt/JNK pathways are both promoting early stages of cardiomyocyte differentiation in human ESCs (Rai *et al.*, 2012), although in mutually exclusive waves. On the other hand, Wnt/β-catenin signaling also indirectly regulates Wnt/JNK signaling in Xenopus heart development (Afouda and Hoppler, 2011; Afouda *et al.*, 2008). With respect to these complex findings, it may well be important to consider Wnt signaling as a network consisting of interwoven rather than isolated signaling entities (Kestler and Kühl, 2008; Chapter 11).

Wnt signaling in heart muscle regeneration

The Wnt signaling mechanisms in heart development described earlier will provide insights into human heart development and possibly the etiology of congenital birth defects. However, translational medicine more recently has focused on the possible lessons to be learned from Wnt signaling mechanisms in embryonic heart development for regeneration of injured heart muscle, for instance, after myocardial infarction (reviewed by Bergmann, 2010). Wnt ligands and extracellular Wnt inhibitors of the sFRP family were found to be re-expressed in damaged heart muscle tissue in experimental animal models of myocardial infarction

(Barandon *et al.*, 2003), suggesting that similar Wnt signaling mechanisms as the ones discovered during embryonic development and ES cell differentiation are recapitulated. Mouse myocardial infarction models with artificial sFRP1 (Barandon *et al.*, 2003) or sFRP2 (Alfaro *et al.*, 2008) overexpression or with compromised levels of β-catenin (Zelarayán *et al.*, 2008) indeed demonstrated improved recovery of cardiomyocyte function. Overexpression of Wnt11 (He *et al.*, 2011) leads to similar improvement. However, regeneration of damaged heart muscle from existing myocardial tissue remains slow and difficult.

There may be other sources of potential cardiomyocytes for repairing damaged adult heart muscle. Particularly *in vitro* ESC experiments and cell lineage tracing in developing embryos provided a detailed understanding of the relationship among different precursor populations. Cardiomyocytes derive from earlier cardiovascular precursor population that share some of the early gene expression and regulatory mechanisms with definitive cardiogenic lineages (e.g., Moretti *et al.*, 2006). Also in the early embryo, there appears to be more of a continuum between myocardium, epicardium, pericardium, and mesocardium (Kruithof *et al.*, 2006; Raffin *et al.*, 2000). Transdifferentiation of some of those cell populations may provide an avenue for recruiting sufficient new cardiomyocytes for effective regeneration of damaged heart muscle. In response to injury, cells from epicardium (Smart *et al.*, 2011) have recently been shown to transdifferentiate into functional cardiomyocytes, even though the epicardium does not feature among sources for cardiomyocytes in the embryo. The role of Wnt signaling mechanisms in this process remains to be investigated. However, Wnt11 overexpression has been shown to enhance transdifferentiation of endothelial cells (Koyanagi *et al.*, 2005) and bone marrow-derived mesenchymal stem cells (He *et al.*, 2011) into cardiomyocytes.

Just coaxing cells from whatever source to differentiate as cardiomyocytes is not yet the complete solution to all of the problems associated with a damaged heart muscle. For a functioning heart, cardiomyocytes need to be integrated into an appropriately shaped heart muscle with the correct anatomical and functional connections to other tissues to ensure, for instance, adequate blood supply and electrical coupling (Mummery, Davis, and Krieger, 2010). However, Wnt signaling is also instrumental for those aspects of heart development (reviewed by Bergmann, 2010; Gessert and Kühl, 2010).

Conclusions

Wnt signaling pathways, both canonical and noncanonical, are clearly central regulators of vertebrate cardiogenesis. However, many questions remain about cross regulation between Wnt signaling branches and how their function is integrated with other signaling pathways (Noseda *et al.*, 2011) and transcriptional regulatory mechanisms (Herrmann *et al.*, 2012) that are necessary for normal heart development. Such a more comprehensive understanding of the regulatory mechanisms of embryonic heart development will also provide valuable insight into novel therapeutic strategies to promote healing of damaged adult heart muscle and reinstate efficient heart function.

References

Abdul-Ghani, M., Dufort, D., Stiles, R. *et al.* (2011) Wnt11 promotes cardiomyocyte development by caspase-mediated suppression of canonical Wnt signals. *Molecular and Cellular Biology*, **31**, 163–178.

Afouda, B.A. and Hoppler, S. (2011) Different requirements for GATA factors in cardiogenesis are mediated by non-canonical Wnt signaling. *Development Dynamics*, **240**, 649–662.

Afouda, B.A., Martin, J., Liu, F. *et al.* (2008) GATA transcription factors integrate Wnt signalling during heart development. *Development*, **135**, 3185–3190.

Alfaro, M.P., Pagni, M., Vincent, A. *et al.* (2008) The Wnt modulator sFRP2 enhances mesenchymal stem cell engraftment, granulation tissue formation and myocardial repair. *Proceedings of the National Academy of Sciences of the United States of America*, **105**, 18366–18371.

Barandon, L., Couffinhal, T., Ezan, J. *et al.* (2003) Reduction of infarct size and prevention of cardiac rupture in transgenic mice overexpressing FrzA. *Circulation*, **108**, 2282–2289.

Bergmann, M.W. (2010) WNT signaling in adult cardiac hypertrophy and remodeling: lessons learned from cardiac development. *Circulation Research*, **107**, 1198–1208.

Bondue, A., Lapouge, G., Paulissen, C. *et al.* (2008) Mesp1 acts as a master regulator of multipotent cardiovascular progenitor specification. *Cell Stem Cell*, **3**, 69–84.

Brown, K., Doss, M.X., Legros, S. *et al.* (2010) eXtra-embryonic ENdoderm (XEN) stem cells produce factors that activate heart formation. *PLoS One*, **5**, e13446.

Chamorro, M.N., Schwartz, D.R., Vonica, A. *et al.* (2005) FGF-20 and DKK1 are transcriptional targets of beta-catenin and FGF-20 is implicated in cancer and development. *The EMBO Journal*, **24**, 73–84.

Chen, V.C., Stull, R., Joo, D. *et al.* (2008) Notch signaling respecifies the hemangioblast to a cardiac fate. *Nature Biotechnology*, **26**, 1169–1178.

Choudhry, P. and Trede, N.S. (2013) DiGeorge syndrome gene tbx1 functions through wnt11r to regulate heart looping and differentiation. *PLoS One*, **8**, e58145.

Cohen, E.D., Miller, M.F., Wang, Z. *et al.* (2012) Wnt5a and Wnt11 are essential for second heart field progenitor development. *Development*, **139**, 1931–1940.

David, R., Brenner, C., Stieber, J. *et al.* (2008) MesP1 drives vertebrate cardiovascular differentiation through Dkk-1-mediated blockade of Wnt-signalling. *Nature Cell Biology*, **10**, 338–345.

Eisenberg, C.A. and Eisenberg, L.M. (1999) WNT11 promotes cardiac tissue formation of early mesoderm. *Developmental Dynamics*, **216**, 45–58.

Foley, A.C. and Mercola, M. (2005) Heart induction by Wnt antagonists depends on the homeodomain transcription factor Hex. *Genes & Development*, **19**, 387–396.

Garriock, R.J., D'Agostino, S.L., Pilcher, K.C., and Krieg, P.A. (2005) Wnt11-R, a protein closely related to mammalian Wnt11, is required for heart morphogenesis in Xenopus. *Developmental Biology*, **279**, 179–192.

Gessert, S. and Kühl, M. (2010) The multiple phases and faces of wnt signaling during cardiac differentiation and development. *Circulation Research*, **107**, 186–199.

Gessert, S., Maurus, D., Brade, T. *et al.* (2008) DM-GRASP/ALCAM/CD166 is required for cardiac morphogenesis and maintenance of cardiac identity in first heart field derived cells. *Developmental Biology*, **321**, 150–161.

Gibb, N., Lavery, D.L., and Hoppler, S. (2013) sfrp1 promotes cardiomyocyte differentiation in Xenopus via negative-feedback regulation of Wnt signalling. *Development*, **140**, 1537–1549.

He, Z., Li, H., Zuo, S. *et al.* (2011) Transduction of Wnt11 promotes mesenchymal stem cell transdifferentiation into cardiac phenotypes. *Stem Cells and Development*, **20**, 1771–1778.

Herrmann, F., Groß, A., Zhou, D. *et al.* (2012) A boolean model of the cardiac gene regulatory network determining first and second heart field identity. *PLoS One*, **7**, e46798.

Kanai-Azuma, M., Kanai, Y., Gad, J.M. *et al.* (2002) Depletion of definitive gut endoderm in Sox17-null mutant mice. *Development*, **129**, 2367–2379.

Kelly, R.G. (2012) The second heart field. *Current Topics in Developmental Biology*, **100**, 33–65.

Kestler, H.A. and Kühl, M. (2008) From individual Wnt pathways towards a Wnt signalling network. *Philosophical Transaction of the Royal Society London. Series B, Biological Sciences*, **363**, 1333–1347.

Kispert, A., Vainio, S., Shen, L. *et al.* (1996) Proteoglycans are required for maintenance of Wnt-11 expression in the ureter tips. *Development*, **122**, 3627–3637.

Klaus, A., Saga, Y., Taketo, M.M. *et al.* (2007) Distinct roles of Wnt/beta-catenin and Bmp signaling during early cardiogenesis. *Proceedings of the National Academy of Sciences of the United States of America*, **104**, 18531–18536.

Korol, O., Gupta, R.W., and Mercola, M. (2008) A novel activity of the Dickkopf-1 amino terminal domain promotes axial and heart development independently of canonical Wnt inhibition. *Developmental Biology*, **324**, 131–138.

Koyanagi, M., Haendeler, J., Badorff, C. *et al.* (2005) Non-canonical Wnt signaling enhances differentiation of human circulating progenitor cells to cardiomyogenic cells. *The Journal of Biological Chemistry*, **280**, 16838–16842.

Kruithof, B.P.T., van Wijk, B., Somi, S. *et al.* (2006) BMP and FGF regulate the differentiation of multipotential pericardial mesoderm into the myocardial or epicardial lineage. *Developmental Biology*, **295**, 507–522.

Ku, M. and Melton, D.A. (1993) Xwnt-11: a maternally expressed Xenopus wnt gene. *Development*, **119**, 1161–1173.

Kwon, C., Arnold, J., Hsiao, E.C. *et al.* (2007) Canonical Wnt signaling is a positive regulator of mammalian cardiac progenitors. *Proceedings of the National Academy of Sciences of the United States of America*, **104**, 10894–10899.

Kwon, C., Qian, L., Cheng, P. *et al.* (2009) A regulatory pathway involving Notch1/β-catenin/Isl1 determines cardiac progenitor cell fate. *Nature Cell Biology*, **11**, 951–957.

Landesman, Y. and Sokol, S.Y. (1997) Xwnt-2b is a novel axis-inducing Xenopus Wnt, which is expressed in embryonic brain. *Mechanisms of Development*, **63**, 199–209.

Lanier, M., Schade, D., Willems, E. *et al.* (2012) Wnt inhibition correlates with human embryonic stem cell cardiomyogenesis: a structure-activity relationship study based on inhibitors for the Wnt response. *Journal of Medical Chemistry*, **55**, 697–708.

Lavery, D.L., Martin, J., Turnbull, Y.D., and Hoppler, S. (2008) Wnt6 signaling regulates heart muscle development during organogenesis. *Developmental Biology*, **323**, 177–188.

Lian, X., Hsiao, C., Wilson, G. *et al.* (2012a) Robust cardiomyocyte differentiation from human pluripotent stem cells via temporal modulation of canonical Wnt signaling. *Proceedings of the National Academy of Sciences of the United States of America*, **109**, E1848–E1857.

Lian, X., Zhang, J., Azarin, S.M. *et al.* (2012b) Directed cardiomyocyte differentiation from human pluripotent stem cells by modulating Wnt/β-catenin signaling under fully defined conditions. *Nature Protocol*, **8**, 162–175.

Lickert, H. (2005) Dissecting Wnt/-catenin signaling during gastrulation using RNA interference in mouse embryos. *Development*, **132**, 2599–2609.

Lin, L., Cui, L., Zhou, W. *et al.* (2007) Beta-catenin directly regulates Islet1 expression in cardiovascular progenitors and is required for multiple aspects of cardiogenesis. *Proceedings of the National Academy of Sciences of the United States of America*, **104**, 9313–9318.

Lindsley, R.C., Gill, J.G., Kyba, M. *et al.* (2006) Canonical Wnt signaling is required for development of embryonic stem cell-derived mesoderm. *Development*, **133**, 3787–3796.

Lindsley, R.C., Gill, J.G., Murphy, T.L. *et al.* (2008) Mesp1 coordinately regulates cardiovascular fate restriction and epithelial-mesenchymal transition in differentiating ESCs. *Cell Stem Cell*, **3**, 55–68.

Liu, Y., Asakura, M., Inoue, H. *et al.* (2007) Sox17 is essential for the specification of cardiac mesoderm in embryonic stem cells. *Proceedings of the National Academy of Sciences of the United States of America*, **104**, 3859–3864.

Lustig, B., Jerchow, B., Sachs, M. *et al.* (2002) Negative feedback loop of Wnt signaling through upregulation of conductin/axin2 in colorectal and liver tumors. *Molecular and Cellular Biology*, **22**, 1184–1193.

Majumdar, A. (2003) Wnt11 and Ret/Gdnf pathways cooperate in regulating ureteric branching during metanephric kidney development. *Development*, **130**, 3175–3185.

Martin, J., Afouda, B.A., and Hoppler, S. (2010) Wnt/beta-catenin signalling regulates cardiomyogenesis via GATA transcription factors. *Journal of Anatomy*, **216**, 92–107.

Marvin, M.J., Di Rocco, G., Gardiner, A. *et al.* (2001) Inhibition of Wnt activity induces heart formation from posterior mesoderm. *Genes & Development*, **15**, 316–327.

Maye, P., Zheng, J., Li, L., and Wu, D. (2004) Multiple mechanisms for Wnt11-mediated repression of the canonical Wnt signaling pathway. *The Journal of Biological Chemistry*, **279**, 24659–24665.

Meilhac, S.M., Esner, M., Kelly, R.G. *et al.* (2004) The clonal origin of myocardial cells in different regions of the embryonic mouse heart. *Developmental Cell*, **6**, 685–698.

Monaghan, A.P., Kioschis, P., Wu, W. *et al.* (1999) Dickkopf genes are co-ordinately expressed in mesodermal lineages. *Mechanisms of Development*, **87**, 45–56.

Monkley, S.J., Delaney, S.J., Pennisi, D.J. *et al.* (1996) Targeted disruption of the Wnt2 gene results in placentation defects. *Development*, **122**, 3343–3353.

Moretti, A., Caron, L., Nakano, A. *et al.* (2006) Multipotent embryonic isl1+ progenitor cells lead to cardiac, smooth muscle, and endothelial cell diversification. *Cell*, **127**, 1151–1165.

Mukhopadhyay, M., Shtrom, S., Rodríguez Esteban, C. *et al.* (2001) Dickkopf1 is required for embryonic head induction and limb morphogenesis in the mouse. *Developmental Cell*, **1**, 423–434.

Mummery, C.L., Davis, R.P., and Krieger, J.E. (2010) Challenges in using stem cells for cardiac repair. *Science Translational Medicine*, **2**, 27ps17.

Mummery, C., Ward-van Oostwaard, D., Doevendans, P. *et al.* (2003) Differentiation of human embryonic stem cells to cardiomyocytes: role of coculture with visceral endoderm-like cells. *Circulation*, **107**, 2733–2740.

Nagy, I.I., Railo, A., Rapila, R. *et al.* (2010) Wnt-11 signalling controls ventricular myocardium development by patterning N-cadherin and beta-catenin expression. *Cardiovascular Research*, **85**, 100–109.

Naito, A.T., Shiojima, I., Akazawa, H. *et al.* (2006) Developmental stage-specific biphasic roles of Wnt/beta-catenin signaling in cardiomyogenesis and hematopoiesis. *Proceedings of the National Academy of Sciences of the United States of America*, **103**, 19812–19817.

Nakamura, T., Sano, M., Songyang, Z., and Schneider, M.D. (2003) A Wnt- and beta-catenin-dependent pathway for mammalian cardiac myogenesis. *Proceedings of the National Academy of Sciences of the United States of America*, **100**, 5834–5839.

Noseda, M., Peterkin, T., Simões, F.C. *et al.* (2011) Cardiopoietic factors: extracellular signals for cardiac lineage commitment. *Circulation Research*, **108**, 129–152.

Onizuka, T., Yuasa, S., Kusumoto, D. *et al.* (2012) Wnt2 accelerates cardiac myocyte differentiation from ES-cell derived mesodermal cells via non-canonical pathway. *Journal of Molecular and Cellular Cardiology*, **52**, 650–659.

Paige, S.L., Thomas, S., Stoick-Cooper, C.L. *et al.* (2012) A temporal chromatin signature in human embryonic stem cells identifies regulators of cardiac development. *Cell*, **151**, 221–232.

Pandur, P., Läsche, M., Eisenberg, L.M., and Kühl, M. (2002) Wnt-11 activation of a non-canonical Wnt signalling pathway is required for cardiogenesis. *Nature*, **418**, 636–641.

Pandur, P., Sirbu, I.O., Kühl, S.J. *et al.* (2013) Islet1-expressing cardiac progenitor cells: a comparison across species. *Development Genes & Evolution*, **223**, 117–129.

Phillips, M.D., Mukhopadhyay, M., Poscablo, C., and Westphal, H. (2011) Dkk1 and Dkk2 regulate epicardial specification during mouse heart development. *International Journal of Cardiology*, **150**, 186–192.

Qyang, Y., Martin-Puig, S., Chiravuri, M. *et al.* (2007) The renewal and differentiation of Isl1+ cardiovascular progenitors are controlled by a Wnt/beta-catenin pathway. *Cell Stem Cell*, **1**, 165–179.

Raffin, M., Leong, L.M., Rones, M.S. *et al.* (2000) Subdivision of the cardiac Nkx2.5 expression domain into myogenic and nonmyogenic compartments. *Developmental Biology*, **218**, 326–340.

Rai, M., Walthall, J.M., Hu, J., and Hatzopoulos, A.K. (2012) Continuous antagonism by Dkk1 counter activates canonical Wnt signaling and promotes cardiomyocyte differentiation of embryonic stem cells. *Stem Cells Development*, **21**, 54–66.

Samuel, L.J. and Latinkic, B.V. (2009) Early activation of FGF and nodal pathways mediates cardiac specification independently of Wnt/beta-catenin signaling. *PLoS One*, **4**, e7650.

Sater, A.K. and Jacobson, A.G. (1990) The restriction of the heart morphogenetic field in Xenopus laevis. *Developmental Biology*, **140**, 328–336.

Schneider, V.A. and Mercola, M. (2001) Wnt antagonism initiates cardiogenesis in Xenopus laevis. *Genes & Development*, **15**, 304–315.

Singh, A.M., Li, F.-Q., Hamazaki, T. *et al.* (2007) Chibby, an antagonist of the Wnt/beta-catenin pathway, facilitates cardiomyocyte differentiation of murine embryonic stem cells. *Circulation*, **115**, 617–626.

Smart, N., Bollini, S., Dubé, K.N. *et al.* (2011) De novo cardiomyocytes from within the activated adult heart after injury. *Nature*, **474**, 640–644.

Terami, H., Hidaka, K., Katsumata, T. *et al.* (2004) Wnt11 facilitates embryonic stem cell differentiation to Nkx2.5-positive cardiomyocytes. *Biochemical and Biophysical Research Communications*, **325**, 968–975.

Tian, Y., Yuan, L., Goss, A.M. *et al.* (2010) Characterization and in vivo pharmacological rescue of a Wnt2-Gata6 pathway required for cardiac inflow tract development. *Developmental Cell*, **18**, 275–287.

Tzahor, E. and Lassar, A.B. (2001) Wnt signals from the neural tube block ectopic cardiogenesis. *Genes & Development*, **15**, 255–260.

Ueno, S., Weidinger, G., Osugi, T. *et al.* (2007) Biphasic role for Wnt/beta-catenin signaling in cardiac specification in zebrafish and embryonic stem cells. *Proceedings of the National Academy of Sciences of the United States of America*, **104**, 9685–9690.

Uosaki, H., Andersen, P., Shenje, L.T. *et al.* (2012) Direct contact with endoderm-like cells efficiently induces cardiac progenitors from mouse and human pluripotent stem cells. *PLoS One*, **7**, e46413.

Verma, A.K., Schmeckpeper, J., Yin, L. *et al.* (2012) Abstract 18934: inhibition of Wnt6 by Sfrp2 regulates CPC differentiation by differential modulation of canonical and non canonical Wnt pathways. *Circulation*, **126**, A18934.

Wang, H., Gilner, J.B., Bautch, V.L. *et al.* (2006) Wnt2 coordinates the commitment of mesoderm to hematopoietic, endothelial, and cardiac lineages in embryoid bodies. *The Journal of Biological Chemistry*, **282**, 782–791.

Xu, Q., D'Amore, P.A., and Sokol, S.Y. (1998) Functional and biochemical interactions of Wnts with FrzA, a secreted Wnt antagonist. *Development*, **125**, 4767–4776.

Zelarayán, L.C., Noack, C., Sekkali, B. *et al.* (2008) β-Catenin downregulation attenuates ischemic cardiac remodeling through enhanced resident precursor cell differentiation. *Proceedings of the National Academy of Sciences of the United States of America*, **105**, 19762–19767.

23 Wnt Signaling in Kidney Organogenesis

Kimmo Halt and Seppo Vainio

The Centre of Excellence in Cell-Extracellular Matrix Research, Biocenter Oulu, Oulu, Finland
Laboratory of Developmental Biology, Department of Medical Biochemistry and Molecular Biology, Institute of Biomedicine, University of Oulu, Oulu, Finland

Introduction

In mammals, the definitive kidney originates from the metanephros, which forms a morphologically distinguishable structure around midgestation (Saxén, 1987). The early metanephros itself consists of the epithelial ureteric bud, an outgrowth from the adjacent Wolffian duct, and the metanephric mesenchyme, which can be divided into the cap mesenchyme directly in contact with the ureteric bud and the cortical interstitial stroma enclosing both the ureteric bud and the cap mesenchyme (Figure 23.1). The metanephric cells originate from the intermediate mesoderm, which also gives rise to the pronephros and mesonephros, which largely regress in mammals but contribute to permanent kidneys in lower vertebrates (Mugford *et al.*, 2008; Saxén, 1987).

The metanephros commences a morphogenic program leading to the mature kidney through an iterative process that includes branching of the ureteric bud and differentiation of nephrons from the cap mesenchyme within the interstitial stroma. The differentiation of the nephron from stem cells begins in the ventral cap mesenchyme adjacent to the tip of the ureteric bud (Figure 23.2). Interplay between the cap mesenchyme and the ureteric bud leads to induction of the ventral part of the cap mesenchyme and subsequent transition of the induced cells into a pretubular aggregate that will further undergo a mesenchyme-to-epithelium transformation to establish a renal vesicle connected to the collecting duct system at the tip of the ureteric bud (Bard *et al.*, 2001; Georgas *et al.*, 2009; Figure 23.3a). Subsequently, the renal vesicle matures into a functional nephron through comma- and s-shaped body stages (Figure 23.3b and c). Eventually, a fully differentiated nephron spanning the corticomedullary axis consists of morphologically distinguishable segments called a Bowman's capsule, a proximal convoluted tube, a Henle's loop, a distal convoluted tube, and a connecting tube (Figure 23.1 and Figure 23.3d). At the vascular pole, the nephron forms a connection with the glomerular tuft – composed of endothelial and mesangial cells – and with the collecting duct at the urinary pole.

Detailed characterization of mouse models depicting morphological and molecular changes in the cap mesenchyme and its derivatives has elucidated the transcriptional program leading

Wnt Signaling in Development and Disease: Molecular Mechanisms and Biological Functions, First Edition.
Edited by Stefan Hoppler and Randall T. Moon.

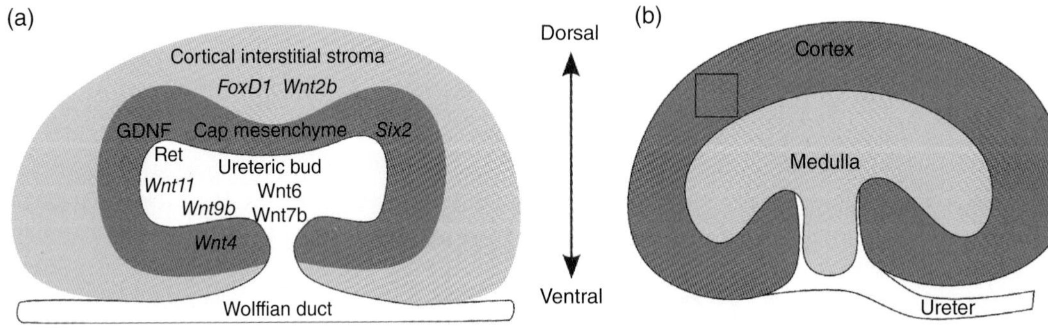

Figure 23.1 (a) Schematic view of a metanephric kidney. A ureteric bud (*white*) – an outgrowth from the Wolffian duct – invades the metanephric mesenchyme, consisting of the cap mesenchyme (*dark gray*) and the cortical interstitial mesenchyme (*light gray*). *Six2* and *FoxD1* transcription factors mark the cap mesenchyme and cortical interstitial mesenchyme, respectively. The primary nephrogenic inductor *Wnt9b* is expressed in the ureteric bud, excluding the tips. Wnt9b induces the expression of *Wnt4* in the ventral cap mesenchyme. *Wnt11* expressed in the tips of the ureteric bud forms a positive regulation loop with the GDNF/Ret signaling pathway to promote branching morphogenesis of the ureteric bud. The role of *Wnt2b* in the cortical interstitial mesenchyme is unknown. (b) The maturing kidney becomes radially organized into the cortex and the medulla. The Bowman's capsules (see Figure 23.3d) connecting with glomerular capillary loops reside in the cortex, where they filter plasma. The medullary region contains both Henle's loops and medullary collecting ducts (see Figure 23.3d), which participate in concentration of urine and thus osmoregulation of the body. The ureter passes urine from the kidney to the bladder. Figure 23.2 shows a close-up of the cortex (squared area), where the tips of the ureteric bud induce a mesenchyme-to-epithelium transformation in the cap mesenchyme.

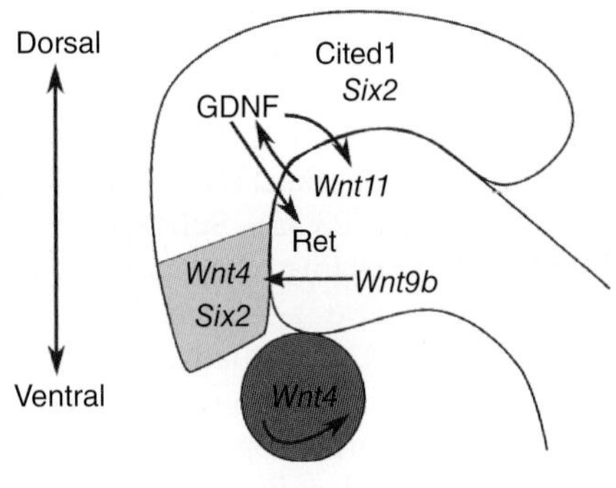

Figure 23.2 Schematic close-up of the ureteric bud tip (*white*) surrounded by the adjacent cap mesenchyme (*white* and *light gray*). *Wnt11* and *Wnt9b* are expressed in the ureteric bud. The cap mesenchyme can be divided into the uninduced and the induced cap mesenchyme based on gene expression pattern and location. *Six2* is expressed throughout the cap mesenchyme, whereas only the uninduced cap mesenchyme (*white*) shows *Cited1* expression, while the induced mesenchyme (*light gray*) commences *Wnt4* expression due to an upstream signal from ureteric bud-derived Wnt9b. The induced cap mesenchyme transforms into a pretubular aggregate (*dark gray*) that can be morphologically distinguished from the cap mesenchyme. In the pretubular aggregate, *Six2* transcription is stopped, whereas *Wnt4* expression is upregulated by autoregulation. A reciprocal positive feedback between GDNF and Wnt11 promotes branching of the ureteric bud. Ret functions as a receptor of the GDNF.

to formation of the mature nephron. The main transcription factor marking the cap mesenchyme, which constitutes the nephron stem cell pool, is the sine oculis-related homeobox 2 (Six2), which is essential for maintenance of self-renewal potential and prevention of premature epithelization of the cap mesenchyme (Kobayashi *et al.*, 2008; Self *et al.*, 2006; Figure 23.1 and Figure 23.2). *Six2* transcription rapidly ceases in the pretubular aggregate (Mugford *et al.*, 2009) but the protein persists at the proximal pole of the renal vesicle (Park *et al.*, 2012). Rapid downregulation of the *Cited1* transcriptional regulator distinguishes the induced cap mesenchyme from the dorsal cap mesenchyme, which is still in the stem cell stage (Mugford *et al.*, 2009).

Wnt signaling has an essential role in multiple aspects of kidney development, from induction of the cap mesenchyme to maturation of the nephron. As we explain in section "Development of the nephron," failures in Wnt signaling during the early stages – comprised of inductive events and mesenchyme-to-epithelium transformation – usually lead to renal agenesis, whereas mutations affecting later nephron maturation steps result in renal dysplasia, most typically including a cystic phenotype. The greatest insight

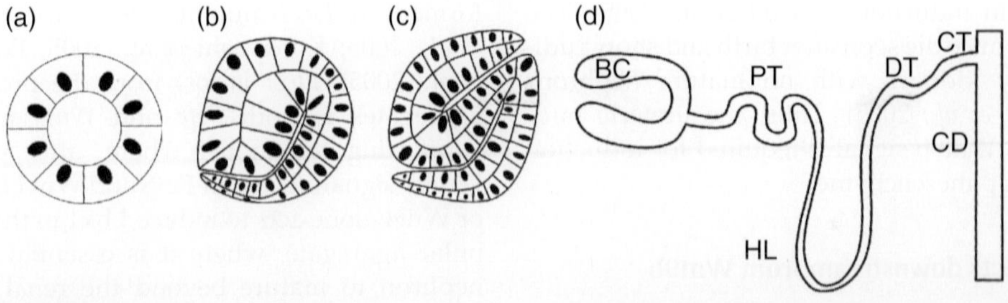

Figure 23.3 Pattern of nephron morphogenesis. As a consequence of mesenchyme-to-epithelium transformation, a renal vesicle (**a**) forms from the pretubular aggregate. A comma-shaped body (**b**) and an s-shaped body (**c**) are the next morphological steps in nephron maturation, and finally the mature nephron (**d**) consists of a Bowman's capsule (BC), a proximal convoluted tubulus (PT), a Henle's loop (HL), a distal convoluted tubulus (DT), and a connecting tube (CT), which joins with the ureteric bud-derived collecting duct (CD). The Bowman's capsule resides in the cortical region of the kidney, whereas the Henle's loop extends to the medullary region (cf. Figure 23.1b).

into vertebrate kidney development has been gained from research done on mice, but experiments in *Xenopus laevis* and the few case reports from humans suggest a conserved role of Wnt signaling in vertebrate kidney development. In this chapter, we sum up the essential literature concerning Wnt signaling in kidney organogenesis.

Development of the nephron

Nephron induction

Wnt9b as a primary signal

The contact between the ureteric bud and the cap mesenchyme is essential to differentiate the nephron from the mesenchyme, since their separation results in regression of each in tissue culture setup, where intact metanephros proceed its development (Kispert, Vainio, and McMahon, 1998; Saxén, 1987). However, the isolated metanephric mesenchyme can be rescued and induced to undergo nephron generation in coculture with various exogenous signal sources. Several embryonic tissues have been shown to have this inductive capacity, the embryonic spinal cord being the most known and used (Grobstein, 1955). Its capability to induce nephrogenesis is presumably due to expression of several Wnts, including Wnt4, which occurs naturally in the induced cap mesenchyme (Parr *et al.*, 1993; Stark *et al.*, 1994). Furthermore, cell lines expressing Wnt1, Wnt3a, Wnt4, Wnt7a, and Wnt7b provoke nephrogenesis in the isolated metanephric mesenchyme in a transfilter assay where the mesenchyme and the signaling cells are separated with a porous polycarbonate filter (Kispert, Vainio, and McMahon, 1998). The search for the ureteric bud-derived signal has led to the discovery of additional synthetic and biological factors that promote nephrogenesis in a cultured metanephric mesenchyme, for example, lithium ions (Davies and Garrod, 1995), leukemia inhibitory factor (LIF), which is secreted by the ureteric bud (Barasch *et al.*, 1999). However, the viability of LIF-deficient mice made LIF an unlikely candidate as an essential endogenous signal (Stewart *et al.*, 1992), and finally Wnt9b was recognized as a ureteric bud-derived factor that was able to induce the cap mesenchyme cells to exit the stem cell stage and permanently commit to the nephron lineage (Carroll *et al.*, 2005). The expression of *Wnt9b* is confined to the Wolffian duct prior to establishment of the metanephros. As the ureteric bud arises from the Wolffian duct, the expression of *Wnt9b* and *Wnt11* establishes a complementary pattern where *Wnt9b* expression dominates in the stalk and *Wnt11* in the tip. *Wnt9b* expression in the collecting duct system is maintained until adulthood in mice (Karner *et al.*, 2009). Primary nephrogenic induction of Wnt9b targets the cap mesenchyme, which in response commences expression of *Wnt4*, *fibroblast growth factor 8 (Fgf8)*, *paired box gene 8 (Pax8)*, and *LIM homeobox protein 1 (Lhx1)* (Carroll *et al.*, 2005; Grieshammer *et al.*, 2005; Kobayashi *et al.*, 2005; Perantoni *et al.*, 2005; Stark *et al.*, 1994). Transcription of these genes remains repressed in the cap mesenchyme of Wnt9b null mice, demonstrating a failure in induction (Carroll *et al.*, 2005). The mutant mice die soon after birth and show rudimentary kidneys with no mature nephrons (Carroll *et al.*, 2005). Thus, the ureteric bud-derived Wnt9b signal is required for induction of the cap mesenchyme.

Wnt4 acts downstream from Wnt9b

The importance of Wnt signaling in kidney organogenesis was originally suggested by the discovery of a requirement of *Wnt4* for the formation of the pretubular aggregate from the cap mesenchyme (Stark *et al.*, 1994). In nephron differentiation cascade, Wnt9b-dependent *Wnt4* expression is detected from the induced cap mesenchyme to the s-shaped body stage (Carroll *et al.*, 2005; Mugford *et al.*, 2009; Stark *et al.*, 1994). Fate mapping studies show that mature nephrons are entirely formed by the progeny of Wnt4-expressing cells (Shan *et al.*, 2010). Wnt4 function is conserved in *X. laevis*, in which *xWnt4* expression is detected in the early pronephric anlage, where it subsequently gets restricted to the tips of the pronephric tubuli that require intact *xWnt4* to develop (Carroll, Wallingford, and Vize, 1999; Saulnier, Ghanbari, and Brändli, 2002). In mice, similarly to *Wnt9b* mutation, homozygous deficiency of *Wnt4* leads to severe failure of differentiation of nephrons and development of only vestigial kidneys (Stark *et al.*, 1994). In addition, *Wnt4* mutations in humans are associated with renal agenesis (Biason-Lauber *et al.*, 2004; Mandel *et al.*, 2008). Specifically, nephron differentiation in mice halts immediately prior to pretubular aggregate formation, consistent with the stage when endogenous *Wnt4* transcription is normally initiated in this tissue. Nonfunctional *Wnt4* mRNA is still transcribed in other tissues, but this expression is lost from the cap mesenchyme in *Wnt4*-deficient mice, pointing to an autoinductive function of Wnt4 in the induced cap mesenchyme (Stark *et al.*, 1994). In the absence of functional *Wnt4*, expression of *Pax8* and *Fgf8* is initially retained in a few rudimentary pretubular aggregates (Perantoni *et al.*, 2005; Stark *et al.*, 1994), suggesting that these factors act upstream of Wnt4. In contrast, *Wnt4* transcripts are not detected in the metanephros of *Fgf8*-deficient mice, whereas *Wnt4* expression is retained in pretubular aggregates formed in *Lhx1*-mutant mice (Grieshammer *et al.*, 2005; Kobayashi *et al.*, 2005; Perantoni *et al.*, 2005). *Lhx1* is lost from the pretubular aggregates in both *Fgf8* and *Wnt4* mutants, supporting the model where, after primary Wnt9b signaling, either Fgf8 and Wnt4 together or Wnt4 alone acts to induce Lhx1 in the pretubular aggregate, where it is essential for the nephron to mature beyond the renal vesicle stage (Kobayashi *et al.*, 2005; Perantoni *et al.*, 2005). In summary, Wnt4 acts downstream of Wnt9b and is necessary for formation of the pretubular aggregate and subsequent epithelial nephron derivatives.

Signal transduction

Wnt9b/β-catenin pathway

Wnt/TCF pathway activity in kidney organogenesis has been analyzed with several TCF reporter mouse lines such as BAT-gal, TCF/Lef-lacZ, BATlacZ, and TCF-gfp (Ferrer-Vaquer *et al.*, 2010; Maretto *et al.*, 2003; Mohamed, Clarke, and Dufort, 2004; Nakaya *et al.*, 2005; for a review, see Barolo, 2006). While these reporters have consistently shown activity in the ureteric bud and its branches, a lack of TCF-driven reporter expression in the cap mesenchyme and its derivatives has been observed (Burn *et al.*, 2011; Ferrer-Vaquer *et al.*, 2010; Maretto *et al.*, 2003; Schmidt-Ott *et al.*, 2007; Tanigawa *et al.*, 2011). In contrast to other reporter lines, TCF/Lef-lacZ (Mohamed, Clarke, and Dufort, 2004) reporter mice activate reporter expression in the distal s-shaped bodies and occasionally in the parietal epithelial layer of the Bowman's capsules (Grouls *et al.*, 2012; Iglesias *et al.*, 2007). Thus, the reporter lines alone do not elucidate Wnt/TCF pathway activity in a fully consistent manner.

In contrast to results provided by the reporter lines, the Wnt/β-catenin/TCF pathway feedback target *Lef1* (see Chapters 4, 5, and 17) is expressed in the induced cap mesenchyme in a Wnt9b-dependent manner (Mugford *et al.*, 2009). Moreover, the β-catenin-stabilizing small molecules – lithium and 5′-bromoindirubin-3′-oxime (BIO) – which function by inhibiting GSK3β, are able to elicit nephrogenesis in the isolated metanephric mesenchyme while also provoking reporter activity in BATlacZ metanephric mesenchymes (Davies and Garrod,

1995; Klein and Melton, 1996; Kuure *et al.*, 2007; Tanigawa *et al.*, 2011). Lithium also induces transcription of the Wnt/β-catenin/TCF feedback genes *Lef1* and *Tcf1* as well as *Wnt4* in the isolated metanephric mesenchyme (Kuure *et al.*, 2007). Thus, activation of the β-catenin/TCF signaling pathway is sufficient to act upstream or at the level of Wnt4 to elicit a nephrogenic response.

A conditional knockout model that removes functional β-catenin from the cap mesenchyme with Six2Cre-mediated recombination showed severe perturbation in nephron differentiation (Park, Valerius, and McMahon, 2007). Molecular analysis of these β-catenin-mutant kidneys show markedly reduced expression of *Fgf8*, *Pax8*, *Wnt4*, and *Lhx1*, suggesting a failure in cap mesenchyme induction (Park, Valerius, and McMahon, 2007). In contrast, constitutive activation of β-catenin in the cap mesenchyme with a degradation-resistant form results in ectopic induction of the cap mesenchyme, leading to depletion of the nephron stem cell pool. The inductive response marked by transcription of the aforementioned genes by degradation-resistant β-catenin occurs independently of both Wnt9b and Wnt4 (Park, Valerius, and McMahon, 2007). However, sustained accumulation of β-catenin prevents the mesenchyme-to-epithelium transformation, suggesting that β-catenin-mediated signaling needs to be suppressed for development to proceed beyond the pretubular aggregate stage.

It was recently demonstrated that β-catenin stabilization with BIO treatment upregulates *Wnt4* and *Fgf8* while suppressing *Six2* in aggregated cap mesenchyme cells (Park *et al.*, 2012). Chromatin immunoprecipitation and sequencing revealed that β-catenin binds cis-regulatory modules of *Wnt4*, *Fgf8*, and *Six2* genes in the cap mesenchyme. These cis-regulatory modules also contain TCF motifs. When introduced as a transgene into mouse embryos, the *Wnt4* and *Fgf8* cis-regulatory modules that drive β-galactosidase expression are able to activate a reporter signal in renal vesicles only when the TCF motif is intact, suggesting that β-catenin/TCF-mediated transcription upregulation of *Wnt4* and *Fgf8* takes place in the cap mesenchyme derivatives. The reporter pattern driven by the *Six2* cis-regulatory module recapitulates endogenous Six2 expression even with a mutated TCF motif. Thus, the suppression of Six2 associated with β-catenin accumulation occurs independently of the TCF motif (Park *et al.*, 2012).

In summary, β-catenin is sufficient to push the cap mesenchyme cells out of the stem cell cycle to start differentiation into mature nephrons via upregulation of *Fgf8* and *Wnt4*. This induction stage is likely controlled by Wnt9b via the β-catenin/TCF/Lef pathway and TCF/Lef-independent yet β-catenin-mediated transcriptional program regulation.

Wnt4/calcium/nuclear factor of activated T-cell pathway

After Wnt9b-mediated induction of the cap mesenchyme and formation of the pretubular aggregate, Wnt4 acts to transform the mesenchymal cells into a polarized epithelium, which is prevented by constitutively active β-catenin signaling (Park, Valerius, and McMahon, 2007; Stark *et al.*, 1994). In the face of negative TCF reporter data and the requirement to suppress β-catenin, the search for alternative pathways for Wnt/β-catenin-mediated signaling revealed a potential role for the Wnt/calcium/nuclear factor of activated T-cells (NFAT) pathway as a downstream effector for Wnt4 in pretubular aggregates (Burn *et al.*, 2011; Tanigawa *et al.*, 2011). Cytoplasmic calcineurin-dependent NFAT genes were found to be expressed in the developing nephrons and calcineurin inhibitor treatment of isolated embryonic kidneys resulted in impaired nephrogenic response, while an ectopic increase in intracellular calcium partly rescued the *Wnt4*-mutant nephron phenotype *in vitro* (Burn *et al.*, 2011). Moreover, Wnt4 was shown to induce a calcium influx and calcium/calmodulin-dependent protein kinase II phosphorylation in primary metanephric mesenchyme cells (Tanigawa *et al.*, 2011).

Thus, Wnt4 induced by Wnt9b/β-catenin acts in the pretubular aggregate via a calcium/NFAT pathway to induce a mesenchyme-to-epithelium transformation. The small molecules – lithium and BIO – probably function upstream of Wnt4, whereas calcium ionophores inducing an increase in intracellular calcium concentration mimic the action of Wnt4 to promote nephrogenesis.

Nephron maturation

Regulation of planar cell polarity

Following the mesenchyme-to-epithelium transformation, the presumptive nephron undergoes differentiation into specialized segments aligned into a unique structure spanning the kidney radially. To achieve such a complex architecture accurately, with correct tube length and diameter, oriented cell division and coordinated cell motility is required. Wnt signaling has been implicated in the regulation of these parameters, thus playing an important role in the pathophysiology of cystic kidney disease (for a review, see Lancaster and Gleeson, 2010).

In addition to its pivotal role in the induction of the cap mesenchyme, Wnt9b acts during nephron maturation to ensure the correct planar cell polarity (PCP) and intercalation of the cells of the developing tubules in order to facilitate convergent extension of the tubular epithelium (Karner *et al.*, 2009; Lienkamp *et al.*, 2012). During the postnatal period, when epithelial cell division becomes oriented in the kidney, Wnt9b is further required to maintain the correct diameter of the tubules (Karner *et al.*, 2009). Hypomorph *Wnt9b* mice exhibit kidney cysts derived from proximal convoluted tubules. The deficient orientation and random division angle of the kidney epithelial cells were attributed to failure in Rho family GTPase (Rho) and Jun kinase activation, which are implicated in the Wnt/PCP pathway.

Consistent with the mouse studies, results from *Xenopus* show that knockdown of Wnt/PCP components, such as *dishevelled-associated activator of morphogenesis 1* (*Daam1*), *Rho guanine nucleotide exchange factor* (*WGEF*), and *Rho*, results in impaired late morphogenesis of the pronephric tubules and duct while still retaining their ability to differentiate, suggesting that the Wnt/PCP pathway acts after mesenchyme-to-epithelium transformation in the kidney tubules (Miller *et al.*, 2011). Furthermore, time-lapse imaging of tubular elongation in *Xenopus* nephron confirmed the role of convergent extension cell movements in nephron maturation (Lienkamp *et al.*, 2012). The convergent extension morphogenesis was shown to include evolutionarily conserved rosette-based mechanism, which is dependent on Wnt9b in mice and dishevelled-2 in *Xenopus* (Lienkamp *et al.*, 2012).

Thus, Wnt/PCP signaling regulates nephron maturation after the initial induction and mesenchyme-to-epithelium transformation stages.

Wnt/β-catenin pathway

As the Wnt/PCP pathway has been shown to regulate the tubular properties of the nephrons, processes that suppress the Wnt/β-catenin branch have also been described. It was noted that overactive β-catenin signaling in the kidney resulted in a cystic phenotype with delayed maturation of the nephrons (Qian *et al.*, 2005; Saadi-Kheddouci *et al.*, 2001). A disease entity called nephronophthisis is associated with a variety of mutations in genes that encode proteins for primary cilium components (for a review, see Hildebrandt, Attanasio, and Otto, 2009) and is characterized by renal cysts exemplified by a mouse model carrying homozygous mutation in *inversin* (Simons *et al.*, 2005). Inversin was shown to suppress β-catenin levels by increasing proteasomal degradation of cytoplasmic dishevelled and to be required for convergent extension movements in *X. laevis* animal caps. Furthermore, cystic pronephros in zebrafish, caused by *inversin* knockdown, was rescued with structurally related *diversin*, which is reported to target β-catenin for degradation. Finally, it was proposed that the flow of urine in the tubules might serve as a signal via the primary cilium to alter the response of the tubular epithelial cells to Wnt signals from the β-catenin pathway to alternative branches. Indeed, flow applied to an inner medullary collecting duct cell culture induced an increase in inversin and a decrease in β-catenin protein levels (Simons *et al.*, 2005). Another mechanism involved in suppression of Wnt/β-catenin signaling after the renal vesicle stage was demonstrated to involve inhibition of protein kinase A (Gallegos *et al.*, 2012).

Likewise, tightly controlled differentiation of the nephron epithelium into highly specialized cell types is critical for kidney function. Specification of proximodistal segmentation begins very early, demonstrated by polarized gene expression patterns in the renal vesicle (Georgas *et al.*, 2009), and already the s-shaped body stage of a presumptive nephron contains distinguishable progenitors for the parietal and visceral layers of the Bowman's capsule (Grouls *et al.*, 2012). Conditional loss-of-function mutation of β-catenin in nephron progenitors beyond the pretubular aggregate stage with Pax8Cre-mediated recombination resulted in an interesting fate switch of parietal epithelial cells into a visceral phenotype with foot processes and connections with capillaries (Grouls *et al.*, 2012).

In summary, the modality of Wnt signaling in later nephron maturation likely includes suppression of the β-catenin branch and domination of the Wnt/PCP pathway. The switch between the different pathways is probably carried out by regulating the responsivity of the tubular cells to existing Wnt ligands.

Development of the collecting duct system

Wnt11/GDNF loop

In addition to successive nephron differentiation from the cap mesenchyme, iterative branching morphogenesis of the ureteric bud is important for the final nephron number and thus the size of the adult kidney. Glial cell line-derived neurotrophic factor and ret proto-oncogene (GDNF/Ret) signaling acts as a critical signaling pathway to promote ureteric bud outgrowth from the Wolffian duct and subsequent branching morphogenesis into the collecting duct system. From the early stages of metanephric specification, the early cap mesenchyme starts to express GDNF, which then signals to the ureteric bud via Ret to initiate bud ingrowth and subsequent branch generation. Mutation in either of these leads to kidney agenesis due to failure in ureteric bud growth (Sánchez *et al.*, 1996; Schuchardt *et al.*, 1994, 1996).

Wnt11 acts together with GDNF/Ret signaling to promote collecting duct development (Majumdar *et al.*, 2003). Unlike Wnt1, Wnt3a, Wnt7a, and Wnt7b, Wnt11 is not competent to induce nephrogenesis in isolated mouse metanephric mesenchyme explants (Kispert, Vainio, and McMahon, 1998). *Wnt11* is expressed in the mouse and human embryonic kidney at the tips of the ureteric bud (Kispert *et al.*, 1996; Lako *et al.*, 1998). In contrast, *Wnt11b*, which is related to *Wnt11* but has been lost from mammals, can act as a nephrogenic inducer in *Xenopus pronephros* (Tételin and Jones, 2010). It has been suggested that the evolutionary loss of *Wnt11b* might correlate with the regression of the pronephros in mammals (Tételin and Jones, 2010). The mouse phenotype of the *Wnt11* knockout is rather subtle, showing reduced kidney size and a diminished number of nephrons (Majumdar *et al.*, 2003). A detailed inspection of the first branching steps beyond the T-stage ureteric bud reveals a delay in the first trifurcation event and a transient reduction in expression of GDNF in the cap mesenchyme. In turn, exogenous GDNF has been shown to promote *Wnt11* expression in the ureteric bud tips of isolated metanephric explants, suggesting that these factors are partly interdependent (Pepicelli *et al.*, 1997). Moreover, the few Ret homozygous mutant embryos that are able to form the T-stage ureteric bud show a complete loss of expression of Wnt11 in the ureteric bud tips. Furthermore, on a *Ret* heterozygous background, *Wnt11* mutation results in a dose-dependent reduction in nephron number. Thus, in mice, Wnt11 and the GDNF/Ret pathway form a reciprocally positively regulated signal loop (Figure 23.2) that participates in the generation of a sufficient number of ureteric bud branches to form the collecting duct system and to induce a corresponding amount of nephrons.

Role of Frizzled receptors

Although the essential role of Wnt ligands in kidney development is well established, the receptors that mediate the signals are still rather poorly known. Frizzled receptors expressed in the mouse embryonic kidney include *Fz2*, *Fz4*, *Fz6*, *Fz7*, *Fz8*, and *Fz10* (McMahon *et al.*, 2008; www.gudmap.org; Ye *et al.*, 2011). Functional analysis done in *Xenopus* with morpholino-mediated inhibition of translation of *Xfz8* expressed in the pronephric anlage impaired epithelial morphogenesis of the pronephric duct and tubules (Satow, Chan, and Asashima, 2004).

Double knockout of *Fz4* and *Fz8* in a mouse preserves nephron segmentation but leads to reduced kidney size, owing to diminished branching of the ureteric bud, and thus resembles the phenotype of *Wnt11* knockout mice (Majumdar *et al.*, 2003; Ye *et al.*, 2011). More specifically, *Fz8* knockout causes dominant kidney size reduction only on a *Fz4*-null background, whereas *Fz4* mutation functions in a recessive manner. The diminished size of the kidney was attributed to reduced proliferation of the mutant ureteric bud tips. Signal transduction analysis shows that Wnt11 together with Fz4 can activate both TFC/Lef-driven reporter and Rho in a cell culture, whereas coexpression of Wnt11 and Fz8 activated only Rho (Ye *et al.*, 2011). Thus, Fz4 and Fz8 may act to transduce Wnt11 signaling to enhance proliferation of the cells of the ureteric bud tips and also to regulate branching morphogenesis via cytoskeletal modification.

Development of the renal medulla

Role of Wnt7b and Wnt4

Filtration of plasma takes place in the kidney cortex, whereas concentration of urine and thus osmoregulation of plasma occurs in the medullary region where the Henle's loops and collecting ducts reside (Figure 23.1). Thus, the division of the kidney into radially defined regions provides means to maintain concentration of plasma in the narrow region compatible with body metabolism.

Wnt7b is expressed in the ureteric bud, excluding the tips (Carroll *et al.*, 2005; Kispert *et al.*, 1996), and was shown to be able to induce nephrogenesis in the isolated metanephric mesenchyme (Kispert, Vainio, and McMahon, 1998). However, epiblast and ureteric bud lineage-specific conditional *Wnt7b* knockout models revealed that the overall nephrogenic zone in the embryonic kidney was unaffected (Yu *et al.*, 2009). Instead, the medullary region of *Wnt7b*-mutant kidneys was completely absent. It is worth noting that *Wnt4-*, *Fgf8-*, and *Lim1*-mutant kidneys, which lack mature nephrons, still form a distinguishable medullary region in the kidney (Grieshammer *et al.*, 2005; Kobayashi *et al.*, 2005; Perantoni *et al.*, 2005; Stark *et al.*, 1994), suggesting that medulla formation is an independent process and not merely a secondary extension of nephron formation. This view is also supported by the fact that the size of the *Wnt7b*-mutant kidney remains similar to the wild type and the ureteric bud tip number is not affected. In *Wnt7b*-mutant mice, nephron segmentation in the cortical region occurs normally with differentiation of podocytes and proximal and distal convoluted tubules. However, the Henle's loops that should reside in the medulla become truncated. In addition, the proximal ureteric bud shows a dilated appearance that was attributed to failure in oriented cell division. Interestingly, mutation of *dickkopf-1* in the ureteric bud and developing nephrons results in an overgrowth of the renal medulla, suggesting that medulla formation requires finely balanced functioning of Wnt7b via negative regulation (Pietilä *et al.*, 2011).

In addition to the induced cap mesenchyme and its subsequent derivatives, *Wnt4* is expressed in the medullary stroma of the developing kidney, where its expression is not dependent on autoinduction, as in the pretubular aggregate (Itäranta *et al.*, 2006; Kispert *et al.*, 1996). A loss of *Wnt4* function leads to reduced stromal *Bmp4* expression and absence of alpha smooth muscle actin-expressing cells, suggesting a failure in pericyte differentiation.

Downstream signaling

One of the first reports of the presence of nuclear translocation β-catenin in the embryonic kidney was actually from human tissue, probably due to more general use of β-catenin antibodies in cancer diagnostics (Eberhart and Argani, 2001). The report showed strong nuclear staining in the medullary stromal cells surrounding the collecting ducts in fetuses with a gestational age of 10–18 weeks. Later, it was shown in mice that these medullary periductal cells express the feedback targets (see Chapter 5) *Axin2* and Lef1, which were dependent on Wnt7b expressed in the collecting duct cells, suggesting that these cells are responding to Wnt signaling (Yu *et al.*, 2009). In that context, also BAT-gal reporter colocalized with Lef1, confirming previous observations that Wnt/β-catenin/TCF signaling takes place in the embryonic kidney medulla. Interestingly, conditional inactivation of β-catenin from the stromal population with FoxD1Cre-mediated recombination phenocopied the *Wnt7b* knockout (Yu *et al.*, 2009). Thus, reciprocal interaction between the collecting ducts and the stroma takes place via Wnt/β-catenin/TCF signaling, coordinating the development of the renal medulla.

Summary and future outlook

Wnt signaling shows fascinating versatility in kidney organogenesis and highlights the context-dependent manner of downstream signaling. While the indispensable role of Wnt signaling in nephron induction and early differentiation is well established, later morphogenic events – including mainly the PCP pathway to ensure the correct architecture of the nephrons – are under intensive research. Instead of often severe and early lethal mutations that affect early kidney organogenesis, genetic diseases involving Wnt signaling that fine-tunes kidney organogenesis form an entity of renal diseases manifested later in life. Particular interest will be in the primary cilium and its role in regulating the responses to Wnt ligands in the tubular cells. Due to often subtle phenotypes of mutations that affect later kidney organogenesis, novel methods such as three-dimensional reconstituting imaging – like optical projection tomography – should be included in the analysis.

Furthermore, Wnt ligands with an unknown role, such as Wnt2b, Wnt5a, and Wnt6, are expressed in the embryonic kidney (Itäranta *et al.*, 2002; Lin *et al.*, 2001; Vainio, unpublished). The role of Wnt2b in kidney development has not been investigated in detail, but apparently, Wnt2b is not a critical factor in induction, since Wnt2/Wnt2b double knockout mice do not show an overt kidney phenotype (Goss *et al.*, 2009). However, Wnt2b may function as an additional stromal player that fine-tunes ureteric bud morphogenesis (Hatini *et al.*, 1996; Lin *et al.*, 2001). Wnt6 expression is detected in the ureteric bud and it has the capacity to induce nephron differentiation in the isolated metanephric mesenchyme (Itäranta *et al.*, 2002). So far, no kidney phenotype resulting from Wnt6 deficiency has been observed (Kispert and Brändli, 2003).

References

Barasch, J., Yang, J., Ware, C.B. *et al.* (1999) Mesenchymal to epithelial conversion in rat metanephros is induced by LIF. *Cell*, **99** (4), 377–386.

Bard, J.B., Gordon, A., Sharp, L., and Sellers, W.I. (2001) Early nephron formation in the developing mouse kidney. *Journal of Anatomy*, **199** (Pt 4), 385–392.

Barolo, S. (2006) Transgenic Wnt/TCF pathway reporters: all you need is Lef? *Oncogene*, **25** (57), 7505–7511.

Biason-Lauber, A., Konrad, D., Navratil, F., and Schoenle, E.J. (2004) A WNT4 mutation associated with Müllerian-duct regression and virilization in a 46,XX woman. *The New England Journal of Medicine*, **351** (8), 792–798.

Burn, S.F., Webb, A., Berry, R.L. *et al.* (2011) Calcium/NFAT signalling promotes early nephrogenesis. *Developmental Biology*, **352** (2), 288–298.

Carroll, T.J., Wallingford, J.B., and Vize, P.D. (1999) Dynamic patterns of gene expression in the developing pronephros of Xenopus laevis. *Developmental Genetics*, **24** (3–4), 199–207.

Carroll, T.J., Park, J.S., Hayashi, S. *et al.* (2005) Wnt9b plays a central role in the regulation of mesenchymal to epithelial transitions underlying organogenesis of the mammalian urogenital system. *Developmental Cell*, **9** (2), 283–292.

Davies, J.A. and Garrod, D.R. (1995) Induction of early stages of kidney tubule differentiation by lithium ions. *Developmental Biology*, **167** (1), 50–60.

Eberhart, C.G. and Argani, P. (2001) Wnt signaling in human development: beta-catenin nuclear translocation in fetal lung, kidney, placenta, capillaries, adrenal, and cartilage. *Pediatric and Developmental Pathology*, **4** (4), 351–357.

Ferrer-Vaquer, A., Piliszek, A., Tian, G. *et al.* (2010) A sensitive and bright single-cell resolution live imaging reporter of Wnt/beta-catenin signaling in the mouse. *BMC Developmental Biology*, **10**, 121.

Gallegos, T.F., Kouznetsova, V., Kudlicka, K. *et al.* (2012) A protein kinase A and Wnt-dependent network regulating an intermediate stage in epithelial tubulogenesis during kidney development. *Developmental Biology*, **364** (1), 11–21.

Georgas, K., Rumballe, B., Valerius, M.T. *et al.* (2009) Analysis of early nephron patterning reveals a role for distal RV proliferation in fusion to the ureteric tip via a cap mesenchyme-derived connecting segment. *Developmental Biology*, **332** (2), 273–286.

Goss, A.M., Tian, Y., Tsukiyama, T. *et al.* (2009) Wnt2/2b and beta-catenin signaling are necessary and sufficient to specify lung progenitors in the foregut. *Developmental Cell*, **17** (2), 290–298.

Grieshammer, U., Cebrián, C., Ilagan, R. *et al.* (2005) FGF8 is required for cell survival at distinct stages of nephrogenesis and for regulation of gene expression in nascent nephrons. *Development*, **132** (17), 3847–3857.

Grobstein, C. (1955) Inductive interaction in the development of the mouse metanephros. *Journal of Experimental Zoology*, **130** (2), 319–340.

Grouls, S., Iglesias, D.M., Wentzensen, N. *et al.* (2012) Lineage specification of parietal epithelial cells requires beta-catenin/Wnt signaling. *Journal of American Society of Nephrology*, **23** (1), 63–72.

Hatini, V., Huh, S.O., Herzlinger, D. *et al.* (1996) Essential role of stromal mesenchyme in kidney morphogenesis revealed by targeted disruption of Winged Helix transcription factor BF-2. *Genes & Development*, **10** (12), 1467–1478.

Hildebrandt, F., Attanasio, M., and Otto, E. (2009) Nephronophthisis: disease mechanisms of a ciliopathy. *Journal of American Society of Nephrology*, **20** (1), 23–35.

Iglesias, D.M., Hueber, P.A., Chu, L. *et al.* (2007) Canonical WNT signaling during kidney development. *American Journal of Physiology Renal Physiology*, **293** (2), F494–F500.

Itäranta, P., Lin, Y., Peräsaari, J. *et al.* (2002) Wnt-6 is expressed in the ureter bud and induces kidney tubule development in vitro. *Genesis*, **32** (4), 259–268.

Itäranta, P., Chi, L., Seppänen, T. *et al.* (2006) Wnt-4 signaling is involved in the control of smooth muscle cell fate via Bmp-4 in the medullary stroma of the developing kidney. *Developmental Biology*, **293** (2), 473–483.

Karner, C.M., Chirumamilla, R., Aoki, S. *et al.* (2009) Wnt9b signaling regulates planar cell polarity and kidney tubule morphogenesis. *Nature Genetics*, **41** (7), 793–799.

Kispert, A. and Brändli, A.W. (2003) Wnts in kidney and genital development, in *Wnt Signaling in Development* (ed M. Kühl), Kluwer Academic/Plenum Publishers, New York, p. 125.

Kispert, A., Vainio, S., and McMahon, A.P. (1998) Wnt-4 is a mesenchymal signal for epithelial transformation of metanephric mesenchyme in the developing kidney. *Development*, **125** (21), 4225–4234.

Kispert, A., Vainio, S., Shen, L. *et al.* (1996) Proteoglycans are required for maintenance of Wnt-11 expression in the ureter tips. *Development*, **122** (11), 3627–3637.

Klein, P.S. and Melton, D.A. (1996) A molecular mechanism for the effect of lithium on development. *Proceedings of the National Academy of Sciences of the United States of America*, **93** (16), 8455–8459.

Kobayashi, A., Kwan, K.M., Carroll, T.J. *et al.* (2005) Distinct and sequential tissue-specific activities of the LIM-class homeobox gene Lim1 for tubular morphogenesis during kidney development. *Development*, **132** (12), 2809–2823.

Kobayashi, A., Valerius, M.T., Mugford, J.W. *et al.* (2008) Six2 defines and regulates a multipotent self-renewing nephron progenitor population throughout mammalian kidney development. *Cell Stem Cell*, **3** (2), 169–181.

Kuure, S., Popsueva, A., Jakobson, M. *et al.* (2007) Glycogen synthase kinase-3 inactivation and stabilization of beta-catenin induce nephron differentiation in isolated mouse and rat kidney mesenchymes. *Journal of American Society of Nephrology*, **18** (4), 1130–1139.

Lako, M., Strachan, T., Bullen, P. *et al.* (1998) Isolation, characterisation and embryonic expression of WNT11, a gene which maps to 11q13.5 and has possible roles in the development of skeleton, kidney and lung. *Gene*, **219** (1–2), 101–110.

Lancaster, M.A. and Gleeson, J.G. (2010) Cystic kidney disease: the role of Wnt signaling. *Trends in Molecular Medicine*, **16** (8), 349–360.

Lienkamp, S.S., Liu, K., Karner, C.M. *et al.* (2012) Vertebrate kidney tubules elongate using a planar cell polarity-dependent, rosette-based mechanism of convergent extension. *Nature Genetics*, **44** (12), 1382–1387.

Lin, Y., Liu, A., Zhang, S. *et al.* (2001) Induction of ureter branching as a response to Wnt-2b signaling during early kidney organogenesis. *Developmental Dynamics*, **222** (1), 26–39.

Majumdar, A., Vainio, S., Kispert, A. *et al.* (2003) Wnt11 and Ret/Gdnf pathways cooperate in regulating ureteric branching during metanephric kidney development. *Development*, **130** (14), 3175–3185.

Mandel, H., Shemer, R., Borochowitz, Z.U. *et al.* (2008) SERKAL syndrome: an autosomal-recessive disorder caused by a loss-of-function mutation in WNT4. *The American Journal of Human Genetics*, **82** (1), 39–47.

Maretto, S., Cordenonsi, M., Dupont, S. *et al.* (2003) Mapping Wnt/beta-catenin signaling during mouse development and in colorectal tumors. *Proceedings of the National Academy of Sciences of the United States of America*, **100** (6), 3299–3304.

McMahon, A.P., Aronow, B.J., Davidson, D.R. *et al.* (2008) GUDMAP: the genitourinary developmental molecular anatomy project. *Journal of American Society of Nephrology*, **19** (4), 667–671.

Miller, R.K., Canny, S.G., Jang, C.W. *et al.* (2011) Pronephric tubulogenesis requires Daam1-mediated planar cell polarity signaling. *Journal of American Society of Nephrology*, **22** (9), 1654–1664.

Mohamed, O.A., Clarke, H.J., and Dufort, D. (2004) Beta-catenin signaling marks the prospective site of primitive streak formation in the mouse embryo. *Developmental Dynamics*, **231** (2), 416–424.

Mugford, J.W., Sipilä, P., McMahon, J.A., and McMahon, A.P. (2008) Osr1 expression demarcates a multi potent population of intermediate mesoderm that undergoes progressive restriction to an Osr1-dependent nephron progenitor compartment within the mammalian kidney. *Developmental Biology*, **324** (1), 88–98.

Mugford, J.W., Yu, J., Kobayashi, A., and McMahon, A.P. (2009) High-resolution gene expression analysis of the developing mouse kidney defines novel cellular compartments within the nephron progenitor population. *Developmental Biology*, **333** (2), 312–323.

Nakaya, M.A., Biris, K., Tsukiyama, T. *et al.* (2005) Wnt3a links left-right determination with segmentation and anteroposterior axis elongation. *Development*, **132** (24), 5425–5436.

Park, J.S., Valerius, M.T., and McMahon, A.P. (2007) Wnt/beta-catenin signaling regulates nephron induction during mouse kidney development. *Development*, **134** (13), 2533–2539.

Park, J.S., Ma, W., O'Brien, L.L. *et al.* (2012) Six2 and Wnt regulate self-renewal and commitment of nephron progenitors through shared gene regulatory networks. *Developmental Cell*, **23** (3), 637–651.

Parr, B.A., Shea, M.J., Vassileva, G., and McMahon, A.P. (1993) Mouse Wnt genes exhibit discrete domains of expression in the early embryonic CNS and limb buds. *Development*, **119** (1), 247–261.

Pepicelli, C.V., Kispert, A., Rowitch, D.H., and McMahon, A.P. (1997) GDNF induces branching and increased cell proliferation in the ureter of the mouse. *Developmental Biology*, **192** (1), 193–198.

Perantoni, A.O., Timofeeva, O., Naillat, F. *et al.* (2005) Inactivation of FGF8 in early mesoderm reveals an essential role in kidney development. *Development*, **132** (17), 3859–3871.

Pietilä, I., Ellwanger, K., Railo, A. *et al.* (2011) Secreted Wnt antagonist Dickkopf-1 controls kidney papilla development coordinated by Wnt-7b signalling. *Developmental Biology*, **353** (1), 50–60.

Qian, C.N., Knol, J., Igarashi, P. *et al.* (2005) Cystic renal neoplasia following conditional inactivation of apc in mouse renal tubular epithelium. *The Journal of Biological Chemistry*, **280** (5), 3938–3945.

Saadi-Kheddouci, S., Berrebi, D., Romagnolo, B. *et al.* (2001) Early development of polycystic kidney disease in transgenic mice expressing an activated mutant of the beta-catenin gene. *Oncogene*, **20** (42), 5972–5981.

Sánchez, M.P., Silos-Santiago, I., Frisén, J. *et al.* (1996) Renal agenesis and the absence of enteric neurons in mice lacking GDNF. *Nature*, **382** (6586), 70–73.

Satow, R., Chan, T.C., and Asashima, M. (2004) The role of Xenopus frizzled-8 in pronephric development. *Biochemical and Biophysical Research Communications*, **321** (2), 487–494.

Saulnier, D.M., Ghanbari, H., and Brändli, A.W. (2002) Essential function of Wnt-4 for tubulogenesis in the Xenopus pronephric kidney. *Developmental Biology*, **248** (1), 13–28.

Saxén, L. (1987) *Organogenesis of the Kidney*, Cambridge University Press, Cambridge, NY.

Schmidt-Ott, K.M., Masckauchan, T.N., Chen, X. *et al.* (2007) β-Catenin/TCF/Lef controls a differentiation-associated transcriptional program in renal epithelial progenitors. *Development*, **134** (17), 3177–3190.

Schuchardt, A., D'Agati, V., Larsson-Blomberg, L. *et al.* (1994) Defects in the kidney and enteric nervous system of mice lacking the tyrosine kinase receptor Ret. *Nature*, **367** (6461), 380–383.

Schuchardt, A., D'Agati, V., Pachnis, V., and Costantini, F. (1996) Renal agenesis and hypodysplasia in ret-k⁻ mutant mice result from defects in ureteric bud development. *Development*, **122** (6), 1919–1929.

Self, M., Lagutin, O.V., Bowling, B. *et al.* (2006) Six2 is required for suppression of nephrogenesis and progenitor renewal in the developing kidney. *The EMBO Journal*, **25** (21), 5214–5228.

Shan, J., Jokela, T., Skovorodkin, I., and Vainio, S. (2010) Mapping of the fate of cell lineages generated from cells that express the Wnt4 gene by time-lapse during kidney development. *Differentiation*, **79** (1), 57–64.

Simons, M., Gloy, J., Ganner, A. *et al.* (2005) Inversin, the gene product mutated in nephronophthisis type II, functions as a molecular switch between Wnt signaling pathways. *Nature Genetics*, **37** (5), 537–543.

Stark, K., Vainio, S., Vassileva, G., and McMahon, A.P. (1994) Epithelial transformation of metanephric mesenchyme in the developing kidney regulated by Wnt-4. *Nature*, **372** (6507), 679–683.

Stewart, C.L., Kaspar, P., Brunet, L.J. *et al.* (1992) Blastocyst implantation depends on maternal expression of leukaemia inhibitory factor. *Nature*, **359** (6390), 76–79.

Tanigawa, S., Wang, H., Yang, Y. *et al.* (2011) Wnt4 induces nephronic tubules in metanephric mesenchyme by a non-canonical mechanism. *Developmental Biology*, **352** (1), 58–69.

Tételin, S. and Jones, E.A. (2010) Xenopus Wnt11b is identified as a potential pronephric inducer. *Developmental Dynamics*, **239** (1), 148–159.

Ye, X., Wang, Y., Rattner, A., and Nathans, J. (2011) Genetic mosaic analysis reveals a major role for frizzled 4 and frizzled 8 in controlling ureteric growth in the developing kidney. *Development*, **138** (6), 1161–1172.

Yu, J., Carroll, T.J., Rajagopal, J. *et al.* (2009) A Wnt7b-dependent pathway regulates the orientation of epithelial cell division and establishes the cortico-medullary axis of the mammalian kidney. *Development*, **136** (1), 161–171.

24 Wnt Signaling Regulation of Tissue Architecture (EMT and MET) and Morphogenesis: Consequences for Colorectal and Liver Cancer

Theodora Fifis[1], Bang M. Tran[2], Renate H.M. Schwab[2], Timothy M. Johanson[2], Nadia Warner[2], Nick Barker[3], and Elizabeth Vincan[2]

[1]Austin Hospital, University of Melbourne, Heidelberg, Melbourne, Victoria, Australia
[2]University of Melbourne and Victorian Infectious Diseases Reference Laboratories, Melbourne, Australia
[3]Institute of Medical Biology, Immunos, Singapore

Introduction

The Wnt signaling pathway is highly conserved through evolution (see Chapter 12), is complex, and is quite unusual in many ways. For example, while standard signal transduction mechanisms involve amplification via "second messengers" in the cytoplasm, in Wnt/β-catenin signaling, there is a stoichiometric relationship between the pathway components leading to transcriptional regulation without cascade amplification. Each step of the pathway is tightly controlled with multiple negative and positive regulators and positive and negative feedback loops (see Chapters 3 and 6). This intricate regulation affords the cell plasticity and the capacity to undergo committed phenotypic transitions, which underlie tissue morphogenesis.

Tissue morphogenesis and cell plasticity are fundamental processes of embryonic, fetal, and postnatal organogenesis and tumorigenesis (Vincan and Barker, 2008; Vincan *et al.*, 2007c). Indeed, our quest to understand the mechanisms of Wnt signaling has seen the inevitable convergence of developmental and cancer biology (Bissell and Radisky, 2001; Egeblad, Nakasone, and Werb, 2010; Radtke, Clevers, and Riccio, 2006). Tissue patterning and remodeling are key events in embryology and are governed by Wnt signaling. These same processes are also essential for solid tumor growth and progression (Kirchner and Brabletz, 2000). For example, for a tumor to increase in size, an increasing demand for nutrients, oxygen, and waste exchange must be met. As in normal development, this demand is met by angiogenesis and lymphogenesis, processes whereby new blood and lymphatic vessels, respectively, sprout from existing vessels (Stacker and Achen, 2008). Ironically, tumor-associated angiogenesis and lymphangiogenesis are the main route by which cancer cells spread to other organs (metastasize) and immune cells infiltrate (Lu, Weaver, and Werb, 2012). Metastasis and inflammation are important aspects of cancer progression and are governed, at least in part, by Wnt signaling, as we will discuss in this review.

Indeed, since the discovery of Wnt (Nusse and Varmus, 1982) and Frizzled (Fz) (Vinson and

Wnt Signaling in Development and Disease: Molecular Mechanisms and Biological Functions, First Edition.
Edited by Stefan Hoppler and Randall T. Moon.

Adler, 1987) and the demonstration that Fz functions as the cell surface Wnt receptor (Bhanot *et al.*, 1996; Yang-Snyder *et al.*, 1996), the Wnt signaling pathway has been shown to govern diverse developmental, homeostatic, and pathological processes. These diverse activities of the Wnt pathway may in part be due to the large family of receptors and ligands (see Chapter 2) but also because of the different branches of the Wnt/Fz pathway. The different Wnt signaling pathways are dealt with in depth in Chapters 1–8 and are only touched upon here to place Wnt signaling in the context of tissue architecture and morphogenesis. Notably, the different branches of the Wnt signaling pathway were first characterized in developmental contexts, but in recent times, each branch has been implicated in tumor growth and progression (Lai, Chien, and Moon, 2009; Vincan, 2004). In this chapter, this link between development and cancer is exemplified by our focus on two Wnt-driven cancers, colorectal cancer (CRC) and hepatocellular carcinoma (HCC). More specifically, we focus on the processes of epithelial-to-mesenchymal transition (EMT) and the reverse transition, mesenchymal-to-epithelial transition (MET), both fundamental mechanisms of tumor metastasis and morphogenesis.

(a)

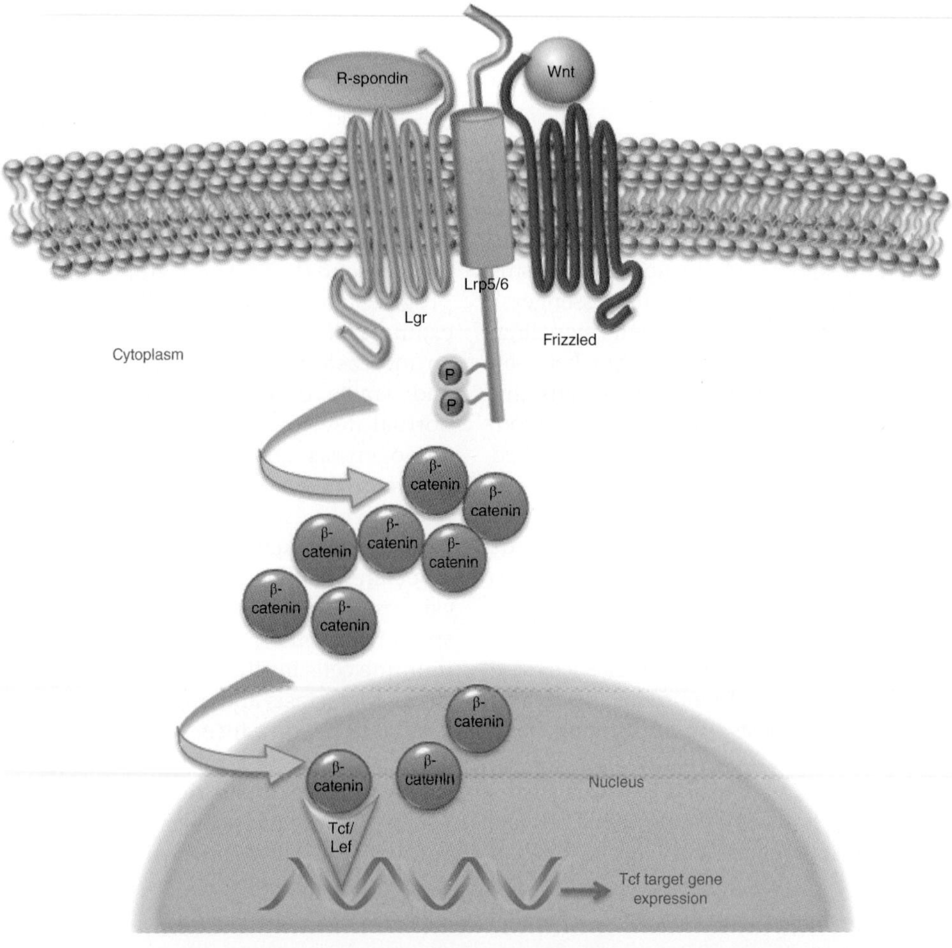

Figure 24.1 Overview of Wnt signaling pathways. (a) Upon binding of Wnt to Fz in a receptor complex that contains LRP and R-spondin bound to LGR, β-catenin accumulates in the cytoplasm and translocates to the nucleus. Once in the nucleus, together with TCF (TCF/lymphoid-enhancer binding protein factor), β-catenin forms a transcriptionally active complex to activate the expression of target genes. *(See insert for color representation of the figure.)*

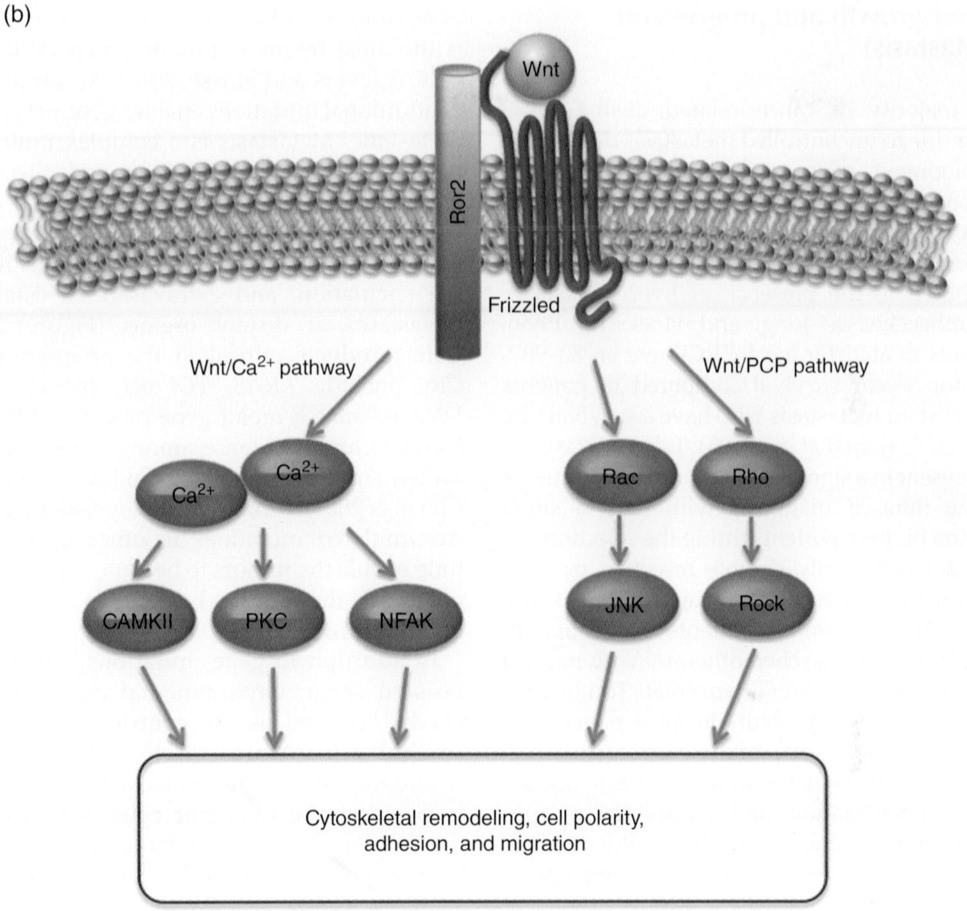

Figure 24.1 (continued) (b) In the β-catenin-independent arm of Wnt signaling, the Wnt signal is relayed into the cell via calcium (Ca^{2+}) and the subsequent activation of CAMKII (calmodulin-dependent protein kinase II), PKC (protein kinase C), NFAK (nuclear factor of activated T-cells) Ca^{2+}-responsive proteins (referred to as Wnt/Ca^{2+} pathway) or Rac/Rho GTPases and JNK/ROCK (referred to as Wnt/PCP pathway). *(See insert for color representation of the figure.)*

Wnt signaling pathways

Wnts are a large family of secreted glycoproteins that activate various intracellular pathways upon binding to seven-pass transmembrane Fz receptor family proteins. Different branches of Wnt signaling have been characterized: the β-catenin-dependent pathway (Wnt/β-catenin) (Clevers and Nusse, 2012) and β-catenin-independent pathways (Wnt/Ca^{2+} and the planar cell polarity (PCP) pathway) (Lai, Chien, and Moon, 2009; Figure 24.1). Apart from multiple pathways, there is also added complexity to Wnt signaling with the involvement of coreceptors such as LDL receptor-related protein (LRP), Knypek, and Ror2; alternative receptors such as Ryk and Norrin; and potentiating receptors such as leucine-rich repeat-containing G protein coupled receptor (LGR) (see Chapters 2 and 7). The latter is particularly important as LGR specifically marks adult stem cells (Barker, van Oudenaarden, and Clevers, 2012; Barker *et al.*, 2007) and binding of R-spondin secreted proteins to LGR potentiates Wnt signaling (Schuijers and Clevers, 2012). Consequently, LGR5 does not inertly mark stem cells but is in the Wnt/Fz receptor complex (Figure 24.1) and plays a critical role in transmitting Wnt signals into stem cells. Moreover, as the $LGR5^+$ intestinal stem cells are the cell of origin of CRC (Barker *et al.*, 2009), targeting the Wnt cell surface receptor complex might provide important anticancer or, on the converse, regenerative treatment.

Tumor growth and progression (metastasis)

The majority of cancer-related deaths (90%) occur due to uncontrolled metastasis despite the development of a wide range of systemic, regional, and local therapies. CRC is the second most frequent cause of cancer death in the Western world with the majority of deaths attributed to the presence of liver metastases (Cromheecke, de Jong, and Hoekstra, 1999). Patients treated for local CRC have an 80–90% rate for 5-year survival compared to patients with distant metastasis who have a survival rate of 10–20% (Jemal *et al.*, 2005). Liver metastases are present in a significant proportion of patients at the time of diagnosis with only a small fraction of these patients fitting the selection criteria for potentially curable resection or local ablation (Cromheecke, de Jong, and Hoekstra, 1999). The remaining patients are generally treated by systemic chemotherapy with modest improvements in overall survival. Tumor progression eventually occurs in most patients by mechanisms that are not fully elucidated.

Tumors arise when mutations enable uncontrolled proliferation and escape from apoptosis. Deregulation of the Wnt pathway leads to uncontrolled growth, causing the formation of adenomas and preinvasive carcinomas, and is the most frequent initiating step in human CRC (Clevers and Nusse, 2012). Accumulation of additional mutations enables CRC to become metastatic. Metastasis is a complex multistep process (Steeg, 2006; Thiery and Sleeman, 2006), requiring cancer cells to invade the adjacent submucosa and muscularis, intravasate the vascular and lymphatic routes, survive in the circulation, and extravasate to establish metastases at distant organs (Figure 24.2). Gene products critical in the progression of CRC include *KRAS*, *TGFBR1*, *BRAF*, *TP53*, *DNA* mismatch repair gene products, FBXW7, NOTCH, and PI3 kinase, among others (Davies, Miller, and Coleman, 2005; Fodde, Smits, and Clevers, 2001; Kinzler and Vogelstein, 1996). Accumulated mutations in other genes over time enable the tumors to become invasive and metastatic through the hematogenous and the lymphatic routes.

In addition to gene mutations, for metastasis to occur, environmental cues are also needed as well as the contribution of host stromal cells. Moreover, the distant organ where metastases are to be established needs to be receptive to tumor cells. Intriguingly, the primary tumor contributes to the establishment of the receptive environment or

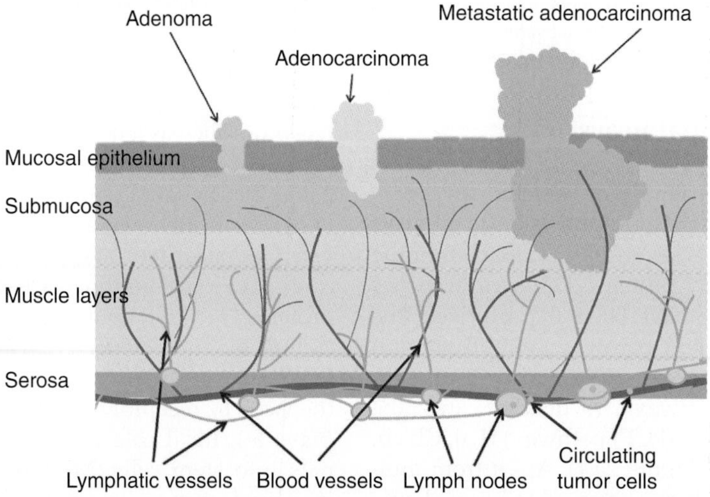

Figure 24.2 Progression in CRC. Initial mutations most frequently involving the APC genes result in benign adenoma growths. Accumulated mutations in other genes over time enable the tumors to become invasive and metastatic through the hematogenous and the lymphatic routes. *(See insert for color representation of the figure.)*

"metastatic niche" of the distant organ through the secretion of molecular factors and the activation of systemically circulating host cells in the tumor microenvironment (Erler and Weaver, 2009; Hiratsuka *et al.*, 2006; Kaplan, Moon, and Vunjak-Novakovic, 2005) well before the arrival of the metastatic cells.

Tumors of different origins preferentially establish in different organs. For example, breast tumors metastasize to the lungs and the bone marrow, while CRC tumors by and large metastasize to the liver. Two factors determine the organ of preference for metastasis. The first receptive site that the hematogenous circulating tumor cell encounters becomes the most frequent metastatic site as is the case for the lung and liver for breast tumor and CRC, respectively (Steeg, 2006). The metastatic site is also determined by the molecular signature of the particular tumor clone. This is supported by evidence that breast cancer clones that colonize the lung have a different molecular signature to clones that colonize the bone marrow (Sihto *et al.*, 2011). Similarly, the distinct roles of Snail and Twist, two transcription factors known to induce EMT, were demonstrated in regional metastasis of CRC (Fan *et al.*, 2013).

Epithelial-to-mesenchymal (EMT) and mesenchymal-to-epithelial (MET) transitions

Tumor metastasis had been explained by two models; the EMT model proposes that a subset of tumor cells in response to local signals undergoes an EMT and acquires metastatic properties (Huber, Kraut, and Beug, 2005). The cancer stem cell (CSC) model holds that tumors are organized in a cellular hierarchy and only a subset of tumor cells with stem cell properties have the capacity of self-renewal and the ability to initiate metastases (Schulenburg *et al.*, 2010). Tumor cells with CSC and EMT morphology are usually seen at the tumor periphery and were proposed to define "migrating cancer cells" capable of metastasis (Brabletz *et al.*, 2005). Indeed, recent studies reported that breast CSCs have a mesenchymal morphology and that induction of EMT results in the upregulation of stem cell markers and the acquisition of metastatic and drug resistance properties (Mani *et al.*, 2008; Morel *et al.*, 2008). Therefore, it now appears that both models describe the same mechanism for tumor progression (Singh and Settleman, 2010).

In any solid tumor, only a small subset of cells at any time displays mesenchymal morphology. Most often, such cells are seen at the tumor invasive front, where, as we and others have shown, the microenvironment is different to the rest of the tumor (Brabletz *et al.*, 1998; Jung *et al.*, 2001; Nguyen *et al.*, 2012; Spaderna *et al.*, 2006). It is, therefore, difficult to study EMT *in vivo*. Most of the current knowledge concerning EMT in cancers has been derived from *in vitro* studies in tumor cell lines. Several different signaling pathways can induce EMT, especially TGF-β signaling in cooperation with oncogenic Ras or the receptor tyrosine kinases (Siegel *et al.*, 2003; Ueda *et al.*, 2004; Xie *et al.*, 2004), Wnt (Eger *et al.*, 2004; Nelson and Nusse, 2004), Notch (Timmerman *et al.*, 2004; Zavadil *et al.*, 2004), Hedgehog (Karhadkar *et al.*, 2004), and hypoxia/HIF1α pathways (Yang and Wu, 2008). Increase in the transcription factors Snail, Slug, ZEB1, ZEB2 E12/E47, Goosecoid, and Twist, among others, also promotes EMT (Yang *et al.*, 2008a).

Clinical studies report higher levels of EMT markers associate with aggressive disease, tumor progression, and worse prognosis. In pancreatic and CRC tumor tissues, gain of mesenchymal characteristics and loss of epithelial characteristics correlated with tumor progression (Buck *et al.*, 2007). Increased ZEB1 expression correlates directly with Gleason grade in prostate tumors (Graham *et al.*, 2008) and aggressive behavior in CRC and uterine cancers (Pena *et al.*, 2006; Spoelstra *et al.*, 2006). An inverse correlation between ZEB1 and E-cadherin was reported in clinical and experimental studies and is a prognostic indicator of poor survival and resistance to several drugs (Arumugam *et al.*, 2009).

Several molecular factors have been reported to be linked to the induction and maintenance of EMT, some of them newly upregulated through newly activated pathways, while others are expressed *de novo* through alternative mRNA splicing (Steeg, 2006; Thiery and Sleeman, 2006). In addition, several miRNA classes are involved in EMT with some of them activating and others repressing EMT (de Krijger *et al.*, 2011; Gregory *et al.*, 2008, 2011; Zhang *et al.*, 2011). More recent studies have shown that regulation also occurs at the translational level where key driver mRNAs are present but prevented from being translated by either miRNAs or inhibitory proteins (Evdokimova *et al.*, 2009). This type of regulation is very important as it enables the cell to respond very fast to environmental changes and to produce key factors without the need for transcription.

For new metastases to be established in the host tissue, it is necessary that the migrating cancer cells revert to an epithelial morphology through MET (Figure 24.3). This entails downregulation of the mesenchymal properties and reacquisition of epithelial properties. This transition is even harder to see *in vivo* than EMT; however, the majority of researchers accept that it happens since metastases by and large display the epithelial morphology of the primary tumor (Peinado, Olmeda, and Cano, 2007). Support for such transition comes from recent publications reporting successful establishment of metastases when constitutively expressed ectopic EMT driver molecules were conditionally downregulated in experimental animal models (Ocana *et al.*, 2012; Tsai *et al.*, 2012). In fact, these authors suggest that tumor dormancy may be due to its failure to revert into the epithelial state at the metastatic site.

Wnt signaling in CRC morphogenesis

The small and large intestine (colon) is lined by a simple epithelium that is continuously replaced. It is estimated that the entire lining is regenerated every 4–5 days. In the mouse, some 300 cells are generated per crypt every day (Marshman, Booth, and Potten, 2002). This remarkable regenerative process is maintained through continuous production of epithelial cells by LGR5$^+$ stem cells that reside at the base of small invaginations or glands (crypts) in the epithelium (Barker *et al.*, 2007). To achieve this, the stem cells undergo symmetric cell division to produce two stem cells, two differentiated cells, or one of each (Snippert *et al.*, 2010). This stochastic process replaces the stem cell (self-renewal) – and thus the number of stem cells is maintained – and provides the precursor cell (progenitor or transit amplifying cell) that gives rise to all the different specialized cell types of the epithelium (Barker, van de Wetering, and Clevers, 2008; Barker *et al.*, 2007; Snippert *et al.*,

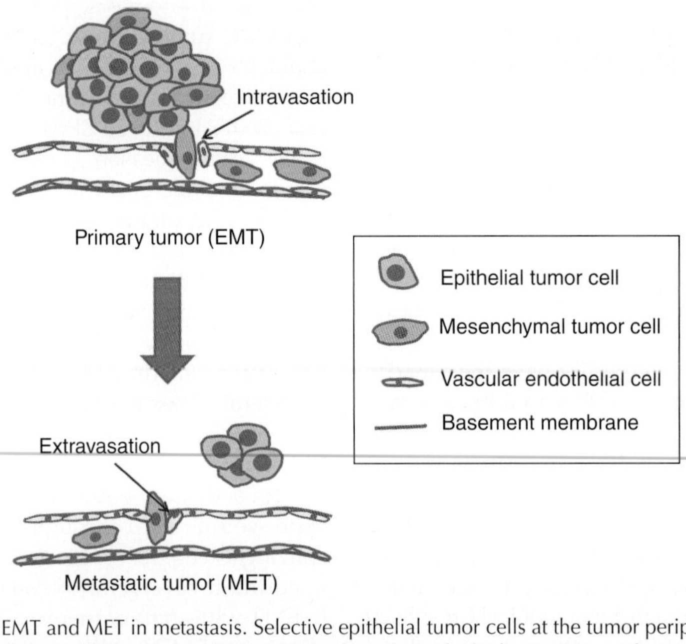

Figure 24.3 Tumor EMT and MET in metastasis. Selective epithelial tumor cells at the tumor periphery acquire a mesenchymal morphology (EMT) and the ability to degrade and invade the basement membrane and enter the circulation. At the metastatic site, the tumor cells undergo the reverse process (MET) to establish micrometastases. *(See insert for color representation of the figure.)*

2010). As the cells differentiate, they migrate up the crypt wall towards the lumen and continue to divide. Once they approach the top third of the crypt, the cells terminally differentiate, stop dividing, and perform their specialized function. The terminally differentiated cells eventually die by apoptosis and are sloughed off into the lumen and excreted. One daughter cell type, the Paneth cell, is excluded from this upward migration and rapid turnover. The progenitors committed to the Paneth cell lineage migrate down to the crypt base where they reside for 3–6 weeks (Sato *et al.*, 2011). The balance between cell production and cell death (crypt homeostasis) is under strict control by Wnt/β-catenin signaling (Clevers and Nusse, 2012). Loss of crypt homeostasis through aberrant activation of Wnt/β-catenin signaling is a necessary initiating step in the vast majority of CRC (see Chapter 27). The Wnt/β-catenin pathway is constitutively activated through truncation of the adenomatous polyposis coli (*APC*) gene product, which impedes destruction complex activity, or, less frequently, through oncogenic mutation of β-catenin or Axin genes (see Chapters 3 and 27).

Once a causative role for constitutive activation of T-cell factor (TCF)/β-catenin transcription in CRC was established, it was assumed that TCF/β-catenin transcription would not be further regulated in this cancer. Consequently, it was thought that any signaling from the upstream components of the pathway would be superfluous as TCF/β-catenin transcription was already activated by genetic mutation of downstream components. Moreover, as genetic mutation is by nature irreversible, it was considered that changes made to a cell by active TCF/β-catenin transcription were similarly irreversible. However, detailed analysis of β-catenin's cellular localization in carcinoma tissues revealed variable patterns of expression. Nuclear localization of β-catenin was primarily confined to invasive areas of the carcinoma, while in the more differentiated central areas, β-catenin was at the plasma membrane. This variable localization of β-catenin was evident even though all cells of the carcinoma harbor truncating mutations in the *APC* gene (Brabletz *et al.*, 1998, 2001).

The distribution of β-catenin in the carcinoma tissues bore many analogies to tissue patterning during embryogenesis (Kirchner and Brabletz, 2000) and indicated that the cells at the invasive front of carcinomas had undergone a partial EMT. Indeed, it was subsequently demonstrated that many of the genes expressed by carcinoma cells at the invasive front are TCF/β-catenin target genes (Brabletz *et al.*, 2005; van Es *et al.*, 2005). Moreover, this EMT is dynamically regulated by many factors in the microenvironment (e.g., extracellular matrix components (Vincan *et al.*, 2010)) and is reversed at the secondary site as most metastases recapitulate the pathology of the primary tumor (Brabletz *et al.*, 2001). Thus, the mesenchymal cells at the invasive front revert back to an epithelial phenotype, that is, undergo MET, to build the tumor mass at the secondary site.

In addition to variable β-catenin subcellular localization, several other factors indicated additional regulation of the Wnt/β-catenin pathway in CRC. TCF/β-catenin transcription can be dampened by ectopic expression of naturally occurring inhibitors of Wnt/FZD or FZD–LRP interaction (reviewed in Vincan and Barker, 2008). Notably, inhibitors of the Wnt/FZD receptor complex are *bona fide* tumor suppressors in CRC and are epigenetically silenced very early in the journey from normal epithelial stem cell to cancer cell, sometimes even before *APC* gene truncation (Caldwell *et al.*, 2004; Suzuki *et al.*, 2004). These studies not only demonstrated *additional* regulation of the Wnt pathway but also indirectly implicated Wnt and FZD *per se* in CRC.

A role for Wnt and FZD in CRC

The fact that inhibitors of the Wnt/FZD receptor complex are *bona fide* tumor suppressors in CRC is a strong indication for a causative role for Wnt and FZD. Indeed, we and others showed that Wnt and FZD are overexpressed in CRC (Dimitriadis *et al.* 2001; Holcombe *et al.*, 2002; Vincan, 2004; Vincan *et al.*, 2007b) and that a dominant-negative FZD7 construct has potent antitumor activity (Vincan *et al.*, 2005). A role for FZD7 in CRC is further supported by findings in a novel CRC morphogenesis model (called LIM1863-*Mph*) (Figure 24.4). LIM1863-*Mph* is a human CRC cell line that expresses LGR5 (Vincan/Barker, unpublished). In this

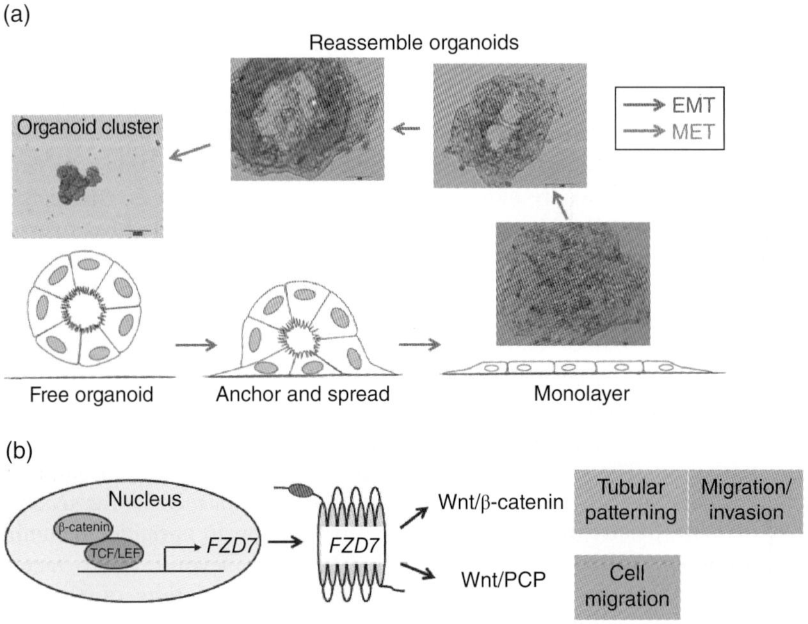

Figure 24.4 Wnt signaling in CRC morphogenesis. (a) The LIM1863-*Mph* model recapitulates many features of CRC tumor morphogenesis. The cells spontaneously undergo transitions between monolayer and organoid states in tissue culture medium. The phenotype change from organoid to monolayer is similar to EMT (*red arrows*), while the reverse is similar to MET (*blue arrows*) observed in carcinoma tissues (shown are DIC images, *blue circle* indicates one organoid in a cluster of organoids, scale bar 100 μM). (b) Experimental evidence indicates that the Wnt receptor FZD7, which is itself a TCF/β-catenin target gene, dictates patterning of the organoids via TCF/β-catenin signaling and monolayer cell migration via Wnt/PCP signaling. *(See insert for color representation of the figure.)*

culture system, the LIM1863-*Mph* cells undergo cyclic transitions between monolayer and organoid states (Vincan *et al.*, 2007a, b). The LIM1863-*Mph* organoids resemble enclosed crypts (or enclosed carcinoma tubules) and contain several epithelial cell types. The LIM1863-*Mph* cells possess some progenitor activity as dispersed organoid cells can be single-cell cloned; subsequently, the single cells expand and again assemble organoids that contain the different cell types (Vincan, unpublished). Thus, the LIM1863-*Mph* culture system recapitulates the organoid self-assembly capacity of the LGR5$^+$ intestinal cells (Sato *et al.*, 2009), which indicates that LGR5$^+$ cells might also "self-assemble" carcinomas. Notably, FZD7, which is itself a TCF/β-catenin target gene (Vincan *et al.*, 2010; Willert *et al.*, 2002), is necessary for LIM1863-*Mph* organoid assembly (Vincan *et al.*, 2007b), implicating an important function for FZD7 in crypt architecture. This prospect and the identity of the Wnts involved in this process are under investigation.

Intriguingly, the *Xenopus* ortholog of FZD7 (Xfz7) can activate different branches of Wnt signaling, depending on the context (Medina, Reintsch, and Steinbeisser, 2000). This versatile signaling capacity might underlie the very important roles Fz7 orthologs have in diverse developmental processes. It might also underlie the frequent overexpression of FZD7 in diverse cancers (Vincan, 2004). Emerging evidence indicates that each branch of Wnt signaling is involved in cancer (Lai, Chien, and Moon, 2009). Initial findings in the LIM1863-*Mph* model indicated that FZD7 was necessary for migration of the monolayer cells and that this was not dependent on Wnt/β-catenin signaling but possibly on the Wnt/PCP pathway (Vincan *et al.*, 2007b). Indeed, this notion was corroborated in further studies by Ueno and colleagues who demonstrated that CRC cancer cell migration involved FZD7-mediated PCP signaling (Ueno *et al.*, 2009). Given FZD7's signaling diversity and frequent overexpression in stem cell and cancer cell contexts (Assou *et al.*, 2007; Vincan, 2004; Vincan and Barker, 2008; Vincan *et al.*, 2007c), mechanisms to block or harness its function are in the not too distant future.

Wnt signaling in liver cancer

Wnt signaling temporally modulates liver development, restricting growth at the early stages (Micsenyi *et al.*, 2004) while inducing liver bud growth and differentiation in later stages (McLin, Rankin, and Zorn, 2007). In adult livers, Wnt signaling is also central to the maintenance of homeostasis. Hepatocytes perform different metabolic functions including gluconeogenesis, glycolysis, glutamine synthesis, and urea formation. These functions are performed by groups of hepatocytes located in different areas along the portocentral axis of the liver lobule, known as the metabolic zonation (Jungermann and Kietzmann, 1996; Torre, Perret, and Colnot, 2010). β-catenin has an important role in maintaining this zonation; it is normally activated in pericentral hepatocytes, while APC expression in the same area is low. Total loss of liver APC results in loss of zonation and in lethal hepatocyte hyperproliferation (Benhamouche *et al.*, 2006). Metabolic enzymes in the pericentral area are induced by β-catenin, while periportal enzymes or transporters are repressed, indicating its central role in liver homeostasis (Torre, Perret, and Colnot, 2010).

Several risk factors are associated with the development of HCC (viral hepatitis B and C, alcohol abuse, metabolic liver disease, α-toxin intoxication). The resulting tumors display heterogeneity and differential mutational profiles (Gougelet and Colnot, 2012). The common thread, however, is chronic liver inflammation leading to injury and hepatocyte necrosis. To replace the injured hepatocytes, liver progenitor stem cells are activated. This is unlike what is seen in normal conditions such as liver regeneration following hepatectomy, where loss of hepatic mass stimulates the proliferation of differentiated hepatocytes. Inflammation causes the activation of fibroblasts, stellate cells, and macrophages, which secrete extracellular matrix proteins (contributing to fibrosis), cytokines, and proinflammatory signaling molecules including Wnt ligands (Chen *et al.*, 2012; Yang *et al.*, 2008b). Increased levels of astrocyte-elevated gene-1 (AEG1) have a major role in chronic inflammation of the liver and activate Wnt/β-catenin and NF-kB signaling in HCC. AEG1 is overexpressed in over 90% of human HCCs (Yoo *et al.*, 2009). Further direct involvement of Wnt signaling in HCC is through interaction with viral antigens such as HBx, which induces the activation of TCF/β-catenin transcription (Keng *et al.*, 2011).

Continuous exposure to unresolved viral infection and viral antigen is thought to result in mutations of the activated liver stem cells giving rise to HCC. Gene signatures of HCC cells include tumor "stemness" molecules, activating Wnt/β-catenin signaling partners, epithelial cell adhesion molecules, hypoxia markers, and proangiogenic factors, indicating a metastatic potential (Chen *et al.*, 2012). Notably, coexpression of β-catenin and HIF1α (indicating hypoxia) in HCCs correlates with shorter patient survival and shorter time to recurrence (Liu *et al.*, 2010).

The role of Wnt signaling pathway in the development and progression of HCC is further exemplified by the frequent mutations in Wnt/β-catenin pathway components (Guichard *et al.*, 2012). However, mutations in the β-catenin gene and nuclear accumulation of the mutated truncated form of β-catenin are not predictive markers of worse progression. HCC tumors with mutated β-catenin displayed low genomic instability and better patient survival (Laurent-Puig *et al.*, 2001). In contrast, nuclear accumulation of wild-type β-catenin, due to mutation or deregulation of one of the Wnt signaling pathway components or other molecules that impact on Wnt signaling such as hepatocyte growth factor (HGF) or transforming growth factor (TGF)-β, is associated with aggressive tumor progression (Hoshida *et al.*, 2009; Yu *et al.*, 2009).

A role for Wnt and FZD in HCC

As in CRC, *additional* regulation of Wnt signaling at the plasma membrane in HCC is indicated by the frequent silencing of inhibitors of the Wnt/Fz receptor complex and the overexpression of ligands and receptors. For example, expression of antagonists such as sFRP1 is decreased by epigenetic silencing, proteolytic degradation, or inactivating mutations (Chan *et al.*, 2006; Shih *et al.*, 2006; Yau *et al.*, 2005). Indeed, repression of sFRP1/sFRP5 correlates with the severity of liver disease and tumor stage (Bengochea *et al.*, 2008). The corollary is that Wnts and FZDs are upregulated. FZD7 is overexpressed in HCC and imparts migratory and invasive properties to HCC cells by interacting with Wnt3 (Kim *et al.*, 2008; Merle *et al.*, 2004), while a dominant-negative form of FZD7 or pharmacological inhibition of FZD7 potently inhibits HCC tumor growth (Nambotin *et al.*, 2011). More recent studies have shown that FZD3/6 and WNT3/4/5a are also upregulated in HCC (Bengochea *et al.*, 2008). Thus, HCC shares many parallels with CRC with respect to the involvement of Wnt signaling in cancer initiation and progression.

Perspectives

Cancer biologists recognized many decades ago that fetal genes are reexpressed in cancer; however, the significance of this observation was realized relatively recently. The now-recognized striking analogies between development and cancer have meant that developmental pathways such as the Wnt pathway have become the subject of intense investigation in the quest for potential anticancer therapies. This has been fuelled by the link between stem cells and Wnt signaling pathways and the gene program overlap between stem cells and cancer cells. The fact that inhibitors of Wnt/FZD interaction are tumor suppressors in various cancers will no doubt result in anticancer therapies targeting the Wnt receptor complex. On the converse, activating the Wnt receptor complex to harness or control stem cell properties has potential for regenerative medicine.

Acknowledgement

This work was supported, in part, by grants from the Cancer Council of Victoria (APP1020716 to E.V. and N.B.) and Melbourne Health (605030 to E.V. and N.B.).

References

Arumugam, T., Ramachandran, V., Fournier, K.F. *et al.* (2009) Epithelial to mesenchymal transition contributes to drug resistance in pancreatic cancer. *Cancer Research*, **69** (14), 5820–5828.

Assou, S., Le Carrour, T., Tondeur, S. *et al.* (2007) A meta-analysis of human embryonic stem cells transcriptome integrated into a web-based expression atlas. *Stem Cells*, **25** (4), 961–973.

Barker, N., van de Wetering, M., and Clevers, H. (2008) The intestinal stem cell. *Genes & Development*, **22** (14), 1856–1864.

Barker, N., van Oudenaarden, A., and Clevers, H. (2012) Identifying the stem cell of the intestinal crypt: strategies and pitfalls. *Cell Stem Cell*, **11** (4), 452–460.

Barker, N., van Es, J.H., Kuipers, J. *et al.* (2007) Identification of stem cells in small intestine and colon by marker gene Lgr5. *Nature*, **449** (7165), 1003–1007.

Barker, N., Ridgway, R.A., van Es, J.H. *et al.* (2009) Crypt stem cells as the cells-of-origin of intestinal cancer. *Nature*, **457** (7229), 608–611.

Bengochea, A., de Souza, M.M., Lefrancois, L. *et al.* (2008) Common dysregulation of Wnt/Frizzled receptor elements in human hepatocellular carcinoma. *British Journal of Cancer*, **99** (1), 143–150.

Benhamouche, S., Decaens, T., Godard, C. *et al.* (2006) Apc tumor suppressor gene is the "zonation-keeper" of mouse liver. *Developmental Cell*, **10** (6), 759–770.

Bhanot, P., Brink, M., Samos, C.H. *et al.* (1996) A new member of the frizzled family from Drosophila functions as a Wingless receptor. *Nature*, **382** (6588), 225–230.

Bissell, M.J. and Radisky, D. (2001) Putting tumours in context. *Nature Reviews Cancer*, **1** (1), 46–54.

Brabletz, T., Jung, A., Hermann, K. *et al.* (1998) Nuclear overexpression of the oncoprotein beta-catenin in colorectal cancer is localized predominantly at the invasion front. *Pathology, Research & Practice*, **194** (10), 701–704.

Brabletz, T., Jung, A., Reu, S. *et al.* (2001) Variable beta-catenin expression in colorectal cancers indicates tumor progression driven by the tumor environment. *Proceedings of the National Academy of Sciences of the United States of America*, **98** (18), 10356–10361.

Brabletz, T., Jung, A., Spaderna, S. *et al.* (2005) Opinion: migrating cancer stem cells – an integrated concept of malignant tumour progression. *Nature Reviews Cancer*, **5** (9), 744–749.

Buck, E., Eyzaguirre, A., Barr, S. *et al.* (2007) Loss of homotypic cell adhesion by epithelial-mesenchymal transition or mutation limits sensitivity to epidermal growth factor receptor inhibition. *Molecular Cancer Therapeutics*, **6** (2), 532–541.

Caldwell, G.M., Jones, C., Gensberg, K. *et al.* (2004) The Wnt antagonist sFRP1 in colorectal tumorigenesis. *Cancer Research*, **64** (3), 883–888.

Chan, D.W., Chan, C.Y., Yam, J.W. *et al.* (2006) Prickle-1 negatively regulates Wnt/beta-catenin pathway by promoting Dishevelled ubiquitination/degradation in liver cancer. *Gastroenterology*, **131** (4), 1218–1227.

Chen, L., Zhang, Q., Chang, W. *et al.* (2012) Viral and host inflammation-related factors that can predict the prognosis of hepatocellular carcinoma. *European Journal of Cancer*, **48** (13), 1977–1987.

Clevers, H. and Nusse, R. (2012) Wnt/beta-catenin signaling and disease. *Cell*, **149** (6), 1192–1205.

Cromheecke, M., de Jong, K.P., and Hoekstra, H.J. (1999) Current treatment for colorectal cancer metastatic to the liver. *European Journal of Surgical Oncology*, **25** (5), 451–463.

Davies, R.J., Miller, R., and Coleman, N. (2005) Colorectal cancer screening: prospects for molecular stool analysis. *Nature Reviews Cancer*, **5** (3), 199–209.

de Krijger, I., Mekenkamp, L.J., Punt, C.J., and Nagtegaal, I.D. (2011) MicroRNAs in colorectal cancer metastasis. *The Journal of Pathology*, **224** (4), 438–447.

Dimitriadis, A., Vincan, E., Mohammed, I.M. *et al.* (2001) Expression of Wnt genes in human colon cancers. *Cancer Letters*, **166** (2), 185–191.

Egeblad, M., Nakasone, E.S., and Werb, Z. (2010) Tumors as organs, complex tissues that interface with the entire organism. *Developmental Cell*, **18** (6), 884–901.

Eger, A., Stockinger, A., Park, J. *et al.* (2004) β-Catenin and TGFbeta signalling cooperate to maintain a mesenchymal phenotype after FosER-induced epithelial to mesenchymal transition. *Oncogene*, **23** (15), 2672–2680.

Erler, J.T. and Weaver, V.M. (2009) Three-dimensional context regulation of metastasis. *Clinical & Experimental Metastasis*, **26** (1), 35–49.

Evdokimova, V., Tognon, C., Ng, T. *et al.* (2009) Translational activation of snail1 and other developmentally regulated transcription factors by YB-1 promotes an epithelial-mesenchymal transition. *Cancer Cell*, **15** (5), 402–415.

Fan, X.J., Wan, X.B., Yang, Z.L. *et al.* (2013) Snail promotes lymph node metastasis and Twist enhances tumor deposit formation through epithelial-mesenchymal transition in colorectal cancer. *Human Pathology*, **44** (2), 173–180.

Fodde, R., Smits, R., and Clevers, H. (2001) APC, signal transduction and genetic instability in colorectal cancer. *Nature Reviews Cancer*, **1** (1), 55–67.

Gougelet, A. and Colnot, S. (2012) A complex interplay between Wnt/beta-catenin signalling and the cell cycle in the adult liver. *International Journal of Hepatology*, **2012**, 816125.

Graham, T.R., Zhau, H.E., Odero-Marah, V.A. *et al.* (2008) Insulin-like growth factor-I-dependent upregulation of ZEB1 drives epithelial-to-mesenchymal transition in human prostate cancer cells. *Cancer Research*, **68** (7), 2479–2488.

Gregory, P.A., Bert, A.G., Paterson, E.L. *et al.* (2008) The miR-200 family and miR-205 regulate epithelial to mesenchymal transition by targeting ZEB1 and SIP1. *Nature Cell Biology*, **10** (5), 593–601.

Gregory, P.A., Bracken, C.P., Smith, E. *et al.* (2011) An autocrine TGF-beta/ZEB/miR-200 signaling network regulates establishment and maintenance of epithelial-mesenchymal transition. *Molecular Biology of the Cell*, **22** (10), 1686–1698.

Guichard, C., Amaddeo, G., Imbeaud, S. *et al.* (2012) Integrated analysis of somatic mutations and focal copy-number changes identifies key genes and pathways in hepatocellular carcinoma. *Nature Genetics*, **44** (6), 694–698.

Hiratsuka, S., Watanabe, A., Aburatani, H., and Maru, Y. (2006) Tumour-mediated upregulation of chemoattractants and recruitment of myeloid cells predetermines lung metastasis. *Nature Cell Biology*, **8** (12), 1369–1375.

Holcombe, R.F., Marsh, J.L., Waterman, M.L. *et al.* (2002) Expression of Wnt ligands and Frizzled receptors in colonic mucosa and in colon carcinoma. *Molecular Pathology*, **55** (4), 220–226.

Hoshida, Y., Nijman, S.M., Kobayashi, M. *et al.* (2009) Integrative transcriptome analysis reveals common molecular subclasses of human hepatocellular carcinoma. *Cancer Research*, **69** (18), 7385–7392.

Huber, M.A., Kraut, N., and Beug, H. (2005) Molecular requirements for epithelial-mesenchymal transition during tumor progression. *Current Opinion in Cell Biology*, **17** (5), 548–558.

Jemal, A., Murray, T., Ward, E. *et al.* (2005) Cancer statistics, 2005. *CA: A Cancer Journal for Clinicians*, **55** (1), 10–30.

Jung, A., Schrauder, M., Oswald, U. *et al.* (2001) The invasion front of human colorectal adenocarcinomas shows co-localization of nuclear beta-catenin, cyclin D1, and p16INK4A and is a region of low proliferation. *The American Journal of Pathology*, **159** (5), 1613–1617.

Jungermann, K. and Kietzmann, T. (1996) Zonation of parenchymal and nonparenchymal metabolism in liver. *Annual Review of Nutrition*, **16**, 179–203.

Kaplan, D.L., Moon, R.T., and Vunjak-Novakovic, G. (2005) It takes a village to grow a tissue. *Nature Biotechnology*, **23** (10), 1237–1239.

Karhadkar, S.S., Bova, G.S., Abdallah, N. *et al.* (2004) Hedgehog signalling in prostate regeneration, neoplasia and metastasis. *Nature*, **431** (7009), 707–712.

Keng, V.W., Tschida, B.R., Bell, J.B., and Largaespada, D.A. (2011) Modeling hepatitis B virus X-induced hepatocellular carcinoma in mice with the Sleeping Beauty transposon system. *Hepatology*, **53** (3), 781–790.

Kim, M., Lee, H.C., Tsedensodnom, O. *et al.* (2008) Functional interaction between Wnt3 and Frizzled-7 leads to activation of the Wnt/beta-catenin signaling pathway in hepatocellular carcinoma cells. *Journal of Hepatology*, **48** (5), 780–791.

Kinzler, K.W. and Vogelstein, B. (1996) Lessons from hereditary colorectal cancer. *Cell*, **87** (2), 159–170.

Kirchner, T. and Brabletz, T. (2000) Patterning and nuclear beta-catenin expression in the colonic adenoma-carcinoma sequence. Analogies with embryonic gastrulation. *The American Journal of Pathology*, **157** (4), 1113–1121.

Lai, S.L., Chien, A.J., and Moon, R.T. (2009) Wnt/Fz signaling and the cytoskeleton: potential roles in tumorigenesis. *Cell Research*, **19** (5), 532–545.

Laurent-Puig, P., Legoix, P., Bluteau, O. *et al.* (2001) Genetic alterations associated with hepatocellular carcinomas define distinct pathways of hepatocarcinogenesis. *Gastroenterology*, **120** (7), 1763–1773.

Liu, L., Zhu, X.D., Wang, W.Q. *et al.* (2010) Activation of beta-catenin by hypoxia in hepatocellular carcinoma contributes to enhanced metastatic potential and poor prognosis. *Clinical Cancer Research*, **16** (10), 2740–2750.

Lu, P., Weaver, V.M., and Werb, Z. (2012) The extracellular matrix: a dynamic niche in cancer progression. *The Journal of Cell Biology*, **196** (4), 395–406.

Mani, S.A., Guo, W., Liao, M.J. *et al.* (2008) The epithelial-mesenchymal transition generates cells with properties of stem cells. *Cell*, **133** (4), 704–715.

Marshman, E., Booth, C., and Potten, C.S. (2002) The intestinal epithelial stem cell. *BioEssays*, **24** (1), 91–98.

McLin, V.A., Rankin, S.A., and Zorn, A.M. (2007) Repression of Wnt/beta-catenin signaling in the anterior endoderm is essential for liver and pancreas development. *Development*, **134** (12), 2207–2217.

Medina, A., Reintsch, W., and Steinbeisser, H. (2000) Xenopus frizzled 7 can act in canonical and non-canonical Wnt signaling pathways: implications on early patterning and morphogenesis. *Mechanisms of Development*, **92** (2), 227–237.

Merle, P., de la Monte, S., Kim, M. *et al.* (2004) Functional consequences of frizzled-7 receptor overexpression in human hepatocellular carcinoma. *Gastroenterology*, **127** (4), 1110–1122.

Micsenyi, A., Tan, X., Sneddon, T. *et al.* (2004) Beta-catenin is temporally regulated during normal liver development. *Gastroenterology*, **126** (4), 1134–1146.

Morel, A.P., Lievre, M., Thomas, C. *et al.* (2008) Generation of breast cancer stem cells through epithelial-mesenchymal transition. *PLoS One*, **3** (8), e2888.

Nambotin, S.B., Lefrancois, L., Sainsily, X. *et al.* (2011) Pharmacological inhibition of Frizzled-7 displays anti-tumor properties in hepatocellular carcinoma. *Journal of Hepatology*, **54** (2), 288–299.

Nelson, W.J. and Nusse, R. (2004) Convergence of Wnt, beta-catenin, and cadherin pathways. *Science*, **303** (5663), 1483–1487.

Nguyen, L., Fifis, T., Malcontenti-Wilson, C. *et al.* (2012) Spatial morphological and molecular differences within solid tumors may contribute to the failure of vascular disruptive agent treatments. *BMC Cancer*, **12** (1), 522.

Nusse, R. and Varmus, H.E. (1982) Many tumors induced by the mouse mammary tumor virus contain a provirus integrated in the same region of the host genome. *Cell*, **31** (1), 99–109.

Ocana, O.H., Corcoles, R., Fabra, A. *et al.* (2012) Metastatic colonization requires the repression of the epithelial-mesenchymal transition inducer Prrx1. *Cancer Cell*, **22**, 709–724.

Peinado, H., Olmeda, D., and Cano, A. (2007) Snail, Zeb and bHLH factors in tumour progression: an alliance against the epithelial phenotype? *Nature Reviews Cancer*, **7** (6), 415–428.

Pena, C., Garcia, J.M., Garcia, V. *et al.* (2006) The expression levels of the transcriptional regulators p300 and CtBP modulate the correlations between SNAIL, ZEB1, E-cadherin and vitamin D receptor in human colon carcinomas. *International Journal of Cancer*, **119** (9), 2098–2104.

Radtke, F., Clevers, H., and Riccio, O. (2006) From gut homeostasis to cancer. *Current Molecular Medicine*, **6** (3), 275–289.

Sato, T., Vries, R.G., Snippert, H.J. *et al.* (2009) Single Lgr5 stem cells build crypt-villus structures in vitro without a mesenchymal niche. *Nature*, **459** (7244), 262–265.

Sato, T., van Es, J.H., Snippert, H.J. *et al.* (2011) Paneth cells constitute the niche for Lgr5 stem cells in intestinal crypts. *Nature*, **469** (7330), 415–418.

Schuijers, J. and Clevers, H. (2012) Adult mammalian stem cells: the role of Wnt, Lgr5 and R-spondins. *The EMBO Journal*, **31** (12), 2685–2696.

Schulenburg, A., Bramswig, K., Herrmann, H. *et al.* (2010) Neoplastic stem cells: current concepts and clinical perspectives. *Critical Reviews in Oncology and Hematology*, **76** (2), 79–98.

Shih, Y.L., Shyu, R.Y., Hsieh, C.B. *et al.* (2006) Promoter methylation of the secreted frizzled-related protein 1 gene SFRP1 is frequent in hepatocellular carcinoma. *Cancer*, **107** (3), 579–590.

Siegel, P.M., Shu, W., Cardiff, R.D. *et al.* (2003) Transforming growth factor beta signaling impairs Neu-induced mammary tumorigenesis while promoting pulmonary metastasis. *Proceedings of the National Academy of Sciences of the United States of America*, **100** (14), 8430–8435.

Sihto, H., Lundin, J., Lundin, M. *et al.* (2011) Breast cancer biological subtypes and protein expression predict for the preferential distant metastasis sites: a nationwide cohort study. *Breast Cancer Research*, **13** (5), R87.

Singh, A. and Settleman, J. (2010) EMT, cancer stem cells and drug resistance: an emerging axis of evil in the war on cancer. *Oncogene*, **29** (34), 4741–4751.

Snippert, H.J., van der Flier, L.G., Sato, T. *et al.* (2010) Intestinal crypt homeostasis results from neutral competition between symmetrically dividing Lgr5 stem cells. *Cell*, **143** (1), 134–144.

Spaderna, S., Schmalhofer, O., Hlubek, F. *et al.* (2006) A transient, EMT-linked loss of basement membranes indicates metastasis and poor survival in colorectal cancer. *Gastroenterology*, **131** (3), 830–840.

Spoelstra, N.S., Manning, N.G., Higashi, Y. *et al.* (2006) The transcription factor ZEB1 is aberrantly expressed in aggressive uterine cancers. *Cancer Research*, **66** (7), 3893–3902.

Stacker, S.A. and Achen, M.G. (2008) From anti-angiogenesis to anti-lymphangiogenesis: emerging trends in cancer therapy. *Lymphatic Research and Biology*, **6** (3–4), 165–172.

Steeg, P.S. (2006) Tumor metastasis: mechanistic insights and clinical challenges. *Nature Medicine*, **12** (8), 895–904.

Suzuki, H., Watkins, D.N., Jair, K.W. *et al.* (2004) Epigenetic inactivation of SFRP genes allows constitutive WNT signaling in colorectal cancer. *Nature Genetics*, **36** (4), 417–422.

Thiery, J.P. and Sleeman, J.P. (2006) Complex networks orchestrate epithelial-mesenchymal transitions. *Nature Reviews Molecular Cell Biology*, **7** (2), 131–142.

Timmerman, L.A., Grego-Bessa, J., Raya, A. *et al.* (2004) Notch promotes epithelial-mesenchymal transition during cardiac development and oncogenic transformation. *Genes & Development*, **18** (1), 99–115.

Torre, C., Perret, C., and Colnot, S. (2010) Molecular determinants of liver zonation. *Progress in Molecular Biology and Translational Science*, **97**, 127–150.

Tsai, J.H., Donaher, J.L., Murphy, D.A. *et al.* (2012) Spatiotemporal regulation of epithelial-mesenchymal transition is essential for squamous cell carcinoma metastasis. *Cancer Cell*, **22**, 725–736.

Ueda, Y., Wang, S., Dumont, N. *et al.* (2004) Overexpression of HER2 (erbB2) in human breast epithelial cells unmasks transforming growth factor beta-induced cell motility. *The Journal of Biological Chemistry*, **279** (23), 24505–24513.

Ueno, K., Hazama, S., Mitomori, S. *et al.* (2009) Down-regulation of frizzled-7 expression decreases survival, invasion and metastatic capabilities of colon cancer cells. *British Journal of Cancer*, **101** (8), 1374–1381.

van Es, J.H., Jay, P., Gregorieff, A. *et al.* (2005) Wnt signalling induces maturation of Paneth cells in intestinal crypts. *Nature Cell Biology*, **7** (4), 381–386.

Vincan, E. (2004) Frizzled/WNT signalling: the insidious promoter of tumour growth and progression. *Frontiers in Bioscience*, **9**, 1023–1034.

Vincan, E. and Barker, N. (2008) The upstream components of the Wnt signalling pathway in the dynamic EMT and MET associated with colorectal cancer progression. *Clinical and Experimental Metastasis*, **25** (6), 657–663.

Vincan, E., Darcy, P.K., Smyth, M.J. *et al.* (2005) Frizzled-7 receptor ectodomain expression in a colon cancer cell line induces morphological change and attenuates tumor growth. *Differentiation*, **73** (4), 142–153.

Vincan, E., Brabletz, T., Faux, M.C., and Ramsay, R.G. (2007a) A human three-dimensional cell line model allows the study of dynamic and reversible epithelial-mesenchymal and mesenchymal-epithelial transition that underpins colorectal carcinogenesis. *Cells Tissues Organs*, **185** (1–3), 20–28.

Vincan, E., Darcy, P.K., Farrelly, C.A. *et al.* (2007b) Frizzled-7 dictates three-dimensional organization of colorectal cancer cell carcinoids. *Oncogene*, **26** (16), 2340–2352.

Vincan, E., Swain, R.K., Brabletz, T., and Steinbeisser, H. (2007c) Frizzled7 dictates embryonic morphogenesis, implications for colorectal cancer progression. *Frontiers in Bioscience*, **12**, 4558–4567.

Vincan, E., Flanagan, D.J., Pouliot, N. *et al.* (2010) Variable FZD7 expression in colorectal cancers indicates regulation by the tumour microenvironment. *Developmental Dynamics*, **239** (1), 311–317.

Vinson, C.R. and Adler, P.N. (1987) Directional non-cell autonomy and the transmission of polarity information by the frizzled gene of Drosophila. *Nature*, **329** (6139), 549–551.

Willert, J., Epping, M., Pollack, J.R. *et al.* (2002) A transcriptional response to Wnt protein in human embryonic carcinoma cells. *BMC Developmental Biology*, **2** (1), 8–15.

Xie, L., Law, B.K., Chytil, A.M. *et al.* (2004) Activation of the Erk pathway is required for TGF-beta1-induced EMT in vitro. *Neoplasia*, **6** (5), 603–610.

Yang, M.H. and Wu, K.J. (2008) TWIST activation by hypoxia inducible factor-1 (HIF-1): implications in metastasis and development. *Cell Cycle*, **7** (14), 2090–2096.

Yang, M.H., Wu, M.Z., Chiou, S.H. *et al.* (2008a) Direct regulation of TWIST by HIF-1alpha promotes metastasis. *Nature Cell Biology*, **10** (3), 295–305.

Yang, W., Yan, H.X., Chen, L. *et al.* (2008b) Wnt/beta-catenin signaling contributes to activation of normal and tumorigenic liver progenitor cells. *Cancer Research*, **68** (11), 4287–4295.

Yang-Snyder, J., Miller, J.R., Brown, J.D. *et al.* (1996) A frizzled homolog functions in a vertebrate Wnt signaling pathway. *Current Biology*, **6** (10), 1302–1306.

Yau, T.O., Chan, C.Y., Chan, K.L. *et al.* (2005) HDPR1, a novel inhibitor of the WNT/beta-catenin signaling, is frequently downregulated in hepatocellular carcinoma: involvement of methylation-mediated gene silencing. *Oncogene*, **24** (9), 1607–1614.

Yoo, B.K., Emdad, L., Su, Z.Z. *et al.* (2009) Astrocyte elevated gene-1 regulates hepatocellular carcinoma development and progression. *The Journal of Clinical Investigation*, **119** (3), 465–477.

Yu, B., Yang, X., Xu, Y. *et al.* (2009) Elevated expression of DKK1 is associated with cytoplasmic/nuclear beta-catenin accumulation and poor prognosis in hepatocellular carcinomas. *Journal of Hepatology*, **50** (5), 948–957.

Zavadil, J., Cermak, L., Soto-Nieves, N., and Bottinger, E.P. (2004) Integration of TGF-beta/Smad and Jagged1/Notch signalling in epithelial-to-mesenchymal transition. *The EMBO Journal*, **23** (5), 1155–1165.

Zhang, Z., Liu, S., Shi, R., and Zhao, G. (2011) miR-27 promotes human gastric cancer cell metastasis by inducing epithelial-to-mesenchymal transition. *Cancer Genetics*, **204** (9), 486–491.

25 Wnt Signaling in Adult Stem Cells: Tissue Homeostasis and Regeneration

Frank J.T. Staal[1] and Riccardo Fodde[2]

[1]Leiden University Medical Center, Leiden, The Netherlands
[2]Josephine Nefkens Institute, Erasmus MC, Rotterdam, The Netherlands

What are adult stem cells?

An adult stem cell is defined as an undifferentiated cell that can be found among specialized, more differentiated cells in a tissue or organ. This type of stem cell has two important characteristics:

(1) It can propagate itself and maintain stem cell properties, a process called **self-renewal**.
(2) It can develop into all of the major specialized cell types of this tissue, a process called **multilineage differentiation** or **multipotency**.

These two properties can be found in stem cells of many different organs. Here, for the sake of simplicity, we will confine our discussion to blood-forming stem cells and intestinal stem cells. Another tissue that displays rapid turnover, the skin, also contains stem cells that are dependent on Wnt signaling. However, for in-depth discussion on skin stem cells, the reader is referred to excellent recent review articles (Arwert, Hoste, and Watt, 2012; Lim and Nusse, 2013).

It is important to distinguish adult stem cells from pluripotent and embryonic stem cells. Pluripotent cells have the ability to differentiate into any of the hundreds of different cell types of the body; embryonic stem cells, derived from early (preimplantation) embryos, are generally considered the prototype pluripotent stem cell, though germ cells and the epiblast of postimplantation embryos can also give rise to pluripotent cell lines in culture. Adult stem cells have a more limited differentiation potential and are able to form only a few different mature cell types. These cells are also referred to as somatic stem cells, where somatic refers to cells of the body and not the germline.

The primary roles of adult stem cells are (i) to maintain the homeostatic equilibrium between self-renewal and differentiation and (ii) to underlie the regenerative response upon tissue insults. Invariably, they constitute only a small fraction of the total pool of cells within an organ or tissue. Possibly, the main reason why stem cells have attracted such great interest from researchers is that they can be transplanted to repair damaged tissues and organs. In fact, adult hematopoietic, or blood-forming, stem cells isolated from bone marrow

Wnt Signaling in Development and Disease: Molecular Mechanisms and Biological Functions, First Edition.
Edited by Stefan Hoppler and Randall T. Moon.

(BM) have been used in clinical transplantation for more than 40 years.

The stem cell niche concept

The existence of a specialized **niche** or microenvironment essential for stem cell maintenance was initially proposed by Schofield (1978). Already by this time, it was suggested that stem cells are seen in association with other fixed-tissue cells that prevent stem cell differentiation and ensure its continuous proliferation. However, only recently advances have been made to define their exact location as well as the molecular mechanisms by which they regulate hematopoietic stem cells (HSCs). It is now clear that other types of stem cells, for instance, in colon, skin, or mammary gland, also reside in niches (Suda and Arai, 2008; Theise, 2006; Yen and Wright, 2006; Zhang and Li, 2008; Zhang *et al.*, 2003).

Within the niche, a unique molecular cross talk takes place between stem cells and niche-constituent cells that, during homeostasis, maintains the finely tuned equilibrium between self-renewal and differentiation but that can also mediate rapid activation of stem cells during the regenerative response to tissue insults such as inflammation or injury. The signals underlying these properties are complex and include **tumor growth factor-β (TGF-β)** and **noncanonical Wnt** signals (Wilson and Trumpp, 2006).

Ample molecular and functional evidence suggests the existence of a complex molecular cross talk between the stem cells and the niche cells in their close vicinity, leading to the definition of one adhesion and signaling unit termed "stem cell–niche synapse," in analogy to the neuronal and immunological synapses. A wide variety of factors are involved in this synapse, and they mediate mainly two types of interactions. They promote both cell–cell and cell–matrix adhesion in order to maintain niche and stem cells in close proximity, and they promote the activation of specific signaling pathways which play key roles in stem cell fate decisions, survival, and proliferation. The activation of signaling pathways can be mediated by direct cell–cell interactions through binding of membrane-associated ligands and receptors, such as Notch signaling (Gering and Patient, 2008), or by binding of soluble factors to specific receptors located both on the stem cell and on the niche cell, such as Wnt, Smad/TGF/BMP/Activin, and Hedgehog signaling (Blank, Karlsson, and Karlsson, 2008) and particularly Wnt signaling, which will be the focus in the following sections.

The prototypical stem cell: The hematopoietic stem cell (HSC)

Although it is currently well-established that HSCs underlie the daily turnover of the blood compartment, it was only in the 1960s that the notion was introduced according to which MPP can be found in the adult BM and are responsible for the continuous production of blood cells throughout life (McCulloch and Till, 1960). Since then, research from many different laboratories contributed to the phenotypic and functional characterization of HSCs, and nowadays, the blood system constitutes the paradigm for understanding stem cell biology. HSCs are located at the top of a **hierarchy** of lineage-specific progenitors that differentiate in an ordered fashion toward fully mature blood cells (Figure 25.1), thereby undergoing a stepwise loss of multilineage potential and becoming progressively committed to a single hematopoietic lineage.

HSCs are functionally defined by their ability to mediate long-term (LT) repopulation after transplantation. The most stringent version of this operational definition requires that HSCs have to be serially transplantable in recipient animals while retaining both self-renewal and multilineage differentiation capacity.

In the mouse, HSCs can be found in the BM within a rare subpopulation defined by the absence of lineage-specific markers, and high expression levels of stem cell antigen 1 (Sca1) and c-Kit (Cheshier *et al.*, 1999). Accordingly, this BM subset is known as **LSK (Lin-Sca1+Kit+)** or sometimes KLS cells, depending on the order of the markers in the abbreviation. Since only a small fraction of the LSK cells encompasses long-term repopulating (LTR) capacity, additional markers have been introduced to subdivide this heterogeneous population. LT-HSC containing LTR activity are

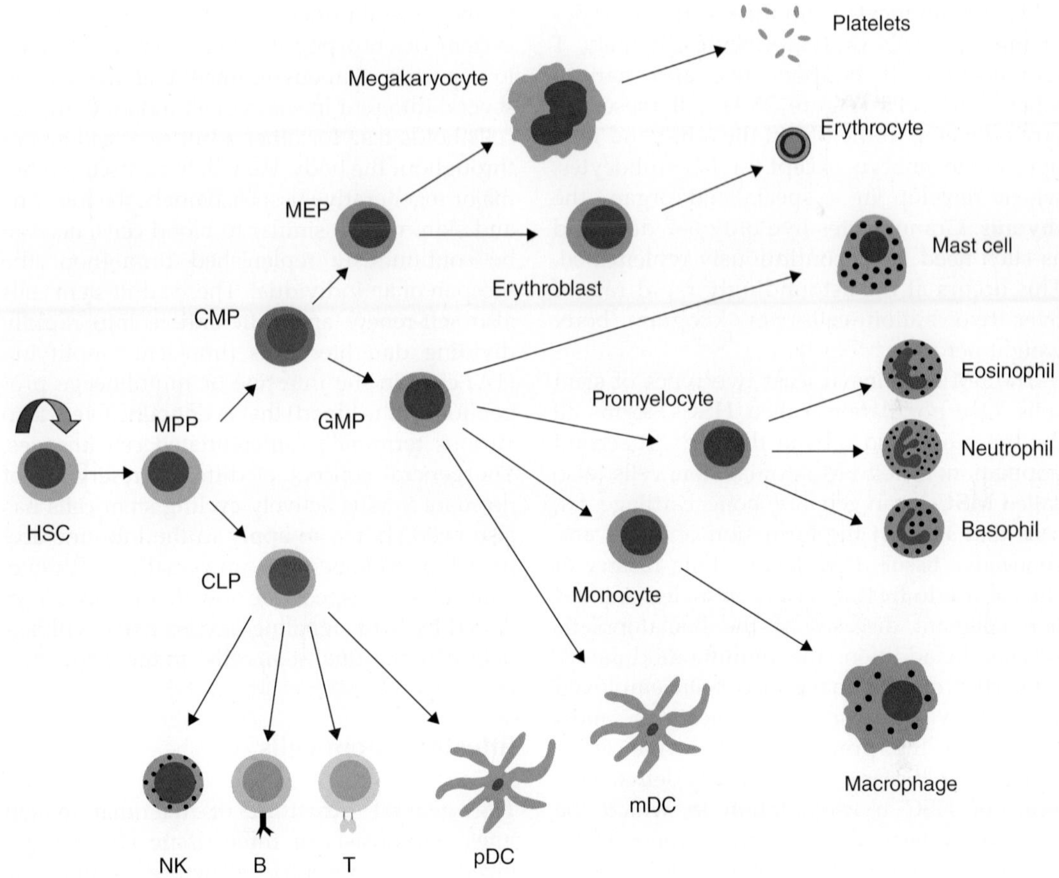

Figure 25.1 Schematic overview of hematopoiesis. HSCs are responsible for blood cell production throughout the lifetime of an individual. Pluripotent HSC can give rise to several different hematopoietic lineages while retaining the capacity for self-renewal. Lineage-committed progenitor cells produce progeny destined to differentiate into red cells, granulocytes, lymphoid cells, and platelets. *(See insert for color representation of the figure.)*

fms-related tyrosine kinase 3 negative ($Flt3^-$), $CD34^-$, $CD48^-$, and $CD150^+$. Short-term (ST-) HSC, characterized by a limited repopulation capacity, are $CD34^+$ and $Flt3^-$, whereas multipotent progenitors (MPP) that lost self-renewal but retain multilineage differentiation potential are $Flt3^+$ and $CD34^+$. In addition to cell surface markers, HSCs can also be identified by their ability to efflux fluorescent dyes such as Hoechst 33342. This property defines a subset named side population (SP) highly enriched for LT-HSCs. The basic characteristics of mouse HSCs seem to also apply for human HSCs. Nevertheless, due to the lack of efficient *in vivo* assays, the correct definition and isolation of human HSCs has been more difficult. Human HSCs are enriched in the $Lin^-CD34^+CD38^-$ subpopulation.

HSCs reside in BM niche that are constituted by a number of stromal cells in the endosteum. A broad spectrum of cell types, including osteoblasts, Cxcl12-abundant reticular (CAR) cells, **mesenchymal stem cells (MSCs)**, perivascular cells, and endothelial cells, have been proposed as integral components of the hematopoietic niche (Wilson and Trumpp, 2006). Moreover, cells from the neural system, nonmyelinating Schwann cells wrapping sympathetic nerve fibers, promote HSC quiescence through activation of TGF-β.

Different types of blood cells, namely, red cells or **erythrocytes**, platelets or **thrombocytes**, and various types of white blood cells or **leukocytes**, represent the progeny of HSCs. The different types of blood cells also include **granulocytes**, the most abundant white blood cells, and **monocytes** and three different types of immune cells or **lymphocytes**, namely, T lymphocytes, B lymphocytes, and natural killer (NK) cells (Figure 25.1). All these cell types develop in the BM in the adult and fetal liver in the embryo, except for T lymphocytes, which develop in a specialized organ, the **thymus**. Granulocytes live only 6–7 days and as such need to be continuously replenished. This occurs at the astoundingly rapid rate of over two billion cells per kilogram body weight per day.

The BM contains at least two types of stem cells. One population, called **HSCs**, forms all the types of blood cells in the body. A second population, called BM stromal stem cells (also called **MSCs**), can generate bone, cartilage, fat, cells that support the formation of blood and connective tissue. HSC have a long history of clinical use to treat a variety of malignant and nonmalignant diseases of the hematopoietic system. In addition, the immunomodulatory properties of MSC have also been employed successfully for clinical applications, most notably in the prevention of severe graft-versus-host disease, a potentially lethal side effect of HSC transplantation in which the immune system generated from donor HSCs recognizes their host as foreign and tries to reject host tissues.

Most researchers think that HSCs actively divide only in case of regeneration of damaged tissues. In steady-state conditions, HSCs divide infrequently and usually do not enter the cell cycle, lending this role to MPP which mostly maintain blood turnover. The ability to maintain this nondividing condition is called **quiescence**. Quiescence prevents stem cell pool exhaustion and protects HSC from acquiring mutations leading to malignant transformation and leukemia. Recent work by Andreas Trumpp and coworkers using modern lineage tracing techniques has elegantly shown (Koch *et al.*, 2008) that two populations of HSCs exist with different division kinetics: a dormant subset (15%) dividing approximately every 145 days in the life of a mouse and an "activated HSC" population (85%) dividing around every 5 weeks. Interestingly, cells may be in transition between the two kinetic states, establishing one subpopulation that is ready to proliferate and another that is a deeply quiescent reserve.

The general principle, as defined in the blood system, of a rare population of stem cells located in a specific microenvironment that gives rise to several different lineages of abundant daughter cells holds true for other adult stem cell niches throughout the body. We will here discuss other major regenerative tissues, namely, the intestine and skin, which, similar to blood cells, need to be continuously replenished throughout the lifespan of an individual. These adult stem cells also self-renew and differentiate into rapidly dividing daughter cells (transient amplifying (TA) cells in the intestine or multilineage progenitors in the blood) that eventually give rise to various terminally differentiated cell lineages. The general concept of different reservoirs of dormant versus actively cycling stem cells has also been shown to apply to the intestine and may, therefore, be more universally applicable. Some of these aspects are now thought to be regulated by Wnt signaling (see Section "Wnt signaling in intestinal stem cells" in the following).

Intestinal stem cells

The general structure of the mammalian intestine consists of three tissue layers organized in a concentric structure around the lumen. The outer layer comprises smooth muscle cells responsible for the rhythmic peristaltic movements of the intestine. The space between the outer muscle and the inner epithelial layer is filled by connective tissue ("**stroma**") encompassing numerous blood and lymph vessels, nerve fibers, and various cells of the immune system. A monolayer of epithelial cells, the so-called mucosa, lines the digestive tract throughout its length.

The intestine is subdivided into two major segments: the small and the large intestine or **colon**. The small intestine is again subdivided into three segments: the duodenum, jejunum, and ileum, whereas the colon is divided into its proximal (ascendens) and distal (descendens) parts, followed by the rectum. The absorptive surface area of the small intestine is dramatically increased by protrusions that point toward the lumen, the so-called villi, and invaginations into the submucosa known as the **crypts** of Lieberkühn. In the colon, only crypts without villi are present.

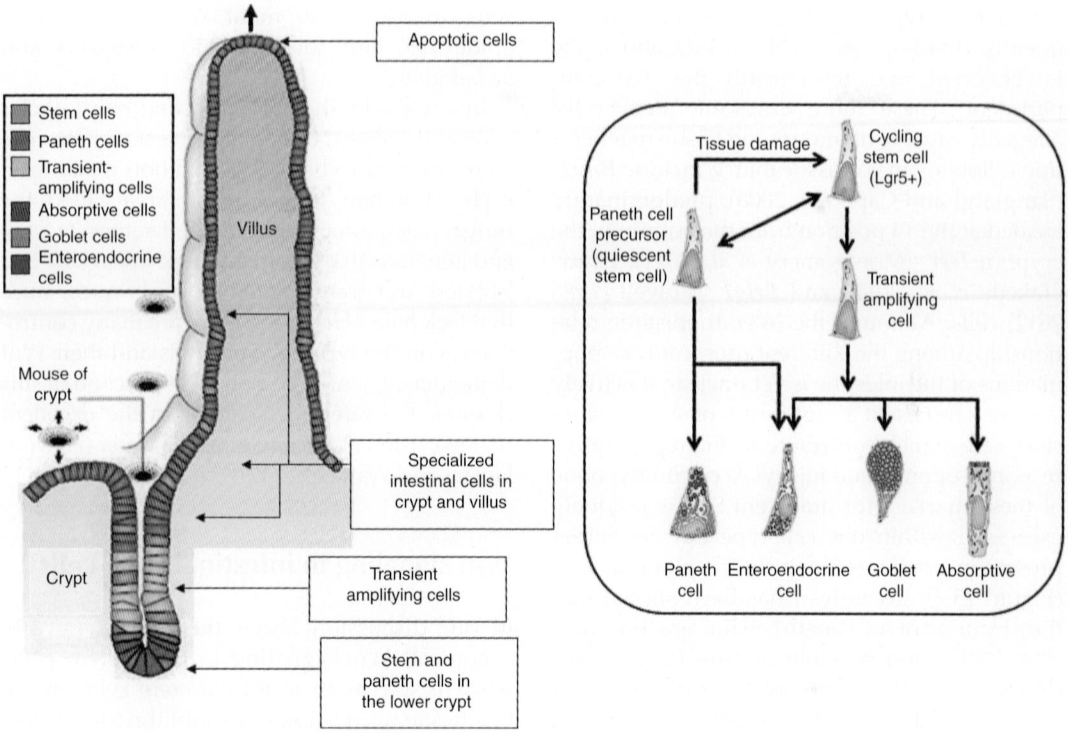

Figure 25.2 Intestinal stem cells in the crypt. The cell population that renews the epithelium of the small intestine occurs in the intestinal crypts, deep invaginations in the lining of the gut. The crypt contains stem cells and other cell types, giving rise to TA cells and differentiating in several other cell types. For details, see main text. *(See insert for color representation of the figure.)*

The small intestinal epithelium renews every 5 days and encompasses five different cell types: **absorptive (enterocytes)**, **enteroendocrine**, **goblet** (mucous-secreting), microfold (**M**), and **Paneth** cells. The relative abundance of each of these five cell types varies markedly within different segments of the intestine (see Figure 25.2). The enterocytes are responsible for the absorption of nutrients from the food and for the secretion of a cocktail of food-digesting enzymes into the lumen. The goblet cell numbers increase from the small intestine to colon and rectum. Enteroendocrine cells constitute a small proportion (<1%) of the intestinal epithelial cells and secrete a number of different hormones including serotonine, substance P, and secretin. Paneth cells reside at the bottommost positions of the crypts of the small intestine but are generally absent in the colon and rectum. They secrete antimicrobial agents such as defensins and lysozyme and provide support to rapidly cycling stem cells (see in the following).

The various types of epithelial cells are derived from stem cells located in the lower third of the crypt that give rise to rapidly dividing (TA) cells that progressively migrate from the crypt to the top of the villus while undergoing differentiation. Paneth cells represent the only terminally differentiated lineage located at the bottom of the crypt in close association with the stem cells. The nature of the intestinal stem cells has been subject to great controversy. Initially, two different stem cell populations have been described with distinct localizations along the crypt–villus axis. The rapidly dividing ***Lgr5***$^{+}$ stem cells are responsible for the daily renewal and high turnover of the intestinal epithelial lining and are located at the crypt bottom intermingled with Paneth cells which provide essential niche signals (Barker *et al.*, 2007; Sato *et al.*, 2011). In the colon, where Paneth cells are absent, a yet poorly characterized secretory lineage is likely to exert a similar niche function (Rothenberg *et al.*, 2012).

Different types of more quiescent (infrequently dividing) stem cells reside above the lower crypt and underneath the TA compartment. These more quiescent stem cells, allegedly capable of underlying tissue regeneration following major tissue injury, include ***Bmi1***$^+$ (Sangiorgi and Capecchi, 2008), predominantly located at the +4 position from the bottom of the crypt; ***mTert***$^+$ (Montgomery *et al.*, 2011); ***Hopx***$^+$ (Takeda *et al.*, 2011); and ***Lrig1***$^+$ (Powell *et al.*, 2012) cells. Although the overall epistatic relationship among the different quiescent SC populations of the intestine is yet unclear, it is likely that they represent a continuum of slow-cycling stem cells which contribute to the regenerative response upon tissue injury. Accordingly, none of these markers for quiescent SCs is uniquely expressed within one cell type but are rather present in gradients along the crypt–villus axis (Figure 25.2). Recently, it has been shown that the precursor of the Paneth cell lineage (Buczacki *et al.*, 2013) and possibly mature Paneth cells (Roth *et al.*, 2012) represent the true quiescent (label-retaining) stem cells of the intestinal crypt. Figure 25.2 depicts an overview of the current unifying model for the intestinal stem cell niche (Clevers, 2013).

The functional dichotomy of adult stem cells observed in the intestine with **quiescent and cycling stem cells** (*"tortoise and hair,"* as it has been referred to with reference to Aesop's fable) (Fuchs, 2009; Roth and Fodde, 2011) appears to be conserved in several adult stem cell niches including the hair follicle, epidermis, and BM. It is generally thought that while cycling stem cells are responsible for the maintenance of the daily homeostasis and tissue self-renewal, their quiescent counterparts are meant to protect "stemness" (stem cell identity) and serve as a backup upon tissue injury. As such, quiescent stem cells are activated only on tissue insults to underlie the regenerative response and thereafter return to their resting state.

Skin stem cells

Skin represents the largest organ in our body and not only forms a critical first line of defense against microbes but also retains our body fluids, preventing lethal dehydration. The skin starts as a single layer of multipotent epithelial cells during development, which give rise to epidermis, hair follicles, and sebaceous and sweat glands.

In adult skin, the epidermis and hair follicles constantly renew. Classically, the skin stem cells were thought to be one population residing in each adult hair follicle in an area called the **bulge**. These stem cells can self-renew *in vitro* and generate all of these different lineages when isolated and transplanted in nude mice, mice that lack hairs. However, there are many controversies on the types of stem cells and their Wnt dependency, which go beyond the scope of this chapter. We refer the readers to the excellent reviews listed at the end of this chapter (Arwert, Hoste, and Watt, 2012; Lim and Nusse, 2013).

Wnt signaling in intestinal stem cells

In our discussion about the role of (mostly canonical) Wnt signaling in adult stem cells, we will start with intestinal stem cells, as for this tissue most is known about the role of Wnt signaling in self-renewal, differentiation, and quiescence.

There is an abundance of evidence that Wnt signaling is essential for the proliferation of progenitor cells and the rapidly cycling stem cells in the crypts of the small intestine and colon. Some investigators postulated that a decreasing Wnt signaling gradient may characterize the crypt–villus axis (Gaspar and Fodde, 2004). The *Lgr5*$^+$ stem cells at the bottom of the intestinal crypt are earmarked by nuclear β-catenin, the hallmark of active Wnt signaling, as reviewed in earlier chapters of this textbook (Chapters 4 and 16). Abrogation of Wnt signaling by removal of Tcf4, the Wnt-responsive transcription factor in the gut, or by intestinal specific removal of β-catenin or by overexpression of the Wnt inhibitor Dickkopf 1 (DKK1) all result in decreased proliferation of stem and TA cells. This indicates that Wnt signaling is an absolute requirement for proliferation of these cells. Moreover, loss of canonical Wnt signaling is thought to directly affect the capacity of stem cells to self-renew and differentiate. Accordingly, loss of *APC* function and constitutive Wnt signaling activation represents the initiating and rate-limiting event in the vast majority of colon cancer in man (Fodde, Smits, and Clevers, 2001), leading to the expansion of the stem and progenitor cell compartment, increased proliferation, and eventually adenoma formation.

The identification of ***Lrg5***$^+$ stem cells by Hans Clevers and coworkers is central to the elucidation of the role of Wnt signaling in the regulation of intestinal homeostasis (Barker *et al.*, 2007). The *Lgr5* gene (also called *Gpr49*) represents a downstream Wnt target and encodes for a G protein-coupled receptor serving as receptor for **R-spondins**. R(oof plate-specific)-spondin 1 was originally identified in the neural tube. Later on, three additional R-spondins were identified. The R-spondin family members all synergize with canonical Wnt signaling and can be blocked by DKK1. Biochemical studies have shown that all four R-spondins can bind to Lgr5 and the related stem cell markers Lgr4 and Lgr6.

As mentioned earlier, Paneth cells are physically associated and intermingled with *Lgr5*$^+$ cells and represent, possibly together with other stromal components, the major sources of Wnt signals thought to play a central role in the regulation of stem cell homeostasis (Sato *et al.*, 2011). Apart from the Wnt ligand Wnt3, Paneth cells also express the Notch ligands Dll1 and Dll4, thus underlying active Notch signaling in *Lgr5*$^+$ cells and thus further contributing to their stem cell function. Apart from their role in self-renewal and maintenance of intestinal stemness, Wnt signals are also crucial for Paneth cell differentiation. Moreover, lack of Wnt signaling inhibits the development of goblet and enteroendocrine lineages. Overall, it appears that a combination of Notch and Wnt signals specifies different cell types in the (mouse) intestinal epithelium: Wnt$^+$ Notch$^-$ crypt cells are forced toward goblet or enteroendocrine differentiation, Wnt$^-$ Notch$^+$ cells are converted to enterocytes, whereas Wnt$^+$ Notch$^+$ cells maintain stemness. Downstream of Notch and Wnt, a broad spectrum of transcription factors, further defines the progeny of stem cells and controls their differentiation. Although the molecular and cellular mechanisms underlying these complex regulatory processes are yet largely unknown, it appears that different Wnt and Notch signals are likely to play crucial roles in determining cell fate decisions in a dosage-, space-, and time-dependent fashion.

Wnt signaling in hematopoietic stem cells

Compared to the convincing studies on the role of Wnt signaling in adult stem cells in skin and gut, a role for Wnt in adult HSCs (the prototypical adult stem cell) has proven much more difficult to demonstrate. In studies reporting an important role for Wnt signaling in blood cells, Wnt seemed to be required for normal HSC self-renewal and, therefore, for efficient reconstitution after transplantation (Reya *et al.*, 2003). Initial attempts to overexpress a constitutively active form of β-catenin in HSCs led to an increase in proliferation and repopulation capacity upon transplantation into lethally irradiated mice. However, later studies using conditional overexpression of a stabilized form of β-catenin led to a block in multilineage differentiation and a transient expansion of the HSC pool, followed by the exhaustion of LT-HSC (Kirstetter *et al.*, 2006). This resulted in anemic mice (fewer red cells) and eventually to lethality, that is, the opposite effect when compared to the improved transplantation setting reported earlier. These and other studies have created confusion concerning the importance of Wnts in maintaining number and integrity of HSCs. Recent findings suggest that this controversy results from differences in levels of Wnt signaling achieved in different experimental circumstances. That is, when Wnt signaling is slightly enhanced over normal levels, HSCs show improved reconstitution capacity. However, when HSCs express high levels of Wnt signaling, they completely fail to reconstitute irradiated recipient mice (Luis *et al.*, 2011). Thus, **different levels of activation** of the pathway can account for discrepancies in previous studies (Luis *et al.*, 2012).

Experimental approaches involving conditional deletion of β-catenin should be complementary to the overexpression studies, but again, the results were confusing. For example, the Mx-Cre system has been used to drive deletion of β-catenin or both β-catenin and its homolog γ-catenin. Mx-Cre is a widely used genetic tool to specifically remove DNA by the Cre DNA-cutting enzyme that is flanked by binding sites (loxP sites).The Mx-promoter is often used because it is inducible by signals given by the investigators that mimic normal inflammation signals. However, no defects were observed in HSC function or cells within lymphopoietic tissues such as the thymus. Surprisingly, *in vivo* reporter assays revealed that canonical Wnt signaling was still active in HSCs despite the absence of both β- and γ-catenin. This could imply the existence of redundant factors capable of transducing Wnt signals in the hematopoietic system. Alternatively, a hypomorphic β-catenin allele may have been generated by the targeting approach. In contrast to these studies targeting β-catenin, other loss-of-function approaches based on loss of Wnt3a, overexpression of the negative regulator DKK1 in osteoblastic stem cell niches, or Vav-Cre-mediated (rather than Mx-Cre) conditional deletion of β-catenin all indicated that a specific threshold of Wnt signaling is necessary for normal HSC function (reviewed in Luis *et al.*, 2012). These studies show that transient inhibition of the Wnt pathway during fetal development or in the adult BM niche irreversibly impairs HSC self-renewal.

In summary, although the strength of its signals in HSCs is possibly lower than in other adult stem cell systems, the Wnt pathway is likely to play an important role in the self-renewal program in hematopoiesis.

Concluding remarks

Canonical Wnt signaling has also been implicated in the self-renewal of a broad spectrum of stem cell compartments in the mammary gland, liver, stomach (Reya and Clevers, 2005), and ES cells (Chapter 20). Notably, Wnt signaling influences the capacity of ES cells to differentiate into the three main germ layers, ectoderm, mesoderm, and definitive endoderm, in a dosage-dependent fashion. These Wnt dosage-dependent effects also seem to hold true for adult self-renewing tissues such as blood, gut, and skin, though the underlying cellular and molecular mechanisms still remain poorly understood (Albuquerque *et al.*, 2002; Gaspar and Fodde, 2004; Luis *et al.*, 2011). In addition, different levels of activation of the pathway confer varying degrees of tumor susceptibility in different tissues, as seen in colon carcinomas versus mammary carcinomas (Gaspar *et al.*, 2009). In addition to the regulation of stem cell fate by means of levels, timing, and duration of Wnt signals, their outcome may be determined by the cellular context in which these signals are received. This may partially be explained by the cross talk between Wnt and other signaling pathways activated at the same time. Indeed, Wnt and Notch, as well as other signaling pathways such as Hedgehog, were previously shown to synergistically or antagonistically regulate each other, as discussed earlier for intestinal stem cells.

Overall, self-renewal, specification, commitment, and differentiation of adult stem cells are complex processes regulated by intricate networks of signals and transcription factors that necessitate a tightly fine-tuning and orchestration. Unraveling these mechanisms will be essential in order to fully explore the potential of stem cells and translate this basic knowledge into clinical applications.

References

Albuquerque, C., Breukel, C., van der Luijit, R. *et al.* (2002) The 'just-right' signaling model: APC somatic mutations are selected based on a specific level of activation of the beta-catenin signaling cascade. *Human Molecular Genetics*, **11**, 1549–1560.

Arwert, E.N., Hoste, E., and Watt, F.M. (2012) Epithelial stem cells, wound healing and cancer. *Nature Reviews Cancer*, **12**, 170–180.

Barker, N., van Es, J.H., Kuipers, J. *et al.* (2007) Identification of stem cells in small intestine and colon by marker gene Lgr5. *Nature*, **449**, 1003–1007.

Blank, U., Karlsson, G., and Karlsson, S. (2008) Signaling pathways governing stem-cell fate. *Blood*, **111**, 492–503.

Buczacki, S.J., Zecchini, H.I., Nicholson, A.M. *et al.* (2013) Intestinal label-retaining cells are secretory precursors expressing Lgr5. *Nature*, **495**, 65–69.

Cheshier, S.H., Morrison, S.J., Liao, X., and Weissman, I.L. (1999) In vivo proliferation and cell cycle kinetics of long-term self-renewing hematopoietic stem cells. *Proceedings of the National Academy of Sciences of the United States of America*, **96**, 3120–3125.

Clevers, H. (2013) Stem cells: a unifying theory for the crypt. *Nature*, **495**, 53–54.

Fodde, R., Smits, R., and Clevers, H. (2001) APC, signal transduction and genetic instability in colorectal cancer. *Nature Reviews Cancer*, **1**, 55–67.

Fuchs, E. (2009) The tortoise and the hair: slow-cycling cells in the stem cell race. *Cell*, **137**, 811–819.

Gaspar, C. and Fodde, R. (2004) APC dosage effects in tumorigenesis and stem cell differentiation.

The International Journal of Developmental Biology, **48**, 377–386.

Gaspar, C., Franken, P., Molenaar, L. *et al.* (2009) A targeted constitutive mutation in the APC tumor suppressor gene underlies mammary but not intestinal tumorigenesis. *PLoS Genetics*, **5**, e1000547.

Gering, M. and Patient, R. (2008) Notch in the niche. *Cell Stem Cell*, **2**, 293–294.

Kirstetter, P., Anderson, K., Porse, B.T. *et al.* (2006) Activation of the canonical Wnt pathway leads to loss of hematopoietic stem cell repopulation and multilineage differentiation block. *Nature Immunology*, **7**, 1048–1056.

Koch, U., Wilson, A., Cobas, M. *et al.* (2008) Simultaneous loss of beta- and gamma-catenin does not perturb hematopoiesis or lymphopoiesis. *Blood*, **111**, 160–164.

Lim, X. and Nusse, R. (2013) Wnt signaling in skin development, homeostasis, and disease. *Cold Spring Harbor Perspectives in Biology*, **5**, a008029.

Luis, T.C., Naber, B.A., Roozen, P.P. *et al.* (2011) Canonical wnt signaling regulates hematopoiesis in a dosage-dependent fashion. *Cell Stem Cell*, **9**, 345–356.

Luis, T.C., Ichii, M., Brugman, M.H. *et al.* (2012) Wnt signaling strength regulates normal hematopoiesis and its deregulation is involved in leukemia development. *Leukemia*, **26**, 414–421.

McCulloch, E.A. and Till, J.E. (1960) The radiation sensitivity of normal mouse bone marrow cells, determined by quantitative marrow transplantation into irradiated mice. *Radiation Research*, **13**, 115–125.

Montgomery, R.K., Carlone, D.L., Richmond, C.A. *et al.* (2011) Mouse telomerase reverse transcriptase (mTert) expression marks slowly cycling intestinal stem cells. *Proceedings of the National Academy of Sciences of the United States of America*, **108**, 179–184.

Powell, A.E., Wang, Y., Li, Y. *et al.* (2012) The pan-ErbB negative regulator Lrig1 is an intestinal stem cell marker that functions as a tumor suppressor. *Cell*, **149**, 146–158.

Reya, T. and Clevers, H. (2005) Wnt signalling in stem cells and cancer. *Nature*, **434**, 843–850.

Reya, T., Duncan, A.W., Ailles, L. *et al.* (2003) A role for Wnt signalling in self-renewal of haematopoietic stem cells. *Nature*, **423**, 409–414.

Roth, S. and Fodde, R. (2011) Quiescent stem cells in intestinal homeostasis and cancer. *Cell Communication & Adhesion*, **18**, 33–44.

Roth, S., Franken, P. Sacchetti, A. *et al.* (2012) Paneth cells in intestinal homeostasis and tissue injury. *PLoS One*, **7**, e38965.

Rothenberg, M.E., Nusse, Y., Kalisky, T. *et al.* (2012) Identification of a cKit(+) colonic crypt base secretory cell that supports Lgr5(+) stem cells in mice. *Gastroenterology*, **142**, 1195–1205.

Sangiorgi, E. and Capecchi, M.R. (2008) Bmi1 is expressed in vivo in intestinal stem cells. *Nature Genetics*, **40**, 915–920.

Sato, T.,van Es, J.H., Snippert, H.J. *et al.* (2011) Paneth cells constitute the niche for Lgr5 stem cells in intestinal crypts. *Nature*, **469**, 415–418.

Schofield, R. (1978) The relationship between the spleen colony-forming cell and the haemopoietic stem cell. *Blood Cells*, **4**, 7–25.

Suda, T. and Arai, F. (2008) Wnt signaling in the niche. *Cell*, **132**, 729–730.

Takeda, N., Jain, R., LeBoeuf, M.R. *et al.* (2011) Interconversion between intestinal stem cell populations in distinct niches. *Science*, **334**, 1420–1424.

Theise, N.D. (2006) The stem cell niche and tissue biology. *Stem Cell Reviews*, **2**, 169–170.

Wilson, A. and Trumpp, A. (2006) Bone-marrow haematopoietic-stem-cell niches. *Nature Reviews Immunology*, **6**, 93–106.

Yen, T.H. and Wright, N.A. (2006) The gastrointestinal tract stem cell niche. *Stem Cell Reviews*, **2**, 203–212.

Zhang, J. and Li, L. (2008) Stem cell niche: microenvironment and beyond. *The Journal of Biological Chemistry*, **283**, 9499–9503.

Zhang, J., Niu, C., Ye, L. *et al.* (2003) Identification of the haematopoietic stem cell niche and control of the niche size. *Nature*, **425**, 836–841.

26 Restoring Tissue Homeostasis: Wnt Signaling in Tissue Regeneration After Acute Injury

Günes Özhan[1] and Gilbert Weidinger[2]

[1]Biotechnology Center, Technische Universität Dresden, Dresden, Germany
[2]Institute for Biochemistry and Molecular Biology, Ulm University, Ulm, Germany

Introduction

All animals have evolved means of dealing with tissue damage due to injury or disease. Integrity of the outer body surface epithelia (skin) can be efficiently restored after insult in most species, but the ability to regenerate damage to other organs and structures varies widely. Regenerative phenomena can be classified by their scale: First, some invertebrates, most prominently the cnidarian hydra and planarian flatworms, are capable of regenerating the entire body from pieces of tissue or even dissociated cells. Secondly, the restoration of the complex architecture of damaged organs or severed appendages is possible in many organisms. In adult mammals, such organ-scale regeneration is only seen for the liver, while many lower vertebrates, in particular fish and urodele amphibia (salamanders and newts), can also completely regenerate lost limbs, fins, and tails; repair damage to the heart, the central nervous system, the eye (lens and retina); and display higher regenerative abilities than mammals in some internal organs including the kidney and the pancreas. Finally, tissue regeneration would describe restoration of a tissue within an otherwise largely intact organ, for example, skeletal muscle regeneration and repair of the epithelia of kidney or lung, all of which efficiently occur also in mammals. Here we will discuss select, fairly well-characterized phenomena from all three scales, namely, planarian whole-body regeneration, lower vertebrate appendage and vertebrate liver regeneration, and finally skeletal muscle and kidney epithelia regeneration.

Cellular mechanisms of regeneration

Much progress has recently been made, largely due to the use of genetic lineage tracing methods, in deciphering the cellular mechanisms of regeneration in several systems. Given the enormous variety of phenomena laid out in the section "Introduction," it is probably not surprising that different systems use very different cellular principles to supply cells replacing the damaged structure (Figure 26.1; Tanaka and Reddien, 2011). Some systems regenerate via activation of stem cells which usually also represent the source of tissue turnover during homeostasis (Figure 26.1a). As described later in this review, such stem cells can either form just

Wnt Signaling in Development and Disease: Molecular Mechanisms and Biological Functions, First Edition.
Edited by Stefan Hoppler and Randall T. Moon.

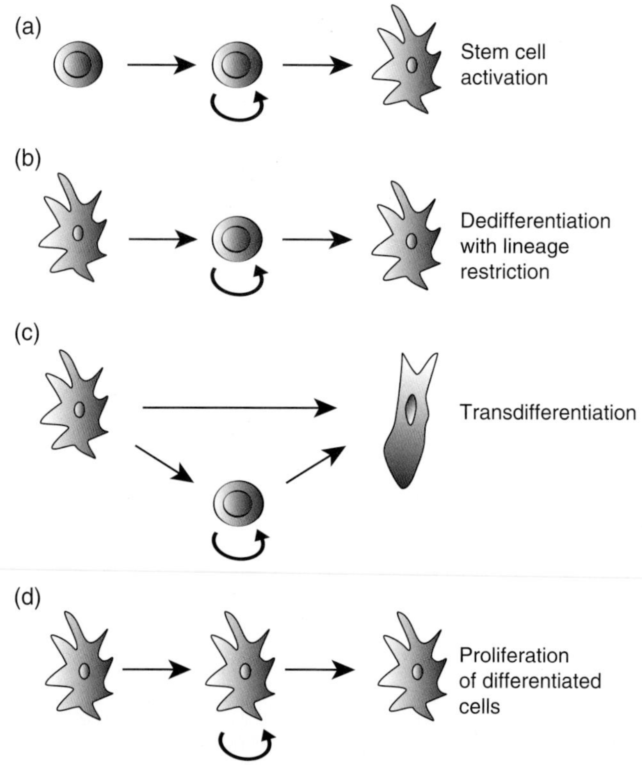

Figure 26.1 Cellular mechanisms of regeneration. (a) Regeneration involving tissue-resident or recruited stem cells that self-renew and produce differentiating daughter cells (e.g., skeletal muscle). (b) Dedifferentiation of mature cells to a progenitor state that self-renews and produces only cells of the same cell type and thus is lineage restricted (e.g., zebrafish fin osteoblasts and likely also heart cardiomyocytes). (c) Transdifferentiation of mature cells into another cell type either directly or via dedifferentiation to a progenitor state (e.g., newt lens). (d) Regeneration via cell cycle reentry of differentiated cells that do not lose their differentiated phenotype (e.g., mammalian liver after partial hepatectomy).

one cell type in the course of regeneration, as in the case of satellite cells of the skeletal muscle; they can be bipotent as seen with oval cells in the liver; or pluripotent as is the case for planarian neoblasts, which can give rise to the entire organism. Intriguingly, mature, differentiated adult cells can also be a source for regenerating tissues, and it appears that regeneration can involve different levels of changes to the differentiation status and cellular potential of mature cells. Some lineages, in particular osteoblasts during zebrafish fin regeneration, have been shown to regenerate from mature cells via dedifferentiation to a progenitor state that is unipotent, thus only gives rise to the same cellular lineage (Figure 26.1b; Knopf *et al.*, 2011; Sousa *et al.*, 2011). Mature cardiomyocytes are the precursors of regenerating heart muscle in zebrafish (Kikuchi *et al.*, 2010), and differentiated cells are likely also the source of other cellular lineages in regenerating fish and salamander appendages (Tanaka and Reddien, 2011). However, genetic lineage tracing of individual cell types will be needed to confirm this and to clarify to which extent these cells dedifferentiate, that is, whether they lose their differentiated character. In rare cases, like regeneration of the newt lens, transdifferentiation of one cell type to another has been proposed to occur (Figure 26.1c, Slack, 2007). Whether this involves dedifferentiation to a proliferating progenitor or direct conversion of cellular phenotype is less clear. Finally, regeneration can occur through proliferation of differentiated cells without a change in their differentiation status, although liver restoration in response to resection is the only known regenerative process thought to involve this mechanism (Figure 26.1d, see succeeding text). Interestingly, liver regeneration in response to toxic injury is achieved via activation of stem cells (Fausto, Campbell, and Riehle, 2006).

It is likely that different tissues within a complex regenerating structure can use different cellular mechanisms to replace themselves (e.g., bone might regenerate via dedifferentiation but muscle from stem cells in an injured appendage). Obviously, the complexity of cellular mechanisms underlying regeneration also will be reflected in a complexity of molecular mechanisms involved. Thus, it appears unlikely that simple common molecular principles of how regeneration is achieved can be deduced; rather we expect to find that regeneration of different systems and cell types within systems will be regulated differently.

Molecular mechanisms of regeneration

While regeneration has fascinated biologists for centuries and its cellular basis has been studied for decades, significant progress in our understanding of the molecular mechanisms of regeneration has only been made quite recently due to technical advances particularly for loss-of-function (LOF) studies in several systems and the introduction of genetically tractable models, like the zebrafish. Thus, many open questions remain, which can be grouped into three complexes: First, which cues trigger regeneration? Second, which mechanisms control regenerative growth, differentiation of cells, and restoration of tissue organization and pattern formation? And finally, how is restoration of the correct adult form achieved, specifically which mechanisms ensure that only the missing pieces are regenerated and that the restored organ reaches its pre-injury size?

Most progress has been made with the second group of questions, and it is here where Wnt signaling has been found to play prominent roles in almost all regenerating structures studied to date. Wnt signaling, in particular the β-catenin pathway, has, for example, been implicated in hydra head regeneration (Galliot and Chera, 2010), newt lens (Hayashi, Mizuno, and Kondoh, 2008) and amniote retina regeneration (Kubo and Nakagawa, 2008), mammalian pancreas regeneration (Figeac *et al.*, 2010) and lung repair (Zhang *et al.*, 2008), and also in injury responses of organs that show compensatory responses but little regeneration like the mammalian adult heart (Duan *et al.*, 2012; Laeremans *et al.*, 2011). However, here, we will concentrate on select systems where more significant progress has been made into a deeper understanding of Wnt signaling function.

Planarian whole-body regeneration

Planarians (flatworms) display some of the most striking examples of regenerative capacity in the animal kingdom. In fact, they are the only known bilateria that can reform missing body parts after almost any type of amputation: When heads are severed from the body, the head will regenerate the missing trunk, and the trunk fragment will reform the missing head; removing one body half in a longitudinal amputation will likewise result in formation of two regenerated animals, and regeneration even occurs from tiny fragments of tissue (Salo, 2006). Planarians also use whole-body regeneration for agametic reproduction via fission, and they display an elevated tissue turnover as adults. These worms continuously grow or shrink in size depending on food availability, and signaling pathways that set up overall body pattern, including Wnt/β-catenin, are required not only during embryogenesis but also to maintain adult form (Gurley, Rink, and Sanchez Alvarado, 2008; Salo, 2006). Thus, morphological plasticity and asexual reproduction are correlated with high regenerative capacity in planaria, but whether the ability to regenerate is the evolutionary basis for the process of fission or whether regeneration is a by-product of agametic reproduction is a matter of debate (Brockes and Kumar, 2008).

Tissue homeostasis and regeneration are achieved via the presence of stem cells, termed neoblasts, which are distributed throughout the body. In response to amputation, neoblasts expand to form a mass of undifferentiated cells, termed blastema, which collectively can form all cell types of the worm (Reddien and Sanchez Alvarado, 2004). Intriguingly, individual neoblasts are pluripotent, since a single transplanted neoblast can rescue regeneration in a host in which all neoblasts were killed by lethal irradiation and a single transplanted neoblast from an asexual strain can convert a sexual host into an asexual genotype (Reddien and Sanchez Alvarado, 2004). Thus, planarian regeneration is driven by an adult population of pluripotent (if not totipotent) stem cells.

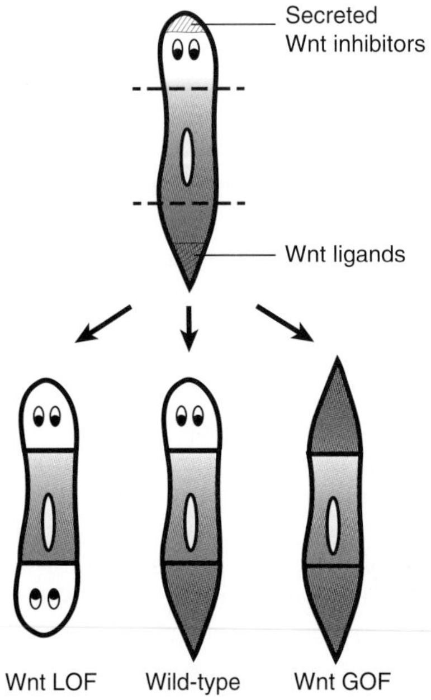

Figure 26.2 Wnt/β-catenin signaling regulates anteroposterior polarity during planaria regeneration. Planaria can regenerate their heads and tails when removed. Heads are enriched with Wnt inhibitors, such as Notum and sFRPs, while tails express Wnt ligands. When Wnt/β-catenin signaling is inhibited systemically (Wnt LOF), a head forms instead of a tail after amputation, while Wnt GOF results in regeneration of two-tailed animals. Thus, β-catenin signaling appears to specify posterior identity, and Wnt signaling needs to be off for heads to form.

Wnt/β-catenin signaling has been found to have essential functions in planarian regeneration. When heads and tails are removed from one animal, both structures regenerate (Figure 26.2). However, when β-catenin (Smed-β-catenin1, see Chapter 16), Dishevelled, or Evi/Wntless are systemically knocked down by RNAi, the tail wound forms a head, resulting in two-headed animals (Figure 26.2; Adell *et al.*, 2009; Gurley, Rink, and Sanchez Alvarado, 2008; Petersen and Reddien, 2008). Conversely, activated Wnt/β-catenin signaling due to the knockdown of APC results in tail regeneration from anterior wounds (Gurley, Rink, and Sanchez Alvarado, 2008). Thus, β-catenin signaling seems to be required for instructing anterior–posterior body polarity in adult planarians, a function that intriguingly is shared with vertebrate embryogenesis where high Wnt/β-catenin activity likewise determines posterior fates and Wnt signaling needs to be off for anterior structures to form (Tanaka and Weidinger, 2008; Reddien, 2011). Astonishingly, after β-catenin knockdown, also noninjured animals slowly convert their tails to heads and form ectopic heads all around the body (Gurley, Rink, and Sanchez Alvarado, 2008; Iglesias *et al.*, 2008; Petersen and Reddien, 2008), indicating that Wnt-dependent polarity needs to be constantly maintained during adult homeostasis.

Several Wnt ligands are expressed in the tail region of uninjured worms, and knockdown of Wnt1 converts regenerating tails to heads (Petersen and Reddien, 2009). Yet, shortly after wounding, Wnt1 expression is induced in all wounds, also the ones that face towards the anterior of the animal and thus regenerate a head (Petersen and Reddien, 2009). However, β-catenin signaling activity seems to be restricted to posterior wounds by pathway inhibitors that are specifically expressed in the anterior, namely, sFRPs, and Notum, whose knockdown results in regeneration of tails in place of heads (Figure 26.2; Petersen and Reddien, 2011; Reddien, 2011). Hedgehog (Hh) signaling appears to act upstream of Wnt/β-catenin in specification of AP polarity during regeneration via regulation of Wnt ligand expression (Rink *et al.*, 2009; Yazawa *et al.*, 2009). Yet, how posterior-specific Hh activity is achieved remains to be elucidated, since Hh pathway components do not show posterior-specific expression.

In conclusion, Wnt/β-catenin signaling plays dramatic roles in establishment and maintenance of tissue patterning during planarian regeneration and homeostasis. Whether Wnt signaling has additional, possibly earlier roles, for example, in processes linking wounding and initiation of a regenerative response, activation, and proliferation of neoblasts or cellular differentiation, has however not been described. Intriguingly, even among planarians, species with limited regenerative capacity exist, and a series of papers has recently shown that failure to downregulate β-catenin signaling in anterior wounds is responsible for the regenerative defects. Strikingly, RNAi-mediated knockdown of β-catenin is sufficient to rescue head regeneration in three planarian species, underscoring the central importance of the Wnt/β-catenin pathway for planarian regeneration (Liu *et al.*, 2013; Sikes and Newmark, 2013; Umesono *et al.*, 2013).

Lower vertebrate appendage regeneration

Cellular mechanisms of appendage regeneration

Some vertebrate groups have a remarkable capacity to fully regenerate lost appendages, that is, limbs, fins, or tails. Most of our knowledge about the mechanisms of appendage regeneration comes from studies of limb regeneration in urodele amphibia (salamanders and newts), fin regeneration in zebrafish, and larval tail regeneration in salamanders and *Xenopus laevis*. In these systems, a blastema forms at the site of injury, which has long been thought to form from mature adult cells via dedifferentiation, but solid evidence for this has only recently been obtained using genetic lineage tracing for osteoblasts during zebrafish fin regeneration. Here, the bone regenerates from mature osteoblasts which dedifferentiate to a lineage-restricted progenitor status and form part of the blastema (Knopf *et al.*, 2011; Sousa *et al.*, 2011). Thus, dedifferentiation of mature cells appears to be an important regenerative mechanism for some cellular lineages in the appendage, but other lineages, in particular muscle, can also regenerate via activation of stem cells (Morrison *et al.*, 2006). Irrespective of their cellular origin, blastema cells have recently been found to be lineage restricted in *Xenopus*, salamander, and zebrafish using lineage tracing techniques (Kragl *et al.*, 2009; Knopf *et al.*, 2011; Slack *et al.*, 2004; Stewart and Stankunas, 2012; Tu and Johnson, 2011). Thus, in contrast to the widely held view that considers the blastema a mass of multipotent cells that arises by dedifferentiation of adult cells back to an embryonic, highly plastic state, new evidence rather suggests a view of the appendage regeneration blastema as a heterogeneous population of cells with restricted potential. Therefore, the term "dedifferentiation" should only be used to describe loss of a molecularly or morphologically defined differentiated phenotype and should not imply a change in a cell's differentiation potential.

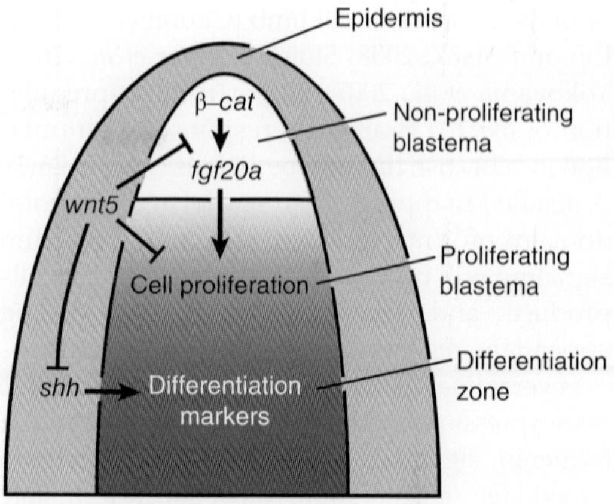

Figure 26.3 Model for Wnt signaling function during zebrafish fin regeneration. During the outgrowth phase of zebrafish fin regeneration, the mesenchymal blastema gets compartmentalized into a small distal domain (not drawn to scale) that is hardly proliferative and a highly proliferating proximal zone. Further proximally, cells start to differentiate. Wnt/β-catenin signaling activity is highest in the distal domain, yet Wnt signaling is essential for blastemal cell proliferation. Likely, signaling in the distal domain regulates expression of secondary diffusible signals, including *fgf20a,* that regulate proliferation in the proximal domain. In the distal wound epidermis, expression of *wnt5b*, which likely activates a β-catenin-independent signaling pathway, is required to restrict expression of *shh* to a more proximal region of the epidermis, which in turn regulates bone differentiation during later stages of fin regeneration. Noncanonical Wnt signaling, activated by *wnt5b*, might also inhibit β-catenin signaling and thus negatively regulate blastema proliferation. *Arrows* do not indicate direct regulation.

Roles of Wnt/β-catenin signaling

In addition to these advances in understanding the cellular mechanisms of blastema formation and potential, quite substantial progress has also been made in uncovering the molecular signals regulating blastema formation and function, in particular in zebrafish tail fin regeneration (Poss, 2010; Stoick-Cooper, Moon, and Weidinger, 2007). Wnt/β-catenin signaling has emerged as an important player in the regulation of appendage regeneration, while β-catenin-independent Wnt pathways appear to be involved as well.

The expression of several Wnt ligands, target genes, and transgenic reporters of β-catenin transcriptional activity is induced in response to amputation of the zebrafish tail fin, of newt tails, or of the *Xenopus* larval limb (Caubit *et al.*, 1997; Lin and Slack, 2008; Stoick-Cooper *et al.*, 2007; Yokoyama *et al.*, 2007). In particular, upregulation of *wnt10a* is an early response to amputation in zebrafish fins and newt tails (Figure 26.3). A detailed mapping of the spatial and temporal domains of Wnt expression and active β-catenin signaling will be required to deduce which cells produce and receive Wnt signals during appendage regeneration in different organisms.

Several lines of evidence, all based on systemic overexpression of pathway inhibitors, show that β-catenin signaling is essential for vertebrate appendage regeneration. Heat-shock-inducible transgenic or adenovirus-mediated overexpression of Dickkopf1 blocks regeneration of the zebrafish caudal fin (Kawakami *et al.*, 2006; Stoick-Cooper *et al.*, 2007), of larval salamander and *Xenopus* limbs (Kawakami *et al.*, 2006; Yokoyama *et al.*, 2007), and of larval *Xenopus* tails (Lin and Slack, 2008). Similarly, overexpression of Axin1 or a dominant-negative form of a TCF transcription factor inhibits zebrafish fin and adult salamander limb regeneration (Kawakami *et al.*, 2006; Stoick-Cooper *et al.*, 2007). The inhibition of β-catenin signaling during early stages of fin or limb regeneration does not impair epithelial migration and wound healing, but results in the formation of a mispatterned wound epidermis that lacks its characteristic stratification and cuboidally shaped basal layer (Kawakami *et al.*, 2006; Stoick-Cooper *et al.*, 2007). Subsequently, accumulation of blastemal cells beyond the amputation plane does not occur in Wnt-inhibited fins (Kawakami *et al.*, 2006; Stoick-Cooper *et al.*, 2007). Appendage blastema formation can be thought of as requiring three cellular processes: cellular dedifferentiation, migration of the dedifferentiated cells beyond the amputation plane, and cell proliferation. Whether Wnt/β-catenin signaling directly regulates any of these primary cellular processes has not yet been tested. The wound epidermis is thought to play a critical role as a source of signals required for blastema formation (Bryant, Endo, and Gardiner, 2002; Tsonis, 1996; Wallace, 1981), and there is also evidence for signaling from the blastemal mesenchyme to the wound epidermis (Chablais and Jazwinska, 2010). Thus it is conceivable, that – similar to limb development – complex reciprocal signaling interactions including Wnt signaling between epidermis and blastemal mesenchyme are necessary for appendage regeneration (Miura, 2013). A future challenge for the field thus will be to dissect the tissue-specific temporal requirements of signaling pathways, including the Wnt/β-catenin pathway, during appendage regeneration. Temporal control over systemic Wnt pathway manipulation has already been possible using heat-shock-regulated zebrafish transgenes and has shown that β-catenin signaling has a second role after blastema formation, namely, in regulating blastemal and epidermal cell proliferation (Stoick-Cooper *et al.*, 2007; Figure 26.3).

Overactivation of β-catenin signaling can enhance regeneration

Intriguingly, experimental activation of β-catenin signaling has been found to be sufficient to augment appendage regeneration. Inducible overexpression of Wnt8, which activates β-catenin signaling in most systems, enhances blastema proliferation during zebrafish fin regeneration, and fish heterozygous for a LOF mutation of the pathway inhibitor Axin1 display faster regenerative growth (Stoick-Cooper *et al.*, 2007). Similarly, adenovirus-mediated overexpression of a constitutively active β-catenin enhances limb regeneration in *Xenopus* larvae at a refractory stage where the regenerative capacity is low (Kawakami *et al.*, 2006). Furthermore, the overexpression of β-catenin or activation of Wnt signaling with the small-molecule GSK3β inhibitor BIO increases cell proliferation and speeds up *Xenopus* larval tail regeneration (Lin and Slack, 2008). Finally, the transplantation of patches of larval limb progenitor cells onto adult *Xenopus* limb stumps, which normally do not regenerate, can promote regeneration if the cells overexpress stabilized β-catenin and the patches are supplemented with sources for Sonic Hedgehog, FGF10, and β-thymosin (Lin, Chen, and Slack, 2013).

Wnt/β-catenin signaling appears to act upstream of FGF signaling during appendage regeneration in several systems. Fish homozygous for a LOF allele of *fgf20a* fail to form a blastema (Whitehead *et al.*, 2005), and *fgf20a* expression in the regenerate is dependent on Wnt signaling (Kawakami *et al.*, 2006; Stoick-Cooper *et al.*, 2007; Figure 26.3b). Likewise, *fgf-8* expression was found to be dependent on β-catenin signaling during *Xenopus* limb regeneration (Yokoyama *et al.*, 2007). Interestingly, epistasis experiments in *Xenopus* larval tail regeneration show that pharmacological inhibition of FGF signaling suppresses regeneration in the presence of drugs that activate Wnt signaling, while FGF activation rescues regeneration inhibited by blockage of β-catenin signaling (Lin and Slack, 2008). Thus, FGF signaling acts downstream of Wnt/β-catenin signaling in *Xenopus* tail regeneration.

Functions of β-catenin-independent Wnt pathways

β-catenin-independent Wnt signaling also has a role in the regulation of appendage regeneration. Expression of *wnt5a* and *wnt5b*, which activate β-catenin-independent pathways in most systems, is upregulated during zebrafish fin and salamander limb regeneration (Ghosh *et al.*, 2008; Stoick-Cooper *et al.*, 2007). Experimental overexpression of *wnt5b* during zebrafish fin regeneration blocks blastema proliferation and regenerative outgrowth, and *wnt5a* overexpression in regenerating salamander limbs likewise interferes with regeneration (Ghosh *et al.*, 2008; Stoick-Cooper *et al.*, 2007). Conversely, zebrafish homozygous for a mutation in *wnt5b* show accelerated fin regeneration (Stoick-Cooper *et al.*, 2007). Thus, it appears that Wnt5-mediated β-catenin-independent Wnt signaling limits appendage regeneration. During zebrafish fin regeneration, *wnt5b* is expressed in the distal wound epidermis, and it was found to repress expression of *sonic hedgehog* (*shh*), which is expressed in a more proximal region of the epidermis and regulates blastemal growth and bone differentiation during later stages of fin regeneration (Lee *et al.*, 2009; Quint *et al.*, 2002). Thus, β-catenin-independent Wnt signaling mediated by Wnt5b is involved in the maintenance of functionally distinct regions of the wound epidermis that are likely essential for blastema proliferation, correct patterning, and outgrowth of the fin (Figure 26.3). Since activation of β-catenin-independent Wnt signaling can interfere with β-catenin signaling (Weidinger and Moon, 2003) and *wnt5b* overexpression in zebrafish fins represses expression of the Wnt/β-catenin target genes *axin2* and *lef1* (Lee *et al.*, 2009; Stoick-Cooper *et al.*, 2007), it is possible that noncanonical Wnt signaling activated by Wnt5 negatively regulates regeneration by counteracting β-catenin signaling.

Surprisingly, in wounded *Xenopus* larval tails, localized overexpression of Wnt5a, but not of β-catenin pathway-activating Wnt ligands, has been found to be able to orchestrate formation of ectopic tails, a process that resembles tail regeneration after amputation (Sugiura *et al.*, 2009). Thus, activation of β-catenin-independent Wnt signaling appears to reprogram a wound healing response to a growth program normally activated during tail regeneration. Whether this Wnt5a-activated signaling has a positive role in normal *Xenopus* tail regeneration as well remains to be tested.

Although much still needs to be learned about the exact function of Wnt signaling pathways in vertebrate appendage regeneration and to which extent its functions vary between different systems, it has become clear that β-catenin signaling is essential for appendage regeneration to occur. Interesting open questions include how Wnt ligand expression is stimulated as an early response to appendage injury, which cellular regenerative processes are directly regulated by β-catenin signaling, whether Wnt has regeneration-specific roles or largely recapitulates functions during appendage development, and where Wnt signaling acts in the hierarchy of signaling pathways regulating regeneration.

Liver regeneration

Cellular mechanisms

In all vertebrates, including mammals, the liver displays a high regenerative capacity. After resection (partial hepatectomy (PH)) of up to 70% of liver mass, the original size of the organ will be restored within 7–10 days in rats and 3–6 months in humans (reviewed in Campbell *et al.*, 2006; Taub, 2004). Similarly, the liver can recover from massive cell death due to toxins or virus infection. Interestingly, depending on the type of liver injury, different cellular mechanisms are being used to regenerate the lost or damaged cells (Figure 26.4). In response to PH, the remaining differentiated hepatocytes enter the cell cycle and go through one to two rounds of cell division until the original size of the organ

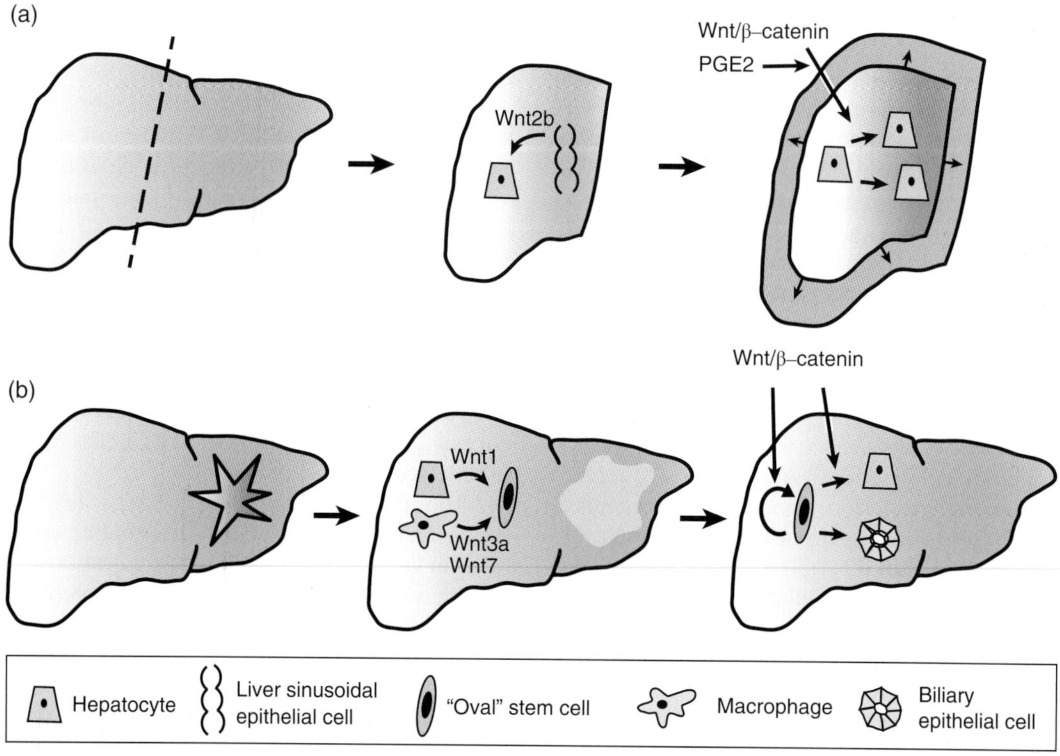

Figure 26.4 Model for Wnt signaling in liver regeneration. (**a**) In response to partial hepatectomy, differentiated hepatocytes proliferate to restore liver mass. Hepatocytes appear to activate Wnt/β-catenin signaling, and this might be induced by Wnt2b secreted from LSECs. Wnt/β-catenin signaling is required for hepatocyte proliferation, and Wnt signaling interacts with PGE2, one of the signals commonly released in response to tissue injury, in the activation of signaling responses. (**b**) In response to toxic injury, stem cells, which include "oval" cells and Lgr5-positive cells, activate β-catenin signaling, possibly in response to Wnt1 produced by hepatocytes and Wnt3a and Wnt7 produced by macrophages. Stem cells self-renew and give rise to hepatocytes and biliary epithelial cells. β-catenin signaling is involved in regulating stem cell renewal and in differentiation into hepatocytes.

has been replaced (reviewed in Fausto, 2004; Fausto, Campbell, and Riehle, 2006; Taub, 2004). Remarkably, the cells do not appear to lose their differentiated phenotype during proliferation, which makes liver regeneration one of the few examples of proliferation of terminally differentiated cells (Figure 26.4a).

In addition to regeneration by proliferation of differentiated cells, progenitor cells can also contribute to restoration of liver mass. In particular, it appears that the so-called oval cells are the main source of regeneration in response to death of hepatic cells caused by virus infection or certain toxic substances (Figure 26.4b; reviewed in Fausto, 2004; Newsome, Hussain, and Theise, 2004; Santoni-Rugiu *et al.*, 2005). Likewise, if hepatocyte proliferation after PH is experimentally blocked by drugs, oval cells are activated and regeneration nevertheless occurs. Oval cells are thought to reside in a defined progenitor cell niche in the biliary ducts, give rise to hepatocytes and biliary cells, and might become positive for the stem marker Lgr5 after liver injury (Evarts *et al.*, 1987, 1989; Huch *et al.*, 2013; Oh, Hatch, and Peterson, 2002; Sell, 2001; Yin *et al.*, 2002; Figure 26.4b). Intriguingly, recent genetic lineage tracing data also suggests that hepatocytes can transdifferentiate into biliary cells in response to toxin-induced liver injury via an intermediate state that expresses markers of both fates (Yanger *et al.*, 2013). However, another study found no evidence for hepatocyte transdifferentiation using genetic lineage tracing (Malato *et al.*, 2011), and the analysis of chimeric livers has indicated that oval cells do not derive from hepatocytes and are not recruited from bone marrow resident cells after liver injury (Wang *et al.*, 2003). Still, if hepatocyte transdifferentiation existed, the liver would be extraordinary among regenerating organs, since it would display enormous versatility of cellular responses to injury. Further studies will be required to clarify the importance of cellular reprogramming for liver regeneration and whether hepatocytes dedifferentiate to the oval stem cell state or whether transdifferentiation represents a third mechanism of regeneration in addition to hepatocyte proliferation and stem cell activation. Wnt/β-catenin signaling has been implicated both in hepatocyte proliferation-dependent and stem cell activation-driven liver regeneration.

Wnt signaling and hepatocyte proliferation-driven liver regeneration

Nuclear β-catenin and the transgenic reporter of β-catenin-dependent transcription TOPGAL can be detected in mouse hepatocytes in periportal regions from 1 to 3 days post PH (Goessling *et al.*, 2008; Stoick-Cooper *et al.*, 2007). In mouse, knockdown of β-catenin with antisense oligos or conditional β-catenin knockout in hepatocytes results in decreased hepatocyte proliferation after PH and an increase in hepatocyte apoptosis (Sodhi *et al.*, 2005; Tan *et al.*, 2006). Similarly, systemic inducible overexpression of a dominant-negative TCF transcription factor completely blocks hepatocyte proliferation and liver regeneration after PH in zebrafish (Goessling *et al.*, 2008). While these experiments indicate that Wnt/β-catenin signaling is required for hepatocyte proliferation after PH, due to the dual role of β-catenin in cell adhesion and Wnt signaling and the systemic nature of the TCF overexpression experiment in zebrafish, additional hepatocyte-specific Wnt LOF experiments are needed to clarify the specific role of Wnt/β-catenin in hepatocyte-driven liver regeneration.

Gain-of-function (GOF) experiments indicate that overactivation of Wnt/β-catenin signaling is sufficient to speed up liver regeneration. Zebrafish and mice heterozygous for a LOF mutation in the Wnt signaling inhibitor APC or fish systemically overexpressing the β-catenin signaling activating ligand Wnt8 display increased levels of liver cell proliferation, particularly in periportal regions (Goessling *et al.*, 2008). Overexpression of a stabilized β-catenin in hepatocytes likewise increases, albeit transiently, hepatocyte proliferation after PH (Nejak-Bowen *et al.*, 2010). A universal response to tissue injury is increased production of prostaglandin 2 (PGE2) (Goldstein *et al.*, 1977). Interestingly, PGE2 signaling was found to be required for the activation of Wnt/β-catenin signaling in wild-type and APC heterozygous zebrafish and mice after PH (Goessling *et al.*, 2009). The inhibition of PGE2 signaling reduces cell proliferation and liver regrowth in both wild-type and APC heterozygotes. Thus, PGE2 and Wnt signaling need to interact to regulate liver regeneration (Figure 26.4a).

Recently, liver sinusoidal endothelial cells (LSECs), which line the dense capillary network in the mammalian liver, were found to regulate proliferation of hepatocytes after PH, and they do so in part by releasing Wnt2b, which usually activates the β-catenin signaling pathway (Ding *et al.*, 2010; Figure 26.4a). Whether LSECs are the only sources of ligands activating β-catenin signaling in hepatocytes is currently unknown.

Wnt signaling- and stem cell-driven liver regeneration

Stem cell-driven liver regeneration likewise appears to require Wnt/β-catenin signaling. PH combined with treatment with the drug 2-acetylaminofluorene (2-AAF), which blocks hepatocyte proliferation, activates oval liver stem cells in the rat (PH/2-AAF model) (Fausto and Campbell, 2003). In addition, liver injury induced by hepatotoxins activates oval cells. In the rat PH/2-AAF liver regeneration model, nuclear levels of β-catenin and expression of the putative direct Wnt target genes *axin2* and *c-myc* are increased after injury with kinetics paralleling the proliferation of oval cells (Apte *et al.*, 2008; Itoh *et al.*, 2009; Yang *et al.*, 2008). In addition, the expression of the Wnt ligands *wnt7a*, *wnt7b*, and *wnt10a* is strongly upregulated in mouse livers injured with hepatotoxin, but the cells expressing these ligands have not been identified (Itoh *et al.*, 2009), whereas *wnt1* has been reported to be upregulated in hepatocytes in the rat PH/2-AAF model (Williams *et al.*, 2010). Toxin-induced liver injury also results in upregulation of the Axin2-lacZ Wnt reporter, of *wnt6*, and of several *r-spondins*, which code for secreted proteins that potentiate Wnt signaling together with their receptor Lgr5 (Huch *et al.*, 2013). Intriguingly, Clevers and colleagues also found that *lgr5* expression, which labels Wnt-dependent stem cells in several adult self-renewing tissues including the gut and hair follicles, is not detectable in the adult noninjured liver, but that Lgr5-positive cells appear near the bile ducts after toxin-induced injury. Lineage tracing showed that these cells give rise to hepatocytes and biliary duct cells *in vivo* and in organoids derived from single-cell cultures, indicating that these Lgr5+ cells represent the bipotent oval stem cells (Huch *et al.*, 2013). Gene expression profiling indicated that these stem cells activate β-catenin signaling (Huch *et al.*, 2013), and nuclear β-catenin has been reported in progenitor cells in diseased human or injured mouse livers (Boulter *et al.*, 2012). Expression of Wnts has also been detected in macrophages which surround progenitor cells in injured mouse livers, and phagocytosis of hepatocyte debris has been found to stimulate expression of *wnt3a* and *wnt7a* in macrophages (Boulter *et al.*, 2012). Together, these data suggest that hepatocytes and/or macrophages produce Wnt ligands in response to injury, while oval stem cells respond to them, yet a more detailed mapping of ligand and Wnt reporter expression will be necessary to confirm this (Figure 26.4b).

An intriguing aspect of the recent identification of Lgr5 as marker for liver stem cells is that – in contrast to the intestine, stomach, or hair follicles – the Lgr5+ cells cannot be detected in the noninjured organ. This raises the possibility that these stem cells only arise after injury, possibly via dedifferentiation from hepatocytes. Further studies will be necessary to clarify whether regeneration of the adult mammalian liver does, surprisingly, involve cellular transdifferentiation and to investigate whether such potential reprogramming events give rise to Lgr5-positive stem cells.

Conditional knockout of β-catenin in both hepatocytes and bile ducts, from which oval cells likely originate, reduces the numbers of oval cells in a mouse stem cell-dependent liver injury model (Apte *et al.*, 2008), while overexpression of stabilized β-catenin in all cells of the mouse liver increases oval cell numbers (Yang *et al.*, 2008). However, the consequences of this for liver regeneration have not been reported. Together, these experiments indicate that Wnt/β-catenin signaling regulates expansion of the oval stem cell pool (Figure 26.4b). This is also supported by the fact that expansion of Lgr5-positive liver stem cells *in vitro* requires R-Spondin (Huch *et al.*, 2013). However, in the rat PH/2-AAF model, systemic knockdown of *wnt1* (which is expressed in hepatocytes, see preceding text) by treatment of animals with a vector expressing a *wnt1* shRNA does not result in an overall failure of liver mass restoration, but causes atypical ductular hyperplasia, indicating that oval cells fail to differentiate into hepatocytes in the absence of Wnt1 but instead form cells of the biliary duct lineage (Williams *et al.*, 2010). Likewise, the inhibition of Wnt signaling using the Wnt inhibitory factor 1 (WIF1) protein reduces differentiation of liver progenitors towards the hepatocyte fate in cocultures with macrophages, and macrophage depletion in liver-injured mice *in vivo* has been reported to cause progenitors to preferentially form biliary cells instead of hepatocytes (Boulter *et al.*, 2012). Thus, further experiments will be required to clarify whether Wnt/β-catenin signaling regulates oval cell proliferation or their commitment towards the hepatocyte lineage or whether it possibly plays a biphasic role, regulating both processes in a temporally distinct manner.

Skeletal muscle regeneration

In vertebrates, including mammals, skeletal muscle has a high ability to renew and regenerate, enabling a lifetime production and maintenance of myofibers. A specific group of mononucleate myogenic precursor cells, termed satellite cells, located between the plasma membrane and the basal lamina of mature myofibers, has been shown to represent the main source of new myotubes during muscle development, regeneration, and physical exercise-mediated growth (Charifi *et al.*, 2003; Mauro, 1961; Wagers and Conboy, 2005). Satellite cells present in isolated single muscle fibers and cultured *in vitro* can self-renew and give rise to differentiated myofibers (Gilbert *et al.*, 2010; Olguin and Olwin, 2004; Wagers and Conboy, 2005). Transplantation of single intact myofibers into radiation-ablated dystrophic muscle has revealed that grafted fibers supply a pool of satellite cells, which are able to produce new myofibers. When an additional muscle injury is induced several weeks after muscle recovery, these newly formed fibers respond by expansion, strongly suggesting that satellite cells are sufficient for proper muscle regeneration (Collins *et al.*, 2005).

β-catenin signaling has been reported to maintain satellite cells in an undifferentiated, proliferative state while inhibiting their myogenic differentiation (Goichberg *et al.*, 2001; Otto *et al.*, 2008; Perez-Ruiz *et al.*, 2008). Surprisingly, Wnt/β-catenin signaling appears to have a second differentiation-promoting role in a later phase of myogenic progression of satellite cells. The inhibition of endogenous Wnt signaling using sFRP3 in isolated mouse muscle fibers prevents myogenic differentiation when added at a time when most satellite cell progeny have progressed to differentiation, while pathway activation with Wnt3a or GSK3 inhibitors after 2 days of satellite cell activation in culture enhances myogenic commitment of the satellite cell progeny and promotes premature muscle differentiation (Brack *et al.*, 2008). Thus, Wnt/β-catenin signaling appears to have a biphasic role in differentiation of satellite cells: While early β-catenin signaling maintains satellite cells in an undifferentiated state, its activation after the myogenic commitment promotes differentiation.

Wnt/β-catenin signaling has been reported to be active in proliferating satellite cells during regeneration of mouse skeletal muscle in response to focal injury in TOPGAL mice (Brack *et al.*, 2008) or to cardiotoxin-induced damage and during repair phases of human muscle in muscular dystrophy (Otto *et al.*, 2008). One mechanism of Wnt signaling activation in response to injury appears to depend on the myogenic transcription factor *mef2a*, which regulates expression of a miRNA cluster that represses expression of sFRPs (Snyder *et al.*, 2013). Overactivation of β-catenin signaling using Wnt3a injection into injured muscle 2 days after injury, that is, towards the end of the early proliferative phase of progenitor cells, promotes myogenic differentiation prematurely while interfering with the proper regeneration of the muscle, suggesting that early induction of differentiation interferes with the regenerative ability of the tissue due to depletion of myogenic progenitor cells (Brack *et al.*, 2008). Loss or reduction of Bcl9, a component of the β-catenin/TCF complex, interferes with the ability of myogenic progenitors to differentiate and regenerate both *in vitro* and *in vivo* (Brack *et al.*, 2009). However, additional LOF experiments will be required to clarify the *in vivo* role of Wnt/β-catenin signaling in muscle regeneration.

Notch signaling has been demonstrated to be crucial for the proliferative expansion of satellite cells and for the inhibition of myogenic differentiation (Conboy and Rando, 2002). Interestingly, inhibition of Notch signaling during the early proliferative phase stimulates a premature activation of Wnt signaling, while activated Notch signaling diminishes nuclear β-catenin levels, suggesting that the two pathways cross talk (Brack *et al.*, 2008). Thus, the late differentiation-promoting role of the β-catenin pathway might be controlled by a transition from early Notch signaling, which is necessary for initial activation and proliferation of satellite cells and inhibition of their premature differentiation. Surprisingly, Wnt signaling activity increases in satellite cells of aging mice and mediates their conversion from myogenic to fibrogenic lineage as they proliferate, most likely due to systemic factors accumulating and circulating in the aged serum (Brack *et al.*, 2007). This conversion can be inhibited by Wnt pathway inhibitors indicating a critical role of Wnt signaling in aging of satellite cells and induction of fibrosis during aging.

Noncanonical Wnt signaling, acting independently of β-catenin, has also been shown to be critical for regulation of the regenerative response of the muscle (Le Grand *et al.*, 2009). The expression of *frizzled-7*, which in many systems is associated with the activation of β-catenin-independent Wnt signaling pathways, and its ligand *wnt7a* are strongly upregulated in proliferating satellite cells and myoblasts during muscle regeneration (Le Grand *et al.*, 2009). Wnt7a-Frizzled7 signaling induces symmetric division of satellite cells, generating a proliferating stem cell pool, which requires *vangl2*, a component of the PCP pathway (Le Grand *et al.*, 2009). Intriguingly, treatment with Wnt7a was found to be sufficient to induce satellite cell expansion and myofiber hypertrophy and to increase muscle strength in a Duchenne muscular dystrophy mouse model (von Maltzahn *et al.*, 2012).

In addition to the satellite cells, other populations of stem cells, in particular blood- and bone marrow-derived side population cells, have myogenic potential (Ferrari *et al.*, 1998; Gussoni *et al.*, 1999; LaBarge and Blau, 2002; Luth *et al.*, 2008). Additionally, muscle-resident stem cells positive for hematopoietic markers (CD45+ Sca1+) appear to exist (Asakura *et al.*, 2002). These latter resident muscle stem cells can undergo myogenic differentiation only in response to muscle damage and display an upregulation of various Wnt pathway components in the regenerating muscle (Polesskaya, Seale, and Rudnicki, 2003). Experimental activation of the Wnt/β-catenin pathway in CD45+ Sca1+ cells induces their myogenic specification, while injection of secreted Wnt antagonists into the regenerating muscles of mice significantly reduces their myogenic differentiation. Although these nonmuscle- and muscle-derived stem cells with hematopoietic character can contribute to the regenerating muscle independently of the satellite cells, the physiological relevance of their myogenic potential remains unclear. So far, it has not been shown that these cells are indispensable, like the satellite cells, for skeletal muscle growth and repair after injury.

Kidney regeneration

The regenerative capacity of the vertebrate kidney lies between that of the appendages, heart, and lens, which regenerate in lower vertebrates but not at all in mammals, and skeletal muscle and liver, which efficiently regenerate in mammals as well. Elasmobranch fish (a subclass of the cartilaginous fish) form new nephrons in response to kidney resection (Elger *et al.*, 2003), and several species of teleost fish likewise display neonephrogenesis in response to kidney injury induced by toxins (Diep *et al.*, 2011; Reimschuessel, 2001; Zhou *et al.*, 2010). In contrast, mammals can only repair damaged nephron epithelia, but not form new ones (Benigni, Morigi, and Remuzzi, 2010). As with other organs, the ability to regenerate new kidney tissue in fish might be associated with their lifelong growth and the addition of new nephrons to the kidney throughout adult life, while the development of the functional kidney units in mammals ceases shortly after birth (Benigni, Morigi, and Remuzzi, 2010; Hartman, Lai, and Patterson, 2007). Neonephrogenesis in response to toxin-induced injury in zebrafish has recently been shown to be achieved by self-renewing nephron progenitor cells (Diep *et al.*, 2011). The molecular mechanisms regulating this progenitor cell-driven kidney regeneration however remain unknown.

While mammals cannot replace lost nephrons, all vertebrate kidneys can repair damaged tubule epithelia even after a severe injury (Benigni, Morigi, and Remuzzi, 2010; Duffield *et al.*, 2005). Injury of the kidney due to ischemia and subsequent reperfusion results in loss of epithelial cells from the tubules of cortex and the medulla, and restoration occurs primarily through a regenerative response initiated by surviving tubular epithelial cells (Duffield *et al.*, 2005; Humphreys *et al.*, 2008). While some evidence has been found for the existence of multipotent progenitor cells in the adult mammalian kidney, it has remained unclear whether such cells contribute to the repair process of tubular epithelia (Benigni, Morigi, and Remuzzi, 2010).

Recent work has shown that macrophages which are recruited to the injury site produce Wnt7b, which activates Wnt/β-catenin signaling and stimulates a regenerative response in epithelial cells (Lin *et al.*, 2010). Ablation of Wnt7b-expressing macrophages by diphtheria toxin injection or deletion of *wnt7b* in macrophages causes a reduction in canonical Wnt signaling in kidney epithelial cells as well as a failure of normal regeneration of the kidney tubular epithelium, strongly suggesting that macrophage-produced Wnt7b is required for the kidney repair process (Lin *et al.*, 2010).

Mammalian kidney podocytes are highly differentiated cells that are thought to have limited renewal capacity. Interestingly, forced overexpression of the telomerase component TERT was found to be sufficient to cause dedifferentiation of mouse kidney podocytes and their entry into the cell cycle, resulting in collapse of glomeruli (Shkreli *et al.*, 2012). This was associated with an upregulation of Wnt/β-catenin signaling, which could be due to a direct effect of TERT as coactivator of β-catenin-dependent transcription (Park *et al.*, 2009). Increases in TERT and β-catenin levels were also found in diseases characterized by glomeruli collapse, and systemic inhibition of Wnt signaling reduces disease progression. Thus, Wnt/β-catenin signaling might be able to induce podocyte dedifferentiation. Whether this is relevant for normal kidney homeostasis or regeneration and whether Wnt signaling plays a role in nephron regeneration in lower vertebrates as well remains to be tested.

Are there common principles of Wnt function in regenerative processes?

The regeneration of the systems discussed here involves quite distinct cellular processes, since some require activation of stem cells, others cellular dedifferentiation, some restore a few, others many different cell types, and some have little requirement for pattern formation and positional information, while others regenerate via redevelopment of intricately patterned complex structures. Nevertheless, all regenerating systems must have molecular mechanisms in place that sense insult and trigger the regenerative response, assure controlled cell proliferation and differentiation, and finally integrate the restored tissue properly into the already existing adult body. Thus, despite the wide spectrum of regenerative phenomena, some commonalities might exist in the molecular regulation of these latter issues. Yet, we are far from a comprehensive understanding of the molecular regulation of any regenerating system, and thus, it is also difficult to assess whether general molecular principles of regeneration exist.

As we have discussed, the known functions of Wnt signaling involve regulation of cell proliferation in several systems; maintenance of the undifferentiated state as in muscle satellite cells but also regulation of cell differentiation, for example in a second role in the muscle and possibly in liver regeneration; pattern formation in planaria and appendages; and possibly also regulation of dedifferentiation in the kidney. At first glance, most of these functions appear to be recapitulations of Wnt function during embryonic development or tissue homeostasis rather than regeneration-specific roles. For example, Wnt is required for establishment of anteroposterior patterning not only during planarian regeneration but also during tissue homeostasis, and it plays essential roles in regulating proliferation and differentiation during vertebrate appendage, liver, and muscle development. However, the potential role in podocyte dedifferentiation in kidney regeneration indicates that Wnt signaling can also be involved in cellular mechanisms specific to regeneration. Clearly, a much more thorough understanding of the exact functions of Wnt signaling in regeneration (and in some systems also development, e.g., of the zebrafish larval fin) will be necessary to conclude which roles are regeneration-specific and whether Wnt signaling has an elevated role in the signaling networks involved, for example serving as a master regulator of regeneration. We do however already know that Wnt/β-catenin signaling appears to play a positive, regeneration-promoting role in most systems, suggesting that therapeutic manipulation of β-catenin signaling might be beneficial in approaches aimed at promoting repair of diverse human organs.

Acknowledgments

We thank Caghan Kizil for help with the figures and discussions and members of the Weidinger lab for helpful comments. Work in the Weidinger lab is supported by the Deutsche Forschungsgemeinschaft via grants WE 4223/3-1 and WE 4223/4-1.

References

Adell, T., Salo, E., Boutros, M., and Bartscherer, K. (2009) Smed-Evi/Wntless is required for beta-catenin-dependent and -independent processes during planarian regeneration. *Development*, **136**, 905–910.

Apte, U., Thompson, M.D., Cui, S. *et al.* (2008) Wnt/beta-catenin signaling mediates oval cell response in rodents. *Hepatology*, **47**, 288–295.

Asakura, A., Seale, P., Girgis-Gabardo, A., and Rudnicki, M.A. (2002) Myogenic specification of side population cells in skeletal muscle. *The Journal of Cell Biology*, **159**, 123–134.

Benigni, A., Morigi, M., and Remuzzi, G. (2010) Kidney regeneration. *Lancet*, **375**, 1310–1317.

Boulter, L., Govaere, O., Bird, T.G. *et al.* (2012) Macrophage-derived Wnt opposes Notch signaling to specify hepatic progenitor cell fate in chronic liver disease. *Nature Medicine*, **18**, 572–579.

Brack, A.S., Conboy, M.J., Roy, S. *et al.* (2007) Increased Wnt signaling during aging alters muscle stem cell fate and increases fibrosis. *Science*, **317**, 807–810.

Brack, A.S., Conboy, I.M., Conboy, M.J. *et al.* (2008) A temporal switch from notch to Wnt signaling in muscle stem cells is necessary for normal adult myogenesis. *Cell Stem Cell*, **2**, 50–59.

Brack, A.S., Murphy-Seiler, F., Hanifi, J. *et al.* (2009) BCL9 is an essential component of canonical Wnt signaling that mediates the differentiation of myogenic progenitors during muscle regeneration. *Developmental Biology*, **335**, 93–105.

Brockes, J.P. and Kumar, A. (2008) Comparative aspects of animal regeneration. *Annual Review of Cell and Developmental Biology*, **24**, 525–549.

Bryant, S.V., Endo, T., and Gardiner, D.M. (2002) Vertebrate limb regeneration and the origin of limb stem cells. *The International Journal of Developmental Biology*, **46**, 887–896.

Campbell, J.S., Riehle, K.J., Brooling, J.T. *et al.* (2006) Proinflammatory cytokine production in liver regeneration is Myd88-dependent, but independent of Cd14, Tlr2, and Tlr4. *Journal of Immunology*, **176**, 2522–2528.

Caubit, X., Nicolas, S., Shi, D.L., and Le Parco, Y. (1997) Reactivation and graded axial expression pattern of Wnt-10a gene during early regeneration stages of adult tail in amphibian urodele Pleurodeles waltl. *Developmental Dynamics*, **208**, 139–148.

Chablais, F. and Jazwinska, A. (2010) IGF signaling between blastema and wound epidermis is required for fin regeneration. *Development*, **137**, 871–879.

Charifi, N., Kadi, F., Feasson, L., and Denis, C. (2003) Effects of endurance training on satellite cell frequency in skeletal muscle of old men. *Muscle Nerve*, **28**, 87–92.

Collins, C.A., Olsen, I., Zammit, P.S. *et al.* (2005) Stem cell function, self-renewal, and behavioral heterogeneity of cells from the adult muscle satellite cell niche. *Cell*, **122**, 289–301.

Conboy, I.M. and Rando, T.A. (2002) The regulation of Notch signaling controls satellite cell activation and cell fate determination in postnatal myogenesis. *Developmental Cell*, **3**, 397–409.

Diep, C.Q., Ma, D., Deo, R.C. *et al.* (2011) Identification of adult nephron progenitors capable of kidney regeneration in zebrafish. *Nature*, **470**, 95–100.

Ding, B.S., Nolan, D.J., Butler, J.M. *et al.* (2010) Inductive angiocrine signals from sinusoidal endothelium are required for liver regeneration. *Nature*, **468**, 310–315.

Duan, J., Gherghe, C., Liu, D. *et al.* (2012) Wnt1/beta-catenin injury response activates the epicardium and cardiac fibroblasts to promote cardiac repair. *The EMBO Journal*, **31**, 429–442.

Duffield, J.S., Park, K.M., Hsiao, L.L. *et al.* (2005) Restoration of tubular epithelial cells during repair of the postischemic kidney occurs independently of bone marrow-derived stem cells. *The Journal of Clinical Investigation*, **115**, 1743–1755.

Elger, M., Hentschel, H., Litteral, J. *et al.* (2003) Nephrogenesis is induced by partial nephrectomy in the elasmobranch Leucoraja erinacea. *Journal of the American Society of Nephrology*, **14**, 1506–1518.

Evarts, R.P., Nagy, P., Marsden, E., and Thorgeirsson, S.S. (1987) A precursor–product relationship exists between oval cells and hepatocytes in rat liver. *Carcinogenesis*, **8**, 1737–1740.

Evarts, R.P., Nagy, P., Nakatsukasa, H. *et al.* (1989) In vivo differentiation of rat liver oval cells into hepatocytes. *Cancer Research*, **49**, 1541–1547.

Fausto, N. (2004) Liver regeneration and repair: hepatocytes, progenitor cells, and stem cells. *Hepatology*, **39**, 1477–1487.

Fausto, N. and Campbell, J.S. (2003) The role of hepatocytes and oval cells in liver regeneration and repopulation. *Mechanisms of Development*, **120**, 117–130.

Fausto, N., Campbell, J.S., and Riehle, K.J. (2006) Liver regeneration. *Hepatology*, **43**, S45–S53.

Ferrari, G., Cusella-De Angelis, G., Coletta, M. *et al.* (1998) Muscle regeneration by bone marrow-derived myogenic progenitors. *Science*, **279**, 1528–1530.

Figeac, F., Uzan, B., Faro, M. *et al.* (2010) Neonatal growth and regeneration of beta-cells are regulated by the Wnt/beta-catenin signaling in normal and diabetic rats. *American Journal of Physiology. Endocrinology and Metabolism*, **298**, E245–E256.

Galliot, B. and Chera, S. (2010) The Hydra model: disclosing an apoptosis-driven generator of Wnt-based regeneration. *Trends in Cell Biology*, **20**, 514–523.

Ghosh, S., Roy, S., Seguin, C. *et al.* (2008) Analysis of the expression and function of Wnt-5a and Wnt-5b in developing and regenerating axolotl (Ambystoma mexicanum) limbs. *Development, Growth & Differentiation*, **50**, 289–297.

Gilbert, P.M., Havenstrite, K.L., Magnusson, K.E. *et al.* (2010) Substrate elasticity regulates skeletal muscle stem cell self-renewal in culture. *Science*, **329**, 1078–1081.

Goessling, W., North, T.E., Lord, A.M. *et al.* (2008) APC mutant zebrafish uncover a changing temporal requirement for wnt signaling in liver development. *Developmental Biology*, **320**, 161–174.

Goessling, W., North, T.E., Loewer, S. *et al.* (2009) Genetic interaction of PGE2 and Wnt signaling regulates developmental specification of stem cells and regeneration. *Cell*, **136**, 1136–1147.

Goichberg, P., Shtutman, M., Ben-Ze'ev, A., and Geiger, B. (2001) Recruitment of beta-catenin to cadherin-mediated intercellular adhesions is involved in myogenic induction. *Journal of Cell Science*, **114**, 1309–1319.

Goldstein, I.M., Malmsten, C.L., Samuelsson, B., and Weissmann, G. (1977) Prostaglandins, thromboxanes, and polymorphonuclear leukocytes: mediation and modulation of inflammation. *Inflammation*, **2**, 309–317.

Gurley, K.A., Rink, J.C., and Sanchez Alvarado, A. (2008) Beta-catenin defines head versus tail identity during planarian regeneration and homeostasis. *Science*, **319**, 323–327.

Gussoni, E., Soneoka, Y., Strickland, C.D. *et al.* (1999) Dystrophin expression in the mdx mouse restored by stem cell transplantation. *Nature*, **401**, 390–394.

Hartman, H.A., Lai, H.L., and Patterson, L.T. (2007) Cessation of renal morphogenesis in mice. *Developmental Biology*, **310**, 379–387.

Hayashi, T., Mizuno, N., and Kondoh, H. (2008) Determinative roles of FGF and Wnt signals in iris-derived lens regeneration in newt eye. *Development, Growth & Differentiation*, **50**, 279–287.

Huch, M., Dorrell, C., Boj, S.F. *et al.* (2013) In vitro expansion of single Lgr5+ liver stem cells induced by Wnt-driven regeneration. *Nature*, **494**, 247–250.

Humphreys, B.D., Valerius, M.T., Kobayashi, A. *et al.* (2008) Intrinsic epithelial cells repair the kidney after injury. *Cell Stem Cell*, **2**, 284–291.

Iglesias, M., Gomez-Skarmeta, J.L., Salo, E., and Adell, T. (2008) Silencing of Smed-betacatenin1 generates radial-like hypercephalized planarians. *Development*, **135**, 1215–1221.

Itoh, T., Kamiya, Y., Okabe, M. *et al.* (2009) Inducible expression of Wnt genes during adult hepatic stem/progenitor cell response. *FEBS Letters*, **583**, 777–781.

Kawakami, Y., Rodriguez Esteban, C., Raya, M. *et al.* (2006) Wnt/beta-catenin signaling regulates vertebrate limb regeneration. *Genes & Development*, **20**, 3232–3237.

Kikuchi, K., Holdway, J.E., Werdich, A.A. *et al.* (2010) Primary contribution to zebrafish heart regeneration by gata4(+) cardiomyocytes. *Nature*, **464**, 601–605.

Knopf, F., Hammond, C., Chekuru, A. *et al.* (2011) Bone regenerates via dedifferentiation of osteoblasts in the zebrafish fin. *Developmental Cell*, **20**, 713–724.

Kragl, M., Knapp, D., Nacu, E. *et al.* (2009) Cells keep a memory of their tissue origin during axolotl limb regeneration. *Nature*, **460**, 60–65.

Kubo, F. and Nakagawa, S. (2008) Wnt signaling in retinal stem cells and regeneration. *Development, Growth & Differentiation*, **50**, 245–251.

LaBarge, M.A. and Blau, H.M. (2002) Biological progression from adult bone marrow to mononucleate muscle stem cell to multinucleate muscle fiber in response to injury. *Cell*, **111**, 589–601.

Laeremans, H., Hackeng, T.M., van Zandvoort, M.A. *et al.* (2011) Blocking of frizzled signaling with a homologous peptide fragment of wnt3a/wnt5a reduces infarct expansion and prevents the development of heart failure after myocardial infarction. *Circulation*, **124**, 1626–1635.

Le Grand, F., Jones, A.E., Seale, V. *et al.* (2009) Wnt7a activates the planar cell polarity pathway to drive the symmetric expansion of satellite stem cells. *Cell Stem Cell*, **4**, 535–547.

Lee, Y., Hami, D., De Val, S. *et al.* (2009) Maintenance of blastemal proliferation by functionally diverse epidermis in regenerating zebrafish fins. *Developmental Biology*, **331**, 270–280.

Lin, G. and Slack, J.M. (2008) Requirement for Wnt and FGF signaling in Xenopus tadpole tail regeneration. *Developmental Biology*, **316**, 323–335.

Lin, G., Chen, Y., and Slack, J.M. (2013) Imparting regenerative capacity to limbs by progenitor cell transplantation. *Developmental Cell*, **24**, 41–51.

Lin, S.L., Li, B., Rao, S. *et al.* (2010) Macrophage Wnt7b is critical for kidney repair and regeneration. *Proceedings of the National Academy of Sciences of the United States of America*, **107**, 4194–4199.

Liu, S.-Y., Selck, C., Friedrich, B. *et al.* (2013) Reactivating head regrowth in a regeneration-deficient planarian species. *Nature*, **500** (7460), 81–84.

Luth, E.S., Jun, S.J., Wessen, M.K. *et al.* (2008) Bone marrow side population cells are enriched for progenitors capable of myogenic differentiation. *Journal of Cell Science*, **121**, 1426–1434.

Malato, Y., Naqvi, S., Schurmann, N. *et al.* (2011) Fate tracing of mature hepatocytes in mouse liver homeostasis and regeneration. *The Journal of Clinical Investigation*, **121**, 4850–4860.

Mauro, A. (1961) Satellite cell of skeletal muscle fibers. *Journal of Biophysical and Biochemical Cytology*, **9**, 493–495.

Miura, T. (2013) Turing and wolpert work together during limb development. *Science Signaling*, **6**, pe14.

Morrison, J.I., Loof, S., He, P., and Simon, A. (2006) Salamander limb regeneration involves the activation of a multipotent skeletal muscle satellite cell population. *The Journal of Cell Biology*, **172**, 433–440.

Nejak-Bowen, K.N., Thompson, M.D., Singh, S. *et al.* (2010) Accelerated liver regeneration and hepatocarcinogenesis in mice overexpressing serine-45 mutant beta-catenin. *Hepatology*, **51**, 1603–1613.

Newsome, P.N., Hussain, M.A., and Theise, N.D. (2004) Hepatic oval cells: helping redefine a paradigm in stem cell biology. *Current Topics in Developmental Biology*, **61**, 1–28.

Oh, S.H., Hatch, H.M., and Petersen, B.E. (2002) Hepatic oval 'stem' cell in liver regeneration. *Seminars in Cell & Developmental Biology*, **13**, 405–409.

Olguin, H.C. and Olwin, B.B. (2004) Pax-7 up-regulation inhibits myogenesis and cell cycle progression in satellite cells: a potential mechanism for self-renewal. *Developmental Biology*, **275**, 375–388.

Otto, A., Schmidt, C., Luke, G. *et al.* (2008) Canonical Wnt signalling induces satellite-cell proliferation during adult skeletal muscle regeneration. *The Journal of Cell Science*, **121**, 2939–2950.

Park, J.I., Venteicher, A.S., Hong, J.Y. *et al.* (2009) Telomerase modulates Wnt signalling by association with target gene chromatin. *Nature*, **460**, 66–72.

Perez-Ruiz, A., Ono, Y., Gnocchi, V.F., and Zammit, P.S. (2008) Beta-catenin promotes self-renewal of skeletal-muscle satellite cells. *Journal of Cell Science*, **121**, 1373–1382.

Petersen, C.P. and Reddien, P.W. (2008) Smed-betacatenin-1 is required for anteroposterior blastema polarity in planarian regeneration. *Science*, **319**, 327–330.

Petersen, C.P. and Reddien, P.W. (2009) A wound-induced Wnt expression program controls planarian regeneration polarity. *Proceedings of the National Academy of Sciences of the United States of America*, **106**, 17061–17066.

Petersen, C.P. and Reddien, P.W. (2011) Polarized notum activation at wounds inhibits Wnt function to promote planarian head regeneration. *Science*, **332**, 852–855.

Polesskaya, A., Seale, P., and Rudnicki, M.A. (2003) Wnt signaling induces the myogenic specification of resident CD45+ adult stem cells during muscle regeneration. *Cell*, **113**, 841–852.

Poss, K.D. (2010) Advances in understanding tissue regenerative capacity and mechanisms in animals. *Nature Reviews Genetics*, **11**, 710–722.

Quint, E., Smith, A., Avaron, F. *et al.* (2002) Bone patterning is altered in the regenerating zebrafish caudal fin after ectopic expression of Sonic Hedgehog and bmp2b or exposure to cyclopamine. *Proceedings of the National Academy of Sciences of the United States of America*, **99**, 8713–8718.

Reddien, P.W. (2011) Constitutive gene expression and the specification of tissue identity in adult planarian biology. *Trends in Genetics*, **27**, 277–285.

Reddien, P.W. and Sanchez Alvarado, A. (2004) Fundamentals of planarian regeneration. *Annual Review of Cell and Developmental Biology*, **20**, 725–757.

Reimschuessel, R. (2001) A fish model of renal regeneration and development. *ILAR Journal*, **42**, 285–291.

Rink, J.C., Gurley, K.A., Elliott, S.A., and Sanchez Alvarado, A. (2009) Planarian Hh signaling regulates regeneration polarity and links Hh pathway evolution to cilia. *Science*, **326**, 1406–1410.

Salo, E. (2006) The power of regeneration and the stem-cell kingdom: freshwater planarians (Platyhelminthes). *BioEssays*, **28**, 546–559.

Santoni-Rugiu, E., Jelnes, P., Thorgeirsson, S.S., and Bisgaard, H.C. (2005) Progenitor cells in liver regeneration: molecular responses controlling their activation and expansion. *APMIS*, **113**, 876–902.

Sell, S. (2001) Heterogeneity and plasticity of hepatocyte lineage cells. *Hepatology*, **33**, 738–750.

Shkreli, M., Sarin, K.Y., Pech, M.F. *et al.* (2012) Reversible cell-cycle entry in adult kidney podocytes through regulated control of telomerase and Wnt signaling. *Nature Medicine*, **18**, 111–119.

Sikes, J.M. and Newmark, P.A. (2013) Restoration of anterior regeneration in a planarian with limited regenerative ability. *Nature*, **500** (7460), 77–80.

Slack, J.M. (2007) Metaplasia and transdifferentiation: from pure biology to the clinic. *Nature Reviews. Molecular Cell Biology*, **8**, 369–378.

Slack, J.M., Beck, C.W., Gargioli, C., and Christen, B. (2004) Cellular and molecular mechanisms of regeneration in Xenopus. *Philosophical Transactions of the Royal Society of London B: Biological Sciences*, **359**, 745–751.

Snyder, C.M., Rice, A.L., Estrella, N.L. *et al.* (2013) MEF2A regulates the Gtl2-Dio3 microRNA mega-cluster to modulate WNT signaling in skeletal muscle regeneration. *Development*, **140**, 31–42.

Sodhi, D., Micsenyi, A., Bowen, W.C. *et al.* (2005) Morpholino oligonucleotide-triggered beta-catenin knockdown compromises normal liver regeneration. *Journal of Hepatology*, **43**, 132–141.

Sousa, S., Afonso, N., Bensimon-Brito, A. *et al.* (2011) Differentiated skeletal cells contribute to blastema formation during zebrafish fin regeneration. *Development*, **138**, 3897–3905.

Stewart, S. and Stankunas, K. (2012) Limited dedifferentiation provides replacement tissue during zebrafish fin regeneration. *Developmental Biology*, **365**, 339–349.

Stoick-Cooper, C.L., Moon, R.T., and Weidinger, G. (2007) Advances in signaling in vertebrate regeneration as a prelude to regenerative medicine. *Genes & Development*, **21**, 1292–1315.

Stoick-Cooper, C.L., Weidinger, G., Riehle, K.J. *et al.* (2007) Distinct Wnt signaling pathways have opposing roles in appendage regeneration. *Development*, **134**, 479–489.

Sugiura, T., Tazaki, A., Ueno, N. *et al.* (2009) Xenopus Wnt-5a induces an ectopic larval tail at injured site, suggesting a crucial role for noncanonical Wnt signal in tail regeneration. *Mechanisms of Development*, **126**, 56–67.

Tan, X., Behari, J., Cieply, B. *et al.* (2006) Conditional deletion of beta-catenin reveals its role in liver growth and regeneration. *Gastroenterology*, **131**, 1561–1572.

Tanaka, E.M. and Weidinger, G. (2008) Heads or tails: can Wnt tell which one is up? *Nature Cell Biology*, **10**, 122–124.

Tanaka, E.M. and Reddien, P.W. (2011) The cellular basis for animal regeneration. *Developmental Cell*, **21**, 172–185.

Taub, R. (2004) Liver regeneration: from myth to mechanism. *Nature Reviews. Molecular Cell Biology*, **5**, 836–847.

Tsonis, P.A. (1996) *Limb Regeneration*, Cambridge University Press, Cambridge.

Tu, S. and Johnson, S.L. (2011) Fate restriction in the growing and regenerating zebrafish fin. *Developmental Cell*, **20**, 725–732.

Umesono, Y., Tasaki, J., Nishimura, Y. *et al.* (2013) The molecular logic for planarian regeneration along the anterior-posterior axis. *Nature*, **500** (7460), 73–76.

von Maltzahn, J., Renaud, J.M., Parise, G., and Rudnicki, M.A. (2012) Wnt7a treatment ameliorates muscular dystrophy. *Proceedings of the National Academy of Sciences of the United States of America*, **109**, 20614–20619.

Wagers, A.J. and Conboy, I.M. (2005) Cellular and molecular signatures of muscle regeneration: current concepts and controversies in adult myogenesis. *Cell*, **122**, 659–667.

Wallace, H. (1981) *Vertebrate Limb Regeneration*, Wiley, Chichester.

Wang, X., Foster, M., Al-Dhalimy, M. *et al.* (2003) The origin and liver repopulating capacity of murine oval cells. *Proceedings of the National Academy of Sciences of the United States of America*, **100** (Suppl 1), 11881–11888.

Weidinger, G. and Moon, R.T. (2003) When Wnts antagonize Wnts. *The Journal of Cell Biology*, **162**, 753–755.

Whitehead, G.G., Makino, S., Lien, C.L., and Keating, M.T. (2005) fgf20 is essential for initiating zebrafish fin regeneration. *Science*, **310**, 1957–1960.

Williams, J.M., Oh, S.H., Jorgensen, M. *et al.* (2010) The role of the Wnt family of secreted proteins in rat oval "stem" cell-based liver regeneration: Wnt1 drives differentiation. *The American Journal of Pathology*, **176**, 2732–2742.

Yang, W., Yan, H.X., Chen, L. *et al.* (2008) Wnt/beta-catenin signaling contributes to activation of normal and tumorigenic liver progenitor cells. *Cancer Research*, **68**, 4287–4295.

Yanger, K., Zong, Y., Maggs, L.R. *et al.* (2013) Robust cellular reprogramming occurs spontaneously during liver regeneration. *Genes & Development*, **27**, 719–724.

Yazawa, S., Umesono, Y., Hayashi, T. *et al.* (2009) Planarian Hedgehog/Patched establishes anterior-posterior polarity by regulating Wnt signaling. *Proceedings of the National Academy of Sciences of the United States of America*, **106**, 22329–22334.

Yin, L., Lynch, D., Ilic, Z., and Sell, S. (2002) Proliferation and differentiation of ductular progenitor cells and littoral cells during the regeneration of the rat liver to CCl4/2-AAF injury. *Histology and Histopathology*, **17**, 65–81.

Yokoyama, H., Ogino, H., Stoick-Cooper, C.L. *et al.* (2007) Wnt/beta-catenin signaling has an essential role in the initiation of limb regeneration. *Developmental Biology*, **306**, 170–178.

Zhang, Y., Goss, A.M., Cohen, E.D. *et al.* (2008) A Gata6-Wnt pathway required for epithelial stem cell development and airway regeneration. *Nature Genetics*, **40**, 862–870.

Zhou, W., Boucher, R.C., Bollig, F. *et al.* (2010) Characterization of mesonephric development and regeneration using transgenic zebrafish. *American Journal of Physiology. Renal Physiology*, **299**, F1040–F1047.

Part 4

Wnt Signaling in Chronic Disease

From the very beginning, Wnt signaling was also associated with disease, and particularly with tumor formation thought to be due to increased cell proliferation. However, the involvement of Wnt signaling in disease in general and in cancer in particular is far more complex. The study of colorectal cancer has provided much insight into how Wnt signaling is implicated in cancer and particularly in cancer progression (see Chapter 27). Similar mechanisms (TCF-mediated regulation of transcription; see Chapter 17) and similar cellular functions (regulation epithelia formation versus migrating mesenchyme; see Chapter 24) as during embryonic development and homeostasis (see Part 3 above) appear to be operating in cancer, at least during colorectal cancer. However, cancer is a complex series of diseases, and new insight from studying Wnt signaling in melanoma (see chapter 28) suggests that the contribution of Wnt signaling to cancer formation and progression is likely to be equally diverse and specific to the particular cancer.

Wnt signaling has a prominent role in normal neural development (see Chapter 21). Recent discoveries suggest that Wnt signaling is also implicated in some psychiatric disease and dementia (see Chapters 29–31). More detailed understanding of the contribution of Wnt signaling will inform future therapeutic strategies to combat these diseases. Drug development of small inhibitory molecules are likely to take center stage (see Chapter 32), but will require detailed understanding not only of the contribution of Wnt signaling to the particular disease (Part 4) and its role in normal development and homeostasis (Part 3), but particularly of the relevant molecules and their protein–protein interactions (Part 2) and how a drug targeting a particular Wnt signaling component will affect the pathway and the whole Wnt signaling network (Part 1). This will lead us full circle to the beginning of the book and gel the different parts of the book together in our shared pursuit to use the knowledge we have gained about Wnt signaling mechanisms and function in order to improve the health and well-being of future generations.

Wnt Signaling in Development and Disease: Molecular Mechanisms and Biological Functions, First Edition.
Edited by Stefan Hoppler and Randall T. Moon.

27 Wnt Signaling and Colorectal Cancer

Kevin Myant and Owen J. Sansom

Beatson Institute for Cancer Research, Garscube Estate, Glasgow, UK

Introduction

Colorectal cancer (CRC) is the third most common human cancer and the second most common cause of cancer death. Of all human cancers types, it is only CRC where there is unquestionable evidence that deregulated Wnt signaling drives tumorigenesis. In many other cancers, important roles for Wnt signaling have been suggested either from murine models (e.g., the Wnt1 mammary model introduced in Chapter 6) or from deregulated expression of pathway components in human disease; however, deregulated Wnt signaling is not sufficient to cause tumorigenesis in those tissues.

In 1991, it was elucidated that germline mutations in the adenomatous polyposis coli (*Apc*) gene were the cause of familial adenomatous polyposis (FAP) (Groden *et al.*, 1991; Kinzler *et al.*, 1991). These patients develop hundreds of colorectal polyps (generally within their 20s), of which a number will progress to adenocarcinoma if left untreated. It was subsequently shown that an *Apc* mutation at codon 850 was responsible for a murine model of intestinal adenoma formation (the "Min" multiple intestinal neoplasia mouse) that mimicked FAP (Su *et al.*, 1992). Importantly, *Apc* mutations were not simply confined to hereditary syndromes, and it was soon discovered that *Apc* mutations were also common in sporadic CRCs (Nishisho *et al.*, 1991; Powell *et al.*, 1992). This work put the *Apc* protein at the center stage of CRC. In 1997, it was shown in CRC cells that *Apc* was a key negative regulator of β-catenin, establishing a direct link between CRC and the Wnt signaling pathway (Korinek *et al.*, 1997; Morin *et al.*, 1997). Recent genome sequencing studies show the true extent of the significance of the Wnt pathway to CRC as over 75% of tumors have *Apc* mutations and ~5% carry activating mutations of β-catenin (TCGA, 2012). Outside of CRC, *Apc* is only rarely mutated, and although β-catenin mutations are more frequent, they are still relatively rare, which argues for a specific function for deregulated Wnt signaling in CRC. In this chapter, we will discuss potential reasons why *Apc* is a key node in intestinal transformation (with reference to intestinal homeostasis, Wnt signaling levels, and other functions of *Apc*) and will conclude with a discussion of Wnt signaling in CRC progression and potential therapeutic targeting.

Wnt Signaling in Development and Disease: Molecular Mechanisms and Biological Functions, First Edition.
Edited by Stefan Hoppler and Randall T. Moon.

Apc loss induces a crypt progenitor cell-like phenotype in an animal model

In Chapter 25, the importance of the Wnt signaling pathway in normal intestinal homeostasis was discussed. This chapter detailed how a reduction in Wnt signaling blocked proliferation and altered differentiation while ablation of Wnt signaling killed the intestinal stem cells (ISCs) and ultimately intestinal crypts. Therefore, if Wnt signaling activation had the opposite effect and induced proliferation and blocked differentiation, then one could envisage why upregulation of Wnt signaling might drive colorectal carcinogenesis. One other important facet of the intestinal tract to consider is the high turnover of intestinal enterocytes. Every 3–5 days, all cells, apart from stem (and in the small intestine paneth) cells, are sloughed off into the intestinal lumen. Therefore, in order to become a cancer cell in the intestine, tumors have to arise from either stem cells or enterocytes with mutations that allow them to escape from the differentiation and sloughing off that would otherwise inevitably occur.

A number of groups have targeted *Apc* gene deletion to the adult murine intestine to assess the specific reason why *Apc* is so frequently mutated in CRC. As complete *Apc* knockout throughout the embryo results in embryonic death, Cre-lox technology has been used to allow *Apc* gene deletion in the adult. Here, exon 14 of *Apc* is flanked by LOXP sites, and when Cre recombinase is expressed, it deletes this exon and induces a frameshift mutation at amino acid 580 (Shibata *et al.*, 1997). This produces a truncated allele that lacks the ability to bind β-catenin or Axin. Mice carrying this conditional *Apc* allele, known as Apc^{580S}, have then been intercrossed to mice carrying inducible intestinal cre transgenes, for example, *AHCre*, *VillinCreER*, and *Lgr5CreER* (Andreu *et al.*, 2005; Barker *et al.*, 2009; Sansom *et al.*, 2004). The phenotype of *Apc* gene deletion is dramatic; there is a rapid localization of β-catenin to the nucleus coincident with a marked phenotypic change within the intestine (Figure 27.1). The intestinal crypts become enlarged, differentiation is perturbed with reduction of goblet and enteroendocrine cell lineages, and proliferation is markedly increased. Moreover, the upward movement and sloughing off of the intestinal enterocytes is stopped. This dramatic phenotype has been termed a crypt progenitor cell-like phenotype (Sansom *et al.*, 2004). Thus, enterocytes lacking *Apc* escape most of the processes that

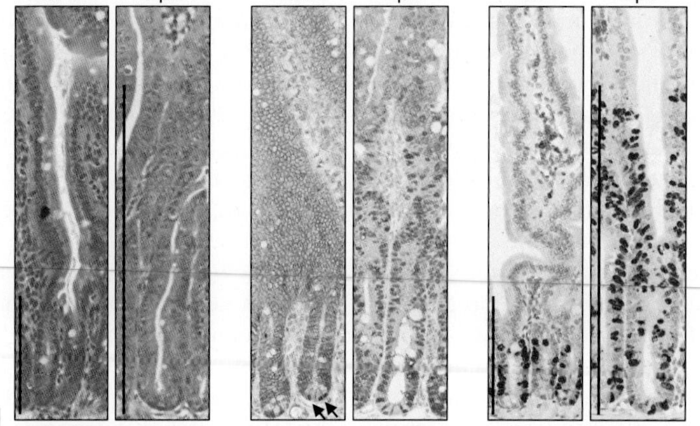

Figure 27.1 Crypt progenitor phenotype following *Apc* loss. (a) Hematoxylin and eosin staining of normal (WT) and *VilCre-ERT* $Apc^{580S/580S}$ (*Apc*) small intestines 4 days following Cre induction with tamoxifen. The expanded *Apc*-deficient crypt is indicated with a black vertical line. (b) Immunohistochemical staining for β-catenin in WT and *Apc*-deficient small intestines. Arrows indicate cells with nuclear localized β-catenin in WT crypt. β-catenin is localized to the nucleus in all *Apc*-deficient cells. (c) Immunohistochemical staining for BrdU in WT and *Apc*-deficient small intestines. The expanded proliferative zone in *Apc*-deficient small intestine is indicated with a black vertical line. *(See insert for color representation of the figure.)*

prevent tumor formation in the intestine. Within the small intestine, there is also a mislocalization and increase in the number of paneth cells. This is pertinent as paneth cells have a high expression of the Wnt signaling pathway (van Es *et al.*, 2005) and they are normally orientated by EphB/Ephrin gradients (Batlle *et al.*, 2002). EphB receptors are expressed at the base of the intestinal crypt and are canonical Wnt target genes; thus, following loss of *Apc*, β-catenin relocalizes to the nucleus and drives expression of EphB2 and EphB3 throughout the enlarged crypt. This removes any gradient, and thus, paneth cells are no longer confined to the base of the crypt. The concurrence of phenotypic onset and β-catenin accumulation in the nucleus suggest that it is the activation of Wnt signaling downstream of *Apc* loss that drives tumor initiation in CRC. This would suggest that genes of the pathway may act as good therapeutic targets. Further work has discovered a number of downstream Wnt targets that can modify the phenotypes following *Apc* loss, for example, TIAM1, Cyclin D2, and MYC (Cole *et al.*, 2010a; He *et al.*, 1998; Malliri *et al.*, 2006). Of these, c-Myc deletion had the most dramatic effect as intestinal enterocytes lacking both c-Myc and *Apc* no longer exhibited a crypt progenitor cell-like phenotype, despite having nuclear β-catenin (Sansom *et al.*, 2007). It is important to note that many CRC cell lines also retain this crypt progenitor cell-like phenotype and inhibition of WNT target genes can also cause arrest and differentiation within these cell lines (van de Wetering *et al.*, 2002). Thus, the dependence on Wnt signaling may be retained in much later stages of the carcinogenic process (see later).

Over the past 5 years, there has been a paradigm shift over the localization and nature of ISCs within the intestine (covered in Chapter 25). The ISC signature refers to a number of genes which mark Lgr5-expressing (Lgr5+) ISCs (van der Flier *et al.*, 2009). Most of these ISC signature genes are targets of the Wnt pathway; however, it is interesting to note that loss of *Apc* does not lead to the deregulation of these targets throughout the intestinal crypt, just an increased number of these cells at the base of the crypt. How the Wnt pathway integrates with other signaling gradients within the intestine to define the ISC is still unknown and will be an area of great interest in the coming years.

The cell of origin of CRC and Wnt signaling

The properties bestowed by loss of both copies of *Apc* could be argued to be sufficient to transform an intestinal enterocyte into an adenoma. One question that remained was whether all cells within the intestinal epithelium respond in the same manner to *Apc* loss and activation of Wnt signaling. To test this concept, a number of groups have aimed to target *Apc* deletion to ISCs or alternatively to differentiated cells. Given that ISCs have a clear identity (e.g., Lgr5+ positivity), using a cre transgene knocked into the Lgr5 locus allowed targeted deletion of *Apc* within ISCs (Barker *et al.*, 2009). This rapidly led to adenoma formation within 6 weeks of *Apc* loss, highlighting that the deletion of both copies of *Apc* is sufficient to initiate tumorigenesis. When *Apc* was deleted outside of stem cells, although small proliferative lesions were formed with nuclear β-catenin, they did not progress quickly to adenoma. More recent studies have shown that additional mutations in KRAS or activation of the NF-κB pathway increases levels of Wnt signaling and ISC marker gene expression within these lesions and accelerates tumor progression (Schwitalla *et al.*, 2013; Figure 27.2). It is interesting to note that in most other epithelia within the mouse when *Apc* is deleted or β-catenin is activated, it does not lead to tumor formation. In the mammary, squamous metaplasia occurs (Gallagher *et al.*, 2002); in the kidney, small lesions occur, which engage the senescence program (Cole *et al.*, 2010b); and in the bladder, small proliferative lesions form that never progress (Ahmad *et al.*, 2011). Thus, activation of Wnt signaling in these tissues has a profound phenotype but is unable to initiate tumorigenesis. In these circumstances, additional oncogenic or tumor suppressor mutations are required for tumor formation, and this may help to explain why CRC is unique in terms of Wnt signaling and carcinogenesis. A deletion of a single gene (*Apc*) in a specific subset of intestinal enterocytes can lead directly to a benign tumor, which then has the capacity to develop further mutations to drive malignant progression. Interestingly, when other common oncogenic or tumor suppressor genes such as *PTEN*, *KRAS*, or *P53* are mutated within the

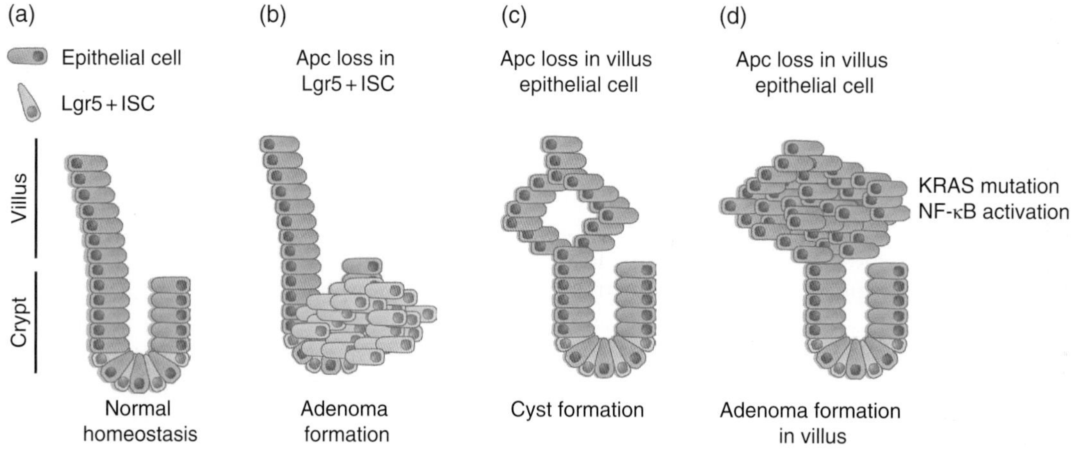

Figure 27.2 Cell of origin of CRC. (a) Graphical representation of a normal intestinal crypt. Lgr5+ ISCs are depicted in green residing at the crypt base; all other intestinal epithelial cells are depicted in blue. (b) *Apc* loss in Lgr5+ ISCs leads to rapid adenoma formation. (c) *Apc* loss in villus epithelial cells gives rise to formation of cyst-like structures that take much longer to progress to adenomas. (d) If *Apc* loss in villus epithelial cells is combined with an activating mutation in *KRAS* or activation of NF-κB signaling, adenoma formation rapidly occurs. *(See insert for color representation of the figure.)*

intestinal epithelium, very few changes to homeostasis are observed, coincident with little or no alteration in candidate downstream pathways (e.g., AKT when *PTEN* is lost, MAPK signaling when *KRAS* in mutation) (Marsh *et al.*, 2008; Sansom *et al.*, 2006). However, in all circumstances, there is marked cooperation when these mutations are targeted alongside *Apc* gene deletion.

Deregulation of Wnt signaling in human CRC and the "just right" hypothesis of signaling model

One question that has persisted is why, given a single mutation of β-catenin should activate Wnt signaling, is mutation of both copies of *Apc* much more common in human CRC. Moreover, most of the mutations in *Apc* occur in the mutation cluster region producing a truncated but expressed fragment that, although able to bind β-catenin, cannot bind Axin. These two observations have led to the "just right" hypothesis of Wnt signaling that proposes there is a window of Wnt signaling that is optimal and selected for transformation. This is not restricted to CRC, as murine models have shown in other tissues that alterations in the levels of *Apc* or mutations that yield only weak Wnt pathway activity promote tumorigenesis (Buchert *et al.*, 2010; Gaspar *et al.*, 2009). Over the last few years, this hypothesis has been tested by the generation of novel knock-in *Apc* alleles, allowing characterization of specific mutations of *Apc* in CRC (Figure 27.3). As a control for these experiments, an *Apc* mutant was produced (Apc^{1638T}) that lacked its C-terminus but retained β-catenin and Axin binding (through three 20 amino acid repeats and one SAMP repeat, respectively). This mouse was not tumor prone, underlining that Wnt activation is required for transformation (Smits *et al.*, 1999). Moreover, another *Apc* mutant mouse was produced where both the N- and C-termini were present, but the β-catenin binding region was lacking (Lewis *et al.*, 2011). These mice were tumor prone, highlighting that the key tumor suppressive function of *Apc* is inhibition of Wnt signaling. To mimic the human *Apc* mutation which retains β-catenin binding but has lost all SAMP repeats, an Apc^{1322T} allele was generated. $Apc^{1322T/+}$ mice developed tumors faster than $Apc^{MIN/+}$ mice, despite tumors from both genotypes losing the wild-type copy of *Apc*. When Wnt signaling activity was examined, Apc^{1322T} tumors exhibited increased Wnt signaling when compared to wild-type intestinal tissue but less than Apc^{MIN} tumors. Moreover, despite the expression of most Wnt target genes being reduced, a subset associated with the ISC signature was increased in these tumors. This

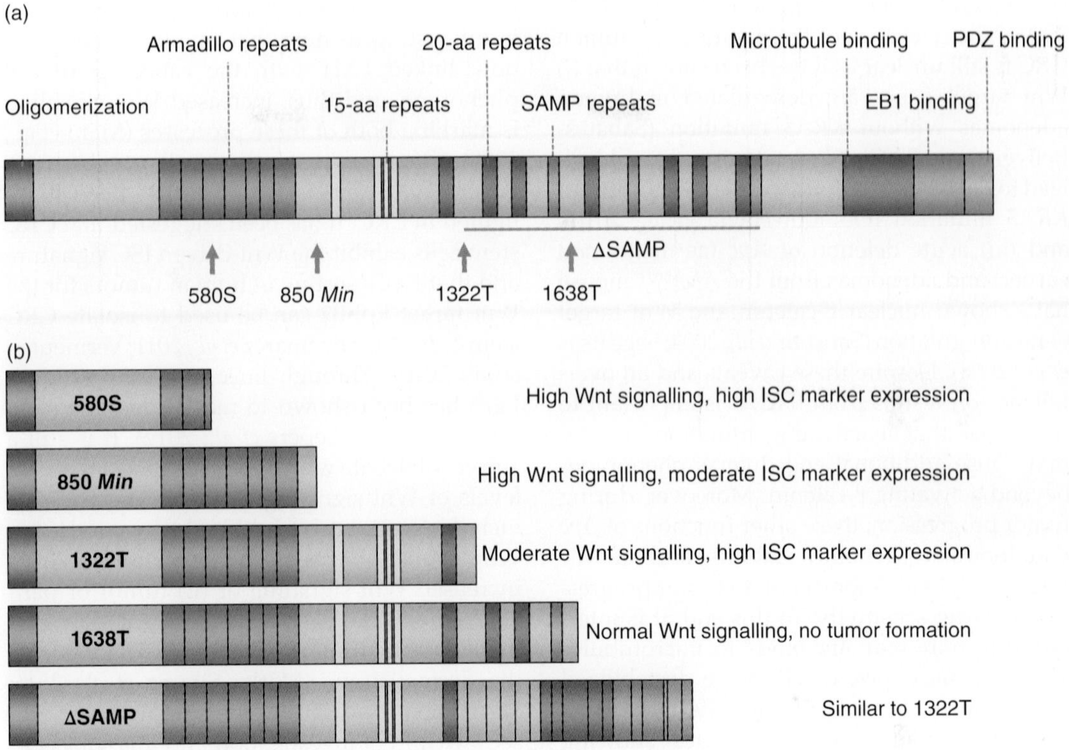

Figure 27.3 Mutant alleles used to study *Apc* function *in vivo*. (a) Graphical representation of the *Apc* protein structure with key domains highlighted. 15-aa and 20-aa repeats are important for β-catenin binding, and SAMP repeats mediate binding to Axin. The positions of mutations creating five mutant *Apc* alleles are shown below the structure. (b) Graphical representations of the truncated protein products produced from these mutations and their effects on Wnt signaling and ISC marker expression. *(See insert for color representation of the figure.)*

suggests that expression of the ISC signature might accelerate tumorigenesis, but more importantly that the precise Wnt signaling outcome, in part determined by the level of signaling, is a very important modifier of tumor initiation in the intestine.

Other functions of *Apc*

Despite these proof-of-principle studies *in vitro* (cell culture models) and *in vivo* (animal models), a number of questions still remain over the role of the Wnt signaling pathway in CRC. In Chapter 24, the role of β-catenin in invasion and EMT was discussed. One concept discussed was the finding that β-catenin expression is much higher at the invasive fronts of tumors compared to the central mass. This is known as the β-catenin paradox as, although the tumor would have *Apc* mutated throughout, intense nuclear β-catenin staining was only observed at the invasive fronts of these tumors. Despite much investigation, the mechanistic basis for this is still unclear. The difficulty in detecting β-catenin by immunohistochemistry (IHC) in the nuclei of human CRC has led some to propose that other functions of *Apc* outside of Wnt signaling may be of relevance for colorectal carcinogenesis. *Apc* is a large protein and interacts with many proteins that are thought to play no role in Wnt signaling. Many of these interactions functionally affect migration or mitosis (McCartney and Nathke, 2008). Of all the studies suggesting that other functions of *Apc* may play important roles in CRC, a study by Phelps *et al.* (2009) demonstrated that in zebrafish *Apc* loss could perturb differentiation via the transcriptional repressor CtBP1 in the absence of nuclear β-catenin. The additional activation of *KRAS* was required for the accumulation of β-catenin in the nucleus and subsequent hyperproliferation (Phelps *et al.*, 2009). The relevance of these studies to human CRC is still unclear as it has been shown that (i) Wnt target genes are deregulated in human adenomas without *KRAS* mutation (Sabates-Bellver *et al.*, 2007), (ii) β-catenin can be localized to the nucleus of human adenomas without *KRAS* mutation (Obrador-Hevia *et al.*, 2010), and (iii) acute deletion of *Apc* (as mentioned earlier) and adenomas from the *Apc*$^{Min/+}$ mouse have shown nuclear β-catenin and Wnt target gene upregulation (Sansom *et al.*, 2004; Segditsas *et al.*, 2008). Despite these caveats and an over-reliance on results from IHC, it is important to remember that inactivating mutations in *Apc* may cause additional and diverse phenotypes beyond activating β-catenin. Moreover, during tumor progression, these other functions of *Apc* that modulate genomic stability and mitotic fidelity might also contribute to tumor progression or drug sensitivity. To this end, it is interesting to note that *Apc* binds to microtubules and that *Apc*-deficient cells have destabilized microtubules. *In vivo*, this manifests itself in *Apc*-deficient intestinal enterocytes showing resistance to microtubule-stabilizing agents such as Taxol (Radulescu *et al.*, 2010).

Wnt signaling and colorectal cancer progression

As mentioned earlier (and in Chapter 24), increased Wnt signaling has been observed during CRC progression. This originally was associated with invasion and EMT; however, a number of recent studies have suggested that increased Wnt signaling during CRC progression drives the "cancer stem cell" phenotype (Merlos-Suarez *et al.*, 2011; Vermeulen *et al.*, 2010). The cancer stem cell concept refers to cells within the tumor that act in a manner analogous to normal stem cells. Thus, a hierarchy within a tumor is retained with cancer stem cells producing proliferative and differentiated cancer cells. One hypothesis is that if chemotherapy does not kill the cancer stem cells, then tumors will regrow. In many cancers, cancer stem cells are thought to be lowly proliferative or in quiescence and thus are thought to be more resistant to treatment. Moreover, if these cancer stem cells seed to new sites, they are proposed to be very efficient in forming metastasis. It is interesting to note that many studies have now linked EMT with the cancer stem cell phenotype, and thus, increased Wnt signaling is marking both of these processes (Mani *et al.*, 2008). Of all cancers, the similarity between normal and cancer stem cells has been highlighted in CRC. It has been suggested that CRC stem cells exhibit the Wnt-driven ISC signature and that FACs sorting of human tumors for the Wnt target EphB2 can be used to isolate CRC stem cells. (Merlos-Suarez *et al.*, 2011; Vermeulen *et al.*, 2010). Through lineage tracing studies, Lgr5 has been shown to mark adenoma stem cells in mice (Schepers *et al.*, 2012). It is still a subject of debate why these cells have increased levels of Wnt signaling; potential mechanisms include (i) mesenchymal fibroblasts producing Wnt ligand or other growth factors to drive increased Wnt signaling or (ii) tumor or paneth/paneth-like cells establishing a Wnt gradient within tumors in a similar manner to the normal tissue (Merlos-Suarez *et al.*, 2011; Vermeulen *et al.*, 2010). The coming years will see a wealth of investigation into this subject as the precise identity and the functional importance of these cells has yet to be fully established (de Sousa *et al.*, 2011).

Wnt signaling as a target in colorectal cancer

The key role that the Wnt signaling pathway plays in CRC has led to the prediction that targeting this pathway should be efficacious in CRC. Proof-of-principle studies have shown that inhibition of β-catenin in established CRC cells through a shRNA approach causes death of CRC cells (Scholer-Dahirel *et al.*, 2011). Moreover, preclinical studies have shown that targets of the Wnt pathway such as c-Myc can have dramatic effects upon *Apc*-deficient cells (Sansom *et al.*, 2007). The final chapter will discuss in much more detail the chemical inhibitors of Wnt signaling that are currently available, but here, we will briefly discuss the current landscape of Wnt inhibition in CRC. One of the biggest hurdles for Wnt inhibitors to overcome in CRC is that of toxicity towards normal intestinal enterocytes. In Chapter 25, the importance of Wnt signaling for the normal intestine and stem cells was discussed, and this may be an important limiting factor. Moreover, given *Apc* is deleted, it remains to be seen if cell surface inhibitors (such as blocking antibodies to Frizzled or Lrp6) or Wnt secretion blockers will suppress Wnt signaling (Ettenberg *et al.*, 2010). Thus, downstream components of the pathway need to be targeted, and so far, there has been some preclinical success in this area. Tankyrase, a member of the poly(ADP-ribose) polymerase (PARP) superfamily, has been shown to be important for the turnover of Axin. Following the inhibition of Tankyrase, Axin levels are increased, and this may stabilize the β-catenin destruction complex even in the absence of *Apc* (Huang *et al.*, 2009). Two recent reports have suggested that in both cell lines and in mice, Tankyrase inhibition can suppress *Apc*-deficient tumor growth (Lau *et al.*, 2013; Waaler *et al.*, 2012). These exciting reports suggest that Tankyrase inhibition may have therapeutic benefit in CRC. However, more research will be required to assess how well this translates clinically, and in particular, concerns regarding toxicity must be addressed. Other approaches that have shown promise *in vivo* include targeting β-catenin shuttling between the cytoplasm and nucleus (de la Roche *et al.*, 2012) or β-catenin interactions with cyclic AMP response element-binding protein (CBP) (Emami *et al.*, 2004). How potent these approaches will be in late-stage cancer, where many other mutations are present, has yet to be defined, but again, they offer promise. One other concern given the roles of Wnt signaling in the regeneration of the intestine and other tissues (Ashton *et al.*, 2010; Chapter 25) is whether Wnt inhibitors might increase gastrointestinal toxicity if combined with conventional chemotherapeutic drugs.

Summary

Despite Wnt signaling first being associated with mammary tumorigenesis (Rijsewijk *et al.*, 1987), the finding that the *Apc* tumor suppressor gene was the most common mutation in CRC and that it negatively regulated Wnt signaling placed the Wnt pathway as the major signaling axis in CRC. Deregulation of β-catenin/TCF4 drives tumor initiation by allowing intestinal enterocytes to escape many of the tumor suppressor pathways in the intestine (namely, growth arrest, differentiation, and clearance). There is an increased expression of oncogenes such as c-Myc and Cyclin D1 that maintains the crypt progenitor cell-like phenotype and of ISC Wnt target genes such ASCL2 and LGR5. As colorectal tumors progress, there is a further elevation of the Wnt pathway driving increased expression of stem cell (and potential cancer stem cell) markers and EMT. Tumors appear to rely on these Wnt targets for maintenance, suggesting that targeting of the Wnt pathway may be an effective strategy in CRC. This is complicated by the absence of *Apc*, but now, 30 years after the identification of Wnt signaling, a number of exciting new inhibitors yield promise for this common and, at late stage, deadly disease.

References

Ahmad, I., Morton, J.P., Singh, L.B. *et al.* (2011) Beta-catenin activation synergizes with PTEN loss to cause bladder cancer formation. *Oncogene*, **30**, 178–189.

Andreu, P., Colnot, S., Godard, C. *et al.* (2005) Crypt-restricted proliferation and commitment to the Paneth cell lineage following Apc loss in the mouse intestine. *Development*, **132**, 1443–1451.

Ashton, G.H., Morton, J.P., Myant, K. *et al.* (2010) Focal adhesion kinase is required for intestinal regeneration and tumorigenesis downstream of Wnt/c-Myc signaling. *Developmental Cell*, **19**, 259–269.

Barker, N., Ridgway, R.A., van Es, J.H. *et al.* (2009) Crypt stem cells as the cells-of-origin of intestinal cancer. *Nature*, **457**, 608–611.

Batlle, E., Henderson, J.T., Beghtel, H. *et al.* (2002) Beta-catenin and TCF mediate cell positioning in the intestinal epithelium by controlling the expression of EphB/ephrinB. *Cell*, **111**, 251–263.

Buchert, M., Athineos, D., Abud, H.E. *et al.* (2010) Genetic dissection of differential signaling threshold requirements for the Wnt/beta-catenin pathway *in vivo*. *PLoS Genetics*, **6**, e1000816.

Cole, A.M., Myant, K., Reed, K.R. *et al.* (2010a) Cyclin D2-cyclin-dependent kinase 4/6 is required for efficient proliferation and tumorigenesis following Apc loss. *Cancer Research*, **70**, 8149–8158.

Cole, A.M., Ridgway, R.A., Derkits, S.E. *et al.* (2010b) p21 loss blocks senescence following Apc loss and provokes tumourigenesis in the renal but not the intestinal epithelium. *EMBO Molecular Medicine*, **2**, 472–486.

de la Roche, M., Rutherford, T.J., Gupta, D. *et al.* (2012) An intrinsically labile alpha-helix abutting the BCL9-binding site of beta-catenin is required for its inhibition by carnosic acid. *Nature Communications*, **3**, 680.

de Sousa, E.M.F., Colak, S., Buikhuisen, J. *et al.* (2011) Methylation of cancer-stem-cell-associated Wnt target genes predicts poor prognosis in colorectal cancer patients. *Cell Stem Cell*, **9**, 476–485.

Emami, K.H., Nguyen, C., Ma, H. *et al.* (2004) A small molecule inhibitor of beta-catenin/CREB-binding protein transcription [corrected]. *Proceedings of the National Academy of Sciences of the United States of America*, **101**, 12682–12687.

Ettenberg, S.A., Charlat, O., Daley, M.P. *et al.* (2010) Inhibition of tumorigenesis driven by different Wnt proteins requires blockade of distinct ligand-binding regions by LRP6 antibodies. *Proceedings of the National Academy of Sciences of the United States of America*, **107**, 15473–15478.

Gallagher, R.C., Hay, T., Meniel, V. *et al.* (2002) Inactivation of Apc perturbs mammary development, but only directly results in acanthoma in the context of Tcf-1 deficiency. *Oncogene*, **21**, 6446–6457.

Gaspar, C., Franken, P., Molenaar, L. *et al.* (2009) A targeted constitutive mutation in the APC tumor suppressor gene underlies mammary but not intestinal tumorigenesis. *PLoS Genetics*, **5**, e1000547.

Groden, J., Thliveris, A., Samowitz, W. *et al.* (1991) Identification and characterization of the familial adenomatous polyposis coli gene. *Cell*, **66**, 589–600.

He, T.C., Sparks, A.B., Rago, C. *et al.* (1998) Identification of c-MYC as a target of the APC pathway. *Science*, **281**, 1509–1512.

Huang, S.M., Mishina, Y.M., Liu, S. *et al.* (2009) Tankyrase inhibition stabilizes axin and antagonizes Wnt signalling. *Nature*, **461**, 614–620.

Kinzler, K.W., Nilbert, M.C., Vogelstein, B. *et al.* (1991) Identification of a gene located at chromosome 5q21 that is mutated in colorectal cancers. *Science*, **251**, 1366–1370.

Korinek, V., Barker, N., Morin, P.J. *et al.* (1997) Constitutive transcriptional activation by a beta-catenin–Tcf complex in $APC^{-/-}$ colon carcinoma. *Science*, **275**, 1784–1787.

Lau, T., Chan, E., Callow, M. *et al.* (2013) A novel Tankyrase small-molecule inhibitor suppresses APC mutation-driven colorectal tumor growth. *Cancer Research*, **73**, 3132–3144.

Lewis, A., Davis, H., Deheragoda, M. *et al.* (2011) The C-terminus of Apc does not influence intestinal adenoma development or progression. *The Journal of Pathology*, **226**, 73–83.

Malliri, A., Rygiel, T.P., van der Kammen, R.A. *et al.* (2006) The rac activator Tiam1 is a Wnt-responsive gene that modifies intestinal tumor development. *The Journal of Biological Chemistry*, **281**, 543–548.

Mani, S.A., Guo, W., Liao, M.J. *et al.* (2008) The epithelial–mesenchymal transition generates cells with properties of stem cells. *Cell*, **133**, 704–715.

Marsh, V., Winton, D.J., Williams, G.T. *et al.* (2008) Epithelial Pten is dispensable for intestinal homeostasis but suppresses adenoma development and progression after Apc mutation. *Nature Genetics*, **40**, 1436–1444.

McCartney, B.M. and Nathke, I.S. (2008) Cell regulation by the Apc protein Apc as master regulator of epithelia. *Current Opinion in Cell Biology*, **20**, 186–193.

Merlos-Suarez, A., Barriga, F.M., Jung, P. *et al.* (2011) The intestinal stem cell signature identifies colorectal cancer stem cells and predicts disease relapse. *Cell Stem Cell*, **8**, 511–524.

Morin, P.J., Sparks, A.B., Korinek, V. *et al.* (1997) Activation of beta-catenin–Tcf signaling in colon cancer by mutations in beta-catenin or APC. *Science*, **275**, 1787–1790.

Nishisho, I., Nakamura, Y., Miyoshi, Y. *et al.* (1991) Mutations of chromosome 5q21 genes in FAP and colorectal cancer patients. *Science*, **253**, 665–669.

Obrador-Hevia, A., Chin, S.F., Gonzalez, S. *et al.* (2010) Oncogenic KRAS is not necessary for Wnt signalling activation in APC-associated FAP adenomas. *The Journal of Pathology*, **221**, 57–67.

Phelps, R.A., Chidester, S., Dehghanizadeh, S. *et al.* (2009) A two-step model for colon adenoma initiation and progression caused by APC loss. *Cell*, **137**, 623–634.

Powell, S.M., Zilz, N., Beazer-Barclay, Y. *et al.* (1992) APC mutations occur early during colorectal tumorigenesis. *Nature*, **359**, 235–237.

Radulescu, S., Ridgway, R.A., Appleton, P. *et al.* (2010) Defining the role of APC in the mitotic spindle checkpoint in vivo: APC-deficient cells are resistant to Taxol. *Oncogene*, **29**, 6418–6427.

Rijsewijk, F., Schuermann, M., Wagenaar, E. *et al.* (1987) The Drosophila homolog of the mouse mammary oncogene int-1 is identical to the segment polarity gene wingless. *Cell*, **50**, 649–657.

Sabates-Bellver, J., Van der Flier, L.G., de Palo, M. *et al.* (2007) Transcriptome profile of human colorectal adenomas. *Molecular Cancer Research*, **5**, 1263–1275.

Sansom, O.J., Reed, K.R., Hayes, A.J. *et al.* (2004) Loss of Apc in vivo immediately perturbs Wnt signaling, differentiation, and migration. *Genes & Development*, **18**, 1385–1390.

Sansom, O.J., Meniel, V., Wilkins, J.A. *et al.* (2006) Loss of Apc allows phenotypic manifestation of the transforming properties of an endogenous K-ras oncogene in vivo. *Proceedings of the National Academy of Sciences of the United States of America*, **103**, 14122–14127.

Sansom, O.J., Meniel, V.S., Muncan, V. *et al.* (2007) Myc deletion rescues Apc deficiency in the small intestine. *Nature*, **446**, 676–679.

Schepers, A.G., Snippert, H.J., Stange, D.E. *et al.* (2012) Lineage tracing reveals Lgr5+ stem cell activity in mouse intestinal adenomas. *Science*, **337**, 730–735.

Scholer-Dahirel, A., Schlabach, M.R., Loo, A. *et al.* (2011) Maintenance of adenomatous polyposis coli (APC)-mutant colorectal cancer is dependent on Wnt/beta-catenin signaling. *Proceedings of the National Academy of Sciences of the United States of America*, **108**, 17135–17140.

Schwitalla, S., Fingerle, A.A., Cammareri, P. *et al.* (2013) Intestinal tumorigenesis initiated by dedifferentiation and acquisition of stem-cell-like properties. *Cell*, **152**, 25–38.

Segditsas, S., Sieber, O., Deheragoda, M. *et al.* (2008) Putative direct and indirect Wnt targets identified through consistent gene expression changes in APC-mutant intestinal adenomas from humans and mice. *Human Molecular Genetics*, **17**, 3864–3875.

Shibata, H., Toyama, K., Shioya, H. *et al.* (1997) Rapid colorectal adenoma formation initiated by conditional targeting of the Apc gene. *Science*, **278**, 120–123.

Smits, R., Kielman, M.F., Breukel, C. *et al.* (1999) Apc1638T: a mouse model delineating critical domains of the adenomatous polyposis coli protein involved in tumorigenesis and development. *Genes & Development*, **13**, 1309–1321.

Su, L.K., Kinzler, K.W., Vogelstein, B. *et al.* (1992) Multiple intestinal neoplasia caused by a mutation in the murine homolog of the APC gene. *Science*, **256**, 668–670.

TCGA. (2012) Comprehensive molecular characterization of human colon and rectal cancer. *Nature*, **487**, 330–337.

van de Wetering, M., Sancho, E., Verweij, C. *et al.* (2002) The beta-catenin/TCF-4 complex imposes a crypt progenitor phenotype on colorectal cancer cells. *Cell*, **111**, 241–250.

van der Flier, L.G., van Gijn, M.E., Hatzis, P. *et al.* (2009) Transcription factor achaete scute-like 2 controls intestinal stem cell fate. *Cell*, **136**, 903–912.

van Es, J.H., Jay, P., Gregorieff, A. *et al.* (2005) Wnt signalling induces maturation of Paneth cells in intestinal crypts. *Nature Cell Biology*, **7**, 381–386.

Vermeulen, L., De Sousa, E.M.F., van der Heijden, M. *et al.* (2010) Wnt activity defines colon cancer stem cells and is regulated by the microenvironment. *Nature Cell Biology*, **12**, 468–476.

Waaler, J., Machon, O., Tumova, L. *et al.* (2012) A novel tankyrase inhibitor decreases canonical Wnt signaling in colon carcinoma cells and reduces tumor growth in conditional APC mutant mice. *Cancer Research*, **72**, 2822–2832.

28 Wnt Signaling in Melanoma

Jamie N. Anastas[1] and Andy J. Chien[1,2]

[1]The University of Washington School of Medicine, Seattle, WA, USA
[2]The Group Health Research Institute, Seattle, WA, USA

An introduction to malignant melanoma

Malignant melanoma is a molecularly diverse cancer that arises from melanocytic cells of the skin and other tissues. Melanoma incidence is increasing at a rate that surpasses that of any other cancer worldwide (Lens and Dawes, 2004). Melanoma accounts for the vast majority of skin cancer-related deaths with a dismally low 5-year survival rate of 5–15% (Siegel *et al.*, 2012). Clinically, melanoma is extremely resistant to available forms of chemotherapy and radiation therapy (Tsao, Atkins, and Sober, 2004). Given the scarcity of effective therapeutic options, research efforts have focused on understanding mechanisms of melanoma initiation and progression in order to develop novel treatment strategies.

Melanomagenesis involves a combination of well-defined genetic and environmental factors, most notably exposure to ultraviolet light (Haluska *et al.*, 2006). One of the ongoing challenges for melanoma research will be to determine if and how changes in the activity of various signal transduction pathways such as those driven by Wnts can be leveraged to improve the diagnosis and treatment of melanomas. Numerous studies now implicate changes in the activity of both β-catenin-dependent Wnt signaling pathways and β-catenin-independent Wnt signaling pathways in melanoma development and progression. Several critical questions remain regarding the Wnt signaling pathways in melanoma:

(1) Can measures of Wnt signaling activity provide any prognostic information in a clinical setting?
(2) How do Wnt signaling pathways functionally contribute to melanoma tumorigenesis, progression, and/or metastasis?
(3) What endogenous and exogenous factors determine the relative activity of Wnt signaling pathways in melanoma?
(4) Can we leverage our emerging understanding of Wnt signaling towards the development of new clinical strategies for therapeutically intractable melanomas?

Wnt Signaling in Development and Disease: Molecular Mechanisms and Biological Functions, First Edition.
Edited by Stefan Hoppler and Randall T. Moon.

Wnt signaling pathways as prognostic indicators in melanoma

The Wnt/β-catenin pathway

Histological detection of cytoplasmic or nuclear β-catenin (encoded by *CTNNB1* in humans and *Ctnnb1* in the mouse) provides a surrogate indicator of active Wnt/β-catenin signaling and is observed in approximately one third of patient melanoma tumors. The detection of increased cytoplasmic or nuclear β-catenin in patient melanoma tumors correlates with improved patient survival (Bachmann *et al.*, 2005; Chien *et al.*, 2009; Gould Rothberg *et al.*, 2009; Kageshita *et al.*, 2001). Consistent with these observations, other studies find that nuclear and cytoplasmic β-catenin abundance gradually decreases in later-stage melanomas compared to primary tumors (Kageshita *et al.*, 2001; Maelandsmo *et al.*, 2003). Another study (often cited incorrectly) finds that increased nuclear phosphorylated β-catenin (phospho-β-catenin) detected using an antibody that recognizes phosphorylation at Ser33, Ser37, and Thr41 correlates with a poorer patient prognosis (Kielhorn *et al.*, 2003). As predicted by the current paradigm of Wnt/β-catenin signaling, tumors with increased *nuclear* phospho-β-catenin exhibit undetectable levels of *total* β-catenin (Kielhorn *et al.*, 2003). In summary, immunohistochemical studies employing either anti-(total) β-catenin or anti-phospho-β-catenin antibodies both support a model in which stabilization of β-catenin correlates with improved clinical outcomes in melanoma.

The prognostic significance of both total and phospho-β-catenin levels in melanoma raises questions regarding the mechanisms underlying the variable activity of the Wnt/β-catenin signaling pathway observed in melanomas. In many cancers, the activation of Wnt/β-catenin signaling occurs primarily through mutations in the pathway components. While an initial report identified activating mutations in *CTNNB1* (β-catenin) in melanomas (Rubinfeld *et al.*, 1997), further studies found that mutations in *CTNNB1* (β-catenin) or genes encoding other Wnt pathway components such as *APC* and *AXIN1* were rare in melanomas (Demunter *et al.*, 2002; Omholt *et al.*, 2001; Pollock and Hayward, 2002; Rimm *et al.*, 1999). Elevated cytoplasmic/nuclear β-catenin staining is sometimes observed in melanoma tumors lacking mutations in Wnt/β-catenin pathway genes, suggesting a model whereby activation of or inactivation of Wnt/β-catenin signaling is regulated not through mutations of known pathway components but rather through the temporal–spatial expression of Wnt ligands.

Expression of WNT5A in metastatic melanoma

Numerous studies now point to possible roles for β-catenin-independent Wnt signaling in melanoma, most notably those signaling pathways activated by WNT5A (Moon *et al.*, 1993; Weeraratna, 2005). Interest in WNT5A-dependent signaling in melanoma was initially generated by large-scale efforts to characterize melanoma phenotypic subtypes by microarray analysis of gene expression (Bittner *et al.*, 2000). After analyzing gene expression in 31 established patient-derived melanoma cell lines, higher levels of *WNT5A* expression correlated with a more invasive phenotype. While the number of clinical samples linked to corresponding patient outcomes in this study was too small to draw definitive conclusions, 4 of 5 patients with a high *WNT5A* signature died, whereas only 3 of 10 patients with the low *WNT5A* signature died. The correlation between increased expression of *WNT5A* and invasive melanoma phenotypes has been independently verified by another large transcriptional profiling study (Hoek *et al.*, 2006).

The first study to report an association between increased cytoplasmic/nuclear β-catenin and increased patient survival also examined WNT5A expression in 312 tumors and noted the presence of cytosolic and nuclear staining for WNT5A in subset of the patient samples (Bachmann *et al.*, 2005). However, WNT5A staining levels did not predict patient outcomes in this dataset (Bachman *et al.*, 2005). In contrast, another immunohistochemical study analyzing 102 patient tumors found that high *cytoplasmic* staining for WNT5A was predictive of reduced time to metastasis and reduced time to patient death (Da Forno *et al.*, 2008). Both of these studies used the same commercially available goat anti-WNT5A antibody (R&D Systems, Minneapolis,

MN) yet employed markedly different antigen retrieval procedures. At present, it remains unclear whether conflicting results from these two studies arise due to technical differences in the WNT5A staining procedures or due to underlying biological differences separating these two cohorts of patient tumors.

Of note, the cytosolic and nuclear staining seen in these studies with WNT5A is somewhat inconsistent with the model of Wnts as strictly secreted ligands (see Chapter 1). Interestingly, intracellular staining of WNT5A by immunohistochemistry is observed with antibodies of varying species and epitopes from different commercial vendors. WNT5A may indeed be localized intracellularly, such as within the secretory machinery or potentially in as yet unidentified signaling structures. Alternatively, the observed intracellular staining of WNT5A may represent an artifact secondary to cross-reactivity of the antibodies with a shared cancer epitope. A better understanding of both WNT5A subcellular localization and the specificity of available antibodies will be an important part of the ongoing effort to study WNT5A in human disease samples, particularly with respect to studies relying on immunohistochemical detection of WNT5A.

Transcriptional profiling of the Wnt pathway in melanoma

Genome-wide transcriptional profiling of melanoma has proven disappointingly inconsistent (Hoek, 2007). Conclusions regarding the activation of Wnt/β-catenin signaling derived from microarray profiling of gene expression often rely on the measurement of downstream transcriptional targets (including well-described β-catenin target genes such as *AXIN2*, *DKK1*, *LEF1*, and *MITF*; see also Chapters 4 and 5). Since these target genes exhibit variable responses to Wnt/β-catenin stimulation in different cell lines and can also act as inhibitors of Wnt/β-catenin signaling via negative feedback loops, it is not clear whether expression of these genes can be interpreted as an accurate readout of Wnt signaling. For example, one study includes *AXIN2*, one of the most reliable Wnt/β-catenin transcriptional targets, in a 45-gene signature that strongly correlates with improved prognosis (Mandruzzato *et al.*, 2006). However, numerous other studies have not associated *AXIN2* expression with either good or bad patient outcomes. One study compared gene expression profiles in a set of >200 archived cutaneous melanomas and used hierarchical clustering of these expression profiles to classify the tumors as either high grade with very poor patient outcomes or low grade with much better patient outcomes (Harbst *et al.*, 2012). Although several genes encoding Wnt signaling pathway proteins such as *FZD4*, *FZD10*, *LEF1*, *TCF4*, *SFRP2*, and *BCL9* were differentially expressed in low-grade compared to high-grade tumors in this dataset, differences in *AXIN2* expression were not observed (Harbst *et al.*, 2012).

Functional relevance of Wnt signaling pathways in melanoma

The role of β-catenin in transgenic mouse melanoma models

Transgenic mouse models play an important role in characterizing the role of Wnt/β-catenin signaling in melanoma initiation and progression. Transgenic mice engineered for melanocyte-specific expression (driven by the tyrosinase promoter) of a nondegradable constitutively active form of β-catenin (β-cat^{STA}; containing Ala substitutions at Ser33, Ser37, and Thr41) have been used to address the effects of β-catenin hyperactivation on melanoma formation and progression. Melanocyte-specific expression of *β-cat*^{STA} alone does not lead to spontaneous melanomas in mice (Delmas *et al.*, 2007). However, transgenic mice expressing both *β-cat*^{STA} and activated mutant *Nras* in melanocytes results in melanoma tumor formation at reduced latency and in enhanced tumor growth (Delmas *et al.*, 2007). A second study similarly finds that melanocyte-specific expression of *β-cat*^{STA} alone fails to induce melanomas, whereas coexpression of *β-cat*^{STA} either in the presence of a constitutively active mutant form of *Braf* or in the presence of inactivating mutations in *Pten* results in melanoma tumor formation (Damsky *et al.*, 2011). Together, these studies suggest a model in which the hyperactivation of β-catenin permits or enhances the ability of the BRAF/NRAS/MAPK and

PTEN/PI3K/AKT signaling pathways to drive melanoma tumorigenesis.

Results from mouse models of melanoma should be interpreted with several caveats in mind. First, these genetic models rely on the expression of a β-cat^{STA} despite the rarity of these mutations in patient tumors. Second, it is uncertain how closely the expression of a constitutively active mutant form of β-catenin (β-cat^{STA}) recapitulates the varied activity of the Wnt/β-catenin signaling pathway observed in patients (reviewed in Lucero *et al.*, 2010). Finally, mice expressing melanocyte-specific β-cat^{STA} and mutant *Nras* develop melanomas arising from the bulge region of the hair follicle rather than from the interfollicular epidermis where most human melanomas originate (Delmas *et al.*, 2007), although species-specific differences in the distribution of melanocytes could account in part for this discrepancy (Walker *et al.*, 2011). Despite these caveats, the data from these mouse models provide important insights, particularly with regard to potentially synergistic effects of hyperactivated β-cat^{STA} on other major signaling networks.

Studies of Wnt/β-catenin signaling in melanoma cell lines

Cellular responses to exogenous Wnt ligands vary widely between different melanoma cell lines (Figure 28.1). The molecular origins of this variability remain unclear but may involve differences in the expression of Wnt receptors. Potential heteromultimerization of various *FZD* isoforms with each other and with other potential coreceptors adds further complexity to the theoretical Wnt ligand and Wnt receptor interactions that may exist in different cell lines. We have observed that melanoma cells exhibiting disparate responses to WNT3A can achieve similar levels of pathway activation using a GSK3 inhibitor (Figure 28.2), implicating differences in cellular determinants upstream of GSK3β. With some of these limitations in mind, the studies to date using established melanoma cell lines have yielded interesting and unexpected results regarding the consequences of Wnt/β-catenin activation.

The B16 murine melanoma cell line represents far and away the most highly utilized and

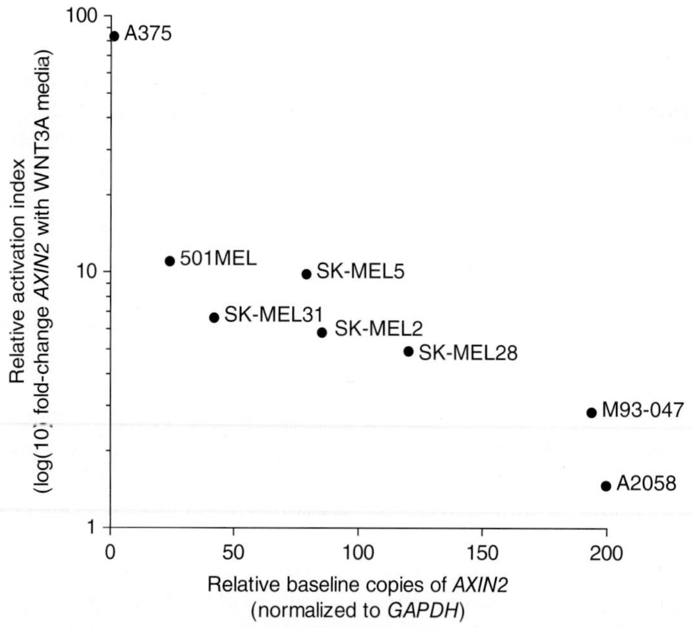

Figure 28.1 Melanoma cells exhibit a wide range of response to WNT3A. This plot reveals variation in both baseline *AXIN2* transcript levels as well as variations in the degree of response to exogenous WNT3A ligand in a panel of established human melanoma cell lines. The relative activation index represents *AXIN2* levels with WNT3A/*AXIN2* levels with control media. Note that the *y*-axis is logarithmic in scale, showing a wide range of BAR activation across melanoma cell lines with WNT3A.

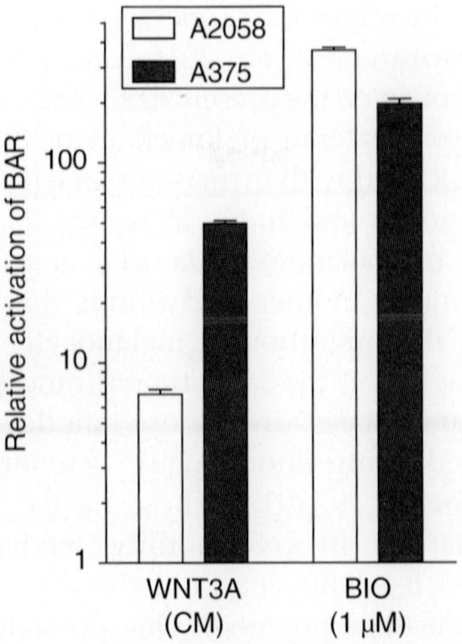

Figure 28.2 Response to WNT ligand may not entirely reflect the status of Wnt/β-catenin signaling potential in melanoma cells. This graph shows two melanoma cell lines that exhibit markedly different responses to WNT3A as measured by a luciferase-based β-catenin-activated reporter (BAR). While the A2058 cells exhibit an almost 10-fold less activation of the reporter by WNT3A, both cell lines exhibit comparable levels of BAR activation with the small-molecule GSK3 inhibitor BIO. These data suggest that cell-specific determinants of Wnt/β-catenin signaling are proximal to GSK3β and likely mediated at the level of ligand-receptor interactions, potentially through variations in the expression of FZD or other receptor isoforms.

published melanoma cell line in the literature. Because they were derived from spontaneous tumors arising in the commonly used C57BL/6 mouse, B16 melanoma cells can be readily engrafted in an isogenic host background for *in vivo* studies of tumor growth and metastasis. Studies of Wnt signaling using this model have found somewhat conflicting results. Forced overexpression of β-catenin enhanced the proliferation of B16 cells, which was inhibited by expression of a dominant negative form of TCF7L2 (Widlund *et al.*, 2002). The enhancement of B16 proliferation induced by overexpressed β-catenin also required *Mitf*, a Wnt/β-catenin target gene in itself (Widlund *et al.*, 2002). These data suggest that β-catenin can promote B16 proliferation by activating a gene expression network driven by the TCF7L2 and MITF transcription factors. Consistent with these initial observations, shRNA-mediated knockdown of *Ctnnb1* (β-catenin) inhibits the proliferation of B16 cells (Takahashi *et al.*, 2008). By contrast, forced expression of WNT3A in the B16 model leads to decreased cell proliferation *in vitro* and decreased tumor growth *in vivo* (Chien *et al.*, 2009). These studies reveal that activation of Wnt/β-catenin signaling either by overexpressed β-catenin or by Wnt ligand stimulation in melanoma can result in distinct cellular responses. One obvious possibility is that the use of overexpressed β-catenin circumvents GSK3β, which may have downstream effects on cellular process that are independent of β-catenin, but still activated by a Wnt ligand (see Chapter 2 for a more detailed discussion of upstream events mediated by Wnt ligand binding). It is also possible that overexpression of β-catenin or Wnts can result in differences in either the amplitude or duration of pathway activation, resulting in distinct signaling outputs.

In contrast to the B16 melanoma model, the experimental assessment of Wnt signaling in human-derived melanoma cell lines *in vivo* requires the use of immunocompromised mouse models for xenograft studies. Studies suggest that forced expression of either a WNT3A or a mutationally stabilized β-catenin leads to decreased tumor size, paralleling results seen *in vitro* (Biechele *et al.*, 2012; Yaguchi *et al.*, 2012). Whether and how these observations relate to the increased survival observed in patients with tumors expressing elevated cytosolic and nuclear β-catenin remains unresolved.

In human melanoma lines, Wnt/β-catenin signaling can enhance apoptosis in response to targeted inhibitors of mutant active BRAF or MEK (Biechele *et al.*, 2012; Conrad *et al.*, 2012). The ability of Wnt/β-catenin signaling to enhance apoptosis in the presence of BRAF inhibitors involves the dynamic regulation of AXIN1 protein abundance. Although some melanoma cell lines do not respond to combined treatment with WNT3A and BRAF inhibitors, knockdown of AXIN1 by siRNA can confer an apoptotic response. This unexpected finding raises the possibility that cross talk between Wnt/β-catenin and BRAF/MAPK signaling may provide an avenue for developing either prognostic or treatment strategies aimed at optimizing the targeted BRAF therapies that are currently used as a first-line therapy for patients with tumors harboring activating *BRAF* mutations (Chapman *et al.*, 2011). We

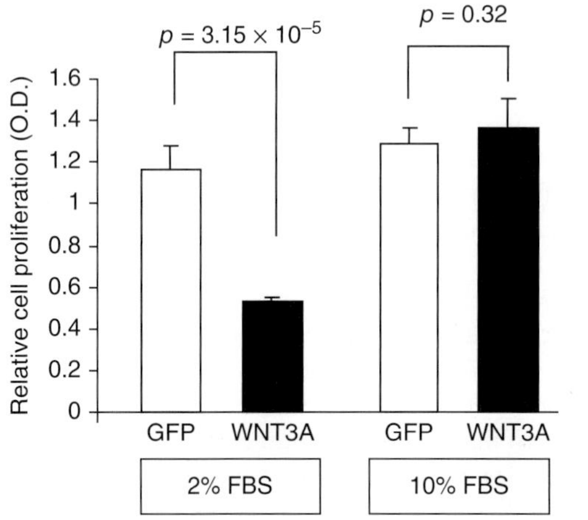

Figure 28.3 Serum levels may help determine functional roles of Wnt/β-catenin activation in cell culture models. This graph shows cellular proliferation measured using the colorimetric tetrazolium dye MTT. Reduction of MTT in proliferating cells is measured by an increase in O.D. Murine B16 cells expressing WNT3A exhibit a reduction in proliferation in 2% serum, paralleling the decrease in tumor growth observed *in vivo* (Chien *et al.*, 2009). Increasing serum to 10% can overcome the effects of WNT3A, potentially due to the negative regulation of Wnt/β-catenin signaling by MAPK signaling. This finding illustrates how variations in cell culture could account for some of the differences in phenotype reported in melanoma cells with Wnt/β-catenin activation.

have also found that variations in serum levels in cell culture media can potentially inhibit or overcome the functional effects of Wnt/β-catenin activation (Figure 28.3). As different serum levels can affect the level of BRAF/MAPK and PI3K/AKT pathway activities, these results argue that potential interplay between Wnt signaling and other kinase-dependent signaling pathways must be carefully considered in order to accurately interpret the results of experiments characterizing Wnt signaling in melanoma cell lines.

Wnt signaling pathways and melanoma metastasis

If cutaneous melanomas are identified and removed prior to metastatic dissemination, patient survival is very high (~90%), which is in stark contrast to the extremely poor patient survival associated with the development of metastatic melanoma (Siegel *et al.*, 2012). Several independent studies find that β-catenin inhibits the migration of melanoma cell lines *in vitro* (Arozarena *et al.*, 2011; Gallagher *et al.*, 2012). However, the presence of β-cat^{STA} (but not wild-type β-catenin or knockout of β-catenin) was associated with increased metastasis to the lymph nodes and lungs in transgenic mouse models of melanoma (Damsky *et al.*, 2011). Interestingly, another study finds that β-cat^{STA} inhibits the migration of melanocytic cells in culture while at the same time promotes melanoma metastasis in mouse models when combined with mutationally activated *Nras* (Gallagher *et al.*, 2012). These results indicate that changes in cell motility mediated by activated β-catenin may not directly correlate with metastatic potential. One possible explanation for these discrepancies is that Wnt/β-catenin signaling mediates complex cell–cell interactions necessary for metastatic progression in animal models but not for the regulation of cell migration and invasiveness *in vitro*.

Studies indicating that *WNT5A* expression is increased in metastatic melanoma cell lines and tumors compared to primary tumors implicate WNT5A-dependent signaling in the regulation of the metastatic process (Bittner *et al.*, 2000; Hoek *et al.*, 2006; Weeraratna *et al.*, 2002). During embryogenesis, WNT5A regulates the establishment of cellular polarity and cell motility in many different tissues and developmental contexts (Andersson *et al.*, 2008; Hardy *et al.*, 2008; Lin *et al.*, 2010; Qian *et al.*, 2007; see also Chapters 6 and 19). Consistent with the elevated expression of *WNT5A* in metastatic melanoma and the functional roles for WNT5A in promoting cell migration, follow-up studies have shown that WNT5A can correspondingly increase melanoma cell motility *in vitro* (Dissanayake *et al.*, 2007; Jenei *et al.*, 2009; Weeraratna *et al.*, 2002). Systemic injections of recombinant WNT5A (rWNT5A) result in the increased engraftment and growth of B16 melanoma cells injected through the tail vein, providing the first evidence that recombinant Wnt proteins could have utility as research tools *in vivo* (Dissanayake *et al.*, 2008).

The molecular events required for WNT5A-dependent induction of cell motility and metastasis are not entirely clear, but likely involve the activation of various kinase-dependent signaling pathways. WNT5A-induced invasiveness of melanoma requires the phosphorylation of PKC (Weeraratna *et al.*, 2002). Recent studies also highlight an important role for the tyrosine kinase-like WNT receptor ROR2 in mediating metastatic cell behaviors. Specifically, siRNA-mediated reduction of Ror2 in B16 mouse melanoma cells reduces the frequency and severity of lung metastases in mice (O'Connell *et al.*, 2010). Collectively, these observations suggest that increased WNT5A can enhance cellular motility and experimental metastasis, providing a possible model to explain why elevated levels of WNT5A are observed in late-stage metastatic melanomas.

Wnt signaling and melanoma heterogeneity

Studies in developmental models find that the activation of Wnt/β-catenin signaling plays a critical role in regulating the differentiation of neural crest stem cells in a time- and cell-type-dependent manner (Ikeya *et al.*, 1997; Hari *et al.*, 2002; Raible, 2006). Interestingly, the activation of Wnt/β-catenin signaling either by WNT3A or by GSK3β inhibition leads to enhanced pigmentation accompanied by the upregulation of cellular genes associated with melanocyte differentiation such as *Silv*, *Mlana*, and *Kit*, suggesting that Wnt/β-catenin pathway activation can result in a fundamental change in the differentiation state of a melanoma cell population (Bellei *et al.*, 2008; Chien *et al.*, 2009). The expression of these differentiation-associated genes in B16 cells is antagonized by WNT5A (Chien *et al.*, 2009; Dissanayake *et al.*, 2008), suggesting that melanoma cell fate may be regulated by a balance between Wnt/β-catenin and β-catenin-independent Wnt signaling (discussed further later).

In heterogeneous tumors, it is likely that the regulation of Wnt/β-catenin and β-catenin-independent signaling is temporally and spatially complex. Several gene expression profiling studies now suggest that melanomas can be categorized into various subtypes (i.e., proliferative, metastatic, melanocyte differentiated, neuronal-like) based on hierarchical clustering of gene expression profiles obtained from patient samples and from melanoma cell lines (Harbst *et al.*, 2012; Jonsson *et al.*, 2010; Widmer *et al.*, 2012). For example, a two-state model as proposed by Hoek and colleagues suggests that melanomas dynamically transition between more proliferative and more invasive phenotypes (Hoek *et al.*, 2008). In this analysis, more invasive cells exhibited a gene signature with relatively higher levels of *WNT5A* and lower levels of genes suggestive of Wnt/β-catenin signaling activation or differentiation. Another study finds that changes in gene expression profiles corresponding to either a more differentiated/proliferative state or a more motile/invasive state are regulated by the abundance of the β-catenin transcriptional coregulators LEF1 and TCF7L2, respectively (Eichhoff *et al.*, 2011; see also Chapter 17). This two-state model predicts that cells with elevated Wnt/β-catenin signaling will be more proliferative, which contrasts with studies showing that activation of Wnt/β-catenin inhibits human melanoma cell proliferation *in vitro* and *in vivo* (Chien *et al.*, 2009; Yaguchi *et al.*, 2012). These studies highlight the difficulty in uniting the existing data on Wnt/β-catenin signaling in melanoma into a viable and consistent model.

More recently, transcriptional profiling studies revealed that melanoma cell resistance to the clinically used targeted BRAF inhibitor PLX4032 (marketed clinically as vemurafenib) correlates with relatively higher levels of *WNT5A* expression, whereas the expression of *CTNNB1* (β-catenin) and several Wnt/β-catenin target genes negatively correlates with PLX4023/vemurafenib resistance (Tap *et al.*, 2010). Given that β-catenin-independent Wnt signaling can antagonize Wnt/β-catenin signaling, the consistent observation of increased *WNT5A* expression in later-stage, more invasive melanomas may reflect a potentially complex interplay between known arms of Wnt signaling (see Chapter 6). In summary, these studies suggest that complex changes in the expression of components of both β-catenin-dependent and β-catenin-independent signaling pathways may result in unpredictable phenotypic changes that are relevant to both melanoma progression and treatment.

Cell nonautonomous effects of Wnt/β-catenin signaling in melanoma

Wnt/β-catenin signaling can also regulate other elements of the tumor microenvironment, with important consequences for tumor formation and host response. Overexpression of Wnt1 in human melanoma cells leads to decreased lymphangiogenesis in mouse xenograft tumors, likely through the regulation of VEGF-C expression (Niederleithner *et al.*, 2012). Interestingly, this regulation is independent of both GSK3β and β-catenin, implying that Wnt ligands may functionally regulate tumor cell phenotypes outside of the existing canonical pathway model. In another study, forced activation of Wnt/β-catenin signaling in human melanoma cells inhibits the ability of antigen-presenting dendritic cells to activate T cells, suggesting that the presence of active Wnt/β-catenin signaling in tumors may be immunosuppressive (Yaguchi *et al.*, 2012). The regulation of the immune-modulating molecule CTLA4 in melanoma cells by Wnt/β-catenin signaling further demonstrates the potential impact that activation of this pathway in melanoma cells can have on the overall tumor environment (Shah *et al.*, 2008). As our models for melanoma become more sophisticated, we will undoubtedly uncover other aspects of tumor biology that help clarify how the regulation of Wnt/β-catenin signaling in laboratory studies relates to the observed associations between Wnt/β-catenin signaling and patient prognosis.

Conclusions

The potential importance of Wnt/β-catenin signaling as well as β-catenin-independent signaling in melanoma is reminiscent of these signaling pathways' context-dependent roles in regulating cell fate and migration during embryogenesis (see Chapters 6 and 19). Accumulating data from mouse and cell line models continues to refine our understanding of how Wnt signaling pathways analogously regulate melanoma cell growth, cell fate, and metastatic spread in a context-dependent manner. Numerous studies indicate that altered expression of Wnt pathway components correlate with patient outcomes in melanoma, raising the possibility that targeting Wnt signaling in the appropriate context could be advantageous in the clinic. However, targeting Wnt signaling in melanoma is still challenging as the functional relevance of Wnt signaling pathways and the mechanisms underlying their activation and inactivation remain unresolved. Overall, current evidence suggests that attempts to target Wnt signaling in melanoma and other disease models will have to take into account the effects of cellular, temporal, and spatial contexts on the regulation of this pathway, as well as the interplay between Wnt signaling pathways and other signaling networks.

Acknowledgements

A.J.C. is currently funded through the University of Washington Provost's Bridge Fund. J.N.A. receives support from the Howard Hughes Medical Institute. The views expressed earlier are not intended to represent the views of these funding agencies, whose contributions to our past and ongoing research endeavors have been greatly appreciated.

References

Andersson, E.R., Prakash, N., Cajanek, L. *et al.* (2008) Wnt5a regulates ventral midbrain morphogenesis and the development of A9–A10 dopaminergic cells in vivo. *PLoS One*, **3**, e3517.

Arozarena, I., Bischof, H., Gilby, D. *et al.* (2011) In melanoma, beta-catenin is a suppressor of invasion. *Oncogene*, **30**, 4531–4543.

Bachmann, I.M., Straume, O., Puntervoll, H.E. *et al.* (2005) Importance of P-cadherin, β-catenin, and Wnt5a/frizzled for progression of melanocytic tumors and prognosis in cutaneous melanoma. *Clinical Cancer Research*, **11**, 8606–8614.

Bellei, B., Flori, E., Izzo, E. *et al.* (2008) GSK3β inhibition promotes melanogenesis in mouse B16 melanoma cells and normal human melanocytes. *Cell Signaling*, **20**, 1750–1761.

Biechele, T.L., Kulikauskas, R.M., Toroni, R.A. *et al.* (2012) Wnt/β-catenin signaling and AXIN1 regulate apoptosis triggered by inhibition of the mutant kinase BRAFV600E in human melanoma. *Science Signaling*, **5**, ra3.

Bittner, M., Meltzer, P., Chen, Y. *et al.* (2000) Molecular classification of cutaneous malignant melanoma by gene expression profiling. *Nature*, **406**, 536–540.

Chapman, P.B., Hauschild, A., Robert, C. *et al.* (2011) Improved survival with vemurafenib in melanoma with BRAF V600E mutation. *The New England Journal of Medicine*, **364**, 9.

Chien, A.J., Moore, E.C., Lonsdorf, A.S. *et al.* (2009) Activated Wnt/β-catenin signaling in melanoma is associated with decreased proliferation in patient tumors and a murine melanoma model. *Proceedings of the National Academy of Sciences of the United States of America*, **106**, 1193–1198.

Conrad, W.H., Swift, R.D., Biechele, T.L. *et al.* (2012) Regulating the response to targeted MEK inhibition in melanoma: enhancing apoptosis in NRAS- and BRAF-mutant melanoma cells with Wnt/β-catenin activation. *Cell Cycle*, **11**, 3724–3730.

Da Forno, P.D., Pringle, J.H., Hutchinson, P. *et al.* (2008) WNT5A expression increases during melanoma progression and correlates with outcome. *Clinical Cancer Research*, **14**, 5825–5832.

Damsky, W.E., Curley, D.P., Santhanakrishnan, M. *et al.* (2011) β-catenin signaling controls metastasis in Braf-activated Pten-deficient melanomas. *Cancer Cell*, **20**, 741–754.

Delmas, V., Beermann, F., Martinozzi, S. *et al.* (2007) β-catenin induces immortalization of melanocytes by suppressing p16INK4a expression and cooperates with N-Ras in melanoma development. *Genes & Development*, **21**, 2923–2935.

Demunter, A., Libbrecht, L., Degreef, H. *et al.* (2002) Loss of membranous expression of beta-catenin is associated with tumor progression in cutaneous melanoma and rarely caused by exon 3 mutations. *Modern Pathology*, **15**, 454–461.

Dissanayake, S.K., Wade, M., Johnson, C.E. *et al.* (2007) The Wnt5A/protein kinase C pathway mediates motility in melanoma cells via the inhibition of metastasis suppressors and initiation of an epithelial to mesenchymal transition. *The Journal of Biological Chemistry*, **282**, 17259–17271.

Dissanayake, S.K., Olkhanud, P.B., O'Connell, M.P. *et al.* (2008) Wnt5A regulates expression of tumor-associated antigens in melanoma via changes in signal transducers and activators of transcription 3 phosphorylation. *Cancer Research*, **68**, 10205–10214.

Eichhoff, O.M., Weeraratna, A., Zipser, M.C. *et al.* (2011) Differential LEF1 and TCF4 expression is involved in melanoma cell phenotype switching. *Pigment Cell and Melanoma Research*, **24**, 631–642.

Gallagher, S.J., Rambow, F., Kumasaka, M. *et al.* (2012) B-catenin inhibits melanocyte migration but induces melanoma metastasis. *Oncogene*, **32**, 2230–2238.

Gould Rothberg, B.E., Berger, A.J., Molinaro, A.M. *et al.* (2009) Melanoma prognostic model using tissue microarrays and genetic algorithms. *Journal of Clinical Oncology*, **27**, 5772–5780.

Haluska, F.G., Tsao, H., Wu, H. *et al.* (2006) Genetic alterations in signaling pathways in melanoma. *Clinical Cancer Research*, **12**, 2301s–2307s.

Harbst, K., Staaf, J., Lauss, M. *et al.* (2012) Molecular profiling reveals low- and high-grade forms of primary melanoma. *Clinical Cancer Research*, **18**, 4026–4036.

Hardy, K.M., Garriock, R.J., Yatskievych, T.A. *et al.* (2008) Non-canonical Wnt signaling through Wnt5a/b and a novel Wnt11 gene, *Wnt11b*, regulates cell migration during avian gastrulation. *Developmental Biology*, **320**, 391–401.

Hari, L., Brault, V., Kléber, M. *et al.* (2002) Lineage-specific requirements of beta-catenin in neural crest development. *The Journal of Cell Biology*, **159**, 867–880.

Hoek, K.S. (2007) DNA microarray analyses of melanoma gene expression: a decade in the mines. *Pigment Cell Research*, **20**, 466–484.

Hoek, K.S., Schlegel, N.C., Brafford, P. *et al.* (2006) Metastatic potential of melanomas defined by specific gene expression profiles with no BRAF signature. *Pigment Cell Research*, **19**, 290–302.

Hoek, K.S., Eichhoff, O.M., Schlegel, N.C. *et al.* (2008) In vivo switching of human melanoma cells between proliferative and invasive states. *Cancer Research*, **68**, 650–656.

Ikeya, M., Lee, S.M., Johnson, J.E. *et al.* (1997) Wnt signalling required for expansion of neural crest and CNS progenitors. *Nature*, **389**, 966–970.

Jenei, V., Sherwood, V., Howlin, J. *et al.* (2009) A t-butyloxycarbonyl-modified Wnt5a-derived hexapeptide functions as a potent antagonist of Wnt5a-dependent melanoma cell invasion. *Proceedings of the National Academy Sciences of the United States of America*, **106**, 19473–19478.

Jonsson, G., Busch, C., Knappskog, S. *et al.* (2010) Gene expression profiling-based identification of molecular subtypes in Stage IV melanomas with different clinical outcome. *Clinical Cancer Research*, **16**, 3356–3367.

Kageshita, T., Hamby, C.V., Ishihara, T. *et al.* (2001) Loss of β-catenin expression associated with disease progression in malignant melanoma. *The British Journal of Dermatology*, **145**, 210–216.

Kielhorn, E., Provost, E., Olsen, D. *et al.* (2003) Tissue microarray-based analysis shows phospho-β-catenin expression in malignant melanoma is associated with poor outcome. *International Journal of Cancer*, **103**, 652–656.

Lens, M.B. and Dawes, M. (2004) Global perspectives of contemporary epidemiological trends of cutaneous malignant melanoma. *The British Journal of Dermatology*, **150**, 179–185.

Lin, S., Baye, L.M., Westfall, T.A., and Slusarski, D.C. (2010) Wnt5b-Ryk pathway provides directional signals to regulate gastrulation movement. *The Journal of Cell Biology*, **190**, 263–278.

Lucero, O.M., Dawson, D.W., Moon, R.T., and Chien, A.J. (2010) A re-evaluation of the "oncogenic" nature of Wnt/β-catenin signaling in melanoma and other cancers. *Current Oncology Report*, **12**, 314–318.

Maelandsmo, G.M., Holm, R., Nesland, J.M. *et al.* (2003) Reduced β-catenin expression in the cytoplasm of advanced-stage superficial spreading malignant melanoma. *Clinical Cancer Research*, **9**, 3383–3388.

Mandruzzato, S., Callegaro, A., Turcatel, G. *et al.* (2006) A gene expression signature associated with survival in metastatic melanoma. *Journal of Translational Medicine*, **4**, 50.

Moon, R.T., Campbell, R.M., Christian, J.L. *et al.* (1993) Xwnt-5A: a maternal Wnt that affects morphogenetic movements after overexpression in embryos of Xenopus laevis. *Development*, **119**, 97–111.

Niederleithner, H., Heinz, M., Tauber, S. *et al.* (2012) Wnt1 is anti-lymphangiogenic in a melanoma mouse model. *The Journal of Investigative Dermatology*, **132**, 2235–2244.

O'Connell, M.P., Fiori, J.L., Xu, M. *et al.* (2010) The orphan tyrosine kinase receptor, ROR2, mediates Wnt5A signaling in metastatic melanoma. *Oncogene*, **29**, 34–44.

Omholt, K., Platz, A., Ringborg, U., and Hansson, J. (2001) Cytoplasmic and nuclear accumulation of β-catenin is rarely caused by CTNNB1 exon 3 mutations in cutaneous malignant melanoma. *International Journal of Cancer*, **92**, 839–842.

Pollock, P.M. and Hayward, N. (2002) Mutations in exon 3 of the β-catenin gene are rare in melanoma cell lines. *Melanoma Research*, **12**, 183–186.

Qian, D., Jones, C., Rzadzinska, A. *et al.* (2007) Wnt5a functions in planar cell polarity regulation in mice. *Developmental Biology*, **306**, 121–133.

Raible, D.W. (2006) Development of the neural crest: achieving specificity in regulatory pathways. *Current Opinion of Cell Biology*, **18**, 698–703.

Rimm, D.L., Caca, K., Hu, G. *et al.* (1999) Frequent nuclear/cytoplasmic localization of β-catenin without exon 3 mutations in malignant melanoma. *The American Journal of Pathology*, **154**, 325–329.

Rubinfeld, B., Robbins, P., El-Gamil, M. *et al.* (1997) Stabilization of β-catenin by genetic defects in melanoma cell lines. *Science*, **275**, 1790–1792.

Shah, K.V., Chien, A.J., Yee, C., and Moon, R.T. (2008) CTLA-4 is a direct target of Wnt/β-catenin signaling and is expressed in human melanoma tumors. *The Journal of Investigative Dermatology*, **128**, 2870–2879.

Siegel, R., DeSantis, C., Virgo, K. *et al.* (2012) Cancer treatment and survivorship statistics, 2012. *CA: A Cancer Journal for Clinicians*, **62**, 220–241.

Takahashi, Y., Nishikawa, M., Suehara, T. *et al.* (2008) Gene silencing of β-catenin in melanoma cells retards their growth but promotes the formation of pulmonary metastasis in mice. *International Journal of Cancer*, **123**, 2315–2320.

Tap, W.D., Gong, K.W., Dering, J. *et al.* (2010) Pharmacodynamic characterization of the efficacy signals due to selective BRAF inhibition with PLX4032 in malignant melanoma. *Neoplasia*, **12**, 637–649.

Tsao, H., Atkins, M.B., and Sober, A.J. (2004) Management of cutaneous melanoma. *The New England Journal of Medicine*, **351**, 998–1012.

Walker, G.J., Soyer, H.P., Terzian, T., and Box, N.F. (2011) Modelling melanoma in mice. *Pigment Cell and Melanoma Research*, **24**, 1158–1176.

Weeraratna, A.T. (2005) A Wnt-er wonderland – the complexity of Wnt signaling in melanoma. *Cancer Metastasis Reviews*, **24**, 237–250.

Weeraratna, A.T., Jiang, Y., Hostetter, G. *et al.* (2002) Wnt5a signaling directly affects cell motility and invasion of metastatic melanoma. *Cancer Cell*, **1**, 279–288.

Widlund, H.R., Horstmann, M.A., Price, E.R. *et al.* (2002) β-catenin-induced melanoma growth requires the downstream target microphthalmia-associated transcription factor. *The Journal of Cell Biology*, **158**, 1079–1087.

Widmer, D.S., Cheng, P.F., Eichhoff, O.M. *et al.* (2012) Systematic classification of melanoma cells by phenotype-specific gene expression mapping. *Pigment Cell and Melanoma Research*, **25**, 343–353.

Yaguchi, T., Goto, Y., Kido, K. *et al.* (2012) Immune suppression and resistance mediated by constitutive activation of Wnt/β-catenin signaling in human melanoma cells. *Journal of Immunology*, **189**, 2110–2117.

29 Wnt Signaling in Mood and Psychotic Disorders

Stephen J. Haggarty[1,2,3], Karun Singh[4], Roy H. Perlis[2,3], and Rakesh Karmacharya[2,3,5,6]

[1]Department of Neurology, Center for Human Genetic Research, Massachusetts General Hospital & Harvard Medical School, Boston, MA, USA
[2]Department of Psychiatry, Center for Human Genetic Research, Massachusetts General Hospital & Harvard Medical School, Boston, MA, USA
[3]MGH Psychiatry Center for Experimental Drugs & Diagnostics, Center for Human Genetic Research, Massachusetts General Hospital & Harvard Medical School, Boston, MA, USA
[4]Department of Biochemistry and Biomedical Sciences, McMaster Stem Cell & Cancer Research Institute, McMaster University, Hamilton, Ontario, Canada
[5]Chemical Biology Program, Broad Institute of Harvard & MIT, Cambridge, MA, USA
[6]Schizophrenia and Bipolar Disorder Program, McLean Hospital, Belmont, MA, USA

Introduction

Mental illness represents one of the areas of greatest unmet medical need in the twenty-first century. Collectively, these illnesses deprive millions of individuals of their mental health, leading to diminished capacity to reason, feel, and remember their life, which presents a tremendous burden to individuals, their families, and society as a whole. The causes of severe mental illnesses are heterogeneous in nature and not fully understood. Despite the immense etiological heterogeneity, affected individuals share common behavioral manifestations that increasingly are being shown to arise due to perturbations of common cellular and molecular mechanisms. In particular, a number of lines of evidence emerging from molecular, cellular, and animal behavioral studies, as well as recent large-scale human genetic studies discussed in the following Chapter 30, have begun to reveal that dysregulation of Wnt signaling pathways provides a mechanistic link between otherwise seemingly disparate forms of neuropsychiatric disorders, including bipolar disorder and schizophrenia. Additionally, as discussed in the following Chapter 31, evidence for potentially convergent mechanisms involving dysregulated Wnt signaling is also found for various dementias, including Alzheimer's disease, frontotemporal dementia, and Lewy body dementia. Consideration of this unifying notion appears to have heuristic value insofar as providing a framework for guiding interpretation of data emerging from efforts to understand the role of genetic variation in determining disease susceptibility as well as for efforts to develop novel targeted, disease-modifying therapeutics that target core aspects of the underlying molecular pathogenesis of these devastating disorders.

Wnt Signaling in Development and Disease: Molecular Mechanisms and Biological Functions, First Edition.
Edited by Stefan Hoppler and Randall T. Moon.

Here, in this chapter, we will summarize particularly salient observations made by investigators in the Wnt field who have been drawn in to the study of this deeply important and fascinating pathway that may hold keys to mental health and disease (Clevers and Nusse, 2012; De Ferrari and Moon, 2006). Given that many neuropsychiatric disorders, particularly schizophrenia and autism, are thought to have a neurodevelopmental component (Rapoport, Giedd, and Gogtay, 2012) and that many of these processes are seen to be compromised in the case of neurodegeneration and more subtly affected in the other cases, the reader is also referred to Chapter 21 for a discussion of the critical role of Wnt signaling in neurogenesis, axon and dendrite development, and synaptogenesis given the pertinence of these processes to the pathophysiology and treatment of mental illnesses.

A theme throughput this chapter, as well as others, is the consideration of the Wnt pathway not in isolation, but rather in the context of numerous other cellular signaling pathways that demonstrate substantial cross talk with Wnt signaling (Figure 29.1). While the extent and specific nature of this cross talk can vary between different organisms and cell types and with time, it is likely that these convergent interactions that are themselves critical for ensuring dynamic regulation of neuroplasticity mechanisms are important for maintaining mental health. As a corollary, it follows that disruption of these convergent pathways in the case of disease-associated mutations or other aspects of disease pathogenesis may lead to dysregulation of Wnt signaling leading to disease pathophysiology. As improved knowledge of the extent and nature of this cross talk emerges at the molecular level, and a new picture of the genetic

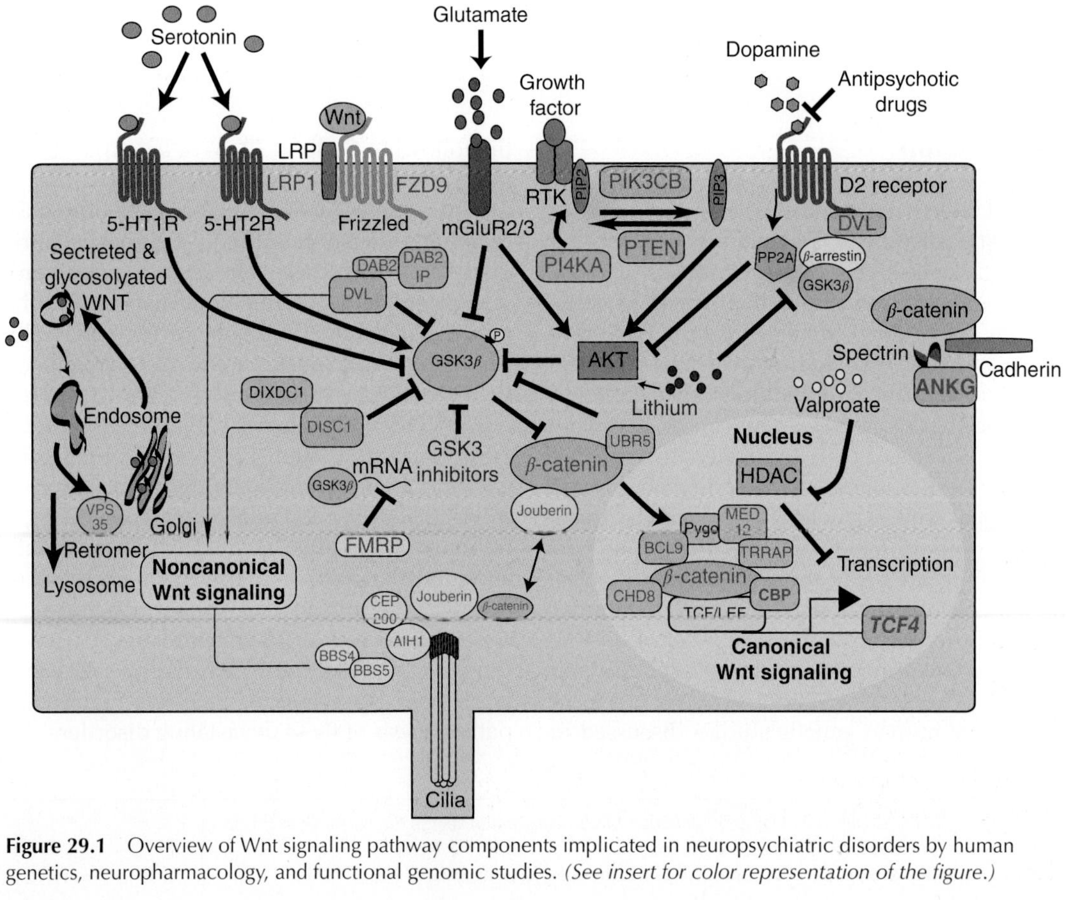

Figure 29.1 Overview of Wnt signaling pathway components implicated in neuropsychiatric disorders by human genetics, neuropharmacology, and functional genomic studies. *(See insert for color representation of the figure.)*

architecture of these disorders comes into better focus, the hypothesis that Wnt signaling is causally involved in the etiology and/or pathophysiology of severe mental illness will become testable, with the outcome having implications for developing next-generation pharmaceuticals and diagnostics (see Chapter 32).

Role of Wnt signaling in mood disorders

Developing novel therapeutics and diagnostic tools based upon an understanding of neuroplasticity is critical in order to improve the treatment and ultimately the prevention of a broad range of nervous system disorders. In the case of mood disorders, such as major depressive disorder and bipolar disorder, where diagnoses are based solely on symptomatology rather than pathophysiology, there exists a clear unmet medical need to advance our understanding of the underlying molecular mechanisms and to develop fundamentally new mechanism-based medicines with improved efficacy. In this context, recent preclinical molecular, cellular, and behavioral findings have begun to reveal the importance of Wnt signaling in the underlying pathophysiology and response to treatments of mood disorders.

The major mood disorders, major depressive disorder and bipolar disorder, are characterized by both episodes of depressed mood and diminished interest and pleasure seeking in conjunction with neurovegetative symptoms, referred to as major depressive episodes. Depressive episodes are typically recurrent and may be chronic as well (Perlis *et al.*, 2006). In bipolar disorder, individuals also experience periods of unusually elevated or irritable mood, referred to as manic episodes. This mood elevation is generally associated with increased impulsive as well as goal-directed activity, decreased need for sleep, increased physical activity, and pressured speech. Manic and depressive symptoms can co-occur as well, a phenomenon known as a mixed state. During mood episodes, individuals with major depressive disorder or bipolar disorder often experience cognitive dysfunction, particularly individuals with bipolar disorder who may experience residual cognitive symptoms between episodes even in the absence of other prominent mood symptoms (Baune *et al.*, 2010; Zarate *et al.*, 2000). In both disorders, the mood of an individual may be influenced by external stimuli, but responses are diminished or exaggerated.

Despite intensive basic research and clinical studies, our understanding of the etiology and pathophysiology of mood disorders is severely limited (Krishnan and Nestler, 2008; Manji and Duman, 2001; Pittenger and Duman, 2008; Quiroz and Manji, 2002). Consequently, mood disorders remain one of the leading causes of disability worldwide, with the World Health Organization projecting that depressive disorders will be the leading cause of disease burden by 2030 (World Health Organization, 2004). However, rational drug development for mood disorders has been hindered by two factors: (i) limited understanding of the fundamental molecular and cellular mechanisms involved in the pathophysiology and (ii) limited understanding of the precise molecular mechanisms of existing therapeutic agents (Berton and Nestler, 2006; Manji and Duman, 2001; Murrough and Charney, 2012; Pittenger and Duman, 2008; Quiroz and Manji, 2002). For these reasons, there has been significant attention paid to the mechanisms of action of existing therapeutics, which, as reviewed in Chapter 32, provided the first and now collectively one of the strongest lines of evidence for the importance of Wnt signaling pathway in mental health.

Lithium and bipolar disorder

Lithium was the first medication found to have mood-stabilizing properties (Cade, 1949) and today remains the first-line treatment for bipolar disorder. Despite the efficacy of lithium in preventing manic episodes and suicide (Bowden, 2000; Tondo, Isacsson, and Baldessarini, 2003), half of bipolar disorder patients do not respond to lithium therapy (Bowden, 2000), and lithium treatment is associated with dose-limiting side effects (Netto and Phutane, 2012). Therefore, there has been much interest in the past half century since its discovery to elucidate the molecular mechanism underlying lithium's modulation of affective behavior. While many targets of lithium have been identified (Beaulieu and Caron, 2008; Beaulieu, Gainetdinov, and Caron, 2009; Beaulieu *et al.*, 2008a; Berridge, Downes, and Hanley, 1989; Klein and Melton, 1996), including compelling evidence for the inhibition of enzymes, such as inositol monophosphatase (IMPase) (Teo *et al.*, 2009), which can impact phosphoinositol signaling and autophagy pathways (Sarkar *et al.*, 2005), growing evidence from both genetic and pharmacological studies implicates the inhibition of glycogen synthase kinase 3 (GSK3) as mediating the effects of lithium relevant to neuropsychiatric disorders (as reviewed in O'Brien and Klein, 2009) and, as discussed in Chapter 31, also in dementia.

Lithium has been shown to inhibit GSK3 kinase activity directly, via competition with magnesium, and indirectly, by increasing inhibitory Ser9/Ser21 phosphorylation of GSK3α and GSK3β, respectively (Beaulieu *et al.*, 2004, 2008a; Chalecka-Franaszek and Chuang, 1999; De Sarno, Li, and Jope, 2002; Klein and Melton, 1996). Furthermore, GSK3α null or GSK3β heterozygous mutant mice phenocopy lithium's effect in mouse models of mania-like and depressive-like behavior (Beaulieu *et al.*, 2004; Kaidanovich-Beilin *et al.*, 2009; O'Brien *et al.*, 2004). Conversely, mice overexpressing GSK3β or carrying mutations preventing inhibitory phosphorylation of GSK3α (Ser21) and GSK3β (Ser9) exhibit behaviors that model mania or psychosis, as do mice with targeted disruption of AKT1, which phosphorylates and inactivates GSK3α (Ser21) and GSK3β (Ser9) (Emamian *et al.*, 2004; Polter *et al.*, 2010; Prickaerts *et al.*, 2006). This inhibition of GSK3 by lithium is sufficient in many cellular contexts to lead to accumulation and nuclear translocation of β-catenin and the transcriptional activation of genes with promoters containing TCF/LEF response elements (Beaulieu *et al.*, 2004; Gould and Manji, 2005; Gould, Chen, and Manji, 2004; Gould *et al.*, 2007; Klein and Melton, 1996; O'Brien *et al.*, 2004). Consistent with the relevance of β-catenin downstream of lithium's inhibition of GSK3 in the context of mood behaviors, the overexpression of β-catenin has been shown to have antidepressant-like properties, similar to lithium treatment, with decreased immobility time in the forced swim test, and antimanic-like properties with an attenuation of D-amphetamine-induced hyperlocomotion (Gould *et al.*, 2007). However, the removal of β-catenin from the forebrain of mice resulted in only mildly affected behaviors with a depressive-like phenotype in the tail suspension test but not other differences in other behavioral tests of mood-related or anxiety-related behaviors pointing to either specific neurocircuits needing to be altered or levels of β-catenin modulation needed to match the effects of lithium (Gould *et al.*, 2008).

Studies in human subjects have shown that bipolar patients treated with lithium, when compared to control subjects, have increased gray matter density, especially in bilateral paralimbic and cingulated cortices (Bearden *et al.*, 2007; Sassi *et al.*, 2002). The same study also showed increased gray matter density in right anterior cingulate in bipolar patients treated with lithium when compared to bipolar patients who were not treated with lithium. Consistent with this result, another longitudinal magnetic resonance imaging study showed that subjects treated with lithium showed increased gray matter volume, which peaked at 10–12 weeks and was maintained at week 16 of treatment (Lyoo *et al.*, 2010). Moreover, the increases in gray matter in the lithium-treated patients correlated with positive clinical response, and these changes in gray matter density were not seen in patients treated with valproic acid or in control subjects. Another study looked at the difference in gray matter density between lithium responders and nonresponders and found that only patients who responded to lithium showed significant increases in gray matter in the prefrontal cortex (Moore *et al.*, 2009). Taken together, these human studies are consistent with lithium's proneurogenic activity through modulation of the Wnt signaling pathway.

Besides being a competitor of magnesium that can directly affect GSK3 kinase activity, lithium has also been recently shown to destabilize a ternary complex consisting of β-arrestin/Akt/protein phosphatase 2A (PP2A) that forms when dopamine or other agonists of the dopamine D2 receptor bind and stimulate phosphorylation by G protein receptor kinases (GRKs), which in turn sterically block coupling to Gi/o proteins that normally regulate cAMP signaling leading instead to the recruitment of β-arrestin

(Beaulieu, Gainetdinov, and Caron, 2007; Beaulieu *et al.*, 2004). As a result of destabilizing this complex, PP2A that is bound to β-arrestin and has sequestered Akt is no longer able to dephosphorylate and inactivate Akt, resulting in GSK3β Ser9 phosphorylation and inactivation by Akt. Additional studies from the Klein laboratory further strengthen the notion that the disruption of the stability of the β-arrestin-2/Akt/PP2A ternary complex occurs upon chronic lithium treatment at physiologically relevant concentrations achieved *in vivo* in humans (O'Brien *et al.*, 2011). In these studies, GSK3 was found to bind to and stabilize the β-arrestin-2/Akt/PP2A complex, leading to PP2A-mediated dephosphorylation and inactivation of Akt. Inhibition of GSK3 through lithium treatment, or through the use of direct ATP-competitive GSK3 inhibitors, led to the disruption of the β-arrestin-2/Akt/PP2A complex, allowing activation of Akt due to the loss of interaction with PP2A, which in turn phosphorylates and inhibits GSK3 on Ser9 (O'Brien *et al.*, 2011). Besides this pharmacological evidence, the conclusion that the disruption of the β-arrestin-2/Akt/PP2A complex occurs through inhibition of GSK3 kinase activity was supported by findings with *Gsk3*+/− heterozygous mice that with the partial loss of function of Gsk3 similarly had decreased levels of the β-arrestin-2/Akt/PP2A ternary complex (O'Brien *et al.*, 2011). Conversely, gain-of-function genetic models with overexpression by ~50% of a GSK3β transgene from the brain-specific prion promoter restored β-arrestin-2/Akt/PP2A ternary complex levels after lithium treatment (O'Brien *et al.*, 2011). Thus, the combined abilities of lithium to directly and indirectly inhibit GSK3β through antagonizing the effects of PP2A on Akt are both likely important mechanisms through which lithium asserts its anti-manic-like activities in rodent models of bipolar disorder. Consistent with this notion, increases in inhibitory Ser9/Ser21 phosphorylation of GSK3 in peripheral blood mononuclear cells of bipolar disorder patients were found to correlate with improvement of symptoms over a 2-month time period of treatment with lithium and antipsychotics (Li *et al.*, 2010).

While the *in vivo* studies described earlier have focused on the regulation of β-arrestin-2/Akt/PP2A complex through lithium and GSK3 inhibition, additional pharmacological loss-of-function studies and viral-mediated gain-of-function studies have demonstrated directly that the kinase activity of Akt is required for lithium to attenuate manic-like behavioral responses *in vivo* in mice (Pan *et al.*, 2011). As part of these studies, inbred strains of mice that were otherwise insensitive to lithium treatment in models of manic-like and depressive-like behaviors that had low levels of Akt activation had their sensitivity to lithium restored through the viral-mediated delivery of constitutively active AKT1 in the striatum, whereas the administration of an Akt kinase inhibitor into the striatum of normally lithium-responsive mice blocked the behavioral effects of lithium. In contrast, a potent and selective GSK3 inhibitor (CHIR99021) that inhibits GSK3 signaling was able to attenuate manic-like and depressive-like behaviors *in vivo* independent of Akt activity in both lithium-sensitive and lithium-insensitive mice strains (Pan *et al.*, 2011). Furthermore, unlike lithium, the effects on behavior of *in vivo* GSK3 inhibition through CHIR99021 treatment were not blocked by cotreatment with AKT kinase inhibitors (Pan *et al.*, 2011). Overall, these findings highlight the therapeutic potential for selective GSK3 inhibitors in bipolar disorder treatment (Beaulieu, Gainetdinov, and Caron, 2009; O'Brien and Klein, 2009), particularly for those patients who do not respond to or tolerate lithium treatment.

Overall, while much progress has been made in terms of supporting the "GSK3 hypothesis" for lithium's therapeutically relevant mechanism of action with the findings highlighting the tight correspondence between the *in vivo* regulation of mood neurocircuitry through the activity of GSK3, additional studies will be needed to clarify the relative contribution of the direct disruption of the β-arrestin-2/Akt/PP2A ternary complex through the effects of lithium on magnesium-sensitive protein–protein interactions versus effects caused by the inhibition of GSK3 altering complex stability. These efforts will benefit from more detailed kinetic studies and by gaining a better understanding of precisely which proteins GSK3 binds to in β-arrestin-2/Akt/PP2A complex and whether the mechanisms through which GSK3 kinase activity is necessary for maintaining complex integrity is through direct phosphorylation or the recruitment of other proteins. Moreover, given the ability of β-arrestin family members to bind to Dvl and form a trimeric complex with Axin that can localize to Frizzled (Fzd) receptors (Bryja *et al.*, 2007; Schulte, Schambony, and Bryja, 2010), as well as for β-arrestin to play a role in noncanonical Wnt signaling pathways (Schulte, Schambony, and Bryja, 2010), it is possible that the regulation of the GSK3β/β-arrestin-2/Akt/PP2A quaternary complex, as discussed later in the context of the regulation of the dopamine D2 receptor by antipsychotics, leads to cross talk with multiple aspects of Wnt signaling on either short-term time scales that are sufficient to regulate mood-related behaviors or over prolonged treatment times that more closely match the clinical paradigm.

Valproate and Wnt signaling

Besides lithium, the mood stabilizer and antiepileptic drug valproate (Atmaca, 2009; Rosenberg, 2007) was shown by Klein and colleagues over a decade ago to enhance canonical Wnt signaling, leading to the activation of TCF/LEF-dependent gene expression (Phiel *et al.*, 2001). While valproate may also have multiple molecular targets, a growing body of evidence suggests that its effects on Wnt signaling are due to its ability to inhibit members of the histone deacetylase (HDAC) family of chromatin-modifying enzymes (Gottlicher *et al.*, 2001; Phiel *et al.*, 2001). Given evidence that β-catenin competes for TCF/LEF binding sites with transducin-like enhancer of Split (TLEs)/Groucho family members that recruit class I HDACs to provide a switch from transcriptional repression to activation during Wnt signaling (Daniels and Weis, 2005), valproate treatment can be considered to mimic the naturally occurring Wnt signal that leads to enhanced histone acetylation through the β-catenin-dependent recruitment of the histone acetyltransferases and other coactivators (Willert and Jones, 2006). More recently, the analysis of the selectivity profile of valproate with purified recombinant HDACs revealed that valproate inhibits class I HDACs (HDAC1, HDAC2, HDAC3, HDAC8) with little or no effect on class IIa HDAC family members (HDAC4, HDAC5, HDAC7, HDAC9) or on the class IIb HDAC6 (Fass *et al.*, 2010). Here, the IC_{50} for valproate towards class I HDACs closely related to the therapeutically achieved doses of valproate in humans that are estimated in brain. Speculatively, these findings suggest that targeting class I HDACs may provide insight into the potential role for chromatin-mediated changes in neuroplasticity in the neurocircuitry involved in bipolar disorder and other neuropsychiatric disorders. In support of this notion (Schroeder *et al.*, 2013), recent findings using more selective HDAC inhibitors that target class I HDACs with a similar selectivity profile as valproate but significantly (~1000×) greater potency are able to mimic the effects of lithium in both the amphetamine-induced hyperactivity model of mania-like behavior and in the forced swim test of antidepressant-like activity. Moreover, the administration of at least one of these selective HDAC inhibitors had similar transcriptional signatures as lithium treatment, suggesting overlapping signaling pathways *in vivo* at the level of chromatin remodeling. As discussed later, recent human genetic findings that implicate Wnt signaling in the etiology of neuropsychiatric disorders at the level of chromatin-modifying and remodeling factors may have particular relevance for efforts to develop novel targeted therapeutics that can ameliorate deficits in Wnt signaling.

Major depression and Wnt signaling

In addition to the ability of lithium and direct GSK3 inhibitors to regulate mood-related behavior and genetic models that implicate GSK3 in the regulation of the neurocircuits involved in mood, there are now additional lines of evidence that point to a critical role for Wnt signaling in the pathophysiology and treatment of depression, as reviewed in Voleti and Duman (2012). Some of the earliest evidence of this comes from treatment paradigms for severe, treatment-resistant depression involving electroconvulsive shock (ECS) treatment that revealed an increased expression of β-catenin and Wnt2 in the subgranular zone of the hippocampus of treated rats (Madsen *et al.*, 2003). Besides ECS, treatment with multiple classes of antidepressants has been demonstrated to alter the expression of a number of

Wnt pathway genes in the hippocampus with evidence for at least one of citalopram, fluoxetine, venlafaxine, and atomoxetine upregulating one or more of genes, including *Wnt2*, *Wnt7b*, *Fz9*, *Frzb*, *Dvl1*, *Ctnnb1*, *Tcf15*, *TsfL1*, *Lef1*, and *Neurod1* (Okamoto *et al.*, 2010). Of these, all of the different classes of antidepressants tested in this study increased *Wnt2*, in line with the findings with ECS. Consistent with these findings of the upregulation of Wnt in response to antidepressant drug treatment, viral-mediated overexpression of Wnt2 in the hippocampus was sufficient to produce antidepressant-like behavioral actions in well-established models of depressive-like behaviors (Okamoto *et al.*, 2010). Additional studies of gene expression changes in response to ECS have revealed upregulation of the Wnt Fzd receptor *Fzd6* mRNA and downregulation of the Wnt antagonist *Dkk2* mRNA in a manner associated with changes to promoter occupancy by CREB (Voleti *et al.*, 2012). *Fzd6* expression was also shown in this study to be decreased in response to exposure to chronic unpredictable stress, which also leads to depressive-like behavior. Consistent with a causal role for these changes in *Fzd6* expression in the regulation of depression-like behavior, viral-mediated silencing of Fzd6 in the hippocampus was sufficient to have a prodepressant-like effect in a model of anhedonia and to increase anxiety in a novelty-suppressed feeding test, again pointing to the importance of Wnt signaling in mood regulation (Voleti *et al.*, 2012).

In agreement with these findings from investigating the response to antidepressant treatments that point to a key role for regulating Wnt signaling in mood behavior regulation, many clinically used drugs that affect serotonin levels, including monoamine oxidase (MAO) inhibitors, tricyclic antidepressants, and selective serotonin reuptake inhibitors (SSRIs), have been shown to elevate the phosphorylation levels of Ser9 and consequently decrease the GSK3β activity *in vivo* in mouse studies (Beaulieu, 2012; Polter and Li, 2011). In further support of a role for serotonergic signaling in the regulation of GSK3 signaling, mice with loss-of-function mutations in the serotonin biosynthesis enzyme, tryptophan hydroxylase 2 (*Tph2*), which results in decreased production of serotonin, display elevated GSK3β activity in the frontal cortex (Beaulieu *et al.*, 2008b). At the biochemical level, there is also evidence that pharmacologically activating serotonin 1A (5-HT1A) receptors, which commonly occurs in response to many classes of antidepressants that increase synaptic levels of serotonin, leads to increased GSK3β phosphorylation and decreased activity, while, conversely, the antagonism of 5-HT2A receptors, which is common to many antipsychotic agents, leads to increased GSK3β phosphorylation and decreased activity *in vivo* in multiple brain regions pointing to the competing effects that serotonin can have on GSK3 activity depending on the neuron type (Polter and Li, 2011). Besides responding to serotonin levels through 5-HT1A and 5-HT2A receptors, GSK3β has also been shown to be a functional selective modulator of 5-hydroxytryptamine-1B (5-HT1B) receptor-regulated signaling with GSK3β directly binding to and phosphorylating the 5-HT1B receptor and its pharmacological inhibition blocking serotonin-induced 5-HT1B receptor coupling and signaling through G proteins, but having no effect on serotonin-induced recruitment of β-arrestin-2 to 5-HT1B receptor (Chen, Salinas, and Li, 2009; Chen *et al.*, 2011). Thus, given evidence for a role of GSK3 in regulating neurogenesis and the ability of lithium and many antidepressant drugs that target the serotonin system to regulate neurogenesis (Samuels and Hen, 2011; Voleti and Duman, 2012), it is highly plausible that the regulation of GSK3 by serotonin neurotransmission and vice versa has an effect on Wnt signaling through potential cross talk of the signaling pathways as described earlier (Chen *et al.*, 2011; Valvezan and Klein, 2012).

Besides evidence emerging from treatment response models, in further support of the notion that the regulation of Wnt signaling and GSK3 activity can alter affective behaviors, several components of Wnt signaling pathways have been shown by Nestler and colleagues to be altered in brain's reward circuits, which includes the ventral striatum, a region of the brain that plays a critical role in mediating stress responses (Wilkinson *et al.*, 2011). These findings included the downregulation of *Dvl1/2/3*, *Ck1*, and *Fzd mRNA levels* and upregulation of *Wnt4*, *Axin2*, and *Groucho* mRNA expression along with decreased levels of phospho-Ser9 GSK3β indicative of elevated GSK3 activity, whereas in resilient mice these changes did not occur but there was an upregulation of β-catenin itself (Wilkinson *et al.*, 2011). Additionally, the mRNA levels of all three *Dvl* genes were significantly decreased in postmortem analysis of the ventral striatum of depressed humans, and changes at the level of protein could be observed for DVL2 (Wilkinson *et al.*, 2011). Consistent with these observations, viral-mediated inhibition of Dvl through dominant-negative form of Dvl1 lacking the PDZ domain (Dvl1-ΔPDZ) and overexpression of GSK3β were both sufficient to promote stress susceptibility; however, augmenting DVL1- or DVL2-mediated inhibition of GSK3 was not found to be sufficient to induce resiliency, but the inhibition of GSK3 with dominant-negative GSK3 (Lys85Ala, Lys86Ala) was sufficient to render mice more resilient to stress (Wilkinson *et al.*, 2011). Taken together, these findings point to an important role of the Wnt signaling pathway in the ventral striatum as a mediator of the reward circuitry and the response to chronic stress.

Wnt pathway and circadian rhythms in mood disorders

Dysregulated circadian rhythms (e.g., in sleep, energy, appetite) have been associated with neuropsychiatric disorders particularly mania and depression since the time of Kraepelin nearly a century ago (Milhiet *et al.*, 2011). One of the master regulators of circadian regulated pathways is CLOCK, a DNA-binding transcription factor with intrinsic histone acetyltransferase activity. Functioning as part of a core transcription–translation feedback mechanisms (Brown, Kowalska, and Dallmann, 2012; Doi, Hirayama, and Sassone-Corsi, 2006), a heterodimer of CLOCK and brain and muscle Arnt-like protein 1 (BMAL1) activates transcription of the genes *PER1–PER3* and *CRY1,2*; the encoded proteins in turn inhibit CLOCK-BMAL1 to repress their own transcription. Mice expressing a deletion mutant of CLOCK lacking exon 19 (CLOCKΔ19) (King *et al.*, 1997) exhibit multiple behavioral abnormalities reminiscent of human mania, including decreased sleep, hyperactivity, lowered anxiety, and reduced depressive-like behavior (Roybal *et al.*, 2007). CLOCKΔ19 mutant mice have also proved to be a useful model for testing pharmacological agents that attenuate manic-like behavioral abnormalities with chronic treatment of the mood stabilizer lithium restoring appropriate responses in tests of anxiety and depressive-like behavior (Roybal *et al.*, 2007). In support that these effects of lithium are mediated through GSK3 inhibition, an ATP-competitive inhibitor of GSK3 kinase activity has been shown to reduce novelty-induced hyperactivity in CLOCKΔ19 mice (Kozikowski *et al.*, 2011). Although not yet demonstrated in the context of the CNS, Bmal1 has been found to localize to the promoter regions of a number of Wnt pathway genes (*Wnt10a, Ctnnb1, Dvl2, Tcf3*), and overexpression of Bmal1 has been observed to activate the canonical Wnt pathway, whereas attenuation of Bmal1 expression downregulated the expression of genes in the canonical Wnt pathway (Guo *et al.*, 2012; Lin *et al.*, 2012). Elevated β-catenin levels due to blockage of its degradation by the APC have been shown to destabilize the clock protein PER2 through the induction of β-TrCP leading to altered circadian rhythms of clock genes (Yang *et al.*, 2009). Taken together, these findings point to a multitude of ways in which alterations of circadian regulators that may be affected in neuropsychiatric disorders can impact Wnt signaling pathways and conversely how alterations in Wnt signaling may affect aspects of the pathophysiology of multiple neuropsychiatric disorders that have altered circadian rhythms and sleep regulation.

Role of Wnt signaling in psychotic disorders

In addition to considerations of the regulation of Wnt signaling in the context of mood disorders and the response to therapeutically used drugs, additional support for a role of Wnt signaling in the pathophysiology and treatment of neuropsychiatric illness comes through consideration of psychotic disorders, such as schizophrenia. This includes, as discussed later, evidence pointing to a role for the protein Disrupted-in-Schizophrenia-1 (DISC1) as direct inhibitor of GSK3 and activator of both canonical and noncanonical Wnt signaling in the brain (Soares *et al.*, 2011), as well as through additional evidence from diverse pharmacological agents (Beaulieu, 2012).

Regulation of dopamine receptors by the Wnt pathway

Consistent with the notion that the destabilization by lithium of the β-arrestin-2/Akt/PP2A/GSK3 quaternary complex formed as a result of dopamine D2 receptor activation has antimanic-like properties, in the context of schizophrenia and rodent models of antipsychotic activity, direct antagonism of the dopamine D2 receptor with antipsychotics also leads to a block of the recruitment of β-arrestin, resulting in a loss of PP2A-mediated dephosphorylation and inactivation of Akt and consequently GSK3 inhibition (Beaulieu, Gainetdinov, and Caron, 2009; Beaulieu *et al.*, 2004; Emamian *et al.*, 2004). Besides these pharmacological effects, while the dopamine D2 receptor has not been traditionally considered as a component of the Wnt signaling pathway, there is growing evidence that this G protein-coupled receptor (GPCR) has properties similar to the Fzd family of GPCRs (see Chapter 14) in terms of regulating canonical Wnt signaling and binding to components of the Wnt pathway. In particular, the dopamine D2 receptor, but not the four other subtypes of dopamine receptors, has been shown to selectively inhibit Wnt signaling at the level of TCF/LEF-mediated transcription (Min *et al.*, 2011). The effects found in this study were independent of stimulation of the dopamine D2 receptor and G protein coupling and occurred without a role for endocytosis of the receptor or its phosphorylation. Instead, the dopamine D2 receptor was shown to directly interact with β-catenin through the second and third intracellular loops, leading to sequestering of β-catenin away from the nucleus and inhibition of TCF/LEF-mediated transcription (Min *et al.*, 2011). Along with binding β-catenin and GSK3, the dopamine D2 receptor has also been shown to directly bind Dvl3 *in vivo* in rat brain, and *in vitro* studies have shown that Dvl3 plays a key role in linking the regulation of the dopamine D2 receptor to both Akt signaling and Wnt signaling (Alimohamad *et al.*, 2005; Sutton *et al.*, 2007). Consistent with these findings, subchronic treatment of rats with the dopamine D2 receptor antagonist raclopride leads to increase phosphorylation of Akt and GSK3 along with elevated Dvl3 levels, whereas the dopamine D2 receptor agonist quinpirole has the opposite response (Sutton and Rushlow, 2012). Thus, while genetic variation in the dopamine D2 receptor has not yet been unambiguously described to increase risk for schizophrenia or determine the response of patients to antipsychotic treatments, its direct interaction with the members of the Dvl family and its interaction with the β-arrestin-2/Akt/PP2A/GSK3 complex point to an important example of potential cross talk between signaling pathways that may have important relevance for the pathophysiology and treatment response in neuropsychiatric disorders.

Interestingly, lithium has not been found to be effective in treating psychosis in schizophrenic patients (Leucht, Kissling, and McGrath, 2004). Consistent with these human findings, whereas antipsychotics (haloperidol and clozapine), mood stabilizers (lithium and valproate), and antidepressants (imipramine and fluoxetine) were all found to lead to activated Akt signaling and inhibition of GSK3 measured through phosphospecific antibodies in the prefrontal cortex, only antipsychotic treatment increased Dvl3 and β-catenin levels and increased binding of GSK3 to the dopamine D2 receptor in the same brain area (Sutton and Rushlow, 2011a). This result suggests that one reason that lithium and other mood stabilizers may not be effective antipsychotics is that they are unable to target GSK3 and Wnt signaling in the prefrontal cortex, which may be a prerequisite for ameliorating psychosis in schizophrenic patients. Thus, it is possible that a potent enough direct GSK3 inhibitor could be an effective antipsychotic. A test of this notion and of the Wnt hypothesis in general in the context of neuropsychiatric disorders awaits the development of highly selective and safe compounds that target GSK3 and other components of Wnt signaling in the CNS, although, as discussed later, there are a number of concerns and challenges that need to be addressed to achieve this goal.

Regulation of glutamate receptors by the Wnt pathway

In addition to the dopaminergic system that is targeted by many antipsychotic treatments, the dysfunction of the glutamatergic system has also been implicated in neuropsychiatric disorders. As an example of linking these two neurotransmitter systems together, using a cortical slice culture model, the elevation of dopamine levels through blocking its uptake leading to GSK3 activation has been shown to mediate the suppression of synaptic *N*-methyl-D-aspartate (NMDA) receptor-mediated currents through the increased internalization of Grin2b subunit of the NMDA receptor and decreased expression of *Grin2b* mRNA levels (Li *et al.*, 2009). The effects in this study were shown to occur due to decreased β-catenin–Grin2b interactions at the synapse and decreased β-catenin in the nucleus with the changes reversible by GSK3 inhibitor treatment and antagonism of the dopamine D2 receptor. These data provide additional evidence for the interaction between the dopamine D2 receptor and GSK3 as well as β-catenin, this time in the regulation of glutamatergic synapse function.

Outside of the serotonergic and dopaminergic neurotransmitter systems that are the target of many of the existing neuropsychiatric therapeutics, one of the additional neurotransmitter systems that has received growing attention both preclinically and clinically, although no drug has yet been approved for human use, is the metabotropic glutamate receptors 2/3 (mGluR2/3) (Vinson and Conn, 2012). Studies by Rushlow and colleagues have shown that, similar to dopamine D2 receptor antagonists, treatment with mGluR2/3 agonists that have antipsychotic-like activity results in both the activation of Akt and inhibitory phosphorylation of GSK3 along with the accumulation of Dvl2 and Dvl3 and β-catenin following repeated treatment (Sutton and Rushlow, 2011b). Moreover, this study demonstrated that the mGluR2/3 complex physically interacts with Dvl2. Taken together, these results provide another example of a class of non-dopamine receptor-targeting pharmacological agents that both modulate psychiatric-relevant behaviors and alter the Wnt pathway and Akt–GSK3 signaling network.

Summary

It is interesting to speculate that the deep evolutionary conservation of Wnt signaling – from flies to humans – is a clue that points to the overall fundamental importance of the Wnt pathway to human biology, such that the subtle forms of its dysregulation will predispose and, in some cases, cause many different types of neuropsychiatric disorders. Gaining a more complete understanding of whether there are shared etiologies of otherwise disparate mental illnesses is of paramount importance to guide our understanding of both health and disease, and the role of Wnt signaling provides a compelling line of investigation to pursue for the foreseeable future. As discussed in Chapter 30, besides the pharmacological evidence and animal data summarized in this chapter, evidence from human genetic studies is accumulating, particularly for schizophrenia and autism, that genetic variation in Wnt signaling pathways may play a fundamentally important role in mental health.

Acknowledgments

Studies in the Haggarty Laboratory on the role of Wnt signaling in neuropsychiatric disorders and dementia have been generously supported by the National Institute of Mental Health, the National Institute of Aging, the National Institute of Drug Addiction, and the Stanley Medical Research Institute. We thank the laboratory of Randall Moon (UW), Li-Huei Tsai (MIT), and the members of the Stanley Center for Psychiatric Research for helpful discussions. We sincerely apologize for not being able to cite due to space limitations a number of additional papers that have contributed to the concepts discussed herein.

References

Alimohamad, H., Sutton, L., Mouyal, J. *et al.* (2005) The effects of antipsychotics on beta-catenin, glycogen synthase kinase-3 and dishevelled in the ventral midbrain of rats. *Journal of Neurochemistry*, **95**, 513–525.

Atmaca, M. (2009) Valproate and neuroprotective effects for bipolar disorder. *International Review of Psychiatry*, **21**, 410–413.

Baune, B.T., Miller, R., McAfoose, J. *et al.* (2010) The role of cognitive impairment in general functioning in major depression. *Psychiatry Research*, **176**, 183–189.

Bearden, C.E., Thompson, P.M., Dalwani, M. *et al.* (2007) Greater cortical gray matter density in lithium-treated patients with bipolar disorder. *Biological Psychiatry*, **62**, 7–16.

Beaulieu, J.M. (2012) A role for Akt and glycogen synthase kinase-3 as integrators of dopamine and serotonin neurotransmission in mental health. *Journal of Psychiatry & Neuroscience*, **37**, 7–16.

Beaulieu, J.M. and Caron, M.G. (2008) Looking at lithium: molecular moods and complex behaviour. *Molecular Interventions*, **8**, 230–241.

Beaulieu, J.M., Gainetdinov, R.R., and Caron, M.G. (2007) The Akt–GSK-3 signaling cascade in the actions of dopamine. *Trends in Pharmacological Sciences*, **28**, 166–172.

Beaulieu, J.M., Gainetdinov, R.R., and Caron, M.G. (2009) Akt/GSK3 signaling in the action of psychotropic drugs. *Annual Review of Pharmacology and Toxicology*, **49**, 327–347.

Beaulieu, J.M., Sotnikova, T.D., Yao, W.D. *et al.* (2004) Lithium antagonizes dopamine-dependent behaviors mediated by an AKT/glycogen synthase kinase 3 signaling cascade. *Proceedings of the National Academy of Sciences of the United States of America*, **101**, 5099–5104.

Beaulieu, J.M., Marion, S., Rodriguiz, R.M. *et al.* (2008a) A beta-arrestin 2 signaling complex mediates lithium action on behavior. *Cell*, **132**, 125–136.

Beaulieu, J.M., Zhang, X., Rodriguiz, R.M. *et al.* (2008b) Role of GSK3 beta in behavioral abnormalities induced by serotonin deficiency. *Proceedings of the National Academy of Sciences of the United States of America*, **105**, 1333–1338.

Berridge, M.J., Downes, C.P., and Hanley, M.R. (1989) Neural and developmental actions of lithium: a unifying hypothesis. *Cell*, **59**, 411–419.

Berton, O. and Nestler, E.J. (2006) New approaches to antidepressant drug discovery: beyond monoamines. *Nature Reviews Neuroscience*, **7**, 137–151.

Bowden, C.L. (2000) The ability of lithium and other mood stabilizers to decrease suicide risk and prevent relapse. *Current Psychiatry Reports*, **2**, 490–494.

Brown, S.A., Kowalska, E., and Dallmann, R. (2012) (Re)inventing the circadian feedback loop. *Developmental Cell*, **22**, 477–487.

Bryja, V., Gradl, D., Schambony, A. *et al.* (2007) Beta-arrestin is a necessary component of Wnt/beta-catenin signaling in vitro and in vivo. *Proceedings of the National Academy of Sciences of the United States of America*, **104**, 6690–6695.

Cade, J.F. (1949) Lithium salts in the treatment of psychotic excitement. *The Medical Journal of Australia*, **2**, 349–352.

Chalecka-Franaszek, E. and Chuang, D.M. (1999) Lithium activates the serine/threonine kinase Akt-1 and suppresses glutamate-induced inhibition of Akt-1 activity in neurons. *Proceedings of the National Academy of Sciences of the United States of America*, **96**, 8745–8750.

Chen, L., Salinas, G.D., and Li, X. (2009) Regulation of serotonin 1B receptor by glycogen synthase kinase-3. *Molecular Pharmacology*, **76**, 1150–1161.

Chen, L., Zhou, W., Chen, P.C. *et al.* (2011) Glycogen synthase kinase-3beta is a functional modulator of serotonin-1B receptors. *Molecular Pharmacology*, **79**, 974–986.

Clevers, H. and Nusse, R. (2012) Wnt/beta-catenin signaling and disease. *Cell*, **149**, 1192–1205.

Daniels, D.L. and Weis, W.I. (2005) Beta-catenin directly displaces Groucho/TLE repressors from Tcf/Lef in Wnt-mediated transcription activation. *Nature Structural & Molecular Biology*, **12**, 364–371.

De Ferrari, G.V. and Moon, R.T. (2006) The ups and downs of Wnt signaling in prevalent neurological disorders. *Oncogene*, **25**, 7545–7553.

De Sarno, P., Li, X., and Jope, R.S. (2002) Regulation of Akt and glycogen synthase kinase-3 beta phosphorylation by sodium valproate and lithium. *Neuropharmacology*, **43**, 1158–1164.

Doi, M., Hirayama, J., and Sassone-Corsi, P. (2006) Circadian regulator CLOCK is a histone acetyltransferase. *Cell*, **125**, 497–508.

Emamian, E.S., Hall, D., Birnbaum, M.J. *et al.* (2004) Convergent evidence for impaired AKT1-GSK3beta signaling in schizophrenia. *Nature Genetics*, **36**, 131–137.

Fass, D.M., Shah, R., Ghosh, B. *et al.* (2010) Effect of inhibiting histone deacetylase with short-chain carboxylic acids and their hydroxamic acid analogs on vertebrate development and neuronal chromatin. *ACS Medicinal Chemistry Letters*, **2**, 39–42.

Gottlicher, M., Minucci, S., Zhu, P. *et al.* (2001) Valproic acid defines a novel class of HDAC inhibitors inducing differentiation of transformed cells. *The EMBO Journal*, **20**, 6969–6978.

Gould, T.D. and Manji, H.K. (2005) Glycogen synthase kinase-3: a putative molecular target for lithium mimetic drugs. *Neuropsychopharmacology*, **30**, 1223–1237.

Gould, T.D., Chen, G., and Manji, H.K. (2004) In vivo evidence in the brain for lithium inhibition of glycogen synthase kinase-3. *Neuropsychopharmacology*, **29**, 32–38.

Gould, T.D., Einat, H., O'Donnell, K.C. *et al.* (2007) Beta-catenin overexpression in the mouse brain phenocopies lithium-sensitive behaviors. *Neuropsychopharmacology*, **32**, 2173–2183.

Gould, T.D., O'Donnell, K.C., Picchini, A.M. *et al.* (2008) Generation and behavioral characterization of beta-catenin forebrain-specific conditional knock-out mice. *Behavioural Brain Research*, **189**, 117–125.

Guo, B., Chatterjee, S., Li, L. *et al.* (2012) The clock gene, brain and muscle Arnt-like 1, regulates adipogenesis via Wnt signaling pathway. *The FASEB Journal*, **26**, 3453–3463.

Kaidanovich-Beilin, O., Lipina, T.V., Takao, K. *et al.* (2009) Abnormalities in brain structure and behavior in GSK-3alpha mutant mice. *Molecular Brain*, **2**, 35.

King, D.P., Zhao, Y., Sangoram, A.M. *et al.* (1997) Positional cloning of the mouse circadian clock gene. *Cell*, **89**, 641–653.

Klein, P.S. and Melton, D.A. (1996) A molecular mechanism for the effect of lithium on development. *Proceedings of the National Academy of Sciences of the United States of America*, **93**, 8455–8459.

Kozikowski, A.P., Gunosewoyo, H., Guo, S. *et al.* (2011) Identification of a glycogen synthase kinase-3beta inhibitor that attenuates hyperactivity in CLOCK mutant mice. *ChemMedChem*, **6**, 1593–1602.

Krishnan, V. and Nestler, E.J. (2008) The molecular neurobiology of depression. *Nature*, **455**, 894–902.

Leucht, S., Kissling, W., and McGrath, J. (2004) Lithium for schizophrenia revisited: a systematic review and meta-analysis of randomized controlled trials. *The Journal of Clinical Psychiatry*, **65**, 177–186.

Li, Y.C., Xi, D., Roman, J. *et al.* (2009) Activation of glycogen synthase kinase-3 beta is required for hyperdopamine and D2 receptor-mediated inhibition of synaptic NMDA receptor function in the rat prefrontal cortex. *The Journal of Neuroscience*, **29**, 15551–15563.

Li, X., Liu, M., Cai, Z. *et al.* (2010) Regulation of glycogen synthase kinase-3 during bipolar mania treatment. *Bipolar Disorders*, **12**, 741–752.

Lin, F., Chen, Y., Li, X. *et al.* (2012) Over-expression of circadian clock gene Bmal1 affects proliferation and the canonical Wnt pathway in NIH-3T3 cells. *Cell Biochemistry and Function*, **31**, 166–172.

Lyoo, I.K., Dager, S.R., Kim, J.E. *et al.* (2010) Lithium-induced gray matter volume increase as a neural correlate of treatment response in bipolar disorder: a longitudinal brain imaging study. *Neuropsychopharmacology*, **35**, 1743–1750.

Madsen, T.M., Newton, S.S., Eaton, M.E. *et al.* (2003) Chronic electroconvulsive seizure up-regulates beta-catenin expression in rat hippocampus: role in adult neurogenesis. *Biological Psychiatry*, **54**, 1006–1014.

Manji, H.K. and Duman, R.S. (2001) Impairments of neuroplasticity and cellular resilience in severe mood disorders: implications for the development of novel therapeutics. *Psychopharmacology Bulletin*, **35**, 5–49.

Milhiet, V., Etain, B., Boudebesse, C., and Bellivier, F. (2011) Circadian biomarkers, circadian genes and bipolar disorders. *Journal of Physiology Paris*, **105**, 183–189.

Min, C., Cho, D.I., Kwon, K.J. *et al.* (2011) Novel regulatory mechanism of canonical Wnt signaling by dopamine D2 receptor through direct interaction with beta-catenin. *Molecular Pharmacology*, **80**, 68–78.

Moore, G.J., Cortese, B.M., Glitz, D.A. *et al.* (2009) A longitudinal study of the effects of lithium treatment on prefrontal and subgenual prefrontal gray matter volume in treatment-responsive bipolar disorder patients. *The Journal of Clinical Psychiatry*, **70**, 699–705.

Murrough, J.W. and Charney, D.S. (2012) Is there anything really novel on the antidepressant horizon? *Current Psychiatry Report*, **14**, 643–649.

Netto, I. and Phutane, V.H. (2012) Reversible lithium neurotoxicity: review of the literature. *The Primary Care Companion to CNS Disorders*, **14**, PCC.11r01197.

O'Brien, W.T. and Klein, P.S. (2009) Validating GSK3 as an in vivo target of lithium action. *Biochemical Society Transactions*, **37**, 1133–1138.

O'Brien, W.T., Harper, A.D., Jove, F. *et al.* (2004) Glycogen synthase kinase-3beta haploinsufficiency mimics the behavioral and molecular effects of lithium. *The Journal of Neuroscience*, **24**, 6791–6798.

O'Brien, W.T., Huang, J., Buccafusca, R. *et al.* (2011) Glycogen synthase kinase-3 is essential for beta-arrestin-2 complex formation and lithium-sensitive behaviors in mice. *The Journal of Clinical Investigation*, **121**, 3756–3762.

Okamoto, H., Voleti, B., Banasr, M. *et al.* (2010) Wnt2 expression and signaling is increased by different classes of antidepressant treatments. *Biological Psychiatry*, **68**, 521–527.

Pan, J.Q., Lewis, M.C., Ketterman, J.K. *et al.* (2011) AKT kinase activity is required for lithium to modulate mood-related behaviors in mice. *Neuropsychopharmacology*, **36**, 1397–1411.

Perlis, R.H., Ostacher, M.J., Patel, J.K. *et al.* (2006) Predictors of recurrence in bipolar disorder: primary outcomes from the Systematic Treatment Enhancement Program for Bipolar Disorder (STEP-BD). *The American Journal of Psychiatry*, **163**, 217–224.

Phiel, C.J., Zhang, F., Huang, E.Y. *et al.* (2001) Histone deacetylase is a direct target of valproic acid, a potent anticonvulsant, mood stabilizer, and teratogen. *The Journal of Biological Chemistry*, **276**, 36734–36741.

Pittenger, C. and Duman, R.S. (2008) Stress, depression, and neuroplasticity: a convergence of mechanisms. *Neuropsychopharmacology*, **33**, 88–109.

Polter, A.M. and Li, X. (2011) Glycogen synthase kinase-3 is an intermediate modulator of serotonin neurotransmission. *Frontiers in Molecular Neuroscience*, **4**, 31.

Polter, A., Beurel, E., Yang, S. *et al.* (2010) Deficiency in the inhibitory serine-phosphorylation of glycogen synthase kinase-3 increases sensitivity to mood disturbances. *Neuropsychopharmacology*, **35**, 1761–1774.

Prickaerts, J., Moechars, D., Cryns, K. *et al.* (2006) Transgenic mice overexpressing glycogen synthase kinase 3beta: a putative model of hyperactivity and mania. *The Journal of Neuroscience*, **26**, 9022–9029.

Quiroz, J.A. and Manji, H.K. (2002) Enhancing synaptic plasticity and cellular resilience to develop novel, improved treatments for mood disorders. *Dialogues in Clinical Neuroscience*, **4**, 73–92.

Rapoport, J.L., Giedd, J.N., and Gogtay, N. (2012) Neurodevelopmental model of schizophrenia: update 2012. *Molecular Psychiatry*, **17**, 1228–1238.

Rosenberg, G. (2007) The mechanisms of action of valproate in neuropsychiatric disorders: can we see the forest for the trees? *Cellular and Molecular Life Sciences*, **64**, 2090–2103.

Roybal, K., Theobold, D., Graham, A. *et al.* (2007) Mania-like behavior induced by disruption of CLOCK. *Proceedings of the National Academy of Sciences of the United States of America*, **104**, 6406–6411.

Samuels, B.A. and Hen, R. (2011) Neurogenesis and affective disorders. *The European Journal of Neuroscience*, **33**, 1152–1159.

Sarkar, S., Floto, R.A., Berger, Z. *et al.* (2005) Lithium induces autophagy by inhibiting inositol monophosphatase. *The Journal of Cell Biology*, **170**, 1101–1111.

Sassi, R.B., Nicoletti, M., Brambilla, P. *et al.* (2002) Increased gray matter volume in lithium-treated bipolar disorder patients. *Neuroscience Letters*, **329**, 243–245.

Schroeder, F.A., Lewis, M.C., Fass, D.M. *et al.* (2013) A selective HDAC 1/2 inhibitor modulates chromatin and gene expression in brain and alters mouse behavior in two mood-related tests. *PLoS One*. **8** (8), e71323.

Schulte, G., Schambony, A., and Bryja, V. (2010) Beta-arrestins – scaffolds and signalling elements essential for WNT/Frizzled signalling pathways? *British Journal of Pharmacology*, **159**, 1051–1058.

Soares, D.C., Carlyle, B.C., Bradshaw, N.J., and Porteous, D.J. (2011) DISC1: structure, function, and therapeutic potential for major mental illness. *ACS Chemical Neuroscience*, **2**, 609–632.

Sutton, L.P. and Rushlow, W.J. (2011a) The effects of neuropsychiatric drugs on glycogen synthase kinase-3 signaling. *Neuroscience*, **199**, 116–124.

Sutton, L.P. and Rushlow, W.J. (2011b) Regulation of Akt and Wnt signaling by the group II metabotropic glutamate receptor antagonist LY341495 and agonist LY379268. *Journal of Neurochemistry*, **117**, 973–983.

Sutton, L.P. and Rushlow, W.J. (2012) The dopamine D2 receptor regulates Akt and GSK-3 via Dvl-3. *The International Journal of Neuropsychopharmacology*, **15**, 965–979.

Sutton, L.P., Honardoust, D., Mouyal, J. *et al.* (2007) Activation of the canonical Wnt pathway by the antipsychotics haloperidol and clozapine involves dishevelled-3. *Journal of Neurochemistry*, **102**, 153–169.

Teo, R., King, J., Dalton, E. *et al.* (2009) PtdIns(3,4,5)P(3) and inositol depletion as a cellular target of mood stabilizers. *Biochemical Society Transactions*, **37**, 1110–1114.

Tondo, L., Isacsson, G., and Baldessarini, R. (2003) Suicidal behaviour in bipolar disorder: risk and prevention. *CNS Drugs*, **17**, 491–511.

Valvezan, A.J. and Klein, P.S. (2012) GSK-3 and Wnt signaling in neurogenesis and bipolar disorder. *Frontiers in Molecular Neuroscience*, **5**, 1.

Vinson, P.N. and Conn, P.J. (2012) Metabotropic glutamate receptors as therapeutic targets for schizophrenia. *Neuropharmacology*, **62**, 1461–1472.

Voleti, B. and Duman, R.S. (2012) The roles of neurotrophic factor and Wnt signaling in depression. *Clinical Pharmacology and Therapeutics*, **91**, 333–338.

Voleti, B., Tanis, K.Q., Newton, S.S., and Duman, R.S. (2012) Analysis of target genes regulated by chronic electroconvulsive therapy reveals role for Fzd6 in depression. *Biological Psychiatry*, **71**, 51–58.

Wilkinson, M.B., Dias, C., Magida, J. *et al.* (2011) A novel role of the WNT-dishevelled-GSK3beta signaling cascade in the mouse nucleus accumbens in a social defeat model of depression. *The Journal of Neuroscience*, **31**, 9084–9092.

Willert, K. and Jones, K.A. (2006) Wnt signaling: is the party in the nucleus? *Genes & Development*, **20**, 1394–1404.

World Health Organization. (2004) The global burden of disease: 2004 update. http://www.who.int/entity/healthinfo/global_burden_disease/GBD_report_2004update_full.pdf (accessed November 18, 2013).

Yang, X., Wood, P.A., Ansell, C.M. *et al.* (2009) Beta-catenin induces beta-TrCP-mediated PER2 degradation altering circadian clock gene expression in intestinal mucosa of ApcMin/+ mice. *Journal of Biochemistry*, **145**, 289–297.

Zarate, C.A., Jr., Tohen, M., Land, M., and Cavanagh, S. (2000) Functional impairment and cognition in bipolar disorder. *The Psychiatric Quarterly*, **71**, 309–329.

30 Neuropsychiatric Disease-Associated Genetic Variation in the Wnt Pathway

Stephen J. Haggarty[1,2,3], Karun Singh[4], Roy H. Perlis[2,3] and Rakesh Karmacharya[2,3,5,6]

[1]Department of Neurology, Center for Human Genetic Research, Massachusetts General Hospital & Harvard Medical School, Boston, MA, USA
[2]Department of Psychiatry, Center for Human Genetic Research, Massachusetts General Hospital & Harvard Medical School, Boston, MA, USA
[3]MGH Psychiatry Center for Experimental Drugs & Diagnostics, Center for Human Genetic Research, Massachusetts General Hospital & Harvard Medical School, Boston, MA, USA
[4]Department of Biochemistry and Biomedical Sciences, McMaster Stem Cell & Cancer Research Institute, McMaster University, Hamilton, Ontario, Canada
[5]Chemical Biology Program, Broad Institute of Harvard & MIT, Cambridge, MA, USA
[6]Schizophrenia and Bipolar Disorder Program, McLean Hospital, Belmont, MA, USA

Introduction

Neuropsychiatric disorders, such as bipolar disorder and schizophrenia, are common disorders affecting ~3% of the adult population, and through multiple lines of evidence, they are known to be highly heritable (Kendler, 2001; Lipina *et al.*, 2012; Psychiatric GWAS Consortium Bipolar Disorder Working Group, 2011; Psychiatric GWAS Consortium Coordinating Committee *et al.*, 2009; Schizophrenia Psychiatric Genome-Wide Association Study Consortium, 2011; Sullivan, Daly, and O'Donovan, 2012). The identification of genes involved in their etiologies has long been sought after as they hold tremendous potential for helping gain insight into the underlying pathogenesis and to aid in the development of improved therapeutics that target core features of the disorders. Owing to advances in our understanding of the human genome and technologies that afford the ability to comprehensively scan the human genome for both single-nucleotide polymorphisms (SNPs) through array-based technologies and now whole exome (coding regions of genes) and even whole genomes (both coding and non-coding regions), the field of neuropsychiatric genetics has experienced a period of rapid growth and excitement in the past decade that is almost certain to continue for the foreseeable future (Sullivan, Daly, and O'Donovan, 2012). Here, we will review in terms of their connections to Wnt signaling and the hypothesis that dysregulation of the Wnt pathway is involved in both the etiology and pathophysiology of neuropsychiatric disorders found in four major areas: (i) classical cytogenetics, (ii) copy number variants (CNVs), (iii) genome-wide association studies (GWAS) findings, and (iv) *de novo* single-nucleotide variants (SNVs). While in certain

Wnt Signaling in Development and Disease: Molecular Mechanisms and Biological Functions, First Edition.
Edited by Stefan Hoppler and Randall T. Moon.

cases previous findings in the literature are not discussed, the rapid nature of expanding knowledge of the genetic architecture of these disorders through the advent of GWAS using SNP analysis (Raychaudhuri, 2011), along with massively parallel DNA sequencing that enables sequencing of exomes and whole genomes (1000 Genomes Project Consortium *et al.*, 2012), requires careful consideration of previous claims of genetic association and causality given the breadth of human genetic data now available.

Cytogenetic findings in schizophrenia and Wnt signaling: DISC1

One of the initial approaches taken to gain insight into the genetic causes of mental illness was the use of classical cytogenetic approaches in combination with family-based linkage studies that sought to identify chromosomal abnormalities that segregated with affected status (Blackwood *et al.*, 2008). This approach, while state-of-the-art at the time, has now been largely replaced by molecular genetics but has nonetheless pointed to the discovery of rare but still highly important examples of genes that regulate the Wnt signaling pathway as being important for mental health. One of the preeminent examples of the success of this approach was the discovery of the gene Disrupted-in-Schizophrenia-1 (*DISC1*), originally discovered using classical cytogenetic approaches as a disrupted open reading frame due to a balanced chromosomal translocation ((1:11) (q42.1; q14.3)) between human chromosomes 1 and 11 in a multigeneration Scottish family that suffered from not only schizophrenia but also bipolar disorder and major depression (Blackwood *et al.*, 2001; Millar *et al.*, 2001; Muir, Pickard, and Blackwood, 2008). Over the past decade, *DISC1* has received much attention as a genetic risk factor associated with schizophrenia. Besides the original Scottish family, an American family with schizophrenia that cosegregates with a small four-base-pair deletion has also been identified (Sachs *et al.*, 2005). What is clear from multiple studies is that the *DISC1* locus encodes a multifunctional protein with multiple protein–protein interactions due to its role as a scaffolding protein and multiple distinct splice forms and subcellular localizations (Soares *et al.*, 2011).

A particularly intriguing finding from the Tsai Laboratory, now replicated in a number of different contexts, has been the demonstration that DISC1 regulates Wnt signaling though a mechanism at least in part involving the direct interaction and inhibition of GSK3β (Mao *et al.*, 2009). In this study, DISC1 overexpression led to decreased GSK3 activity and increased stability of β-catenin leading to stimulation of TCF/LEF-mediated transcription, whereas the silencing of DISC1 through RNA interference had the opposite effect. These biological activities of DISC1 in regulating neurogenesis and behavior are reminiscent of the effects of lithium itself prompting the reference to DISC1 as an "endogenous lithium." Several studies have now sought to determine if common variants in DISC1 is involved in schizophrenia risk through targeted DNA sequencing in large numbers of schizophrenia cases and unaffected controls leading to the identification of a number of common and also ultrarare DISC1 variants. A number of these DISC1 variants (Ala83Val, Arg264Glu, Leu607Phe), but not others (e.g., Ser704Cys), were found to be unable to activate Wnt signaling compared with wild-type DISC1 resulting in decreased neural progenitor proliferation (Singh *et al.*, 2011).

Building on these observations of a role for DISC1 in regulating multiple aspects of brain development, a developmentally regulated switch involving the phosphorylation of DISC1 has been identified as a key determinant of the regulation of Wnt signaling and neurogenesis by DISC1 versus its functions in controlling neuronal migration (Ishizuka *et al.*, 2011). While nonphosphorylated DISC1 was found to be capable of regulating canonical Wnt signaling via its interaction with GSK3β, phosphorylation of DISC1 at serine 710 (Ser710) triggered the recruitment of Bardet–Biedl syndrome (BBS) proteins to the centrosome to regulate migration (Ishizuka *et al.*, 2011). These findings point to the dynamic nature of signaling pathways that can affect Wnt signaling that may be relevant to neurodevelopment and to the etiology of neuropsychiatric disorders.

Paralleling these mouse studies performed with mouse Disc1 and pointing to an important evolutionarily conserved function of DISC1, a zebrafish Disc1-derived peptide was found to directly bind GSK3β *in vitro*, and loss of function of zebrafish Disc1 through morpholino-based gene silencing led to aberrant early forebrain development and impaired axonogenesis. Disc1 loss of function also was found to decrease expression of a β-catenin-responsive transcriptional reporter in a manner that could be rescued by systemic administration of a potent and selective direct GSK3 inhibitor (CHIR-99021) (De Rienzo *et al.*, 2011). Thus, like its mouse and human ortholog, zebrafish Disc1 promotes β-catenin-mediated Wnt signaling and inhibits GSK3β activity. Interestingly, besides affecting canonical Wnt signaling, an important finding in these studies was the observation that loss of Disc1 in zebrafish also led to alterations in the noncanonical Wnt pathway through the formin-domain protein Daam and RhoA GTPase signaling resulting in a convergence and extension phenotype that could be rescued by a dominant-negative GSK3β construct (De Rienzo *et al.*, 2011).

Besides its interactions with GSK3 and further strengthening of the connection between Disc1 and Wnt signaling, Disc1 has recently been shown to interact with another Wnt signaling protein, DIX domain containing 1 (Dixdc1), which has been shown to be involved in the regulation of neurogenesis through the Wnt pathway (Kivimae *et al.*, 2011; Singh *et al.*, 2010). Deletion of *Dixdc1* in mice led to abnormal locomotory activity, anxiety-like activity, and startle reactivity, all indicative of changes to neurocircuitry important to pathophysiology of neuropsychiatric disorders.

Finally, beyond neurodevelopment pathways, further evidence for the importance of the interaction of DISC1 and GSK3 comes from the observation that the inhibition of GSK3 was sufficient to reverse a variety of behavioral phenotypes in mutant Disc1 mice that have a leucine to proline at amino acid position 100 (Disc1–Leu100Pro), including sensorimotor deficits measured through assessing prepulse inhibition, latent inhibition deficits, and hyperactivity (Lipina *et al.*, 2011; Yan *et al.*, 2005). Biochemically, this Leu100Pro missense mutation was shown to be sufficient to reduce the interaction between DISC1 and both GSK3α and GSK3β providing a powerful example of how a single amino acid change is sufficient to alter complex neurocircuitry relevant to the pathophysiology of neuropsychiatric disorders when the variant alters an important protein–protein interaction or biochemical function.

Taken together, given the importance of both the canonical and noncanonical Wnt signaling in synaptic plasticity and dendrite development as described in Chapter 21, these findings on role of DISC1-mediated signaling in brain development and Wnt signaling suggest that loss of DISC1 may affect multiple pathways relevant to the etiology of neuropsychiatric disorders.

Copy number variations (CNVs) in schizophrenia and Wnt signaling

The advent of large-scale human genetics projects in the past decade, such as the Human Haplotype Mapping project (Manolio, Brooks, and Collins, 2008) and 1000 Genomes project (1000 Genomes Project Consortium *et al.*, 2012), has enabled probing of the entire genome with increasingly high-resolution maps of over a million SNPs spaced across the genome so as to maximally capture common genetic variation in the human population (Raychaudhuri, 2011). Using this technology, a number of GWAS have identified that rare or *de novo* (i.e., not transmitted from unaffected parents to the affected proband) CNVs consisting of either microdeletions or microduplications of chromosomal regions have an elevated frequency in individuals with neuropsychiatric disorders compared to unaffected individuals (Girirajan, Campbell, and Eichler, 2011). However, while the genetic evidence clearly supports that CNVs are associated with an increase risk for developing a neuropsychiatric disorder in a population, understanding the precise relationship of a given CNV to the etiology of a particular neuropsychiatric disorder and specific disease symptoms in any one individual is complicated by the variable expressivity and reduced penetrance of the psychiatric phenotypes (Girirajan, Campbell, and Eichler, 2011). In many cases, the same CNVs are identified as susceptibility factors for a range of neurodevelopmental disorders, including but not limited to schizophrenia, autism, and neurodevelopmental delay, as well as for other neurological disorders, such as epilepsy. Nonetheless, this feature of CNVs may reflect as of yet poorly understood shared aspects to the underlying molecular etiology and pathophysiology of neuropsychiatric disorders. That these CNVs cross traditional diagnostic boundaries may reflect more on our diagnostic categories of mental illness. Furthermore, given the clear directionality of the hypothesis of the consequence of the perturbation (i.e., loss of function due to deletions or gain of function due to duplications), this class of human genetic variation is instructive in terms of pointing to potentially critical Wnt pathway genes that are within the critical regions of CNVs that may be involved in the etiology of neuropsychiatric disorders.

Most notably, microdeletion of 1q21.1, which has been associated with schizophrenia, autism, and neurodevelopmental delay, contains the human homolog of *Drosophila* legless gene (*lgs*), *BCL9* (Kirov, 2010). BCL9 encodes an adaptor protein that binds to both β-catenin and Pygopus, the latter of which has been shown to anchor BCL9 within the nucleus where it is constitutively localized and bound to chromatin at least in part through interactions of its PHD finger with histones methylated (mono–tri) at Lys4 (Kessler, Hausmann, and Basler, 2009; Kramps *et al.*, 2002). Thus, the loss of BCL9 leads to a decrease in Wnt signal transduction due to an attenuation of β-catenin-dependent transcriptional activity. Potentially consistent with the notion of decreased Wnt signaling upon loss of *BCL9*, patients with 1q21.1 microdeletions characteristically have microcephaly, whereas patients with microduplications have macrocephaly. While a number of genes within the locus may be involved, given evidence for the ability of a stabilized form of β-catenin that is capable of hyperactivating canonical Wnt signaling through increasing Tcf/Lef-mediated transcription to cause increased brain size (Chenn and Walsh, 2002), it is tempting to speculate that the attenuation of neural progenitor cell proliferation in response to a loss of *BCL9*, or conversely elevated levels of proliferation in response to a duplication of *BCL9*, is mechanistically involved in the observed microcephaly and macrocephaly, respectively. Understanding whether this is the case or not will require more precise investigation of whether a 50% change in BCL9 levels caused by a CNV is sufficient to alter Wnt signaling in human neural progenitor cells. Additionally, more fine-scale mapping of the locus and consideration of rare variants identified by exome and genome sequencing may also shed light on these questions. Further support for the relevance of *BCL9* within the recurrent microdeletion locus at 1q21.1 for schizophrenia susceptibility comes from the discovery of common variants of *BCL9* that confer risk for schizophrenia through analysis of patients with schizophrenia in the Chinese Han population (Li *et al.*, 2011). However, the discovery of common variants significantly associating with schizophrenia risk in *BCL9* has so far not been replicated as part of the large meta-analysis of schizophrenia possibly due to differences of patient populations investigated and increased heterogeneity of larger sample sizes.

Another CNV associated with neurodevelopmental disorders that has been discovered as a *de novo* CNV in an autistic patient, and in an individual with bipolar disorder, is that of 1p36 deletions (Kirov, 2010). 1p36 deletion syndrome is one of the most common subtelomeric microdeletion syndromes (~1 in 5000 live births) that is characterized by intellectual disability, seizures, reduced speech capacity, visual and auditory deficits, developmental delay, limited speech ability, and distinctive facial features, including microcephaly. Within the 1p36 interval commonly (but not always) deleted in this syndrome is the human homolog of the *Drosophila* gene Dishevelled (Dsh), *DVL1*. While deletion of the mouse homolog of *DVL1* does not show a phenotype as severe as the 1q36.6 deletion syndrome in humans, these mice nonetheless show deficits in the recognition of social hierarchy and dominance, indicative of an alternation of neurocircuitry important for social behavior (Long *et al.*, 2004).

Certain CNVs associated with neurodevelopmental disorders also encompass multiple genes that may interact with the Wnt pathway. For example, microdeletions in the chromosomal region 3q29 have been associated with both autism spectrum disorders and schizophrenia (Mulle *et al.*, 2010; Willatt *et al.*, 2005). This interval includes the *PAK2* gene encoding the RAC effector protein p21 protein (Cdc42/Rac)-activated kinase 2 that has been demonstrated to be essential for canonical Wnt signaling through the phosphorylation and regulation of β-catenin-mediated transcription (Zhou *et al.*, 2011). Also within this same interval in 3q29 is *DLG1*, which encodes a protein known to bind to multiple Frizzled receptors (Fzd1, 2, 4, 7) through an interaction mediated by its PDZ domain (Hering and Sheng, 2002). Similarly, *de novo* microdeletions of *DLG2*, which can also interact with Frizzled receptors (Hering and Sheng, 2002), have been reported in multiple cases with schizophrenia (Kirov *et al.*, 2012).

Finally, one of the most prevalent and strongest genetic risk factors for schizophrenia is the 22q11.2 deletion syndrome, with roughly 25% of individuals who are hemizygous for the deletion having schizophrenia along with other physical abnormalities (Bassett *et al.*, 2008; Rodriguez-Murillo, Gogos, and Karayiorgou, 2012). The 1.5 – 4 Mb genomic interval spanned by the 22q11.2 deletion is known to contain multiple genes important for neurodevelopment, neurotransmission, and microRNA processing. This includes *TBX1*, which is known to be a transcriptional target of TCF/LEF-mediated transcription, that has been shown to be important for multiple aspects of the physical abnormalities as well as behavioral abnormalities using mice that have lost one copy of the gene (Hiramoto *et al.*, 2011; Huh and Ornitz, 2010; Paylor *et al.*, 2006). Furthermore, loss-of-function mutations in *TBX1* cosegregated with multiple neuropsychiatric phenotypes in a large family, including features of autism spectrum disorder (Paylor *et al.*, 2006). The 22q11.12 deletion also includes *PI4KA* (Vorstman *et al.*, 2009), a gene encoding a phosphatidylinositol 4-kinase catalytic subunit that phosphorylates phosphatidylinositol to produce phosphatidylinositol 4-phosphate in the first step of the biosynthesis of phosphatidylinositol 4,5-bisphosphate (PIP_2). PIP_2 is later acted upon by phosphatidylinositol-4,5-bisphosphate 3-kinases (PI3Ks), such as that encoded by the *PIK3CB* gene that, as described in the succeeding text, has been identified as containing *de novo* variants in a schizophrenia trios, to produce phosphatidylinositol-3,4,5-triphosphate (PIP_3) that as a second messenger can activate AKT in turn leading to alteration of GSK3 and potential cross talk with canonical Wnt signaling at multiple levels. Lastly, the 22q11.12 deletion causes a loss of one of the two copies of *MED15*, a gene encoding a component of the Mediator complex involved in the regulation of RNA polymerase II-dependent transcription (Napoli *et al.*, 2012). MED15 is also known to interact with MED12, which as described earlier regulates canonical Wnt signaling and is implicated in intellectual disability syndromes along with other Mediator complex members (Napoli *et al.*, 2012), and, as described in the succeeding text, binds to TRRAP (Finkbeiner *et al.*, 2008), a gene which has been identified as having a *de novo* variant in a schizophrenia family and is known to be involved in the regulation of the Wnt pathway transcriptional target gene *TCF4 (aka ITF-2).*

Common variants of Wnt pathway genes in neuropsychiatric disorders

Building on the approach that led to the identification of CNVs associated with increased risk for schizophrenia and other neurodevelopmental disorders such as autism, similar GWAS seeking to identify common variants (>1% in the population) in genes in neuropsychiatric disorders have been completed. While the early nature of these studies, and the relatively small sample sizes considering the potential heterogeneity and polygenecity of the disorders, necessitates follow-up validation of the functional effects of variants identified, we comment here on the ongoing studies due to the likely importance of the results in terms of shaping research on the role of Wnt signaling in neuropsychiatric disorders in the years to come.

Multiple recent human genetic studies using large samples of DNA from cases and controls have identified genome-wide significant associations between schizophrenia susceptibility and SNP variants in the intronic region of the *TCF4* gene (Entrez Gene ID: 6925) on chromosome 18q21.1 encoding a member of the basic helix–loop–helix (bHLH) family of transcription factors (Blake *et al.*, 2010; Li *et al.*, 2010; Purcell *et al.*, 2009; Stefansson *et al.*, 2009). *TCF4* (aka *ITF-2*), and not to be confused with the transcription factor 7-like 2 (T-cell specific, HMG-box) gene *TCF7L2* (Entrez Gene ID: 6934) that is a well-known regulator of Wnt-mediated transcription, has been shown to be a transcriptional target of Wnt signaling due to canonical

LEF family transcription factor-binding sites in its promoter that are capable of recruiting SWI/SNF family chromatin-modifying complexes family in a β-catenin-dependent manner (Feng, Lee, and Fearon, 2003; Kolligs *et al.*, 2002). The locus encoding *TCF4* is complex, with recent studies demonstrating that it encompasses 41 annotated exons with 21 alternative transcriptional start sites, and encodes multiple alternatively spliced isoforms with 18 potential alternative translational start sites and corresponding N-termini (Sepp *et al.*, 2011). *TCF4* is highly, but not exclusively, expressed in the brain particularly within the telencephalic regions in early development and the hippocampus, a region known to be critical to learning and memory, in adults (Brzozka *et al.*, 2010; Sepp *et al.*, 2011; Sepp, Pruunsild, and Timmusk, 2012). Consistent with this expression pattern, the overexpression of one isoform of Tcf4 in mice postnatally leads to deficits in contextual and cued fear conditioning along with sensorimotor gating measured operationally through assessment of prepulse inhibition (Brzozka *et al.*, 2010), a well-known endophenotype of neuropsychiatric disorders (Powell, Weber, and Geyer, 2012). Consistent with these findings in a rodent model, genetic variation in the *TCF4* locus in humans is associated with deficits in sensorimotor gating measured through prepulse inhibition as well, suggesting that TCF4 may have a gain of function of expression in schizophrenia (Lennertz *et al.*, 2011). However, the basis for the association of the *TCF4* locus with SCZ and the functionally relevant, causal SNP variant(s) has not yet been identified, leading to the need for further investigation of how the associated genetic variation may affect TCF4 function and if there are alterations in Wnt-mediated *TCF4* expression.

In further support of the relevance of the Wnt pathway target gene *TCF4* in schizophrenia pathogenesis, *TCF4* has recently been shown to be a functional target of the microRNA, miR-137, another gene implicated in the etiology of schizophrenia by recent GWAS (Kwon, Wang, Tsai, 2011; Ripke *et al.*, 2011). Furthermore, as discussed in the succeeding text, massively parallel exome sequencing in schizophrenia trios has identified *de novo* variants in two genes (*TRRAP*, *RUVBL1*) that have been shown to be involved in the transcriptional regulation of *TCF4*. The observation that two genome-wide significant loci and rare variation in schizophrenia appear to converge on a common pathway that is known to be under control of Wnt signaling strongly suggests that additional genes from loci discovered by GWAS and from massively parallel DNA sequencing will also implicate Wnt signaling further – a speculation that the rapid nature of advances in the field of neuropsychiatric disease genetics will undoubtedly soon be able to support or refute.

Although not yet as advanced as the genetic analysis in schizophrenia and autism in terms of exome sequencing, recently performed GWAS of bipolar disorder on over 10,000 individuals have identified genome-wide significant association between DNA variants in the Ankyrin 3 gene (*ANK3*) and susceptibility to bipolar disorder (Baum *et al.*, 2008; Ferreira *et al.*, 2008; Psychiatric GWAS Consortium Bipolar Disorder Working Group, 2011; Rueckert *et al.*, 2012; Schulze *et al.*, 2009). *ANK3* shows tissue and cell-specific variation in expression and encodes multiple protein isoforms (ankyrin-G) composed of several highly conserved domains with multiple functions (Bennett and Chen, 2001). Ankyrin-G is known to play a critical function as part of the axon initial segment and nodes of Ranvier that regulate diverse aspects of neurophysiology (Mohler and Bennett, 2005). Besides roles in ion channel regulation, ankyrin-G has also been shown to form a complex with the actin cytoskeleton binding protein β2-spectrin that binds to E-cadherin/β-catenin complexes at the cell membrane (Kizhatil *et al.*, 2007). Since E-cadherin recruits β-catenin to the cell membrane in a manner mutually exclusive with binding of LEF1, thereby preventing its nuclear localization and regulation of TCF/LEF-mediated transcription (Orsulic *et al.*, 1999), along with evidence for feedback loops between E-cadherin glycosylation and Wnt signaling (Sengupta *et al.*, 2012), as well as a role for other cadherins in the regulation of β-catenin-mediated Wnt signaling (Zhang *et al.*, 2010), the disruption of the ankyrin-G/cadherin complex due to *ANK3* genetic variation may lead to dysregulation of β-catenin and alteration of Wnt signaling that increase an individual's susceptibility to developing a neuropsychiatric disorder.

Single-nucleotide variants (SNVs) of Wnt pathway genes in neuropsychiatric disorders

In addition to rare CNVs and common polymorphic loci that have been identified by GWAS, much recent attention of late has turned to the use of high-throughput DNA sequencing approaches that allow the analysis of both exomes (full complement of coding regions of genes) and whole genomes (coding and noncoding regions) (1000 Genomes Project Consortium *et al.*, 2012). These studies generally follow one of two designs. The first design uses a family-based design where the affected proband is analyzed in addition to family members (ideally parents) in order to identify *de novo* variants that only occur in the affected proband and therefore can be reasoned to be more likely to be relevant to disease pathogenesis. Being able to identify and analyze such *de novo* genetic variation is an important departure from the investigation of inherited genetic variation in human diseases with the realization that the average human germline genome has 74 *de novo* SNVs, 3 *de novo* indels, 0.02 *de novo* CNVs, and 1 *de novo* mutation in the protein-coding region of the genome (Veltman and Brunner, 2012). The second design follows that of the association studies and relies on the analysis of the distribution of variants in a sufficiently large sample of cases and controls. Here, the sample size needs to be much larger in order to have sufficient statistical power to identify variants truly overrepresented in the context of disease. Overall, the nature of the design of these studies provides the potential for the implication of rare variation in specific genes known to play a role in the Wnt signaling pathway. This stands in contrast to GWAS that are capable of identifying common variation in genes such as the TCF/LEF target gene *TCF4*.

Excitingly, using exome sequencing, studies in schizophrenia and autism spectrum disorders have indeed begun to identify variants in multiple components of Wnt signaling pathway (Girard *et al.*, 2011; Need *et al.*, 2012; Xu *et al.*, 2012). Although the biological function of most of these variants has not been characterized to date, nonetheless, under the hypothesis that variants in the Wnt signaling pathway may be of critical importance to mental health, the investigation of these variants is warranted and will almost certainly be clarified in the years to come. We will summarize a few of the notable gene that appear to implicate specific aspects of Wnt signaling. As summarized in Figure 30.1, these variants when grouped together strongly implicate a role for chromatin remodeling and transcription under the control of Wnt signaling and regulated by Wnt signaling.

Schizophrenia exome variants in Wnt signaling pathway

In schizophrenia, *de novo* variants identified from the work of the Karayiorgou and Rouleau groups to date include the following (Girard *et al.*, 2011; Xu *et al.*, 2012; Table 30.1): (i) low-density lipoprotein 1 (*LRP1*), a member of the low-density lipoprotein receptor family that interacts with Frizzled receptors and, unlike the Wnt coreceptors LRP5/6, potentially represses Wnt signaling due to the sequestration of the Frizzled receptor (Zilberberg, Yaniv, and Gazit, 2004); (ii) *DAB2IP*, a protein that interacts with disabled 2 (DAB2), which also has a *de novo* variant found in an autism spectrum disorder patient, to antagonize Wnt/β-catenin signaling through activation of GSK3β by reducing S9 phosphorylation (Xie *et al.*, 2010); (iii) *UBR5* (also known as EDD) encoding an E3 ubiquitin ligase that ubiquitinates β-catenin leading to increased stability of β-catenin and upon interaction with β-catenin and GSK3β leads to their enhanced nuclear accumulation and increased transcriptional activity (Hay-Koren *et al.*, 2011); (iv) *VPS35* encoding a subunit of the retromer complex that functions to regulate the retrieval and reutilization of a key sorting receptor for Wnt, Wntless (*WLS*, G protein-coupled receptor 177), that is required for the recycling of Wnt from the membrane through the endosome to the *trans*-Golgi network, the loss of which leads to accumulation of Wntless in lysosomes and decreased Wnt signaling (Hay-Koren *et al.*, 2011; Port *et al.*, 2008); and (v) *TRRAP* encoding the transformation/transcription domain-associated protein, a component of multiple histone modifying and chromatin remodeling complexes that regulate Wnt target gene expression (Finkbeiner *et al.*, 2008; Sierra *et al.*, 2006), including that of the schizophrenia

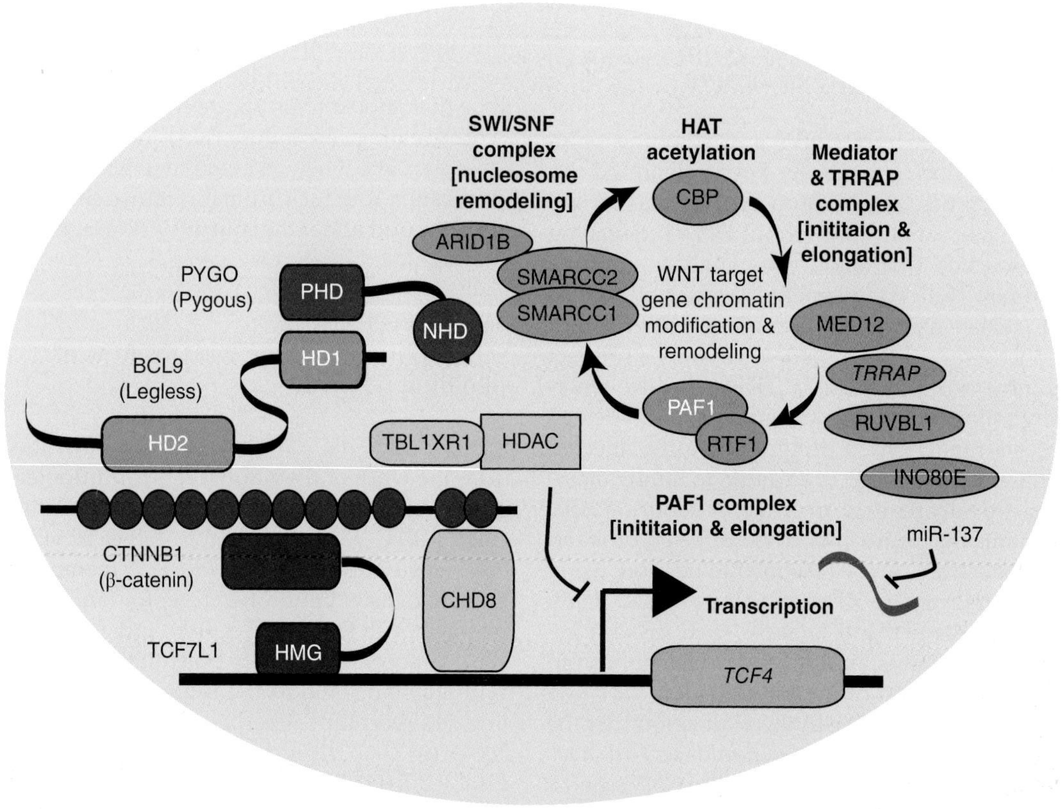

Figure 30.1 Wnt signaling pathway components involved in chromatin remodeling and transcription that have been implicated in neuropsychiatric disorders by human genetics, neuropharmacology, and functional genomic studies. *(See insert for color representation of the figure.)*

Table 30.1 *De novo* variants in Wnt pathway components identified in schizophrenia exome sequencing as of end of 2013

Gene symbol	Name	Entrez Gene ID
DAB2IP	DAB2 interacting protein	153090
LRP1	Low-density lipoprotein-related protein 1 (alpha-2-macroglobulin receptor)	4035
PIK3CB	Phosphoinositide-3-kinase, catalytic, beta polypeptide	5291
TRRAP	Transformation/transcription domain-associated protein	8295
UBR5	Ubiquitin protein ligase E3 component n-recognin 5	51366
VPS35	Vacuolar protein sorting 35 homolog	55737

susceptibility gene *TCF4* (Feng, Lee, and Fearon, 2003). TRRAP is also known to interact with the Skp–Cullin–F-box ubiquitin ligase complexes that meditate the ubiquination of β-catenin bound to chromatin such that its loss leads to abnormal retention of β-catenin at the chromatin and concomitant hyperactivation of the canonical Wnt pathway, as well as to interact with MED15, encoded by a gene within the 22q11.12 deletion interval, and MED12, encoded by a gene that, as discussed in the succeeding text, mutations cause an X-linked intellectual disability syndrome (Napoli *et al.*, 2012); and (vi) *PIK3CB* encoding a PI3K catalytic subunit that phosphorylates the 3′- hydroxyl group of the inositol ring of phosphatidylinositol-containing lipids, leading to elevated PIP_3 that as mentioned earlier, leads to activation of AKT that

Table 30.2 *De novo* variants in Wnt pathway components identified in autism spectrum disorder exome sequencing as of end of 2013

Gene symbol	Name	Entrez Gene ID
ARID1B	AT-rich interactive domain 1B (SWI1-like)	57492
CHD8	Chromodomain helicase DNA-binding protein 8	57680
CTNNB1	Catenin (cadherin-associated protein), beta 1, 88 kDa	1499
DAB2	Disabled homolog 2, mitogen-responsive phosphoprotein	1601
DYRK1A	Dual-specificity tyrosine-(Y)-phosphorylation-regulated kinase 1A	1859
LGR4	Leucine-rich repeat-containing G protein-coupled receptor 4	55366
LRP1	Low-density lipoprotein-related protein 1 (alpha-2-macroglobulin receptor)	4035
PTEN	Phosphatase and tensin homolog; phosphatase and tensin homolog	5728
PTK7	PTK7 protein tyrosine kinase 7	5754
RTF1	Rtf1, Paf1/RNA polymerase II complex component, homolog	23168
RUVBL1	RuvB-like 1	8607
SMARCC1	SWI/SNF-related, matrix-associated, actin-dependent regulator of chromatin, subfamily c, member 1	6599
SMARCC2	SWI/SNF-related, matrix-associated, actin-dependent regulator of chromatin, subfamily c, member 2	6601
TBL1XR1	Transducin (beta)-like 1 X-linked receptor 1	79718
TCF4	Transcription factor 4	6925
TCF7L1	Transcription factor 7-like 1 (T-cell specific, HMG-box)	83439
TRRAP	Transformation/transcription domain-associated protein	8295
TSC2	Tuberous sclerosis 2	7249
WNT5A	Wingless-type MMTV integration site family, member 5A	7474

can in turn phosphorylate and inhibit GSK3 and thereby provide cross talk between PI3K signaling pathway and canonical Wnt signaling at multiple levels.

Autism exome variants in Wnt signaling pathway

In the case of autism spectrum disorders, which, similar to schizophrenia, are considered complex neurodevelopmental disorders for which there is evidence of a significant genetic component to susceptibility (Rapoport, Giedd, and Gogtay, 2012), a larger sample size of trios has been investigated for *de novo* variants to date (Iossifov *et al.*, 2012; Neale *et al.*, 2012; O'Roak *et al.*, 2011, 2012a, b; Sanders *et al.*, 2012), with a number of compelling observations made that include well-known components, including (Table 30.2) (i) *WNT5A* encoding a Wnt ligand that activates the non-canonical arm of Wnt signaling through Frizzled receptors and DVL family members (Kikuchi *et al.*, 2012), (ii) *CTNNB1* encoding β-catenin (Clevers and Nusse, 2012), and (iii) *TCF7L1* (transcription factor 7-like 1) encoding a member of the TCF/LEF family of transcription factors (Clevers and Nusse, 2012). In addition, other components, particularly those involved in chromatin regulation (Figure 30.1), include the following: (i) *CHD8*, encoding the chromodomain helicase DNA-binding protein 8 that binds to β-catenin and histone H1 and suppresses transcription at TCF/LEF sites (Nishiyama *et al.*, 2009; Nishiyama, Skoultchi, and Nakayama, 2012); (ii) *TBL1XR1* (transducin beta-like 1 X-linked receptor 1, also known as *TBLR1*), which encodes a multifunctional protein on the X-chromosome that is known to be required for Wnt/β-catenin-mediated transcription (Choi *et al.*, 2011; Li and Wang, 2008), for which there is also evidence of common genetic variation associated with increased risk for developing an autism spectrum disorder in a closely related family member transducin (β)-like 1X-linked (*TBL1X*) (Chung *et al.*, 2011); (iii) *DYRK1A* encoding a dual-specificity tyrosine phosphorylation-regulated kinase that plays a positive role in the regulation of p120-catenin protein levels, which can bind and prevent inhibition of the β-catenin/TCF transcriptional complex by BTB/POZ zinc finger family member of transcription factors Kaiso (Hong *et al.*, 2012); and (iv) *TRRAP*, as mentioned earlier in schizophrenia, along with (v) *ARID1B* encoding a protein that forms a complex with proteins encoded by the *SMARCC1* and *SMARCC2* genes, both of which also have *de novo* variants and are part of the SWI/SNF chromatin remodeling complex that is important for Wnt-mediated transcription (Barker *et al.*, 2001). Taken together, these findings are beginning to outline an important role for the Wnt signaling pathway suggesting that alteration of the balance of both activation and repression of the pathway is important in the etiology of autism spectrum disorders and schizophrenia.

Neuropsychiatric disorders with a single gene cause and Wnt signaling

Besides complex polygenic neuropsychiatric disorders, such as schizophrenia and autism described earlier, where the role of Wnt signaling in disease etiology and pathogenesis requires further clarification, a number of single genes, Mendelian disorders with altered cognition, and behavioral features have been identified that point more clearly to a critical role of Wnt signaling in the function of the nervous system. As examples discussed in the succeeding text these include (i) Pitt–Hopkins syndrome (PTHS), (ii) fragile X syndrome (FXS), (iii) Rubinstein–Taybi syndrome (RTS), and (iv) Opitz–Kaveggia/Lujan syndrome.

(1) **Pitt–Hopkins syndrome and *TCF4*.** In addition to genetic variation affecting the noncoding intronic region of the Wnt pathway target gene *TCF4* that is, as discussed above, associated with schizophrenia, hemizygosity of this same *TCF4* gene causes PTHS (OMIM ID #610042) (Amiel *et al.*, 2007; Pitt and Hopkins, 1978; Zweier *et al.*, 2007), which is characterized by severe intellectual disability and autistic-like behavior (Hasi *et al.*, 2011; O'Donnell *et al.*, 2010; Talkowski *et al.*, 2012). Besides the originally identified deletions, nonsense (stop) mutations, splice-site mutations, and missense mutations have been found to cause PTHS with the mode of inheritance being that of autosomal dominant (Peippo and Ignatius, 2012). While the phenotypes are generally thought to be caused by haploinsufficiency, certain truncating and missense mutations have a dominant-negative mode of interaction given that TCF4 functions as both a homodimer and a heterodimer with other bHLH family transcription factors and interacts with other transcriptional regulatory machinery. The observed differences in severity, age of onset, and penetrance of PTHS versus schizophrenia may represent a human allelic series for the *TCF4* gene. The investigation of this overlap between PTHS and schizophrenia and determining whether Wnt signaling is disrupted in either or both cases will be an important avenue for research on human neurodevelopmental disorders that result in intellectual disability, given the potential to reveal shared cellular pathways and molecular mechanisms that underlie cognitive deficits and other common behavioral features of these disorders.

(2) **Fragile X syndrome, GSK3 dysregulation, and lithium.** FXS is the most common known genetic cause of autism and the most frequent known cause of intellectual disability that occurs due to an expanded CGG trinucleotide repeat in the 5′ untranslated region of the fragile X mental retardation (*FMR1*) gene on the X chromosome leading to epigenetic silencing and loss of expression of the fragile X mental retardation protein (FMRP) (Martin and Bell, 1943; Santoro, Bray, and Warren, 2012). Clinical symptoms of FXS patients include cognitive deficits ranging from mild to severely impaired intellectual abilities, hyperactivity, attention deficits, social difficulties, anxiety, and other autistic-like behaviors (Santoro, Bray, and Warren, 2012). Despite the known relationship between *FMR1* CGG-repeat expansion, *FMR1* silencing, and the role of FMRP as an RNA-binding protein involved in translational regulation, the consequences of the loss of FMRP on neurodevelopment and neuronal function still remain poorly understood but are beginning to come to light (Santoro, Bray, and Warren, 2012). In particular, evidence linking dysregulation of Wnt

signaling to the loss of FMRP has emerged from multiple sources. First, GSK3β mRNA was identified as one of the top FMRP targets through CLIP-Seq studies by Darnell *et al.* (2011). Consistent with these findings, Fmrp has been shown to directly control the translation of GSK3β such that in the absence of Fmrp-mediated translational repression, elevated GSK3 activity decreases the activation state of the Wnt signaling network due to downregulation of β-catenin levels (Luo *et al.*, 2010). As shown in this study, the decreased TCF/LEF-dependent transcription, the balance between proliferation and subsequent fate specification into differentiated neurons or astrocytes is altered. These findings are consistent with the observed adult neurogenesis phenotype seen in FXS knockout mice and may be relevant for the learning and neurogenesis deficits also seen when Fmrp was selectively ablated in adult neural stem cells (Guo *et al.*, 2012). Furthermore, these findings are highly reminiscent of observations made using recently described human-induced pluripotent stem cell model of FXS which was reported to exhibit aberrant neurodevelopment upon differentiation, including elevated GFAP levels and decreased neuronal differentiation (Sheridan *et al.*, 2011), where aberrant response to Wnt pathway activation has also been observed (S.J.H., unpublished observation).

In further support of the notion of dysregulated Wnt signaling as being relevant to FXS, elevated GSK3 activity has been observed upon loss of FMRP as measured by decreased inhibitory serine phosphorylation of GSK3 in multiple brain regions (Min *et al.*, 2009; Mines *et al.*, 2010; Yuskaitis *et al.*, 2010). In terms of a potential causal role for this dysregulated GSK3 activity in FXS pathogenesis, lithium treatment was first shown to improve behavioral dysfunction in a *Drosophila* model of FXS (McBride *et al.*, 2005), as well as now in multiple FXS mouse models where lithium has been shown to attenuate hyperactivity, decrease sensitivity to audiogenic seizures, improve sociability, and increase learning in tests of hippocampus-dependent learning and memory (Guo *et al.*, 2012; Liu, Chuang, and Smith, 2011; Min *et al.*, 2009; Mines *et al.*, 2010; Yuskaitis *et al.*, 2010). Besides normalizing behavioral deficits, lithium treatment was able to normalize dendritic spine abnormalities (Guo *et al.*, 2012), a hallmark of FXS pathology. Interestingly, while elevated metabotropic glutamate receptor 5 (mGluR5) activity has been shown to be caused by the loss of Fmrp (Dolen *et al.*, 2007), acute treatment with an mGluR5 antagonist, MPEP, was found to increase inhibitory serine phosphorylation of GSK3 in multiple brain regions of *Fmr1* knockout mice but not in wild-type mice (Yuskaitis *et al.*, 2010). This provides an example of cross talk between glutamatergic signaling and GSK3 signaling, similar to what has been observed for another metabotropic glutamate receptor, mGluR1 in other cellular contexts. Finally, an open-label, clinical trial of lithium in FXS patients showed that it had beneficial effects (Berry-Kravis *et al.*, 2008), a result that awaits replication with additional larger and placebo-controlled trials, and potentially with other Wnt pathway targeting agents. Taken together, these data from multiple model systems suggest that enhancing Wnt signaling may be beneficial for FXS and other autism spectrum disorders.

(3) **Wnt signaling and Rubinstein–Taybi syndrome**. The loss of function of one or more copies of the *CREBBP* gene on chromosome 16p13.3 encoding CREB-binding protein (CBP) leads to RTS, a rare, childhood onset, neurodevelopmental disorder. Patients with RTS have characteristic broad thumbs and toes, short stature, and facial features along with varying degrees of intellectual disability. CBP has been shown to have histone acetyltransferase activity and to function as a transcriptional coactivator that is critically involved in the regulation of Wnt signaling pathway (Hecht *et al.*, 2000; Takemaru and Moon, 2000). CBP directly associates with β-catenin that upon stabilization and nuclear accumulation assembles with LEF/TCF transcription factors and other transcriptional cofactors to

activate Wnt target genes through the acetylation of histone and potentially other factors both locally at promoters and at a distance (Hecht *et al.*, 2000; Takemaru and Moon, 2000). In doing so, β-catenin/CBP overcomes repressive chromatin states established by repression of Groucho/TLE proteins that function to recruit antagonizing corepressor activities, such as members of the histone deacetylase (HDAC) family (Willert and Jones, 2006). CBP is also known to directly acetylate β-catenin at Lys49 leading to an increased ability of β-catenin to activate certain promoters but not others (Wolf *et al.*, 2002). Taken as a whole, while not directly demonstrated to date in the context of the CNS, the loss of CBP function in RTS may presumably lead to defects in Wnt signaling by disrupting the normal balance of transcriptional coactivation and repressive pathways.

(4) **Opitz–Kaveggia/Lujan syndrome and Wnt signaling.** In addition to the transcriptional coactivator CBP, other transcriptional regulatory proteins involved in Wnt signaling have been found implicated in neuropsychiatric disorders. One example includes the *MED12* gene encoding a subunit of the Mediator complex that functions to bridge sequence-specific DNA-binding transcription factors to the RNA polymerase II transcriptional machinery (Napoli *et al.*, 2012), which interacts with Pygopus and has been shown to be essential for both canonical and noncanonical Wnt signaling (Carrera *et al.*, 2008; Kim *et al.*, 2006; Rocha *et al.*, 2010). Recurrent mutations in *MED12* at Xq13 have been found to be associated with an X linked intellectual disability syndrome with multiple craniofacial, musculoskeletal, and behavioral abnormalities known as Opitz–Kaveggia or Lujan syndrome (Napoli *et al.*, 2012). Additionally, a 12 bp insertional polymorphism in one of the C-terminal coding exons of MED12 has been found associated with an increased risk for schizophrenia in subjects of European ancestry (Philibert *et al.*, 2007). However, perhaps owing to the nature of the polymorphism, there has not been a consistent association of this 12 bp polymorphisms to MED12 to schizophrenia in recent GWAS studies or in exome sequencing studies. However, as mentioned earlier, variants in a MED12-binding protein TRRAP that also binds β-catenin have been identified by whole-exome sequencing in schizophrenia, and MED12 also binds to MED15, a gene within the 22q11.12 deletion interval that has a strong association with increased schizophrenia susceptibility. Finally, mutations in another Mediator complex subunit, MED23, have also been identified to cause intellectual disability (Hashimoto *et al.*, 2011; Napoli *et al.*, 2012), pointing to a key role of the Mediator complex in human cognition, although whether there is dysregulated Wnt signaling in this case is presently unknown.

Summary

As the genetic architecture of individual neuropsychiatric disorders comes into clearer focus in the coming years through the use of massively parallel, next-generation sequencing technology (Sullivan, Daly, and O'Donovan, 2012), this will afford the opportunity to assess in an unbiased manner whether there is a statistical enrichment of variants in Wnt pathway genes in the context of disease risk or prevention. This will need to follow with considerable effort to assess the functional consequence of the variants and the resulting phenotypes. In doing so, this will answer the intriguing question of whether the pharmacology of lithium in terms of its regulation of Wnt signaling in the nervous system is at all related to the etiology of the disorders in which it is therapeutically active. Although lithium is not an antipsychotic for humans, its ability to modulate the neurocircuitry regulated by the D2/β-arrestin/AKT pathway and to affect GSK3 activity leads the speculation that additional components of Wnt signaling will be identified by human genetic studies that point to a clear etiological role Wnt signaling and additional examples of cross talk between canonical and noncanonical branches of the Wnt pathway.

Acknowledgements

Studies in the Haggarty Laboratory on the role of Wnt signaling in neuropsychiatric disorders and dementia have been generously supported by the National Institute of Mental Health, the National Institute of Aging, the National Institute of Drug Addiction, the Stanley Medical Research Institute, and the Tau Consortium. We thank the laboratory of Randall Moon (UW), Li-Huei Tsai (MIT), and members of the Stanley Center for Psychiatric Research for helpful discussions. We sincerely apologize for not being able to cite due to space limitations a number of additional papers that have contributed to the concepts discussed herein.

References

Amiel, J., Rio, M., de Pontual, L. *et al.* (2007) Mutations in TCF4, encoding a class I basic helix–loop–helix transcription factor, are responsible for Pitt-Hopkins syndrome, a severe epileptic encephalopathy associated with autonomic dysfunction. *American Journal of Human Genetics*, **80**, 988–993.

Barker, N., Hurlstone, A., Musisi, H. *et al.* (2001) The chromatin remodelling factor Brg-1 interacts with beta-catenin to promote target gene activation. *The EMBO Journal*, **20**, 4935–4943.

Bassett, A.S., Marshall, C.R., Lionel, A.C. *et al.* (2008) Copy number variations and risk for schizophrenia in 22q11.2 deletion syndrome. *Human Molecular Genetics*, **17**, 4045–4053.

Baum, A.E., Akula, N., Cabanero, M. *et al.* (2008) A genome-wide association study implicates diacylglycerol kinase eta (DGKH) and several other genes in the etiology of bipolar disorder. *Molecular Psychiatry*, **13**, 197–207.

Bennett, V. and Chen, L. (2001) Ankyrins and cellular targeting of diverse membrane proteins to physiological sites. *Current Opinion in Cell Biology*, **13**, 61–67.

Berry-Kravis, E., Sumis, A., Hervey, C. *et al.* (2008) Open-label treatment trial of lithium to target the underlying defect in fragile X syndrome. *Journal of Developmental and Behavioral Pediatrics*, **29**, 293–302.

Blackwood, D.H., Fordyce, A., Walker, M.T. *et al.* (2001) Schizophrenia and affective disorders – cosegregation with a translocation at chromosome 1q42 that directly disrupts brain-expressed genes: clinical and P300 findings in a family. *American Journal of Human Genetics*, **69**, 428–433.

Blackwood, D.H., Thiagarajah, T., Malloy, P. *et al.* (2008) Chromosome abnormalities, mental retardation and the search for genes in bipolar disorder and schizophrenia. *Neurotoxicity Research*, **14**, 113–120.

Blake, D.J., Forrest, M., Chapman, R.M. *et al.* (2010) TCF4, schizophrenia, and Pitt-Hopkins syndrome. *Schizophrenia Bulletin*, **36**, 443–447.

Brzozka, M.M., Radyushkin, K., Wichert, S.P. *et al.* (2010) Cognitive and sensorimotor gating impairments in transgenic mice overexpressing the schizophrenia susceptibility gene Tcf4 in the brain. *Biological Psychiatry*, **68**, 33–40.

Carrera, I., Janody, F., Leeds, N. *et al.* (2008) Pygopus activates Wingless target gene transcription through the mediator complex subunits Med12 and Med13. *Proceedings of the National Academy of Sciences of the United States of America*, **105**, 6644–6649.

Chenn, A. and Walsh, C.A. (2002) Regulation of cerebral cortical size by control of cell cycle exit in neural precursors. *Science*, **297**, 365–369.

Choi, H.K., Choi, K.C., Yoo, J.Y. *et al.* (2011) Reversible SUMOylation of TBL1-TBLR1 regulates beta-catenin-mediated Wnt signaling. *Molecular Cell*, **43**, 203–216.

Chung, R.H., Ma, D., Wang, K. *et al.* (2011) An X chromosome-wide association study in autism families identifies TBL1X as a novel autism spectrum disorder candidate gene in males. *Molecular Autism*, **2**, 18.

Clevers, H. and Nusse, R. (2012) Wnt/beta-catenin signaling and disease. *Cell*, **149**, 1192–1205.

Darnell, J.C., Van Driesche, S.J., Zhang, C. *et al.* (2011) FMRP stalls ribosomal translocation on mRNAs linked to synaptic function and autism. *Cell*, **146**, 247–261.

De Rienzo, G., Bishop, J.A., Mao, Y. *et al.* (2011) Disc1 regulates both beta-catenin-mediated and noncanonical Wnt signaling during vertebrate embryogenesis. *The FASEB Journal*, **25**, 4184–4197.

Dolen, G., Osterweil, E., Rao, B.S. *et al.* (2007) Correction of fragile X syndrome in mice. *Neuron*, **56**, 955–962.

Feng, Y., Lee, N., and Fearon, E.R. (2003) TIP49 regulates beta-catenin-mediated neoplastic transformation and T-cell factor target gene induction via effects on chromatin remodeling. *Cancer Research*, **63**, 8726–8734.

Ferreira, M.A., O'Donovan, M.C., Meng, Y.A. *et al.* (2008) Collaborative genome-wide association analysis supports a role for ANK3 and CACNA1C in bipolar disorder. *Nature Genetics*, **40**, 1056–1058.

Finkbeiner, M.G., Sawan, C., Ouzounova, M. *et al.* (2008) HAT cofactor TRRAP mediates beta-catenin ubiquitination on the chromatin and the regulation of the canonical Wnt pathway. *Cell Cycle*, **7**, 3908–3914.

1000 Genomes Project Consortium, Abecasis, G.R., Auton, A. *et al.* (2012) An integrated map of genetic variation from 1,092 human genomes. *Nature*, **491**, 56–65.

Girard, S.L., Gauthier, J., Noreau, A. *et al.* (2011) Increased exonic de novo mutation rate in individuals with schizophrenia. *Nature Genetics*, **43**, 860–863.

Girirajan, S., Campbell, C.D., and Eichler, E.E. (2011) Human copy number variation and complex genetic disease. *Annual Review of Genetics*, **45**, 203–226.

Guo, W., Murthy, A.C., Zhang, L. *et al.* (2012) Inhibition of GSK3beta improves hippocampus-dependent learning and rescues neurogenesis in a mouse model of fragile X syndrome. *Human Molecular Genetics*, **21**, 681–691.

Hashimoto, S., Boissel, S., Zarhrate, M. *et al.* (2011) MED23 mutation links intellectual disability to dysregulation of immediate early gene expression. *Science*, **333**, 1161–1163.

Hasi, M., Soileau, B., Sebold, C. *et al.* (2011) The role of the TCF4 gene in the phenotype of individuals with 18q segmental deletions. *Human Genetics*, **130**, 777–787.

Hay-Koren, A., Caspi, M., Zilberberg, A., and Rosin-Arbesfeld, R. (2011) The EDD E3 ubiquitin ligase ubiquitinates and up-regulates beta-catenin. *Molecular Biology of the Cell*, **22**, 399–411.

Hecht, A., Vleminckx, K., Stemmler, M.P. *et al.* (2000) The p300/CBP acetyltransferases function as transcriptional coactivators of beta-catenin in vertebrates. *The EMBO Journal*, **19**, 1839–1850.

Hering, H. and Sheng, M. (2002) Direct interaction of Frizzled-1, -2, -4, and -7 with PDZ domains of PSD-95. *FEBS Letters*, **521**, 185–189.

Hiramoto, T., Kang, G., Suzuki, G. *et al.* (2011) Tbx1: identification of a 22q11.2 gene as a risk factor for autism spectrum disorder in a mouse model. *Human Molecular Genetics*, **20**, 4775–4785.

Hong, J.Y., Park, J.I., Lee, M. *et al.* (2012) Down's-syndrome-related kinase Dyrk1A modulates the p120-catenin-Kaiso trajectory of the Wnt signaling pathway. *Journal of Cell Science*, **125**, 561–569.

Huh, S.H. and Ornitz, D.M. (2010) Beta-catenin deficiency causes DiGeorge syndrome-like phenotypes through regulation of Tbx1. *Development*, **137**, 1137–1147.

Iossifov, I., Ronemus, M., Levy, D. *et al.* (2012) De novo gene disruptions in children on the autistic spectrum. *Neuron*, **74**, 285–299.

Ishizuka, K., Kamiya, A., Oh, E.C. *et al.* (2011) DISC1-dependent switch from progenitor proliferation to migration in the developing cortex. *Nature*, **473**, 92–96.

Kendler, K.S. (2001) Twin studies of psychiatric illness: an update. *Archives of General Psychiatry*, **58**, 1005–1014.

Kessler, R., Hausmann, G., and Basler, K. (2009) The PHD domain is required to link Drosophila pygopus to legless/beta-catenin and not to histone H3. *Mechanisms of Development*, **126**, 752–759.

Kikuchi, A., Yamamoto, H., Sato, A., and Matsumoto, S. (2012) Wnt5a: its signalling, functions and implication in diseases. *Acta Physiologica*, **204**, 17–33.

Kim, S., Xu, X., Hecht, A., and Boyer, T.G. (2006) Mediator is a transducer of Wnt/beta-catenin signaling. *The Journal of Biological Chemistry*, **281**, 14066–14075.

Kirov, G. (2010) The role of copy number variation in schizophrenia. *Expert Review of Neurotherapeutics*, **10**, 25–32.

Kirov, G., Pocklington, A.J., Holmans, P. *et al.* (2012) De novo CNV analysis implicates specific abnormalities of postsynaptic signalling complexes in the pathogenesis of schizophrenia. *Molecular Psychiatry*, **17**, 142–153.

Kivimae, S., Martin, P.M., Kapfhamer, D. *et al.* (2011) Abnormal behavior in mice mutant for the Disc1 binding partner, Dixdc1. *Translational Psychiatry*, **1**, e43.

Kizhatil, K., Davis, J.Q., Davis, L. *et al.* (2007) Ankyrin-G is a molecular partner of E-cadherin in epithelial cells and early embryos. *The Journal of Biological Chemistry*, **282**, 26552–26561.

Kolligs, F.T., Nieman, M.T., Winer, I. *et al.* (2002) ITF-2, a downstream target of the Wnt/TCF pathway, is activated in human cancers with beta-catenin defects and promotes neoplastic transformation. *Cancer Cell*, **1**, 145–155.

Kramps, T., Peter, O., Brunner, E. *et al.* (2002) Wnt/wingless signaling requires BCL9/legless-mediated recruitment of pygopus to the nuclear beta-catenin-TCF complex. *Cell*, **109**, 47–60.

Kwon, E., Wang, W., and Tsai, L.H. (2011) Validation of schizophrenia-associated genes CSMD1, C10orf26, CACNA1C and TCF4 as miR-137 targets. *Molecular Psychiatry*, **18**, 11–12.

Lennertz, L., Quednow, B.B., Benninghoff, J. *et al.* (2011). Impact of TCF4 on the genetics of schizophrenia. *European Archives of Psychiatry and Clinical Neuroscience*, **261** (Suppl 2), S161–S165.

Li, J. and Wang, C.Y. (2008) TBL1-TBLR1 and beta-catenin recruit each other to Wnt target-gene promoter for transcription activation and oncogenesis. *Nature Cell Biology*, **10**, 160–169.

Li, T., Li, Z., Chen, P. *et al.* (2010) Common variants in major histocompatibility complex region and TCF4 gene are significantly associated with schizophrenia in Han Chinese. *Biological Psychiatry*, **68**, 671 673.

Li, J., Zhou, G., Ji, W. *et al.* (2011) Common variants in the BCL9 gene conferring risk of schizophrenia. *Archives of General Psychiatry*, **68**, 232–240.

Lipina, T.V., Kaidanovich-Beilin, O., Patel, S. *et al.* (2011) Genetic and pharmacological evidence for schizophrenia-related Disc1 interaction with GSK-3. *Synapse*, **65**, 234–248.

Lipina, T.V., Wang, M., Liu, F., and Roder, J.C. (2012) Synergistic interactions between PDE4B and

GSK-3: DISC1 mutant mice. *Neuropharmacology*, **62**, 1252–1262.

Liu, Z.H., Chuang, D.M., and Smith, C.B. (2011) Lithium ameliorates phenotypic deficits in a mouse model of fragile X syndrome. *The International Journal of Neuropsychopharmacology*, **14**, 618–630.

Long, J.M., LaPorte, P., Paylor, R., and Wynshaw-Boris, A. (2004) Expanded characterization of the social interaction abnormalities in mice lacking Dvl1. *Genes, Brain, and Behavior*, **3**, 51–62.

Luo, Y., Shan, G., Guo, W. *et al.* (2010) Fragile x mental retardation protein regulates proliferation and differentiation of adult neural stem/progenitor cells. *PLoS Genetics*, **6**, e1000898.

Manolio, T.A., Brooks, L.D., and Collins, F.S. (2008) A HapMap harvest of insights into the genetics of common disease. *The Journal of Clinical Investigation*, **118**, 1590–1605.

Mao, Y., Ge, X., Frank, C.L. *et al.* (2009) Disrupted in schizophrenia 1 regulates neuronal progenitor proliferation via modulation of GSK3beta/beta-catenin signaling. *Cell*, **136**, 1017–1031.

Martin, J.P. and Bell, J. (1943) A pedigree of mental defect showing sex-linkage. *Journal of Neurology and Psychiatry*, **6**, 154–157.

McBride, S.M., Choi, C.H., Wang, Y. *et al.* (2005) Pharmacological rescue of synaptic plasticity, courtship behavior, and mushroom body defects in a Drosophila model of fragile X syndrome. *Neuron*, **45**, 753–764.

Millar, J.K., Christie, S., Anderson, S. *et al.* (2001) Genomic structure and localisation within a linkage hotspot of Disrupted In Schizophrenia 1, a gene disrupted by a translocation segregating with schizophrenia. *Molecular Psychiatry*, **6**, 173–178.

Min, W.W., Yuskaitis, C.J., Yan, Q. *et al.* (2009) Elevated glycogen synthase kinase-3 activity in Fragile X mice: key metabolic regulator with evidence for treatment potential. *Neuropharmacology*, **56**, 463–472.

Mines, M.A., Yuskaitis, C.J., King, M.K. *et al.* (2010) GSK3 influences social preference and anxiety-related behaviors during social interaction in a mouse model of fragile X syndrome and autism. *PLoS One*, **5**, e9706.

Mohler, P.J. and Bennett, V. (2005) Defects in ankyrin-based cellular pathways in metazoan physiology. *Frontiers in Bioscience*, **10**, 2832–2840.

Muir, W.J., Pickard, B.S., and Blackwood, D.H. (2008) Disrupted-in-schizophrenia-1. *Current Psychiatry Reports*, **10**, 140–147.

Mulle, J.G., Dodd, A.F., McGrath, J.A. *et al.* (2010) Microdeletions of 3q29 confer high risk for schizophrenia. *American Journal of Human Genetics*, **87**, 229–236.

Napoli, C., Sessa, M., Infante, T., and Casamassimi, A. (2012) Unraveling framework of the ancestral Mediator complex in human diseases. *Biochimie*, **94**, 579–587.

Neale, B.M., Kou, Y., Liu, L. *et al.* (2012) Patterns and rates of exonic de novo mutations in autism spectrum disorders. *Nature*, **485**, 242–245.

Need, A.C., McEvoy, J.P., Gennarelli, M. *et al.* (2012) Exome sequencing followed by large-scale genotyping suggests a limited role for moderately rare risk factors of strong effect in schizophrenia. *American Journal of Human Genetics*, **91**, 303–312.

Nishiyama, M., Skoultchi, A.I., and Nakayama, K.I. (2012) Histone H1 recruitment by CHD8 is essential for suppression of the Wnt-beta-catenin signaling pathway. *Molecular and Cellular Biology*, **32**, 501–512.

Nishiyama, M., Oshikawa, K., Tsukada, Y. *et al.* (2009) CHD8 suppresses p53-mediated apoptosis through histone H1 recruitment during early embryogenesis. *Nature Cell Biology*, **11**, 172–182.

O'Donnell, L., Soileau, B., Heard, P. *et al.* (2010) Genetic determinants of autism in individuals with deletions of 18q. *Human Genetics*, **128**, 155–164.

O'Roak, B.J., Deriziotis, P., Lee, C. *et al.* (2011) Exome sequencing in sporadic autism spectrum disorders identifies severe de novo mutations. *Nature Genetics*, **43**, 585–589.

O'Roak, B.J., Vives, L., Fu, W. *et al.* (2012a) Multiplex targeted sequencing identifies recurrently mutated genes in autism spectrum disorders. *Science*, **338**, 1619–1622.

O'Roak, B.J., Vives, L., Girirajan, S. *et al.* (2012b) Sporadic autism exomes reveal a highly interconnected protein network of de novo mutations. *Nature*, **485**, 246–250.

Orsulic, S., Huber, O., Aberle, H. *et al.* (1999) E-cadherin binding prevents beta-catenin nuclear localization and beta-catenin/LEF-1-mediated transactivation. *Journal of Cell Science*, **112** (Pt 8), 1237–1245.

Paylor, R., Glaser, B., Mupo, A. *et al.* (2006) Tbx1 haploinsufficiency is linked to behavioral disorders in mice and humans: implications for 22q11 deletion syndrome. *Proceedings of the National Academy of Sciences of the United States of America*, **103**, 7729–7734.

Peippo, M. and Ignatius, J. (2012) Pitt-Hopkins syndrome. *Molecular Syndromology*, **2**, 171–180.

Philibert, R.A., Bohle, P., Secrest, D. *et al.* (2007) The association of the HOPA(12bp) polymorphism with schizophrenia in the NIMH Genetics Initiative for Schizophrenia sample. *American Journal of Medical Genetics Part B*, **144B**, 743–747.

Pitt, D. and Hopkins, I. (1978) A syndrome of mental retardation, wide mouth and intermittent overbreathing. *Australian Paediatric Journal*, **14**, 182–184.

Port, F., Kuster, M., Herr, P. *et al.* (2008) Wingless secretion promotes and requires retromer-dependent cycling of Wntless. *Nature Cell Biology*, **10**, 178–185.

Powell, S.B., Weber, M., and Geyer, M.A. (2012) Genetic models of sensorimotor gating: relevance to neuropsychiatric disorders. *Current Topics in Behavioural Neuroscience*, **12**, 251–318.

Psychiatric GWAS Consortium Bipolar Disorder Working Group (2011) Large-scale genome-wide association analysis of bipolar disorder identifies a new susceptibility locus near ODZ4. *Nature Genetics*, **43**, 977–983.

Psychiatric GWAS Consortium Coordinating Committee, Cichon, S., Craddock, N. et al. (2009) Genomewide association studies: history, rationale, and prospects for psychiatric disorders. *The American Journal of Psychiatry*, **166**, 540–556.

Purcell, S.M., Wray, N.R., Stone, J.L. *et al.* (2009) Common polygenic variation contributes to risk of schizophrenia and bipolar disorder. *Nature*, **460**, 748–752.

Rapoport, J.L., Giedd, J.N., and Gogtay, N. (2012) Neurodevelopmental model of schizophrenia: update 2012. *Molecular Psychiatry*, **17**, 1228–1238.

Raychaudhuri, S. (2011) Mapping rare and common causal alleles for complex human diseases. *Cell*, **147**, 57–69.

Ripke, S., Sanders, A.R., Kendler, K.S. *et al.* (2011) Genome-wide association study identifies five new schizophrenia loci. *Nature Genetics*, **43**, 969–976.

Rocha, P.P., Scholze, M., Bleiss, W., and Schrewe, H. (2010) Med12 is essential for early mouse development and for canonical Wnt and Wnt/PCP signaling. *Development*, **137**, 2723–2731.

Rodriguez-Murillo, L., Gogos, J.A., and Karayiorgou, M. (2012) The genetic architecture of schizophrenia: new mutations and emerging paradigms. *Annual Review of Medicine*, **63**, 63–80.

Rueckert, E.H., Barker, D., Ruderfer, D. *et al.* (2012) Cis-acting regulation of brain-specific ANK3 gene expression by a genetic variant associated with bipolar disorder. *Molecular Psychiatry*, **18**, 922–929.

Sachs, N.A., Sawa, A., Holmes, S.E. *et al.* (2005) A frameshift mutation in Disrupted in Schizophrenia 1 in an American family with schizophrenia and schizoaffective disorder. *Molecular Psychiatry*, **10**, 758–764.

Sanders, S.J., Murtha, M.T., Gupta, A.R. *et al.* (2012) De novo mutations revealed by whole-exome sequencing are strongly associated with autism. *Nature*, **485**, 237–241.

Santoro, M.R., Bray, S.M., and Warren, S.T. (2012) Molecular mechanisms of fragile X syndrome: a twenty-year perspective. *Annual Review of Pathology*, **7**, 219–245.

Schizophrenia Psychiatric Genome-Wide Association Study Consortium (2011) Genome-wide association study identifies five new schizophrenia loci. *Nature Genetics*, **43**, 969–976.

Schulze, T.G., Detera-Wadleigh, S.D., Akula, N. *et al.* (2009) Two variants in Ankyrin 3 (ANK3) are independent genetic risk factors for bipolar disorder. *Molecular Psychiatry*, **14**, 487–491.

Sengupta, P.K., Bouchie, M.P., Nita-Lazar, M. *et al.* (2012) Coordinate regulation of N-glycosylation gene DPAGT1, canonical Wnt signaling and E-cadherin adhesion. *Journal of Cell Science*, **126**, 484–496.

Sepp, M., Pruunsild, P., and Timmusk, T. (2012) Pitt-Hopkins syndrome-associated mutations in TCF4 lead to variable impairment of the transcription factor function ranging from hypomorphic to dominant-negative effects. *Human Molecular Genetics*, **21**, 2873–2888.

Sepp, M., Kannike, K., Eesmaa, A. *et al.* (2011) Functional diversity of human basic helix–loop–helix transcription factor TCF4 isoforms generated by alternative 5′ exon usage and splicing. *PLoS One*, **6**, e22138.

Sheridan, S.D., Theriault, K.M., Reis, S.A. *et al.* (2011) Epigenetic characterization of the FMR1 gene and aberrant neurodevelopment in human induced pluripotent stem cell models of fragile X syndrome. *PLoS One*, **6**, e26203.

Sierra, J., Yoshida, T., Joazeiro, C.A., and Jones, K.A. (2006) The APC tumor suppressor counteracts beta-catenin activation and H3K4 methylation at Wnt target genes. *Genes & Development*, **20**, 586–600.

Singh, K.K., Ge, X., Mao, Y. *et al.* (2010) Dixdc1 is a critical regulator of DISC1 and embryonic cortical development. *Neuron*, **67**, 33–48.

Singh, K.K., De Rienzo, G., Drane, L. *et al.* (2011) Common DISC1 polymorphisms disrupt Wnt/GSK3beta signaling and brain development. *Neuron*, **72**, 545–558.

Soares, D.C., Carlyle, B.C., Bradshaw, N.J., and Porteous, D.J. (2011) DISC1: structure, function, and therapeutic potential for major mental illness. *ACS Chemical Neuroscience*, **2**, 609–632.

Stefansson, H., Ophoff, R.A., Steinberg, S. *et al.* (2009) Common variants conferring risk of schizophrenia. *Nature*, **460**, 744–747.

Sullivan, P.F., Daly, M.J., and O'Donovan, M. (2012) Genetic architectures of psychiatric disorders: the emerging picture and its implications. *Nature Reviews Genetics*, **13**, 537–551.

Takemaru, K.I. and Moon, R.T. (2000) The transcriptional coactivator CBP interacts with beta-catenin to activate gene expression. *The Journal of Cell Biology*, **149**, 249–254.

Talkowski, M.E., Rosenfeld, J.A., Blumenthal, I. *et al.* (2012) Sequencing chromosomal abnormalities reveals neurodevelopmental loci that confer risk across diagnostic boundaries. *Cell*, **149**, 525–537.

Veltman, J.A. and Brunner, H.G. (2012) De novo mutations in human genetic disease. *Nature Reviews Genetics*, **13**, 565–575.

Vorstman, J.A., Chow, E.W., Ophoff, R.A. *et al.* (2009) Association of the PIK4CA schizophrenia-susceptibility gene in adults with the 22q11.2 deletion syndrome. *American Journal of Medical Genetics Part B*, **150B**, 430–433.

Willatt, L., Cox, J., Barber, J. *et al.* (2005) 3q29 microdeletion syndrome: clinical and molecular characterization of a new syndrome. *American Journal of Human Genetics*, **77**, 154–160.

Willert, K. and Jones, K.A. (2006) Wnt signaling: is the party in the nucleus? *Genes & Development*, **20**, 1394–1404.

Wolf, D., Rodova, M., Miska, E.A. *et al.* (2002) Acetylation of beta-catenin by CREB-binding protein (CBP). *The Journal of Biological Chemistry*, **277**, 25562–25567.

Xie, D., Gore, C., Liu, J. *et al.* (2010) Role of DAB2IP in modulating epithelial-to-mesenchymal transition and prostate cancer metastasis. *Proceedings of the National Academy of Sciences of the United States of America*, **107**, 2485–2490.

Xu, B., Ionita-Laza, I., Roos, J.L. *et al.* (2012) De novo gene mutations highlight patterns of genetic and neural complexity in schizophrenia. *Nature Genetics*, **44**, 1365–1369.

Yan, Q.J., Rammal, M., Tranfaglia, M., and Bauchwitz, R.P. (2005) Suppression of two major Fragile X Syndrome mouse model phenotypes by the mGluR5 antagonist MPEP. *Neuropharmacology*, **49**, 1053–1066.

Yuskaitis, C.J., Mines, M.A., King, M.K. *et al.* (2010) Lithium ameliorates altered glycogen synthase kinase-3 and behavior in a mouse model of fragile X syndrome. *Biochemical Pharmacology*, **79**, 632–646.

Zhang, J., Woodhead, G.J., Swaminathan, S.K. *et al.* (2010) Cortical neural precursors inhibit their own differentiation via N-cadherin maintenance of beta-catenin signaling. *Developmental Cell*, **18**, 472–479.

Zhou, L., Ercolano, E., Ammoun, S. *et al.* (2011) Merlin-deficient human tumors show loss of contact inhibition and activation of Wnt/beta-catenin signaling linked to the PDGFR/Src and Rac/PAK pathways. *Neoplasia*, **13**, 1101–1112.

Zilberberg, A., Yaniv, A., and Gazit, A. (2004) The low density lipoprotein receptor-1, LRP1, interacts with the human frizzled-1 (HFz1) and down-regulates the canonical Wnt signaling pathway. *The Journal of Biological Chemistry*, **279**, 17535–17542.

Zweier, C., Peippo, M.M., Hoyer, J. *et al.* (2007) Haploinsufficiency of TCF4 causes syndromal mental retardation with intermittent hyperventilation (Pitt-Hopkins syndrome). *American Journal of Human Genetics*, **80**, 994–1001.

31 Wnt Signaling in Dementia

Stephen J. Haggarty

Department of Neurology, Center for Human Genetic Research, Massachusetts General Hospital & Harvard Medical School, Boston, MA, USA
Department of Psychiatry, Center for Human Genetic Research, Massachusetts General Hospital & Harvard Medical School, Boston, MA, USA
MGH Psychiatry Center for Experimental Drugs & Diagnostics, Center for Human Genetic Research, Massachusetts General Hospital & Harvard Medical School, Boston, MA, USA

Introduction

Paralleling the case of the neuropsychiatric disorders described in the previous two chapters, numerous lines of evidence in the past decade point to a key role for dysregulation of Wnt signaling in the pathogenesis of dementia and response to potential therapeutic agents (De Ferrari and Moon, 2006). In the following sections, we will review seminal findings that collectively demonstrate that important aspects of both the core age-related pathologies of Alzheimer's disease (AD) originally described by Lois Alzheimer over a century ago, namely, extracellular amyloid plaques composed of peptides derived from the amyloid precursor protein (APP) and intracellular neurofibrillary tangles composed of phosphorylated Tau proteins (Mandelkow and Mandelkow, 2012; Mucke and Selkoe, 2012), have intimate connections to aberrant Wnt signaling. Overall, these findings have the potential for deep implications for the development of novel classes of disease-modifying therapeutics for AD and other forms of dementia (Figure 31.1).

The term dementia, originally derived from the Latin word "demens" meaning "without a mind," refers to a category of neuropsychiatric disorders characterized by impairment of memory and thought processes (cognition) and often also disturbances of emotion and behavior. Historically, this term encompassed a much larger set of neuropsychiatric disorders, including "dementia praecox" (precocious or early-onset dementia) as was referred to by Dr Emil Kraepelin in the late nineteenth century for the symptoms that are now referred to as schizophrenia and other forms of intellectual disability. In more recent times, the category of dementia has been refined and is distinguished from other forms of mental illness by characteristic clinical and histopathological phenotypes (Iglewicz, Meeks, and Jeste, 2011). While there are many different types and causes of dementia, outside of vascular dementia caused by stroke, the most common forms include AD, frontotemporal dementia (FTD), and dementia with Lewy bodies, all of which have connections to Wnt signaling (De Ferrari and Moon, 2006).

Wnt Signaling in Development and Disease: Molecular Mechanisms and Biological Functions, First Edition.
Edited by Stefan Hoppler and Randall T. Moon.

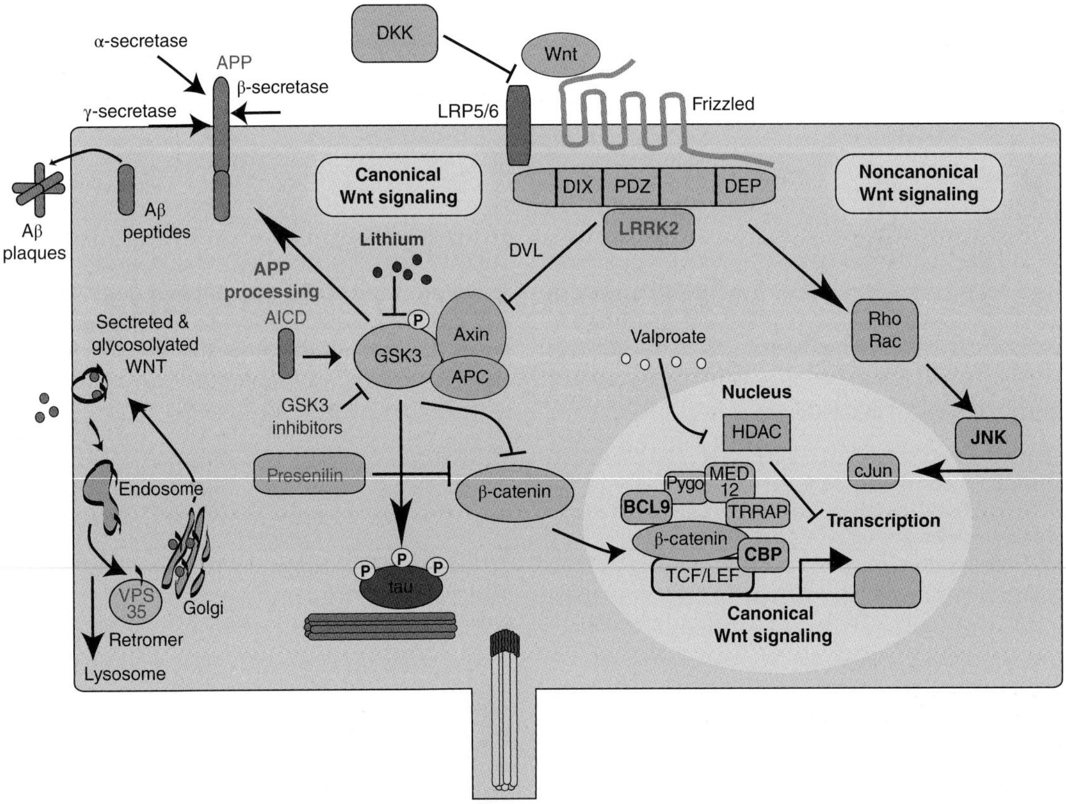

Figure 31.1 Wnt signaling pathway components implicated in dementias by human genetics and functional genomic studies. *(See insert for color representation of the figure.)*

Wnt signaling and Alzheimer's disease

AD is one of the most common of all dementias and neurodegenerative disorders, with an estimated 5.3 million individuals affected in the United States alone and over 25 million affected individuals worldwide. AD progressively attenuates cognition in those afflicted, ultimately leading to severe and irreversible neurodegeneration and death. The characteristic neuropathology of AD that has been known for over a century includes neuronal loss, extracellular amyloid plaques, and neurofibrillary tangles throughout the neocortex and hippocampus proteins (Mandelkow and Mandelkow, 2012; Mucke and Selkoe, 2012). Amyloid plaques are extracellular proteinaceous aggregates composed mainly of the amyloid β (Aβ) peptides, which are usually between 39 and 42 amino acids in length. Neurofibrillary tangles are insoluble intracellular aggregates of straight filaments and paired helical filaments found in neuronal cell bodies. The main components of neurofibrillary tangles are abnormally hyperphosphorylated forms of the microtubule-binding protein Tau. As discussed in more detail later, hyperphosphorylated tau-associated neurofibrillary pathology is not unique to AD, as other dementias including FTD and certain forms of Parkinson's disease, all of which share commonalities that may implicate aberrant Wnt signaling.

Role of Wnt signaling in amyloid precursor protein processing in Alzheimer's disease

One of the earliest studies to directly link Wnt signaling to pathways relevant to the neuropathology observed in AD came from the demonstration that overexpression of Dishevelled 1 (Dvl1) alters the processing of the APP (Mudher *et al.*, 2001). APP is normally cleaved at three main sites either by β-secretase (BACE1) and γ-secretase resulting in the release of amyloidogenic Aβ peptides and the APP intracellular domain (AICD) or by α-secretase resulting in the cleavage of APP within the Aβ sequence, creating a nonamyloidogenic, secreted fragment (sAPPα). Dvl1 overexpression was found to activate noncanonical Wnt pathways involving both the c-Jun N-terminal kinase (JNK) and protein kinase C (PKC)/mitogen-activated protein (MAP) kinase leading to enhanced cleavage of APP by α-secretase and therefore favoring sAPPα production over the amyloidogenic Aβ peptides (Mudher *et al.*, 2001). Furthermore, the AICD was recently found to inhibit Wnt/β-catenin-mediated transcription through a direct interaction with GSK3β resulting in activation of its kinase activity and enhanced Axin–GSK3β complex-mediated β-catenin polyubiquitination and consequently degradation (Zhou *et al.*, 2012). Besides Dvl1, GSK3α has been demonstrated to regulate the processing of APP with increased production of Aβ occurring in response to elevated GSK3 activity (Phiel *et al.*, 2003), although conflicting results have also been obtained (Jaworski *et al.*, 2011). Other studies have also observed that GSK3 inhibition leads to an induction of lysosomal and autophagy-mediated degradation of APP (Parr *et al.*, 2012).

In addition to mutations in APP, early-onset familial AD is also known to be caused by mutations in the *PSEN1* and *PSEN2* genes encoding the presenilins, which are responsible along with γ-secretase for the intramembranous cleavage of β-APP to generate the more amyloidogenic Aβ1–Aβ42 peptides. Besides regulating Aβ peptide production, presenilin 1 has also been shown to bind to β-catenin and attenuate the stabilization and transcriptional activation induced by Wnt and Dvl family members with familial AD mutations generally having an attenuated response (Killick *et al.*, 2001; Soriano *et al.*, 2001).

VPS35, which as described in the previous two chapters regulates Wnt recycling through its function as a component of the retromer complex and regulation of Wntless (Belenkaya *et al.*, 2008; Port *et al.*, 2008), has also been implicated in the suppression of Alzheimer's neuropathology as the hemizygous deletion of Vps35 in the Tg2576 AD mouse model shortened the age of onset of neuropathology and cognitive deficits and deficits in long-term potentiation and postsynaptic glutamatergic neurotransmission (Wen *et al.*, 2011). VPS35 loss of function also led to increased BACE1 activity and Aβ production by decreasing retrieval of BACE1 from the endosome (Wen *et al.*, 2011).

Taken together, these findings point to multiple mechanisms through which APP metabolism implicated in the pathogenesis of AD directly intersects with Wnt signaling pathway. Accordingly, maintaining the proper balance between DVL1-stimulated α-secretase cleavage and β-/γ-secretase cleavage as well as GSK3 activity may be an important part of proper APP metabolism in the brain and a point of aberration in the amyloid cascade.

Role of Wnt signaling in synaptic dysfunction and neuronal loss in Alzheimer's disease

One of the first genetic implications of aberrant Wnt signaling in association with late-onset AD came from pioneering work by the Moon Laboratory that identified a variant in the low-density lipoprotein receptor-related protein 6 (LRP6) that interacted genetically with the apolipoprotein E-epsilon4 (APOE-epsilon4) carrier status of individuals (De Ferrari *et al.*, 2007). LRP6 normally functions along lipoprotein receptor-related protein 5 (LRP5) and Frizzled receptors as a coreceptor that binds Wnt proteins to activate the canonical Wnt signaling pathway (Bafico *et al.*, 2001; Semenov *et al.*, 2001). The AD-associated variant demonstrated in LRP6 reduced Wnt coreceptor function leading to diminished Wnt/β-catenin signaling (De Ferrari *et al.*, 2007). Additionally, further studies of LRP6 have identified an alternative splice form of *LRP6* that lacks exon 3 with reduced Wnt/β-catenin signaling that is significantly unregulated in AD patients' brain compared to controls (Alarcon *et al.*, 2012).

Consistent with these findings implicating decreased Wnt/β-catenin signaling in AD, levels of the secreted Wnt antagonist Dickkopf-1 (Dkk1), which blocks canonical Wnt signaling through binding of the ectodomains of LRP5/6 (Ahn *et al.*, 2011; Cheng *et al.*, 2011), have been shown to be elevated both in AD patients' brains and in an AD transgenic (Tg) mouse model (TgCRND8) overexpressing the Swedish and Indiana mutations in the human APP (Caricasole *et al.*, 2004; Purro, Dickins, and Salinas, 2012; Rosi *et al.*, 2010), as well as in response to acute treatment with Aβ oligomers (Purro, Dickins, and Salinas, 2012). In agreement with these *in vivo* observations, using *in vitro* slice cultures and dissociated neurons, blockage of Wnt signaling through Dkk1 induced disassembly of both presynaptic and postsynaptic components as well as reduced the overall size of synapses in mature neurons (Purro, Dickins, and Salinas, 2012). Besides affecting synaptic disassembly, signaling through LRP5/6 normally functions to inhibit GSK3, a key mediator of phosphorylation of Tau, such that elevating Dkk1 levels leads to activation of GSK3 and increased Tau phosphorylation (Caricasole *et al.*, 2004). In agreement with these *in vitro* observations, direct administration of Dkk1 itself into the hippocampus and basal ganglia of rats was found to be sufficient to cause neuronal cell death and astrocytosis in a manner that could be reversed by systemic administration of lithium as an inhibitor of GSK3 (Rosi *et al.*, 2010; Scali *et al.*, 2006).

Another risk factor for AD that has been recently identified by GWAS is *CLU* encoding clusterin (also known as apolipoprotein J), with the minor allele within the *CLU* gene strongly associated with reduced risk for developing AD (Harold *et al.*, 2009; Lambert *et al.*, 2009). While the mechanism for the protective effects of CLU is not fully understood, *CLU* has been shown to be downregulated at the level of mRNA expression by the Wnt signaling pathway through a TCF7-dependent mechanism (Schepeler *et al.*, 2007). The loss of clusterin expression in cultured neurons using RNAi-mediated gene silencing has been shown to reduce Aβ toxicity and to reduce the upregulation of the LRP5/6 antagonist Dkk1 that occurs upon Aβ treatment; in contrast, treatment of neurons with Aβ led to increased intracellular clusterin and a p53-dependent induction of Dkk1 that led to the activation of the noncanonical Wnt planar cell polarity pathway resulting in JNK signaling that activates c-Jun-mediated transcription, which could also be observed *in vivo* in Tg mice with overexpression of Dkk1 that mimicked the effects of Aβ (Killick *et al.*, 2012). Furthermore, this same study identified evidence for activation of the Wnt planar cell polarity pathway in postmortem brain from AD patients (Killick *et al.*, 2012). Thus, besides the ability of elevated DKK1 to antagonize canonical Wnt signaling leading to loss of synapses (Purro, Dickins, and Salinas, 2012), Aβ elevation in AD may also lead to activation of noncanonical Wnt signaling through the clusterin/p53/Dkk1/Wnt–PCP–JNK pathway (Killick *et al.*, 2012). Since one of the hallmark features of AD thought to be responsible for cognitive decline is the loss of synaptic proteins (Mandelkow and Mandelkow, 2012; Mucke and Selkoe, 2012), these findings of the consequence of decreased Wnt signaling as a result of elevated Dkk1 and compromised Wnt–LRP5/6 signaling through GSK3 suggest new approaches for preventing synaptic loss in the context of AD, other dementias, and possibly other neuropsychiatric disorders where there may be loss of synapses, such as schizophrenia.

Role of Wnt/β-catenin-mediated neurogenesis and neuroprotection in Alzheimer's disease

In addition to neuronal loss and synaptic dysfunction that accompany cognitive deficits in AD mouse models, a loss of hippocampal neurogenesis has also been observed that may be important for the underlying pathogenesis of AD. The treatment of young (2-month old) double Tg CRND8 mice overexpressing the Swedish and Indiana mutations in human APP with lithium rescued the deficit in hippocampal neurogenesis and ameliorated cognitive deficits in a manner correlated with inhibition of GSK3β and activation of the Wnt/β-catenin signaling pathway measured by increased phospho-Ser9GSK3β and β-catenin immunoreactivity in multiple brain regions (Fiorentini *et al.*, 2010), and similar results with Wnt treatment have been observed to rescue neurogenesis deficits caused by Aβ1–Aβ42 exposure (Shruster *et al.*, 2011). The existence of a therapeutic window for rescuing AD pathogenesis with stimulation of the Wnt signaling pathway is suggested by these studies as the efficacy of lithium was lost in aged (6-month) Tg mice (Fiorentini *et al.*, 2010). These observations have important implications for efforts to target Wnt signaling in the context of varying stages of disease progression. Besides enhancing neurogenesis, activation of Wnt signaling has been shown in multiple systems to provide a more general neuroprotection that may be relevant to AD and other dementia, including preventing apoptosis and other forms of cell death in response to toxic oligomeric Aβ species (Chacon, Varela-Nallar, and Inestrosa, 2008). Consistent with this notion, the inhibition of GSK3 activity either genetically or pharmacologically with lithium or small-molecule inhibitors of GSK3 has been shown to prevent Aβ-induced neurodegeneration both *in vitro* and *in vivo* in a variety of animal models of AD leading to the GSK3 hypothesis of AD (Hong *et al.*, 1997; Hooper, Killick, and Lovestone, 2008; Lovestone *et al.*, 1999; Takashima *et al.*, 1993).

Use of lithium to treat Alzheimer's disease: Early tests of GSK3 hypothesis

Although, as discussed earlier, the polypharmacology of lithium ultimately limits the ability to make a precise statement about the relationship between Wnt pathway modulation and its therapeutically relevant targets, its approval and widespread use as therapeutic for bipolar disorder and the preclinical animal work in the past decade strongly support the hypothesis that GSK3 inhibition would be beneficial in the context of AD pathogenesis. A number of clinical trials with lithium have been recently conducted. For example, the effect of short-term lithium treatment in patients with AD has been investigated with a 10-week trial that included a 6-week titration phase to reach a lithium serum concentration level of 0.5–0.8 mmol/l (Hampel *et al.*, 2009). However, in this study, neither changes in pharmacodynamic markers of GSK3 activity in cerebrospinal fluid (CSF)- or blood-derived lymphocytes were observed nor were changes observed in cognitive performance or in depressive symptoms, leaving open the question of whether insufficient levels of GSK3 inhibition occurred or whether there is a lack of efficacy of GSK3 inhibition. Alternatively, because the patients being treated in this case already had mild to moderate dementia, it is also possible that the appropriate window for preventing cognitive dysfunction had already passed and that early intervention may still yield a beneficial, disease-modifying effect.

A more recent double-blind, long-term (12-month) clinical trial in the context of amnesic mild cognitive impairment (aMCI), which is thought to be antecedent in many cases to later onset of more severe AD symptoms, was able to demonstrate both a significant decrease in concentration of phospho-tau in the CSF and improved cognitive performance along with the overall tolerability with doses of lithium between 150 and 600 mg/daily with plasma levels of 0.25 and 0.5 mmol/l (Forlenza *et al.*, 2011). Additionally, to avoid concerns of toxicity of lithium in AD patients, a recent microdose study of 300 µg administered once daily for 18 months was performed and found that lithium was efficacious at preventing cognitive loss (Nunes, Viel, and Buck, 2012). In further support of the notion that lithium may be effective in preventing symptoms of dementia, long-term treatment of patients with mania or bipolar disorder with lithium was found to be associated with a decreased rate of subsequently developing dementia, while other treatments, such as anticonvulsants, antidepressants, or antipsychotics, did not (Kessing, Forman, and Andersen, 2010).

Taken as whole, these intervention and epidemiological studies with lithium, while not directly proving one way or another that it is the Wnt pathway *per se* that is involved, provide a glimmer of hope that the inexpensive and widely available lithium salts may have disease-modifying properties that may be relevant to the treatment or prevention of AD. Moreover, based on the Wnt pathway hypothesis, these encouraging findings provide a strong impetus for further consideration of additional therapeutic agents that can modulate Wnt signaling for a more direct test of the hypothesis of its relevance to disease pathogenesis.

To attempt to overcome the limitations of lithium in terms of its narrow therapeutic index, multiple targets, and weak inhibition of GSK3, there has been significant effort in the past decade to develop potent and selective GSK3 inhibitors that could be used to test the GSK3 hypothesis of AD pathogenesis. As reviewed (Cohen and Goedert, 2004; Kramer, Schmidt, and Lo Monte, 2012; Selenica *et al.*, 2007), multiple structural classes of small-molecule GSK3 inhibitors have been tested and demonstrated to have efficacy in Tg mouse models of AD and related tauopathies. Despite the apparent promise of direct GSK3 inhibition for ameliorating AD pathogenesis, the ultimate proof of concept of this therapeutic strategy remains to be demonstrated. One compound, tideglusib (NP-12; NP031112), a reported non-ATP-competitive inhibitor that is a covalent and irreversible GSK3 inhibitor (Dominguez *et al.*, 2012), has reached clinical trials; however, the potency of this compound is limited (low μM), and the compound acts as a covalent and irreversible inhibitor, which may limit its tolerability and efficacy (Del Ser *et al.*, 2012).

Efforts by a number of major pharmaceutical companies to develop GSK3 inhibitors for advancing into clinical trials in AD have purposefully focused on mild GSK3 inhibition to avoid potential mechanism-based toxicity. As a result, a number of pharmaceutical companies appear to have counterscreened in their programs against significantly elevating β-catenin levels so as to minimize concerns of tumorigenesis. However, if dysregulation of Wnt signaling is of paramount importance to AD pathogenesis, with hyperphosphorylation of Tau as a marker of disease pathogenesis but not a sufficient endpoint to have ameliorated, the strategies used to date to develop GSK3 inhibitors for AD may need to be revised. Along these lines, a number of strategies are being taken to try to overcome limitations of the available chemical matter for targeting GSK3. This include efforts to target GSK3 in regions other than the highly conserved ATP binding site as most of the existing inhibitors are ATP-competitive type I inhibitors that bind to the same site as ATP within the catalytic domain that is highly evolutionarily conserved and therefore challenging to achieve selectivity that translates over into robust cellular and *in vivo* efficacy. The development of next-generation GSK3 inhibitors that are capable of occupying pockets adjacent to the ATP binding site (i.e., type II inhibitors) and thereby trapping the kinase in the inactive, so-called DFG-out state, or novel types of non-ATP-competitive inhibitors that are not covalent binders may provide further traction. GSK3 is also known to play important roles in multiple pathways outside of Wnt signaling and the regulation of Tau phosphorylation, and APP processing, which may limit the effective doses that can be administered without obtaining toxicity. Here, an attractive feature may be to develop substrate-competitive inhibitors that could, for example, selectively block the phosphorylation of Tau or substrates or protein–protein interactions that are key to the Wnt signaling pathway. As discussed in Chapter 32, future efforts to target Wnt signaling with small molecules may lead to the discovery of cell-type or other specific mechanisms for enhancing Wnt signaling, for example, blocking DKK1's antagonism of LRP5/6, that do not afford elevated risk of tumorigenesis or other undesired side effects.

Wnt signaling and frontotemporal dementia

In addition to AD, aberrant Wnt signaling has been implicated in other highly related forms of dementia referred to as FTD (Rabinovici and Miller, 2010). One of the known genetic causes of FTD is mutations in the gene Progranulin (*GRN*) (Ward and Miller, 2011). The analysis of the global gene expression profile in human brains from FTD patients with *GRN* haploinsufficiency compared to other forms of sporadic FTD not caused by mutations in *GRN* revealed differential expression of multiple components of the Wnt signaling pathway (Rosen *et al.*, 2011). These expression differences included Wnt ligands and receptors and members of the TCF/LEF family of transcription factors with Wnt signaling overall increased in *GRN*-haploinsufficient human brains (Rosen *et al.*, 2011). Similar to these observations of elevated Wnt signaling in postmortem human brain of FTD patients with mutations in *GRN*, the analysis of brain tissue from the JNPL3 mouse model of FTD that expresses the human P301L mutant Tau protein revealed accumulation of β-catenin and activation of Wnt signaling pathway as an early feature of the neurodegenerative process (Wiedau-Pazos *et al.*, 2009). Further *in vitro* studies with cultured hNPCs demonstrated a reciprocal relationship between Wnt signaling and *GRN* levels with loss of *GRN* expression induced by RNAi-mediated silencing leading to increased *WNT1* expression and conversely overexpression of WNT1 leading to decreased *GRN* expression. Furthermore, whereas the Wnt receptor Fzd2 was upregulated in the *Grn* knockout mouse brains, FZD2 knockdown *in vitro* in hNPCs increased apoptotic death while overexpressing it was neuroprotective (Wexler *et al.*, 2011). Finally, Geschwind and colleagues also characterized the time-dependent changes in the global expression in response to Wnt1 treatment of cultured hNPCs (Wexler *et al.*, 2011). These analyses revealed an oscillatory pattern of expression of many known and novel Wnt-responsive genes along with a higher-order, network-based relationship between Wnt-responsive genes and the FTD gene *GRN* and the *PSEN1* gene that causes AD (Wexler *et al.*, 2011). Taken together, these findings point to the existence of potential negative feedback loop regulating Wnt signaling in FTD and another potentially overall important role of Wnt signaling in the context of dementia.

Dementia with Lewy bodies and Wnt signaling

Next to AD, one of next most frequent cause of dementia is dementia with Lewy bodies, which shares overlapping clinical features with AD and Parkinson's disease, including cognitive dysfunction and behavioral disturbances. Characteristic of this type of dementia are fibrillar aggregates referred to as Lewy bodies that are composed of a number of proteins, including α-synuclein. Although the etiological basis of dementia with Lewy bodies is varied, one of the most common known genetic causes is that of mutations in the leucine-rich repeat kinase 2 (*LRRK2*) gene that along with the α-synuclein gene (*SCNA*) are the most prevalent causes of autosomal dominant familial and sporadic Parkinson's disease (Ross *et al.*, 2006). Recently, LRRK2 has been shown to be a key regulator of Wnt signaling with direct physical interactions between the membrane-bound Wnt coreceptor LRP6 and cytoplasmic components consisting of DVL1 and the β-catenin destruction complex (Sancho, Law, and Harvey, 2009). Whereas elevating LRRK2 levels enhanced DVL1 and LRP6-driven Wnt signaling, pathogenic familial mutant forms of LRRK2 (R1441C, Y1699C, and G2019S) were found to exhibit decreased DVL1-induced activation of canonical Wnt signaling and to have reduced interaction with the Wnt coreceptor LRP6 (Sancho, Law, and Harvey, 2009). These data point to a potentially important role for aberrant Wnt-mediated signaling in dementia with Lewy bodies and in Parkinson's disease in general.

Additionally, aberrant Tau phosphorylation has also been found as part of the neuropathology of patients with Parkinson's disease caused by mutations in LRKK2, as well as in mouse models expressing pathogenic forms of LRRK2 (G2019S) (Melrose *et al.*, 2010; Rajput *et al.*, 2006; Ujiie *et al.*, 2012). While the precise mechanistic basis for the elevated levels of Tau phosphorylation in human and mouse brain remains unknown, in a *Drosophila* model of Parkinson's disease, expressing a pathogenic human LRRK2 (G2019S) mutant was found to enhance Tau phosphorylation by the *Drosophila* GSK3β homolog Shaggy (Sgg) (Lin *et al.*, 2010). Taken together, these findings suggest that dysregulation of GSK3 leading to Tau hyperphosphorylation in response to altered Wnt signaling may be a core feature of the neuropathology of multiple types of dementia in addition to AD.

Etiological overlap between dementia and other neuropsychiatric disorders

While we have discussed neuropsychiatric disorders in Chapters 29 and 30 and this chapter as distinct entities with separate etiologies with potentially convergent pathways of molecular pathophysiology involving dysregulation of Wnt signaling, there is also emerging epidemiological evidence suggesting there may exist crossover at the level of etiology. For example, patients with dementia have been demonstrated to have an increased risk of developing mania and depression (Nilsson *et al.*, 2002); conversely, based upon the analysis of the Danish psychiatric case registry, patients with bipolar disorder or major depression are at increased risk for developing dementia, and lithium treatment can reduce the prevalence to the same level as the normal elderly population (Kessing *et al.*, 1999; Nunes, Forlenza, and Gattaz, 2007). These observations provide evidence at a population level that there are shared risk factors between neuropsychiatric disorders and dementias.

Summary

Despite the potential for shared risk factors between neuropsychiatric disorders and dementias, the observations of dysregulation of Wnt signaling firmly rooted in the observed neuropathology and Tg mouse models of degenerative dementias stand in stark contrast to the case of bipolar disorder, schizophrenia, and neurodevelopment disorders where there is a relative paucity of overt phenotypes observable at the level of postmortem neuropathology and through neuroimaging methods. Yet, at a finer scale of resolution, a number of commonalities exist, including evidence for dysregulation of the process of neurogenesis and synaptic function, both of which Wnt signaling play a key role. While the true relationship between lithium's ability to modulate Wnt signaling and its therapeutic effects to treat neuropsychiatric disorders and dementias may never be fully understood, it is nonetheless intriguing that lithium, which arguably was the beginning of the development of modern neuropsychopharmacology, has provided at least in part the impetus for consideration of the role of Wnt signaling in general, and GSK3 in particular, in the molecular pathophysiology of these disorders, which points to the core aspects of a shared pathophysiology between multiple different neuropsychiatric disorders and dementias. Research in the Wnt field in the next decade will undoubtedly clarify the question of this relationship through the development of new targeted pharmacology that will enable the selective and conditional modulation of Wnt signaling in animal models and ultimately as experimental therapeutics in humans. Given the tremendous burden of these disorders to individuals, their families, and society as a whole, these anticipated future developments hold tremendous promise for improving mental health.

Acknowledgments

Studies in the Haggarty Laboratory on the role of Wnt signaling in neuropsychiatric disorders and dementia have been generously supported by the National Institute of Mental Health, the National Institute of Aging, the National Institute of Drug Addiction, the Stanley Medical Research Institute, and the Tau Consortium. We thank the laboratory of Randall Moon (UW), Li-Huei Tsai (MIT), and members of the Stanley Center for Psychiatric Research for helpful discussions. We sincerely apologize for not being able to cite due to space limitations a number of additional papers that have contributed to the concepts discussed herein.

References

Ahn, V.E., Chu, M.L., Choi, H.J. *et al.* (2011) Structural basis of Wnt signaling inhibition by Dickkopf binding to LRP5/6. *Developmental Cell*, **21**, 862–873.

Alarcon, M.A., Medina, M.A., Hu, Q. *et al.* (2012) A novel functional low-density lipoprotein receptor-related protein 6 gene alternative splice variant is associated with Alzheimer's disease. *Neurobiology of Aging*, **34**, e9–e18.

Bafico, A., Liu, G., Yaniv, A. *et al.* (2001) Novel mechanism of Wnt signalling inhibition mediated by Dickkopf-1 interaction with LRP6/Arrow. *Nature Cell Biology*, **3**, 683–686.

Belenkaya, T.Y., Wu, Y., Tang, X. *et al.* (2008) The retromer complex influences Wnt secretion by recycling wntless from endosomes to the trans-Golgi network. *Developmental Cell*, **14**, 120–131.

Caricasole, A., Copani, A., Caraci, F. *et al.* (2004) Induction of Dickkopf-1, a negative modulator of the Wnt pathway, is associated with neuronal degeneration in Alzheimer's brain. *The Journal of Neuroscience*, **24**, 6021–6027.

Chacon, M.A., Varela-Nallar, L., and Inestrosa, N.C. (2008) Frizzled-1 is involved in the neuroprotective effect of Wnt3a against Abeta oligomers. *Journal of Cellular Physiology*, **217**, 215–227.

Cheng, Z., Biechele, T., Wei, Z. *et al.* (2011) Crystal structures of the extracellular domain of LRP6 and its complex with DKK1. *Nature Structural & Molecular Biology*, **18**, 1204–1210.

Cohen, P. and Goedert, M. (2004) GSK3 inhibitors: development and therapeutic potential. *Nature Reviews Drug Discovery*, **3**, 479–487.

De Ferrari, G.V. and Moon, R.T. (2006) The ups and downs of Wnt signaling in prevalent neurological disorders. *Oncogene*, **25**, 7545–7553.

De Ferrari, G.V., Papassotiropoulos, A., Biechele, T. *et al.* (2007) Common genetic variation within the low-density lipoprotein receptor-related protein 6 and late-onset Alzheimer's disease. *Proceedings of the National Academy of Sciences of the United States of America*, **1104**, 9434–9439.

Del Ser, T., Steinwachs, K.C., Gertz, H.J. *et al.* (2012) Treatment of Alzheimer's disease with the GSK-3 inhibitor Tideglusib: a pilot study. *Journal of Alzheimer's Disease*, **33**, 205–215.

Dominguez, J.M., Fuertes, A., Orozco, L. *et al.* (2012) Evidence for irreversible inhibition of glycogen synthase kinase-3beta by tideglusib. *The Journal of Biological Chemistry*, **287**, 893–904.

Fiorentini, A., Rosi, M.C., Grossi, C. *et al.* (2010) Lithium improves hippocampal neurogenesis, neuropathology and cognitive functions in APP mutant mice. *PLoS One*, **5**, e14382.

Forlenza, O.V., Diniz, B.S., Radanovic, M. *et al.* (2011) Disease-modifying properties of long-term lithium treatment for amnestic mild cognitive impairment: randomised controlled trial. *The British Journal of Psychiatry*, **198**, 351–356.

Hampel, H., Ewers, M., Burger, K. *et al.* (2009) Lithium trial in Alzheimer's disease: a randomized, single-blind, placebo-controlled, multicenter 10-week study. *The Journal of Clinical Psychiatry*, **70**, 922–931.

Harold, D., Abraham, R., Hollingworth, P. *et al.* (2009) Genome-wide association study identifies variants at CLU and PICALM associated with Alzheimer's disease. *Nature Genetics*, **41**, 1088–1093.

Hong, M., Chen, D.C., Klein, P.S., and Lee, V.M. (1997) Lithium reduces tau phosphorylation by inhibition of glycogen synthase kinase-3. *The Journal of Biological Chemistry*, **272**, 25326–25332.

Hooper, C., Killick, R., and Lovestone, S. (2008) The GSK3 hypothesis of Alzheimer's disease. *Journal of Neurochemistry*, **104**, 1433–1439.

Iglewicz, A., Meeks, T.W., and Jeste, D.V. (2011) New wine in old bottle: late-life psychosis. *The Psychiatric Clinics of North America*, **34**, 295–318, vii.

Jaworski, T., Dewachter, I., Lechat, B. *et al.* (2011) GSK-3alpha/beta kinases and amyloid production in vivo. *Nature*, **480**, E4–E5, discussion E6.

Kessing, L.V., Forman, J.L., and Andersen, P.K. (2010) Does lithium protect against dementia? *Bipolar Disorders*, **12**, 87–94.

Kessing, L.V., Olsen, E.W., Mortensen, P.B., and Andersen, P.K. (1999) Dementia in affective disorder: a case-register study. *Acta Psychiatrica Scandinavica*, **100**, 176–185.

Killick, R., Pollard, C.C., Asuni, A.A. *et al.* (2001) Presenilin 1 independently regulates beta-catenin stability and transcriptional activity. *The Journal of Biological Chemistry*, **276**, 48554–48561.

Killick, R., Ribe, E.M., Al-Shawi, R. *et al.* (2012) Clusterin regulates β-amyloid toxicity via Dickkopf-1-driven induction of the wnt-PCP-JNK pathway. *Molecular Psychiatry*. doi: 10.1038/mp.2012.163. [Epub ahead of print].

Kramer, T., Schmidt, B., and Lo Monte, F. (2012) Small-molecule inhibitors of GSK-3: structural insights and their application to Alzheimer's disease models. *International Journal of Alzheimer's Disease*, **2012**, 381029.

Lambert, J.C., Heath, S., Even, G. *et al.* (2009) Genome-wide association study identifies variants at CLU and CR1 associated with Alzheimer's disease. *Nature Genetics*, **41**, 1094–1099.

Lin, C.H., Tsai, P.I., Wu, R.M., and Chien, C.T. (2010) LRRK2 G2019S mutation induces dendrite degeneration through mislocalization and phosphorylation of tau by recruiting autoactivated GSK3ss. *The Journal of Neuroscience*, **30**, 13138–13149.

Lovestone, S., Davis, D.R., Webster, M.T. *et al.* (1999) Lithium reduces tau phosphorylation: effects in living cells and in neurons at therapeutic concentrations. *Biological Psychiatry*, **45**, 995–1003.

Mandelkow, E.M. and Mandelkow, E. (2012) Biochemistry and cell biology of tau protein in neurofibrillary degeneration. *Cold Spring Harbor Perspectives in Medicine*, **2**, a006247.

Melrose, H.L., Dachsel, J.C., Behrouz, B. *et al.* (2010) Impaired dopaminergic neurotransmission and microtubule-associated protein tau alterations in human LRRK2 transgenic mice. *Neurobiology of Disease*, **40**, 503–517.

Mucke, L. and Selkoe, D.J. (2012) Neurotoxicity of amyloid beta-protein: synaptic and network dysfunction. *Cold Spring Harbor Perspectives in Medicine*, **2**, a006338.

Mudher, A., Chapman, S., Richardson, J. *et al.* (2001) Dishevelled regulates the metabolism of amyloid precursor protein via protein kinase C/mitogen-activated protein kinase and c-Jun terminal kinase. *The Journal of Neuroscience*, **21**, 4987–4995.

Nilsson, F.M., Kessing, L.V., Sorensen, T.M. *et al.* (2002) Enduring increased risk of developing depression and mania in patients with dementia. *Journal of Neurology, Neurosurgery, and Psychiatry*, **73**, 40–44.

Nunes, P.V., Forlenza, O.V., and Gattaz, W.F. (2007) Lithium and risk for Alzheimer's disease in elderly patients with bipolar disorder. *The British Journal of Psychiatry*, **190**, 359–360.

Nunes, M.A., Viel, T.A., and Buck, H.S. (2012) Microdose lithium treatment stabilized cognitive impairment in patients with Alzheimer's disease. *Current Alzheimer Research*, **10**, 104–107.

Parr, C., Carzaniga, R., Gentleman, S.M. *et al.* (2012) Glycogen synthase kinase 3 inhibition promotes lysosomal biogenesis and autophagic degradation of the amyloid-beta precursor protein. *Molecular and Cellular Biology*, **32**, 4410–4418.

Phiel, C.J., Wilson, C.A., Lee, V.M., and Klein, P.S. (2003) GSK-3alpha regulates production of Alzheimer's disease amyloid-beta peptides. *Nature*, **423**, 435–439.

Port, F., Kuster, M., Herr, P. *et al.* (2008) Wingless secretion promotes and requires retromer-dependent cycling of Wntless. *Nature Cell Biology*, **10**, 178–185.

Purro, S.A., Dickins, E.M., and Salinas, P.C. (2012) The secreted Wnt antagonist Dickkopf-1 is required for amyloid beta-mediated synaptic loss. *The Journal of Neuroscience*, **32**, 3492–3498.

Rabinovici, G.D. and Miller, B.L. (2010) Frontotemporal lobar degeneration: epidemiology, pathophysiology, diagnosis and management. *CNS Drugs*, **24**, 375–398.

Rajput, A., Dickson, D.W., Robinson, C.A. *et al.* (2006) Parkinsonism, Lrrk2 G2019S, and tau neuropathology. *Neurology*, **67**, 1506–1508.

Rosen, E.Y., Wexler, E.M., Versano, R. *et al.* (2011) Functional genomic analyses identify pathways dysregulated by progranulin deficiency, implicating Wnt signaling. *Neuron*, **71**, 1030–1042.

Rosi, M.C., Luccarini, I., Grossi, C. *et al.* (2010) Increased Dickkopf-1 expression in transgenic mouse models of neurodegenerative disease. *Journal of Neurochemistry*, **112**, 1539–1551.

Ross, O.A., Toft, M., Whittle, A.J. *et al.* (2006) Lrrk2 and Lewy body disease. *Annals of Neurology*, **59**, 388–393.

Sancho, R.M., Law, B.M., and Harvey, K. (2009) Mutations in the LRRK2 Roc-COR tandem domain link Parkinson's disease to Wnt signalling pathways. *Human Molecular Genetics*, **18**, 3955–3968.

Scali, C., Caraci, F., Gianfriddo, M. *et al.* (2006) Inhibition of Wnt signaling, modulation of Tau phosphorylation and induction of neuronal cell death by DKK1. *Neurobiology of Disease*, **24**, 254–265.

Schepeler, T., Mansilla, F., Christensen, L.L. *et al.* (2007) Clusterin expression can be modulated by changes in TCF1-mediated Wnt signaling. *Journal of Molecular Signaling*, **2**, 6.

Selenica, M.L., Jensen, H.S., Larsen, A.K. *et al.* (2007) Efficacy of small-molecule glycogen synthase kinase-3 inhibitors in the postnatal rat model of tau hyperphosphorylation. *British Journal of Pharmacology*, **152**, 959–979.

Semenov, M.V., Tamai, K., Brott, B.K. *et al.* (2001) Head inducer Dickkopf-1 is a ligand for Wnt coreceptor LRP6. *Current Biology*, **11**, 951–961.

Shruster, A., Eldar-Finkelman, H., Melamed, E., and Offen, D. (2011) Wnt signaling pathway overcomes the disruption of neuronal differentiation of neural progenitor cells induced by oligomeric amyloid beta-peptide. *Journal of Neurochemistry*, **116**, 522–529.

Soriano, S., Kang, D.E., Fu, M. *et al.* (2001) Presenilin 1 negatively regulates beta-catenin/T cell factor/lymphoid enhancer factor-1 signaling independently of beta-amyloid precursor protein and notch processing. *The Journal of Cell Biology*, **152**, 785–794.

Takashima, A., Noguchi, K., Sato, K. *et al.* (1993) Tau protein kinase I is essential for amyloid beta-protein-induced neurotoxicity. *Proceedings of the National Academy of Sciences of the United States of America*, **90**, 7789–7793.

Ujiie, S., Hatano, T., Kubo, S. *et al.* (2012) LRRK2 I2020T mutation is associated with tau pathology. *Parkinsonism & Related Disorders*, **18**, 819–823.

Ward, M.E. and Miller, B.L. (2011) Potential mechanisms of progranulin-deficient FTLD. *Journal of Molecular Neuroscience*, **45**, 574–582.

Wen, L., Tang, F.L., Hong, Y. *et al.* (2011) VPS35 haploinsufficiency increases Alzheimer's disease neuropathology. *The Journal of Cell Biology*, **195**, 765–779.

Wexler, E.M., Rosen, E., Lu, D. *et al.* (2011) Genome-wide analysis of a Wnt1-regulated transcriptional network implicates neurodegenerative pathways. *Science Signalling*, **4**, ra65.

Wiedau-Pazos, M., Wong, E., Solomon, E. *et al.* (2009) Wnt-pathway activation during the early stage of neurodegeneration in FTDP-17 mice. *Neurobiology of Aging*, **30**, 14–21.

Zhou, F., Gong, K., Song, B. *et al.* (2012) The APP intracellular domain (AICD) inhibits Wnt signalling and promotes neurite outgrowth. *Biochimica et Biophysica Acta*, **1823**, 1233–1241.

32 Therapeutic Targeting of the Wnt Signaling Network

Felicity Rudge and Trevor Dale

School of Biosciences, Cardiff University, Cardiff, UK

Introduction

In the last few years, the Wnt pathway has become a major focus for drug discovery, as mechanisms leading to its deregulation have been uncovered. Cancer therapeutic development has historically driven Wnt drug discovery based on the biology of pathway mutations and because clinical trials can be rapidly progressed for patients with late-stage terminal disease (Cancer Genome Atlas Network, 2012). However, therapeutic opportunities are abound for the treatment of osteoporosis (Regard *et al.*, 2013) and vascular (Daskalopoulos *et al.*, 2012), psychiatric (Klein, 2012), and neurodegenerative diseases (Inestrosa, Montecinos-Oliva, and Fuenzalida, 2012), together with metabolic and inflammatory conditions (Whyte, Smith, and Helms, 2013). Mechanistic details regularly spur the development of new therapeutic agents, while *in vitro* control of the pathway is emerging as a key tool for the growth of replacement tissues (Lian *et al.*, 2012; Merrill, 2013).

This review considers small-molecule and biological therapeutics that modulate the canonical Wnt signaling pathway upstream of β-catenin/TCF-dependent transcription. The initial section is structured according to a linear order of "core" Wnt pathway components, Wnt → Fzd/LRP → Dvl–| Axin/APC/GSK3/CK1–| β-catenin → TCF, and briefly discusses how and where novel therapeutics function (with a comprehensive list of Wnt "canonical pathway" therapeutics detailed in Table 32.1). The latter section uses the concept of a "Wnt network" to frame a discussion of Wnt inhibitor toxicity and the use of Wnt inhibitors in combinatorial therapy. Related recent reviews discuss disease-specific opportunities (Zimmerman, Moon, and Chien, 2012), the targeting of noncanonical Wnt pathways, and the importance of target validation (Polakis, 2012).

Extracellular

The most upstream approach to the inhibition of Wnt signaling has been the prevention of Wnt ligand secretion and receptor binding. Cell-based screening identified inhibitors of PORCN, a member of the membrane-bound O-acyltransferase (MBOAT) gene family that acyl-modify Wnt proteins, an essential step for Wnt secretion and binding to the extracellular

Wnt Signaling in Development and Disease: Molecular Mechanisms and Biological Functions, First Edition.
Edited by Stefan Hoppler and Randall T. Moon.

Table 32.1 Wnt "canonical pathway" therapeutics.

Loc[n]	Target	Therapeutic	Effect	Type	Disease	Stage	Supplier	Formatted
Ex	PORCN	**LGK-974 (C57)**	I	SM	Melanoma, breast, pancreatic cancer	P I	Miltenyi, Novartis, Cellagen, BioVision	Proffitt *et al.* (2012) NCT01351103
Ex	PORCN	3 IWP series I, II, III	I	SM	Developmental system tests	P I	Dodge *et al.* (2012)	Dodge *et al.* (2012)
Ex	PORCN	IWP2	I	SM	Developmental system tests	P I	Multiple	Chen *et al.* (2009a)
Ex	Wnt1	**Wnt1 Ab.**	I	B	Lung, breast cancer	D	He *et al.* (2004)	He *et al.* (2004), Wei *et al.* (2009)
Ex	Wnt1	WIF1	I	B	Bladder, osteosarcoma, prostate	T	Tang *et al.* (2009)	Tang *et al.* (2009)
Ex	Wnt2	**Wnt2 Ab.**	I	B	Melanoma	D	You *et al.* (2004)	You *et al.* (2004)
Ex	Soluble Wnt3A	Ant1.4Br, Ant1.4Cl	I	SM	Not tested	T	Morrell *et al.* (2008)	Morrell *et al.* (2008)
Ex	Wnt3	**sFZD7**	I	B	HCC	D	Wei *et al.* (2011)	Wei *et al.* (2011)
Ex	Wnt	**sFRP4**	I	B	Renal fibrosis	Pre	Surendran, Schiavi, and Hruska (2005)	Surendran, Schiavi, and Hruska (2005)
Ex	sFRP1	WAY-316606	A	SM	Osteoporosis	T	Multiple	Bodine *et al.* (2009), Moore *et al.* (2009)
Ex	sFRP1	Iminooxothiazolidines no. 5	A	SM	Osteoporosis	D	Wyeth Research	Shi *et al.* (2009)
Ex	Fzd1,2-*Wnt3,5A*	**Wnt peptide UM206**	I	P	Myocardial infarction	D	Laeremans *et al.* (2011)	Laeremans *et al.* (2011)
Ex	Fzd-5-*Wnt5A*	**Wnt peptide Foxy-5**	I	P	Prostate cancer	Pre	WntResearch	Säfholm *et al.* (2008)
Ex	Fzd7	Fzd7 Ab.	I	B	Wilms' tumor	D	Pode-Shakked *et al.* (2011)	Pode-Shakked *et al.* (2011)
Ex	Fzd1, 2, 5, 7, 8	**Fzd Ab OMP-18R5 vantictumab**	I	B	Breast, pancreatic, colon, lung cancer	P I	OncoMed	Gurney *et al.* (2012)
Ex	Wnt1	**F8CRDhFc OMP-54F28**	I	B	Teratocarcinoma, breast, solid tumors	P I	OncoMed	DeAlmeida *et al.* (2007), Liu *et al.* (2012)
Ex	Fzd10	**Fzd10 Ab. (^{90}Y)-MAb 92-13**	I	B	Synovial sarcoma	D	Fukukawa *et al.* (2008)	Fukukawa *et al.* (2008)
Ex	SOST	**Sclerostin Ab. AMG-785**	A	B	Osteoporosis	P III	Amgen	Padhi *et al.* (2010)
Ex	SOST	**Sclerostin Ab.**	A	B	Osteoporosis	Pre	Novartis	US Patent 7,879,322
Ex	SOST	**Sclerostin Ab.**	A	B	Osteoporosis, osteopenia, osteoarthritis	Pre	Eli Lilly	US Patent 20,100,221,263
Ex	Dkk1	**Dkk1 Ab. BHQ880, RH2-18LC01**	A	B	MM	P II	Novartis, Merck	Fulciniti *et al.* (2009), Glantschnig *et al.* (2010)
Ex	LRP5/6	**IIIC3**	A	SM	Diabetes	D	Enzo Life Sciences	Li *et al.* (2012)
Ex	ZNFR3 + Lgr5	**R-spondin1 Nu206**	A	B	Inflammatory colitis	P I	Nuvelo	Zhao *et al.* (2007)

Ex	LRP6 – *Dkk1*	**Mesd peptide**	I	P	Breast cancer	Pre	Abgent	Liu *et al*. (2010)
Ex	LRP6 – *Wnt3/3A*	**LRP6 Ab, mAb. A7-IgG**	I	B	Breast cancer	Pre	Novartis	Ettenberg *et al*. (2010)
Ex	LRP6 – *Wnt1, 2, 6, 7a, 7b, 9a, 10a, 10b*	**LRP6 Ab, mAb. B2 -IgG**	I	B	Breast cancer	Pre	Novartis	Ettenberg *et al*. (2010)
Ex	LRP6 – *Wnt3/3A*	**LRP6 Ab. YW211.31**	I	B	Teratocarcinoma	Pre	Genentech	Gong *et al*. (2010)
Ex	LRP6 –*Wnt1, 2, 2b, 6, 8a, 9a, 9b, 10b*	**LRP6 Ab. YW210.09**	A	B	Breast cancer	Pre	Genentech	Gong *et al*. (2010)
Ex	LRP6 – *Dkk1*	LRP6 Ab. MAb135	A	B	Osteolysis in MM	D	Nuvelo	Binnerts *et al*. (2009)
Cy	LRP6/Fzd1	**Niclosamide**	I	SM	Colon, ovarian cancer	Pre	Multiple	Chen *et al*. (2009b), Fonseca *et al*. (2012), Lu *et al*. (2011b), Osada *et al*. (2011), Yo *et al*. (2012)
Cy	LRP6	**Salinomycin**	I	SM	CLL, breast, MM, prostate cancer	Pre	Multiple	Gupta *et al*. (2009), Lu *et al*. (2011a)
Cy	Dvl *Fzd7*	**Fzd peptide RHPD**	I	P	HCC	D	Nambotin *et al*. (2011)	Nambotin *et al*. (2011)
Cy	Gαo	BIM-46174	I	SM	Lung prostate cancer	D	Prevost (2006)	Prevost (2006)
Cy	ARFGAP1	QS11	A	SM	Breast cancer cell migration	T	Multiple	Zhang *et al*. (2007)
Cy	Dvl – *Fzd*	3289–8625	I	SM	Prostate cancer	T	Multiple	Grandy *et al*. (2009)
Cy	Dvl – *Fzd7*	**FJ9**	I	SM	Lung cancer, melanoma	D	Fujii *et al*. (2007)	Fujii *et al*. (2007)
Cy	Dvl – *Fzd*	NSC668036, J01-017a	I	SM	Cancer	T, D	Multiple	Shan *et al*. (2012)
Cy	CK1δ/ε	PF670462	I	SM	Breast cancer, fibrosarcoma cells	T	Multiple	Cheong *et al*. (2011)
Cy	CK1α activator	**Pyrvinium**	I	SM	Colon cancer, cardiomyogenesis	Pre	Multiple	Basu, Reyes-Mugica, and Rebbaa (2012), Saraswati *et al*. (2010, 2012), Thorne *et al*. (2010)
Cy	PKCα activator	**CGK062**	I	SM	Prostate tumors	Pre	Glixx Labs	Gwak *et al*. (2012)
Cy	TNKS1/TNKS2	**XAV-939**	I	SM	Colon cancer, fibrosis, cardiomyogenesis, multiple sclerosis	T	Novartis, Multiple	Distler *et al*. (2012), Fancy *et al*. (2011), Huang *et al*. (2009), Ulsamer *et al*. (2012), Wang, Hao, and Hong (2011)
Cy	TNKS1/TNKS2	IWR-1	I	SM	Colon, prostate cancer	T	Multiple	Chen *et al*. (2009a)
Cy	TNKS1/TNKS2	JW67, **JW74**, and **JW55**	I	SM	Colon cancer	Pre	ChemBioNet	Shultz *et al*. (2012), Waaler *et al*. (2011, 2012)
Cy	TNKS2	WIKI4	I	SM	Colon cancer	D	ChemBridge	James *et al*. (2012)
Cy	GSK3	**AZD2858**	A	SM	Osteoporosis	T	AstraZeneca	Marsell *et al*. (2012)
Cy	GSK3	**SB-216763**	A	SM	Type 2 diabetes, colon cancer	T	Multiple	Coghlan *et al*. (2000)
Cy	GSK3	**Tideglusib (NP031112)**	A	SM	Alzheimer's, progressive supranuclear palsy	P II	Multiple	Luna-Medina *et al*. (2007), NCT01049399, NCT01350362

(*continued*)

Table 32.1 (continued)

Loc[n]	Target	Therapeutic	Effect	Type	Disease	Stage	Supplier	Formatted
Cy	Axin/β-catenin	SKL2001	A	SM	Osteoporosis	T	Merck/Millipore	Gwak *et al.* (2011)
Cy	(β-catenin)	**AV-65**	I	SM	MM	D	Yao *et al.* (2011)	Yao *et al.* (2011)
Cy	(β-catenin)	CCT031374	I	SM	Colon cancer	T	Multiple	Ewan *et al.* (2010)
Cy	(β-catenin)	Hexachlorophene	I	SM	Colon cancer	T	Multiple	Park *et al.* (2006)
Cy	(β-catenin)	DCA bile acid	A	SM	Colon cancer (diet mechanism)	T	Multiple	Pai, Tarnawaski, and Tran (2004)
N	β-catenin-*TCF4*	iCRT3, iCRT5, iCRT14	I	SM	Breast, colon cancer, MM	D	Sigma Aldrich	Gonsalves *et al.* (2011), Narayanan *et al.* (2012)
N	β-catenin-*TCF4*	BC21	I	SM	Colon cancer	D	Calbiochem	Tian *et al.* (2012)
N	β-catenin-*TCF4*	PNU-74654	I	SM	None tested	T	Multiple	Trosset *et al.* (2006)
N	β-catenin-*TCF4*	fStAx-35	I	P	Colon cancer	D	Grossmann and Yeh (2012)	Grossmann and Yeh (2012)
N	β-catenin-*TCF4*	Stapled BCL9 peptide	I	P	None tested	D	Kawamoto *et al.* (2012)	Kawamoto *et al.* (2012)
N	TCF4 – β-catenin	**PKF115-584 (calphostin), CGP049090, PKF118-310, PKF118-774 (haloquinone)**	I	SM	Colon cancer HCC, CLL, melanoma	T	Multiple	Chen, Ding, and McCormick (2000), Gandhirajan *et al.* (2010), Halbedl *et al.* (2013), Lepourcelet *et al.* (2004), Minke *et al.* (2009), Sinnberg *et al.* (2011), Wei *et al.* (2010)
N	(TCF4 – β-catenin)	**NC043**	I	SM	Colon cancer	Pre	Wang *et al.* (2011)	Wang *et al.* (2011)
N	(TCF4 – β-catenin)	**HQBA**	I	SM	Colon cancer	Pre	Coombs *et al.* (2011)	Coombs *et al.* (2011)
N	PP2A	**IQ-1**	I	SM	Inflammatory lung disease	T	Calbiochem	Miyabayashi *et al.* (2007), Zemans *et al.* (2011)
N	CBP – β-*catenin*	**ICG-001**	I	SM	Colon cancer, pulmonary fibrosis	T	Multiple	Emami *et al.* (2004), Henderson *et al.* (2010)
N	CBP – β-*catenin*	**PRI-724**	I	SM	Advanced solid tumors	P I	Prism Pharma Co.	Garber (2009)
N	CBP – β-*catenin*	**CWP232291**	I	SM	AML	P I	JW Pharma.	Garber (2009)
N	β-catenin – *BCL9*	Carnosic acid	I	SM	Colon cancer	D	Multiple	Barni *et al.* (2012), de la Roche *et al.* (2012)
N	LEF-1	**Ethacrynic acid**	I	SM	CLL, MM	Pre	Multiple	Kim *et al.* (2012b), Lu *et al.* (2009)
N	CDK8	**Senexin A**	I	SM	Lung cancer stroma	D	Senex Biotech.	Porter, Farmaki, and Altilia (2012)
MU								
MU	Unknown	KY02111	I	SM	Cardiomyocyte differentiation	T	Glixx Labs	Minami *et al.* (2012)
MU	Unknown	GDK-100017	I	SM	NSCLC, colon	D	Lee *et al.* (2013)	Lee *et al.* (2013)
MU	Unknown	WAY-262611	A	SM	Osteoporosis	D	Wyeth, Calbiochem	Pelletier *et al.* (2009)
MU	Unknown	3,3′-Bisindoles	I	SM	Not tested	D	Arai *et al.* (2013)	Arai *et al.* (2013)

MU	Unknown	Cpd1, Cpd2	A	SM	Not tested	T	Merck	Verkaar *et al.* (2011)
MU	Unknown	2-amino-4-[3,4-(methylenedioxy) benzyl-amino]-6-(3-methoxyphenyl) pyrimidine	A	SM	Not tested	T	Liu *et al.* (2005)	Liu *et al.* (2005)
MU	Unknown	Trifluridine, kaempferol, FG7142, eseroline, ethyl 9H-pyrido[3,4-b] indole-3-carboxylate	A	SM	Not tested	D	Multiple	Zhao *et al.* (2012b)
MU	Unknown	Quinoxaline derivatives	I	SM	NSCLC	D	Gong *et al.* (2011)	Gong *et al.* (2011)
MU	Unknown	Diaminoquinazolines	I	SM	Colorectal cancer	D	Pfizer	Chen *et al.* (2009c), Dehnhardt *et al.* (2010), Mao *et al.* (2012)
MU	Various	**NSAIDS, for example, sulindac, celecoxib**	I		Colon	N/A	N/A	Lee *et al.* (2009), Grosch *et al.* (2001), Takahashi-Yanaga, Yoshihara, and Jingushi (2008)
MU	Various	Flavonoids, for example, quercetin, curcumin, flavone, dihydroxy flavones	I		Various	N/A	N/A	Amado *et al.* (2011), Thorne *et al.* (2011)
MU	RAR, RXR, PPARγ, AR, GR, VDR	NHRs, for example, FH535	I		Various	N/A	N/A	Beildeck, Gelmann, and Stephen (2010), Handeli and Simon (2008), Wang *et al.* (2009)
MU	Various	**GPCR agonist/ antagonists, for example, clozapine**	A		For example, schizophrenia	N/A	N/A	Kang *et al.* (2004), Roh *et al.* (2007)

Wnt therapeutics are listed according to their approximate order of action from the extracellular space (Ex) via the cytoplasm (Cy) to the nucleus (N). MU, molecules with multilevel or undefined action. Targets are indicated together with specific interactions (in italics) that are blocked if known. Therapeutics in **bold** have been demonstrated to have action in mammalian systems *in vivo*. Inhibitory (I) or activating (A) effects are noted by small molecule (SM), peptide (P), or biological agents (B). Stage of development of the therapeutic is noted or estimated (preclinical, Pre; phase I, P I; tool compound, T; discovery phase, D).

cysteine-rich domain (CRD) of Frizzled (Fzd) receptors (Chen *et al.*, 2009a; Janda *et al.*, 2012). The PORCN inhibitor LGK-974 has now entered phase I clinical trials for solid tumors including breast cancer and pancreatic adenocarcinoma in part predicated on its ability to prevent MMTV-Wnt1 tumor growth *in vivo* (Proffitt *et al.*, 2012). Diseases in which suppression of Wnt ligand expression is predicted to have therapeutic benefit include rheumatoid arthritis (de Sousa Rabelo *et al.*, 2010) and cancers that are dependent on autocrine or paracrine Wnt signaling from the stroma (Zong, Huang, and Sankarasharma, 2012).

Anti-Wnt antibodies and Wnt-binding proteins titrate Wnt ligands and have been used as biological therapeutics (Cruciat and Niehrs, 2013). Anti-Wnt1 and anti-Wnt2 antibodies blocked tumor growth in xenograft models of melanoma and hepatocellular carcinoma that express corresponding Wnt ligands (You *et al.*, 2004; Wei *et al.*, 2009). Peptides based on Wnt3A and Wnt5A ligand sequences (Foxy-5 and UM206) blocked both canonical and noncanonical Wnt signaling by competing with endogenous Wnt ligands, presumably for binding to Fzd/LRP5, and were shown to block breast cancer metastasis or to prevent infarct-induced heart failure, respectively (Laeremans *et al.*, 2011; Säfholm *et al.*, 2008). Secreted frizzled-related proteins (sFRPs) and the structurally unrelated Wnt inhibitory factor (WIF) sequester and inactivate Wnts and are frequently silenced in cancer, even in the presence of downstream APC or β-catenin mutations. Restoration of sFRP expression blocked tumor growth, presumably by binding Wnt ligands through its CRD (Ying and Tao, 2009). A recombinant CRD from Fzd7 (sFZD7) inhibited the growth of hepatocellular carcinoma tumors *in vivo* (Wei *et al.*, 2011), while recombinant sFRP4 reduced obstruction-associated renal fibrosis, possibly by inhibiting Wnt/β-catenin-dependent epithelial–mesenchymal transition (EMT) (Surendran, Schiavi, and Hruska, 2005). One of the most advanced "Wnt-titration" therapeutics is a soluble Fzd8 CRD fused to a humanized immunoglobulin Fc domain (F8CRDhFc). This reagent was initially shown to be effective against teratocarcinomas and MMTV-Wnt1-driven breast cancer (DeAlmeida *et al.*, 2007; Liu *et al.*, 2012) and has now been developed (as OMP-54F28) for phase I clinical trials against a range of solid tumors. By contrast, two classes of small-molecule inhibitors of sFRP1 binding to Wnt were shown to increase bone formation in organ culture (Bodine *et al.*, 2009; Shi *et al.*, 2009), a system that requires activation of Wnt signaling (reviewed in Regard *et al.*, 2013).

A number of biological agents have been developed that inhibit binding to Fzd and LRP5/6 receptors. One of the furthest developed is an antibody (OMP-18R5) that binds 5/10 Fzd receptors (Fzd1, Fzd2, Fzd5, Fzd7, Fzd8) and is active as a single agent against a subset of breast, pancreatic, colon, and lung cancers (Gurney *et al.*, 2012). This antibody, now known as vantictumab, entered phase I clinical trials in 2011. A number of molecules inhibit Wnt signaling via interactions with the Wnt coreceptors LRP5 and LRP6. A peptide derived from the ER chaperone Mesd bound cell surface LRP5/6 and inhibited growth of MMTV-Wnt1-driven tumors (Li, 2005; Liu *et al.*, 2010). Two groups showed that antibodies developed by phage display blocked multiple Wnt interactions with LRP6 and were active against MMTV-Wnt1, MMTV-Wnt3, and Ntera-2 tumors (Ettenberg *et al.*, 2010; Gong *et al.*, 2010). Antibody combinations or the use of biparatopic antibodies that recognize at least two YWTD Wnt-binding, beta-propeller repeat domains in LRP6 will be required to fully block signaling since distinct YWTD domains bind Wnt subsets. For example, the antibody YW210.09 blocked signaling by Wnt1, Wnt2, Wnt2b, Wnt6, Wnt8a, Wnt9a, Wnt9b, and Wnt10b, while YW211.31 blocked Wnt3 and Wnt3A, and neither antibody blocked Wnt4, Wnt7a, Wnt7b, and Wnt10a. Antibodies directed against LRP6 domains not actively involved in Wnt binding potentiated signaling by dimerizing the LRP6 receptors and were shown to promote bone mineralization (Ettenberg *et al.*, 2010; Gong *et al.*, 2010). By contrast, the anti-LRP antibody mAb135 acted as a Wnt activator by blocking the Wnt inhibitory effects of Dkk1 binding to LRP6 (Binnerts *et al.*, 2009).

Secreted proteins of the Dkk family (Dkk1-3) bind LRP5/6 and interfere with Wnt/β-catenin signaling by preventing the formation of a Wnt/LRP6/Fzd receptor complex and by enhancing Kremen-dependent LRP6 endocytosis (Cruciat and Niehrs, 2013). Adenoviral-expressed

Dkk1 prevented kidney fibrosis through Wnt-dependent but β-catenin-independent pathways induced by PDGF/PDGFR/LRP6 and TGFβ/TGFβR/LRP6 complexes (Ren *et al.*, 2013). Repression of Dkk1 function through the use of blocking antibodies enhanced canonical Wnt signaling in models of Alzheimer's disease, brain injury, osteoporosis, osteoarthritis, fracture healing, and multiple myeloma (reviewed in Zimmerman, Moon, and Chien, 2012). Of particular note, the anti-Dkk1 antibodies (BHQ880 and RH2-18LC01) prevented osteolytic bone lesions induced by Dkk1-secreting, multiple myeloma cells by restoring Wnt signaling in osteoblasts and by improving bone volume, structure, and density, respectively (Fulciniti *et al.*, 2009; Glantschnig *et al.*, 2010). A small molecule IIIc3a, which bound the third extracellular YWTD repeat of LRP5/6, prevented Dkk2 inhibition of Wnt/β-catenin signaling and reduced blood glucose levels, suggesting a possible therapeutic role in the treatment of diabetes (Li *et al.*, 2012). One of the most advanced therapeutics is an antisclerostin antibody, AMG-785, that enhances the formation of productive Wnt/Fzd/LRP5/6 signaling, by preventing the competitive binding of Sclerostin to LRP5/6. AMG-785 is currently in a phase III trial as a treatment for osteoporosis in postmenopausal women (Padhi *et al.*, 2010). One major potential advantage for this therapeutic is that Sclerostin is selectively expressed in osteoblasts and osteocytes and its action is therefore restricted to the bone.

Two families of growth factors have been shown to activate Wnt signaling in addition to Wnt ligands: R-spondins and Norrin (Cruciat and Niehrs: 2013). R-spondins function by binding the Lgr5 and ZNRF3 family transmembrane proteins and inhibiting ZNFR3-mediated degradation of Fzd, leading to raised Fzd receptor levels and signaling (MacDonald and He, 2012; see Chapters 2 and 8). Purified R-spondin1 (NU206) has been shown to boost signaling by endogenous Wnt ligands and to protect animals against intestinal damage induced by radiation, chemotherapy, and inflammation (Takashima *et al.*, 2011; Zhao *et al.*, 2007, 2009). There is huge potential for the further development of extracellular biological Wnt pathway modulators that is guided by structural characterization of Wnt, LRP, and Fzd interactions. Definition of the ligand binding specificity of further Wnt receptors (e.g., LRP5, Ror, Ryk) and inhibitor binding domains (e.g., sFRP, WIF, Dkk, Cerberus) should offer additional opportunities to target distinct Wnt subsets and explain some currently unpredictable effects, such as the enhancement of Wnt signaling by Dkk2 and sFRP2 in a subset of cellular contexts (Cruciat and Niehrs, 2013; von Marschall and Fisher, 2010).

Cytoplasm

Wnt/Fzd-initiated G protein-coupled receptor (GPCR) signaling has historically been regarded as separate from canonical Wnt/β-catenin signaling. However, an accumulating body of evidence has linked G protein (particularly Gαo) activation to key events including the recruitment of Axin to the plasma membrane (reviewed in Koval *et al.*, 2011; Malbon, 2011; see Chapter 14). In an attempt to reconcile GPCR-dependent and independent views, it has been suggested that GPCR signaling is necessary for the expression of "low-threshold" β-catenin/TCF-dependent target genes via signaling from the plasma membrane, while "high-threshold" target genes require the formation of signaling endosomes (Dobrowolski and De Robertis, 2012). From a drug discovery perspective, the recent demonstration that Fzd receptors biochemically activate Gα nucleotide exchange in Wnt3A-treated, cell-free membrane preparations should allow systematic screening for agonists and antagonists what may be a key step in signaling without the confounding effects of context-dependent feedback (Koval and Katanaev, 2011). The potential for further drug development is highlighted by the observation that the Gαo inhibitor BIM-46174 blocked the migration of Fzd2-transformed or Wnt3A-stimulated colorectal cancer cells (Prevost, 2006). Immediately distal to the Wnt receptors, small-molecule inhibitors of Dvl's PDZ domain interaction with Fzd's C-terminus have been developed using structure-based modeling (FJ9, 3289-8625, NSC668036, J01-017a) (Fujii *et al.*, 2007; Grandy *et al.*, 2009; Shan *et al.*, 2012). FJ9 was shown to be active *in vivo* against H460 non-small-cell lung cancers (NSCLC), a tumor type that overexpresses Dvl (Zhao *et al.*,

2010). Cell-penetrating peptides based on the C-terminal tail (KTLQSW) of Fzd7 were also shown to directly interfere with signaling and to inhibit HCC growth *in vivo* (Nambotin *et al.*, 2011).

The antihelminthic compound niclosamide was identified as a Wnt signaling inhibitor in a cell imaging-based assay for Fzd1 endocytosis (Chen *et al.*, 2009b) and was shown to reduce Fzd and LRP6 levels together with TCF-dependent transcription (Lu *et al.*, 2011b). Niclosamide inhibited the growth of colorectal and ovarian tumors in both preventative and therapeutic settings *in vivo* (Osada *et al.*, 2011; Yo *et al.*, 2012). Cell-based screening for molecules that synergized with Wnt3A-induced TCF-dependent transcription identified QS11 as an ArfGAP1 binding protein and suggested that it may also function by altering LRP phosphorylation and endocytosis (Kim *et al.*, 2012a; Zhang *et al.*, 2007). Salinomycin, a potassium anticoccidial ionophore, was identified as a selective inhibitor of breast cancer stem cells and was subsequently shown to block LRP6 phosphorylation, β-catenin/TCF-dependent transcription, and the growth of chronic lymphocytic leukemia (CLL) cells (Gupta *et al.*, 2009; Lu *et al.*, 2011a). Salinomycin's activity was suggested to be due to its ability to inhibit the generation of a proton gradient following Wnt-dependent endocytosis of vesicles containing complexes of LRP6, the prorenin receptor and the vacuolar H^+-adenosine triphosphatase (V-ATPase) (Cruciat *et al.*, 2010). Interestingly, an alteration of intracellular pH was also suggested to be required for the activity of niclosamide (Fonseca *et al.*, 2012).

By binding β-catenin, CK1α, and GSK3, Axin acts as a scaffold that enhances β-catenin phosphorylation and ubiquitin-dependent degradation (Clevers and Nusse, 2012). An inhibitor of tankyrase (XAV939) was identified in a cell-based screen for repressors of TCF-dependent transcription (Huang *et al.*, 2009). TNKS1 and TNKS2 ADP-ribosylate Axin (and Axin2), marking it for ubiquitylation/degradation by the RNF146 E3 ubiquitin ligase. The increased Axin levels that result from tankyrase inhibition enhance β-catenin degradation (reviewed in Riffell, Lord, and Ashworth, 2012). Compounds that inhibited TNKS1/TNKS2 but not the related PARP1/PARP2 were also identified by Chen *et al.* (Iwr-1, Chen *et al.*, 2009a), Waaler *et al.* (JW55/JW67/JW74; Waaler *et al.*, 2011, 2012), and James *et al.* (WIKI4; James *et al.*, 2012). Tankyrase inhibitors reduced breast and colorectal cancer cell growth under conditions of low serum (Casas-Selves *et al.*, 2012), reduced breast cancer cell migration and rates, and adenoma formation in the mouse intestine following APC deletion. Roles for tankyrase inhibitors outside cancer therapy are starting to emerge in areas including the prevention of fibrosis and the acceleration of remyelination in disorders such as multiple sclerosis (Distler *et al.*, 2012; Fancy *et al.*, 2011; Ulsamer *et al.*, 2012). By contrast with the tankyrase stabilization, interfering with the Axin/β-catenin interaction was suggested to account for the small molecule SKL2001's ability to activate TCF-dependent transcription (Gwak *et al.*, 2011).

Therapeutic regulation of GSK3 has been explored in many disorders including Alzheimer's disease, diabetes, heart disease, and cancer (reviewed in Cheng *et al.*, 2011; Mills *et al.*, 2011). Inhibition of GSK3 activates β-catenin/TCF-dependent transcription but also alters a wide range of additional cellular processes including glucose metabolism. Some GSK3 inhibitors have entered the clinic, particularly based on their activity in Alzheimer's disease, although it is currently unclear whether they will progress since they have been described as having a narrow therapeutic window (reviewed in Kramer, Schmidt, and Lo Monte, 2012). In the treatment of bipolar disorder, it has been suggested that partial induction of β-catenin/TCF-dependent transcription is central to the efficacy of lithium based on the effective therapeutic concentrations *in vivo* (Cheng *et al.*, 2011). By contrast, in cell culture, GSK3 inhibitors are central to the activation of Wnt signaling for tissue regeneration where the careful timing of Wnt activation/inhibition is well suited to the use of small-molecule inhibitors (Blair, Wray, and Smith, 2011, Lian *et al.*, 2012).

In a biochemical screen for regulators of β-catenin stability, Thorne *et al.* (2010) identified the antihelminthic compound pyrvinium as a pan-CK1 binding molecule that showed selective allosteric activation of purified CK1α (Thorne *et al.*, 2010). However, recent biochemical studies have suggested that pyrvinium may not function by binding CK1, but instead functions through an AKT-dependent mechanism leading to GSK3 activation (Venerando *et al.*, 2013). In addition to promoting β-catenin turnover, pyrvinium promoted Axin stability and the degradation of Pygopus, a β-catenin transcriptional coactivator. Pyrvinium inhibited the growth of colon cancer cells *in vitro* and enhanced cardiac wound repair and mesenchymal stem cell engraftment *in vivo* (Saraswati *et al.*, 2010, 2012). A similar activation of kinase activity, in this case of PKCα by CGK062, was proposed to explain compound-dependent increases in β-catenin phosphorylation and degradation leading to the inhibition of prostate tumor growth (Gwak *et al.*, 2012). By contrast, PF670462 reduced Wnt signaling and fibrosarcoma growth by inhibiting CK1ε and CK1δ, presumably by reducing the positive signaling activity of Dvl/CK1δ/ε complexes (Cheong and Virshup, 2011).

Nucleus

One of the most direct approaches to interfere with β-catenin/TCF-dependent transcription is to block the interaction between β-catenin and TCF transcription factors. Lepourcelet *et al.* (2004) identified natural products that blocked β-catenin's binding to TCF in biochemical assays, colon cancer cell proliferation *in vitro*, and β-catenin-driven *Xenopus* axis duplication (Lepourcelet *et al.*, 2004). Some of these compounds (PKF115-584, PKF118-310, and CGP049090) were subsequently shown to have *in vivo* activity against hepatocellular and hematological cancers (Gandhirajan *et al.*, 2010; Minke *et al.*, 2009; Wei *et al.*, 2010). *In silico* virtual screening for compounds that bound the TCF-binding surface of β-catenin identified two small molecules, NU-74654 and BC21 (Tian *et al.*, 2012; Trosset *et al.*, 2006). BC21 prevented TCF binding, TCF-dependent transcription, and colorectal cancer growth in cell culture. Structure-based modeling was also central to the design of "stapled" alpha-helical peptides that blocked β-catenin's interactions with TCF4 or BCL9, a transcriptional coactivator (Grossmann and Yeh, 2012; Kawamoto *et al.*, 2012). In this technically challenging approach, cross-linking "staples" were used to stabilize and increase the β-catenin affinity of short alpha-helical peptides derived from Axin (fStAx-35) and BCL9. fStAx-35 blocked TCF-dependent transcription without altering levels of β-catenin and inhibited proliferation of colorectal cancer cells at 10–20 μM concentrations.

Cell-based screening for inhibitors of TCF-dependent transcription (induced by siRNA-mediated depletion of Axin) identified a series of oxazole ligands (iCRT3, iCRT5, iCRT14) that bound β-catenin, blocking its interaction with TCF4. These compounds increased colorectal cancer cell cycle arrest in the G1/S phase and blocked Wnt-dependent morphological transformation of mammary epithelial cells (Gonsalves *et al.*, 2011). The natural product carnosic acid was identified in a biochemical screen for inhibitors of the β-catenin/BCL9 interaction and was shown to inhibit Wnt target gene expression in colorectal cancer cells (de la Roche *et al.*, 2012). Carnosic acid was suggested to have an unusual mechanism of action, in that it induced the aggregation-dependent proteolysis of free β-catenin by destabilizing an N-terminal helix that abuts the BCL9 binding site. The TCF family member LEF-1 was identified among the cross-linked targets of ethacrynic acid, a small-molecule that blocked β-catenin/TCF-dependent transcription. Ethacrynic acid induced apoptosis in CLL cells *in vivo* and reduced levels of LEF-1–β-catenin complexes (Kim *et al.*, 2012b; Lu *et al.*, 2009).

Once β-catenin has formed a complex with DNA-bound TCF factors, it activates transcription through the recruitment of a range of coactivating factors (reviewed in Cadigan and Waterman, 2013). The C-terminal transactivation domain of β-catenin interacts with the histone acetyltransferase CBP, contributing to changes in histone H3 and H4 modification and chromatin structure. In a cell-based screen, Emami *et al.* (2004) identified ICG-001 as an inhibitor of β-catenin/TCF-dependent transcription and showed it bound to CBP and blocked the β-catenin/CBP interaction (Emami *et al.*, 2004). Two related inhibitors (PRI-724 and CWP232291) have entered phase I clinical studies for the treatment of advanced solid tumors and AML (Garber, 2009). In addition to inhibiting the growth of intestinal tumors in APC mutant min mice, ICG-001 reduced idiopathic pulmonary fibrosis (Henderson *et al.*,

2010). Cell-based screening for regulators of ES cell differentiation identified IQ-1 as an inhibitor of Wnt/β-catenin signaling, and subsequent assays showed it reduced epithelial proliferation in a mouse model of inflammatory lung disease (Miyabayashi *et al.*, 2007; Zemans *et al.*, 2011). IQ-1, an inhibitor of the protein phosphatase PP2A/PR72, was thought to function indirectly by blocking β-catenin's interaction with the CBP family member, p300.

One of the key links between the β-catenin/TCF/DNA complex and transcriptional initiation/extension by RNA polymerase II is the multiprotein mediator complex (Xu and Ji, 2011). β-catenin binds to Med12 within the mediator "kinase module" that comprises CDK8, cyclin C, Med12, and Med13. β-catenin is also linked to the kinase module via the β-catenin coactivators, BCL9, and Pygopus which in turn bind Med12 and Med13 (Carrera *et al.*, 2008). CDK8 is amplified in a subset of colorectal, ovarian, and breast cancers and has been shown to be required for colorectal tumor growth *in vivo* and for the maintenance of an undifferentiated state (Adler *et al.*, 2012; Firestein *et al.*, 2008). CDK8 phosphorylates a number of nuclear targets including the C-terminus of RNA polymerase II. CDK8 also activates TCF-dependent transcription through the inhibitory phosphorylation of E2F1, interfering with E2F1's ability to repress β-catenin/TCF-dependent transcription (Morris *et al.*, 2008; Zhao, Ramos, and Demma, 2012). Compounds that target CDK8 are under development by Selvita (Sel-120) and were identified indirectly in a cell-based screen for inhibitors of p21-induced transcription (Senexin A; Porter, Farmaki, and Altilia, 2012). Senexin A blocked β-catenin/TCF-dependent transcription in colon cancer cells and cooperated with the chemotherapeutic doxorubicin in preventing lung cancer growth.

Multilevel/undefined mechanisms

Cell-based screening for small-molecule regulators of β-catenin/TCF-dependent transcription identified a number of additional pathway regulators without characterizing their molecular targets. AV-65 increased β-catenin degradation and binding to the E3 ubiquitin ligase βTrCP even in the absence of APC and prolonged the survival of mice with multiple myeloma (Yao *et al.*, 2011). Two groups identified inhibitors, CCT031374 and KY02111, that reduced levels of β-catenin- and TCF-dependent transcription, even in the presence of inhibitors of GSK3 (Ewan *et al.*, 2010; Minami *et al.*, 2012). The therapeutic potential of targeting alternative β-catenin degradation pathways was further supported by the finding that hexachlorophene promoted β-catenin degradation through a Siah1/APC but GSK3-independent pathway (Park *et al.*, 2006). A series of quinoxaline derivatives reduced levels of nuclear β-catenin- and TCF-dependent transcription in NSCLC cells (Gong *et al.*, 2011), while a series of diaminoquinazolines inhibited transcription at an undefined level in colon cancer cells (Mao *et al.*, 2012). Both the diterpenoid NC043 and the Fe^{2+} binding compound HQBA blocked signaling downstream of β-catenin accumulation and inhibited tumor growth *in vivo* (Coombs *et al.*, 2011; Wang, Hao, and Hong, 2011; Wang *et al.*, 2011). Interestingly, reducing luminal iron levels in the gut was also shown to lower rates of tumorigenesis in an APC mutant mouse model (Radulescu *et al.*, 2012). By contrast, two cell-based screens identified several activators of β-catenin/TCF-dependent transcription in U2OS cells and iPSC-derived neural progenitors (Verkaar *et al.*, 2011; Zhao *et al.*, 2012).

Many small-molecule inhibitors of non-Wnt pathway components interfere with β-catenin/TCF-dependent transcription in specific cellular contexts (e.g., Src, PKA, PI3K; reviewed in Voronkov and Krauss, 2012). This raises a question as to what should be considered a "Wnt inhibitor," particularly as some responses may be secondary to cellular transcription changes induced by engagement of non-Wnt targets. Nonetheless, mechanistic details support direct action on the Wnt pathway for some compound classes.

Nonsteroidal anti-inflammatory drugs (NSAIDS)

Nonsteroidal anti-inflammatory drugs (NSAIDS) including sulindac, aspirin, and celecoxib have been used in the clinic to prevent colon cancer and are thought to act in part by inhibiting cyclooxygenase enzymes (COX) leading to a reduction in the levels of the bioactive lipid prostaglandin E2 (PGE2) (Smalley and DuBois, 1997). Raised levels of PGE2 in cancer bind the GPCR EP2 and activate β-catenin/TCF-dependent transcription (Castellone *et al.*, 2005), while celecoxib lowers PGE2 levels and blocks β-catenin/TCF-dependent transcription (Takahashi-Yanaga, Yoshihara, and Jingushi, 2008). However, celecoxib also acts through COX-independent pathways (Grosch *et al.*, 2001). Some NSAIDs including indomethacin bind to peroxisome proliferator-activated receptor gamma (PPARγ) nuclear receptors and, by acting as partial agonists, block the action of strong agonists (Bishop-Bailey and Warner, 2003). As PPARγ forms a ligand-dependent complex with β-catenin/TCF, this offers an alternative route mechanism for NSAID action against Wnt signaling. Interestingly, the PPARγ/PPARδ antagonist FH535 was isolated in a screen for inhibitors of β-catenin/TCF-dependent transcription and was shown to interfere with the PPARγ–β-catenin interaction (Handeli and Simon, 2008). Surprisingly, the NSAID sulindac bound the Dvl PDZ domain with a Ki of 10 μM and inhibited β-catenin/TCF-dependent target gene expression in *Xenopus* embryos (Lee *et al.*, 2009). Although NSAIDS have clear effects *in vivo* and have been linked to multiple mechanisms, careful studies will be required to link effect to mechanism since many NSAIDS do not achieve the concentrations and exposures that are frequently studied *in vitro* (Ettarh, Cullen, and Calamai, 2010).

Flavonoids

Flavonoids are a broad family of polyphenolic plant compounds that target a range of cellular pathways and processes (reviewed in Havsteen, 2002). Several flavonoids are active against Wnt signaling, including genistein, quercetin, isoquercitrin, and curcumin (reviewed in Amado *et al.*, 2011). With the exception of flavone activity against tankyrase (Yashiroda *et al.*, 2010), little evidence has so far identified a direct Wnt molecular target that could account for the array of biochemical changes observed, including reductions in β-catenin or Dvl protein levels and the prevention of DNA binding by β-catenin/TCF protein complexes. Part of the difficulty in identifying a direct mechanism may be due to the effects many flavonoids have on pathways including PI3K, MAPK, and Notch that may indirectly modulate the Wnt pathway activity.

Ligands for distinct molecular pathways

Nuclear hormone receptors (NHR)

Extensive reciprocal inhibitory interactions between nuclear hormone receptor (NHR; e.g., RAR, RXR, PPARγ, AR, GR, VDR) and Wnt signaling are central to ES, mesenchymal, and keratinocyte stem cell differentiation (reviewed in Beildeck, Gelmann, and Stephen, 2010). Much cross talk occurs indirectly through alterations in the expression of ligands and antagonists. However, multiple direct interactions have been described. In summary, β-catenin or TCF binding to DNA-bound NHRs usually activates NHR-dependent transcription, while binding of liganded NHRs to DNA-bound TCFs inhibits TCF-dependent transcription, although there are context-dependent exceptions to this rule (Beildeck, Gelmann, and Stephen, 2010; Cadigan and Waterman, 2013). Distinct mechanisms have been implicated. For example, liganded NHR receptors were suggested to compete free β-catenin or CBP away from TCF (Chen *et al.*, 2006; Xiao *et al.*, 2003), while liganded AR was shown to bind TCF recognition sites in the Myc promoter (Amir *et al.*, 2003). From a therapeutic perspective, treatments such as antiandrogen therapy in prostate cancer may in part operate through the modulation of Wnt signaling.

G protein-coupled receptors

Multiple GPCR agonists including parathyroid hormone (PTH), lipopolysaccharide, FP prostanoid ligands, and GnRH activate β-catenin/TCF-dependent transcription (Castellone *et al.*, 2005; Fujino and Regan, 2001; Gardner *et al.*, 2007; Yang *et al.*, 2005). In addition, LRP6 has been suggested to act as a general GPCR accessory protein since it was required for GPCR/cAMP coupling in response to isoproterenol, adenosine, and glucagon (Jernigan *et al.*, 2010; Wan *et al.*, 2008, 2011). The context-dependent expression of non-Fzd GPCRs, their well-established pharmacopoeia, and the potential for cooperativity with Wnt/Fzd receptor signaling offer opportunities for modulating levels of Wnt signaling. As a potential example of this, it has been shown that the antipsychotic, serotonergic/dopaminergic receptor binding drug clozapine induces Dvl phosphorylation, GSK3 inactivation, and the accumulation of nuclear β-catenin (Kang *et al.*, 2004; Roh *et al.*, 2007).

Drugging a Wnt network

The conventional, linear view of the Wnt signaling underemphasizes key aspects of the pathway. Firstly, signaling is not either "on" or "off"; key outputs such as the level of β-catenin/TCF-dependent transcription can be modulated over four orders of magnitude, and the level of activity determines biological outcomes (Buchert *et al.*, 2010; Luis *et al.*, 2011). Secondly, hundreds of context-specific "Wnt pathway regulators" have been identified, particularly in the last few years with the onset of high-throughput siRNA and proteomic screens (Major *et al.*, 2008; Tang *et al.*, 2008). Thirdly, the absolute level of pathway activity likely reflects the integrated output of multiple regulators in patterns that are not simply additive. For example, R-spondin does not itself signal but alters the output from a fixed level of Wnt ligand (MacDonald and He, 2012). Lastly, non-Wnt signals such as EGF can directly modulate the activity of β-catenin/TCF-dependent transcription independent of upstream Wnt pathway components (Yang *et al.*, 2011). The range of potential functional interactions, when fully described, may best be represented as a graph network (Kestler and Kuhl, 2008; see also Chapter 11). The network view can be used to highlight aspects of therapeutic targeting including compound dosing, dynamics, and the use of compound combinations that target a distinct pathway, a subset of which are alluded to in Figure 32.1.

Doses and dynamics

"Just right" levels of Wnt signaling are required for homeostasis within normal and, at altered levels, diseased tissue. For example, liver and melanoma oncogenesis is associated with lower levels of β-catenin/TCF-dependent transcription than intestinal tumorigenesis (Buchert *et al.*, 2010; Lucero *et al.*, 2010). Therapeutic molecules that partially alter signaling levels may therefore reach an efficacy threshold in one tissue but may be ineffective in another, even if the molecular target is equally expressed. Partial efficacy may result from the partial inhibition of a strong pathway regulator (well-connected network node) or full inhibition of a molecule that contributes a smaller quantum to total pathway activity (peripheral node). For example, the partial inhibition of GSK3 (a well-connected node) by lithium may be needed to generate a "just right" level of β-catenin/TCF-dependent transcription in bipolar patients (Klein, 2012). A therapeutic level of Wnt signaling may be either lower or higher than the disease state, even in cancer. For example, riluzole, an FDA-approved drug for treating amyotrophic lateral sclerosis, may act on a melanoma-specific network by superactivating the pathway above a "just right" level (Biechele *et al.*, 2010).

Different levels of pathway activity distinguish stem and "transit-amplifying" cells within one tissue. In the intestinal epithelium, a high level of mutant β-catenin induced supernumerary stem cell formation and was accompanied by low levels of proliferation, while a lower level led to fewer stem cells but increased progenitor cell proliferation (Hirata *et al.*, 2012). Within individual tumors, heterogeneous levels of nuclear β-catenin- and TCF-dependent transcription are driven by microenvironmental factors that in turn couple to distinct cell phenotypes including stem cell, migratory and proliferative (Vermeulen *et al.*, 2010). Each tumor cell subpopulation (e.g., stem or progenitor) will likely have cellular networks that respond differentially to therapeutics. Inhibitors such as the anti-Fzd antibody OMP18-R5 have been shown to be active against tumor-initiating/cancer stem cells (Gurney *et al.*, 2012), but it is not currently clear whether they are directly active against the Wnt-driven proliferative compartment. These hidden details of

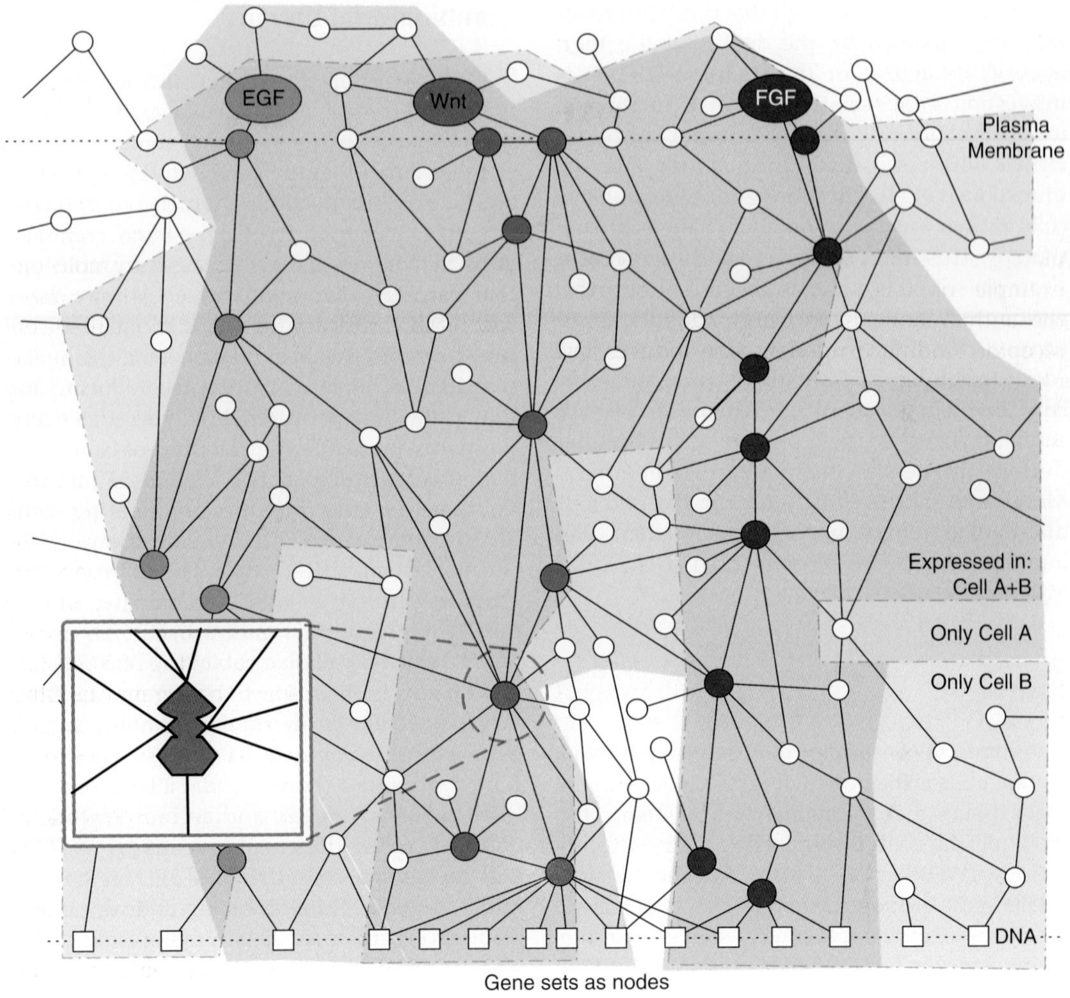

Figure 32.1 Network nodes represent functional units (proteins or protein complexes; see inset), and vertices represent functional interactions (+ve or –ve). While the total network is complex, individual cell contexts express a simpler subset of nodes, each of which would contribute towards total pathway activity. Targeting highly connected core components of pathways that are expressed in all cell types (shaded nodes; e.g., Wnt, MAPK, Notch) would be most effective at blocking the corresponding pathway but would be predicted to maximize toxicity. Targeting noncore nodes or individual vertices (e.g., a signal-transducing protein–protein interaction) might lack single-agent efficacy due to a partial effect on Wnt pathway activity but would minimize toxicity. By contrast, inhibiting molecules such as CBP (e.g., ICG-001) or CK1α (e.g., pyrvinium) that can be represented as highly connected nodes or as components of multiple nodes would enhance the probability of single-agent efficacy but would increase the probability of toxicity via action at unintended nodes. Combinations of single node-specific inhibitors should maximize the cell context specificity. Drug resistance to single-agent inhibitors may develop through mutation to target nodes or by expression of nodes from outside the cell context.

therapeutic mechanism may make the development of biomarkers of drug response (e.g., modified forms of β-catenin) difficult, if the target cell subpopulation is not first purified.

On a longer time scale, immediate-early responses to Wnt inhibitors will feed through to changes in target gene expression, including alterations to cell differentiation. Analysis of the timing of "Wnt-off" responses in an intestinal hyperplasia model showed that responses (changes in apoptosis/differentiation) were complete within only 48 hours (Jarde *et al.*, 2012). By contrast, tumor regression frequently took 2–4 weeks for therapeutics described earlier. Although this time difference may be explained by the details of the therapeutic, its access, or the tumor model, it is interesting to speculate that initial treatments induce a rapid response from tumor subpopulations with a sensitive Wnt network and that unresponsive cells later convert into responsive cells as their microenvironment changes during therapy. This interpretation would predict that short pulses of Wnt pathway inhibitor combinations that target distinct cells/network structures would induce very rapid responses and could minimize long-exposure-associated toxicity. Interestingly, short-period pulses of Wnt signaling have been suggested to be optimal during tissue regeneration (Zimmerman, Moon, and Chien, 2012) and for the efficient differentiation of human pluripotent stem cells into cardiomyocytes *in vitro* (Lian *et al.*, 2012; Minami *et al.*, 2012).

Toxicity

One common concern for the use Wnt pathway therapeutics is the potential for acute toxicity in adult tissues that are maintained by stem cells, based on the central role for Wnts in stem cell biology (Wend *et al.*, 2010). Additional side effects may include metabolic changes, based, for example, on the role of Wnt signaling in the maintenance of liver zonation, and the potential for neurological effects, based on the action of Wnts on synapse formation in the CNS and PNS (Benhamouche *et al.*, 2006; Harrison-Uy and Pleasure, 2013; Klein, 2012; Koles and Budnik, 2013; Liu *et al.*, 2011; Zarnescu and Zinsmaier, 2009). Toxicity may also be associated with off-target effects of inhibitors. For example, pyrvinium has been suggested to have alkylating activity in addition to its effects on CK1α (Saraswati *et al.*, 2010). Longer-term treatment with Wnt inhibitor therapeutics might be expected to promote the onset of diseases for which Wnt activators are being developed and vice versa. This could include aging-related diseases (Naito *et al.*, 2012). Nonetheless, an understanding of the "Wnt network" in diseased tissue may be able to maximize on-target specificity by exploiting unique dependencies/network structures that are not likely to be present within other adult tissues.

Combinatorial therapy

Standard-of-care (SoC) chemotherapy in cancer

Combinatorial therapy can be divided into two types: combinations in which compounds target distinct tumor cell types and combinations that target distinct processes within one cell. Targeting Fzd receptors in solid tumors with OMP-18R5 synergized with chemotherapeutic agents including taxol, irinotecan and gemcitabine, at least in part by reducing the slow-growing, tumor-initiating/cancer stem cell compartment, while the standard-of-care (SoC) chemotherapeutics agents "debulked" tumors by targeting the rapidly proliferating cells (Curtin and Lorenzi, 2010; de Sousa *et al.*, 2011; Gurney *et al.*, 2012; Malanchi *et al.*, 2011). Providing further evidence for this hypothesis, the CK1α inhibitor pyrvinium and sFZD7 potentiated the activity of doxorubicin against prostate and HCC tumors, respectively (Wei *et al.*, 2011; Yu *et al.*, 2008). Furthermore, salinomycin and gemcitabine combined to repress pancreatic tumor growth (Zhang *et al.*, 2011). *In vitro*, PKF115-584, Quercetin and an anti-Wnt1 antibody increased the chemosensitivity of colon and melanoma cells towards 5FU, temozolomide, cisplatin, doxorubicin, and docetaxel, respectively (He *et al.*, 2005; Sinnberg *et al.*, 2011; Xavier, 2011), while GDK100017 enhanced lung cancer radiosensitivity (Lee *et al.*, 2013). By contrast, the tankyrase inhibitor XAV-939 failed to synergize with 5FU or oxaliplatin in the killing of primary colorectal cancer sphere cultures (Tenbaum *et al.*, 2012). *In vivo*, Wnt inhibitors may also work indirectly by reducing a side effect of standard chemotherapy – the ability to promote tumor relapse. The CDK8 inhibitor Senexin A and pyrvinium were shown to prevent doxorubicin-induced stromal phenotypes that supported tumor progression, while pyrvinium also enhanced SoC efficacy by reducing the Wnt ligand-dependent expression of the drug export protein, mdr-1 (Basu, Reyes-Mugica, and Rebbaa, 2012; Porter, Farmaki, and Altilia, 2012). The tankyrase inhibitor XAV939 reduced paracrine stromal Wnt signaling to tumor cells and synergized with ara-C to increase survival in acute lymphoblastic leukemia (Yang *et al.*, 2013). Similarly, DNA damage-induced

Wnt16 expression in fibroblasts promoted prostate cancer resistance to chemotherapeutics through a pathway that was inhibited by XAV939 (Sun *et al.*, 2012). Far less explored are the mechanisms by which chemotherapeutic agents synergize with Wnt inhibitors within a single cell type. Studies in ES cells showed that the DNA-damaging agent cisplatin induced β-catenin/TCF-dependent transcription, which in turn blocked apoptosis, suggesting that the inhibition of Wnt signaling may enhance the cell-killing efficacy of DNA-damaging SoC agents (Carreras Puigvert *et al.*, 2013).

Wnt and non-Wnt combinations

In the context of a network, the definition of which components are "Wnt specific" is somewhat fuzzy. Components of pathways such as PI3K, Ras/MAPK, and Notch interact to affect the levels of β-catenin/TCF-dependent transcription as well effecting well-studied non-Wnt outcomes (reviewed in Bertrand *et al.*, 2012; Hu and Li, 2010; Itasaki and Hoppler, 2009; Voronkov and Krauss, 2012). For example, activation of Ras or PI3K signaling increased β-catenin nuclear accumulation and tumor initiation and progression in the intestine (He *et al.*, 2007; Marsh *et al.*, 2008; Phelps *et al.*, 2009; Sansom *et al.*, 2006). Unexpectedly however, reduction of PI3K signaling through treatment of colorectal cancers with PI3K inhibitors did not revert cancers to a less advanced tumor phenotype as might have been predicted, but instead led to the activation of a metastatic phenotype that was dependent on high levels of nuclear β-catenin (Tenbaum *et al.*, 2012). Encouragingly though, treatment of primary colorectal cancers expressing high levels of nuclear β-catenin in sphere culture with the tankyrase inhibitor XAV-939 reduced β-catenin levels, allowing PI3K inhibition to induce apoptosis rather than metastasis through a FOXO3a-dependent pathway (Tenbaum *et al.*, 2012). More in line with expectation, combinations of the Wnt inhibitors pyrvinium and PKF115-584 with a Ras inhibitor (FTS) were found to synergize in the *in vitro* killing of colorectal cancer cell lines with mutant KRAS and APC or β-catenin (Mologni *et al.*, 2012). The Wnt inhibitors XAV939 and pyrvinium also synergistically inhibited NSCLC cell line growth in combination with the EGFR inhibitor gefitinib (Casas-Selves *et al.*, 2012).

Wnt and Wnt inhibitor combinations

Although combinations involving different Wnt inhibitors have not yet been described, they should offer therapeutic advantages. Firstly, they should allow greater control over the absolute level of pathway activity than single agents. Secondly, they should allow the tailoring of inhibitor combinations (and therefore maximal effect) to pathway branches that are selectively active in the disease setting, thereby reducing toxicity. Thirdly, in the cancer context, they should reduce the chances of resistance developing through the activation of alternative branches of the network (Figure 32.1). A useful initial combination would likely involve both extracellular and intracellular Wnt pathway inhibitors in colorectal cancer since APC/β-catenin mutations are frequently accompanied by reduced expression of extracellular Wnt repressors such as Dkk1 (Ying and Tao, 2009). In addition, combinations should allow a greater range of disease-associated phenotypes to be targeted since activation at the Wnt ligand level can induce a greater range of cancer hallmark changes than downstream changes induced by, for example, mutant β-catenin alone (Collu, Meurette, and Brennan, 2009).

A major future task will be the identification of efficacious therapeutic combinations and corresponding susceptible patient populations from the huge numbers of potential drug combinations and genotypes. One approach to this would require the systematic mapping of functional dependencies among "Wnt pathway regulators" combined with mutation and expression analyses of diseased tissues to allow the definition of "patient stratification functional networks" that would represent each cell type within the diseased tissue (Figure 32.1). A second more empirical approach would be to identify efficacious combinations through direct experiment using patient tissues (Tenbaum *et al.*, 2012). Here, the recent identification of R-spondin-dependent organoid growth conditions for normal and diseased tissues has been a major advance (Schuijers and Clevers, 2012).

However, high-throughput implementations of these technologies would be needed to maximize their potential in drug combination studies. A third empirical approach is to use "synthetic lethal" genome-scale screens to identify disease-specific, druggable molecular targets that synergize. A suitable combination for a screen may be a tankyrase inhibitor and breast cancer cells since tankyrase inhibition was able to reduce β-catenin/TCF-dependent transcription but was unable to induce cell death under normal cell growth conditions (Bao *et al.*, 2012). Lastly, genome-scale siRNA-based approaches that focus on the overlap between signaling pathways can uncover conditional dependencies that emerge following deregulation of non-Wnt pathways. For example, the loss of Lkb1 in cancer would be predicted to render cells susceptible to porcupine or tankyrase inhibitors (Jacob *et al.*, 2011).

Conclusion

A great deal of activity in the development of Wnt pathway therapeutics is now leading to compounds entering the early stages of clinical development. These therapies and the many tool compounds that have been developed are helping develop a better understanding of the Wnt signaling network that will be needed to direct specific inhibitors and combinations to defined patient subsets.

References

Adler, A.S., McCleland, M.L., Truong, T. *et al.* (2012) CDK8 maintains tumor de-differentiation and embryonic stem cell pluripotency. *Cancer Research*, **72**, 2129–2139.

Amado, N.G., Fonseca, B.F., Cerqueira, D.M. *et al.* (2011) Flavonoids: potential Wnt/beta-catenin signaling modulators in cancer. *Life Science*, **89**, 545–554.

Amir, A.L., Barua, M., McKnight, N.C. *et al.* (2003) A direct beta-catenin-independent interaction between androgen receptor and T cell factor 4. *The Journal of Biological Chemistry*, **278**, 30828–30834.

Arai, T., Yamamoto, Y., Awata, A. *et al.* (2013) Catalytic asymmetric synthesis of mixed 3,3′-bisindoles and their evaluation as Wnt signaling inhibitors. *Angewandte Chemie International Edition*, **52**, 2486–2490.

Bao, R., Christova, T., Song, S. *et al.* (2012) Inhibition of tankyrases induces axin stabilization and blocks Wnt signalling in breast cancer cells. *PLoS One*, **7**, e48670.

Barni, M.V., Carlini, M.J., Cafferata, E.G. *et al.* (2012) Carnosic acid inhibits the proliferation and migration capacity of human colorectal cancer cells. *Oncology Reports*, **27**, 1041–1048.

Basu, D., Reyes-Mugica, M., and Rebbaa, A. (2012) Role of the Beta catenin destruction complex in mediating chemotherapy-induced senescence-associated secretory phenotype. *PLoS One*, **7**, e52188.

Beildeck, M.E., Gelmann, E.P., and Stephen, W.B. (2010) Cross-regulation of signaling pathways: an example of nuclear hormone receptors and the canonical Wnt pathway. *Experimental Cell Research*, **316**, 1763–1772.

Benhamouche, S., Decaens, T., Godard, C. *et al.* (2006) Apc tumor suppressor gene is the "zonation-keeper" of mouse liver. *Developmental Cell*, **10**, 759–770.

Bertrand, F.E., Angus, C.W., Partis, W.J., and Sigounas, G. (2012) Developmental pathways in colon cancer: crosstalk between WNT, BMP, Hedgehog and Notch. *Cell Cycle*, **11**, 4344–4351.

Biechele, T.L., Camp, N.D., Fass, D.M. *et al.* (2010) Chemical-genetic screen identifies riluzole as an enhancer of Wnt/β-catenin signaling in melanoma. *Chemistry & Biology*, **17**, 1177–1182.

Binnerts, M.E., Tomasevic, N., Bright, J.M. *et al.* (2009) The first propeller domain of LRP6 regulates sensitivity to DKK1. *Molecular Biology of the Cell*, **20**, 3552–3560.

Bishop-Bailey, D. and Warner, T.D. (2003) PPARγ ligands induce prostaglandin production in vascular smooth muscle cells: indomethacin acts as a peroxisome proliferator-activated receptor-γ antagonist. *The FASEB Journal*, **13** 1925–1927.

Blair, K., Wray, J., and Smith, A. (2011) The liberation of embryonic stem cells. *PLoS Genetics*, **7**, e1002019.

Bodine, P.V.N., Stauffer, B., Ponce-de-Leon, H. *et al.* (2009) A small molecule inhibitor of the Wnt antagonist secreted frizzled-related protein-1 stimulates bone formation. *Bone*, **44**, 1063–1068.

Buchert, M., Athineos, D., Abud, H.E. *et al.* (2010) Genetic dissection of differential signaling threshold requirements for the Wnt/beta catenin pathway in vivo. *PLoS Genetics*, **6**, e1000816.

Cadigan, K.M. and Waterman, M.L. (2013) TCF/LEFs and Wnt signaling in the nucleus, in *Wnt Signaling* (eds X.H. Roel Nusse and R. van Amerongen), Cold Spring Harbor Laboratory Press, Cold Spring Harbor, NY, pp. 105–126.

Cancer Genome Atlas Network (2012) Comprehensive molecular characterization of human colon and rectal cancer. *Nature*, **487**, 330–337.

Carrera, I., Janody, F., Leeds, N. *et al.* (2008) Pygopus activates wingless target gene transcription through the mediator complex subunits Med12 and Med13. *Proceedings of the National Academy of Sciences*, **105**, 6644–6699.

Carreras Puigvert, J., von Stechow, L., Siddappa, R. *et al.* (2013) Systems biology approach identifies the kinase Csnk1a1 as a regulator of the DNA damage response in embryonic stem cells. *Science Signaling*, **6**, ra5.

Casas-Selves, M., Kim, J., Zhang, Z. *et al.* (2012) Tankyrase and the canonical Wnt pathway protect lung cancer cells from EGFR inhibition. *Cancer Research*, **72**, 4154–4164.

Castellone, M.D., Teramoto, H., Williams, B.O. *et al.* (2005) Prostaglandin E2 promotes colon cancer cell growth through a novel Gs-Axin-β-catenin signaling axis. *Science*, **310**, 1504–1510.

Chen, R.H., Ding, W.V., and McCormick, F. (2000) Wnt signaling to β-catenin involves two interactive components. *The Journal of Biological Chemistry*, **275**, 17894–17899.

Chen, S.Y., Wulf, G., Zhou, X.Z. *et al.* (2006) Activation of beta-catenin signaling in prostate cancer by peptidyl-prolyl isomerase Pin1-mediated abrogation of the androgen receptor-beta-catenin interaction. *Molecular and Cellular Biology*, **26**, 929–939.

Chen, B., Dodge, M.E., Tang, W. *et al.* (2009a) Small molecule-mediated disruption of Wnt-dependent signaling in tissue regeneration and cancer. *Nature Chemical Biology*, **5** 100–107.

Chen, M., Wang, J., Lu, J. *et al.* (2009b) The anti-helminthic niclosamide inhibits Wnt/Frizzled1 signaling. *Biochemistry*, **48**, 10267–10274.

Chen, Z., Venkatesan, A.M., Dehnhardt, C.M. *et al.* (2009c) 2,4-Diamino-quinazolines as inhibitors of β-catenin/Tcf-4 pathway: potential treatment for colorectal cancer. *Bioorganic & Medicinal Chemistry Letters*, **19**, 4980–4983.

Cheng, H., Woodgett, J., Maamari, M., and Force, T. (2011) Targeting GSK-3 family members in the heart: a very sharp double-edged sword. *Journal of Molecular & Cellular Cardiology*, **51**, 607–613.

Cheong, J.K. and Virshup, D.M. (2011) Casein kinase 1: complexity in the family. *The International Journal of Biochemistry & Cell Biology*, **43**, 465–469.

Cheong, J.K., Hung, N.T., Wang, H. *et al.* (2011) IC261 induces cell cycle arrest and apoptosis of human cancer cells via CK1δ/ε and Wnt/β-catenin independent inhibition of mitotic spindle formation. *Oncogene*, **30**, 2558–2569.

Clevers, H. and Nusse, R. (2012) Wnt/β-catenin signaling and disease. *Cell*, **149**, 1192–1205.

Coghlan, M.P., Culbert, A.A., Cross, D.A. *et al.* (2000) Selective small molecule inhibitors of glycogen synthase kinase-3 modulate glycogen metabolism and gene transcription. *Chemistry & Biology*, **7**, 793–803.

Collu, G.M., Meurette, O., and Brennan, K. (2009) Is there more to Wnt signalling in breast cancer than stabilisation of β-catenin? *Breast Cancer Research*, **11**, 105–106.

Coombs, G.S., Schmitt, A.A., Canning, C.A. *et al.* (2011) Modulation of Wnt and beta-catenin signaling and proliferation by a ferrous iron chelator with therapeutic efficacy in genetically engineered mouse models of cancer. *Oncogene*, **31**, 213–225.

Cruciat, C.M. and Niehrs, C. (2013) Secreted and transmembrane Wnt inhibitors and activators, in *Wnt Signaling* (eds X.H. Roel Nusse and R. van Amerongen), Cold Spring Harbor Laboratory Press, Cold Spring Harbor, NY, pp. 39–64.

Cruciat, C.M., Ohkawara, B., Acebron, S.P. *et al.* (2010) Requirement of prorenin receptor and vacuolar H^+-ATPase-mediated acidification for Wnt signaling. *Science*, **327**, 459–463.

Curtin, J.C. and Lorenzi, M.V. (2010) Drug discovery approaches to target Wnt signaling in cancer stem cells. *Oncotarget*, **1**, 563–577.

Daskalopoulos, E.P., Hermans, K.C.M., Janssen, B.J., and Blankesteijn, W.M. (2012) Targeting the Wnt/frizzled signaling pathway after myocardial infarction. A new tool in the therapeutic toolbox? *Trends in Cardiovascular Medicine*, **23**, 121–127.

de la Roche, M., Rutherford, T.J., Gupta, D. *et al.* (2012) An intrinsically labile α-helix abutting the BCL9-binding site of β-catenin is required for its inhibition by carnosic acid. *Nature Communications*, **3**, 680–610.

de Sousa, E.M.F., Vermeulen, L., Richel, D., and Medema, J.P. (2011) Targeting Wnt signaling in colon cancer stem cells. *Clinical Cancer Research*, **17**, 647–653.

de Sousa Rabelo, F., da Mota, L.M.H., Lima, R.A.C. *et al.* (2010) The Wnt signaling pathway and rheumatoid arthritis. *Autoimmunity Reviews*, **9**, 207–210.

DeAlmeida, V.I., Miao, L., Ernst, J.A. *et al.* (2007) The soluble wnt receptor Frizzled8CRD-hFc inhibits the growth of teratocarcinomas in vivo. *Cancer Research*, **67**, 5371–5379.

Dehnhardt, C.M., Venkatesan, A.M., Chen, Z. *et al.* (2010) Design and synthesis of novel diaminoquinazolines with *in vivo* efficacy for β-catenin/T-cell transcriptional factor 4 pathway inhibition. *Journal of Medicinal Chemistry*, **53**, 897–910.

Distler, A., Deloch, L., Huang, J. *et al.* (2012) Inactivation of tankyrases reduces experimental fibrosis by inhibiting canonical Wnt signalling. *Annals of Rheumatic Diseases*, **72**, 1575–1580.

Dobrowolski, R. and De Robertis, E.M. (2012) Endocytic control of growth factor signalling: multivesicular bodies as signalling organelles. *Nature Reviews Molecular Cell Biology*, **13**, 53–60.

Dodge, M.E., Moon, J., Tuladhar, R. *et al.* (2012) Diverse chemical scaffolds support direct inhibition of the membrane-bound O-acyltransferase Porcupine. *The Journal of Biological Chemistry*, **287**, 23246–23254.

Emami, K.H., Nguyen, C., Ma, H. *et al.* (2004) A small molecule inhibitor of beta-catenin/cyclic AMP response element-binding protein transcription. *Proceedings of the National Academy of Sciences*, **101**, 12682–12687.

Ettarh, R., Cullen, A., and Calamai, A. (2010) NSAIDs and cell proliferation in colorectal cancer. *Pharmaceuticals*, **3**, 2007–2021.

Ettenberg, S.A., Charlat, O., Daley, M.P. *et al.* (2010) Inhibition of tumorigenesis driven by different Wnt proteins requires blockade of distinct ligand-binding regions by LRP6 antibodies. *Proceedings of the National Academy of Sciences*, **107**, 15473–15478.

Ewan, K., Pajak, B., Stubbs, M. *et al.* (2010) A useful approach to identify novel small-molecule inhibitors of Wnt-dependent transcription. *Cancer Research*, **70**, 5963–5973.

Fancy, S.P.J., Harrington, E.P., Yuen, T.J. *et al.* (2011) Axin2 as regulatory and therapeutic target in newborn brain injury and remyelination. *Nature Neuroscience*, **14**, 1009–1016.

Firestein, R., Bass, A.J., Kim, S.Y. *et al.* (2008) CDK8 is a colorectal cancer oncogene that regulates beta-catenin activity. *Nature*, **455**, 547–551.

Fonseca, B.D., Diering, G.H., Bidinosti, M.A. *et al.* (2012) Structure–activity analysis of niclosamide reveals potential role for cytoplasmic pH in control of mammalian target of rapamycin complex 1 (mTORC1) signaling. *The Journal of Biological Chemistry*, **287**, 17530–17545.

Fujii, N., You, L., Xu, Z. *et al.* (2007) An antagonist of dishevelled protein–protein interaction suppresses beta-catenin-dependent tumor cell growth. *Cancer Research*, **67**, 573–579.

Fujino, H. and Regan, J.W. (2001) FP prostanoid receptor activation of a T-cell factor/beta-catenin signaling pathway. *The Journal of Biological Chemistry*, **276**, 12489–12492.

Fukukawa, C., Hanaoka, H., Nagayama, S. *et al.* (2008) Radioimmunotherapy of human synovial sarcoma using a monoclonal antibody against FZD10. *Cancer Science*, **99**, 432–440.

Fulciniti, M., Tassone, P., Hideshima, T. *et al.* (2009) Anti-DKK1 mAb (BHQ880) as a potential therapeutic agent for multiple myeloma. *Blood*, **114**, 371–379.

Gandhirajan, R.K., Staib, P.A., Minke, K. *et al.* (2010) Small molecule inhibitors of Wnt/beta-catenin/lef-1 signaling induces apoptosis in chronic lymphocytic leukemia cells in vitro and in vivo. *Neoplasia*, **12**, 326–335.

Garber, K. (2009) Drugging the Wnt pathway: problems and progress. *Journal of the National Cancer Institute*, **101**, 548–550.

Gardner, S., Maudsley, S., Millar, R.P., and Pawson, A.J. (2007) Nuclear stabilization of beta-catenin and inactivation of glycogen synthase kinase-3beta by gonadotropin-releasing hormone: targeting Wnt signaling in the pituitary gonadotrope. *Molecular Endocrinology*, **21**, 3028–3038.

Glantschnig, H., Hampton, R.A., Lu, P. *et al.* (2010) Generation and selection of novel fully human monoclonal antibodies that neutralize Dickkopf-1 (DKK1) inhibitory function in vitro and increase bone mass in vivo. *The Journal of Biological Chemistry*, **285**, 40135–40147.

Gong, Y., Bourhis, E., Chiu, C. *et al.* (2010) Wnt isoform-specific interactions with coreceptor specify inhibition or potentiation of signaling by LRP6 antibodies. *PLoS One*, **5**, e12682.

Gong, Y.D., Dong, M.S., Lee, S.B. *et al.* (2011) A novel 3-arylethynyl-substituted pyrido[2,3,-b]pyrazine derivatives and pharmacophore model as Wnt2/β-catenin pathway inhibitors in non-small-cell lung cancer cell lines. *Bioorganic & Medicinal Chemistry Letters*, **19**, 5639–5647.

Gonsalves, F.C., Klein, K., Carson, B.B. *et al.* (2011) An RNAi-based chemical genetic screen identifies three small-molecule inhibitors of the Wnt/wingless signaling pathway. *Proceedings of the National Academy of Sciences*, **108**, 5954–5963.

Grandy, D., Shan, J., Zhang, X. *et al.* (2009) Discovery and characterization of a small molecule inhibitor of the PDZ domain of dishevelled. *The Journal of Biological Chemistry*, **284**, 16256–16263.

Grosch, S., Tegeder, I., Niederberger, E. *et al.* (2001) COX-2 independent induction of cell cycle arrest and apoptosis in colon cancer cells by the selective COX-2 inhibitor celecoxib. *The FASEB Journal*, **15**, 2742–2744.

Grossmann, T.N. and Yeh, J. (2012) Inhibition of oncogenic Wnt signaling through direct targeting of β-catenin. *Proceedings of the National Academy of Sciences*, **109**, 17942–17947.

Gupta, P.B., Onder, T.T., Jiang, G. *et al.* (2009) Identification of selective inhibitors of cancer stem cells by high-throughput screening. *Cell*, **138**, 645–659.

Gurney, A., Axelrod, F., Bond, C.J. *et al.* (2012) Wnt pathway inhibition via the targeting of Frizzled receptors results in decreased growth and tumorigenicity of human tumors. *Proceedings of the National Academy of Sciences*, **109**, 11717–11722.

Gwak, J., Hwang, S.G., Park, H.S. *et al.* (2011) Small molecule-based disruption of the Axin/β-catenin protein complex regulates mesenchymal stem cell differentiation. *Cell Research*, **22**, 237–247.

Gwak, J., Lee, J.H., Chung, Y.H. *et al.* (2012) Small molecule-based promotion of PKCα-mediated β-catenin degradation suppresses the proliferation of CRT-positive cancer cells. *PLoS One*, **7**, e46697.

Halbedl, S., Kratzer, M.C., Rahm, K. *et al.* (2013) Synthesis of novel inhibitors blocking Wnt signaling downstream of beta-catenin. *FEBS Letters*, **586**, 522–527.

Handeli, S. and Simon, J.A. (2008) A small-molecule inhibitor of Tcf/beta-catenin signaling down-regulates PPARgamma and PPARdelta activities. *Molecular Cancer Therapy*, **7**, 521–529.

Harrison-Uy, S.J. and Pleasure, S.J. (2013) Wnt signaling and forebrain development, in *Wnt Signaling* (eds X.H. Roel Nusse and R. van Amerongen), Cold Spring Harbor Laboratory Press, Cold Spring Harbor, NY, pp. 297–307.

Havsteen, B.H. (2002) The biochemistry and medical significance of the flavonoids. *Pharmacology & Therapeutics*, **96**, 67–202.

He, B., You, L., Uematsu, K. *et al.* (2004) A monoclonal antibody against Wnt-1 induces apoptosis in human cancer cells. *Neoplasia*, **6**, 7–14.

He, B., Reguart, N., You, L. *et al.* (2005) Blockade of Wnt-1 signaling induces apoptosis in human colorectal cancer cells containing downstream mutations. *Oncogene*, **24**, 3054–3058.

He, X.C., Yin, T., Grindley, J.C. *et al.* (2007) PTEN-deficient intestinal stem cells initiate intestinal polyposis. *Nature Genetics*, **39**, 189–198.

Henderson, W.R., Chi, E.Y., Ye, X. *et al.* (2010) Inhibition of Wnt/β-catenin/CREB binding protein (CBP) signaling reverses pulmonary fibrosis. *Proceedings of the National Academy of Sciences*, **107**, 14309–14314.

Hirata, A., Utikal, J., Yamashita, S. *et al.* (2012) Dose-dependent roles for canonical Wnt signalling in de novo crypt formation and cell cycle properties of the colonic epithelium. *Development*, **140**, 66–75.

Hu, T. and Li, C. (2010) Convergence between Wnt-β-catenin and EGFR signaling in cancer. *Molecular Cancer*, **9**, 236.

Huang, S.M., Mishina, Y.M., Liu, S. *et al.* (2009) Tankyrase inhibition stabilizes axin and antagonizes Wnt signalling. *Nature*, **461**, 614–620.

Inestrosa, N.C., Montecinos-Oliva, C., and Fuenzalida, M. (2012) Wnt signaling: role in Alzheimer disease and schizophrenia. *Journal of Neuroimmune Pharmacology*, **7**, 788–807.

Itasaki, N. and Hoppler, S. (2009) Crosstalk between Wnt and bone morphogenic protein signaling: a turbulent relationship. *Developmental Dynamics*, **239**, 16–33.

Jacob, L.S., Wu, X., Dodge, M.E. *et al.* (2011) Genome-wide RNAi screen reveals disease-associated genes that are common to Hedgehog and Wnt signaling. *Science Signaling*, **4**, ra4.

James, R.G., Davidson, K.C., Bosch, K.A. *et al.* (2012) WIKI4, a novel inhibitor of tankyrase and Wnt/β-catenin signaling. *PLoS One*, **7**, e50457.

Janda, C.Y., Waghray, D., Levin, A.M. *et al.* (2012) Structural basis of Wnt recognition by Frizzled. *Science*, **337**, 59–64.

Jarde, T., Evans, R.J., McQuillan, K.L. *et al.* (2012) In vivo and in vitro models for the therapeutic targeting of Wnt signaling using a Tet-OΔN89β-catenin system. *Oncogene*, **32**, 883–893.

Jernigan, K.K., Cselenyi, C.S., Thorne, C.A. *et al.* (2010) Gβγ activates GSK3 to promote LRP6-mediated β-catenin transcriptional activity. *Science Signaling*, **3**, ra37.

Kang, U.G., Seo, M.S., Roh, M.S. *et al.* (2004) The effects of clozapine on the GSK-3-mediated signaling pathway. *FEBS Letters*, **560**, 115–119.

Kawamoto, S.A., Coleska, A., Ran, X. *et al.* (2012) Design of triazole-stapled BCL9 α-helical peptides to target the β-catenin/B-cell CLL/lymphoma 9 (BCL9) protein–protein interaction. *The Journal of Medicinal Chemistry*, **55**, 1137–1146.

Kestler, H.A. and Kuhl, M. (2008) From individual Wnt pathways towards a Wnt signalling network. *Philosophical Transactions of the Royal Society of London B: Biological Sciences*, **363**, 1333–1347.

Kim, W., Kim, S.Y., Kim, T. *et al.* (2012a) ADP-ribosylation factors 1 and 6 regulate Wnt/β-catenin signaling via control of LRP6 phosphorylation. *Oncogene*, **32**, 3390–3396.

Kim, Y., Gast, S.M., Endo, T. *et al.* (2012b) In vivo efficacy of the diuretic agent ethacrynic acid against multiple myeloma. *Leukemia Research*, **36**, 598–600.

Klein, P.S. (2012) GSK-3 and Wnt signaling in neurogenesis and bipolar disorder. *Frontiers in Molecular Neuroscience*, **5**, 1–13.

Koles, K. and Budnik, V. (2013) Wnt signaling in neuromuscular junction development, in *Wnt Signaling* (eds X.H. Roel Nusse and R. van Amerongen), Cold Spring Harbor Laboratory Press, Cold Spring Harbor, NY, pp. 373–394.

Koval, A. and Katanaev, V.L. (2011) Wnt3a stimulation elicits G-protein-coupled receptor properties of mammalian Frizzled proteins. *Biochemical Journal*, **433**, 435–440.

Koval, A., Purvanov, V., Egger-Adam, D., and Katanaev, V.L. (2011) Yellow submarine of the Wnt/Frizzled signaling: submerging from the G protein harbor to the targets. *Biochemical Pharmacology*, **82**, 1311–1319.

Kramer, T., Schmidt, B., and Lo Monte, F. (2012) Small-molecule inhibitors of GSK-3: structural insights and their application to Alzheimer's disease models. *International Journal of Alzheimer's Disease*, **2012**, 1–32.

Laeremans, H., Hackeng, T.M., van Zandvoort, M.A.M.J. *et al.* (2011) Blocking of Frizzled signaling with a homologous peptide fragment of Wnt3a/Wnt5a reduces infarct expansion and prevents the development of heart failure after myocardial infarction. *Circulation*, **124**, 1626–1635.

Lee, H.J., Wang, N.X., Shi, D.L., and Zheng, J.J. (2009) Sulindac inhibits canonical Wnt signaling by blocking the PDZ domain of the protein Dishevelled. *Angewandte Chemie International Edition*, **48**, 6448–6452.

Lee, S.B., Gong, Y.D., Park, Y.I., and Dong, M.S. (2013) 2,3,6-Trisubstituted quinoxaline derivative, a small molecule inhibitor of the Wnt/beta-catenin signaling pathway, suppresses cell proliferation and enhances radiosensitivity in A549/Wnt2 cells. *Biochemical and Biophysical Research Communications*, **431**, 746–752.

Lepourcelet, M., Chen, Y.N., France, D.S. *et al.* (2004) Small-molecule antagonists of the oncogenic Tcf/beta-catenin protein complex. *Cancer Cell*, **5**, 91–102.

Li, Y. (2005) Mesd binds to mature LDL-receptor-related protein-6 and antagonizes ligand binding. *Journal of Cell Science*, **118**, 5305–5314.

Li, X., Shan, J., Chang, W. *et al.* (2012) Chemical and genetic evidence for the involvement of Wnt antagonist Dickkopf2 in regulation of glucose metabolism. *Proceedings of the National Academy of Sciences*, **109**, 11402–11407.

Lian, X., Hsiao, C., Wilson, G. *et al.* (2012) Robust cardiomyocyte differentiation from human pluripotent stem cells via temporal modulation of canonical Wnt signaling. *Proceedings of the National Academy of Sciences*, **109**, E1848–E1857.

Liu, J., Wu, X., Mitchell, B. *et al.* (2005) A small-molecule agonist of the Wnt signaling pathway. *Angewandte Chemie International Edition*, **44**, 1987–1990.

Liu, C.C., Prior, J., Piwnica-Worms, D., and Bu, G. (2010) LRP6 overexpression defines a class of breast cancer subtype and is a target for therapy. *Proceedings of the National Academy of Sciences*, **107**, 5136–5141.

Liu, H., Fergusson, M.M., Wu, J.J. *et al.* (2011) Wnt signaling regulates hepatic metabolism. *Science Signaling*, **4**, ra6.

Liu, B.Y., Soloviev, I., Huang, X. *et al.* (2012) Mammary tumor regression elicited by Wnt signaling inhibitor requires IGFBP5. *Cancer Research*, **72**, 1568–1578.

Lu, D., Liu, J.X., Endo, T. *et al.* (2009) Ethacrynic acid exhibits selective toxicity to chronic lymphocytic leukemia cells by inhibition of the Wnt/β-Catenin pathway. *PLoS One*, **4**, e8294.

Lu, D., Choi, M.Y., Yu, J. *et al.* (2011a) Salinomycin inhibits Wnt signaling and selectively induces apoptosis in chronic lymphocytic leukemia cells. *Proceedings of the National Academy of Sciences*, **108**, 13253–13257.

Lu, W., Lin, C., Roberts, M.J. *et al.* (2011b) Niclosamide suppresses cancer cell growth by inducing Wnt co-receptor LRP6 degradation and inhibiting the Wnt/β-catenin pathway. *PLoS One*, **6**, e29290.

Lucero, O.M., Dawson, D.W., Moon, R.T., and Chien, A.J. (2010) A re-evaluation of the "oncogenic" nature of Wnt/beta-catenin signaling in melanoma and other cancers. *Current Oncology Reports*, **12**, 314–318.

Luis, T.C., Naber, B., Roozen, P., and Brugman, M.H. (2011) Canonical Wnt signaling regulates hematopoiesis in a dosage-dependent fashion. *Cell Stem Cell*, **9**, 345–356.

Luna-Medina, R., Cortes-Canteli, M., Sanchez-Galiano, S. *et al.* (2007) NP031112, a thiadiazolidinone compound, prevents inflammation and neurodegeneration under excitotoxic conditions: potential therapeutic role in brain disorders. *The Journal of Neuroscience*, **27**, 5766–5776.

MacDonald, B.T. and He, X. (2012) A finger on the pulse of Wnt receptor signaling. *Cell Research*, **22**, 1410–1412.

Major, M.B., Roberts, B.S., Berndt, J.D. *et al.* (2008) New regulators of Wnt/beta-catenin signaling revealed by integrative molecular screening. *Science Signaling*, **1**, ra12.

Malanchi, I., Santamaria-Martínez, A., Susanto, E. *et al.* (2011) Interactions between cancer stem cells and their niche govern metastatic colonization. *Nature*, **481**, 85–89.

Malbon, C.C. (2011) Wnt signalling: the case of the "missing" G-protein. *The Biochemical Journal*, **433**, e3–e5.

Mao, Y., Lin, N., Tian, W. *et al.* (2012) Design, synthesis, and biological evaluation of new diaminoquinazolines as β-catenin/Tcf4 pathway inhibitors. *Journal of Medicinal Chemistry*, **55**, 1346–1359.

Marsell, R., Sisask, G., Nilsson, Y. *et al.* (2012) GSK-3 inhibition by an orally active small molecule increases bone mass in rats. *Bone*, **50**, 619–627.

Marsh, V., Winton, D.J., Williams, G.T. *et al.* (2008) Epithelial Pten is dispensable for intestinal homeostasis but suppresses adenoma development and progression after Apc mutation. *Nature Genetics*, **40**, 1436–1444.

Merrill, B.J. (2013) Wnt pathway regulation of embryonic stem cell self-renewal, in *Wnt Signaling* (eds X.H. Roel Nusse and R. van Amerongen), Cold Spring Harbor Laboratory Press, Cold Spring Harbor, NY, pp. 233–249.

Mills, C.N., Nowsheen, S., Bonner, J.A., and Yang, E.S. (2011) Emerging roles of glycogen synthase kinase 3 in the treatment of brain tumors. *Frontiers in Molecular Neuroscience*, **4**, 1–8

Minami, I., Yamada, K., Otsuji, T.G. *et al.* (2012) A small molecule that promotes cardiac differentiation of human pluripotent stem cells under defined, cytokine- and xeno-free conditions. *Cell Reports*, **2**, 1448–1460.

Minke, K.S., Staib, P., Puetter, A. *et al.* (2009) Small molecule inhibitors of WNT signaling effectively induce apoptosis in acute myeloid leukemia cells. *European Journal of Haematology*, **82**, 165–175.

Miyabayashi, T., Teo, J.L., Yamamoto, M. *et al.* (2007) Wnt/beta-catenin/CBP signaling maintains long-term murine embryonic stem cell pluripotency. *Proceedings of the National Academy of Sciences*, **104**, 5668–5673.

Mologni, L., Brussolo, S., Ceccon, M., and Gambacorti-Passerini, C. (2012) Synergistic effects of combined Wnt/KRAS inhibition in colorectal cancer cells. *PLoS One*, **7**, e51449.

Moore, W.J., Kern, J.C., Bhat, R. *et al.* (2009) Modulation of Wnt signaling through inhibition of secreted frizzled-related protein I (sFRP-1) with N-substituted piperidinyl diphenylsulfonyl sulfonamides. *Journal of Medicinal Chemistry*, **52**, 105–116.

Morrell, N.T., Leucht, P., Zhao, L. *et al.* (2008) Liposomal packaging generates Wnt protein with in vivo biological activity. *PLoS One*, **3**, e2930.

Morris, E.J., Ji, J.Y., Yang, F. *et al.* (2008) E2F1 represses beta-catenin transcription and is antagonized by both pRB and CDK8. *Nature*, **455**, 552–556.

Naito, A.T., Sumida, T., Nomura, S. *et al.* (2012) Complement C1q activates canonical Wnt signaling and promotes aging-related phenotypes. *Cell*, **149**, 1298–1313.

Nambotin, S.B., Lefrancois, L., Sainsily, X. *et al.* (2011) Pharmacological inhibition of Frizzled-7 displays anti-tumor properties in hepatocellular carcinoma. *Journal of Hepatology*, **54**, 288–299.

Narayanan, B.A., Doudican, N.A., Park, J. *et al.* (2012) Antagonistic effect of small-molecule inhibitors of Wnt/β-catenin in multiple myeloma. *Anticancer Research*, **32**, 4697–4707.

Osada, T., Chen, M., Yang, X.Y. *et al.* (2011) Antihelminth compound niclosamide downregulates Wnt signaling and elicits antitumor responses in tumors with activating APC mutations. *Cancer Research*, **71**, 4172–4182.

Padhi, D., Jang, G., Stouch, B. *et al.* (2010) Single-dose, placebo-controlled, randomized study of AMG 785, a sclerostin monoclonal antibody. *Journal of Bone and Mineral Research*, **26**, 19–26.

Pai, R., Tarnawski, A.S., and Tran, T. (2004) Deoxycholic acid activates β-catenin signaling pathway and increases colon cell cancer growth and invasiveness. *Molecular Biology of the Cell*, **15**, 2156–2163.

Park, S., Gwak, J., Cho, M. *et al.* (2006) Hexachlorophene inhibits Wnt/beta-catenin pathway by promoting Siah-mediated beta-catenin degradation. *Molecular Pharmacology*, **70**, 960–966.

Pelletier, J.C., Lundquist, J.T., Gilbert, A.M. *et al.* (2009) (1-(4-(Naphthalen-2-yl)pyrimidin-2-yl)piperidin-4-yl)methanamine: a wingless beta-catenin agonist that increases bone formation rate. *Journal of Medicinal Chemistry*, **52**, 6962–6965.

Phelps, R.A., Chidester, S., Dehghanizadeh, S. *et al.* (2009) A two-step model for colon adenoma initiation and progression caused by APC loss. *Cell*, **137**, 623–634.

Pode-Shakked, N., Harari-Steinberg, O., Haberman-Ziv, Y. *et al.* (2011) Resistance or sensitivity of Wilms' tumor to anti-FZD7 antibody highlights the Wnt pathway as a possible therapeutic target. *Oncogene*, **30**, 1664–1680.

Polakis, P. (2012) Drugging Wnt signalling in cancer. *The EMBO Journal*, **31**, 2737–2746.

Porter, D.C., Farmaki, E., and Altilia, S. (2012) Cyclin-dependent kinase 8 mediates chemotherapy-induced tumor-promoting paracrine activities. *Proceedings of the National Academy of Sciences*, **109**, 13799–13804.

Prevost, G.P. (2006) Anticancer activity of BIM-46174, a new inhibitor of the heterotrimeric Gα/Gβ protein complex. *Cancer Research*, **66**, 9227–9234.

Proffitt, K.D., Madan, B., Ke, Z. *et al.* (2012) Pharmacological inhibition of the Wnt acyltransferase PORCN prevents growth of Wnt-driven mammary cancer. *Cancer Research*, **73**, 502–507.

Radulescu, S., Brookes, M.J., Salgueiro, P. *et al.* (2012) Luminal iron levels govern intestinal tumorigenesis after Apc loss in vivo. *Cell Reports*, **2**, 270–282.

Regard, J.B., Zhong, Z., Williams, B.O., and Yang, Y. (2013) Wnt signaling in bone development and disease: making stronger bone with Wnts, in *Wnt Signaling* (eds X.H. Roel Nusse and R. van Amerongen), Cold Spring Harbor Laboratory Press, Cold Spring Harbor, NY, pp. 265–281.

Ren, S., Johnson, B.G., Kida, Y. *et al.* (2013) LRP-6 is a coreceptor for multiple fibrogenic signaling pathways in pericytes and myofibroblasts that are inhibited by DKK-1. *Proceedings of the National Academy of Sciences*, **110**, 1440–1445.

Riffell, J.L., Lord, C.J., and Ashworth, A. (2012) Tankyrase-targeted therapeutics: expanding opportunities in the PARP family. *Nature Reviews Drug Discovery*, **11**, 923–936.

Roh, M-S., Seo, M.S., Kim, Y. *et al.* (2007) Haloperidol and clozapine differentially regulate signals upstream of glycogen synthase kinase 3 in the rat frontal cortex. *Experimental and Molecular Medicine*, **39**, 353–360.

Säfholm, A., Tuomela, J., Rosenkvist, J. *et al.* (2008) The Wnt-5a-derived hexapeptide Foxy-5 inhibits breast cancer metastasis in vivo by targeting cell motility. *Clinical Cancer Research*, **14**, 6556–6563.

Sansom, O.J., Meniel, V., Wilkins, J.A. *et al.* (2006) Loss of Apc allows phenotypic manifestation of the transforming properties of an endogenous K-ras oncogene in vivo. *Proceedings of the National Academy of Sciences*, **103**, 14122–14127.

Saraswati, S., Alfaro, M.P., Thorne, C.A. *et al.* (2010) Pyrvinium, a potent small molecule Wnt inhibitor, promotes wound repair and post-MI cardiac remodeling. *PLoS One*, **5**, e15521.
Saraswati, S., Deskins, D.L., Holt, G.E., and Young, P.P. (2012) Pyrvinium, a potent small molecule Wnt inhibitor, increases engraftment and inhibits lineage commitment of mesenchymal stem cells (MSCs). *Wound Repair and Regeneration*, **20**, 185–193.
Schuijers, J. and Clevers, H. (2012) Adult mammalian stem cells: the role of Wnt, Lgr5 and R-spondins. *The EMBO Journal*, **31**, 2685–2696.
Shan, J., Zhang, X., Bao, J. *et al.* (2012) Synthesis of potent dishevelled PDZ domain inhibitors guided by virtual screening and NMR studies. *Chemical Biology & Drug Design*, **79**, 376–383.
Shi, M., Stauffer, B., Bhat, R., and Billiard, J. (2009) Identification of iminooxothiazolidines as secreted frizzled related protein-1 inhibitors. *Bioorganic & Medicinal Chemistry Letters*, **19**, 6337–6339.
Shultz, M.D., Kirby, C.A., Stams, T. *et al.* (2012) [1,2,4] Triazol-3-ylsulfanylmethyl)-3-phenyl-[1,2,4]oxadiazoles: antagonists of the Wnt pathway that inhibit tankyrases 1 and 2 via novel adenosine pocket binding. *Journal of Medicinal Chemistry*, **55**, 1127–1136.
Sinnberg, T., Menzel, M., Ewerth, D. *et al.* (2011) β-catenin signaling increases during melanoma progression and promotes tumor cell survival and chemoresistance. *PLoS One*, **6**, e23429.
Smalley, W.E. and DuBois, R.N. (1997) Colorectal cancer and nonsteroidal anti-inflammatory drugs. *Advances in Pharmacology*, **39**, 1–20.
Sun, Y., Campisi, J., Higano, C. *et al.* (2012) Treatment-induced damage to the tumor microenvironment promotes prostate cancer therapy resistance through Wnt16B. *Nature Medicine*, **18**, 1359–1368.
Surendran, K., Schiavi, S., and Hruska, K.A. (2005) Wnt-dependent beta-catenin signaling is activated after unilateral ureteral obstruction, and recombinant secreted frizzled-related protein 4 alters the progression of renal fibrosis. *Journal of the American Society of Nephrology*, **16**, 2373–2384.
Takahashi-Yanaga, F., Yoshihara, T., and Jingushi, K. (2008) Celecoxib-induced degradation of T-cell factors-1 and -4 in human colon cancer cells. *Biochemical and Biophysical Research Communications*, **377**, 1185–1190.
Takashima, S., Kadowaki, M., Aoyama, K. *et al.* (2011) The Wnt agonist R-spondin1 regulates systemic graft-versus-host disease by protecting intestinal stem cells. *Journal of Experimental Medicine*, **208**, 285–294.
Tang, W., Dodge, M., Gundapaneni, D. *et al.* (2008) A genome-wide RNAi screen for Wnt/beta-catenin pathway components identifies unexpected roles for TCF transcription factors in cancer. *Proceedings of the National Academy of Sciences*, **105**, 9697–9702.
Tang, Y., Simoneau, A.R., Liao, W.X. *et al.* (2009) WIF1, a Wnt pathway inhibitor, regulates SKP2 and c-myc expression leading to G1 arrest and growth inhibition of human invasive urinary bladder cancer cells. *Molecular Cancer Therapeutics*, **8**, 458–468.
Tenbaum, S.P., Ordóñez-Morán, P., Puig, I. *et al.* (2012) β-catenin confers resistance to PI3K and AKT inhibitors and subverts FOXO3a to promote metastasis in colon cancer. *Nature Medicine*, **18**, 892–901.
Thorne, C.A., Hanson, A.J., Schneider, J. *et al.* (2010) Small-molecule inhibition of Wnt signaling through activation of casein kinase 1α. *Nature Chemical Biology*, **6**, 829–836.
Thorne, C.A., Lafleur, B., Lewis, M. *et al.* (2011) A biochemical screen for identification of small-molecule regulators of the Wnt pathway using Xenopus egg extracts. *Journal of Biomolecular Screening*, **16**, 995–1006.
Tian, W., Han, X., Yan, M. *et al.* (2012) Structure-based discovery of a novel inhibitor targeting the β-catenin/Tcf4 interaction. *Biochemistry*, **51**, 724–731.
Trosset, J.Y., Dalvit, C., Knapp, S. *et al.* (2006) Inhibition of protein–protein interactions: the discovery of druglike β-catenin inhibitors by combining virtual and biophysical screening. *Proteins*, **64**, 60–67.
Ulsamer, A., Wei, Y., Kim, K.K. *et al.* (2012) Axin pathway activity regulates in vivo pY654-β-catenin accumulation and pulmonary fibrosis. *The Journal of Biological Chemistry*, **287**, 5164–5172.
Venerando, A., Girardi, C., Ruzzene, M., and Pinna, L. (2013) Pyrvinium pamoate does not activate protein kinase CK1, but promotes Akt/PKB down-regulation and GSK3 activation. *Biochemical Journal*. **452**, 131–137.
Verkaar, F., van der Stelt, M., Blankesteijn, W.M. *et al.* (2011) Discovery of novel small molecule activators of β-catenin signaling. *PLoS One*, **6**, e19185.
Vermeulen, L., de Sousa, E.M.F., Van Der Heijden, M. *et al.* (2010) Wnt activity defines colon cancer stem cells and is regulated by the microenvironment. *Nature Cell Biology*, **12**, 468–476.
von Marschall, Z., and Fisher, L.W. (2010). Secreted Frizzled-related protein-2 (sFRP2) augments canonical Wnt3a-induced signaling. *Biochemical and Biophysical Research Communications* **400**, 299--304.
Voronkov, A. and Krauss, S. (2012) Wnt/beta-catenin signaling and small molecule inhibitors. *Current Pharmaceutical Design*, **19**, 634–664.
Waaler, J., Machon, O., von Kries, J.P. *et al.* (2011) Novel synthetic antagonists of canonical Wnt

signaling inhibit colorectal cancer cell growth. *Cancer Research*, **71**, 197–205.

Waaler, J., Machon, O., Tumova, L. *et al.* (2012) A novel tankyrase inhibitor decreases canonical Wnt signaling in colon carcinoma cells and reduces tumor growth in conditional APC mutant mice. *Cancer Research*, **72**, 2822–2832.

Wan, M., Yang, C., Li, J. *et al.* (2008) Parathyroid hormone signaling through low-density lipoprotein-related protein 6. *Genes & Development*, **22**, 2968–2979.

Wan, M., Li, J., Herbst, K. *et al.* (2011) LRP6 mediates cAMP generation by G protein-coupled receptors through regulating the membrane targeting of Gαs. *Science Signaling*, **4**, ra15.

Wang, H., Hao, J., and Hong, C.C. (2011) Cardiac induction of embryonic stem cells by a small molecule inhibitor of Wnt/β-catenin signaling. *ACS Chemical Biology*, **6**, 192–197.

Wang, P.S., Chou, F.S., Bloomston, M. *et al.* (2009) Thiazolidinediones downregulate Wnt/beta-catenin signaling via multiple mechanisms in breast cancer cells. *Journal of Surgical Research*, **153**, 210–216.

Wang, W., Liu, H., Wang, S. *et al.* (2011) A diterpenoid derivative 15-oxospiramilactone inhibits Wnt/β-catenin signaling and colon cancer cell tumorigenesis. *Cell Research*, **21**, 730–740.

Wei, W., Chua, M.S., Grepper, S., and So, S.K. (2009) Blockade of Wnt-1 signaling leads to anti-tumor effects in hepatocellular carcinoma cells. *Molecular Cancer*, **8**, 76.

Wei, W., Chua, M.S., Grepper, S., and So, S. (2010) Small molecule antagonists of Tcf4/β-catenin complex inhibit the growth of HCC cells in vitro and in vivo. *International Journal of Cancer*, **126**, 2426–2436.

Wei, W., Chua, M.S., Grepper, S., and So, S.K. (2011) Soluble Frizzled-7 receptor inhibits Wnt signaling and sensitizes hepatocellular carcinoma cells towards doxorubicin. *Molecular Cancer*, **10**, 1–12.

Wend, P., Holland, J.D., Ziebold, U., and Birchmeier, W. (2010) Wnt signaling in stem and cancer stem cells. *Seminars in Cell & Development Biology*, **21**, 855–863.

Whyte, J.L., Smith, A.A., and Helms, J.A. (2013) Wnt signaling and injury repair, in *Wnt Signaling* (eds X.H. Roel Nusse and R. van Amerongen), Cold Spring Harbor Laboratory Press, Cold Spring Harbor, NY, pp. 409–421.

Xavier, C.P., Lima, C.F., Rohde, M., and Pereira-Wilson, C. (2011) Quercetin enhances 5-fluorouracil-induced apoptosis in MSI colorectal cancer cells through p53 modulation. *Cancer Chemotherapy and Pharmacology*, **68**, 1449–1457.

Xiao, J.H., Ghosn, C., Hinchman, C. *et al.* (2003) Adenomatous polyposis coli (APC)-independent regulation of beta-catenin degradation via a retinoid X receptor-mediated pathway. *The Journal of Biological Chemistry*, **278**, 29954–29962.

Xu, W. and Ji, J.Y. (2011) Dysregulation of CDK8 and cyclin C in tumorigenesis. *Journal of Genetics & Genomics*, **38**, 439–452.

Yang, M., Zhong, W.W., Srivastava, N. *et al.* (2005) G protein-coupled lysophosphatidic acid receptors stimulate proliferation of colon cancer cells through the β-catenin pathway. *Proceedings of the National Academy of Sciences*, **102**, 6027–6032.

Yang, W., Xia, Y., Ji, H. *et al.* (2011) Nuclear PKM2 regulates β-catenin transactivation upon EGFR activation. *Nature*, **480**, 118–122.

Yang, Y., Mallampati, S., Sun, B. *et al.* (2013) Wnt pathway contributes to the protection by bone marrow stromal cells of acute lymphoblastic leukemia cells and is a potential therapeutic target. *Cancer Letters*, **333**, 9–17.

Yao, H., Ashihara, E., Strovel, J.W. *et al.* (2011) AV-65, a novel Wnt/β-catenin signal inhibitor, successfully suppresses progression of multiple myeloma in a mouse model. *Blood Cancer Journal*, **1**, e43.

Yashiroda, Y., Okamoto, R., Hatsugai, K. *et al.* (2010) A novel yeast cell-based screen identifies flavone as a tankyrase inhibitor. *Biochemical and Biophysical Research Communications*, **394**, 569–573.

Ying, Y. and Tao, Q. (2009) Epigenetic disruption of the WNT/β-catenin signaling pathway in human cancers. *Epigenetics*, **4**, 307–312.

Yo, Y.T., Lin, Y.W., Wang, Y.C. *et al.* (2012) Growth inhibition of ovarian tumor-initiating cells by niclosamide. *Molecular Cancer Therapeutics*, **11**, 1703–1712.

You, L., He, B., Xu, Z. *et al.* (2004) An anti-Wnt-2 monoclonal antibody induces apoptosis in malignant melanoma cells and inhibits tumor growth. *Cancer Research*, **64**, 5385–5389.

Yu, D.H., Macdonald, J., Liu, G. *et al.* (2008) Pyrvinium targets the unfolded protein response to hypoglycemia and its anti-tumor activity is enhanced by combination therapy. *PLoS One*, **3**, e3951.

Zarnescu, D.C. and Zinsmaier, K.E. (2009) Ferrying wingless across the synaptic cleft. *Cell*, **139**, 229–231.

Zemans, R.L., Briones, N., Campbell, M. *et al.* (2011) Neutrophil transmigration triggers repair of the lung epithelium via β-catenin signaling. *Proceedings of the National Academy of Sciences*, **108**, 15990–15995.

Zhang, Q., Major, M.B., Takanashi, S. *et al.* (2007) Small-molecule synergist of the Wnt/beta-catenin signaling pathway. *Proceedings of the National Academy of Sciences*, **104**, 7444–7448.

Zhang, G.N., Liang, Y., Zhou, L.J. *et al.* (2011) Combination of salinomycin and gemcitabine eliminates pancreatic cancer cells. *Cancer Letters*, **313**, 137–144.

Zhao, J., Ramos, R., and Demma, M. (2012) CDK8 regulates E2F1 transcriptional activity through S375 phosphorylation. *Oncogene*, **2012**, 1–11.

Zhao, J., de Vera, J., Narushima, S. *et al.* (2007) R-spondin1, a novel intestinotrophic mitogen, ameliorates experimental colitis in mice. *Gastroenterology*, **132**, 1331–1343.

Zhao, J., Kim, K.A., De Vera, J. *et al.* (2009) R-spondin1 protects mice from chemotherapy or radiation-induced oral mucositis through the canonical Wnt/β-catenin pathway. *Proceedings of the National Academy of Sciences*, **106**, 2331–2336.

Zhao, Y., Yang, Z.Q., Wang, Y. *et al.* (2010) Dishevelled-1 and dishevelled-3 affect cell invasion mainly through canonical and noncanonical Wnt pathway, respectively, and associate with poor prognosis in nonsmall cell lung cancer. *Molecular Carcinogenesis*, **49**, 760–770.

Zhao, W.N., Cheng, C., Theriault, K.M. *et al.* (2012) A high-throughput screen for Wnt/β-catenin signaling pathway modulators in human iPSC-derived neural progenitors. *Journal of Biomolecular Screening*, **17**, 1252–1263.

Zimmerman, Z.F., Moon, R.T., and Chien, A.J. (2012) Targeting Wnt pathways in disease. *Cold Spring Harbor Perspectives in Biology*, **4**, a008086.

Zong, Y., Huang, J., and Sankarasharma, D. (2012) Stromal epigenetic dysregulation is sufficient to initiate mouse prostate cancer via paracrine Wnt signaling. *Proceedings of the National Academy of Sciences*, **109**, E3395–E3404.

Index

Page references followed by f denote figures. Page references followed by t denote tables.

2-acetylaminofluorene (2-AAF), 347–348
Activity-based proteomics, 126
Acylation, of Wnt proteins, 4
Adenlyl cyclase, regulation of, 200
Adenomatous polyposis coli (APC)
  β-catenin destruction complex and, 33, 34f, 36–37, 39, 138, 244, 245–246
  in cancers, 321, 323, 359–367, 360f, 362f–363f
  familial adenomatous polyposis (FAP), 36, 359
  functions of, 363–364
  mathematical modeling and, 154–155
  MS analysis, 130
  structure of, 245
  tissue regeneration and, 342
Adult stem cells, 329–336
  cycling, 334
  hematopoietic stem cell (HSC)
    overview, 330–332, 331f
    Wnt signaling, 335–336
  intestinal stem cells
    colorectal cancer and, 360–361, 362f, 364–365
    overview, 332–334, 333f
    Wnt signaling, 334–335
  mesenchymal stem cell (MSC), 331–332
  overview of, 329–330
  quiescence, 332, 334
  roles of, 329
  skin stem cells, 334
  stem cell niche concept, 330
AEG1 (astrocyte-elevated gene-1), 323
Affinity purification-mass spectrometry (APMS), 126–133, 128t
AKT (AKT1), 382–384, 387
α-catenin, 217–218
α-synuclein (*SCNA*) gene, 417
Alternative splicing, in vertebrate TCF/LEF proteins, 230–234
Alzheimer's disease, 412–416
  amyloid precursor protein processing, 412–413
  GSK3 hypothesis, 415–416
  lithium for, 415–416
  neurogenesis and neuroprotection, 414–415
  synaptic dysfunction and neuronal loss, 413–414
Amer1/WTX, 37, 130, 244
Amnesic mild cognitive impairment, 415
Amniote vertebrates, Wnt signaling in, 164–166
Amphibians, Wnt signaling in, 166–167
Amyloid precursor protein processing, 412–413
Angiogenesis, 323
Ankyrin 3 (ANK3), 398
Antagonists, 16
AP-2, 115, 116f, 117
APC. *See* Adenomatous polyposis coli (APC)
APMS (affinity purification-mass spectrometry), 126–133, 128t

Wnt Signaling in Development and Disease: Molecular Mechanisms and Biological Functions, First Edition.
Edited by Stefan Hoppler and Randall T. Moon.

Appendage regeneration, lower vertebrate, 342–345
β-catenin-independent Wnt pathways, 345
cellular mechanisms, 342–345
overactivation of Wnt/β-catenin signaling, 344–345
roles of Wnt/β-catenin signaling, 343–344, 343f
ARID1B, 402
Arm, 140, 143, 218, 220
Armadillo, 120
Astrocyte-elevated gene-1 (AEG1), 323
Autism, single-nucleotide variants (SNVs) in, 401–402, 401t
Axial patterning
early, 280–281
of mesoderm, 256–277
Axin
APC competition with, 245
β-catenin destruction complex and, 33–35, 34f–35f, 38, 41, 138, 244–247
binding of, 243
caveolin-mediated endocytosis and, 115
in colorectal cancer, 360, 362, 365
degradation
in response to Wnt stimulation, 40f, 41
by TNK, 144, 365
displacement from β-catenin, 140, 142
GSK3 binding, 245
heterotrimeric G protein, interaction with, 200t, 201
in liver regeneration, 347
LRP6 and, 23
mathematical modeling and, 155–156
melanoma and, 370–371, 373
membrane recruitment of, 244
MS analysis, 130
polyubiquitination of, 247
posttranslational control of, 144–145
role of turnover in β-catenin turnover, 38–39
stability, regulation of, 144
as target for developing Wnt pathway inhibitors, 246–247
Axis induction by maternal Wnt/β-catenin signaling, 254–256

BACE1, 413
Bardet-Biedl syndrome (BBS), 394
B-cell lymphoma 9/*legless* (BCL-9/Lgs), 138, 220, 217–218, 396
β-arrestin, 116f, 117–118, 382–384, 387
Beta-catenin activated reporter (BAR), 144, 221
β-catenin
in β-catenin destruction complex, 33–34, 34f–35f
binding partners of, 219f
in colorectal cancer, 360, 362–365
discovery of, 217
evolution of, 220–222
function
cell adhesion, 217–218
relating structure to, 218–219
subfunctionalization of, 220–221
Wnt signaling, 218
gene duplication, 220–221
in melanoma, 371–372
models for inhibition of degradation, 39–43, 40f
nuclear, 131, 138, 247–248, 254, 347, 364, 370
phosphorylation of, 138, 244, 246, 370
polyubiquitination, 138
structure, 219f
subcellular localization, 219–220
therapeutic targeting, 424t, 429–430
transcription factors that bind β-catenin, 61–63, 62t
turnover
as a futile cycle, 43
structural basis of, 244–247
ubiquitination, 138, 246
Wnt5a activation/inhibition of Wnt/β-catenin signaling, 102
β-catenin-dependent Wnt signaling pathway
mass spectrometry (MS) analysis of, 127–131
β-catenin destruction complex, 129–131
dishevelled (Dvl), 129
nuclear β-catenin, 131
proximal components of Wnt signaling, 127–129
overview of, 316f, 317
receptor-mediated endocytosis, 115–119, 116f
modulation by HSPGs, 116f, 118–119
negative regulation by Dickkopf (Dkk) and Disabled-2 (Dab2), 116f, 118
selective activation of Wnt signaling pathways by, 115–117, 116f
β-catenin destruction complex, 33–44
β-catenin degradation by, 38
components of, 33–38, 34f–35f
adenomatous polyposis coli (APC), 33, 34f, 36–37, 39
Axin, 33–35, 34f–35f, 38, 41
β-catenin, 33–34, 34f–35f
CK1α, 33, 34f, 36, 40f, 41
GSK3, 33, 34f, 35–36, 38, 40f, 41–42
formation of, 38
future directions for study of, 43–44
models for inhibition of β-catenin degradation, 39–43, 40f
Axin degradation in response to Wnt stimulation, 40f, 41
complex dissociation, 41
GSK3 inhibition by LRP6, 42
inhibition of β-catenin ubiquitination, 42–43
phosphorylation of GSK3, 40f, 41–42
sequestration of GSK3, 42
MS analysis, 129–131
structure of, 244–247

β-catenin-independent Wnt signaling pathway, 89–96
calcium signaling, 89–91, 90f
cilia formation and function, 92–93
lower vertebrate appendage regeneration, 345
overview, 317, 317f
planar cell polarity (PCP) signaling, 91–93, 92f
stem cell niche concept and, 330
Wnt/cAMP/protein kinase A signaling, 94f, 95
Wnt/rac/JNK signaling, 93–94
Wnt/rho/ROCK signaling, 93–94
Wnt/Ror signaling, 94
Wnt/Ryk signaling, 94–95
Wnt/TOR signaling, 95–96, 95f
β-catenin paradox, 363
β-transducin repeat containing protein (β-TrCP), 138
Bipolar disorder, lithium and, 381–384
Birds, Wnt signaling in, 164–166
BMP signaling
cross talk with Wnt, 156
in nonamniote vertebrates, 166–167
BRAF, 144–145, 371, 373–375
Brain and muscle Arnt-like protein-1 (BMAL1), 386
Bruton's tyrosine kinase (BTK), 131
Bulge, hair follicle, 334

C1q, 25
Cadherin-catenin complex, 217–220
Calcium signaling, 89–91, 90f
G protein-dependent, 199–201
Wnt4/calcium/nuclear factor of activated T-cell pathway in nephron development, 307
Wnt5a and, 101–102, 103
CamKII, 103
cAMP
adenlyl cyclase, regulation of, 200
Wnt/cAMP/protein kinase A signaling, 94f, 95
Cancer
chemotherapy, 434–435
colorectal, 316–323, 318f, 322f, 359–367
hepatocellular, 323–324
melanoma, 369–376, 372f–374f
tumor growth and progression, 318–320
Cancer genome, screening of, 145–146
Cancer stem cell (CSC) model, 319, 364
Canonical Wnt signaling pathway. *See also* β-catenin-dependent Wnt signaling pathway
cross-regulatory interactions with other signaling pathways, 139f, 140–141
Dishevelled in, 243
evolution of, 161–172, 163f
overview of, 138, 139f
schematic representation of, 139f
therapeutic targeting, 422t–425t
Carnosic acid, 424t, 429
Casein kinase 1α (CK1α), 138
Casein kinase 1ε (CK1ε), 107, 210
Casein kinase 2 (CK2), 107, 210
Caudal-related homeodomain protein 2 (CDX2), 81
CBP (CREB-binding protein), 57t, 58t, 61, 365, 403–404
C-clamp, in T-cell factors, 75
CDK8, 248, 424t, 430
CDK14, 22, 24t
Cell adhesion function of β-catenin, 217–218
Cell fate, 137–138, 253
Central nervous system formation, 279–285, 280f, 283f–284f
cell-type specificity in CNS, 283–284
CNS induction, 280–281
early axial patterning, 280–281
maintenance of continuous neurogenesis, 284–285, 284f
neural plate formation, 280–281, 280f
organizer establishment, 281–282
overview, 279–280
regulation of neurogenesis, 282–283, 283f
Cephalochordates, Wnt signaling in, 167–168
Cerberus (Cer), 182–183
CGG trinucleotide repeat, 402
CHD8, 401
Chemotherapy in cancer, 434–435
Chibby (Cby), 56t, 60
ChIP (chromatin immunoprecipitation), for identification of Wnt target genes, 76
Cholera toxin (CTX), 195, 200
Chronic lymphocytic leukemia, 121
CID (collision-induced dissociation), 125
Cilia formation and function, 92–93
Circadian rhythms in mood disorders, 386
Cited1, 304, 304f
c-Jun, 81
CK1α
in β-catenin destruction complex, 33, 34f, 36, 40f, 41
therapeutic targeting, 423t, 428–429
CK1γ
phosphorylation in LRP6, 23, 24t
therapeutic targeting, 423t, 429
CK1 site phosphorylation in LRP6, 23, 24t
Classification of Wnt proteins, 101
CLOCK, 386
CLU gene, 414
c-Myc, as Wnt target gene, 77, 157, 364–365
Cnidaria, Wnt signaling in, 170–171
Coated pits, 114–115, 114f
Coats disease, 188
Coco, 182–183
Collecting duct system development, 309–310
Frizzled receptors and, 309–310
Wnt11/GDNF loop, 309
Collision-induced dissociation (CID), 125

Colon, 332
Colorectal cancer, 316–323, 359–365
  APC (adematous polyposis coli) and, 321, 323, 359–367, 360f, 362f–363f
  cell of origin and, 361–362, 362f
  crypt progenitor phenotype, 360–361, 360f
  epithelial-to-mesenchymal transition (EMT), 319, 364, 365
  tumor growth and progression, 318–319, 318f, 364
  Wnt signaling, 320–323, 359–365
    cancer progression, 364
    cell of origin and, 361–362, 362f
    deregulation, 362–363
    "just right" hypothesis, 362–363
    in morphogenesis of, 320–323, 322f
    as target in CRC, 364
Connective tissue growth factor (CTGF), 185–186
Coop (Corepressor of Pangolin), 55, 56t
Copy number variations (CNVs) in schizophrenia, 395–397
Coreceptors
  for Frizzled signaling and specification, 198–199
  heparan sulfate proteoglycans (HSPGs), 26–27
  LRP family, 20–26
  MuSK, 26
  PTK7, 26
  ROR1/2, 26
  RYK, 26
CREB-binding protein (CBP), 57t, 58t, 61, 365, 403–404
Cross talk between signaling pathways, 139f, 140–141
Crypts of Lieberkühn, 332–334, 333f, 360–361, 360f, 362f
CSC (cancer stem cell) model, 319, 364
Ctenophora, Wnt signaling in, 170–171
C-terminal binding protein (CtBP), 55, 57t, 59t, 60, 143, 363
CTGF (connective tissue growth factor), 185–186
CTNNB1, 370, 375, 401, 401t
CTX (cholera toxin), 195, 200
Cyclin D1, 157, 365
Cycloheximide, 76
Cysteine, conserved in Wnt proteins, 3
Cytogenetic findings in schizophrenia, 394–395

Dab2. *See* Disabled-2 (Dab2)
DAB2IP, 399, 400t
Dally, as coreceptor for Wg, 10
Dapper (DpR), 243
Dementia, 411–418, 412f
  Alzheimer's disease, 412–416
    amyloid precursor protein processing, 412–413
    GSK3 hypothesis, 415–416
    lithium for, 415–416
    neurogenesis and neuroprotection, 414–415
    synaptic dysfunction and neuronal loss, 413–414
  etiological overlap with other neuropsychiatric disorders, 417
  frontotemporal, 416–417
  with Lewy bodies, 417
Depression and Wnt signaling, 384–386
Desmosome, 221
Deubiquitination of Frizzled by UBPY, 120–121, 120f
Development. *See* Vertebrate development
Dickkopf (Dkk)
  in Alzheimer's disease, 413–414
  glypicans and, 119
  LRP6 complex structure, 240f, 241–242
  mathematical modeling and, 155–156
  negative regulation of β-catenin-dependent Wnt signaling pathway in receptor-mediated endocytosis, 116f, 118
  in regulation of canonical Wnt/β-catenin signaling in heart development, 295–296
  regulation of LRP6, 24–25
  secreted Wnt modulator, 184
  therapeutics targeting, 422t–423t, 426–427
  Wnt signaling in hematopoietic stem cells, 336
  Wnt signaling in intestinal stem cells, 334–335
Diego, 157
Differential expression of TCF genes, 228–229
Diffusion of Wnt, mechanisms promoting and controlling, 8–11
  HSPG roles in gradient formation, 9–11
  long-range signaling, 8–9
  short-range signaling, 8
  Tiki and, 11
Disabled-2 (Dab2)
  negative regulation of β-catenin-dependent Wnt signaling pathway in receptor-mediated endocytosis, 116f, 118
  schizophrenia and, 399
Dishevelled (Dvl), 207–214
  amyloid precursor protein processing and, 412–413
  β-arrestin and, 117
  clathrin-mediated endocytosis in β-catenin-independent pathway, 116f, 117–118
  copy number variations (CNVs), 396
  dopamine receptors and, 387
  Frizzled interaction with, 196
  glutamate receptors and, 388
  heterotrimeric G protein, interaction with, 200–201, 200t
  kinases associated with, 209–211
  major depression and, 385–386
  overview, 207–208
  PCP signaling, 157
  phosphorylation, 209–211
  posttranslational modification, 211f
  proteomics, 129
  proteosomal degradation of, 211
  secrets of, 213–214

stability, 211–212
structure, 208–209, 208f, 243–244
DEP domain, 209
DIX domain, 209
PDZ domain, 209
TAZ and, 146
ubiquitination, 211–212
in Wnt/β-catenin pathway, 212–213, 212f
Wnt/PCP pathway, 212f, 213
Wnt signaling specificity and, 107
Disrupted-in-Schizophrenia-1 (DISC1), 387, 394–395
Disulfide bonds, in Wnt proteins, 3
Diversin, 37
Dkk. *See* Dickkopf (Dkk)
DLG1/2, 397
Dlp, Wg flow and, 10–11
Dopamine receptors, regulation by Wnt pathway, 387
Double-stranded RNA (dsRNA), 141
DpR (Dapper), 243
Dvl. *See* Dishevelled (Dvl)
DYRK1A, 401, 401t

Ecdysozoa, Wnt signaling in, 168–170
Echinoderms, Wnt signaling in, 167–168
Éclair, 5
EGF (epidermal growth factor), endocytosis and, 114
EGFR (epidermal growth factor receptor), 114, 119–120, 142
Electroconvulsive shock treatment, 384–385
Embryonic axis induction by maternal Wnt/β-catenin signaling, 254–256
Embryonic stem cells
core pluripotency transcriptional network, 272–274
markers of, 268, 268t
naïve *versus* prime state pluripotency, 269
as pluripotent stem cells, 267–268
Wnt/β-catenin signaling in, 267–274
acquisition of pluripotency, 272
differentiation of primed ESCs, 270–272
regulation of self-renewal in naïve ESCs, 269–270
Emp24, 5
EMT. *See* Epithelial-mesenchymal transition
Endocytosis, 113–122
caveolin-mediated endocytosis
β-catenin-dependent Wnt signaling pathway, 115–117, 116f
overview, 114f, 115
clathrin-mediated endocytosis
β-catenin-independent Wnt signaling pathway, 116f, 117–118
LRP5/6 internalization, 118
overview, 114–115, 114f
overview, 113–114
receptor ubiquitination and, 119–121, 120f
selective activation of Wnt signaling pathways by receptor-mediated endocytosis, 115–119, 116f
activation in β-catenin-dependent Wnt signaling pathway, 115–117, 116f
activation in β-catenin-independent Wnt signaling pathway, 116f, 117–118
modulation by HSPGs, 116f, 118–119
negative regulation by Dickkopf (Dkk) and Disabled-2 (Dab2), 116f, 118
Wnt signaling specificity and, 107
Endoplasmic reticulum (ER)
p24 protein mediation of ER to Golgi transport of Wnt, 5
Wnt proteins in, 3–4, 4f
Endosomal sorting complex required for transport (ESCORT) complex, 117
Endosomes, recycling of Wntless (Wls) and, 6–7
Enhancer Element Locator (EEL), 77
Enterocytes, 333, 333f, 360–361
Enteroendocrine cells, 333, 333f
Epidermal growth factor (EGF), endocytosis and, 114
Epidermal growth factor receptor (EGFR), 114, 119–120, 142
Epithelial-mesenchymal transition (EMT), 140, 142, 220, 258, 319–321, 320f
colorectal cancer and, 319, 364, 365
metastasis and, 319–321, 320f
ERK signaling cross talk with Wnt, 156, 156f, 158
Erythrocytes, 331, 331f
ESRRB, 273–274
Ethacrynic acid, 424t, 429
evenness interrupted (*evi*), 143
Evolution
of vertebrate TCF/LEF, 225–234
altered and alternative splicing, 230–234
conservation and innovation in exon structure and functional domains, 226f
differential transcriptional regulation, 228–230
genome duplications, 225
shared features of invertebrate and vertebrate TCF structure and function, 226–228
variety in context-dependent regulatory domain, 231–232
variety in C-terminal tail, 232–234
of Wnt signaling, 161–172
β-catenin, 220–222
distribution of *Wnt* and *Fzd* genes in metazoans, 163f, 164
gene duplication of β-catenin, 220–221
Wnt signaling in amniote vertebrates, 164–166
Wnt signaling in nonamniote vertebrates, 166–167
Wnt signaling in nondeuterostome bilaterians, 168–170
Wnt signaling in nonvertebrate deuterostomes, 167–168
Wnt signaling in prebilaterians, 170
Exosomes, secretion of Wnt and, 7–8

Factor X hypothesis, 157
Familial adenomatous polyposis (FAP), 36, 359
Familial exudative vitreoretinopathy (FEVR), 188, 197
FGF, 344–345
Fishes, Wnt signaling in, 166–167
Flamingo, 157
Flavonoids, therapeutic targeting and, 425t, 431
FOXO, 248
Fragile X syndrome, 402–403
Frizzled (FZD,Fz) family of Wnt receptors, 16–20
  activating ligand of, 197
  biochemical regulation of FZD proteins, 19
  calcium signaling, 90–91, 90f
  clathrin-mediated endocytosis in β-catenin-independent pathway, 116f, 117–118
  in collecting duct system development, 309–310
  in colorectal cancer, 321–323
  coreceptors for signaling and specification, 198–199
  CRD (cysteine-rich domain), 16f, 18–19
  deubiquitination of Frizzled by UBPY, 120–121, 120f
  dimerization of, 196–197
  distribution of *Wnt* and *Fzd* genes in metazoans, 163f, 164–165, 168–172
  as G protein-coupled receptors, 195–203
  G protein signaling, 199–203, 202f
  in hepatocellular cancer, 323– 324
  heterotrimeric G proteins, 199
  interaction partners, 18f
  ligand-receptor selectivity, 197–199, 198f
  ligand trafficking, 197–199, 198f
  major depression and, 385
  molecular pharmacology, 196
  multiple signaling routes through single family members, 103–104
  overview, 16–17
  PCP signaling, 157
  regulated maturation of, 20
  R-Spondin derepression of, 27
  secreted FZD-related proteins (sFRP1-5), 18–19
  in skeletal muscle regeneration, 349
  specificity of Wnt signaling and, 104–107, 105f
  structure of, 16f, 17–18, 17f, 196
  structure of Wnt-Frizzled complex, 239–241, 240f
  therapeutics targeting, 422t, 426
  ubiquitination of Frizzled by ZNRF3/RNF43, 120f, 121
  Wnt5a and, 101–102
Frontotemporal dementia, 416–417
fused gene, 34

GAG (glycosaminoglycan), 9–11
GATA1/2, 81
GDNF (glial cell line-derived neurotrophic factor), 309
Gefitinib, 435
Gemcitabine, 434
Gene duplication
  β-catenin, 220–221
  TCF/LEF, 225
Gene regulation by Wnt/β-catenin signaling, 51–64
  coregulators of β-catenin/TCF transcription, 55–61, 56t–59t
    activation of Wnt target genes, 60–61
    repression of Wnt target genes, 55, 60
  T-cell factor (TCF) transcriptional switch, 52–55, 52f–54f
  transcription factors that bind β-catenin, 61–63, 62t
Genome duplications, 225
Genome-wide association studies of neuropsychiatric disorders, 393–395, 397–399
Genome-wide studies of T-cell factor (TCF) binding, 77–82, 80t, 82t
  colocalization of TCFs with other transcription factors, 81–82, 82
  functional faction of TCF-bound regions, 79–81, 80t
  general features, 78–79
Girdin, 130
Glial cell line-derived neurotrophic factor (GDNF), 309
Glutamate receptors, regulation by Wnt pathway, 387–388
Glycogen synthase kinase 3. *See* GSK3
Glycosaminoglycan (GAG), 9–11
Glycosylation-deficient mutants, 3
Glypicans, 106, 118–119
Goblet cells, 333, 333f
Golgi
  p24 protein mediation of ER to Golgi transport of Wnt, 5
  recycling of Wntless (Wls), 6–7
GPC. *See* Glypicans
G protein-coupled receptors (GPCRs)
  Frizzleds as, 195–203, 387
  therapeutic targeting and, 425t, 431–432
G protein-dependent signaling, 199–201, 427
  adenlyl cyclase, regulation of, 200
  biological relevance of Wnt-induced, 201–202
  calcium, 199–201
  Frizzled and, 199–203, 202f
  heterotrimeric G proteins, 199, 200–201, 200t, 202f
G proteins, heterotrimeric, 199, 200–201, 200t, 202f
Gradient formation, HSPG roles in, 9–11
Granulocytes, 331–332, 331f
Grin2b, 388
GRK5/6 (G protein-coupled receptor kinases 5/6), 22, 24t
Gro/TLE, 55, 56t
Groucho, 220
GSK3 (glycogen synthase kinase 3)
  Axin binding of, 245
  β-catenin destruction complex and, 33, 34f, 35–36, 38, 40f, 41–42, 244, 245

β-catenin phosphorylation, 138
cross-regulation between signaling pathways, 138
Disrupted-in-Schizophrenia-1 (DISC1) and, 394–395
dopamine receptors and, 387
fragile X syndrome and, 402–403
glutamate receptors and, 388
heterotrimeric G protein, interaction with, 200t
inhibition by lithium, 382, 414
inhibition by LRP6, 42
LRP6 phosphorylation, 22, 24t
major depression and, 384–386
mood disorders and, 382–384
phosphorylation of, 40f, 41–42
sequestration of, 42, 117
therapeutic regulation of, 423t, 428
GSK3β, 95f, 96, 138
GSK3 hypothesis, 383, 415–416

HDAC (histone deacetylase), 384, 404
Heart organogenesis, 293–298
canonical Wnt/β-catenin signaling, 293–296, 294f
gene targets, 296
importance in cardiogenic endoderm, 296
multiphasic roles in heart muscle development, 293–296
regulation of first and second heart field development, 294–295
Wnt signals activating signaling, 295
heart muscle regeneration, 297–298
noncanonical Wnt/JNK signaling, 294f, 296–297
Hematopoietic stem cell (HSC)
overview, 330–332, 331f
Wnt signaling, 335–336
Hemichordates, Wnt signaling in, 167–168
Heparan sulfate proteoglycans (HSPG)
modulation of receptor-mediated endocytosis by, 116f, 118–119
roles in Wnt gradient formation, 9–11
as Wnt coreceptors, 26–27
Wnt signaling specificity and, 106
Hepatocellular cancer, 323–324
Heterotrimeric G proteins, 199, 200–201, 200t, 202f
HIF1, 248, 323
High-Mobility Group (HMG) family of transcription factors, 73, 74f
Hippo pathway, cross talk with Wnt signaling, 146
Histone deacetylase (HDAC), 384, 404
HMP-2, 221
HSC. *See* Hematopoietic stem cell
HSPG. *See* Heparan sulfate proteoglycans
5-hydroxytryptamine-1 (5-HT1), 385

IGF-binding protein 4 (IGFBP-4), 186
IGFBP-4, 25
Incoherent type 1 feedforward loop (iFFL-1), 154–156, 156f
Induced pluripotent stem cells (iPSCs), 267, 272
Inhibitors
regulation canonical Wnt/β-catenin signaling in heart development, 295–296
therapeutic targeting and, 435–436
Intestinal crypts, 332–334, 333f, 360–361, 360f, 362f
Intestinal stem cells
overview, 332–334, 333f
Wnt signaling, 334–335
inversion, 308
Invertebrates
evolution of Wnt signaling
Wnt signaling in nondeuterostome bilaterians, 168–170
Wnt signaling in nonvertebrate deuterostomes, 167–168
Wnt signaling in prebilaterians, 170
shared features of invertebrate and vertebrate TCF structure and function, 226–228

JNK
noncanonical Wnt/JNK signaling in heart organogenesis, 294f, 296–297
ROR2-dependent JNK signaling, 103
Wnt5a activation of, 119
Wnt/rac/JNK signaling, 93–94
"Just right" hypothesis of Wnt signaling, 362–363, 432

Kidney organogenesis, 303–311, 304f–305f
collecting duct system development, 309–310
Frizzled receptors and, 309–310
Wnt11/GDNF loop, 309
metanephric kidney, schematic of, 304f
nephron development, 305–309, 305f
nephron induction, 305–307
nephron maturation, 308–309
primary signal, 305–306
signal transduction, 306
Wnt4 and, 306, 307
Wnt9b and, 305–307
overview, 303–305
renal medulla development, 310
downstream signaling, 310
Wnt7b and Wnt4, 310
uretic bud tip, 304f
Kidney regeneration, 350–351
Kinases
in β-catenin destruction complex, 34–38, 40f
Dishevelled (Dvl) associated, 209–211
of LRP6, 22–24, 24t
Wnt5a activation of, 103
KRAS, 361–362, 435

LEF1. *See* Lymphoid enhancer factor 1
Leucine-rich repeat kinase 2 (LRRK2), 417
Leukocytes, 331–332, 331f

Lewy bodies, 417
LGR5, 317, 320, 322, 333–335, 348, 361, 362f, 364
Lhx1, 306
LIF, 269–270
Lipidation, in Wnt proteins, 3–4, 4f
Lipophorin, 7
Lipoproteins
  association of secreted Wnt proteins with, 7
  long-range Wnt signaling and, 8
Lithium
  Alzheimer's disease and, 415–416
  bipolar disorder and, 381–384
  Fragile X syndrome and, 402–403
Liver cancer, 323–324
Liver metastasis, 318
Liver regeneration, 345–348, 346f
  cellular mechanisms, 345–347
  hepatocyte proliferation-driven, 347
  stem cell-driven, 347–348
Long-range signaling, 8–9
Lophotrochozoa, Wnt signaling in, 168–170
LRG4/5/6, 121
LRP1, 399, 400t
LRP6
  in Alzheimer's disease, 413–414
  cell surface localization by Mesd, 25–26
  Dickkopf and, 24–25, 118
  disabled-2 (Dab2) and, 118
  DKK complex structure, 240f, 241–242
  endocytosis, 115, 118
  as GPRC accessory protein, 431–432
  GSK3 inhibition by, 42
  heterotrimeric G protein, interaction with, 200t, 201
  interaction partners, 18f
  kinases, 22–24, 24t
    CDK14, 22, 24t
    CK1 site phosphorylation, 23, 24t
    GRK5/6, 22, 24t
    GSK3, 22, 24t
    MAPK, 23, 24t
    PKA, 23, 24t
    Wnt-dependent *versus* Wnt-independent phosphorylation, 23–24, 24t
  phosphorylation of, 115
  PPSPxS motifs, 21–22
  regulation by Wise/SOST, IGFDBP-4, Waif1, and C1q, 25
  regulatory phosphorylation of intracellular domain, 21–24, 24t
  signalosome formation by, 115
  SOST binding to, 242
  structure of, 21, 240f, 241–242
  therapeutics targeting, 422t–423t, 426
  Wnt pathway specificity and, 104–107, 105f
LRP (low-density lipoprotein receptor-related proteins) family of Wnt coreceptors, 20–26
  endocytosis, 115, 118
  heterotrimeric G protein, interaction with, 200t, 201
  overview of, 20–21
  PPSPxS motifs, 21–24, 24t
  regulatory phosphorylation of intracellular domain, 21–24, 24t
  schizophrenia, 399
  specificity of Wnt signaling and, 104–107, 105f
  structure of, 21, 240f, 241–242
  therapeutics targeting, 422t–423t, 426
LSK (Lin-Sca1+Kit+) cells, 330
Lymphocytes, 331f, 332
Lymphogenesis, 323
Lymphoid enhancer factor 1 (LEF1), 73, 81
  axial patterning of mesoderm and, 256–257
  evolution of the TCF/LEF family, 225–234
  phosphorylation by CamKII, 103

Major depression and Wnt signaling, 384–386
Mammalian target of Rapamycin (mTOR), 95–96, 95f, 145
Mammals
  evolution of Wnt signaling in, 164–166
  Wnt/β-catenin signaling in early development, 257–258
MAPK (mitogen-activated protein kinase), 23, 24t, 140, 373–374
MAPKKK (mitogen-activated protein kinase kinase kinase), 140
Mass spectrometry, 125–133
  activity-based proteomics, 126
  affinity purification (APMS), 126–133, 128t
  integration of previously described APMS experiments, 131–133, 132f
  network analysis of β-catenin-dependent Wnt signaling, 127–131
    β-catenin destruction complex, 129–131
    dishevelled (Dvl), 129
    nuclear β-catenin, 131
    proximal components of Wnt signaling, 127–129
  overview, 125
  quantitative proteomics, 126–127
  selected reaction monitoring (SRM), 126–127
  tandem affinity purification with (TAP-MS), 146
Mass to charge ration (m/z) ratio, 125
mastermind (*mam*), 140
Mastermind-like 1 (Mam-l1), 140
Mathematical models, 153–158, 155f–156f
  incoherent type 1 feedforward loop (iFFL-1), 154–156, 156f
  modeling approaches, 153–154
  PCP signaling, 157
  Wnt/β-catenin signaling, 154–157
MED12, 400, 404, 430
MED13, 430
MED15, 397, 400, 404

MED23, 404
Melanoma, Wnt signaling in, 369–376, 372f–374f
  cell nonautonomous effects, 375–376
  functional relevance of pathway, 371–376
  heterogeneity, 375
  metastasis, 374–375
  prognostic indicators, 370–371
  questions regarding, 369
  role of β-catenin in transgenic mouse models, 371–372
  studies in cell lines, 372–374, 372f–374f
  transcriptional profiling, 371
  Wnt5a expression, 370–371
Mesd, LRP6 cell surface localization by, 25–26
Mesenchymal-epithelial transition (MET), 319–320, 320f
Mesenchymal stem cell (MSC), 331–332
Metanephros, 303, 304f
Metastasis, 318–320, 318f, 320f
  epithelial-mesenchymal transition (EMT) and, 319–321, 320f
  liver, 318
  in melanoma, 374–375
Metazoans
  distribution of *Wnt* and *Fzd* genes, 163f, 164
  molecular phylogeny of, 162f
Microarrays, for identification of Wnt target genes, 76
Microfold (M) cells, 333
Mindbomb 1 (MIB1), ubiquitination of Ryk by, 120f, 121
miRNAs, 145
Mitf, 285, 373
Mitogen-activated protein kinase (MAPK), 23, 24t, 140, 373–374
Mitogen-activated protein kinase kinase kinase (MAPKKK), 140
Modulators of Wnt signaling
  context-dependent modulation of Wnt/β-catenin signaling pathway, 143–144
  grouping of, 179
  interacting with Wnt pathway components, 184–187
    connective tissue growth factor (CTGF), 185–186
    Dickkopf (Dkk), 184
    IGF-binding protein 4 (IGFBP-4), 186
    Norrin, 187
    Notum, 187
    Sclerostin/SOST, 185
    Serpina3k, 186–187
    Shisa, 185
    Tsukushi (TSK), 186
    Wnt modulator in surface ectoderm (Wise), 185
  interacting with Wnt proteins, 181–184
    Cerberus (Cer), 182–183
    Coco, 182–183
    secreted Frizzled-related proteins (sFRPs), 181–182
    secreted wingless interacting molecule (Swim), 184
    Wnt-inhibitory factor 1 (WIF-1), 183–184
  pathological implications of malfunction, 187–188
  RNAi-based whole-genome screens for modulators of Wnt/wg pathway, 141–143
  secreted, 179–189, 183f
  structure of, 179, 180f
Molecular pharmacology, Frizzled structure, 196
Molecular phylogeny of metazoans, 162f
Monocytes, 331f, 332
Mood disorders, 381–386
  circadian rhythms in, 386
  lithium and bipolar disorder, 381–384
  major depression and Wnt signaling, 384–386
  valproate and Wnt signaling, 384
Morphogenesis, 258–260
  Wnt signaling in morphogenesis of colorectal cancer, 320–323, 322f
  during *Xenopus* gastrulation, 258f
MS2 spectrum, 125
MSC (mesenchymal stem cell), 331–332
MTG (Myeloid Translocation Gene), 55, 56t
mTOR signaling, 95–96, 95f, 145
MuSK, 26
Mx-Cre system, 335
Myocardial infarction, 297–298
Myotubularian lipid phosphatases, 7
m/*z* ratio, 125

NANOG, 272–273
Nemo-like kinase (NLK), 103, 140
Nephron development, 305–309, 305f
  nephron induction, 305–307
    primary signal, 305–306
    signal transduction, 306
    Wnt4 and, 306, 307
    Wnt9b and, 305–307
  nephron maturation, 308–309
    regulation of planar cell polarity, 308
    Wnt/β-catenin pathway, 308–309
Nervous system formation, 279–287, 280f, 283f–284f
  cell-type specificity in CNS, 283–284
  CNS induction, 280–281
  CNS organizer establishment, 281–282
  early axial patterning, 280–281
  maintenance of continuous CNS neurogenesis, 284–285, 284f
  neural connectivity, 286–287, 286f
  neural crest formation, 285
  neural crest migration, control of, 285
  neural plate formation, 280–281, 280f
  overview, 279–280
  regulation of CNS neurogenesis, 282–283, 283f
  regulation of PNS neurogenesis, 285–286

Neural crest
  formation, 285
  migration, control of, 285
Neuropsychiatric disorders, 379–388, 380f, 411–418
  dementia, 411–418, 412f
    Alzheimer's disease, 412–416
    frontotemporal, 416–417
    with Lewy bodies, 417
  genetic variation in Wnt pathway associated with, 393–404
    common variants, 397–399
    copy number variations (CNVs) in schizophrenia, 395–397
    cytogenetic findings in schizophrenia, 394–395
    Fragile X syndrome, 402–403
    Opitz-Kaveggia/Lujan syndrome, 404
    Pitt-Hopkins syndrome, 402
    Rubinstein-Taybi syndrome, 403–404
    single-nucleotide variants (SNVs), 399–402, 400f
    autism, 401–402, 401t
    schizophrenia, 399–401, 400t
  mood disorders, 381–386
  psychotic disorders, 386–388
  schizophrenia, 386–387
    copy number variations (CNVs) in, 395–397
    cytogenetic findings in, 394–395
    single-nucleotide variants (SNVs), 399–401, 400t
Nexins, recycling of Wntless (Wls) and, 7
NFAT (nuclear factor of activated T-cells), 102, 307
N-glycosylation, in Wnt proteins, 3
Niche, stem cell, 330
Niclosamide, 423t, 428
NLK (Nemo-like kinase), 103, 140
NMDA receptor, 388
Nonamniote vertebrates, Wnt signaling in, 166–167
Noncanonical Wnt/JNK signaling in heart organogenesis, 294f, 296–297
Noncanonical Wnt signaling pathway. *See* β-catenin-independent Wnt signaling pathway
Nonsteroidal anti-inflammatory drugs (NSAIDs), therapeutic targeting and, 425t, 430–431
Norrie disease, 188, 197
Norrin, 16, 17, 187, 197
Notum, 106, 187, 342
Nuclear factor of activated T-cells (NFAT), 102, 307
Nuclear hormone receptors, 431
Nuclear protein complexes and Wnt signaling, 247–248

OCT4, 272–273
Opitz-Kaveggia/Lujan syndrome, 404
Opossum, 5
Ordinary differential equation (ODE), 154–158
Organogenesis
  heart organogenesis, 293–298
    canonical Wnt/β-catenin signaling, 293–296, 294f
    heart muscle regeneration, 297–298
    noncanonical Wnt/JNK signaling, 294f, 296–297
  kidney organogenesis, 303–311, 304f–305f
    collecting duct system development, 309–310
    metanephric kidney, schematic of, 304f
    nephron development, 305–309, 305f
    overview, 303–305
    renal medulla development, 310
    uretic bud tip, 304f

p24 protein mediation of ER to Golgi transport of Wnt, 5
PAK2 gene, 396
Palmitate modification, of Wnt proteins, 3–4
Palmitoleic acid modification, of Wnt proteins, 4
Paneth cells, 333–335, 333f, 361
Parkinson's disease, 417
Partial differential equation (PDE), 154, 158
PCP. *See* Planar cell polarity (PCP) signaling
PDGF (platelet derived growth factor), 140, 142
Peripheral nervous system
  neural crest formation, 285
  neural crest migration, control of, 285
  regulation of PNS neurogenesis, 285–286
Pertussis toxin (PTX), 195, 199–201
Phosphatidylinositol 4,5-bisphosphate (PIP_2), 397, 400
Phosphatidylinositol 4,5-bisphosphate 3-kinases (PI3Ks), 397, 400–401, 435
Phosphorylation. *See specific target molecules*
PIK3CB, 400–401, 400t
Pitt-Hopkins syndrome, 402
PKA, 23, 24t, 94t, 95, 200
PKC, 102, 103
Planar cell polarity (PCP) signaling, 91–93, 92f
  architecture of pathway, 91–92, 92f
  cilia formation and function, 92–93
  CNS development, 281
  Dishevelled (Dvl) in, 212f, 213
  early vertebrate development and, 260–261
  in nephron development, 308
  Wnt5a/Ror2 and, 102
Planarian whole-body regeneration, 341–343, 342f
Platelet derived growth factor (PDGF), 140, 142
Pluripotent stem cells
  acquisition of pluripotency, 272
  core pluripotency transcriptional network, 272–274
  culturing of, 268
  differentiation of primed ESCs, 270–272
  markers of, 268, 268t
  naïve *versus* prime state pluripotency, 269
  overview of, 267–268, 329
PLX4032, 375
Polyubiquitination, 120, 138, 247
POP1, 53, 53f
PORCN, 421, 422t, 426

Porcupine (Porc), 4, 4f, 243
Porifera, Wnt signaling in, 170–171
Posttranslational modification, 3–5, 4f
of Dvl, 211f
mass spectrometry and, 125–127, 129
ubiquitination, 119–121, 120f
PP2A (protein phosphatase 2A), 244
in β-catenin destruction complex, 37
dopamine receptors and, 387
mood disorders and, 382–383
PPAR-γ, 140, 431
PPAR-γ response element (PPRE), 140
PPSPxS motifs, 21–24, 24tr
Prebilaterians, Wnt signaling in, 170
Prickle, 157
Prickle-Spiny legs, 157
Progranulin (*GRN*), 416–417
Prostaglandin E2 (PGE2), 347, 431
Protein kinase A (PKA), 23, 24t, 94f, 95, 200
Protein kinase C (PKC), 102, 103
Protein-protein interactions by Wnt pathway components, 239–248
Proteomic analysis of Wnt signaling, 125–133
activity-based proteomics, 126
integration of previously described APMS experiments, 131–133, 132f
integration of RNAi screens with genomic and proteomic approaches, 146–147
mass spectrometry (MS), 125–133
network analysis of β-catenin-dependent Wnt signaling, 127–131
β-catenin destruction complex, 129–131
dishevelled (Dvl), 129
nuclear β-catenin, 131
proximal components of Wnt signaling, 127–129
protein-protein interaction networks, 126
quantitative proteomics, 126–127
selected reaction monitoring (SRM), 126–127
PS1, 37–38
PSEN1/2, 413
Psychotic disorders, 386–388
dopamine receptors, regulation of, 387
glutamate receptors, regulation of, 387–388
PTK7, 26
Pygopus (Pygo) proteins, 58t, 60, 138, 220, 248
Pyrvinium, 423t, 428–429, 434, 435

Quantitative proteomics, 126–127
Quiescence, stem cell, 332, 334

Rac, in Wnt/rac/JNK signaling, 93–94
RAR-γ, 140
Ras, 435
Ras-related GTP-binding protein (Rab), 114f, 115
Receptor-mediated endocytosis. *See* Endocytosis
Receptors, 15–27, 317. *See also* Coreceptors; *specific receptors*
clustering, 27
Frizzed (FZD) family, 16–20
LRP family of Wnt coreceptors, 20–26
overview of, 15–16, 16f
signalosomes, 27
ubiquitination of, 119–121, 120f
ubiquitinylation of, 16, 19–20
Recycling of Wntless (Wls), 6–7
Regeneration, 339–351
cellular mechanisms of, 339–341, 340f
common principles of Wnt function in, 351
kidney, 350–351
liver, 345–348, 346f
cellular mechanisms, 345–347
hepatocyte proliferation-driven, 347
stem cell-driven, 347–348
lower vertebrate appendage regeneration, 343–345
β-catenin-independent Wnt pathways, 345
cellular mechanisms, 343–345
overactivation of Wnt/β-catenin signaling, 344–345
roles of Wnt/β-catenin signaling, 343–344, 343f
molecular mechanisms of, 341
planarian whole-body, 341–343, 342f
skeletal muscle, 348–350
Reggie-1/flotillin-2, 8–9
Release of Wnt from producing cells, 7–8
Renal medulla development
downstream signaling, 310
Wnt7b and Wnt4, 310
Reptin/TIP49b, 55, 57t
Ret, 309
Retinoic acid receptor (RAR), 140
Retinopathy of prematurity, 188
Retromer, recycling of Wntless (Wls) and, 6–7
Rho, in Wnt/rho/ROCK signaling, 93–94
Riluzole, 432
RING finger protein 146 (RNF146), 144
RNAi (RNA interference), 141–147
cancer genome screening, 145–146
context-dependent modulation of Wnt/β-catenin signaling pathway, 143–144
functionally targeted RNAi screens, 144–145
integration of RNAi screens with genomic and proteomic approaches, 146–147
whole-genome screens for modulators of Wnt/wg pathway, 141–143
RNF43, 120f, 121, 242
Robinow syndrome, 102
ROCK, in Wnt/rho/ROCK signaling, 93–94
Ror, 26
clathrin-mediated endocytosis in β-catenin-independent pathway, 116f, 117–118
in melanoma metastasis, 375
Wnt5a and Ror2, 102, 103
Wnt/Ror signaling, 94, 102

R-spondins, 16, 27, 243
liver regeneration and, 348
Wnt signaling in intestinal stem cells and, 335
ZNRF3 and, 121, 243
Rubinstein-Taybi syndrome, 403–404
Ryk, 26
clathrin-mediated endocytosis in β-catenin-independent pathway, 116f, 117–118
MS analysis of, 127, 129
ubiquitination by Mindbomb 1 (MIB1), 120f, 121
Wnt/Ryk signaling, 94–95

Salinomycin, 423t, 428, 434
Schizophrenia, 386–387
copy number variations (CNVs) in, 395–397
cytogenetic findings in, 394–395
single-nucleotide variants (SNVs), 399–401, 400t
Sclerostin. *See* SOST
SDC. *See* Syndecans
Secreted Frizzled-related proteins (sFRPs)
heart muscle regeneration and, 297–298
in regulation of canonical Wnt/β-catenin signaling in heart development, 295–296
repression and liver disease, 324
as secreted Wnt modulator, 181–182
structure of, 243
therapeutics targeting, 422t, 426
Wnt sequestration by, 197
Secreted wingless interacting molecule (Swim), 9, 184
Secretion pathway, 5–8
p24 protein mediation of ER to Golgi transport of Wnt, 5
release of Wnt from cells, 7–8
Wntless and, 5–7
Selected reaction monitoring, 126–127
Senexin, 424t, 430, 434
Sense, 120
Serpina3k, 186–187
Shaggy (Sgg), 417
Shisa, 185
Short hairpin RNA (shRNA), 145
Short interfering RNA (siRNA), 141, 144–146
Short-range signaling, 8
Siah2, 102
Signalosomes, 27, 115, 117, 243
Sine oculis-related homeobox 2 (Six2), 304, 304f
Single-nucleotide variants (SNVs), 399–402, 400f
autism, 401–402, 401t
schizophrenia, 399–401, 400t
Skeletal muscle regeneration, 348–350
Skin stem cells, 334
Small intestine, 332–334, 333f
Smoothened, 195
SNX-BAR nexins, recycling of Wntless (Wls) and, 7
SOST (Sclerostin), 25
binding to LRP6, 242
secreted Wnt modulator, 185
structure of, 242
therapeutics targeting, 422t
SOX2, 272–273
Sox17, 248
Specificity of Wnt signaling pathway, 101–107
determinants of, 104–107, 105f
multiple layers of, 105f
multiple signaling routes through single Frizzled family members, 103–104
Wnt5a activation of multiple signaling cascades, 101–103
Sponges, Wnt signaling in, 170–171
Stem cell niche concept, 330
Stem cells. *See also* Adult stem cells; Embryonic stem cells; Pluripotent stem cells
cancer stem cell (CSC) model, 319
characteristics of, 329
cycling, 334
hematopoietic stem cell (HSC)
overview, 330–332, 331f
Wnt signaling, 335–336
intestinal stem cells
colorectal cancer and, 360–361, 362f, 364–365
overview, 332–334, 333f
Wnt signaling, 334–335
liver regeneration and, 347–348
mesenchymal stem cell (MSC), 331–332
quiescence, 332, 334
skin stem cells, 334
Stroma, connective tissue, 332
Structural analysis of protein-protein interactions by Wnt pathway components, 239–248
SWIM (secreted wingless interacting molecule), 9, 184
Synaptogenesis, 286–287, 286f
Syndecans, 106, 119
SYS-1, 221

TAK1 (TGF-β-activated kinase 1), 103, 140
Tandem affinity purification with mass spectrometry (TAP-MS), 146
Tankyrase (TNK), 144, 365, 423t, 428, 431, 434
Target genes, identification of Wnt, 75–77
computational approaches, 77
direct and indirect targets, 76–77
TAZ (transcriptional activator with PDZ-binding motif), 63, 146
TBL1/TBLR1, 59t, 60
TBX1, 397
T-cell factors (TCFs)
alternative transcription start sites with TCF genes, 229–230
in autism, 401, 401t
axial patterning of mesoderm and, 256–257

β-catenin/TCF complex, 247–248
in colorectal cancer, 321–322
differential expression of TCF genes, 228–229
DNA binding by, 73–75, 74f
binding site flexibility, 73–75, 74f
bipartite binding, 75
dopamine receptors and, 387
evolution of vertebrate TCF/LEF family, 225–234
altered and alternative splicing, 230–234
conservation and innovation in exon structure and functional domains, 226f
differential transcriptional regulation, 228–230
genome duplications, 225
shared features of invertebrate and vertebrate TCF structure and function, 226–228
variety in context-dependent regulatory domain, 231–232
variety in C-terminal tail, 232–234
genome-wide studies of binding, 77–82, 80t, 82t
colocalization of TCFs with other transcription factors, 81–82, 82
functional faction of TCF-bound regions, 79–81, 80t
general features, 78–79
in nephron development, 306–307
in neuropsychiatric disorders, 397–402
structure of, 73, 74f
therapeutic targeting, 424t, 429–430
transcriptional switch, 52–55, 52f–54f, 227f
Wnt Response Element (WRE) and, 73, 75, 77–79, 81–83, 226–227
TCF7L1, 273–274, 401, 401t
TERT, 350–351
TGF-β
activation of TAK1 by, 140
stem cell niche concept and, 330
TGF-β-activated kinase 1 (TAK1), 103, 140
TGF-β receptor, 114
Therapeutic targeting of Wnt signaling network, 421–436, 433
canonical pathway therapeutics, 422t–425t
combinatorial therapy, 434–436
chemotherapy in cancer, 434–435
Wnt and non-Wnt combinations, 435
Wnt and Wnt inhibitor combinations, 435–436
cytoplasm, 423t–424t, 427–429
doses and dynamics, 432–434
drugging a Wnt network, 432
extracellular, 421, 422t–423t, 426–427
flavonoids, 425t, 431
G protein-coupled receptors, 425t, 431–432
nonsteroidal anti-inflammatory drugs (NSAIDs), 425t, 430–431
nucleus, 424t, 429–430
toxicity, 434
Thrombocytes, 331, 331f
Thymus, 332
Tiki, 11
Tissue homeostasis, restoring, 339–351
Tissue regeneration. *See* Regeneration
TNK. *See* Tankyrase
TNKS, 247
TOPGAL, 347, 349
TOR
shRNA RNAi screen against mTOR, 145
Wnt/TOR signaling, 95–96, 95f
Toxicity, therapeutic targeting and, 434
Transcriptional network, core pluripotency, 272–274
Transcriptional profiling of Wnt pathway in melanoma, 371
Transcriptional switch
β-catenin-dependent, 227f
T-cell factor (TCF), 52–55, 52f–54f, 227f
Transcription factors. *See also specific transcription factors*
colocalization of TCFs with other transcription factors, 81–82, 82
in Wnt/β-catenin signaling, 51–64
coregulators of β-catenin/TCF transcription, 55–61, 56t–59t
T-cell factor (TCF) transcriptional switch, 52–55, 52f–54f
transcription factors that bind β-catenin, 61–63, 62t
Transducin beta-like 1 X-linked receptor 1 (TBL1XR1), 401, 401t
TRAPP, 399–400, 402
TRRAP, 397, 400t, 404
Tsukushi (TSK), 186
Tumor growth and progression, 318–320

Ubiquitination
β-catenin, 138, 246
deubiquitination of Frizzled by UBPY, 120–121, 120f
of Dishevelled (Dvl), 211–212
of Frizzled by ZNRF3/RNF43, 120f, 121
inhibition of β-catenin, 42–43
polyubiquitination, 120, 138, 247
of receptors, 16, 19–20, 119–121, 120f
of Ryk by MIB1, 120f, 121
Ubiquitin-specific protease Y (UBPY), 120–121, 120f
UBR5, 399, 400t
Uretic bud tip, 304f
Urochordates, Wnt signaling in, 167–168

Valproate and Wnt signaling, 384
Van Gogh, 157
Vertebrate(s)
alternative transcription start sites with TCF genes, 229–230
differential expression of TCF genes, 228–229

Vertebrate(s) (*cont'd*)
evolution of vertebrate TCF/LEF family, 225–234
altered and alternative splicing, 230–234
conservation and innovation in exon structure and functional domains, 226f
differential transcriptional regulation, 228–230
genome duplications, 225
shared features of invertebrate and vertebrate TCF structure and function, 226–228
variety in context-dependent regulatory domain, 231–232
variety in C-terminal tail, 232–234
evolution of Wnt signaling
Wnt signaling in amniote vertebrates, 164–166
Wnt signaling in nonamniote vertebrates, 166–167
lower vertebrate appendage regeneration, 343–345
β-catenin-independent Wnt pathways, 345
cellular mechanisms, 343–345
overactivation of Wnt/β-catenin signaling, 344–345
roles of Wnt/β-catenin signaling, 343–344, 343f
Vertebrate development
early, 253–261, 255f
axial patterning of mesoderm, 256–277
core planar cell polarity (PCP) proteins, 260–261
embryonic axis induction by maternal Wnt/β-catenin signaling, 254–256
morphogenesis, 258–260, 258f
schematic diagram of early mouse and *Xenopus* development, 255f
Wnt/β-catenin signaling in mammalian development, 257–258
heart organogenesis, 293–298, 294f
canonical Wnt/β-catenin signaling, 293–296, 294f
heart muscle regeneration, 297–298
noncanonical Wnt/JNK signaling, 294f, 296–297
kidney organogenesis, 303–311, 304f–305f
collecting duct system development, 309–310
metanephric kidney, schematic of, 304f
nephron development, 305–309, 305f
overview, 303–305
renal medulla development, 310
uretic bud tip, 304f
nervous system formation, 279–287, 280f, 283f–284f
cell-type specificity in CNS, 283–284
CNS induction, 280–281
CNS organizer establishment, 281–282
early axial patterning, 280–281
maintenance of continuous CNS neurogenesis, 284–285, 284f
neural connectivity, 286–287, 286f
neural crest formation, 285
neural crest migration, control of, 285
neural plate formation, 280–281, 280f
overview, 279–280
regulation of CNS neurogenesis, 282–283, 283f
regulation of PNS neurogenesis, 285–286
VPS35
Alzheimer's disease and, 413
schizophrenia and, 399, 400t

WAIF1 (Wnt-activated inhibitory factor 1), 25, 115
Wg
diffusion of, mechanisms controlling, 8–11
release from producing cells, 7–8
Whole-genome screens for modulators of Wnt/wg pathway, RNAi-based, 141–143
WIF-1. *See* Wnt-inhibitory factor 1
Wise (Wnt modulator in surface ectoderm), 25, 185
Wnt
in colorectal cancer, 321–323
in hepatocellular cancer, 323– 324
structure of, 239
Wnt1
liver regeneration and, 348
in nervous system formation, 279–281
overexpression in melanoma, 376
therapeutics targeting, 422t, 426
Wnt2
in heart development, 297
major depression and, 384–385
therapeutics targeting, 422t, 426
Wnt3
in development, 257–258
therapeutics targeting, 422t, 426
Wnt3a
in development, 257
embryonic stem cells and, 272
in liver regeneration, 348
in melanoma, 372–373, 372f–374f
MS analysis of, 127
phosphorylation of LRP6, 115
in skeletal muscle regeneration, 349
therapeutics targeting, 422t, 426
Wnt4
in nephron development, 306, 307
in renal medulla development, 310
Wnt4/calcium/nuclear factor of activated T-cell pathway, 307
Wnt5a
activation of signaling cascades, 101–103
autism and, 401, 401t
clathrin-mediated endocytosis in β-catenin-independent pathway, 116f, 117–118
expression in metastatic melanoma, 370–371, 374–375
in heart development, 297
JNK activation, 119
in tissue regeneration, 345
Wnt5b, in tissue regeneration, 345

Wnt7a
in liver regeneration, 347–348
in skeletal muscle regeneration, 349
Wnt7b
in kidney development, 310
in kidney regeneration, 350
in liver regeneration, 347
Wnt8
axial patterning of mesoderm and, 256
in development, 256–258
posttranslational modification of *Xenopus* Wnt8 (XWnt8), 3–4
structure of, 17–18, 17f, 239–241, 240f
tissue regeneration and, 344
Wnt9b
as primary signal in nephron induction, 305–306
Wnt9b/β-catenin pathway, 306–307
Wnt10a, in liver regeneration, 347
Wnt11
in collecting duct system development, 309–310
in heart development, 297
heart muscle regeneration and, 298
mRNA localization, 254, 256
Wnt-activated inhibitory factor 1 (WAIF1), 25, 115
Wnt/β-catenin signaling. *See also* β-catenin-dependent Wnt signaling pathway
context-dependent modulation of, 143–144
Dishevelled (Dvl) in, 212–213, 212f
in early vertebrate development, 253–261, 255f
in embryonic stem cells, 267–274
gene regulation by, 51–64
mathematical models, 154–157
in melanoma, 369–376, 372f–374f
in nephron development, 306, 308–309
regeneration and
common principles of Wnt function in, 351
kidney, 350–351
liver, 345–348, 346f
lower vertebrate appendage regeneration, 343–345, 343f
planarian whole-body, 341–343, 342f
skeletal muscle, 348–350
stoichiometric relationships in, 315
Wnt/cAMP/protein kinase A signaling, 94f, 95
WntD, 5
Wnt-Frizzled complex, structure of, 239–241, 240f
Wnt inhibitors. *See* Inhibitors
Wnt-inhibitory factor 1 (WIF-1)
liver regeneration and, 348
secreted Wnt modulator, 183–184
structure of, 240f, 242–243
Wntless (Wls)
recycling of, 6–7
required for Wnt secretion, 5–6
structure of, 243
Wnt modulator in surface ectoderm (Wise), 25, 185
Wnt/rac/JNK signaling, 93–94
Wnt Response Element (WRE), 52, 55, 60–61, 73-83
T-cell factors and, 73, 75, 77–79, 81–83, 226–227
Wnt/rho/ROCK signaling, 93–94
Wnt/Ror signaling, 94, 102
Wnt/Ryk signaling, 94–95
Wnt/TOR signaling, 95–96, 95f
WRE. *See* Wnt Response Element
WRM-1, 221
WTX, 37, 130, 244

XAV-939, 423t, 428, 434
Xenopus
morphogenesis, 258–260, 258f
schematic, 255f
Spemann's organizer, 253
tissue regeneration, 343–345, 343f
Wnt signaling in early development, 253–261
Xenopus Wnt8 (XWnt8)
posttranslational modification of, 3–4
structure of, 239–241

Yan, 143
YES1-associated protein (YAP1), 146

ZEB1, 319
Zebrafish
fin regeneration, 343–345, 343f
Wnt signaling in early development, 253–261
ZNRF3, 120f, 121, 242, 422t, 427